Inland Fisheries Management in North America
Third Edition

INLAND FISHERIES MANAGEMENT IN NORTH AMERICA, 3RD EDITION
is a special project of and partially funded by the
**Education Section and Fisheries Management Section,
American Fisheries Society**

INLAND FISHERIES MANAGEMENT IN NORTH AMERICA
THIRD EDITION

Edited by

Wayne A. Hubert
Hubert Fisheries Consulting, LLC
1063 Colina Drive
Laramie, Wyoming 82072

and

Michael C. Quist
U.S. Geological Survey
Idaho Cooperative Fish and Wildlife Research Unit
Department of Fish and Wildlife Resources, University of Idaho
P.O. Box 441141, Moscow, Idaho 83844

American Fisheries Society
Bethesda, Maryland
2010

A suggested citation format for this book follows.

Entire Book

Hubert, W. A. and M. C. Quist, editors. 2010. Inland fisheries management in North America, 3rd edition. American Fisheries Society, Bethesda, Maryland.

Chapter within the Book

Kolar, C. S., W. R. Courtenay, Jr., and L. G. Nico. 2010. Managing undesired and invading fishes. Pages 213–259 *in* W. A. Hubert and M. C. Quist, editors. Inland fisheries management in North America, 3rd edition. American Fisheries Society, Bethesda, Maryland.

© Copyright 2010 by the American Fisheries Society

All rights reserved. Photocopying for internal or personal use, or for the internal or personal use of specific clients, is permitted by AFS provided that the appropriate fee is paid directly to Copyright Clearance Center (CCC), 222 Rosewood Drive, Danvers, Massachusetts 01923, USA; phone 978-750-8400. Request authorization to make multiple copies for classroom use from CCC. These permissions do not extend to electronic distribution or long-term storage of articles or to copying for resale, promotion, advertising, general distribution, or creation of new collective works. For such uses, permission or license must be obtained from AFS.

Printed in the United States of America on acid free paper

Library of Congress Control Number 2010934465
ISBN 978-1-934874-16-5

American Fisheries Society Web site address: *www.fisheries.org*

American Fisheries Society
5410 Grosvenor Lane, Suite 100
Bethesda, Maryland 20814
USA

Table of Contents

Contributors...xi
List of Species..xv
Preface..xxi

1   Historical Perspectives on Inland Fisheries Management in North America
    Christine M. Moffitt, Gary Whelan, and Randy Jackson
    1.1   Importance of Historic Accounts..1
    1.2   Fisheries Before European Settlement...2
    1.3   Early Ichthyological Surveys in North America..4
    1.4   Changing Concepts of Management...5
    1.5   Use of Historic Information in Contemporary Management........................31
    1.6   References..34

2   Fish Population Dynamics: Mortality, Growth, and Recruitment
    Micheal S. Allen and Joseph E. Hightower
    2.1   Introduction...43
    2.2   Overview of Dynamics in Inland Fish Populations....................................44
    2.3   Methods of Estimating Population Parameters..46
    2.4   Use of Basic Fisheries Models..62
    2.5   Conclusions...77
    2.6   References...77

3   Scale and Fisheries Management
    James T. Peterson and Jason Dunham
    3.1   Introduction...81
    3.2   Defining Scale and Importance...81
    3.3   Working with Scale...86
    3.4   Identifying the Appropriate Scales..86
    3.5   Incorporating Scales into Management...97
    3.6   Conclusions...101
    3.7   References...102

4   The Legal Process and Fisheries Management
    Jeffery A. Ballweber and Harold L. Schramm, Jr.
    4.1   Introduction...107
    4.2   Overview of North American Governments..108
    4.3   Historical Basis of Governments' Inland Fisheries Management
          Authority...113
    4.4   Primary Inland Fisheries Management Structures in North
          America...121
    4.5   Inland Fisheries Management within a Watershed or Ecosystem
          Management Framework...124
    4.6   Water...125
    4.7   Land..129

	4.8	Conclusions	131
	4.9	References	131

5 The Process of Fisheries Management
STEVE L. MCMULLIN AND EDMUND PERT

	5.1	Introduction	133
	5.2	Roles of Fisheries Professionals, Stakeholders, and Policy Makers	137
	5.3	A Contrast of Marine and Inland Fisheries Management	139
	5.4	Adaptive Fisheries Management	142
	5.5	Conclusions	149
	5.6	References	131

6 Communication Techniques for Fisheries Scientists
SCOTT A. BONAR AND MICHAEL E. FRAIDENBURG

	6.1	Introduction	157
	6.2	Influence Techniques	158
	6.3	Verbal Judo and Conflict Resolution	161
	6.4	Principles of Negotiation	166
	6.5	Management of a Group Project	168
	6.6	The Successful Presentation	173
	6.7	The Successful Publication	174
	6.8	Effective Meetings	178
	6.9	Community Change through Community Participation	180
	6.10	Conclusions	183
	6.11	References	183

7 Regulating Harvest
DANIEL A. ISERMANN AND CRAIG P. PAUKERT

	7.1	Introduction	185
	7.2	Regulatory Authority, Implementation, and Enforcement	187
	7.3	Types of Harvest Regulation	189
	7.4	Selection of Harvest Regulation	198
	7.5	Evaluation of Harvest Regulations	201
	7.6	Factors Confounding Evaluation and Effectiveness of Harvest Regulations	202
	7.7	Conclusions	205
	7.8	References	206

8 Managing Undesired and Invading Fishes
CINDY S. KOLAR, WALTER R. COURTENAY, JR., AND LEO G. NICO

	8.1	Introduction	213
	8.2	History of Fish Introductions	214
	8.3	Prevention of Unintended Introductions and Reduction of Risks of Deliberate Introductions	221
	8.4	Eradication and Population Control	228
	8.5	Integrated Pest Management	243
	8.6	Conclusions	245
	8.7	References	249

9 Use of Hatchery Fish for Conservation, Restoration, and Enhancement of Fisheries
Jesse Trushenski, Thomas Flagg, and Christopher Kohler

9.1	Introduction	261
9.2	Stocking Philosophy	263
9.3	Hatcheries and Approaches to Fish Culture	264
9.4	Genetic Considerations	267
9.5	Distinctions between Wild and Hatchery Fish	269
9.6	Best Management Practices for Propagation and Stocking Programs	270
9.7	Criticisms of Fish Culture and Hatchery Fish	283
9.8	The Future of Fish Culture as a Management Tool	286
9.9	Conclusions	287
9.10	References	287

10 Habitat Improvement in Altered Systems
Mark A. Pegg and John H. Chick

10.1	Introduction	295
10.2	Habitat Concepts	296
10.3	Assessment of Habitat for Fishes	296
10.4	Habitat Improvement Process	298
10.5	Agency Roles	300
10.6	Unique Systems and Issues	304
10.7	Evaluation of Habitat Improvements	312
10.8	Conclusions	316
10.9	References	321

11 Methods for Assessing Fish Populations
Kevin L. Pope, Steve E. Lochmann, and Michael K. Young

11.1	Introduction	325
11.2	Need for Assessment	325
11.3	Sampling Considerations	326
11.4	Characteristics, Statistics, Parameters, and Indices	329
11.5	Distinction Between Assessment and Monitoring	344
11.6	Conclusions	344
11.7	References	345

12 Assessment and Management of Ecological Integrity
Thomas J. Kwak and Mary C. Freeman

12.1	Introduction	353
12.2	Ecological Integrity Assessment	361
12.3	Management for Ecological Integrity	373
12.4	Politics and Environmental Legislation	380
12.5	Ecological Integrity in Practice	386
12.6	References	387

13 Ecology and Management of Lake Food Webs
Steven R. Chipps and Brian D. S. Graeb

13.1	Introduction	395

13.2	Nutrients and Productivity	396
13.3	Lake Food Webs	399
13.4	Food Web Theory	410
13.5	Food Web Management	412
13.6	Case Histories: Food Web Dynamics	415
13.7	Conclusions	419
13.8	References	419

14 Use of Social and Economic Information in Fisheries Assessments
Kevin M. Hunt and Stephen C. Grado

14.1	Introduction	425
14.2	Social and Economic Research	426
14.3	Conclusions	444
14.4	References	444

15 Natural Lakes
Michael J. Hansen, Nigel P. Lester, and Charles C. Krueger

15.1	Introduction	449
15.2	Types of Fisheries	455
15.3	Fishery Potential of Natural Lakes	457
15.4	Management of Goals and Objectives	464
15.5	Stock Assessment Designs	472
15.6	Management Strategies	481
15.7	Conclusions	490
15.8	References	490

16 Farm Ponds and Small Impoundments
David W. Willis, Robert D. Lusk, and Jeffrey W. Slipke

16.1	Introduction	501
16.2	Habitat Development	504
16.3	Common Habitat Management Techniques	511
16.4	Traditional Management Strategies	519
16.5	Innovative Management Strategies	526
16.6	Population and Assemblage Analysis	530
16.7	Conclusions	537
16.8	References	537

17 Large Reservoirs
Leandro E. Miranda and Phillip W. Bettoli

17.1	Introduction	545
17.2	The Diversity of Reservoirs	546
17.3	The Reservoir	550
17.4	Tributaries	568
17.5	The Watershed	570
17.6	The River Basin	574
17.7	Conclusions	576
17.8	References	579

18 Coldwater Streams
ROBERT E. GRESSWELL AND BRUCE VONDRACEK
- 18.1 Introduction..587
- 18.2 Characteristics of Coldwater Streams..588
- 18.3 Management of Coldwater Streams in the 21st Century..........................594
- 18.4 Conclusions..609
- 18.5 References..609

19 Coldwater Rivers
DARIN G. SIMPKINS AND JESSICA L. MISTAK
- 19.1 Introduction..619
- 19.2 Coldwater River Ecology..620
- 19.3 Management of Coldwater Rivers..631
- 19.4 Conclusions..648
- 19.5 References..648

20 Warmwater Streams
DANIEL C. DAUWALTER, WILLIAM L. FISHER, AND FRANK J. RAHEL
- 20.1 Introduction..657
- 20.2 Factors Influencing Warmwater Stream Fishes..................................658
- 20.3 Warmwater Streams: Issues and Management..................................665
- 20.4 Conclusions..688
- 20.5 References..690

21 Warmwater Rivers
CRAIG P. PAUKERT AND DAVID L. GALAT
- 21.1 Introduction..699
- 21.2 Characteristics of Large Warmwater Rivers......................................700
- 21.3 River Ecology Concepts..701
- 21.4 Major Issues..706
- 21.5 Ecological Integrity..716
- 21.6 Conclusions..718
- 21.7 References..722
- Appendix..731

Symbols and Abbreviations..737
Index..739

Contributors

Micheal S. Allen (Chapter 2): Department of Fisheries and Aquatic Science, University of Florida, 7922 NW 71st Street, Gainesville, Florida 32653, USA

Jeffery A. Ballweber (Chapter 4): Agriculture and Resource Economics, Colorado State University, B308 Clark Building, Fort Collins, Colorado 80523, USA

Phillip W. Bettoli (Chapter 17): U.S. Geological Survey, Tennessee Cooperative Fish and Wildlife Research Unit, Box 5114, Tennessee Tech University, Cookeville, Tennessee 38505, USA

Scott A. Bonar (Chapter 6): U.S. Geological Survey, Arizona Cooperative Fish and Wildlife Research Unit, University of Arizona, 104 Bio Sci E, Tucson, Arizona 8572, USA

John H. Chick (Chapter 10): Illinois Natural History Survey, 8450 Montclair Avenue, Brighton, Illinois 62012, USA

Steven R. Chipps (Chapter 13): U.S. Geological Survey, South Dakota Cooperative Fish and Wildlife Research Unit, Department of Wildlife and Fisheries Sciences, South Dakota State University, Box 2140B NPB, Brookings, South Dakota 57007, USA

Walter R. Courtenay (Chapter 8): 5005 NW 59th Terrace, Gainesville, Florida 32653, USA

Daniel C. Dauwalter (Chapter 20): Trout Unlimited, Inc., 910 Main St., Suite 342, Boise, Idaho 83702, USA

Jason Dunham (Chapter 3): U.S. Geological Survey, Forest and Rangeland Ecosystem Science Center, 3200 SE Jefferson Way, Corvallis, Oregon 97331, USA

William L. Fisher (Chapter 20): U.S. Geological Survey, New York Cooperative Fish and Wildlife Research Unit, 206F Fernow Hall, Cornell University, Ithaca, New York 14853, USA

Thomas Flagg (Chapter 9): National Oceanic and Atmospheric Administration, Fisheries Service, Northwest Fisheries Science Center, Manchester Research Station, P.O. Box 130, Manchester, Washington 98353, USA

Michael E. Fraidenburg (Chapter 6): Dynamic Solutions Group, LLC, 5432 Keating Road Northwest, Olympia, Washington 98502, USA

Mary C. Freeman (Chapter 12): U.S. Geological Survey, Patuxent Wildlife Research Center, Institute of Ecology, University of Georgia, Athens, Georgia, 30602, USA

David L. Galat (Chapter 21): U.S. Geological Survey, Missouri Cooperative Fish and Wildlife Research Unit, Department of Fisheries and Wildlife Sciences, University of Missouri, 302 Anheuser-Busch Natural Resources Building, Columbia, Missouri 65211, USA

Brian D. S. Graeb (Chapter 13): Department of Wildlife and Fisheries Science, South Dakota State University, P.O. Box 2140B, Brookings, South Dakota 57007, USA

Robert E. Gresswell (Chapter 18):U.S. Geological Survey, Northern Rocky Mountain Science Center, 2327 University Way, Suite 2, Bozeman, Montana 59715, USA

Michael J. Hansen (Chapter 15): University of Wisconsin—Stevens Point, 800 Reserve Street, Stevens Point, Wisconsin 54481, USA

Joseph E. Hightower (Chapter 2): U.S. Geological Survey, North Carolina Cooperative Fish and Wildlife Research Unit, North Carolina State University, Box 7617, Raleigh, North Carolina 27695, USA

Wayne A. Hubert (Coeditor): Hubert Fisheries Consulting, LLC, 1063 Colina Drive, Laramie, Wyoming 82072, USA

Kevin M. Hunt (Chapter 14): Department of Wildlife and Fisheries, Mississippi State University, Box 9690, Mississippi State, Mississippi 39762, USA

Daniel A. Isermann (Chapter 7): University of Wisconsin—Stevens Point, 800 Reserve Street, Stevens Point, Wisconsin 54481, USA

Randy Jackson (Chapter 1): Cornell Biological Field Station, 900 Shackelton Point Road, Bridgeport, New York 13030, USA

Christopher Kohler (Chapter 9): Fisheries and Illinois Aquaculture Center, Southern Illinois University, Carbondale, Illinois 62901, USA

Cindy S. Kolar (Chapter 8): U.S. Geological Survey, 12201 Sunrise Valley Drive, Reston, Virginia 20192, USA

Charles C. Krueger (Chapter 15): Great Lakes Fishery Commission, 2100 Commonwealth Blvd., Suite 100, Ann Arbor, Michigan 48105, USA

Thomas J. Kwak (Chapter 12): U.S. Geological Survey, North Carolina Cooperative Fish and Wildlife Research Unit, North Carolina State University, Box 7617, Raleigh, North Carolina 27695, USA

Nigel P. Lester (Chapter 15): Ontario Ministry of Natural Resources, DNA Building, 2140 East Bank Drive, Peterborough, Ontario K9J 7B8, Canada

Steve E. Lochmann (Chapter 11): University of Arkansas at Pine Bluff, 1200 N. University Drive, Mail Slot 4912, Pine Bluff, Arkansas 71601, USA

Robert D. Lusk (Chapter 16): *Pond Boss* Magazine, P.O. Box 12, Sadler, Texas 76264, USA

Steve L. McMullin (Chapter 5): Department of Fisheries and Wildlife Sciences, Virginia Polytechnic Institute and State University, 100 Cheatham Hall, Blacksburg, Virginia 24061, USA

Leandro E. (Steve) Miranda (Chapter 17): U.S. Geological Survey, Mississippi Cooperative Fish and Wildlife Research Unit, Mississippi State University, Mail Stop 9691, Mississippi State, Mississippi 39762, USA

Jessica L. Mistak (Chapter 19): Michigan Department of Natural Resources, Marquette Fisheries Station, 484 Cherry Creek Road, Marquette, Michigan 49855, USA

Christine M. Moffitt (Chapter 1): U.S. Geological Survey, Idaho Cooperative Fish and Wildlife Research Unit, Department of Fish and Wildlife Resources, University of Idaho, P.O. Box 441141, Moscow, Idaho 83844, USA

Craig P. Paukert (Chapter 7 and 21): U.S. Geological Survey, Missouri Cooperative Fish and Wildlife Research Unit, Department of Fisheries and Wildlife Sciences, University of Missouri, 302 Anheuser-Busch Natural Resources Building, Columbia, Missouri 65211, USA

Mark A. Pegg (Chapter 10): University of Nebraska, 402 Hardin Hall, Lincolns, Nebraska 68583, USA

Edmund Pert (Chapter 5): California Department of Fish and Game, 1416 Ninth Street, 12th floor, Sacramento, California 95814, USA

James T. Peterson (Chapter 3): U.S. Geological Survey, Georgia Cooperative Fish and Wildlife Research Unit, Warnell School of Forestry and Natural Resources, University of Georgia, Athens, Georgia 30602, USA

Kevin L. Pope (Chapter 11): U.S. Geological Survey, Nebraska Cooperative Fish and Wildlife Research Unit, University of Nebraska, 402 Hardin Hall, Lincoln, Nebraska 68583, USA

Michael C. Quist (Coeditor): U.S. Geological Survey, Idaho Cooperative Fish and Wildlife Research Unit, Department of Fish and Wildlife Resources, University of Idaho, P.O. Box 441141, Moscow, Idaho 83844, USA

Frank J. Rahel (Chapter 20): Department of Zoology and Physiology, University of Wyoming, Department 3166, 1000 East University Avenue, Laramie, Wyoming 82071, USA

Harold L. Schramm, Jr. (Chapter 4): U.S. Geological Survey, Mississippi Cooperative Fish and Wildlife Research Unit, Mississippi State University, Mail Stop 9691, Mississippi State, Mississippi 39762, USA

Darin G. Simpkins (Chapter 19): U.S. Fish and Wildlife Service, Green Bay Ecological Services Office, 2661 Scott Tower Drive, New Franken, Wisconsin 54229, USA

Jeffrey W. Slipke (Chapter 16): Southeastern Pond Management, 104 Gingham Drive, LaVergne, Tennessee 37086

Jesse Trushenski (Chapter 9): Fisheries and Illinois Aquaculture Center, Southern Illinois University, Carbondale, Illinois 62901, USA

Bruce Vondracek (Chapter 18): U.S. Geological Survey, Minnesota Cooperative Fish and Wildlife Research Unit, Department of Fisheries, Wildlife, and Conservation Biology, University of Minnesota, St. Paul, Minnesota 55108, USA

Gary Whelan (Chapter 1): Michigan Department of Natural Resources and Environment, P.O. Box 30446, Lansing, Michigan 55108, USA

David W. Willis (Chapter 16): Department of Wildlife and Fisheries Science, South Dakota State University, P.O. Box 2140B, Brookings, South Dakota 57007, USA

Michael K. Young (Chapter 11): U.S. Department of Agriculture, U.S. Forest Service, Forest Science Laboratory, 800 E. Beckwith Ave., Missoula, Montana 59801, USA

List of Species

Fish Taxa

Alewife	*Alosa pseudoharengus*
American eel	*Anguilla rostrata*
American shad	*Alosa sapidissima*
Arctic char	*Salvelinus alpinus*
Arctic grayling	*Thymallus arcticus*
Asian carps	Family Cyprinidae
Asian swamp eel	*Monopterus albus*
Atlantic salmon	*Salmo salar*
Bighead carp	*Hypophthalmichthys nobilis*
Bigmouth buffalo	*Ictiobus cyprinellus*
Black bass	*Micropterus* spp.
Black bullhead	*Ameiurus melas*
Black carp	*Mylopharyngodon piceus*
Black crappie	*Pomoxis nigromaculatus*
Black piranha	*Serrasalmus rhombeus*
Blue catfish	*Ictalurus furcatus*
Blue sucker	*Cycleptus elongates*
Blue tilapia	*Oreochromis aureus*
Bluegill	*Lepomis macrochirus*
Bluntnose minnow	*Pimephales notatus*
Bonneville cutthroat trout	*Oncorhynchus clarkii utah*
Brook trout	*Salvelinus fontinalis*
Brown trout	*Salmo trutta*
Buffaloes	*Ictiobus* spp.
Bull trout	*Salvelinus confluentus*
Bullheads	*Ameiurus* spp.
Bullseye snakehead	*Channa marulius*
Butterfly peacock bass	*Cichla ocellaris*
Carps	Family Cyprinidae
Carpsuckers	*Carpiodes* spp.
Catfishes	Family Ictaluridae
Channel catfish	*Ictalurus punctatus*
Chinook salmon	*Oncorhynchus tshawytscha*
Chubs	Family Cyprinidae
Chum salmon	*Oncorhynchus keta*
Cichlids	Family Cichlidae
Cisco	*Coregonus artedi*
Ciscoes	Family Salmonidae

Clear Creek gambusia..*Gambusia heterochir*
Cod...Family Gadidae
Coho salmon..*Oncorhynchus kisutch*
Colorado pikeminnow..*Ptychocheilus lucius*
Common carp..*Cyprinus carpio*
Convict cichlid...*Cichlasoma nigrofasciatum*
Coppernose bluegill...*Lepomis macrochirus purpurescens*
Crappies..*Pomoxis* spp.
Creek chub..*Semotilus atromaculatus*
Cutthroat trout..*Oncorhynchus clarkii*
Darters..Family Percidae
Delta smelt..*Hypomesus transpacificus*
Devils Hole pupfish...*Cyprinodon diabolis*
Eels..*Anguilla* spp.
Emerald shiner...*Notropis atherinoides*
Fathead minnow..*Pimephales promelas*
Flannelmouth sucker..*Catostomus latipinnis*
Flathead catfish..*Pylodictis olivaris*
Florida largemouth bass..*Micropterus salmoides floridanus*
Freshwater drum..*Aplodinotus grunniens*
Gars...Family Lepisosteidae
Gila topminnows...*Poeciliopsis occidentalis*
Gila trout..*Oncorhynchus gilae*
Gizzard shad...*Dorosoma cepedianum*
Gobies..Family Gobiidae
Golden redhorse...*Moxostoma erythrurum*
Golden shiner...*Notemigonus crysoleucas*
Golden trout..*Oncorhynchus mykiss aguabonita*
Goldfish...*Carassius auratus*
Grass carp...*Ctenopharyngodon idella*
Green sunfish..*Lepomis cyanellus*
Guadalupe bass..*Micropterus treculii*
Haddock..*Melanogrammus aeglefinus*
Hickory shad...*Alosa mediocris*
Humpback chub..*Gila cypha*
Hybrid striped bass...*Morone* spp.
Ide...*Leuciscus idus*
Inconnu...*Stenodus leucichthys*
Inland silverside...*Menidia beryllina*
Kokanee..*Oncorhynchus nerka*
Lahontan cutthroat trout..*Oncorhynchus clarkii henshawi*
Lake sturgeon..*Acipenser fulvescens*
Lake trout...*Salvelinus namaycush*
Lake whitefish..*Coregonus clupeaformis*
Largemouth bass...*Micropterus salmoides*
Largescale silver carp..*Hypophthalmichthys harmandi*

Little Colorado spinedace	*Lepidomeda vittata*
Longnose dace	*Rhinichthys cataractae*
Longnose gar	*Lepisosteus osseus*
Menhaden	*Brevoortia* spp.
Mimic shiner	*Notropis volucellus*
Minnows	Family Cyprinidae
Mosquitofishes	*Gambusia* spp.
Mottled sculpin	*Cottus bairdii*
Mountain whitefish	*Prosopium williamsoni*
Muskellunge	*Esox masquinongy*
Neosho madtom	*Noturus placidus*
Northern pike	*Esox lucius*
Northern pikeminnow	*Ptychocheilus oregonensis*
Northern snakehead	*Channa argus*
Pacific salmon	*Oncorhynchus* spp.
Paddlefish	*Polyodon spathula*
Pallid sturgeon	*Scaphirhynchus albus*
Palmetto bass	*Morone saxatilis* × *M. chrysops*
Pecos gambusia	*Gambusia nobilis*
Perches	Family Percidae
Pink salmon	*Oncorhynchus gorbuscha*
Plains topminnow	*Fundulus sciadicus*
Pumpkinseed	*Lepomis gibbosus*
Rainbow darter	*Etheostoma caeruleum*
Rainbow smelt	*Osmerus mordax*
Rainbow trout	*Oncorhynchus mykiss*
Razorback sucker	*Xyrauchen texamus*
Red shiner	*Cyprinella lutrensis*
Redear sunfish	*Lepomis microlophus*
Reticulate sculpin	*Cottus perplexus*
Riffle sculpin	*Cottus gulosus*
Rio Grande cutthroat trout	*Oncorhynchus clarkii virginalis*
River redhorse	*Moxostoma carinatum*
Robust redhorse	*Moxostoma robustum*
Rock bass	*Ambloplites rupestris*
Round goby	*Neogobius melanostomus*
Rudd	*Scardinius erythrophthalmus*
Ruffe	*Gymnocephalus cernuus*
Sailfin catfishes	*Pterygoplichthys* spp.
Sailfin molly	*Poecilia latipinna*
Salmon	Family Salmonidae
Sauger	*Sander canadensis*
Saugeye	*Sander vitreus* × *S. canadensis*
Sculpins	Family Cottidae
Sea lamprey	*Petromyzon marinus*
Shads	Family Clupeidae

Sheepshead minnow	*Cyprinodon variegatus*
Shimofuri goby	*Tridentiger bifasciatus*
Shoal bass	*Micropterus cataractae*
Shovelnose sturgeon	*Scaphirhynchus platorynchus*
Silver carp	*Hypophthalmichthys molitrix*
Silversides	*Chirostoma* spp.
Slimy sculpin	*Cottus cognatus*
Smallmouth bass	*Micropterus dolomieu*
Smoky madtom	*Noturus baileyi*
Snakeheads	Family Channidae
Sockeye salmon	*Oncorhynchus nerka*
Speckled dace	*Rhinichthys osculus*
Spotted bass	*Micropterus punctulatus*
Spotted jewelfish	*Hemichromis guttatus*
Spotted tilapia	*Tilapia mariae*
Steelhead	*Oncorhynchus mykiss*
Striped bass	*Morone saxatilis*
Sturgeons	Family Acipenseridae
Suckermouth armored catfishes	Family Loricariidae
Suckers	Family Catostomidae
Sunfishes	Family Centrarchidae, *Lepomis* spp.
Sunshine bass	*Morone chrysops* × *M. saxatilis*
Tench	*Tinca tinca*
Threadfin shad	*Dorosoma petenense*
Tiger muskellunge	*Esox lucius* × *E. masquinongy*
Tilapias	*Tilapia* spp.
Topminnows	Family Fundulidae
Trout	Family Salmonidae
Tubenose goby	*Proterorhinus marmoratus*
Tui chub	*Gila bicolor*
Tunas	*Thunnus* spp.
Utah chub	*Gila atraria*
Utah sucker	*Catostomus ardens*
Walleye	*Sander vitreus*
Western mosquitofish	*Gambusia affinis*
Westslope cutthroat trout	*Oncorhynchus clarkia lewisi*
White bass	*Morone chrysops*
White crappie	*Pomoxis annularis*
White perch	*Morone americana*
White sturgeon	*Acipenser transmontanus*
White sucker	*Catostomus commersonii*
Whitefishes	Family Salmonidae
Yellow bass	*Morone mississippiensis*
Yellow bullhead	*Ameiurus natalis*
Yellow perch	*Perca flavescens*
Yellowstone cutthroat trout	*Oncorhynchus clarkii bouvieri*

Zander..*Stizostedium lucioperca*

Nonfish Taxa

American mink..*Neovison vison*
American pondweed..*Potamogeton nodosus*
American water willow...*Justicia americana*
American white pelican..*Pelecanus erythrorhynchos*
Amphipods..Order Amphipoda
Asian clam...*Corbicula fluminea*
Bald eagles...*Haliaeetus leucocephalus*
Baldcypress..*Taxodium distichum*
Bears..Family Ursidae
Blackflies..Family Simuliidae
Bulrushes...*Scirpus* spp.
Caddisflies..Family Trichoptera
Caspian tern..*Sterna caspia*
Cattails..*Typha* spp.
Common merganser...*Mergus merganser*
Coontail..*Ceratophyllum demersum*
Cormorants (North American)..*Phalacrocorax* spp.
Cottonwood...*Populus* spp.
Crayfishes...Families Astacidae and Cambaridae
Curly-leaf pondweed..*Potamogeton crispus*
Double-crested cormorant..*Phalacrocorax auritus*
Dragonflies...Order Odonata
Duckweeds..Family Lemnaceae
Eastern swamp privet...*Forestiera acuminata*
Eurasian watermilfoil..*Myriophyllum spicatum*
Eurasian zebra mussel..*Dreissena polymorpha*
Freshwater mussels..Order Unionoida
Glaucous-winged western gull...*Larus glaucescens* × *L. occidentalis*
Great blue heron..*Ardea herodias*
Grizzly bear..*Ursus arctos horribilis*
Hydrilla..*Hydrilla verticillata*
Mayflies...Order Ephmeroptera
Midges..Family Chironomidae
Milfoils..*Myriophyllum* spp.
Millet...*Echinochola* spp.
Muskrat..*Ondatra zibethicus*
Northern crayfish...*Orconectes virilis*
Nutria..*Myocastor coypus*
Opossum shrimp..*Mysis relicta*
Oysters...Family Ostreidae
Pacific treefrog..*Pseudacris regilla*
Pig...*Sus* spp.

Common name	Scientific name
Pondweeds	*Potamogeton* spp.
Quagga mussel	*Dreissena rostriformis bugensis*
Red-eared slider	*Trachemys scripta elegans*
Red mulberry	*Morus rubra*
Reed canarygrass	*Phalaris arundinacea*
River otter	*Lontra canadensis*
Rusty crayfish	*Orconectes rusticus*
Ryegrass	*Lolium* spp.
Sago pondweed	*Potamogeton pectinatus*
Saw grass	*Cladium jamaicense*
Sea lice	*Lepeophtheirus* spp. and *Caligus* spp.
Sierra Madre mountain yellow-legged frog	*Rana muscosa*
Sierra Nevada yellow-legged frog	*Rana sierrae*
Sitka spruce	*Picea sitchensis*
Spotted owl	*Strix occidentalis*
Water hyacinth	*Eichhornia crassipes*
Waterdogs (tiger salamanders)	*Ambystoma tigrinum*
Western toad	*Bufo boreas*
White-tailed deer	*Odocoileus virginianus*
Wild celery	*Vallisneria americana*
Willows	*Salix* spp.
Winter wheat	*Triticum aestivum*
Yosemite toad	*Bufo canorus*
Zebra mussel	*Dreissena polymorpha*

Preface

The first and second editions of *Inland Fisheries Management in North America* became standard texts and references used by educators, students, and practicing fisheries managers. However, they became outdated as the scientific literature grew and philosophy and practices for fisheries management evolved. This third edition of *Inland Fisheries Management in North America* represents a full revision of the book, not a simple update of literature. This edition was developed by fisheries scientists associated with the American Fisheries Society, particularly the Education and Fisheries Management sections, with the intent of providing information on the past and emerging conceptual basis and evolving practices for managing freshwater fisheries in North America. This is a text for use in introductory university courses for juniors, seniors, or graduate students and assumes that readers will have a general foundation in ecology, limnology, ichthyology, and college-level mathematics. The book is written as an educational tool, not a reference work. Nonetheless, recent advances in fisheries management make this a valuable reference for practicing fisheries professionals.

Preparation of the third edition of *Inland Fisheries Management in North America* began in 2006 when it was recognized by educators and fisheries managers that the second edition, published in 1999, was becoming outdated. A process was initiated to identify structure and content for the third edition that would be informative and useful to educators, students, and managers. The process involved formation of a steering committee with representatives from the Fisheries Management and Education sections to assist in the identification of structure, content, and authors for specific chapters. A survey of the membership of the American Fisheries Society was conducted to identify strengths and weaknesses of the previous editions and appropriate structure and topics for inclusion in the third edition. The steering committee was composed of Joseph Larscheid, President of the Fisheries Management Section; Thomas Kwak, President of the Education Section; Fred Harris; Jeffrey Boxrucker; Jason Vokoun; and the coeditors, Wayne Hubert and Michael Quist. The editors are indebted to the steering committee for their assistance in developing the text outline and identifying authors. An online survey was conducted during January and February 2007. Most questions were posed with multiple response options that could be tabulated. Several additional open-ended questions were asked that provided qualitative data. The outcome of the survey and deliberations of the steering committee led to an outline for the third edition that differed substantially from the previous editions, all with the intent of making the third edition a pertinent text that will be applicable for many years into the future.

The structure and content of the third edition differs from the previous editions in several ways. This edition has 21 chapters whereas the second edition had 24 chapters. New and innovative chapters were added on topics such as Scale and Fishery Management; Communication Techniques for Fisheries Scientists; Use of Hatchery Fish for Conservation, Restoration, and Enhancement of Fisheries; Assessment and Management of Ecological Integrity; Ecology and Management of Lake Food Webs; and Use of Social and Economic Information in Fisheries Assessments. Most chapters were reorganized with both merging and division of chapters to create a structure that is more pertinent to current management. Because the book is written for use in North America, attempts were made to include information for the United States, Canada, and, when possible, Mexico. Additionally, most chapters begin by providing a conceptual framework focused on dominant management paradigms or important ecological concepts. This framework is then used to discuss current management theories and practices.

The array of authors contributing to the third edition is almost completely different from that of the second edition, with only five individuals contributing to both. The authors who contributed to the third edition represent the best of the current generation of educators and scientists focusing on inland fisheries management in North America.

Each chapter was drafted by the authors, reviewed by the editors, revised, and then an anonymous external peer review was conducted. The criticisms of the peer reviewers were thoroughly considered by the authors and editors in the development of final chapter manuscripts. We are sincerely appreciative to the following peer reviewers: P. Angermeier, P. Bailey, D. Beauchamp, C. Berry, P. Braaten, M. Brown, D. Chapman, M. Colvin, N. Cook, P. Dey, R. Essig, B. Finlayson, J. Fischer, K. Gelwicks, C. Guy, E. Hansen, F. Harris, T. Hartley, J. Hebdon, R. Hughes, C. Jennings, J. Kershner, C. Kruse, T. Lang, J. Larscheid, J. Lott, J. Lyons, M. Maceina, J. Margraf, P. Martinez, K. Meyer, J. Morris, J. Mulhollem, R. Neumann, N. Nibbelink, J. Nickum, R. Noble, B. Parsons, D. Periera, J. Petty, Q. Phelps, R. Schneidervin, C. Schoenebeck, R. Schultz, A. Sindt, J. Slipke, W. Smith, J. Vokoun, D. Wahl, L. Wang, M. Weullner, G. Wilde, R. Wiley, K. Wolf, M. Wolter, and D. Zafft.

Several authors acknowledged help they received as they created their respective chapters. The authors of Chapter 2 acknowledge B. Pine and C. Walters for discussions that improved the chapter. Chapter 4 authors recognize P. Hartfield for providing text for Box 4.1; W. Hubbard for providing text for Box 4.4 and review of Box 4.3; S. Cutts and J. Odenkirk for information used in Box 4.2; C. Williams for providing text for Box 4.5 and reviews of the manuscript; J. Thompson for review of the manuscript and input for Canada; and J. Williams for review of early drafts. Chapter 5 authors thank C. Milliron for his contribution to Box 5.1 and J. Newman for her contribution to Box 5.5. The authors of Chapter 6 recognize individuals inside and outside the fisheries profession who taught them communications techniques through personal contacts or their writings, including B. Bolding, A. Bonar, D. Bonar, J. Bonar, D. Burns, R. Cialdini, M. Divens, P. Ehrlich, R. Gibbons, K. Hamel, W. Hubert, P. Krausmann, P. Mongillo, W. Shaw, L. Susskind, D. Willis, E. O. Wilson, and others from the University of Washington, University of Arizona, Washington Department of Fish and Wildlife, and U.S. Geological Survey Cooperative Fish and Wildlife Research Units Program; they also thank T. Schmidt of ManagementPro.com and the Organization of Wildlife Planners for the best materials on fisheries planning. Authors of Chapter 7 thank J. Sorensen for material regarding the Gavins Point paddlefish fishery; J. Hennessy for reviewing material regarding walleye management in the ceded territory of Wisconsin; M. Sullivan and R. Jobin for providing photos of tagged walleye; D. Stein for the image of a trout regulation sign; and B. Borkholder for aid in describing harvest regulations within tribal fisheries. The authors of Chapter 12 acknowledge P. Angermeier for an excellent review of the manuscript; J. Hauxwell and B. Weigel for helpful comments on an early manuscript; and the Wisconsin Department of Natural Resources and Federal Aid in Sport Fish Restoration Project F-95-P for support of J. Lyon's contribution of Box 12.1. For Chapter 15, the authors thank S. Miehls, H. Patrick, and M. Zimmerman for helpful comments on an early draft; S. Miehls for library research and preparation of draft figures; and N. Nate for assistance with references. Chapter 16 authors gratefully acknowledge S. Flickinger and F. Bulow, who authored this chapter for the first and second editions of *Inland Fisheries Management in North America*, for their willingness to allow copying of multiple sections of their chapter into the third edition. Authors of Chapter 18 thank J. Blagen, R. Bronk, R. Dodd, L. Fujishin, J. Loomis, S. Mackenthun, N. Rudh, D. Vanderbosch, B. Vondra, and V. Were who provided the basis for much of the text and literature through a class project. Authors of Chapter 20 thank the New York Department of Environmental Conservation for permission to use the fish images in Figure 20.1; the American Institute of Biological Sciences for permission to use Figure 20.5; and the Oklahoma Department of Wildlife Conservation for use of photos in Figure 20.7. Authors employed by the U.S. Geological Survey Cooperative Research Units Program acknowledge the cooperation of their respective universities, state resource management agencies, the Wildlife Management Institute, and the U.S. Fish and Wildlife Service. Authors who are employees of the U.S. Government note that the use of trade, product, or firm names in their chapters is for descriptive purposes only and does not imply endorsement by the U.S. Government.

All figures in the book were drawn by Elizabeth Ono-Rahel to achieve consistency in appearance; she also designed the cover. We deeply appreciate her assistance throughout this project. Funding of her work was provided by the Fisheries Management and Education sections of the American Fisheries Society.

We acknowledge the assistance provided by Eva Silverfine who did the copy editing. Her detailed comments and editorial suggestions greatly improved the book.

We also appreciate the help provided by Aaron Lerner and Kurt West of the American Fisheries Society's Book Department for their management of the production of the book.

Wayne A. Hubert
Michael C. Quist

Chapter 1

Historical Perspectives on Inland Fisheries Management in North America

CHRISTINE M. MOFFITT, GARY WHELAN, AND RANDY JACKSON

1.1 IMPORTANCE OF HISTORIC ACCOUNTS

The management of inland fisheries and ecosystems is impossible without full appreciation and understanding of the actions and conditions of the past. The material in this chapter provides the reader with a history of the science of and philosophical approaches to inland fisheries management in North America. Contemporary ecosystems are a reflection of historic habitat alterations, fisheries exploitation, and management actions. Therefore, selection of different reference baselines for understanding the historic conditions of populations and ecosystems can lead to various correlations of causative factors (Humphries and Winemiller 2009). The social context of historic fisheries management is also a critical component to understanding and dealing with change. Records of fish, wildlife, landscape management agencies, and historic scientific literature provide important information and insight needed by contemporary managers and scientists. Equally important to understanding our history is recognizing the importance of recording and documenting contemporary management actions for future generations. Steedman et al. (1996) provided a comprehensive review of reasons why fisheries professionals need to understand the historic context of aquatic systems.

- Those who ignore history are doomed to repeat the worst of it.
- Historical information is frequently required during the specification of targets for ecosystem restoration.
- Historical information can be used to modify values and beliefs as they relate to habitat.
- By its very nature, information about natural ecosystem processes needs to be interpreted in a context that is long term and retrospective.
- Humans are not very good at perceiving slow processes or rare events without the help of scientists or historians.
- In culture, religion, and science, humans have often preferred to seek out and preserve stability in the natural world and to filter human perceptions through models grounded on stability or steady-state dynamics.
- The present is the "history of the future."

Early fisheries scientists and naturalists did not have the breadth of information and tools that are available today, but they understood many of the challenges of fisheries management, observed natural systems, and provided accurate documentation. The contemporary student of fisheries management may overlook important historical publications when searching the scientific literature because of the wealth of digital data and contemporary publications are often more easily available. Scientists must be encouraged to pursue historic data and images not yet available through digital resources. Fortunately, historic resources are increasingly being made more accessible via the internet through efforts by public and private entities worldwide, but methods and access have yet to be fully harmonized (Shepherd 2006; Colati et al. 2009; Seadle 2009).

1.2 FISHERIES BEFORE EUROPEAN SETTLEMENT

Native Americans exploited fisheries resources of North America prior to settlement by Europeans, and in many regions fish were central to the culture and economy of aboriginal inhabitants. North America was characterized by large regional variability in fish production, dependent upon factors such as quantity and types of waters, climate, and availability of nutrients, and the extent of use of fish populations by Native Americans varied. Rostlund (1952) synthesized descriptions from early explorers, post-settlement records of commercial catches, and early scientific estimates of fish yields to characterize the regional differences in the availability of fishes in North America. Pacific salmon provided food resources for native peoples from northwestern California to Alaska along the Pacific coast, and in many areas the annual salmon runs were a central element in native cultures. Similarly, fish represented a staple food for many tribes in the Great Lakes region, who used a variety of species and developed fishing cultures analogous to those on the Pacific coast. The fish fauna of the Atlantic coast and Mississippi basin were diverse and abundant, but their importance to Native Americans was not on the scale observed in the Pacific coast and Great Lakes regions. Historians have speculated that this was a result of a higher level of importance of hunting and agricultural activities, with fishing serving to supplement other food sources. Native American fishers used most of the gears familiar to modern fisheries workers, including nets, traps, weirs, spears, fishhooks, and even poisons, although preferred techniques and the degree of technological development varied across the continent.

Even in areas such as the Pacific coast, where fishing represented a primary source of food acquisition, little evidence exists that Native American fisheries exceeded sustainable levels. The apparently limited impact that Native American fisheries had on fish populations is often attributed to low human population densities and inefficient fishing and food storage techniques. More recent evidence suggests that Natives Americans in some areas were technologically capable of overexploiting fish populations, but complex social and cultural traditions tempered harvests and maintained sustainability (Taylor 1999). Native American belief systems typically involved more spiritual connections to nature, particularly with resources important for subsistence. In Native American cultures with strong ties to fisheries resources, fish played a central role in myths and seasonal ceremonies, and traditions developed that acted to prevent overfishing (Taylor 1999; Bogue 2000). European immigrants encountered fisheries resources in an essentially unexploited state but brought a value system that necessitated a more formal system of fisheries management.

Prior to large-scale immigration of Europeans to North America, European travelers to the New World sent back seemingly fantastical tales of the richness of the natural resources. Fisheries resources were no exception, and reports suggested a limitless supply of food fishes. Salmon runs on both coasts were so large they sometimes inspired complaints. Nicolas Denys, French governor of Acadia, said this of the Atlantic salmon in the Miramichi River in 1672 (quoted in Montgomery 2003): "If the [passenger] pigeons plagued us by their abundance, the salmon gave us even more trouble. So large a quantity of them enters into this river that at night one is unable to sleep, so great is the noise they make in falling upon the water after having thrown themselves into the air."

Salmon runs on the Pacific coast were no less spectacular. Ezra Meeker, an early pioneer to the Washington territory, described the abundance of salmon in the Puyallup River in the 1850s (quoted in Montgomery 2003): "I have seen the salmon so numerous on the shoal water of the channel as to literally touch each other. It was utterly impossible to wade across without touching the fish. At certain seasons I have sent my team, accompanied by two men with pitchforks, to load up from the riffle for fertilizing hop fields."

Reports from interior waters suggested that fish of incredible diversity and abundance could be found throughout North America. In the late 1800s, reports came from North Dakota lakes of net hauls "so large that four horses on each side were required to land the catch" (Eastgate 1918). Vanderkemp (1880) described the fish resources of New York's Oneida Lake in 1792 in this way: "Never did I see yet a country, where all kind of fish was so abundant and good: It may be equalled [sic], it cannot be excelled.…It is enough to set out a few lines at evening, to make now and then an excursion to the woods, without sacrificing much of his time, that a settler may supply his family with meat and fish during five or six months." Reports from Hernando de Soto's travels through the lower Mississippi River valley in the early 1500s included accounts of the easy access to fish enjoyed by the Native American, who "everyday…brought fish until they come to be in such plenty the town was covered in them" (quoted in Pearson 1972). Samuel Williams, writing from Vermont in 1794, captured the overall tone of European explorers and settlers across much of North America (quoted in Pearson 1972): "In the production of fish, nature seems to have been extremely prolific, in every part of America. Their species, their multiplying power, and the ages at which they become prolific, are beyond our knowledge and computation. The brooks, rivers, ponds and lakes are everwhere [sic] stored with them. The sea coasts are one continued range of fishing banks, covered with cod, haddock, and other animals of the ocean."

Many of the reports from the New World came at a time when Europe was experiencing rapid population growth. The early stages of the industrial revolution were resulting in increased urbanization, and the demand on natural resources in Europe was beginning to take its toll, resulting in resource depletion near large population centers (Goudie 2005). News of a vast continent of seemingly unlimited natural riches provided a strong attraction. In fact, Fagen (2006) argues that demand for fish fueled by both a growing European population and the spread of Christianity (and its popularity of fish on abstinence days, which accounted for more than half the calendar year in the 13th century) brought largely secretive commercial fisheries to the shores of North America long before John Cabot and the Pilgrims, particularly those operated by Basque fishermen. European immigrants would bring a philosophy that natural resources were a fuel for economic development and religious fulfillment, and this view would ultimately change the nature of the continent's aquatic resources (e.g., Worster 1992).

1.3 EARLY ICHTHYOLOGICAL SURVEYS IN NORTH AMERICA

The North American freshwater fish fauna comprises about 1,060 species, 50 families, and 200 genera (Burr and Mayden 1992). The North American fauna encompasses wide phylogenetic diversity, including representatives from ancient pre-teleostean lineages as well as more modern teleosts; 128 genera are found only in North America. Most of the adaptive radiations contributing to the diversity of freshwater fishes in North America were centered in the Mississippi River basin, which includes the majority of the continent's species. The native fish fauna on the Pacific coast contained only about a quarter of the number of species found in the eastern part of the continent, but a high proportion of species were endemic to the region (Briggs 1986). North American freshwater fishes are among the best understood faunas in the world. The zoogeography and distribution of freshwater fishes in North America are well described in three monumental works: *Atlas of North American Freshwater Fishes* (Lee et al. 1980); *The Zoogeography of North American Freshwater Fishes* (Hocutt and Wiley 1986); and *Systematics, Historical Ecology and North American Freshwater Fishes* (Mayden 1992).

While only a small fraction of North America's fish species would figure in the development of inland commercial and sport fisheries, early scientific activities on the continent focused on describing and cataloging the continent's fauna. Prior to 1800, North American fishes were little studied, and those species that had been described appeared primarily in publications emanating from European museums (Dymond 1964; Myers 1964). The first significant work of American ichthyology published by an American came from New York, where Samuel Latham Mitchell's studies culminated in his 1814 paper on the fishes of New York (Mitchell 1814). The formation of the Academy of Natural Sciences of Philadelphia in 1812 began to attract workers to North America intent on cataloging the continent's wealth of natural resources. Constantine Samuel Rafinesque was one of the more eccentric scientists drawn to the New World. His work included a survey of the fishes of the Ohio River, *Ichthyologia Ohiensis*, which was published in 1820. The monograph included descriptions of fish species drawn from completely fictitious drawings provided to him by John James Audubon as repayment for Rafinesque having destroyed his violin in an effort to collect bats while a guest in Audubon's cabin (Rafinesque 1820; Myers 1964). Charles Alexandre LeSueur arrived in North America about the same time as Rafinesque and concentrated his efforts on Atlantic coast fishes.

The expansion of ichthyological surveys into the western waters of the USA was accomplished in large part through the activities of Spencer Fullerton Baird at the Smithsonian Institution, which was founded in 1846. Baird, who would later become the first commissioner of the U.S. Commission of Fish and Fisheries, served as the Smithsonian's first Assistant Secretary. Baird's efforts at the Smithsonian reflected his desire that the institution become home to a comprehensive collection of natural history specimens. Among Baird's first actions as Assistant Secretary were efforts to attach naturalists to Army exploration crews surveying the boundary with Mexico and various routes for railroads to the Pacific coast between 1848 and 1855 (Goetzmann 1959). The resulting flood of fish specimens returning to the Smithsonian from the various surveys was assigned to Charles Frederic Girard, a former assistant of Louis Agassiz at Harvard University. Agassiz was perhaps the most respected naturalist in North America at the time, with ambitions of writing the definitive work on North American fishes.

Agassiz apparently resented both Girard's departure for the Smithsonian and his access to specimens that Agassiz wanted for his newly formed Museum of Comparative Zoology, and Agassiz openly criticized Girard's publications on new species of western fishes (Jackson and Kimler 1999). Despite criticism by Agassiz, however, Girard's analysis of fish specimens sent to the Smithsonian resulted in the discovery of 146 new species, ranking him second among all ichthyologists working on North American fishes for number of valid species described. Girard's work during the 1850s accounted in large part for the first great pulse of discovery of North American fish species and provided important early documentation of the fauna of the West prior to large-scale European settlement.

Another worker with early attachments to Agassiz and Baird was David Starr Jordan, whose career spanned much of the last half of the 1800s and into the early 20th century. Jordan's accomplishments earned him the title, "father of North American ichthyology" (Hubbs 1964), and his work included the initial descriptions of over 200 fish species currently recognized in North America. He also trained many of the leading ichthyologists that followed him, including Carl Leavitt Hubbs; many contemporary ichthyologists can trace their academic lineages to Jordan. Jordan's monumental work with Barton Warren Evermann in 1896, *The Fishes of North and Middle America*, provided a comprehensive and accurate catalog of North American fishes. By the early 1900s, the golden era of ichthyological discovery had ended. New fish species would continue to be discovered as more intensive local surveys were undertaken, but the overall pattern of fish species distributions had been described. While ichthyological research would continue, a new area of fish-related research was developing with efforts to identify causes and remedies for impacts on fish populations resulting from human activities.

Descriptive surveys of fishes, habitats, and pollution were conducted by many state commissions during the 1920s and 1930s (e.g., Belding et al. 1924), mostly with the goal of better understanding how and where to stock fish. Particularly noteworthy was a survey led by the first woman biologist working for the New York State Department of Conservation, Emmeline Moore. She first studied the fish and limnology of Lake George and then went on to conduct surveys throughout state waters. These biological surveys of the surface waters and reports on the watershed characteristics were landmarks in management for their logical and careful sequence of study (Moore 1926, 1927). Moore became the first woman president of the American Fisheries Society in 1927 (Moffitt 2001) and also contributed to understanding fish health (Moore 1923; Michell 2001; Figure 1.1).

1.4 CHANGING CONCEPTS OF MANAGEMENT

Establishing authority for managing inland fisheries was difficult in the early years of both the USA and Canada. During colonial times there was confusion regarding ownership and access to inland aquatic resources. Aristocracy and nobility held property rights, and there was considerable disregard for Native American tribes and their claims to natural resources. At the conclusion of the American Revolution, and establishment of a sovereign representative democracy in the USA, many terrestrial, riparian, and freshwater resources were opened to public use under the public trust doctrine (see Chapter 4, this volume). Laws governing fisheries were put in place (particularly concerning fish passage at dams), but the management paradigm and infrastructures for these laws were neither understood nor clearly defined.

Figure 1.1. Emmeline Moore led many efforts to raise awareness of the importance of science-based management. She was elected the first woman president of the American Fisheries Society in 1927. She is shown at her desk in 1963 (photo courtesy of New York State Department of Environmental Conservation, from the *New York State Conservationist*).

English law regarding ownership of wildlife and fisheries provided the basis for these laws, but the laws regarding fisheries were confusing (Goble 1999, 2005). Many laws in the eastern USA defined riparian zones to fall under private ownership. The failure to manage fisheries resources and access to aquatic systems properly was likely due, in part, to the small European population, which was distracted by the many opportunities for commerce and had a general lack of biological understanding of the vast territory and its resources.

In Canada, ties to European-based systems of nobility and far-away governments made it difficult to initiate fisheries management. Each province had laws regulating fisheries that were held in common. Most fishing regulations targeted anadromous fishes, marine fishes, or estuarine finfish and shellfish resources or fish passage at dams. Canadian citizens recognized that habitat degradation and overfishing affected inland fisheries, but governments were ineffective in stopping exploitation. For example, the Province of New Brunswick authorized strict Atlantic salmon laws in 1845, and local associations of fishers were established to manage harvests; however, there was little compliance with the laws (Johnstone 1977).

Central control of Canadian fisheries began after the Dominion of Canada was established with the provinces of New Brunswick, Ontario, Quebec and Nova Scotia. In 1868, the Canadian Confederation passed the federal Fisheries Act to create a Department of Marine and Fisheries (Box 1.1). Canada's first Minister of Marine and Fisheries was Peter Mitchell, and the Fisheries Act mandated the appointment of federal fisheries officers, creation of federal fishing licenses, closed seasons for some species, and passage for fishes named in the act. These provisions included free passage of fish on Sundays and prohibition of Sunday fishing. The Fisheries Act prohibited pollution in waters frequented by fishes, allowed for creation of fish sanctuaries or fish reserves, and included controls on oyster and shellfish fisheries. The act continues as the policy directive for fisheries in Canada.

> **Box 1.1. Historical Names of Canada's Federal Fisheries Agencies**
>
> The names and missions of federal fisheries agencies in Canada changed over time as a result of governmental reorganizations.
>
> - 1867–1884   Department of Marine and Fisheries
> - 1884–1892   Department of Fisheries
> - 1892–1914   Department of Marine and Fisheries
> - 1914–1920   Department of Naval Services
> - 1920–1930   Department of Marine and Fisheries
> - 1930–1969   Department of Fisheries
> - 1930–1936   Department of Marine[1]
> - 1969–1971   Department of Fisheries and Forestry
> - 1971–1976   Department of the Environment
> - 1976–1979   Department of Fisheries and the Environment
> - 1979–present   Department of Fisheries and Oceans
>
> [1]The Department of Marine was merged with the Civil Aviation Branch of the Department of National Defense in 1936 to form the Department of Transport.

A federal mandate for fisheries management in the USA began in 1871 when the U. S. Congress authorized the U.S. Commission on Fish and Fisheries (Fish Commission) in response to the decline in fisheries. Similarly, many states established fish commissions for the same reasons at about that time. The first U.S. Commissioner of Fisheries, Spencer F. Baird, was discussed previously (section 1.3) for his contributions to ichthyological surveys. Baird was based in Woods Hole Massachusetts, and the Fish Commission and its succeeding authorities (Box 1.2) established laboratories with field operations at several locations on both coasts, on the Great Lakes, and in the interior of the USA. The primary mission of the Fish Commission was to determine the reasons for declines of fisheries in New England and the Great Lakes and to develop methods for fish culture. The early legislation also provided the Fish Commission with the right to collect specimens from all states and territories and to enlarge the collections of the Smithsonian Institution.

1.4.1 Early Management

Well before federal mandates for fisheries management and fish culture were established, lay people were interested in fish culture as a way to enhance fish production. Entrepreneurial efforts in fish culture included those of Seth Green, who established a fish hatchery in Caledonia, New York, in 1870 (Bowen 1970). Equally enthusiastic about fish culture, residents of Canada developed techniques for fish culture. In 1868 Samuel Wilmot built a fish hatchery on his farm near Newcastle, Ontario, and in 1876 he became the Superintendent of Fish Breeding for the federal government in Ottawa. Wilmot subsequently established hatcheries in Quebec, Ontario, and the maritime provinces of Canada, and a division for hatcheries was retained following his tenure (Huntsman 1938).

> **Box 1.2. Historical Names of Federal Agencies Responsible for Fisheries Management in the USA**
>
> The names and missions of federal fisheries agencies in the USA changed over time as a result of governmental reorganizations.
>
> - 1871 U.S. Commission on Fish and Fisheries (independent)
> - 1879 Division of Economic Ornithology and Mammalogy (U.S. Department of Agriculture)
> - 1885 Bureau of Biological Survey (U.S. Department of Agriculture)
> - 1902 Bureau of Fisheries (U.S. Department of Commerce)
> - 1939–1940 U.S. Fish and Wildlife Service (U.S. Department of the Interior)—consolidated the Bureau of Biological Survey and Bureau of Fisheries into one agency
> - 1956 U.S. Fish and Wildlife Service contained Bureau of Sport Fisheries and Bureau of Commercial Fisheries
> - 1970 Bureau of Commercial Fisheries moved to National Oceanic and Atmospheric Administration, National Marine Fisheries Service

Edward E. Prince was the second Commissioner of Fisheries in Canada and provided strong leadership for providing scientific information for fisheries. Prince (1923) supported the development of laboratories and field explorations. The Canadian government conducted a complete survey of fisheries of boundary waters from the Bay of Fundy to the Puget Sound and in 1899 established the first marine biology laboratory at St. Andrews, New Brunswick. Soon thereafter, the Canadian government started the Pacific Biological Station in Nanaimo, British Columbia (Johnstone 1977). Additionally, the Canadian Fisheries and Biological Board began the Georgian Bay Station on Lake Huron and continued research in fish culture techniques (Clemens 1932).

Fish culture was also a priority for the new U.S. government fisheries agency. Among the congressional trade-offs used to gain approval for the U.S. Fish Commission was an agreement to stock American shad in the Mississippi River drainage (Moffitt 2001). The Fish Commission succeeded as an institution, but American shad did not become established in the Mississippi River drainage. The Fish Commission also built the USS *Fish Hawk* to serve as a floating hatchery and distribution system, capturing fish such as American shad and distributing their spawn (Figure 1.2). Fish culture operations were robust, and millions of small fish were released into the waters each year. The research and survey work used several vessels, including the U.S. Bureau of Fisheries' ship, the *Albatross*, that sailed on exploration cruises and collecting trips along both coasts and into the tropics.

Because of the value of fisheries as a commodity, politics have been intertwined in the development of fisheries management. The American Fish Culturists Association, a group of citizens and professionals interested in fish culture, began in 1870 and discussed fish culture, resource management, and politics. One of the association's first actions was to write letters

Figure 1.2. The USS *Fish Hawk* was used for distributing eggs of many fish species in an effort to replenish stocks affected by fishing. Photo taken in 1896 by archival photographer Stefan Claesson as part of the Gulf of Maine Cod Project, National Oceanic and Atmospheric Administration, National Marine Sanctuaries (photo courtesy of National Archives).

to the U.S. Congress to urge elevation of federal funding for fisheries because of the clear declines in U.S. fisheries. By 1878, the American Fish Cultural Association provided exhibits at international fisheries exhibitions. In 1885, the American Fish Cultural Association adopted the name of the American Fisheries Society (AFS) with a mission to "promote the cause of fish culture, gather and diffuse information bearing upon its practical successes, and upon all matters relating to the fisheries; the uniting and encouraging of the interests of fish culture and the fisheries, and treatment of all questions regarding fish of scientific and economic character" (Bower 1911).

Records of fisheries activities were published in the *Transactions of the American Fisheries Society* and *Bulletin of the U.S. Fish Commission*. Many states also formed commissions or surveys and published bulletins or special publications. Outreach activities for the general public were a part of meetings of fisheries scientists, and the political leaders at national, state, and provincial levels were enthusiastic about the potential for fisheries development. As early as 1873, *Forest and Stream* magazine published the papers and much of the proceedings of AFS meetings (still the American Fish Cultural Association) before the *Transactions of the American Fisheries Society* appeared in print. This magazine was officially connected to AFS until the magazine moved to New York City in 1875 and eventually changed its name to *Field and Stream* (Moffitt 2001). In addition, the American Association for the Advancement of Science and *The New York Times* published reports of the AFS and other fisheries organizations.

Use of cultured fishes and other aquatic organisms was seen as a way to rehabilitate degraded fisheries and re-establish extirpated fisheries to reverse the results of overharvest of fish stocks and destruction of stream habitats from timber removal, mining, industrial development, and introduced species (Whitaker 1892; Farley 1957; Beeton 2002). The Fish Commission published a pivotal book in 1897, *A Manual of Fish-Culture* (U.S. Commission on Fish and Fisheries 1897). This was a compendium of methods to culture more than 30 species or groups of fish, shellfish, and frogs, with descriptions of the general biology of each species, details on incubation of embryos, and advice on choice of food for rearing. The compendium also included information on methods for transportation and described the operation of 25 stations or hatcheries established across the nation for the purpose of propagation and stocking of fishes. Livingston Stone, the first secretary of the AFS, was a major pioneer and lobbyist for fish culture and was Deputy Commissioner of the Fish Commission. He was sent to California to develop the first federal hatchery, which began operations in 1872 on the McCloud River. Development of the hatchery was assisted by McCloud Wintu tribal members (Yoshiyama and Fisher 2001). Chinook salmon eggs or fry from this site were sent to at least 37 states and 14 countries, including destinations as far away as Italy, Japan, Australia, and New Zealand. During the late 1800s, fish harvest and stocking statistics for inland waters were reported in the *Bulletin of the U.S. Fish Commission* and as frequent notes in *Transactions of the American Fisheries Society*. These volumes included combinations of reports from stations and surveys and information on international fisheries (Ito 1886; Weber 1886).

The newly developed railroads provided access to waters for stocking fish. Special railroad cars were developed for transport of fish and eggs (Figure 1.3). In a report for the Fish Commission, McDonald (1896) stated, "Of the above transportation, 26,212 mile were furnished by the railroads gratuitously, and 48,593 mile were paid for at the rate of 20 cents per mile. The Commission is indebted to the personnel and management of the railroads for much courtesy, consideration, and dispatch."

Figure 1.3. The "fish car," a specialized railroad car, was used by the U.S. Department of the Interior (USDI) and also by many states for transporting fishes for stocking in waters across the USA (photo courtesy of USDI).

Development of canning and freezing of fisheries products increased the demand for harvest and subsequently products from Canada and the USA were shipped throughout the world (McArthur 1947; Clark 1985). Specialized fisheries resources developed in different geographic regions. For example, the inland fisheries of the Mississippi River started as subsistence operations, but by the late 1870s the fisheries had developed into organized commercial operations. One of the earliest commercial industries on the Mississippi River was the catfish and buffalo fishery. In 1894 the commercial catch was more than 1.4 million kg (3.75 million pounds) of catfish and more than 2.7 million kg (7.24 million pounds) of buffalo from the upper Mississippi River (Carlander 1954).

Harvests from inland waters were not limited to fishes. Widespread harvest of freshwater mussels from rivers for the button industry, for food, and later for freshwater pearls decimated populations and altered population dynamics of many aquatic species dependent on the mussels (Anthony and Downing 2001). Freshwater mussels had been important protein sources for native peoples living along these rivers. By as early as 1860, mussel harvests were extensive in Arkansas, Florida, Iowa, Kentucky, Michigan, Nebraska, New Jersey, Ohio, Tennessee, Texas, Vermont, Washington, and Wisconsin (Kunz 1893; Kunz 1898; Claassen 1994).

The U.S. Bureau of Fisheries (see Box 1.2) established the Fairport Biological Station in Iowa (Coker 1914) with the mission of learning how to culture freshwater mussels. Between 1914 and 1919 many of the upper Mississippi River and some Great Lakes states adopted harvest regulations, but regulations were too late to save the mussel fisheries and the industry (Figure 1.4). Curiously, interest in freshwater mussels has increased in recent times due to the threatened and endangered status of many species, concerns regarding host fish species that support the glochidial stage of mussels, use of mussel shells as seed material for cultured pearls, and their utility as biological indicators (Neves et al. 1985; Williams et al. 1993).

Figure 1.4. The Fairport Fisheries Station was developed to propagate freshwater mollusks to replace stocks depleted by the button industry and by harvest for food. Photo of station, southwest portion of grounds and principal buildings in 1914 (photo courtesy of USDI).

As the areas along the Mississippi and Missouri rivers developed commercially, "fish rescue" became a management activity. These efforts involved the salvage of sport and commercial species from isolated flood pools after spring floods and the placement of these fishes back into rivers or their transport to other locations. Before completion of extensive flood control efforts, large numbers of fish would be "stranded" each summer in flooded backwater areas of floodplain rivers. Iowa led this program, and other state and federal agencies followed during the 1880s (Box 1.3). Fish rescue activities resulted in introductions of fishes into waters outside their native distributions, but philosophically managers considered sport fish to be important resources to propagate for harvest. The economic advantage of allowing fish to propagate naturally and then moving the juveniles to other locations was highly attractive, and in the 1920s the U.S. federal government adopted fish rescue programs enthusiastically. Near the Fairport Biological Station in Iowa, workers engaged in both fish rescue and mussel propagation. Before rescued fish were released they were exposed to mussel glochidia. By the 1920s, fisheries stations throughout the Mississippi River basin released millions of fish, to which were attached billions of glochidia of commercially-valuable mussel species (Anfinson 2003).

These efforts were designed to sustain commercial mussel fisheries. Unfortunately, overharvest of both fish and mussel populations, municipal and industrial pollution, and siltation from farming and timber practices all contributed to the decline of these resources. Mussel propagation efforts ended by the early 1930s, and at the same time the great era of dam building began to alter fish and mussel habitat across North America. Over the next 40 years dams impounded thousands of river kilometers, fragmented fish and mussel populations, and changed habitat needed by both fishes and mussels.

1.4.2 Regulations and Fish Stocking

The first management actions taken by states and provinces were to enact regulations, mostly regarding access, and to form commissions for management oversight and fish production in hatcheries. The general philosophical approach for management at the time was that fisheries resources could be sustained if they were regulated through harvest or access control and, furthermore, fisheries resources could be enhanced or recovered through stocking of cultured fishes (Bowen 1970). Fish were valued first as food and direct service to humans and secondarily as recreational resources (Viosca 1945). Much of this philosophy was built on agrarian principles by which crops are grown and harvested for human use. The crop could easily be fish, corn, or cattle. With this philosophy, the idea that some species were superior to others was fostered; hence the importation and culture of selected species of commercial or sport fishes and the movement of fishes familiar to European settlers to new locations being settled.

With the development of the concept of conservation of special land resources, the U.S. national park system began with Yellowstone National Park in 1872. For several years Yellowstone was the only park under federal management, and fishing and hunting were allowed because fish and wildlife provided both food and recreation. A fish-stocking program was established for "fishless waters" of the park, and rainbow trout, brown trout, brook trout, and lake trout were stocked. At the same time, eggs from native cutthroat trout from within the park were collected and shipped to many locations across North America. With the expansion of the national parks, similar programs of stocking were begun. The federal management authority for Yellowstone National Park and national monuments rested with the U.S. Army until the National Park Service was established in 1916.

Box 1.3. Locations of Fish Rescue Operations in the Mississippi River System

Fish rescue involved the salvage of sport and commercial species from isolated flood pools after spring floods and the placement of these fishes back into rivers or their transport to other locations. Below are locations and dates of fish operations within the Mississippi River system based on data from Carlander (1954).

State and station name	Dates of operation
Iowa	
Bellevue	1903–1938
Fairport	1917–1938
Gordon's Ferry	1922
Gutterberg	1921–1923, 1939
Montpelier	1923
North McGregor (renamed Marquette)	1904–1939
Illinois	
Andalusia	1928–1930
Cairo	1919–1922
Galena	1917
Lake Cooper	1917
Meredosia	1894–1904, 1918–1922
New Boston	1918
Quincy	1889–1921
Rock Island	1922–1928
Louisiana and Mississippi	
Various	1917–1930
Minnesota	
Brownsville	1921–22
Dakota	1922
Hastings	1924
Homer	1911–1938
Lake City	1917
Latsch Estate	1921–1922
Minneiska	1917, 1922
Minnesota City	1921–1922
Red Wing	1918
Richmond	1917
Winona	1917, 1922
Missouri	
Candon	1919
Clarksville	1919–1920
Hannibal	1920

(Box continues)

> **Box 1.3. Continued.**
>
> | State and station name | Dates of operation |
> |---|---|
> | Wisconsin | |
> | Ferryville | 1921–1923 |
> | Fountain City | 1917–1921 |
> | La Crosse | 1904–1938 |
> | Lake Pepin | 1917–1918 |
> | Lynxville | 1917–1918 |
> | Genoa | 1917, 1922–1923, 1931, 1938 |
> | Prescott | 1921–1922 |
> | Trempealeau | 1917 |

In Canada, conflicts regarding authority between national and provincial governments led to the Imperial Fisheries Judgment of 1898. This judgment ruled that jurisdiction and making of laws regarding fisheries were vested in the federal government, and the federal government was the ultimate agent for conservation. The governmental policy was to manage resources to allow maximum production for use and guarantee the supply in perpetuity. Public harbors and fisheries in these waters were also vested in federal control. The judgment declared that property rights, leases, and licenses were vested in the provinces, but the federal government had the right to impose a tax on fisheries licensed by provinces (Young 1952).

Establishment of national parks occurred in Canada with Banff National Park in 1887 via the Rocky Mountain Park Act. As was the case in the USA, from the early 1900s to 1980 fishless lakes in Canadian parks throughout the Rocky Mountains were stocked with fish. Fishes stocked included cutthroat trout, rainbow trout, and brook trout (Donald 1987), and stocking peaked in the 1960s (Solman et al. 1952). The value of fish stocking in the USA and Canada began to be questioned in the late 1800s (Bahls 1992), as in many cases few fish survived and the transfer of nonendemic species to new areas caused alterations of the aquatic systems.

Livingston Stone called for the formation of a National Salmon Park in 1889 as he recognized the importance of preserving a natural environment for native fish populations to reproduce. As a result of Stone's work, Afognak Island, Alaska, was set aside in 1892 as a forest and fish cultural reserve. Stone (1892) wrote "artificial breeding can do a great deal, and has done a great deal, but it cannot be relied upon for a certainty."

1.4.3 Public Trust Doctrine

Conflicts between private and public interests in fisheries have been dynamic and continue to affect fisheries management. Early fisheries management practices were to limit access during certain times of the year and protect breeding areas, but controversy existed regarding these practices given a debate over the benefits of fish culture versus wild spawning of fishes. Dickenson (1898), Commissioner of Fisheries for the state of Michigan, wrote the following regarding the public trust doctrine and conflicts in the Great Lakes between commercial interests and public rights: "While the catching and marketing of commercial fish should for the most part be left to private enterprises…The public should be empowered to say, through its authorized agents, whether the title to public property should pass."

Early fisheries managers clearly recognized the tremendous effects from lack of regulation of pollution, dam building, and gear types used for fishing, and many doubted the role that hatchery production could play in rehabilitation. Many of these individuals understood the lack of political will to intervene in the degradation of aquatic habitat effectively and realized that their only tool was fish stocking, even with its limitations. Spangler (1893) described the decimation of fisheries and fish habitats: "What adds to the incomprehensibility is the fact that within the memory of many now living, those streams, lakes, and coasts, almost without exception, teemed with food-fishes. Some of them are still prolific in that respect, but it is a deplorable truth that a very large proportion of them—those inland especially—have been either almost entirely depleted, or their productiveness so diminished as to practically amount to depopulation." He commented on the success of hatchery stocking with some disappointment: "These well-meant endeavors to arrest further diminution have, unfortunately, been only partially successful. This failure has been largely disappointing, for great results were expected from the carefully framed and very stringent statutes, as well as from the distribution of millions of young fish annually from the state hatcheries and from the national hatcheries under the control of the U. S. Fish Commission." Finally, Spangler recommended directives to reclaim control of these waters and increase protection through public education of lawmakers (Box 1.4).

Titcomb (1917) wrote about the need for a permanent stocking policy (Box 1.5) and the conflicting options for increasing yields of fish for consumption. Some of his comments provide an interesting reflection of the role of self-interest and the conflicts of common property versus individual resources.

> Just at this crisis in our history the fishery resources of the country are receiving especial attention and many impractical recommendations have been made by well intentioned persons who are not familiar with the subject, as well as by some who, from selfish motives, want to let down the bars to conservation and to disregard the laws of nature which are the basis for regulations in regard to the methods and seasons for taking fish.
>
> College professors have come forward with recommendations to loosen up on the laws for the protection of fish during the present war. Incidentally I may say that similar appeals are being made to those in authority with reference to the taking of all kinds of game. (Titcomb 1917)

The debate on how far the interaction between the public trust doctrine should encroach on the private rights is something that continues to this day. In 1933 Gordon wrote: "How far should we go in acquiring fishing rights by lease and purchase as Connecticut has done? How far can we go in the improvement of fishing on privately-owned waters? And how far should we go toward encouraging private initiative in the production of fishing for the angler? These are all matters upon which we can agree if we try, but we can't do it without an acceptable policy."

Even within federal authority, there was confusion about who had the responsibility to manage within national forests and where the regulative authority resided. Davis (1935) reported: "The Bureau of Fisheries assumes responsibility for conducting research necessary

Box 1.4. Spangler's Policy Directives regarding Fisheries Management in Public Waters

The following eight policy directives were aimed at reclaiming authority for management over public waters, increasing protection of public waters through education of lawmakers, and defining responsibilities (Spangler 1893).

1. The inculcation, to the extent of a full comprehension, of the truth that the fish in the public waters of a State are the property of that State, and the taking of them, by any means, a privilege.

2. That the guardianship of such waters and their finny inhabitants is the sworn duty of the people's representatives, just as is the guardianship of any other kind of public property.

3. That the laws enacted in order to make that guardianship effective are binding upon and demand implicit obedience from all.

4. That it is the sworn duty of sheriffs, magistrates, constables, and fish wardens, as far as they have cognizance and jurisdiction, to arrest or cause to be arrested and tried, and without fear or favor, any and all offenders against the restrictive statutes.

5. That it is a patriotic obligation resting upon all citizens to aid the authorities in their endeavors to restore the original fecundity of American waters, for the reason that such restoration would benefit the country annually to the extent of millions of dollars.

6. That it is the duty of the people's representatives in Congress to enact laws that will place the menhaden and other coast fisheries under such restrictions as will prevent the edible fishes from being so largely and wastefully diminished in numbers as they have been for years past, and still are.

7. That artificial propagation, judicious distribution, and the thereafter protection of edible fishes should be prosecuted to the fullest needed extent by every State and Territory.

8. That fish-protective associations, being potent helpers in the work of restoring edible fish fruitfulness to our waters, should be warmly encouraged in every State, and the powerful aid of the newspaper press of the entire country evoked in its behalf.

Box 1.5. Recommendations for Fish Stocking Policy

Titicomb (1917) saw the need for a permanent stocking policy and recommended the following.

1. List the waters under your jurisdiction and establish a permanent policy as to the species with which you will specialize in each.

2. Prohibit the introduction of any species of fish foreign to the waters, unless approved by the commission and also the introduction of any species contrary to the established policy; this to include connecting privately stocked and controlled waters.

3. Co-operate with the United States Bureau of Fisheries in the adherence to a permanent policy as to the selection of species for restocking waters in which both authorities are interested.

4. Give the commissioners power to exterminate and market rough fish at any time and by any means, either directly or by the issuance of licenses. By rough fish is meant those kinds which are antagonistic to the maintenance of a successful permanent policy already decided upon. It is immaterial for statistical results whether fish are taken by nets or by hook and line. How fish are taken should be regulated according to local conditions and effects upon property values. In this connection the value of recreation, as an asset in its effect upon property values, must not be under rated.

5. The selection of species for planting in any water system should be determined with reference to a permanent policy, in the hope that our successor will continue the policy which we have established.

6. A campaign of education as to the importance of care in planting should be waged in order to save wastage.

7. Educate the public also to appreciate private property rights in fish when privately propagated entirely under the control of the owner.

8. In the present crisis, conservation of our resources is of greater importance than at any previous time in our history, and proper regulations for the protection of fishes is as important as fish propagation.

9. The farming of our waters should be with a view to maximum annual production of the kind of fish crop that, after careful investigation, it is decided to specialize in.

10. The retention in office of men conversant with the propagation and protection of fishes is essential to insure the best results.

(Box continues)

> **Box 1.5. Continued.**
>
> 11. Finally, build for the future generations regardless of present political conditions. Leave a monument for yourselves by setting an example for your successors whether of the same political faith or not. Set a pace for them and turn over to them such complete records of your work that there can be no excuse for not following the permanent policy which you have established.

to lay the foundations for fishery management throughout the national forests and will also provide the fish required for stocking forest waters. On the other hand the Forest Service assumes responsibility for the administration and operation of management plans and will also undertake stream and lake improvement and all stocking work under instructions and recommendation provided by the Bureau of Fisheries." This agreement ceded public trust doctrine authority for regulating the fishery in the forests to the respective states.

1.4.4 A Critical Turning Point in Defining Management

Between 1910 and 1970 many dams were constructed on major rivers of North America for power, flood control, transportation, and irrigation with little consideration to the environmental effects of this development. Federal power development was instrumental in providing efficient and inexpensive power for rural areas (Figure 1.5). The combination of river development and overfishing was particularly pronounced in the Columbia River, and attitudes regarding fisheries and other uses for the river were mixed. As early as the 1870s, the U.S. Army Corps of Engineers altered the Columbia River to make it more navigable. The development of a large industry of salmon canneries reached peak harvest in the 1880s and again during World War I. Chapman (1986) estimated the likely catches at the peak of the fishery (mostly 1880s) included 1.7 million summer Chinook salmon; 382,000 steelhead; 1.1 million fall Chinook salmon; 400,000 spring Chinook salmon; 476,000 coho salmon; 1.9 million sockeye salmon; and 359,000 chum salmon. During this time, a general lack of regulation of the fisheries resulted in overfishing of stocks, as gas engines and refrigeration made vast areas of the world's ocean possible for fisheries exploitation. More federal assistance for fisheries was requested from both fishers and states.

In 1930, the Mitchell Act in the USA provided funding for management improvement, fisheries engineering, and fish culture in the Columbia River basin. This legislation approved construction of more than 25 fish culture stations, three new laboratories, and two fish distribution railroad cars over the next 5 years. The results from these fish culture operations were mixed, and fisheries recovery was complicated by increased habitat alterations associated with river development, particularly water withdrawal for agriculture and urban areas and new dams for hydropower, irrigation, transportation, and flood control. For many years, the development of the river for human use rather than natural fish production was favored (Committee on Protection and Management of Pacific Northwest Anadromous Salmonids 1999). Initially, two major dams were built, Bonneville and Grand Coulee dams. In 1937, the Bonneville Power Administration was established within the Department of the Interior to manage these projects. From 1932 to 1975, 19 major dams were built on the Columbia and

Figure 1.5. Building of dams has occurred throughout history, but during the early to mid 1900s in the USA the efforts were increased to provide rural electrification, flood control, irrigation water supply, and navigation. These two dams altered major river systems of the west: **(top)** Boulder Dam (renamed Hoover Dam) on the Colorado River between Arizona and Nevada (photo by Ansel Adams, courtesy of the National Park Service, USDI) and **(bottom)** excavation behind coffer dam during construction of the Grand Coulee Dam on the Columbia River in Washington in 1936 (photo courtesy of Bureau of Reclamation, USDI).

Snake rivers, and many more dams were built on tributaries. These dams not only caused problems for fish passage but also altered instream flows and biological productivity. Adult salmon passage via fishways was provided at dams on the main stem of the Columbia River downstream from Chief Joseph and Hells Canyon dams, but the problem of juvenile salmon downstream passage was not addressed, except with more hatcheries as mitigation (Committee on Protection and Management of Pacific Northwest Anadromous Salmonids 1999).

In contrast, development of the nearby Frazier River, British Columbia, was debated, and plans to erect dams throughout the drainage were not implemented. These fish populations remained intact and self-sustaining, and the economy of the region moved in a different direction (Evenden 2004).

In the first half of the 20th century, the most influential act to protect fisheries in the USA was the Fish and Wildlife Coordination Act of 1934. This act provided the first basic authority for the secretaries of Agriculture and Commerce to provide assistance to federal and state agencies to protect game and fur-bearing animals and to study the effects of pollution on wildlife. The definition of wildlife was formally defined to include birds, fish, mammals, and other classes of wild animals and aquatic and terrestrial vegetation upon which wildlife depends. The act authorized protecting and surveying wildlife resources on public lands. It also directed the Bureau of Fisheries to use water resources for fish culture stations and migratory bird resting and nesting areas and required federal agencies to consult with the Bureau of Fisheries prior to the construction of any new dams and to provide for fish migration.

The first half of the 20th century was a critical turning point in defining management priorities and values, and the AFS took a lead to help establish the first North American Fish Policy, adopted by the AFS in 1938 and embraced by states and provinces through their governmental associations. This policy outlined state, provincial, national, and international relations; the roles of administration and research; and the need for management. The policy recognized that change was inevitable, and practices would evolve. The policy document also contained the wording that "fish are crops, capable of being conserved, restored, and increased through sound management practices" (*Transactions of the American Fisheries Society* 68:40–51). This established the economic and social roles of fish and fisheries: "fishery resources were important elements of national wealth and not a minor incident in the development of power, flood control, drainage, irrigation, reclamation, and recreational projects, as has been done in the past." A further component of this document was the "objective of fisheries research," which included guidance for lake and stream surveys, fisheries statistics, and other standard practices. It included suggestions that stocking should not occur in waters that had good fishing opportunities for native fish species.

In 1940, the Bureau of Fisheries and Bureau of Biological Survey were consolidated into the U.S. Fish and Wildlife Service and placed in the Department of the Interior. Through the authority of the Fish and Wildlife Coordination Act the new agency was required to consult regarding water resource projects and their effects on fish and wildlife resources. In 1946, this act was amended to require consultation with the U.S. Fish and Wildlife Service and the fish and wildlife agencies of states where any body of water was controlled or modified by any federal agency in order to prevent loss and damage of wildlife resources. However, the amendments specifically exempted the Tennessee Valley Authority from these provisions. Additional amendments were added in 1958 to define and require equal consideration and coordination of wildlife conservation with other water resource development programs. At this time, the act also authorized the Secretary of the Interior to provide public fishing areas and accept donations of lands and funds.

In Canada prior to 1930, the federal government controlled all Crown lands in Manitoba, Saskatchewan, and Alberta. In 1930, the Constitution Act transferred control of Crown lands and public trust resources via the Natural Resource Transfer Act to each of the three prairie provinces, with the exception of remaining tracts of Crown lands such as First Nation reserve lands and national parks. Wording from the Canadian Constitution Act affirms this move and states "Except as herein otherwise provided, all rights of fishery shall, after the coming into force of this agreement, belong to and be administered by the Province, and the Province shall have the right to dispose of all such rights of fishery by sale, license or otherwise, subject to the exercise by the Parliament of Canada of its legislative jurisdiction over sea-coast and inland fisheries."

1.4.5 Interjurisdictional Management

There have been many debates about how to manage shared resources such as the Great Lakes. Contributions written by Joslyn (1905) reflect some of the dilemma that the states faced: "[E]fforts of a single state, no matter how well directed, were wholly inadequate to meet the demands and accomplish practical results." He later referenced the take of lake sturgeon in unregulated fashion and commented on "imported Russian caviar" that was made and put up at Grand Haven, Michigan, from lake sturgeon harvested from the Great Lakes. This industry carried on to such an extent that lake sturgeon was almost exterminated from these waters.

In 1909, the Boundary Waters Treaty established the International Joint Commission of Canada and the USA. The treaty created a process for cooperation in the use of all the waterways that crossed the border between the two nations, including the Great Lakes. However, it was nearly 40 years later that a bi-national fisheries management agreement was negotiated. The 1955 Convention on Great Lakes Fisheries created the Great Lakes Fishery Commission, one of the most successful models for joint fisheries management in the world.

The USA and Canada have had several interjurisdictional disputes regarding fishing rights. In 1870, the Canadians forbade foreign fishermen from fishing in Canadian waters. The Washington Treaty was drawn in 1873 to allow U.S. fishermen access to inshore waters of Canadian fisheries in return for Canadian access in the USA, including U.S. fishing rights on the Grand Banks and free entry of Canadian fish to U.S. markets. The Great Lakes was a site of conflict regarding harvest, and limited markets in Canada were a problem and source of misunderstanding between the countries (Bogue 2000).

The concept of reciprocity was established early on with a mutual reduction of duties charged on goods exchanged between Canada and the USA. The movement toward reciprocity began in 1846–50 in Canada's west and the maritime colonies, particularly New Brunswick. British diplomats negotiated in Washington, D.C., without success, when a dispute developed over the rights of American fishers in British coastal waters in North America. Both governments became anxious for a comprehensive settlement to dispose of the reciprocity and the fisheries issues. Even today, many fisheries negotiations include issues of reciprocity.

1.4.6 Contemporary Management Goals

The environmental, social, and economic value of different fishes, and whether they were considered native or introduced species, has changed over time (Lucas 1939; Dill and Cordone 1997; Fuller et al. 1999; Rahel 2002). For many years, the nongame or noncommercial native fishes were considered "rough or trash fish" and were removed from systems to enhance desirable game fish. Reports such as counts of the number of trout eggs found in the stomachs of bullheads and suckers (Atkinson 1931) and the observations of predation on fish by piscivorous birds and reptiles (Salyer 1933; Huntsman 1938) supported predator removals.

The consequences of the wide introductions and selective removal of different aquatic species were not considered critically until more recent times. Management programs organized in many locations to reduce or eliminate specific fish populations produced various outcomes (Meronek et al. 1996; Clarkson et al. 2005; Chapter 8, this volume). Selective fish removal activities continue to be used and evaluated as tools for different management objectives, with some directed at removing nonnative species to restore ecosystems and others focused on native species that have increased in numbers due to habitat alterations (e.g., Beamesderfer 2000; Weidel et al. 2007; Herbst et al. 2009). Today over 200 nonindigenous fish species have been stocked in waters of North America, and the management goals for fisheries have evolved (Nelson 1965; Benson 1970; Leach and Lewis 1991).

Unfortunately, fisheries and water resources were not considered at more holistic community and ecosystem levels until the modern conservation movement. During the late 1960s to 1970s, increased awareness of threats to environmental and human health resulted in substantial national legislation in the USA, including the Endangered Species Act, National Environmental Policy Act, Clean Air Act, Coastal Zone Act, and Fisheries Conservation and Management Act. These active U.S. public laws are organized under 49 titles of the U.S. Code, with Title 16 focused on conservation. However, legislation appears elsewhere; for example Title 33 contains laws regarding navigable waters, including the Federal Water Pollution Control Act Amendments of 1972, thereafter called the Clean Water Act. Title 50 concerns most specific codes written for fish and wildlife.

The Canadian Environmental Protection Act was passed in 1999 and regulates toxic substances and ocean dumping (Boyd 2003). According to Boyd (2003), federal and provincial governments have nonbinding water quality guidelines that establish the maximum allowable concentrations of substances in water for particular uses. Under the Fisheries Act, the federal government of Canada can call for minimum flows for fishes and fish passage and regulate substances that can be harmful to fishes. Water rights in western Canada are allocated on a first-come, first-served basis; in eastern Canada, water rights are based on property ownership, meaning property owners enjoy riparian rights to use adjoining waters. These regional differences are similar to those in the USA (see Chapter 4). The Species at Risk law was passed in 2002 and protects all aquatic and terrestrial species on federal lands. The provinces are directed to protect species that are on provincial and private lands.

1.4.7 Aboriginal Rights

Many native peoples' rights to healthy lives, their territory, and their fishing resources were disregarded during the European settlement until some key social factors and legal decisions mandated new ways of thinking (e.g., Scott 1923; Lurie 1957; Landeen and Pinkham 1999). A representative example of the attitude toward development of land and disregard for Native American rights in the USA is provided in the wording of the Pacific Railway Act of 1862 (12 U.S. Statutes at Large 489, section 2):

> That the right of way through the public lands be…granted to said company for the construction of said railroad and telegraph line; and the right…is hereby given to said company to take from the public lands adjacent to the line of said road, earth, stone, timber, and other materials for the construction thereof; said right of way is granted to said railroad to the extent

of two hundred feet in width on each side of said railroad when it may pass over the public lands, including all necessary grounds, for stations, buildings, workshops, and depots, machine shops, switches, side tracks, turn tables, and water stations. The United States shall extinguish as rapidly as may be the Indian titles to all lands falling under the operation of this act....

An active role for indigenous American people in fisheries management emerged in the 1970s and challenged existing paradigms for fisheries management. In the USA, because of formal signed treaties with many Indian nations, rights for fish and wildlife resources have been slowly recovered and defined through the courts. As a result of this process, values of traditional knowledge and cooperative management techniques have been recognized as key parts of management decisions, and aboriginal Americans have emerged as important forces in inland fisheries management.

Landmark Native American rights decisions in the USA occurred in association with Columbia River and Puget Sound Indian tribal challenges. The basis for these decisions came from the fact that Isaac Ingalls Stevens, the first governor and Superintendent of Indian Affairs of Washington Territory, had summoned the tribes to a series of meetings in 1854 and 1855 at which they were invited to sell their lands to the USA at a price of something less than half a cent per acre. Governor Stevens was provided this authority to negotiate treaties because of the Donation Land Law Act of 1850 for homesteading. In treaties negotiated with tribes, wording was provided as follows for Indian fishing rights: "The right of taking fish, at all usual and accustomed grounds and stations, is further secured to said Indians, in common with all citizens...." Since the tribal members did not speak English, all negotiations for these treaties were translated into Chinook (a language used by many tribes for trade) by an interpreter and translated again into the various languages of the tribes. These few words quoted above have been debated by lawyers, and throughout this debate courts agreed that only a general meaning of these words could have been conveyed to the Indians by the two-stage translation process from English through the Chinook trade language. Also important in understanding the context of the negotiations for native claims at this time is to recognize that nearly three-quarters of the population in the region were indigenous peoples (Clark 1985). Therefore these rights were not minority issues.

The challenges by Native American tribes for fishing rights in states included *Sohappy v. Smith*, filed in U.S. District Court for the District of Oregon to secure the rights of the Columbia River Indian tribes to harvest salmon. This challenge was based on the Nez Perce Treaty of 1855, signed by Washington Territorial Governor Isaac Stevens, General Palmer, and Chief Looking Glass in Walla Walla, Washington. Governor Stevens also signed three other treaties with Columbia Basin tribes during 1855: Umatilla, Warm Springs, and Yakima. Federal Circuit Court Judge Robert Belloni ruled in 1969 that states could not restrict Indian fishing except for clearly defined conservation reasons and Indians were entitled to a fair share of the fishery. *Sohappy v. Smith* was the first of a series of cases known as *U.S. v. Oregon* (Marsh and Johnson 1985; Landeen and Pinkham 1999).

In challenges brought by treaty tribes from Puget Sound, also Stevens' Treaty tribes, Federal Circuit Court Judge George Boldt in *U.S. v. Washington* ruled in 1974 that "fair share" was half the allowable catch destined for usual and accustomed fishing places. Currently, each year of the season, negotiations with tribal harvest biologists of the Stevens Treaty tribes and state and federal agencies determine the precise number of fish allowed for the 50% take in these waters. This decision on allocation has been upheld in many challenges and has been extended to the rights to harvest shellfishes by the Puget Sound tribes (Combs 1999).

In the 1955 Convention on Great Lakes Fisheries that resulted in the establishment of the Great Lakes Fishery Commission, the concerns of aboriginal American and Canadian tribes were not considered. Native American rights were originally established in lakes Superior, Huron, and Michigan in 1836 through a treaty between the U.S. government with five Chippewa and Ottawa tribes. However, it was not until 1985, through the challenge of the *U.S. v. Michigan* (Western District Court of Michigan 1985), that fishing rights of Bay Mills Indian Community, Sault Ste. Marie Tribe of Chippewa Indians, Grand Traverse Band of Ottawa and Chippewa Indians, Little River Band of Ottawa Indians, and Little Traverse Bands of Odawa Indians were established. Recently, these rights were renegotiated via an extensively mediated process in a 2000 consent decree to establish fishing allocation, management, and regulation in lakes Michigan, Huron, and Superior ceded waters (Western District Court of Michigan 2000). The consent decree provided for joint fisheries management on Great Lakes areas within the 1836 treaty-ceded waters. In 2007, 1836 treaty-ceded fisheries harvest rights were extended to inland waters with another consent decree. Both of these consent decrees follow similar allocation rules of the Pacific coast adjudications, as do other court cases regarding Indian treaty areas in Wisconsin and Minnesota.

Although the tribal rights of indigenous peoples in the USA are better recognized, the First Nations rights in Canada are still under negotiation and evolution, and Canadian federal control allows provinces to retain public trust ownership of lake and riverbeds and riparian rights to fishes. In Canada, resolution of First Nations' rights has been more complex as treaties were not established, and documents and authorities are in continual deliberations. In British Columbia, First Nations fishing rights were summarized by Jones et al. (2004). They detailed the three different categories of fisheries in aboriginal settings: (1) a fishery for food, (2) social and ceremonial fishery (aboriginal food fishery), and (3) commercial fishery. The aboriginal food fishery is recognized by the Supreme Court of Canada as a right enshrined in the constitution and thus has priority over all other fishing rights. The commercial aboriginal fishery allocations are negotiated on a case-by-case basis, depending on the respective stock assessment and allocation.

The concept of co-management of fisheries with aboriginal Americans is now fully recognized as the proper approach for managing most tribal claims; however, the methods and approaches are varied and in development both in Canada and the USA. Management approaches include cooperative management, collaborative management, and management by community (Busiahn 1989; Tipa and Welch 2006). In 2007, the U.S. Bureau of Indian Affairs listed 561 tribal entities within the contiguous 48 states that have status as Indian tribes (U.S. Federal Register 2007), and each has a stake in co-management of fisheries.

1.4.8 Sport Fish Management

Fishing as sport was brought to North America from Europe. Even though authority for sport fish management was established for the states and provinces, there was little money available for inland fisheries research and monitoring early in the 20th century. The financing was generally derived from license sales, and there was difficulty in establishing infrastructure in both states and provinces. Palmer (1912) wrote that licenses for hunting game for sport, as distinguished from market hunting licenses, were gradually adopted after the beginning of the 20th century. By 1912, fishing licenses were required of residents in 34 states and 6 provinces, and nonresidents were required to purchase licenses in all states and provinces

(Palmer 1912). At that time, several of the states exempted women and children from fishing license requirements.

During the days of the state and provincial fish commissions in the early 20th century, fisheries surveys were conducted within each state or province, but few of these governments had fisheries management plans or cohesive strategies. A pivotal point for developing infrastructure in states for fisheries management was the passage of the Federal Aid in Sport Fish Restoration Act of 1950, also known as the Dingell–Johnson Act. The Wildlife Restoration Program had begun in 1938 following the passage of the Federal Aid in Wildlife Restoration Act, often called the Pittman–Robertson Act. As a result of the Pittman–Robertson Act, states began to develop wildlife management programs to restore, conserve, manage, and enhance wildlife resources and to provide for public use and benefits from these resources. After World War II, this philosophy was expanded to include fisheries restoration and enhancement via the Dingell–Johnson Act. As a result of this funding, the staffs of state agencies dealing with freshwater fisheries increased from a few hatchery workers to include fisheries managers and researchers by the mid-1950s. The act was a tremendous success, and by 1979 the total budgets for 50 state fisheries agencies were US$143 million dollars (Sullivan 1979). The expansion of the federal aid in sport fish restoration program occurred in 1984 with the passage of the Wallop–Breaux Act, which increased revenue even further by including excise taxes on additional fishing equipment and federal taxes from small boat fuels. This program today provides funds to restore and manage sport fishery resources and to provide public use and benefits from these resources. In a survey conducted in 2001, inland fisheries management programs in individual states employed an average of 106 full-time permanent employees, varying from 6 in Delaware to 416 in Minnesota, and states spent an average of $9,994,571 annually on their inland fisheries programs, varying from $432,000 in North Dakota to $39,276,052 in Minnesota (Gabelhouse 2005).

Federal aid for fisheries was patterned after the Pittman–Robertson Act for wildlife and used the 10% excise tax initiated during World War II on fishing rods, reels, lures, baits, and flies for dispersal to state fisheries agencies and required matching funds from states to support all aspects of recreational fishing. Allocation to states was based 60% on the number of licensed sport anglers and 40% on the land and water area of the state. The expansion of money for fisheries brought with it a dilemma of success. At the outset, there was the question of whether or not a public conservation agency should employ only management technicians and "farm out" its research problems to colleges and universities. In the 1940s and 1950s there were debates in the AFS as to the best role of fisheries agencies and whether research should be more removed from management agencies. Universities were happy to invite research into their infrastructure (Harkness et al. 1950), but most state agencies chose to develop their own research infrastructure with support from universities. Federal support for research and training for fisheries biologists came with the addition of Cooperative Fishery Research Units to the successful Wildlife Research Units. The Cooperative Units Act (P.L. 86–686) was passed by the U.S. Congress in 1960 and authorized the unit program as a separate budget item within the U.S. Fish and Wildlife Service. Starting in 1961, the fishery units increased opportunities for training of fisheries professionals. These and other training programs have successfully trained fisheries and aquatic biologists that are now in private, public, and tribal agencies and institutions throughout the world. Today, recreational and commercial fisheries are assessed and evaluated by state fisheries agencies that have extensive research infrastructure and receive external funding for additional research.

1.4.9 Evolution of Scale and Complexity in Management

As a result of growing demands for natural resources by expanding human populations, changing human values toward the environment, and accompanying legislative mandates and regulations (Box 1.6), the complexity of fisheries management changed rapidly from the 1960s to the turn of the 21st century and beyond. Marine and inland fisheries management moved from its traditional single-species focus on maximum sustainable yield and optimum sustainable yield to more holistic science-based approaches mandated by legislative authorities to consider the linkages between terrestrial and aquatic systems. These new paradigms included incorporation of ecosystem considerations, environmental fluctuations, and socioeconomic factors (Caddy 1999). The stock–recruitment tools for exploited stocks provided by the deterministic models of Beverton and Holt (1957) and Ricker (1975) were improved and modified with a suite of models and multidimensional approaches (Walters and Korman 1999; Quinn 2003; Walters and Martel 2004). Since the 1990s, inclusion of Bayesian and time series methods in stock–recruit models have allowed for explicit specification of uncertainty (Quinn 2003; Koen-Alonso 2009). These new approaches in management include tools to understand and integrate differences in genetic stock structures, changing species structures and predator–prey dynamics, bioenergetics, ecosystem dynamics, and human values (Walters et al. 1997; Caddy 1999; Rothschild and Beamish 2009).

Because of species introductions, xenobiotics, and trophic changes, fish stocks in the Great Lakes and other systems that were depleted from overharvest have not recovered by simply reducing fishing mortality or stocking (Coble et al. 1990; Holey et al. 1995; Mercado-Silva 2006). The increased pressures from human alterations of habitat, nonpoint and point source pollution, and species introductions have led to restructured habitats and altered ecosystem dynamics with enormous consequences (Hatch et al. 2001; Anderson 2009). Conflicts over freshwater resources have emerged as key driving forces in inland fisheries management (Reisner 1989; Postel et al. 1996; Postel 2000). In the USA, the Clean Water Act of 1972 called for improved water quality and restoration of ecosystem services such as recreation and fish habitat (Brown et al. 2009). In Canada, Fisheries and Oceans Canada (DFO 1986) began enforcing a principle of no net loss of habitat under its authorization via the Fisheries Act, section 35(2) (Harper and Quigley 2005). The results of these mandates and economic and social conflicts led to major restoration and mitigation programs across the inland and coastal landscapes, and such programs continue to emerge to restore the ecological functions of aquatic systems (Poff et al. 1997; Naiman et al. 2000; Palmer et al. 2009). Prominent programs established with interagency and public agreements include the Columbia River Basin Fish and Wildlife Program, developed in the early 1980s (Williams et al. 1999); the California Central Valley Project and CALFED Bay–Delta Program, which evolved in the late 1980s (Schick and Lindley 2007; Brown et al. 2009); Mississippi Interstate Cooperative Resource Agreement, established in 1989 (Montgomery 1991); the Colorado River Basin Restoration in the mid-1990s (Gloss et al. 2005; Adler 2007); and the Klamath River Basin mitigations currently underway (Committee on Endangered and Threatened Fishes in the Klamath River Basin 2004; Committee on Hydrology, Ecology, and Fishes of the Klamath River 2008).

In addition to recognition of the complex biological and hydrological cycles, restoration projects expanded the role for social sciences in management and increasingly acknowledged the footprint of the human environment on the greater ecosystem (Stevens et al. 1997; Van

Box 1.6. Selected U.S. Legislative Acts

Below are summaries of selected U.S. legislative acts that provided authority for conserving or managing fish, fish habitat, or related environmental components. The acts are organized in chronological order.

The Rivers and Harbors Act (1899). Passage of this legislation aimed at prohibiting the obstruction of navigable waters gave the U.S. Army Corps of Engineers (ACE) increased authority to regulate activities in the U.S. rivers. The construction of bridges, dams, or dikes across navigable waters required approval by the Chief of Engineers, the Secretary of the Army, and the consent of the U.S. Congress, and the law outlawed the deposit of refuse in these waters. In 1905, ACE established a permit system to implement this congressional act. Anyone who wished to change the course, location, condition, or capacity of a water body now had to apply for permission from the local ACE district office.

The Antiquities Act (1906). This bill set to preserve all objects of historic or cultural interest that are situated upon lands owned or controlled by the government of the USA. As with natural parks such as Yosemite and Yellowstone, the governing prerequisite behind the preservation of what this bill called "national monuments" was that the land in question offered no economic value beyond that of scenic interest. The Antiquities Act granted exclusive decision-making power to the President, and it was through this piece of legislation that Theodore Roosevelt earned the lasting admiration of the preservationists.

The National Park Service Organic Act (1916). This act was a historic departure from previously unregulated land development activities in the West. This legislation established the National Park Service, and stewards were charged with the duty to conserve the scenery, the natural and historic objects, and the wildlife therein and to provide for the enjoyment of the same in such manner and by such means as would leave them unimpaired for the enjoyment of future generations.

Federal Food, Drug, and Cosmetic Act (1938). This is the nation's major law regulating contaminants in food, including pesticides. The Food and Drug Administration (FDA) implements most of this law; the Environmental Protection Agency (EPA) carries out its pesticide standard setting provisions (with FDA enforcement). See also Food Quality Protection Act.

Federal Insecticide, Fungicide, and Rodenticide Act (1947). This law controls the sale, distribution, and application of pesticides; it was amended in 1972, 1988, and 1996. See also Food Quality Protection Act.

(Box continues)

Box 1.6. Continued.

Atomic Energy Act (1954). This legislation was passed because of the government's keen interest in monitoring the commercial and national defense uses of atomic energy. Government concerns included radiation hazards and the disposal of radioactive waste. The act established a general regulatory structure for construction and use of nuclear power plants and nuclear weapons facilities. Unlike most environmental statutes, it does not permit citizen suits and affords only limited opportunities for suits by public interest groups.

The Wilderness Act (1964). In this act, Congress recognized that the expansion of human activities posed a threat to the existence of natural lands and gave a legal definition to wilderness and protection to lands so designated.

Wild and Scenic Rivers Act (1968). The act established the policy that certain rivers of the nation which, with their immediate environments, possess outstandingly remarkable scenic, recreational, geologic, fish and wildlife, historic, cultural, or other similar values, shall be preserved in free-flowing condition and that they and their immediate environments shall be protected for the benefit and enjoyment of present and future generations. The act both identifies specific river reaches for designation as wild or scenic and provides criteria to be used for classifying additional river reaches.

National Environmental Policy Act (1970). The first of the modern environmental statutes, this act became effective 1 January 1970. The National Environmental Policy Act created environmental policies and goals for the country and established the President's Council on Environmental Quality. Its most important feature is its requirement that federal agencies conduct thorough assessments of the environmental impacts of all major activities undertaken or funded by the federal government. Many states have enacted similar laws governing state activities.

Clean Air Act (1970). This legislation sets goals and standards for the quality and purity of air in the USA. By law, it is periodically reviewed. A significant set of amendments in 1990 toughened air quality standards and placed new emphasis on market forces to control air pollution.

Clean Water Act (1972). This legislation establishes and maintains goals and standards for U.S. water quality and purity. It has been amended several times, most prominently in 1987 to increase controls on toxic pollutants, and in 1990, to address more effectively the hazard of oil spills.

Coastal Zone Management Act (1972). This act provides a partnership structure allowing states and the federal government to work together for the protection of U.S. coastal zones from environmentally harmful overdevelopment. The program provides federal funding to participating coastal states and territories for the implementation of measures that conserve coastal areas.

(Box continues)

Box 1.6. Continued.

Marine Mammal Protection Act (1972). This law seeks to protect whales, dolphins, sea lions, seals, manatees, and other species of marine mammals, many of which remain threatened or endangered. The law requires wildlife agencies to review any activity—for example, the use of underwater explosives or high-intensity active sonar—that has the potential to "harass" or kill these animals in the wild. The law is our nation's leading instrument for the conservation of these species and is an international model for such laws.

Endangered Species Act (1973). This legislation is designed to protect and recover endangered and threatened species of fish, wildlife, and plants in the USA and beyond. The law works in part by protecting species habitats.

Safe Drinking Water Act (1974). This act establishes drinking water standards for tap water safety and requires rules for groundwater protection from underground injection; it was amended in 1986 and 1996. The 1996 amendments added a fund to pay for water system upgrades, revised standard-setting requirements, required new standards for common contaminants and included public "right to know" requirements to inform consumers about their tap water.

Federal Land Policy and Management Act (1976). This act provides for protection of the scenic, scientific, historic, and ecological values of federal lands and for public involvement in their management.

Fisheries Conservation and Management Act (1976). Better known as the Magnuson–Stevens Act, this legislation governs the management and control of U.S. marine fish populations and is intended to maintain and restore healthy levels of fish stocks and prevent overharvesting.

Resource Conservation and Recovery Act (1976). This legislation seeks to prevent the creation of toxic waste dumps by setting standards for the management of hazardous waste. Like the Comprehensive Environmental Response, Compensation and Liability Act (see below), this law also includes some provisions for cleanup of existing contaminated sites.

Toxic Substances Control Act (1976). This law authorizes the EPA to regulate the manufacture, distribution, import, and processing of certain toxic chemicals.

Surface Mining Control and Reclamation Act (1977). This act is intended to ensure that coal mining activity is conducted with sufficient protections of the public and the environment and provides for the restoration of abandoned mining areas to beneficial use.

(Box continues)

Box 1.6. Continued.

Comprehensive Environmental Response, Compensation and Liability Act (1980). Commonly referred to as "Superfund," this law requires the cleanup of sites contaminated with toxic waste. In 1986, major amendments were made in order to clarify the level of cleanup required and degrees of liability. This legislation is retroactive, which means it can be used to hold liable those responsible for disposal of hazardous wastes before the law was enacted in 1980.

Coastal Barrier Resources Act (1982). The act (reauthorized and amended in 1990) established a policy that coastal barriers, in certain geographic areas of the USA, and their adjacent inlets, waterways, and wetland resources are to be protected by restricting federal expenditures that have the effect of encouraging development of coastal barriers. The act provided for a Coastal Barrier Resources System, which identified undeveloped coastal barriers along the Atlantic and Gulf coasts, including islands, spits, and bay barriers that are subject to wind, waves, and tides, such as estuaries and nearshore waters. These areas were outlined on a set of maps dated 30 September 1982 and approved by the Congress.

Emergency Planning and Community Right-to-Know Act (1986). This law requires companies to disclose information about toxic chemicals they release into the air and water and dispose of on land.

Oil Pollution Act (1990). Enacted a year after the disastrous Exxon Valdez oil spill in Alaska's Prince William Sound, this law streamlines federal response to oil spills by requiring oil storage facilities and vessels to prepare spill-response plans and provide for their rapid implementation. The law also increases polluters' liability for cleanup costs and damage to natural resources and imposes measures—including a phase out of single-hulled tankers—designed to improve tanker safety and prevent spills. This law will be prominent in litigations following the BP oil spill in the Gulf of Mexico in 2010.

Food Quality Protection Act (1996). This legislation is designed to ensure that levels of pesticide residues in food meet strict standards for public health protection. Under this law, which overhauled the Federal Food, Drug, and Cosmetic Act and the Federal Insecticide, Fungicide, and Rodenticide Act, the EPA is required to protect infants and children better from pesticides in food and water and from indoor exposure to pesticides.

Winkle et al. 1997; Adler 2007). New modeling approaches used spatially explicit approaches and attributes and incorporated variation in models of complex phenomena (Burke et al. 2008; Cressie et al. 2009; Sharma et al. 2009). Recognition of the consequences and challenges of global climate change on inland lake and river systems has been well documented in studies of the Great Lakes and Canadian lakes (Magnuson et al. 2000; Casselman 2002; Latifovic and Pouliot 2007) and more recently in reviews of river ecosystems and associated fish populations (Reist et al. 2006; Palmer et al. 2009; Williams et al. 2009). The increasing extent of harmful algal blooms has been shown to be associated with many human activities, especially with invasive species that are transported in ballast water and toxic compounds that are released with industrial, agricultural, and sewage effluents and transported into rivers and coastal waters (Anderson 2009).

1.5 USE OF HISTORIC INFORMATION IN CONTEMPORARY MANAGEMENT

A variety of historic resources should be considered by contemporary fisheries biologists and managers to provide inferences on past conditions. A number of historic fisheries information sources are obvious and include agency management and research reports, including fishery management plans. Frequently overlooked sources of information are the early reports of various state, provincial, and federal fish commissions. These reports frequently contain detailed observations and data on the condition of fisheries resources starting around 1870. Annual or biannual reports provide information on the location of early fish stocking, stocking success, habitat impairments, and fisheries surveys. Most of the early fish commissions sponsored special reports on specific aspects of fisheries resources. County and local histories along with early plat and survey maps, while more difficult to access, can provide a landscape context along with the specific changes that European colonization made to the landscape. Frequently, county and local histories along with surveyor notes have detailed accounts of fisheries and landscapes that are particularly valuable for unique and easily-identified fishes such as lake sturgeon. Other overlooked sources of historic fisheries information are tax ledgers, commercial records, or other community record keeping of fish harvests. This is particularly true in the original 13 U.S. colonies, and this information is frequently found in state or county historical society archives. Another source of information on fish harvest can be obtained from cannery records, often available in local archives.

Key sources of historic habitat condition of waters and their riparian zones can be found in early surveyors' journals. Many early journals are very detailed and include information on plant species, observed fishes, lake and wetland locations, and stream widths. Surveyors were keen observers of the natural world, and one of their tasks was to inventory qualitatively natural resources along survey lines. An easy way to access historic data is to contact state, provincial, or federal archives to determine their current fisheries-related holdings along with historic photos of waters. County and local historical societies are rich and inexpensive sources of information and photos that are often overlooked.

Fisheries workers also can overlook the holdings of many museums that have data from archaeological middens and archived samples. Environmental historians and archaeologists can provide assistance with archaeological information through contacts with state or provincial historical societies. A number of museums have large shell collections that can provide information on historic conditions in lakes or river systems. Museum collections of archived fish samples can provide information on what fishes were found historically in a system.

Some cautions on the use of historic information are warranted and include understanding that historic information can often be biased by values held at the time. Knowing the cultural context can inform the user of reasons for decisions made through time. An excellent example of documentation of the evolution of social values is provided by Reuss (2005), who followed water management and development framed by social reasons over time. Reuss maintained that only after World War II did public attention shift in favor of a more inclusive ecological approach to water development, partly because large dam projects had forced basin inhabitants from their homes, chemical and nuclear pollutants threatened the environment, and urban populations sought opportunities for recreation.

Historic data cannot be used as direct replacement for experimentation, but they are valuable sources for generating hypotheses and providing complementary information. However, there are some situations in which historic information may be the only source of information, such as processes that can be examined only over very long time periods; unusual events that require historic context; unplanned experiments where systems are affected by unplanned events and require historic information to interpret; chronicles of historic patterns in a water or system; or determination of past conditions for modeling waters or systems (Gould 1986; Steedman et al. 1996). No one tool will be sufficient for most historic interpretation, and combinations of tools will be needed to maximize success. The most powerful analyses use multiple tools to collaborate and verify multiple data sources.

1.5.1 Records of Habitat Conditions, Species Population Size, and Distribution

Hooke (1997) provides an excellent checklist on the use of historic fluvial geomorphologic data that can be adapted to assist users of historic information desiring "replicate" data sets. This process includes obtaining all historic information from complimentary data sources to allow cross checking. Background information about the reliability of source data should be pursued, especially investigating document quality, accuracy, and applicability, with verification through secondary data sources. If historic data are going to be used, assessment of data accuracy can be accomplished through comparisons with other data in the same area and time frame. Once data are considered accurate, a time sequence of conditions should be developed using both qualitative and quantitative methods, and finally a field analysis and check of the data should be conducted if possible.

Historical fisheries and landscape information have been used in many ways for establishing contemporary management goals, such as understanding historic river and landscape conditions. Archaeological and human artifacts can be used to examine the geomorphology of river basins (Brown et al. 2003). Many resources can be used to date locations and provide information on how river dynamics have changed over time including pottery, coins, hearths, bones, earthworks, middens, stonework, structures (e.g., buildings, bridges, wharves and jetties, wells, or aqueducts), and mining debris. Mining debris can also provide specific mineral tracers to help locate and date sediment deposits. Brown et al. (2003) provided four case studies to illustrate the use of residue information obtained from excavations: (1) reconstructing river channels from bridge structures (2) using mining slag, bed load, and hydraulic sorting to examine river movement; (3) using artifacts to describe floodplain deposition and erosion; and (4) using metal mining and their residues to examine fluvial responses over time.

Many resources are useful in understanding river dynamics. Potential tools include land surveys, botanical collections, general surveys and travel accounts, bridge surveys, channel surveys, building locations, historic ground photos, topographic records, navigation surveys, lake sediments, reservoir storage changes, diaries and journals, and water level and flood records on buildings (Gurnell et al. 2003).

To develop rehabilitation options adequately for aquatic species whose populations are threatened or badly degraded in numbers, information is needed regarding historic population sizes or harvests so targets for rehabilitation can be determined. An example of the use of historic information to estimate historic population size from replicate information is found in Holzkamm and McCarty (1998), who used isinglass records from Hudson's Bay Company Lac la Pluie District to estimate Obijway harvest of lake sturgeon from the Rainy River and Lake of the Woods. Reconstructions of Pacific salmon populations in the Columbia River system have been estimated using cannery data (Gresh et al. 2000) and a diverse number of historic data sets (Northwest Power Planning Council 1986).

Historic population and production data have been key data in the U.S. Endangered Species Act listing processes. As part of the processes dictated by the Endangered Species Act, historic stock status and distribution are pivotal pieces of information required for species rehabilitation actions and legal proceedings. There are a number of examples of such studies, including Hamilton et al. (2005) on the distribution of salmon in the Klamath River system and Kaczynski and Alverado (2006) and Adams et al. (2007) on the southern extent of coho salmon in California. In these cases, historic information including residue and replicate information was used to develop best estimates of historic stock status and species distributions.

Some states in the USA are currently developing databases that will catalog historic stocking of fishes in their jurisdictions to facilitate analyzes of historic genetic stock structures. Historic fish stocking data provide information on the locations where fish have not been stocked, so wild fish presumably still have the historic fish genetic structure and could be used for future broodstocks for rehabilitation efforts. Other analyses will combine the fish stocking database with broodstock source information to determine where unique genetic strains of fish have been stocked. For example, reef-spawning populations of walleye in Saginaw Bay of Lake Huron are believed to be extinct at this time, but an analysis of historic fish stocking information, citizen accounts, and broodstock source references in Michigan Fish Commission reports indicate this walleye strain was stocked into Lake Gogebic in the western Upper Peninsula and no other walleye have been stocked. The self-sustaining population of reef-spawning walleye in Lake Gogebic may be a broodstock source for future rehabilitation efforts in Saginaw Bay.

1.5.2 Preservation for the Future

Resource agency personnel often get placed into interagency relationships that are confrontational without knowledge of the history of the interactions among agencies. This history can frequently color interactions among agencies for decades and examinations of their earliest interactions can provide insight into current relations. The history of fisheries on the Great Lakes is replete with interactions among different resource agencies that illustrate the conflicts between interests and agencies supporting commercial fishing, recreational fishing, transportation infrastructure, agriculture, natural resource extraction, and industrial development of the watershed (Bogue 2000). These historic interactions help explain how agencies may act and react to positions and assist fisheries managers to a better job.

While much of this chapter details the importance of historic fisheries information, it is critical for current fisheries managers to record in detail current conditions and the rationale and processes for management decisions for use by managers and scientists in the future. Written information on fisheries projects, management decisions, and condition of systems should be provided to federal, state, or provincial archives in paper copies for long-term storage in consultation with professional archivists. Digital records systems are increasing in prevalence, but to date, potential loss of information from lack of redundancy and adequate storage systems provides many challenges. In addition, photographs and video data are important resources that should be preserved, and documentation of the date, time, location, and subject as well as geo-reference data should be included to allow future placement on the landscape. One method to document habitat conditions is to take time series photographs or videos from fixed locations. Equally important to preservation of fisheries information is the preservation of individual fisheries workers' materials. Most fisheries workers have information that likely has not been placed in agency files, archives, or publications. Personal photographs, videos, field log books, work diaries, papers, and other media could be critical in understanding the context of decisions, system conditions, or how work was conducted. Fisheries workers should either provide their materials to archival locations upon their retirement or give directions to family members for the long-term storage of their materials in their wills. Potential archival locations include federal, state, or provincial archives along with archives managed by the U.S. Fish and Wildlife Service and National Oceanic and Atmospheric Administration. In addition, the AFS Fisheries History Section has begun a process to provide a series of recommendations for archiving of information.

1.6 REFERENCES

Adams, P. B., L. W. Botsford, K. W. Gobalet, R. A. Letdy, D. R. McEwan, P. B. Moyle, J. J. Smith, J. G. Williams, and R. M. Yoshiyama. 2007. Coho salmon are native south of San Francisco Bay: a reexamination of North American coho salmon's southern range limit. Fisheries 32(9):441–451.

Adler, R. W. 2007. Restoring Colorado River ecosystems: a troubled sense of immensity. Island Press, Washington, D.C.

Anderson, D. M. 2009. Approaches to monitoring, control and management of harmful algal blooms (HABs). Ocean and Coastal Management 52:342–347.

Anfinson, J. O. 2003. The river we have wrought: a history of the upper Mississippi. University of Minnesota Press, Minneapolis.

Anthony, J. L., and J. A. Downing. 2001. Exploitation trajectory of a declining fauna: a century of freshwater mussel fisheries in North America. Canadian Journal of Fisheries and Aquatic Sciences 58:2071–2090.

Atkinson, N. J. 1931. The destruction of gray trout eggs by suckers and bullheads. Transactions of the American Fisheries Society 61:183–188.

Bahls, P. 1992. The status of fish populations and management of high mountain lakes in the western United States. Northwest Science 66:183–193.

Beamesderfer, R. C. 2000. Managing fish predators and competitors: deciding when intervention is effective and appropriate. Fisheries 25(6):18–23.

Beeton, A. M. 2002. Large freshwater lakes: present state, trends, and future. Environmental Conservation 29:21–38.

Belding, D. L., A. Merrill, and J. Kitson. 1924. Fisheries investigations in Massachusetts. Transactions of the American Fisheries Society 54: 29–47.

Benson, N. G., editor. 1970. A century of fisheries in North America. American Fisheries Society, Special Publication 7, Bethesda, Maryland.

Beverton, R. J. H., and S. J. Holt. 1957. On the dynamics of exploited fish populations. Chapman and Hall, London.

Bogue, M. B. 2000. Fishing the Great Lakes: an environmental history 1783–1933. University of Wisconsin Press, Madison.

Bowen, J. T. 1970. A history of fish culture as related to the development of fishery programs. Pages.71–94 in N. G. Benson, editor. A century of fisheries in North America. American Fisheries Society, Special Publication 7, Bethesda, Maryland.

Bower, W. T. 1911. History of the American Fisheries Society. Transactions of the American Fisheries Society 40:323–358.

Boyd, D. R. 2003. Unnatural law: rethinking Canadian environmental law and policy. University of British Columbia Press, Vancouver.

Briggs, J. C. 1986. Introduction to the zoogeography of North American fishes. Pages 1–16 in C. H. Hocutt and E. O. Wiley, editors. The zoogeography of North American freshwater fishes. John Wiley and Sons, New York.

Brown, A. G., F. Petit, and A. James. 2003. Archaeology and human artifacts. Pages 59–76 in G. M. Kondoff and H. Piegay, editors. Tools in fluvial geomorphology. John Wiley and Sons, Chichester, UK.

Brown, L. R., W. Kimmerer, and R. Brown. 2009. Managing water to protect fish: a review of California's environmental water account, 2001–2005. Environmental Management 43:357–36.

Burke, M., K. Jorde, J. M. Buffington. 2008. Application of a hierarchical framework for assessing environmental impacts of dam operation: changes in streamflow, bed mobility and recruitment of riparian trees in a western North American river. Journal of Environmental Management 90:S224–S236.

Burr, B. M., and R. L. Mayden. 1992. Phylogenetics and North American freshwater fishes. Pages 18–75 in R. L. Mayden, editor. Systematics, historical ecology, and North American freshwater fishes. Stanford University Press, Stanford, California.

Busiahn, T. R. 1989. The development of state/tribal co-management of Wisconsin fisheries. Pages 170–185 in E. Pinkerton, editor. Co-operative management of local fisheries. University of British Columbia Press, Vancouver.

Caddy, J. F., 1999. Fisheries management in the twenty-first century: will new paradigms apply? Reviews in Fish Biology and Fisheries 9:1–43.

Carlander, H. B. 1954. A history of fish and fishing in the upper Mississippi River. Mississippi River Conservation Commission Report, Onalaska, Wisconsin.

Casselman J. M. 2002. Effects of temperature, global extremes, and climate change on year-class production of warmwater, coolwater, and coldwater fishes in the Great Lakes Basin. Pages 39–60 in N. A. McGinn, editor. Fisheries in a changing climate. American Fisheries Society, Symposium 32, Bethesda, Maryland.

Chapman, D. W. 1986. Salmon and steelhead abundance in the Columbia River in the 19th century. Transactions of the American Fisheries Society 115:662–670.

Claassen, C. 1994. Washboards, pigtoes, and muckets: historic musseling in the Mississippi watershed. Historical Archaeology 28:1–145.

Clark, A. H. 1886. History of the iced fish and frozen fish trade of the United States. Transactions of the American Fisheries Society 15:68–83.

Clark, W. G. 1985. Fishing in a sea of court orders: Puget Sound salmon management 10 years after the Boldt Decision. North American Journal of Fisheries Management 5:417–434.

Clarkson, R. F., P. C. Marsh, S. E. Stefferud, and J. A. Stefferud. 2005. Conflicts between native fish and nonnative sport fish management in the southwestern United States. Fisheries 30(9):20–27.

Clemens, W. A. 1932. The aim of research in fish culture in Canada. Transactions of the American Fisheries Society 62:261–266.

Coble, D. W., R. E. Bruesewitz, T. W. Fratt, and J. W. Scheirer. 1990. Lake trout, sea lamprey, and overfishing in the upper Great Lakes: a review and reanalysis. Transactions of the American Fisheries Society 119:985–995.

Coker, R. E. 1914. The Fairport Fisheries Biological Station: its equipment, organization and functions. U.S. Bureau of Fisheries Bulletin 34:383–405.

Colati, G. C., K. M. Crowe, and E. S. Meagher. 2009. Better, faster, stronger integrating archives processing and technical services. Library Resources and Technical Services 53(4):261–270.

Combs, M. J. 1999. United States v. Washington: the Boldt decision reincarnated. Environmental Law 29:683–720.

Committee on Endangered and Threatened Fishes in the Klamath River Basin. 2004. Endangered and threatened fishes in the Klamath River basin: causes of decline and strategies for recovery. National Research Council, National Academy Press, Washington, D.C.

Committee on Hydrology, Ecology, and Fishes of the Klamath River. 2008. Hydrology, ecology, and fishes of the Klamath River basin. National Research Council, National Academy Press, Washington, D.C.

Committee on Protection and Management of Pacific Northwest Anadromous Salmonids. 1999. Upstream: salmon and society in the Pacific Northwest. National Research Council, National Academy Press, Washington, D.C.

Cressie, N., C. A. Calder, J. S. Clark, J. M. Ver Hoef, and C. K. Wikle. 2009. Accounting for uncertainty in ecological analysis: the strengths and limitations of hierarchical statistical modeling. Ecological Applications 19: 553–570.

Davis, H. S. 1935. Stream management in the national forests. Transactions of the American Fisheries Society 65:234–239.

DFO (Department of Fisheries and Oceans). 1986. Policy for the management of fish habitat. Department of Fisheries and Oceans, Ottawa.

Dickenson, F. B. 1898. The protection of fish and a closed season. Transactions of the American Fisheries Society 27:32–46.

Dill, W. A., and A. J. Cordone. 1997. History and status of introduced fishes in California, 1871–1996. California Department of Fish and Game, Fish Bulletin 178.

Ditton, R. B., S. M. Holland, and D. K. Anderson. 2002. Recreational fishing as tourism. Fisheries 27(3):17–24.

Donald, D. B. 1987. Assessment of the outcome of eight decades of trout stocking in the mountain national parks, Canada. North American Journal of Fisheries Management 7:545–553.

Dymond, J. R. 1964. A history of ichthyology in Canada. Copeia 1964:2–33.

Eastgate, A. 1918. Planting fish in an alkali lake. Transactions of the American Fisheries Society 47:89–91.

Evenden, M. D. 2004. Fish versus power: an environmental history of the Fraser River. Cambridge University Press, Cambridge, UK.

Fagen, B. 2006. Fish on Friday: feasting, fasting and the discovery of the New World. Basic Books, New York.

Farley, J. L. 1957. The role of the Great Lakes Fishery Commission in the solution of Great Lakes problems. Transactions of the American Fisheries Society 86:424–429.

Fuller, P. L., L. G. Nico, and J. D. Williams. 1999. Nonindigenous fishes introduced into inland waters of the United States. American Fisheries Society, Special Publication 27, Bethesda, Maryland.

Gabelhouse, D. W., Jr. 2005. Staffing, spending, and funding of state inland fisheries programs. Fisheries 30(2):10–17.

Gloss, S. P., J. E. Lovich, and T. S. Melis, editors, 2005. The state of the Colorado River ecosystem in Grand Canyon. U.S. Geological Survey Circular 1282.

Goble, D. D.1999. Salmon in the Columbia basin: from abundance to extinction. Pages 229–263 in D. Goble and P. W. Hirt, editors. Northwest lands and peoples: readings in environmental history. University of Washington Press, Seattle.

Goble, D. D. 2005. Three cases/four tales: commons, capture, the public trust, and property in land. Environmental Law 35(4): 807–853.

Goetzmann, W. H. 1959. Army exploration in the American west 1803–1863. Yale University Press, New Haven, Connecticut.

Goudie, A. S. 2005. The human impact on the natural environment: past, present and future, 6th edition. Wiley–Blackwell, Hoboken, New Jersey.

Gould, S. J. 1986. Evolution and the triumph of homology, or why history matters. American Scientist 74:60–69.

Gresh, T., J. Lichatowich, and P. Schoolmaker. 2000. An estimation of historic and current levels of salmon production in the northwest Pacific ecosystem: evidence of nutrient deficit in the freshwater systems of the Pacific Northwest. Fisheries 25(1):15–21.

Gurnell, A. M., J. Peiry, and G. E. Petts. 2003. Using historical data in fluvial geomorphology. Pages 77–101 in G. M. Kondoff, and H. Piegay, editors. Tools in fluvial geomorphology. John Wiley and Sons, Chichester, UK.

Hamilton, J. B., G. L. Curtis, S. M. Snedaker, and D. K. White. 2005. Distribution of anadromous fishes in the upper Klamath River watershed prior to hydropower dams—a synthesis of the historical evidence. Fisheries 30(4):10-20.

Harkness, W. J., K., J. W. Leonard, and P. R. Needham. 1954. Fishery research at mid-century. Transactions of the American Fisheries Society 83:212–216.

Harper, D. J., and J. T. Quigley. 2005. No net loss of fish habitat: a review and analysis of habitat compensation in Canada. Environmental Management 36:343–355.

Hatch, L. K., A. Mallawatantri, D. Wheeler, A. Gleason, D. Mulla, J. Perry, K. W. Easter, R. Smith, L. Gerlach, and P. Brezonik. 2001. Land management at the major watershed–agroecoregion intersection. Journal of Soil and Water Conservation 56: 44–51.

Herbst, D. B., E. L. Silldorff, and S. D. Cooper. 2009. The influence of introduced trout on the benthic communities of paired headwater streams in the Sierra Nevada of California. Freshwater Biology 54:1324–1342.

Hocutt, C. H., and E. O. Wiley, editors. 1986. The zoogeography of North American freshwater fishes. John Wiley and Sons, New York.

Holey, M. E., R. W. Rybicki, G. W. Eck, E. H. Brown, J. E. Marsden, D. S. Lavis, M. L Toneys, T. N. Trudeau, and R. M. Horrall. 1995. Progress toward lake trout restoration in Lake Michigan. Journal of Great Lakes Research 21 (Supplement 1):128–151.

Holzkamm, T., and M. McCarthy. 1988. Potential fishery for lake sturgeon (*Acipenser fulvescens*) as indicated by the returns of the Hudson's Bay Company Lac Le Pluie District. Canadian Journal of Fisheries and Aquatic Sciences 45:921–923.

Hooke, J. M. 1997. Style of channel change. Pages 237–268 in C. R. Thorne, R. D. Hey, and M. D. Newson, editors. Applied fluvial geormorphology for river engineering and management. John Wiley and Sons, Chichester, UK.

Hubbs, C. L. 1964. History of ichthyology in the United States after 1850. Copeia 1964:42–60.

Humphries, P. L., and K. O. Winemiller. 2009. Historical impacts on river fauna, shifting baselines and challenges for restoration. BioScience 59:673-684.

Huntsman, A. G. 1938. Fish culture–past and future. Transactions of the American Fisheries Society 67:87–93.

Ito, K. 1886. Fishery industries of the Island of Hokkaido, Japan. Report 105. U.S. Fish Commission Bulletin 342.

Jackson, J. R., and W. C. Kimler. 1999. Taxonomy and the personal equation: the historical fates of Charles Girard and Louis Agassiz. Journal of the History of Biology 32:509–555.

Johnstone, K. 1977. The aquatic explorers. A history of the Fisheries Research Board of Canada. University of Toronto Press, Toronto.

Jones, R., M. Shepert, and N. J. Sterritt. 2004. Our place at the table: First Nations in the B.C. fishery. First Nation Panel on Fisheries, Vancouver.

Jordan, D. S., and B. W. Evermann. 1896. The fishes of North and Middle America: a descriptive catalog of the species of fish-like vertebrates found in the waters of North America, north of the Isthmus of Panama. Bulletin of the U.S. National Museum 47.

Joslyn, C. D. 1905. The policy of ceding the control of the Great Lakes from state to national supervision. Transactions of the American Fisheries Society 34:217–222.

Kaczynski, V. W., and F. Alverado. 2006. Assessment of the southern range limit of North American coho salmon: difficulties in establishing natural range boundaries. Fisheries 31(8):374–391.

Koen-Alonso, M. 2009. Some observations on the role of trophodynamic models for ecosystem approaches to fisheries. Pages 185–208 *in* R. J. Beamish and B. J. Rothschild, editors. The future of fisheries science in North America. Fish and Fisheries Series 31, Springer, Netherlands.

Kunz, G. F. 1893. On the occurrence of pearls in the United States, and shall we legislate to preserve the fisheries. Transactions of the American Fisheries Society 22:16–34.

Kunz, G. F. 1898. A brief history of the gathering of freshwater pearls in the United States. U.S. Fish Commission Bulletin 17:321–330.

Landeen, D., and A. Pinkham. 1999. Salmon and his people: fish and fishing in Nez Perce culture. Confluence Press, Lewiston, Idaho.

Latifovic, R., and D. Pouliot. 2007. Analysis of climate change impacts on lake ice phenology in Canada using the historical satellite data record. Remote Sensing of Environment 106:492–507.

Leach, J. H., and C. A. Lewis. 1991. Fish introductions in Canada: provincial laws and regulations. Canadian Journal of Fisheries and Aquatic Sciences 48 (Supplement 1):156–161.

Lee, D. S., C. R. Gilbert, C. H. Hocutt, R. E. Jenkins, D. E. McAllister, and J. R. Stauffer Jr. 1980. Atlas of North American freshwater fishes. North Carolina State Museum of Natural History, Raleigh.

Lucas, C. R. 1939. Game fish management. Transactions of the American Fisheries Society 68: 67–75.

Lurie, N. O. 1957. The Indian Claims Commission Act. Annals of the American Academy of Political and Social Science 311:56–70.

Magnuson, J. J., D. M. Robertson, B. J. Benson, R. H. Wayne, D. M. Livingstone, T. Arai, R. A. Assel, R. G. Barry, V. Card, E. Kuusisto, N. G. Granin, T. D. Prowse, K. M. Stewart, and V. S. Vuglinski. 2000. Historical trends in lake and river ice cover in the northern hemisphere. Science 289:1743–1746.

Marsh, J. H., and J. H. Johnson. 1985. The role of Stevens Treaty tribes in the management of anadromous fish runs in the Columbia basin. Fisheries 10(4):2–5.

Mayden, R. L., editor. 1992. Systematics, historical ecology, and North American freshwater fishes. Stanford University Press, Stanford, California.

McArthur, I. S. 1947. The fisheries of Canada. Annals of the American Academy of Political and Social Science 253:59–65.

McDonald, M. 1886. XXVII. Report on the distribution of fish and eggs by the U. S. Fish Commission from January 1, 1886, to June 30, 1887. Part XIV, Report to the Commissioner for 1886. U.S. Commission of Fish and Fisheries, Government Printing Office, Washington, D.C.

Mercado-Silva, N., J. D. Olden, J. T. Maxted, T. R. Hrabik, and M. J. Vander Zanden. 2006. Forecasting the spread of invasive rainbow smelt in the Laurentian Great Lakes region of North America. Conservation Biology 20:1740–1749.

Meronek, T. G., R. M. Bouchard, E. R. Buckner, T. M. Burri, K. K. Demmerly, D. C. Hatleli, R. A.

Klumb, S. H. Schmidt, and D. W. Coble. 1996. A review of fish control projects. North American Journal of Fisheries Management 16:63–74.

Michell, A. J. 2001. Finfish health in the United States (1609–1969): historical perspective, pioneering researchers and fish health workers, and annotated bibliography. Aquaculture 196:347–438.

Mitchell, S. L. 1814. The fishes of New York described and arranged. Transactions of the Literary and Philosophical Society of New York 1:355–492.

Moffitt, C. M. 2001. Reflections: a photographic history of fisheries and the American Fisheries Society in North America. American Fisheries Society, Bethesda, Maryland.

Montgomery, D. R. 2003. King of fish: the thousand year run of salmon. Westview Press, Boulder, Colorado.

Montgomery, R. 1991. Restoring large river fishery resources: the Mississippi Interstate Cooperative Research Agreement. Fisheries 16(5):44–47.

Moore, E. 1923. Octomitus salmonis, a new species of intestinal parasite in trout. Transactions of the American Fisheries Society 52:74–97.

Moore, E. 1926. Some features of the stream survey undertaken in New York State. Transactions of the American Fisheries Society 56:108–121.

Moore, E. 1927. Progress of the biological survey in New York State. Transactions of the American Fisheries Society 57:65–72.

Myers, G. S. 1964. A brief sketch of the history of ichthyology in America to the year 1850. Copeia 1964:33–41.

Naiman, R. J., S. R. Elliot, J. M. Helfield, and T. C. O'Keefe. 2000. Biophysical interactions and the structure and dynamics of riverine ecosystems: the importance of biotic feedbacks. Hydrobiologia 410:79–86.

Nelson, J. S. 1965. Effects of fish introductions and hydroelectric development on fishes in the Kananaskis River system, Alberta. Journal of the Fisheries Research Board of Canada 22:721–753.

Neves, R. J., S. N. Moyer, L. R. Weaver, and A. V. Zale. 1985. An evaluation of host fish suitability for glochidia of *Villosa Vanuxemi* and *V. nebulosa* (Pelecypoda: Unionidae). American Midland Naturalist 113:13–19.

Northwest Power Planning Council. 1986. Compilation of information on salmon and steelhead losses in the Columbia River basin. Appendix D of the 1987 Columbia River Basin Fish and Wildlife Program, Northwest Power Planning Council, Portland, Oregon.

Palmer, M. A., D. P. Lettenmaier, N. L. Poff, S. L. Postel, B. R. Richter, and R. Warner. 2009. Climate change and river ecosystems: protection and adaptation options. Environmental Management 44:1053–1068.

Palmer, T. S. 1912. Licenses for hook and line fishing. Transactions of the American Fisheries Society 41:91–97.

Pearson, J. C. 1972. The fish and fisheries of colonial North America: a documentary history of the fishery resources on the United States and Canada. Part VII. The inland states. U.S. Department of Commerce, National Marine Fisheries Service Report NOAA (National Oceanic and Atmospheric Administration) 72100305.

Poff, N. L., J. D. Allan, M. B. Bain, J. R. Karr, K. L. Prestegaard, B.D. Richter, R. E. Sparks, and J. C. Stromberg. 1997. The natural flow regime: a paradigm for river conservation and restoration. BioScience 47:769-784.

Postel, S. L. 2000. Entering an era of water scarcity: the challenges ahead. Ecological Applications 10(4):941–948.

Postel, S. L., G. C. Daily, and P. R. Ehrlich. 1996. Human appropriation of renewable freshwater. Science 271:785–787.

Prince, E. E. 1923. The fisheries of Canada. Annals of the American Academy of Political and Social Science 107:88–94.

Quinn, T. J., II. 2003. Ruminations of the development and future of population dynamics models in fisheries. Natural Resource Modeling 16:341–392.

Rahel, F. J. 2002. Homogenization of freshwater faunas. Annual Review of Ecology and Systematics 33:291–315.

Reisner, M. 1989. The next water war—cities versus agriculture. Issues in Science and Technology 5(2):98–102.

Reist, J. D., F. J. Wrona, T. D. Prowse, M. Power, J. B. Dempson, J. R. King, and R. J. Beamish. 2006. An overview of effects of climate change on selected Arctic freshwater and anadromous fishes. Ambio 35:381–6387.

Reuss, M. 2005. Ecology, planning, and river management in the United States: some historical reflections. Ecology and Society 10(1):34. Available online: http://www.ecologyandsociety.org/vol10/iss1/art34/.

Ricker, W. E. 1975. Computation and interpretation of biological statistics of fish populations. Fisheries Research Board of Canada Bulletin 191.

Rostlund, E. 1952. Freshwater fish and fishing in Native North America. University of California Press, Berkeley.

Rothschild, B. J., and R. J. Beamish 2009. On the future of fisheries science. Pages 1–11 *in* R. J. Beamish and B. J. Rothschild, editors. The future of fisheries science in North America. Fish and Fisheries Series 31, Springer, Netherlands.

Salyer, J. C. 1933. Predator studies in Michigan waters. Transactions of the American Fisheries Society 63:229–239.

Schick, R. S., and S. T. Lindley. 2007. Directed connectivity among fish populations in a riverine network. Journal of Applied Ecology 44:1116–1126.

Scott, D. C. 1923. The aboriginal races. The Annals of the American Academy of Political and Social Science 107:63–66.

Seadle, M. 2009. Archiving in the networked world: betting on the future. Library Hi Tech 27:319–325.

Sharma, S., L.-M. Herborg, and T. W. Therriault. 2009. Predicting introduction, establishment and potential impacts of smallmouth bass. Diversity and Distributions 15: 831–840.

Shepherd, E. 2006. Developing a new academic discipline: UCL's contribution to the research and teaching of archives and records management. Aslib (Association for Information Management) Proceedings: New Information Perspectives 58:10–19.

Solman, V. E. F., J.-P. Cuerrier, and W. C. Cable. 1952. Why have fish hatcheries in Canada's national parks? Transactions of the North American Wildlife Conference 17:226–234.

Spangler, A. M. 1893. The decrease of food fishes in American waters and some of the causes. U.S. Fish Commission Bulletin 13(1893):21–35.

Steedman, R. J., T. H. Whillans, A. P. Behm, K. E. Bray, K. I. Cullis, M. M. Holland, S. J. Stoddart, and R. J. White. 1996. Use of historical information for conservation and restoration of Great Lakes aquatic habitat. Canadian Journal of Fisheries and Aquatic Sciences 53 (Supplement 1):415–423.

Stevens, L. E., J. P. Shannon, and D. W. Blinn. 1997. Colorado River benthic ecology in Grand Canyon, Arizona, USA: dam, tributary, and geomorphological influences. Regulated Rivers and Research Management 13:129–149.

Sullivan, C. R. 1979. Dingell–Johnson—an increasing role in the future. Fisheries 4(3):5, 26.

Taylor, J. E. 1999. Making salmon: an environmental history of the northwest fisheries crisis. University of Washington Press, Seattle.

Titcomb, J. W. 1917. Importance of a permanent policy in stocking inland waters. Transactions of the American Fisheries Society 47:11–21.

Tipa, G., and R. Welch. 2006. Co-management of natural resources: issues of definition from an indigenous community perspective. Journal of Applied Behavioral Science 42:373–391.

U. S. Commission of Fish and Fisheries. 1897. A manual of fish-culture, based on the methods of the U.S. Commission of Fish and Fisheries, with chapters on the cultivation of oysters and frogs. U.S. Commission of Fish and Fisheries, Government Printing Office, Washington, D.C.

U. S. Federal Register. 2007. Indian entities recognized and eligible to receive services from the U.S. Bureau of Indian Affairs. Federal Register 72:55 (22 March 2007):13648–13652.

Van Winkle, W., C. C. Coutant, H. I. Jager, J. S. Mattice, D. J. Orth, R. G. Otto, S. F. Railsback, and M. J. Sale. 1997. Uncertainty and instream flow standards: perspectives based on hydropower research and assessment. Fisheries 22(7):21–22.

Vanderkemp, A. F. 1880. Extracts from the Vanderkemp papers: from the Hudson to Lake Ontario in 1792. Publications of the Buffalo Historical Society 2(2):49–80.

Viosca, P., Jr. 1945. A critical analysis of practices in the management of warm-water fish with a view to greater food production. Transactions of the American Fisheries Society 73:274–283.

Walters, C., and J. Korman. 1999. Linking recruitment to trophic factors: revisiting the Beverton–Holt recruitment model from a life history and multispecies perspective. Reviews in Fish Biology and Fisheries 9:187–202.

Walters, C. J., V. Christensen, and D. Pauly. 1997. Structuring dynamic models of exploited ecosystems from trophic mass-balance assessments. Review in Fish Biology and Fisheries 7:139–172.

Walters, C. J., and S. Martell. 2004. Fisheries ecology and management. Princeton University Press, Princeton, New Jersey.

Weber, M. 1886. Report 99. Pearls and pearl fisheries. U.S. Fish Commission Bulletin 6:321–328.

Weidel, B. C., D. C. Josephson, and C. E. Kraft. 2007. Littoral fish community response to smallmouth bass removal from an Adirondack Lake. Transactions of the American Fisheries Society 136:778–789.

Western District Court of Michigan. 1979. United States v. State of Michigan. Federal Supplement 471:192.

Western District Court of Michigan. 1985. United States v. State of Michigan (Consent Order 1985 Settlement Agreement). Indian Law Reporter 12:3079.

Western District Court of Michigan. 2000. United States v. State of Michigan (Consent Decree). Case 2:73-cv-00026 (M26 73).

Whitaker, H. 1892. Early history of the fisheries on the Great Lakes. Transactions of the American Fisheries Society 21:163–179

Williams, J. D., M. L. Warren Jr., K. S. Cummings, J. L. Harris, and R. J. Neves. 1993. Conservation status of freshwater mussels of the United States and Canada. Fisheries 18(9):6–22.

Williams, J. E., A. L. Haak, H. M. Neville, and W. T. Colyer. 2009. Potential consequences of climate change to persistence of cutthroat trout populations. North American Journal of Fisheries Management 29:533–548.

Williams, R. N., P. A. Bisson, D. L. Bottom, L. D. Calvin, C. C. Coutant, M. W. Erho Jr., C. A. Frissell, J. A. Lichatowich, W. J. Liss, W. E. McConnaha, P. R. Mundy, J. A. Stanford, and R. R. Whitney. 1999. Scientific issues in the restoration of salmonid fishes in the Columbia River. Fisheries 24(3):10–19.

Worster, D. 1992. Under western skies: nature and history in the American west. Oxford University Press, New York.

Yoshiyama, R. M., and F. W. Fisher. 2001. Long time past: Baird Station and the McCloud Wintu. Fisheries 26(3):6–22.

Young, H. A. 1952. Conservation and wise utilization of natural resources in Canada. Annals of the American Academy of Political and Social Science 281:196–202.

Chapter 2

Fish Population Dynamics: Mortality, Growth, and Recruitment

MICHEAL S. ALLEN AND JOSEPH E. HIGHTOWER

2.1 INTRODUCTION

Fisheries management is a rewarding career because it is challenging and fun and, most importantly, has a real impact on the quality of people's lives. Decisions made by fisheries managers about a commercial fishery directly affect the income of fishers. Decisions about a recreational fishery can influence angler satisfaction and the level of participation, which has direct economic effects on tackle shops, motel and restaurant owners, and fishing guides. Because these decisions can have an impact on a community or region, it is critical to have the best available information about fisheries resources, including habitat quality and species interactions, as well as the needs of human users of a resource. Methods to evaluate many of these factors are described in other chapters of this text.

The focus of this chapter is the use of quantitative methods to evaluate how management actions regarding harvest may influence fish abundance, the size of fish in a population, angler catch, and total yield (i.e., biomass of fish removed from a population). Assessment of these basic population characteristics enables a fisheries manager to detect changes occurring in a population in response to fishing. Diagnosing the condition of overfishing is an important step in fisheries management, and identifying management actions that can improve fish abundance and angler catches is obviously required for sustaining and improving fisheries. Thus, fish population dynamics and assessment are literally where "the rubber meets the road" in fisheries management.

Assessment of fish populations usually contains much uncertainty. John Shepherd's adage that "fish are like trees, except they are invisible and they move," provides a first look at the difficulties in evaluating fish populations. Fish are not typically visible, and thus our "view" of a fish population usually comes from a variety of sources, including anglers, commercial fisheries, and different sampling gears. All sampling gears have inherent sampling biases, and fisheries managers almost always work with incomplete information about fish stocks.

The literature contains a wide range of complex methods to analyze fish populations from sampling data, and reading through the latest journal articles can be discouraging to students trying to gain a basic understanding of fish population dynamics. Although some facets of fisheries assessment require highly quantitative methods and sophisticated software programs, we contend that most fish population assessments can be relatively straightforward and require only simple mathematics and practice with spreadsheet software. Canned software packages are useful, but they are not as helpful in learning how methods work and are often inflexible in their analysis options. In contrast, the spreadsheet methods presented here can be used to tailor population models to specific needs of investigators. The objective of this chapter is to summarize the basics of fish population dynamics and the skills needed to evaluate fisheries management scenarios. Spreadsheet examples are provided so the reader can learn by doing the analyses and apply them (see http://fishweb.ifas.ufl.edu/allenlab/courses.html for spreadsheets).

2.2 OVERVIEW OF DYNAMICS IN INLAND FISH POPULATIONS

The abundance of fish (or any animal) is limited by available resources. Fish populations in new reservoirs and farm ponds previously devoid of fish will exhibit a brief period of unlimited exponential population growth. This occurs because resources are initially unlimited, but as the fish population expands, food and space resources become limiting and fish abundance is then regulated by density-dependent growth and (or) survival. A common representation of this process is the logistic growth model:

$$B_{t+1} = B_t + rB_t(1 - \frac{B_t}{K}), \tag{2.1}$$

where B is the biomass at time t, r is the maximum population growth rate, and K represents the carrying capacity, the maximum equilibrium biomass that can be supported by the resources available in the system. Notice from equation (2.1) that if B_t is low, the population will grow rapidly because resources are not limited (i.e., $1 - B_t/K$ is close to 1). This period of unlimited population growth is considered the exponential growth phase (Figure 2.1), and the rate of increase is determined by the parameter r. As B_t approaches K, population growth is slowed by density dependent processes until the population reaches K (Figure 2.1).

However, the biomass of fish populations does not remain static but fluctuates around an average abundance due to changes in environmental conditions, habitat quality and quantity, fishing mortality, and interactions with other species such as predators or competitors (Figure 2.1). Fishing can hold average fish biomass well below the carrying capacity for the system, but in these cases random fluctuations in biomass still occur due to variation in fish recruitment. Most fisheries managers attempt to manage fish stocks that are varying around some average abundance, which may or may not be close to K.

The specific factors that influence fish abundance and biomass are typically described by three dynamic rate functions: mortality, growth, and recruitment. Mortality is usually divided into two categories: death due to fishing and death due to natural causes. Fishing mortality is often the focus of fisheries managers because it can be controlled with management actions. Natural mortality is almost always unobserved and is often outside managers' control. Growth is the increase in size of individual fish and can be measured in terms of length or weight. Growth affects a fish's vulnerability to predation and fishing, as well as the food resources available to each individual fish. Recruitment refers to young fish entering the population, and from a management perspective it usually means recruitment to the fishable stock. If new recruits are not replacing losses due to mortality, then the population will eventually decline to zero. Recruitment and growth both increase the biomass of a cohort (year-class), whereas mortality causes both the number of fish and total biomass of a cohort to decline.

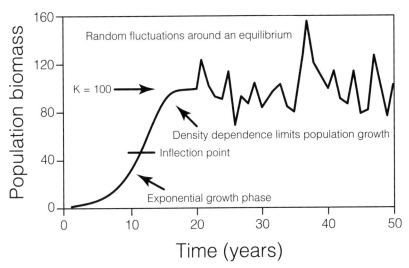

Figure 2.1. Example of a logistic population growth model with maximum growth rate, r, equal to 0.5 and carrying capacity, K, equal to 100. The inflection point indicates the change from exponential to decreasing incremental population growth. Random fluctuations after 20 years demonstrate a fish population whose abundance fluctuates around an equilibrium biomass value.

A good understanding of fish population dynamics (i.e., how mortality, growth, and recruitment interact to affect abundance) is required for informed fisheries management. An example of how important these factors can be is illustrated by the case of the endangered Kootenai River white sturgeon (Paragamian et al. 2005). Surveys have shown that this population has had essentially no recruitment since the early 1970s. Harvest was prohibited in 1984 to help protect the remaining adult stock. The fishery closure substantially reduced annual mortality, but the population still declined by about 9% each year. Field studies have been conducted to learn where and when spawning occurs and what habitat changes may be causing the lack of recruitment. Population models have been developed to predict future abundance of the population and to understand how releases of small fish produced in hatcheries contribute to the overall population. The hope is to use all these sources of information to bring about the population's recovery.

Studies of population dynamics usually involve the development of models. Some fisheries managers are skeptical about the use of models to inform management decisions; however, an experienced manager generally has an idea of how the population or fishery operates. Constructing a model with numbers and equations forces the investigator to be explicit about the hypothesized processes that influence fish population size. Thus, population models should be viewed as hypotheses for management and future research needs. For example, rather than speculating that low recruitment is limiting the abundance of fish in a population, constructing a model can highlight the need to estimate recruitment trends to evaluate the impacts. The model can help identify data gaps and can guide future research toward the areas of greatest uncertainty. When managers do have a good understanding of a stock's dynamics or how a fishery operates, the model serves as a repository for knowledge and experience that may have been gained over many years (Hilborn et al. 1984). Many fish population models are fairly simple because managers rarely have enough data to justify elaborate population models. In many cases, relatively simple population models perform better for management than do complex models because of the high uncertainty associated with numerous parameters required by complex models (Walters and Martell 2004).

Population models are usually developed for exploited species. One reason for this pattern is that it is easier to justify the cost of conducting surveys and catch sampling programs for a species that supports an important fishery. Modeled scenarios can be used to examine potential effects of fishing and to predict how harvest regulations may increase population size or fishery yield (i.e., biomass of fish harvested). Often the purpose of a population model is to determine whether overfishing is occurring. Overfishing is generally defined as a fishing mortality rate above some target level; for example, the rate that is estimated to result in the maximum long-term (sustainable) yield. Overfishing may occur in two ways—growth overfishing or recruitment overfishing. Growth overfishing results in reduced yield because fishing mortality is too high on young or small fish. If the rate of fishing is reduced or the size at first harvest is increased, fish would have additional time to grow before being harvested and long-term average yields would increase. Recruitment overfishing means that fishing has reduced the spawning stock (large adult fish) to a level at which recruitment is limiting population abundance. If recruitment overfishing continues, the population will decline to very low levels, eventually causing the collapse of the fishery. If the rate of fishing can be reduced, the spawning stock will increase, resulting in higher recruitment and sustainable long-term yield.

Population models also play an important role in the study of rare or threatened species. As in the case of exploited populations, when managing for threatened fishes the need is to understand which factors regulate abundance. The difference is that the information is used not to regulate harvest but to aid in rebuilding the population to a viable level. There is no fishery in the case of rare species, so biological data generally come from research or management surveys rather than commercial or recreational harvest.

2.3 METHODS OF ESTIMATING POPULATION PARAMETERS

2.3.1 Expressions of Mortality

Estimates of mortality are an essential part of assessing fish populations. Fish populations typically exhibit very high mortality during larval and juvenile life stages (often exceeding 99%), followed by lower mortality rates during adult life. Most fisheries investigations are concerned with fish mortality during adulthood, and thus managers emphasize adult fish mortality rates. Estimates of mortality are required to understand how fishing influences fish abundance, angler harvest (numbers of fish), and yield (weight of fish). We begin with some expressions of mortality rates that are commonly used in fisheries investigations, describe how estimates of mortality rates are obtained, and then discuss the advantages and disadvantages of various methods. Definitions of all symbols are found in Table 2.1.

Mortality is typically separated into two components: fishing mortality and natural mortality. Fishing mortality can be controlled via length and creel limits, closed seasons, closed areas, or restrictions on fishing effort (see Chapter 7). Thus, managing fishing mortality is one of the most common practices of fisheries managers. Natural mortality occurs due to predation, disease, parasitism, and any other natural cause. Natural mortality of adult fish is not typically controlled by fisheries management actions, but the level of natural mortality is very

Table 2.1. Population parameters, their definitions, and common methods for estimation of each parameter.

Parameter symbol	Definition	Estimation methods
Z	Instantaneous total mortality	Catch curve, tagging study, $F + M$
M	Instantaneous natural mortality	Tagging study, surrogate methods, subtraction ($M = Z - F$)
F	Instantaneous fishing mortality	Tagging study (angler reported or telemetry), catch or population size estimates, catch-at-age methods
A	Annual total mortality	As above for Z, $A = u + v$
S	Annual total survival	$e^{-Z}, 1 - A, S = N_{t+1} / N_t$
S_0	Annual natural survival	e^{-M}
u	Annual exploitation rate	As above for F
v	Annual natural mortality rate	As above for M
cf	Conditional fishing mortality	As above for F
cm	Conditional natural mortality	As above for M
L_∞	Asymptotic length	Age-growth, tagging study
k	Growth rate	Age-growth, tagging study
t_0	Age at zero length	Age-growth, tagging study

important for establishing harvest criteria. We provide some basic mortality expressions here, and detailed analysis methods are described by Miranda and Bettoli (2007).

First, we describe finite and instantaneous mortality rates. Finite mortality rates are the fraction of the fish stock that dies in a finite time period (e.g., a year). Instantaneous mortality rates can be calculated from finite rates and are useful for estimating the number of fish at any continuous time interval (e.g., fractions of a year). For a cohort (year-class) of fish, the decline in numbers over time usually follows an exponential pattern (Figure 2.2).

The change in cohort size (N) per unit time (t) depends on the total instantaneous mortality rate (Z) and population size (more deaths per unit time when N is large):

$$\frac{dN}{dt} = -ZN, \qquad (2.2)$$

After integration, we obtain an exponential decline in N per unit time as:

$$N_t = N_0 e^{-Zt}, \qquad (2.3)$$

where N_t = number alive at time t, N_0 = number alive initially (at t_0), Z = instantaneous total mortality rate, and t = time units since t_0.

Fisheries managers frequently work in time step units of a year. Given a finite time step of 1 year, the annual total survival rate is: $S = N_{t+1}/N_t$. The value of Z can be determined by $S = e^{-Z}$ and thus $Z = -\log_e(S)$. Box 2.1 shows an example of how to work between finite and instantaneous rates.

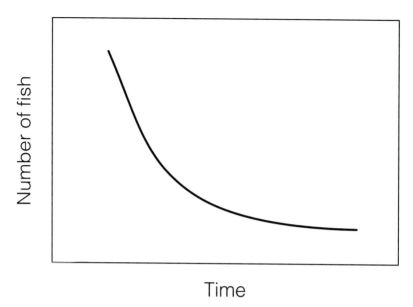

Figure 2.2. Depiction of an exponential decline in numbers of fish with time. The finite mortality rate is a constant proportion of the population at each incremental time step.

Discrete (type I) fisheries.—A discrete fishery is one in which fishing mortality and natural mortality occur separately within the year. Examples would be highly-seasonal fisheries in which most fishing mortality occurs in a short time period, and natural mortality can be assumed to be separated in time. The mortality expressions for discrete fisheries are

$$u = cf = 1 - e^{-F}, \qquad (2.4)$$

$$v = cm = 1 - e^{-M}, \qquad (2.5)$$

where u is the annual exploitation rate (the fraction of the fish stock harvested within a year), cf is the conditional fishing mortality rate, F is the instantaneous fishing mortality rate, v is the annual natural mortality rate (the fraction of a fish stock that dies due to natural mortality within a year), cm is the conditional natural mortality rate, and M is the instantaneous natural mortality rate (Table 2.1). Instantaneous total mortality (Z) is defined as $Z = F + M$. In discrete fisheries, because M and F are assumed to operate separately, the conditional rates (cf and cm) are the annual fishing (u) and natural mortality (v) rates.

Although it may seem that truly discrete fisheries are relatively rare, many fisheries have highly seasonal rates of fishing mortality with most of the harvest occurring during relatively short periods of the year. In practice, many fisheries models use $S_0 = e^{-M} = (1 - v)$ to approximate the survival rate from natural mortality (S_0), and u to represent annual exploitation even if the fishery is continuous. The F in this case of discrete fisheries can be found by $F = -\log_e(1 - u)$ (as per equation 2.4 above). Box 2.2 shows mortality expressions used for discrete fisheries and in other applications. Recent fish stock assessment textbooks have utilized the discrete mortality equations for fisheries where F and M occur together because the approximation is

> **Box 2.1. Finite and Instantaneous Mortality Rates**
>
> Suppose you start with 1,000 fish at time zero 0 and lose 9% per year due to total mortality. The number of fish in each yearly step is simply the number the year before times 0.91.
>
> | Year | 0 | 1 | 2 | 3 | 4 | 5 |
> |---|---|---|---|---|---|---|
> | Number of fish | 1,000 | 910 | 828 | 753 | 686 | 624 |
>
> Let's define the following:
>
> A = annual total mortality rate,
> S = annual total survival rate, and
> Z = instantaneous total mortality.
>
> In this example, $A = 0.09$, and
>
> $S = 1 - A = 0.91$, given that
>
> $S = e^{-Z}$,
>
> $Z = -\log_e(S)$.
>
> Therefore
>
> $Z = -\log_e(0.91) = 0.0943$.
>
> Now use the instantaneous rate, $N_t = N_0 e^{-Zt}$, to predict the number of fish at age 5 (N_5).
>
> $N_5 = 1,000 e^{-0.0943(5)} = 624$.
>
> Thus, any mortality rate can be described either as a finite or instantaneous rate, and this example shows the ease of transferring between the rate types.

typically satisfactory even when fishing and natural mortality occur simultaneously throughout the year (Walters and Martell 2004).

Continuous (type II) fisheries.—A continuous fishery is one in which fishing mortality and natural mortality operate concurrently, so use of instantaneous rates F and M is required to model this relationship. Annual survival is indicated by $S = e^{-Z} = e^{-(F+M)}$. The relationships

> **Box 2.2. Discrete Fishery Mortality**
>
> Suppose a fish stock is determined to have the following mortality rates:
>
> $$u = 0.3, \text{ and}$$
>
> $$M = 0.2.$$
>
> Here we transform this instantaneous mortality rate to a natural survival rate as
>
> $$S_0 = e^{-0.2} = 0.82.$$
>
> Using this format and starting with 1,000 age-2 fish, we calculate the number of fish surviving from age 2 to age 3:
>
> $$N_3 = 1,000 \times (0.82) \times (1 - 0.3) = 574.$$
>
> Natural deaths are: $1,000 \times 0.18 = 180$, and the total catch would be $1,000 \times 0.82 \times 0.3 = 246$, because we have assumed that fishing took place after natural mortality occurred. So, the total deaths are 426 fish (i.e., 180 from natural causes and 246 from fishing). Notice that the total deaths (426) plus survivors (574) is the original value of 1,000 fish. This is because we assumed that fishing and natural mortality operate separately within the year.

between instantaneous and finite rates are proportional, such that the exploitation rate can be obtained as the fraction of total annual mortality (A) that is due to F:

$$u = \frac{F}{Z} \times A, \qquad (2.6)$$

In continuous fisheries, the total annual mortality rate A is found by

$$A = cf + cm - (cf \times cm), \qquad (2.7)$$

where the quantity $cf \times cm$ signifies that some fish that die due to fishing mortality would have died due to natural mortality, and vice versa. This is a key difference between the discrete fishery model and the continuous fishery model because the equations accounting for a discrete fishery do not impart any interaction between fishing mortality and natural mortality: they are assumed to occur separately in time. The equations accounting for a continuous fishery explicitly model the fact that fish dying from one cause of mortality (e.g., fishing) are no longer available to die from the other cause of mortality (e.g., natural mortality) and vice versa.

2.3.2 Estimation of mortality rates

Total mortality.—Fisheries managers seek to estimate mortality to understand how fishing mortality and natural mortality rates are influencing fish populations. The most basic approach is to estimate total annual mortality, which is frequently evaluated as the change in fish abundance with age. Catch curves are a regression of the natural log of the number of fish at age on fish age, and the slope of the relationship is an estimate of Z (see Box 2.3). The assumptions of a catch curve are that (1) mortality rate is constant across ages, (2) recruitment is constant, and (3) the age sample is a random sample of fish abundance with age. Although these assumptions are not strictly met in most applications, catch curves usually provide general estimates of Z. Assumption 1 is usually addressed by including only fish that are expected to have similar mortality rates. Assumption 2 is often not a major problem provided that recruitment has not exhibited an increasing or decreasing trend through time. Random recruitment variation tends to make a catch curve bumpy but does not bias the slope (Ricker 1975). Selectivity of the sample gear must be considered relative to assumption 3 and the youngest fish not fully vulnerable to the gear are typically excluded from catch curves (see Box 2.3).

Passive tagging estimates of fishing mortality.—Estimates of fishing mortality may be obtained using passive tagging or active tagging methods. Passive tagging involves tagging fish with external tags and obtaining anglers' reports of harvesting tagged fish. The estimate of annual exploitation rate is obtained by:

$$u = \frac{C}{T}, \tag{2.8}$$

where u is the exploitation rate, C is the corrected number of tagged fish caught, and T is the corrected number of tagged fish in the population. Values of C must be corrected for nonreporting of tags, and values of T should be corrected for short term tag loss and tagging-associated mortality. If long term tag loss is substantial, T can also be adjusted downwards to account for chronic tag loss. Angler reporting rates are the most difficult issue with this approach. The most common method of estimating reporting rates is to use some high-reward tags for which it can be assumed that reporting rate is 100%, then adjust the number of standard tags returned based on the assumption that capture rate of fish by anglers is not influenced by reward value (Pollock et al. 2002):

$$\hat{\lambda} = \frac{\left(\dfrac{C_S}{T_S}\right)}{\left(\dfrac{C_H}{T_H}\right)}, \tag{2.9}$$

where $\hat{\lambda}$ is the estimated reporting rate for standard tags, C_S is the number of standard-tag fish reported by anglers, T_S is the number of fish tagged with standard tags, C_H is the number of high-value-tag fish reported by anglers, and T_H is the number of fish tagged with high-value-reward tags. Once an estimate of $\hat{\lambda}$ is obtained, it can be used to correct C in equation (2.8) for the standard-tag fish.

Correcting estimates of C and T for tag loss and tagging mortality is typically required. Tag loss is frequently evaluated through double tagging a subset of fish to estimate the tag loss

Box 2.3. Catch Curve Analysis

Age structure was determined by means of an age–length key. Ten fish per centimeter-group were aged and the aged fish applied back to the total length sample (see DeVries and Frie 1996). The age structure data are shown below.

Table. Age structure data for largemouth bass from the Apalachicola River, Florida.

Age	N	$(\log_e N)$
0	155	5.04
1	283	5.65
2	128	4.85
3	285	5.65
4	73	4.29
5	31	3.43
6	22	3.09
7	4	1.39
8	2	0.69
9	5	1.61
10	0	
11	2	0.69

If abundance at time t is defined as $N_t = N_0 e^{-Z t}$, then the log-transformed equation is linear ($\log_e N_t = \log_e N_0 + -Z_t$, with intercept $\log_e N_0$ and slope $-Z$). Results of the regression show that $Z = 0.64$. Annual total survival and total mortality can be obtained as: $S = e^{-Z} = 0.53$, and $A = 1 - S = 0.47$. So, total annual mortality in this population is around 47%. Notice that we did not use ages 0, 1, 10, or 11 in the regression. Ages 0 and 1 were excluded because they were apparently not fully vulnerable to the gear. Ages 10 and 11 were excluded because of low sample size. Older ages with less than five fish are often removed from a catch curve to reduce their influence on the overall estimate of Z. Although ages 10 and 11 were removed, techniques are available that allow their inclusion in the catch curve analysis (i.e., weighted catch curves; Miranda and Bettoli 2007). Note that the catch curve exhibits some bumpiness, likely due to both variation in recruit-

(Box continues)

Box 2.3. Continued.

ment and sampling variability. Age-3 fish appeared to be from a relatively strong year-class, whereas fish ages 7 and 8 were from relatively weak year-classes. See Maceina (1997) for more discussion of how residuals around a catch curve can provide an index of past recruitment.

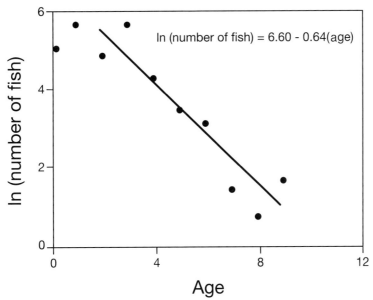

Figure. Example of a catch curve for largemouth bass from the Apalachicola River, Florida. Fish were collected with electrofishing by Florida Fish and Wildlife Conservation Commission biologist Rich Cailteux.

rate for single-tagged fish. Tagging mortality is usually conducted with cage experiments to evaluate short-term mortality from the tagging process (Box 2.4). Pine et al. (2003) provided a detailed discussion of the assumptions and methods of estimating angler reporting rates, tag loss, and tagging mortality (see Box 2.4). A detailed example of how to deal with these biases through time during a tagging study is shown by Smith et al. (2009).

If a multiyear tagging program is established, both fishing and natural mortality can be estimated (Hoenig et al. 1998; Jiang et al. 2007). As in the above short-term approach, addressing practical issues such as tagging mortality, tag loss, and nonreporting of tags is critical. An auxiliary estimate of the reporting rate (e.g., through the use of high-reward tags) is generally necessary to separate total mortality into F and M.

Active tagging methods for estimating F *and* M.—Active tagging involves the use of telemetry techniques to estimate fishing mortality and natural mortality (Hightower et al. 2001). By means of this method, fish tagged with sonic or radio transmitters are located at regular intervals. The status of each fish is based on movement (or lack of movement) between searches. Fish that move between successive locations are obviously still alive, whereas fish that stop moving are classified as natural mortalities. Live fish are sometimes found in the same location on consecutive searches,

Box 2.4. Estimate of Fishing Mortality from a Passive-Tagging Study

Here we describe the steps for estimating fishing mortality for redear sunfish from Lake Panasoffkee, Florida. Data were obtained from Crawford and Allen (2006). Estimates of fishing mortality were obtained in 1998 and 1999, but in this example we focus on the 1999 estimates. In January 1999, 753 redear sunfish greater than 15 cm in total length (TL) were collected by means of electrofishing, tagged with passive dart tags, and released into Lake Panasoffkee.

To estimate tag loss, 163 of the fish were double tagged. The estimate of annual tag loss from reported fish was 25% (in 1999, two of the eight double-tagged fish that were returned had only a single tag). Anglers were contacted to verify that harvested fish contained only one tag at time of capture.

Short-term tagging mortality was estimated with a cage experiment in both 1998 ($N = 2$ cage treatments for 3 d) and 1999 ($N = 1$ cage treatment for 6 d). Fish mortality through 3 d was 0 for all replicates, indicating that short-term tagging associated mortality was nil.

Nonreporting was estimated directly via a creel survey in both years. The creel clerk recorded tagged fish numbers in angler creels, and the reporting rate (λ) was estimated directly in this case as the proportion of tagged fish observed in angler creels that were reported. In 1999, the reporting rate was 83%. See Crawford and Allen (2006) for discussion of a second method for estimating reporting rates in this study by use of a variable reward system.

Anglers returned a total of 55 tagged redear sunfish. Therefore, the approximate estimate of the number of fish available for capture by anglers (T) was

$$T = 753 \times 0.75 \times 1 = 564,$$

where 753 is the number tagged, 0.75 is the tag retention rate (1 minus tag loss) and 1 is the tag survival because short-term tagging mortality was estimated as 0. The estimate of the number of fish caught by anglers (C) was

$$C = 55 / 0.83 = 66,$$

where 0.83 is the reporting rate.

The estimate of annual exploitation (u; equation 2.8) was

$$u = 66 / 564 = 0.12.$$

Therefore the annual fishing mortality for fish greater than 15 cm at Lake Panasoffkee was 12%.

so classifying a fish as a natural mortality should be done only after several searches indicate no movements. Fish that are harvested disappear from the system, so those fish provide an indirect estimate of fishing mortality. However, not every fish with a transmitter is located on each search occasion, so the probability that a fish was harvested depends on how many consecutive searches it has not been found. The study area should be closed to emigration to avoid the risk of confusing emigration and fishing mortality. If emigration can occur (e.g., in a section of river), then an array of receivers should be set up to detect emigrating fish. Those fish are then censored from the tagged population so that they are not incorrectly classified as fishing mortalities.

Advantages of this approach are that it does not rely on angler-reported tags, natural mortality is directly estimated, and the study can be carried out over a shorter interval than can the multiyear tagging approach described above. Mortality estimates can be made over a fine time scale (e.g., monthly or quarterly), and estimates of fishing mortality can be highly seasonal (Hightower et al. 2001; Waters et al. 2005; Thompson et al. 2007). Another benefit of this approach is that periodic searches provide valuable information about fish movements and habitat use, in addition to estimates of fishing mortality and natural mortality.

2.3.3 Growth

One of the first steps in developing a management plan for a fishery is to characterize growth. The growth rate of a fish determines various aspects of its ecology (e.g., vulnerability to predation and sexual maturation) as well as its recruitment into a fishery. For recreational fisheries, the growth rate determines when fish reach a size that would be considered desirable, either for harvest or as trophy fish.

Harvest regulations for recreational fisheries are usually defined in terms of fish length. For example, a minimum length of 356 mm (14 in) might be established for a largemouth bass fishery to protect age-3 and younger fish from harvest and thereby increase the number of adults. Growth rates strongly influence the potential for a minimum length limit to improve the abundance of large fish in a population. Enacting minimum length regulations on fish populations with slow growth could exacerbate management problems (see Chapter 7).

Although growth is usually discussed in terms of length, fish growth in weight is used in some analyses. For example, weight can be used as a surrogate for fecundity or the contribution of females to the spawning population. Harvest regulations in some fisheries are set to allow the average weight of fish to increase, with the expectation that protecting large, highly fecund females will improve recruitment.

Information about growth also indicates the "health" of a population relative to its food resources and the quality of the aquatic environment. Fast growth suggests that fish density is in balance with food resources and that habitat quality is adequate. If fish are growing slowly (e.g., small mean length at a given age), it could indicate that the density is too high (relative to the food supply) or that habitat is not suitable to support an adequate prey base.

One way to summarize growth is to fit a model relating age to length or weight. For example, age in years and total length in centimeters could be determined for a random sample of fish. A model can be fitted to the data to describe the relationship. The model is typically a curve because the rate of growth usually decreases with age (e.g., Figure 2.3). A growth curve is convenient for modeling because the pattern developed from many data points can be described using a single curve with only two or three parameters.

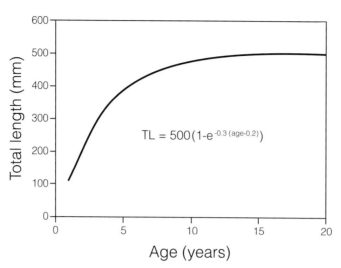

Figure 2.3. Example of a von Bertalanffy growth curve relating total length (TL) to age. Notice that incremental growth between ages declines with age.

A single model is typically appropriate for only one life stage (e.g., adult fish) because fish growth usually varies with age. Rapid growth occurs at the larval and juvenile stages, and then growth slows for adult fish as energy is diverted from somatic growth to gonadal development. The adult life stage is generally of greatest interest to fisheries managers because it is the life stage of harvestable fish.

Another way of estimating growth is through tagging (e.g., Smith and McFarlane 1990; Smith and Botsford 1998; Quinn and Deriso 1999). Ideally, fish of a range of lengths are tagged and released. When tagged fish are encountered either through surveys or by anglers, length is again determined. The time at large and the change in length between tagging and recapture are used to estimate growth rate and maximum length. For instance, the change in length might be considerable for a small fish that is growing rapidly or for a fish that is at large for many months. A fish that is close to its maximum length when tagged will not grow substantially regardless of the time at large. An advantage of this approach is that it is not necessary to estimate fish ages. This means that the method can be used on fish for which aging is not possible or can be used as an independent check on a growth curve derived from age data. It is important that the length at recapture be measured accurately, so lengths reported by fishers must be used with caution. Detailed analyses of growth data are shown in Isely and Grabowski (2007) and Quist et al. (in press).

Weight–length relationships.—In many fisheries management applications, it is useful to predict fish weight from length or vice versa. The relationship between fish weight and length is typically exponential and can be described by

$$W = aL^b, \tag{2.10}$$

where W is fish weight, a is the intercept of the weight–length relationship, L is fish length, and b is the exponent that describes the steepness of the change in weight as fish grow in length. Most fish exhibit b near 3, which is commonly called isometric growth. Allometric growth occurs when *b* is lower or higher than 3, which means that the fish changes shape (in terms of weight) with an incremental change in length. Equation (2.10) has historically been estimated with $\log_{10}$ transformations of both length and weight data followed by computation by linear regression, but the ease of computer optimization routines in spreadsheets means that least-squares fits of the nonlinear equation (2.10) are simple and obtain nearly identical parameter estimates (see "Model fitting" below).

Models for length and weight at age.—The model most commonly used for length-at-age data is the von Bertalanffy growth curve:

$$L_t = L_\infty \left(1 - e^{-k(t-t_0)}\right), \qquad (2.11)$$

where L_t is length at age t (usually in years), L_∞ is the asymptotic length, k is the growth rate, and t_0 is the theoretical age at which the fish would have a length of 0 (Table 2.1). The von Bertalanffy growth curve describes growth that slows with age, as fish approach their maximum length (Figure 2.3)—a relationship that frequently occurs with adult fish. Note that individual fish may be larger than L_∞, because L_∞ simply represents the expected average maximum length. The growth rate k, sometimes referred to as the growth completion rate, is the rate at which fish approach L_∞. The parameter t_0 will generally be close to 0 if the growth data include young fish. If the growth data are from fish of harvestable length (so that t_0 represents a considerable extrapolation), the estimate of t_0 may be far from 0 and have no real biological meaning. Other growth curves have been put forward (e.g., Schnute 1981), but the von Bertalanffy curve continues to be widely used because it fits observed data for a variety of organisms (Cushing 1981), the parameters have a biological interpretation, and many published parameter estimates are available for comparison across populations of a given species.

Inspection of a von Bertalanffy curve (Figure 2.3) shows that the growth rate (tangent to the curve) is continuously decreasing towards a growth rate of 0 at $L_t = L_\infty$, such that the incremental growth (e.g., growth per year) declines with age. This assumption appears to work well to model the adult (i.e., age 1 and older) phase for most fish species when the growth measure of interest is change in length with age.

When growth is described in terms of increasing weight, the growth rate typically increases at younger ages, reaches a maximum at some intermediate ages, and then decreases at older ages. For this situation, a curve with an inflection point is needed. The Gompertz curve can be used to model growth in size (Quinn and Deriso 1999):

$$Y_t = Y_\infty e^{\left(-\frac{1}{k} e^{-k(t-t_0)}\right)}, \qquad (2.12)$$

where *Y* is length, weight, or some other measure of size, and the other parameters are as defined for equation (2.11).

Model fitting.—Growth curves can be fitted using least-squares methods where the squared differences between the observations and the fitted curve are minimized. In linear regression there are exact formulas for calculating the slope and intercept. In fitting a nonlinear model, parameter estimates are obtained iteratively (over a series of steps) by making small changes in the parameter estimates until no further improvement in the error sum of squares can be made. The calculations are readily done in a spreadsheet (see Box 2.5). Starting values must be chosen for each parameter. For a von Bertalanffy growth equation, the maximum observed length in the data is usually a good starting value for L_∞. A value of 0.2 or 0.5 tends to work well for k, and a starting value of 0 should be sufficient for t_0. The same final estimates should be obtained from any reasonable starting values if the model fits well and the data cover a reasonable range of ages and sizes.

Several factors affect the reliability of the fitted growth model. The most important factor is having a range of lengths, including older fish with lengths approaching L_∞. It is common to overestimate L_∞ if the growth data include only young, fast-growing fish. Small sample sizes at older ages can be a source of bias, depending on whether the curve is fitted using means by age or individual observations. It is generally more convenient to fit the curve using means, but the disadvantage is that each mean is given the same weight regardless of the number of observations. Means at older ages are often quite variable because of the small sample sizes. There are techniques for weighting each observation based on its sample size or inverse of the variance at each age, but a simpler approach is just to fit the curve to the individual observations. Aging error can introduce bias depending on whether the errors are random and unbiased or systematic (e.g., consistent underaging). In most instances where aging bias is a problem, the age of older fish is often underestimated, which results in an overestimate of k (Leaman and Beamish 1984).

2.3.4 Recruitment

Fish recruitment is typically defined as the number of fish that survive to a specific age or size in a given year. Although recruitment can be defined in a number of ways, it is most commonly specified as either the number of fish that reach age 1 each year or the number of fish that survive to the first age at which they may be captured in a fishery. Thus, the term recruitment can be used to denote the number of fish at various life stages (age 0, age 1, or an older age), and it is important to specify the life stage at which fish are considered recruits. In most cases, fish are considered recruited to the population after they reach a size or age at which the very high larval mortality rates (see section 2.3.1) have already occurred and the fish can be considered part of the adult population.

Measures of fish recruitment are vital to assessing fish stocks. Recruitment can vary from year to year by orders of magnitude. This high variation in recruitment influences population abundance, age structure, and number of large fish, and it can influence fish growth rates when large year-classes cause density-dependent interactions. Thus, variation in recruitment strongly influences adult fish abundance as strong or weak year-classes move through a fishery, and understanding recruitment variation among years is an important consideration when evaluating harvest policies.

Fish recruitment is influenced by a variety of density-dependent and density-independent effects. Traditional fisheries management has used stock–recruitment relationships to predict recruits from spawning stock abundance (Ricker 1975). Stock abundance undoubtedly contributes to variation in recruitment, although it has been widely noted that recruitment tends to remain about the same (with high variation around an average value) across a distribution of stock abundances (Walters and Martell 2004). This infers that recruitment exhibits density dependence because recruits produced per spawner increase with declines in spawner abundance for nearly all fish stocks (see Myers et al. 1999). However, in many freshwater fisheries applications, the threat of recruitment overfishing via recreational fisheries is not as large as con-

Box 2.5. Fitting a von Bertalanffy Growth Curve in Excel

Fitting a least-squares estimate of growth parameters is easy in Microsoft® Office Excel 2007. In the table below are mean total-length-at-age estimates for black crappie *Pomoxis nigromaculatus* from Lake Dora, Florida. Data were obtained from the Florida Fish and Wildlife Conservation Commission and represent angler-caught black crappie at the lake in 2006. Ten fish per centimeter-group were selected for aging, and a larger sample of fish was measured for total length. Mean total-length-at-age estimates were obtained by means of the fixed-length subsampling methods of DeVries and Frie (1996). Copies of this and the other spreadsheets can be found at: http://fishweb.ifas.ufl.edu/allenlab/courses.html.

Table. Excel spreadsheet of mean total-length-at-age (TL, mm) estimates for black crappie from Lake Dora, Florida, used to fit a von Bertalanffy growth curve. See Table 2.1 for explanation of symbols; SSE is the sum of squared residuals.

A	B	C	D	E	F	G
2	L_∞	350				
3	k	0.41				
4	t_0	−0.49				
5						Predicted	
6				Age	TL	TL	Residuals2
7				2	226	224	5
8				3	262	267	18
9				4	295	295	0
10				5	311	313	4
11				6	329	326	11
12				7	345	334	132
13				8	328	339	119
14							
15							SSE = 289

The following steps describe how a spreadsheet is used to fit a von Bertalanffy growth model. First input reasonable starting values, then create a column of predicted values based on those hypothesized growth parameters. Create a column that calculates the observed minus predicted values squared (i.e., the squared residuals). Now sum the squared residuals in cell labeled SSE. To obtain the least-squares parameter estimates, click "Data," then "Analysis," then "Solver." Choose the SSE cell G15 as the target cell, choose the option to minimize this cell, and then in the box "By Changing Cells," select cells C2 to C4. Now click "Solve" and notice that the parameter estimates change as Solver's optimization routine finds the least-squares parameter estimates. Solver tends to perform better when using the following options: "Automatic Scaling," "Quadratic Estimates," and "Central Derivatives." It is also useful to run Solver from more than one set of initial values to make sure the optimization routine converges on the same parameter values. The solved solution is shown below along with a graph of the observed and predicted values. *(Box continues)*

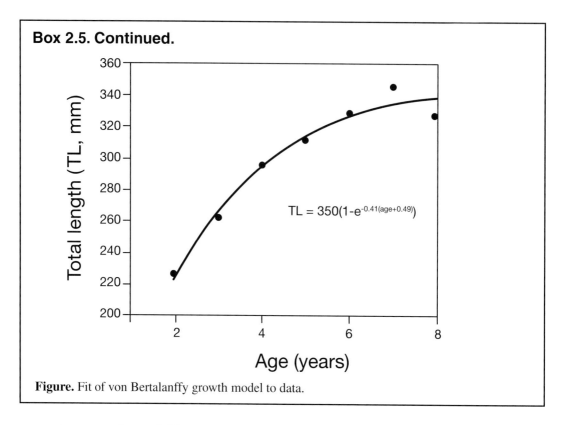

Box 2.5. Continued.

Figure. Fit of von Bertalanffy growth model to data.

cerns about growth overfishing. It may not even be feasible to estimate the stock–recruitment relationship because of the lack of information about recruitment at very low stock abundances at which recruitment would presumably be limited by stock size. Thus, in many freshwater applications, understanding stock–recruitment relationships is not as important as understanding the magnitude of recruitment variation and the factors that cause strong or weak year-classes for a particular water body. Fish recruitment in freshwater systems is often influenced by flow changes in rivers (Bain et al. 1988), water levels in reservoirs (Aggus and Elliot 1975; Ploskey 1986), aquatic plant abundance and species composition (Bettoli et al. 1993), and water temperatures (Cargnelli and Gross 1996). Here we present some common methods for measuring fish recruitment and show how these estimates can be used as part of freshwater fisheries stock assessments. For a detailed analysis of stock–recruit relationships and fitting methods, we refer the reader to Walters and Martell (2004) and Maceina and Pereira (2007).

Measures of fish recruitment and variability.—Most fisheries managers measure recruitment through catch per unit effort (C/f, also known as CPUE) indices. Electrofishing, trawls, trap nets, and hoop nets have been used to measure C/f of small fish. The "recruit" C/f is usually designated by the lengths of fish (e.g., first mode of a length-frequency distribution) or through aging fish to verify catches as recruits (e.g., age-1 fish). Use of C/f indices to measure fish abundance makes the implicit assumption that the relationship between catch rate and population abundance is

$$CPUE = \frac{C}{f} = q \times N, \quad (2.13)$$

where *C* = is catch, *f* = is fishing effort (e.g., trawl time or net night), *q* = is the catchability coefficient (the fraction of population caught per unit of effort), and *N* = is fish abundance (Ricker 1975). This equation infers a linear relationship between *C*/*f* and abundance, with a constant slope *q*. However, studies that evaluate relationships between *C*/*f* and *N* show substantial variation due to environmental conditions, changes in fish distribution and behavior, fish size, and gear selectivity (Hilborn and Walters 1992; Bayley and Austen 2002; Rogers et al. 2003). Thus, use of *C*/*f* data as an index of abundance should be accompanied by other methods that validate whether changes in *C*/*f* reflect changes in *N*. Alternately, managers could verify whether trends in *C*/*f* correspond to changes in recruitment using annual age structure estimates to verify strong and weak year classes moving through the population, or catch-at-age models (see below).

As an example, bottom trawl *C*/*f* data have been used to assess recruitment of black crappie at Lake Okeechobee, Florida (Figure 2.4). Substantial recruitment variability was indicated with mean annual age-1 catch per minute varying from near zero to nearly eight fish per minute over the period of record. Very strong year-classes were produced in 1981, 1987, and 1998 (caught in trawls a year later at age 1), whereas very weak year-classes were evident in other years. Weak year-classes in 2004 and 2005 were associated with hurricane events that caused lakewide changes in aquatic plant abundance (Rogers and Allen 2008). This time series shows a typical scenario of highly variable recruitment. Obviously, the quality of the black crappie fishery would be expected to vary with the large variation in recruitment, and it did! Angler catch per hour showed a lag effect where high angler catch rates occurred 1–2 years after high age-1 catches in the bottom trawl (Figure 2.4). This example shows how monitoring recruitment trends can allow fisheries managers to anticipate the quality of the fishery in the future.

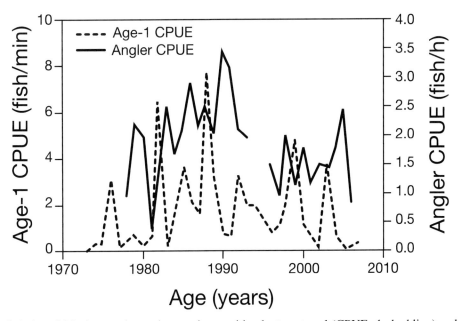

Figure 2.4. Age-1 black crappie catch per minute with a bottom trawl (CPUE, dashed line) and angler catch per hour (fish harvested/angler-hour, solid line) plotted on year for samples from Lake Okeechobee, Florida. Data were collected by Don Fox, Florida Fish and Wildlife Conservation Commission.

Estimating the magnitude of recruitment variation is important for fisheries management. Several authors have proposed indices to measure variation in recruitment through time. Maceina (1997) showed that residuals around a catch curve (see example Box 2.3) can serve as an index of recruitment variability. Isermann et al. (2002) and Quist (2007) compared a number of published methods for measuring recruitment variability. We have not reiterated all those methods, but the most straightforward method is to evaluate the coefficient of variation (CV) in recruit-size fish C/f across years. The CV is a standard measure of variability:

$$CV = \frac{s}{\bar{x}} \times 100\%, \tag{2.14}$$

where CV is the coefficient of variation, s is the standard deviation of the sample, and $\bar{x}$ is the mean. Knowledge of the across-year CV in recruits is useful for population modeling exercises, because it allows the investigator to explore how variable adult fish abundance and angler catches are likely to be due to variation in recruitment. Allen and Pine (2000) found that recruitment variability could influence the ability of managers to detect fish population responses to changes in minimum length limits. Thus, understanding recruitment variability is a key component in managing recreational fisheries (Maceina and Pereira 2007).

2.4 USE OF BASIC FISHERIES MODELS

Fisheries managers can use population models to predict how estimates of growth, mortality, and recruitment will interact to determine fish yield, fish population size, and angler catch rates. Fish population models provide a conceptualization of how a fish population changes in abundance and age structure in response to harvest. Many fisheries textbooks provide complex model formulations that are mathematically challenging. There are cases in which such models are essential, but a key objective of this chapter is to show that building simple fish population models can be both easy and effective for exploring harvest policy options.

Several key points should be considered before beginning a modeling exercise. First, all models are a simplified version of reality, and no model considers all factors that influence a fish population. Models should not attempt to capture all dynamics influencing fisheries, but they should consider the major factors that influence abundance, such as fishing mortality and recruitment. Population models should not be used to make specific predictions but to compare the relative responses to a range of management actions (Hilborn et al. 1984; Johnson 1995). A good way to think about a modeling exercise is that the model is a hypothesis about how a fish population may respond to management actions. Models typically compile much of the existing data into one framework, which allows investigators to identify data gaps clearly. Thus, a modeling exercise may have as much value in guiding future sampling efforts to reduce uncertainty as in evaluating the relative response to a variety of management actions. In this section, we show how a simple age-structured yield-per-recruit model and a catch-at-age approach can be used to assess freshwater fish populations.

2.4.1 Yield-per-Recruit Models

Yield-per-recruit models are typically used to evaluate the potential for growth overfishing (i.e., fishing at a level that reduces the maximum yield per recruit). A number of formulations have been proposed for these models. Here, we used a simple formulation by Botsford and Wickham (1979) and Botsford (1981a, b) that was summarized by Walters and Martell (2004). The approach uses Botsford incidence functions, which estimate the abundance and biomass per recruit for a fish population. Consider a fish population with only instantaneous natural mortality (M) of 0.2. The S_0 (annual survival from natural mortality) for this population is $e^{-M} = 0.82$. The number of fish alive at each age starting with 1,000 recruits to age-1 would therefore be found as follows.

Age	Number alive	Survival per recruit (lx_a)
1	1,000	1
2	820	0.820
3	670	0.670
4	549	0.549
5	449	0.449
6	368	0.368

Simply, the survivors to any age $a + 1$ is equal to the survivors to age a times 0.82. This example describes the survivorship schedule, lx_a, on a per recruit basis, which is useful for the calculations below. Now let us consider the same population with an annual exploitation rate (u) of 0.2. We assume that fish become vulnerable to fishing at age 3. The Botsford incidence functions can be used to predict survivors to each age. Survivorship at age 1, the youngest age in our simulated population, is

$$lx_1 = 1.$$

Survivorship at any older age a can be depicted as:

$$lx_a = lx_{a-1} \times S_0 \times (1 - u \times V_{a-1}), \qquad (2.15)$$

where lx_{a-1} is the survivorship from the previous age, S_0 is survival from natural mortality, u is the annual exploitation rate, and V is a vulnerability parameter that determines whether fish are vulnerable ($V = 1$) to exploitation or not ($V = 0$). In this case, V is 0 for ages 1 and 2, and V is 1 for all older ages. Thus, the survivorship to age 2 would be: $1 \times 0.82 \times (1 - 0.2 \times 0) = 0.82$. because this age group is not yet vulnerable to fishing. Following through with our example, the number of fish at any age can simply be determined by $R \times lx_a$, where R is the number of simulated recruits at age 1.

We now have estimates of the number of survivors per recruit and can compute the vulnerable biomass per recruit as:

$$\phi_{VB} = w_1 V_1 + lx_2 w_2 V_2 + lx_3 w_3 V_3 ..., \qquad (2.16)$$

where ϕ_{VB} is the vulnerable biomass per recruit, lx_a is the age-specific survivorship as defined above, w is the average weight of fish at each age, and V is the vulnerability schedule. Total vulnerable population biomass (B) is then simply:

$$B = R \times \phi_{VB}, \qquad (2.17)$$

where R is the number of recruits at age 1. This per-recruit formulation of the population model makes then next step easy. The equilibrium yield is estimated as:

$$Y = u \times R \times \phi_{VB}, \qquad (2.18)$$

where Y is yield in biomass expressed in the same units as fish weight (w). Botsford incidence functions (such as ϕ_{VB}) are easy to set up in a spreadsheet and they allow relatively complex population models to be expressed in simple terms. Box 2.6 shows a demonstration of the Botsford incidence functions and a yield-per-recruit model for black crappie at Lake Dora, Florida.

In some fisheries total yield (or weight of fish harvested) is not the variable of interest because anglers place little value on harvesting fish but high value on catching large numbers of fish, large-sized fish, or both. For instance many black bass *Micropterus* spp. anglers release their catch even if the fish is legal to harvest, and managers of black bass fisheries seldom consider yield as an important aspect of the fishery. The same could be said for some trout fisheries where the catch of large fish is a more important management objective than is total yield. In these cases, total catch or the total catch of large fish may be a more useful model output. As illustrated in Box 2.6, a change in one spreadsheet formula is all that is required to reflect numbers, rather than biomass, of fish caught under each management alternative. Additional relationships, such as discard mortality of fish caught and released and voluntary release of fish by anglers, can also be incorporated. Thus, spreadsheet models can be built to address the specific needs of the investigator, and their flexibility to handle a wide range of modeling approaches is an advantage to learning this approach.

Software.—There are a number of software packages that can be used for yield-per-recruit and other age-structured population models as well as simple model fitting, such as for growth curves and mortality estimation. Perhaps the most popular is the program FAST (Fisheries Analysis and Simulation Tools; Slipke and Maceina 2001), which is a user-friendly software package that can perform all these analyses in a straightforward Windows platform. The FAST model has been used by many state management agencies, and the user can fit growth curves and catch curves and conduct analyses of yield-per-recruit and more complex age-structured models.

2.4.2 Catch-at-Age Methods

Management of a fish population is greatly enhanced if population size can be estimated. An estimate of absolute abundance can be compared with the catch to evaluate the impact of fishing on the population. Spawning stock abundance can be estimated to determine whether the stock abundance is low enough to limit recruitment. Population levels for prey and predator species can be used to estimate food resources for predators or the impact of predation on prey species. In combination, analyses that predict population abundance provide a wealth of information about mortality, recruitment, and growth that cannot be obtained from relative abundance (i.e., CPUE) data.

Box 2.6. Yield-per-Recruit Model

This yield-per-recruit (YPR) model for black crappie at Lake Dora, Florida, was based on data collected by Florida Fish and Wildlife Conservation Commission and University of Florida personnel. The purpose of the model was to evaluate whether growth overfishing was occurring in this fishery. Copies of the spreadsheet can be found at http://fishweb.ifas.ufl.edu/allenlab/courses.html.

Annual exploitation (M) was estimated with a variable-reward passive-tagging study during 2006; u was estimated at 0.42 (Dotson 2007). The first three cells contain parameters for a von Bertalanffy growth equation (see Box 2.5) used to predict mean total length at age (TL):

$$TL = 350[1 - e^{-0.41(\text{age} + 0.49)}].$$

Natural mortality ($M = 0.40$) was obtained from a literature review for black crappie, and thus $S_0 = e^{-M} = 0.67$. "Recruits" was used to designate the total number of recruits for this population and was set arbitrarily as 1,000. The next cell, "Reg," designated the fish length at entry to the fishery. These parameters were placed into the left column of the spreadsheet and were named as indicated below (see spreadsheet for how to name cells).

Table 1. Parameters for Excel spreadsheet for yield-per-recruit model for black crappie at Lake Dora, Florida. Symbols defined here and in Table 2.1.

L_∞	350
k	0.412
t_0	−0.49
M	0.40
S_0	0.67
u	0.42
a	6.310E-06[a]
b	3.32
Recruits	1000
Reg	250

[a] 6.31×10^{-6}

(Box continues)

Box 2.6. Continued.

Next we construct the length, weight, and mortality structure of the population.

Table 2. Values of each variable for eight ages of black crappie. Explanation of variables follow table.

Age	1	2	3	4	5	6	7	8
TL	161	225	267	295	314	326	334	339
wt	0.13	0.40	0.72	1.00	1.22	1.39	1.51	1.59
V	0	0	1	1	1	1	1	1
lxfished	1.000	0.670	0.449	0.175	0.068	0.026	0.010	0.004

Total length in millimeters was estimated with the von Bertalanffy growth model, and weight (wt) in kilograms was estimated by the standard weight–length relationship for black crappie as $W = a \times TL^b$ (Anderson and Neumann 1996). The V (vulnerability) schedule was used to set the length and age at which fish become vulnerable to the fishery. In this case, we used an IF statement to set V equal to 0 if the mean length at that age was less than Reg and 1 if the mean length was equal to or larger than Reg. The row for "lxfished" is the survivorship per recruit in the fished condition, found by $lx_a = lx_{a-1} \times S_0 \times (1 - u \times V_{a-1})$, where lx_{a-1} was the survivorship from the previous age, S_0 is annual survival from natural mortality, u was the annual exploitation rate, and V is a vulnerability parameter that determines whether fish are vulnerable to u or not, per equation (2.15).

The Botsford incidence function of vulnerable biomass per recruit ($ø_{VB}$) was calculated as SUMPRODUCT(wt,V,lxfished). Yield per recruit was then found by $Y = u \times R \times ø_{VB}$, where R was the number of recruits. A second incidence function was set up as vulnerable number of fish per recruit, $ø_n$, by SUMPRODUCT(V,lxfished). Thus, total angler catch in numbers of fish was estimated as $C = u \times R \times ø_n$.

Table 3. Summary of Botsford incidence function values.

Vulnerable biomass per recruit ($ø_{VB}$)	0.64
Yield per recruit (YPR)	268.11
Vulnerable number per recruit ($ø_n$)	0.73
Catch per recruit (CPR)	307.69

To finish the analysis we simulated a range of exploitation rates and potential sizes of harvest (i.e., minimum length at harvest). We used the "Table" function in Excel to iterate the spreadsheet across a wide range of both values and show how equilibrium yield was predicted to change. See spreadsheet for instructions. The yield isopleth curve is shown below.

(Box continues)

Box 2.6. Continued.

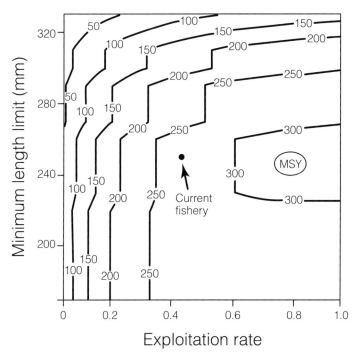

Figure. Yield isopleths (i.e., numbers in the plot represent yield in kilograms) for minimum length limit (y-axis) and exploitation rate (u, x-axis) combinations (MSY = maximum sustainable yield).

These yield isopleths (i.e., numbers within the plot represent yield in kilograms) show that the maximum sustainable yield (MSY) occurred at a minimum length limit of about 240–250 mm with annual exploitation rates of about 0.8. Growth overfishing was predicted to occur at u over about 0.6 if the minimum length at harvest was below 240 mm (notice the decline in yield if u was high [>0.80] and the minimum length limit dropped from 250 mm to 200 mm). This resulted because fish would be harvested before they reached the size that would maximize the yield. In this case the exploitation rate at Lake Dora was 0.42 (gray circle on plot). If the management objective was to maximize yield, increasing the exploitation rate would be recommended with about the same minimum length limit (250 mm TL). Establishing a minimum length limit above about 270 mm TL would cause declines in yield because many fish would die from natural mortality before reaching harvestable size. This example provides a way to construct YPR models to evaluate harvest policies in common recreational fisheries scenarios.

There are a variety of ways to estimate population abundance. For example, if a trawl is assumed to catch 100% of the fish in its path, the area swept by the trawl provides a measure of catch per unit area that can be used to estimate total abundance. Hydroacoustic sampling can provide estimates of total population abundance for certain species, depending on their vertical distribution (Brandt 1996). Capture–recapture methods can be effective in small systems such as streams or small lakes but are harder to apply in large lakes and rivers because of the difficulty in tagging and recapturing a sufficient fraction of the population. Overviews of capture–recapture methods can be found in Pine et al. (2003) and Hayes et al. (2007). One related approach that can be effective on larger systems is to use tagging in combination with a creel survey. Total harvest from the creel survey divided by the exploitation rate from the tagging program provides an estimate of absolute abundance.

An approach commonly used in large systems (e.g., large lakes) is to estimate the total harvest by age and then reconstruct the population from the catch-at-age matrix. This matrix of catches by age and year provides a record of removals from each cohort or year-class. The total catch from a cohort over its lifetime in the fishery is a minimum estimate of the initial size of that cohort. Correcting for natural deaths provides a better estimate of initial cohort size. Methods that attempt to recreate the stock abundance using historical catches are usually termed virtual population analysis (VPA).

An exceptional example of a catch-at-age dataset exists for the walleye *Sander vitreus* fishery at Lake Escanaba, Wisconsin, for the years 1956 to 1997 (Box 2.7). The entire catch-at-age matrix includes a few fish that were age 0 and older than age 12, but those have been omitted for this example. This lake is unique because anglers are required to report their entire catch when leaving the lake. Mandatory reporting has resulted in a high-quality dataset compared with the typical situation in which total harvest is estimated from a small subsample of catches. Walleye ages were determined from jaw tags and by examining scales. Catch sampling began in 1956, so age-1 fish caught in that year would be from the 1955 cohort (age 0 in 1955). Age-2 fish in 1956 are from the 1954 cohort. The earliest cohort in that year is the 1946 cohort at age 10. The most recent cohort that has completed its lifetime in the fishery is the 1985 cohort, which is age 12 in 1997. These completed cohorts are the simplest to analyze because it can be assumed that no fish from those cohorts remain (Hilborn and Walters 1992).

Estimates of age-1 abundance for the completed cohorts showed that recruitment has varied widely over time, from about 2,000 to 18,000 fish per year (Box 2.7). Strong and weak year-classes were apparent and can be tracked across years (e.g., the weak 1960 year-class is evident through at least age 6 in 1966). The ability to track strong and weak year-classes across years is a sign that the age data are reliable. The occasional strong year-classes (e.g., 1955, 1973, and 1981) can have a big impact on the population and result in several years of high catches. Slight modifications of the method shown here provide estimates for incomplete cohorts (Hilborn and Walters 1992), so that the catch-at-age matrix can be transformed into estimates of population size for every age and year. There are also statistically-based catch-at-age analyses that use the same information and produce similar results but provide estimates of the uncertainty in estimating population abundance and fishing mortality (Hilborn and Walters 1992; Quinn and Deriso 1999). These methods are beyond the scope of this chapter but are recommended for carrying out catch-at-age analyses.

Catch-at-age methods are dependent on an assumed value of natural mortality. Changing M produces a new set of population estimates that will be consistently higher or lower, depending on whether M is decreased or increased. Although absolute abundance will differ,

Box 2.7. Virtual Population Analysis

Table 1. Data based on catches of walleye by age (1–12) and year (1956–1997) from Escanaba Lake, Wisconsin (unpublished data, M. Hansen, University of Wisconsin–Stevens Point).

Year	1	2	3	4	5	Age 6	7	8	9	10	11	12
1956	702	2,247	448	309	492	97	129	23	11	1	0	0
1957	9	1,330	1,543	186	147	293	79	45	11	2	0	0
1958	1	26	452	462	49	36	108	29	13	5	0	0
1959	210	35	17	366	284	43	24	14	4	7	0	0
1960	736	553	58	28	581	336	38	7	14	1	1	7
1961	6	2,750	233	33	15	265	229	16	3	2	0	0
1962	27	34	1,869	111	20	61	134	225	8	0	0	0
1963	475	169	3	368	34	7	4	69	117	2	0	0
1964	428	963	122	6	112	11	23	25	28	34	1	0
1965	164	497	695	55	3	50	8	21	6	20	0	0
1966	73	1,739	389	328	35	5	61	96	7	5	0	0
1967	0	35	2,130	247	137	65	26	36	25	10	6	27
1968	2	175	220	371	141	37	26	14	12	12	0	33
1969	27	201	352	180	221	34	15	9	3	3	2	0
1970	164	682	430	454	181	198	33	9	6	2	2	0
1971	85	579	872	325	301	129	164	26	9	3	3	7
1972	41	131	171	223	157	25	16	16	12	5	1	3
1973	67	271	381	278	99	54	30	24	7	3	3	2
1974	112	121	239	193	213	71	61	38	13	12	3	0
1975	4	2,846	278	382	370	277	177	88	37	8	3	3
1976	38	789	1,801	345	133	171	116	60	32	16	3	0
1977	97	387	1,519	866	65	34	21	8	3	0	7	2
1978	120	625	749	1,178	468	93	38	24	9	3	0	0
1979	6	716	766	393	418	213	81	55	33	31	9	5

(Box continues)

Box 2.7. Continued.

Table 1. Continued.

Year	1	2	3	4	5	Age 6	7	8	9	10	11	12
1980	9	140	2,040	335	129	183	116	47	38	26	9	4
1981	77	496	144	539	80	22	24	13	2	5	0	0
1982	124	442	971	139	251	54	24	17	7	5	0	0
1983	8	1,495	283	450	101	241	37	18	15	6	1	1
1984	6	107	2,172	129	126	19	103	29	20	4	3	1
1985	17	101	348	1,960	54	31	21	43	12	6	3	2
1986	4	336	374	109	370	17	14	3	14	2	3	0
1987	64	567	1,734	370	61	90	16	9	0	1	0	0
1988	148	1,788	1,469	754	117	15	30	4	3	0	3	1
1989	37	622	2,804	577	165	27	10	12	3	0	0	0
1990	5	354	811	1,188	220	48	10	3	4	1	0	2
1991	8	52	415	300	208	23	10	2	2	1	0	0
1992	21	1,068	107	245	136	87	29	7	1	2	4	2
1993	8	137	998	138	174	97	68	14	10	8	1	1
1994	5	171	315	498	51	71	31	47	7	9	2	2
1995	40	135	525	277	216	31	28	28	17	1	1	0
1996	0	0	362	220	102	55	13	8	6	2	1	1
1997	0	0	1,952	298	111	50	30	4	5	6	6	1

Box 2.7. Continued.

Each cohort can be reconstructed by summing the catches and adjusting upward for natural mortality. Because natural deaths are not observed, the instantaneous rate of natural mortality (M) is often an assumed value based on the life history characteristics of that species. Here, a value of 0.4 is assumed.

The abundance estimate for each cohort begins at the oldest age and works backward. This is very convenient for cohorts that have completed their life in the fishery because it can be assumed that no fish from that cohort remain in the population. Abundance at the start of age 11 is the population abundance at age 12 (assumed to be the catch), adjusted upward for a year of natural mortality (simply divided by S_0 [e^{-M}]), plus the catch of age-11 fish that year. For the 1955 cohort, the expression would be

$$N_{11,\,1966} = N_{12,\,1967}/S_0 + C_{11,\,1966}.$$

The equation for the number at age 10 is

$$N_{10,\,1965} = N_{11,\,1966}/S_0 + C_{10,\,1965}.$$

A similar calculation is made for each age, working backwards up the diagonal to age 1.

(Box continues)

Box 2.7. Continued.

Table 2. Abundance estimates for each cohort of walleye from Escanaba Lake, Wisconsin.

Year	1	2	3	4	5	6	7	8	9	10	11	12
1956	16,462											
1957	1,817	10,564										
1958	1,579	1,212	6,190									
1959	5,570	1,058	795	3,846								
1960	13,166	3,593	686	521	2,333							
1961	1,125	8,332	2,038	421	331	1,174						
1962	2,133	750	3,742	1,210	260	212	609					
1963	6,625	1,412	480	1,255	737	161	101	319				
1964	5,164	4,122	833	320	595	471	103	65	167			
1965	14,297	3,175	2,118	477	210	324	308	54	27	93		
1966	4,627	9,473	1,795	954	283	139	183	201	22	14	49	
1967	6,676	3,053	5,185	942	419	166	90	82	71	10	6	33
1968	6,826	4,475	2,023	2,047	466	189	68	43	31	31	0	0
1969	8,679	4,574	2,883	1,208	1,124	218	102	28	19	13	12	0
1970	8,690	5,799	2,932	1,696	689	605	123	58	13	11	6	7
1971	7,253	5,715	3,430	1,677	833	341	273	61	33	4	6	3
1972	5,675	4,805	3,443	1,715	906	356	142	73	23	16	1	2
1973	5,055	3,777	3,133	2,193	1,000	502	222	84	38	7	7	0
1974	17,344	3,343	2,350	1,845	1,284	604	300	129	41	21	3	3
1975	13,118	11,551	2,160	1,415	1,107	718	357	161	61	18	6	0
1976	5,823	8,790	5,835	1,262	693	494	295	121	49	16	7	2
1977	5,975	3,878	5,363	2,704	614	375	217	120	41	11	0	0
1978	12,928	3,940	2,340	2,577	1,232	368	229	131	75	25	7	0
1979	1,918	8,586	2,222	1,066	938	512	185	128	72	44	15	5
1980	5,796	1,282	5,275	976	451	348	201	69	49	26	9	4

(Box continues)

Box 2.7. Continued.

Table 2. Continued.

Year	1	2	3	4	Age 5	6	7	8	9	10	11	12
1981	2,540	3,879	765	2,169	430	216	111	57	15	7	0	0
1982	17,237	1,651	2,268	417	1,092	235	130	58	29	9	1	0
1983	2,135	11,471	810	869	186	564	121	71	28	15	2	1
1984	3,566	1,426	6,687	354	281	57	216	56	36	8	6	1
1985	8,989	2,386	884	3,027	150	104	25	76	18	10	3	2
1986	8,189	6,014	1,532	359	715	65	49	3	22	4	3	0
1987		5,486	3,806	776	168	231	32	23	0	5	1	0
1988			3,298	1,389	272	72	95	11	10	0	3	1
1989				1,226	426	104	38	43	4	4	0	0
1990					435	175	52	19	21	1	3	0
1991						144	85	28	11	11	0	2
1992							81	50	17	6	7	0
1993								35	29	11	2	2
1994									14	13	2	1
1995										5	2	0
1996											2	1
1997												1

The exploitation rate (u) can be estimated as the ratio of catch to population abundance (e.g., 702 / 16,462 for age-1 walleye in 1956). The instantaneous fishing mortality rate (F) is $-\log_e(1 - u)$, resulting in the following matrix of F estimates.

(Box continues)

Box 2.7. Continued.

Table 3. Instantaneous fishing mortality rate estimates for walleye from Escanaba Lake, Wisconsin.

Year	1	2	3	4	5	Age 6	7	8	9	10	11
1956	0.04										
1957	0.00	0.13									
1958	0.00	0.02	0.08								
1959	0.04	0.03	0.02	0.10							
1960	0.06	0.17	0.09	0.06	0.29						
1961	0.01	0.40	0.12	0.08	0.05	0.26					
1962	0.01	0.05	0.69	0.10	0.08	0.34	0.25				
1963	0.07	0.13	0.01	0.35	0.05	0.04	0.04	0.24			
1964	0.09	0.27	0.16	0.02	0.21	0.02	0.25	0.49	0.18		
1965	0.01	0.17	0.40	0.12	0.01	0.17	0.03	0.50	0.25	0.24	
1966	0.02	0.20	0.24	0.42	0.13	0.04	0.40	0.65	0.38	0.44	0.00
1967	0.00	0.01	0.53	0.30	0.40	0.50	0.34	0.58	0.44		
1968	0.00	0.04	0.12	0.20	0.36	0.22	0.48	0.40	0.49	0.50	0.18
1969	0.00	0.04	0.13	0.16	0.22	0.17	0.16	0.39	0.17	0.27	0.37
1970	0.02	0.13	0.16	0.31	0.30	0.40	0.31	0.17	0.64	0.20	0.70
1971	0.01	0.11	0.29	0.22	0.45	0.48	0.92	0.56	0.32	1.10	
1972	0.01	0.03	0.05	0.14	0.19	0.07	0.12	0.25	0.73	0.37	
1973	0.01	0.07	0.13	0.14	0.10	0.11	0.15	0.33	0.20	0.51	0.51
1974	0.01	0.04	0.11	0.11	0.18	0.13	0.23	0.35	0.39	0.85	
1975	0.00	0.28	0.14	0.31	0.41	0.49	0.68	0.79	0.94	0.57	0.70
1976	0.01	0.09	0.37	0.32	0.21	0.42	0.50	0.69	1.07		
1977	0.02	0.11	0.33	0.39	0.11	0.10	0.10	0.07	0.08	0.00	
1978	0.01	0.17	0.39	0.61	0.48	0.29	0.18	0.20	0.13	0.13	0.00
1979	0.00	0.09	0.42	0.46	0.59	0.54	0.58	0.56	0.62	1.20	0.92
1980	0.00	0.12	0.49	0.42	0.34	0.74	0.86	1.13	1.51		

(Box continues)

Box 2.7. Continued.

Table 3. Continued.

Year	1	2	3	4	5	Age 6	7	8	9	10	11
1981	0.03	0.14	0.21	0.29	0.21	0.11	0.24	0.26	0.14	1.18	
1982	0.01	0.31	0.56	0.41	0.26	0.26	0.20	0.35	0.27	0.85	0.00
1983	0.00	0.14	0.43	0.73	0.78	0.56	0.36	0.29	0.78	0.51	0.51
1984	0.00	0.08	0.39	0.45	0.59	0.41	0.65	0.72	0.82	0.64	0.70
1985	0.00	0.04	0.50	1.04	0.44	0.35	1.74	0.83	1.07	0.85	
1986	0.00	0.06	0.28	0.36	0.73	0.30	0.34		1.00	0.64	
1987		0.11	0.61	0.65	0.45	0.49	0.69	0.49		0.20	0.00
1988			0.59	0.78	0.56	0.24	0.38	0.47	0.37		
1989				0.64	0.49	0.30	0.31	0.32	1.10		
1990					0.71	0.32	0.22	0.17	0.21	0.00	0.00
1991						0.17	0.13	0.07	0.21	0.09	
1992							0.44	0.15	0.06	0.43	0.85
1993								0.51	0.42	1.30	0.51
1994									0.69	1.23	
1995										0.24	0.51
1996											0.51

It is not possible to estimate F for the last nonzero catch for a cohort. Notice that values of F increase across ages 1–4 for each year as fish grow and become more vulnerable to angling. The fishing mortality rates can be viewed across years, as desired for management (see figure below). This analysis suggested that fishing mortality varied widely from 1965 to 1986 but also showed a general increase through time.

(Box continues)

Box 2.7. Continued.

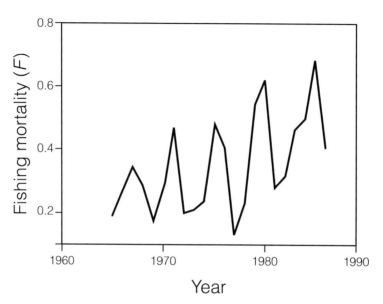

Figure A. Fishing mortality of walleye from 1965 to 1986 in Escanaba Lake, Wisconsin.

Other useful results from a virtual population analysis (VPA) include population abundance (totaled across all ages) and annual recruitment to age-1 (shown below). Notice how the strong and weak year-classes are evident simply by reconstructing the cohorts in the VPA.

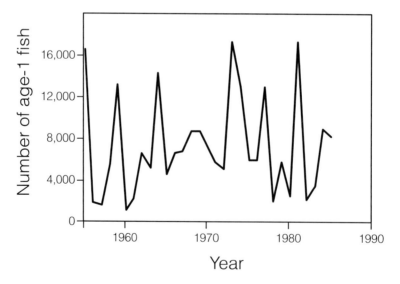

Figure B. Annual recruitment of age-1 walleye at Escanaba Lake, Wisconsin.

Population biomass can also be calculated by multiplying each abundance estimate by the associated average weight-at-age value.

the trend and year-to-year variability in year-class strength will be similar for different assumed values of *M*. Field studies to estimate *M* (e.g., a tagging study) can be used to reduce this source of uncertainty.

Catch-at-age models are routinely used in marine fisheries, but they have not been commonly used in freshwater systems other than the Great Lakes. They require more effort than do relative abundance surveys, but they can be derived from creel survey data if accompanied by estimates of age composition of the angler catch. The abundance estimates provide a strong foundation for single-species or multispecies models and are superior to relative abundance data for selecting an appropriate harvest rate.

2.5 CONCLUSIONS

Fisheries management requires making choices about harvest regulations, fish stocking programs, and habitat restoration Those choices influence fisheries resources and the human users who benefit from those resources. Estimating fish population parameters including mortality, growth, and recruitment and integrating those estimates into simple population models improves understanding of the factors influencing fish abundance and angler harvest. The methods outlined in this chapter serve as a first step towards proficiency in assessment of fish populations. Quantitative analysis of fish populations will always be a critical element for effective management, and the purpose of this chapter has been to show that most analyses are not difficult to draw basic fishery conclusions needed in most instances. When combined with effective use of harvest restrictions and other management strategies related to fish habitat and species composition, quantitative fisheries assessment methods will inform management decisions and improve fisheries in the future.

2.6 REFERENCES

Aggus, L. R., and G. V. Elliot. 1975. Effects of cover and food on year class strength of largemouth bass. Pages 317–322 *in* R. H. Stroud and H. Clepper, editors. Black bass biology and management. Sport Fishing Institute, Washington, D.C.

Allen, M. S., and W. E. Pine, III. 2000. Detecting fish population responses to a minimum length limit: effects of variable recruitment and duration of evaluation. North American Journal of Fisheries Management 20:672–682.

Anderson, R. O., and R. M. Neumann. 1996. Length, weight, and structural indices. Pages 447–482 *in* B. R. Murphy and D. W. Willis, editors. Fisheries techniques, 2nd edition. American Fisheries Society, Bethesda, Maryland.

Bain, M. B., J. T. Finn, and H. E. Booke. 1988. Streamflow regulation and fish community structure. Ecology 69:382–392.

Bayley, P. B., and D. J. Austen. 2002. Capture efficiency of a boat electrofisher. Transactions of the American Fisheries Society 131:435–451.

Bettoli, P. W., M. J. Maceina, R. L. Noble, and R. K. Betsill. 1993. Response of a reservoir fish community to aquatic vegetation removal. North American Journal of Fisheries Management 13:110–124.

Botsford, L. W., and D. E. Wickham. 1979. Population cycles caused by inter-age, density-dependent mortality in young fish and crustaceans. Pages 73–82 *in* E. Naybr and R. Hartnoll, editors. Cyclic phenomena in marine plants and animals. Proceedings of the 13th European marine biology symposium. Permagon Press, New York.

Botsford, L. 1981a. Optimal fishery policy for size-specific density dependent population models. Journal of Mathematical Biology 12:265–293.

Botsford, L. 1981b. The effects of increased growth rates on depressed population size. American Naturalist 117:38–63.

Brandt, S. B. 1996. Acoustic assessment of fish abundance and distribution. Pages 385–432 in B. R. Murphy and D. W. Willis, editors. Fisheries techniques, 2nd edition. American Fisheries Society, Bethesda, Maryland.

Cargnelli, L. M., and M. R. Gross. 1996. The temporal dimension in fish recruitment: birth date, body size, and size-dependent survival in a sunfish (bluegill: *Lepomis macrochirus*). Canadian Journal of Fisheries and Aquatic Sciences 53:360–367.

Crawford, S., and M. S. Allen. 2006. Fishing and natural mortality of bluegills and redear sunfish at Lake Panasoffkee, Florida: implications for size limits North American Journal of Fisheries Management 26:42–51.

Cushing, D. L. 1981. Fisheries biology: a study in population dynamics. The University of Wisconsin Press, Madison.

DeVries, D. R., and R. V. Frie. 1996. Determination of age and growth. Pages 483–512 in B. R. Murphy and D. W. Willis, editors. Fisheries techniques, 2nd edition. American Fisheries Society, Bethesda, Maryland.

Dotson, J. R. 2007. Effects of commercial gill net bycatch on the black crappie fishery at Lake Dora, Florida. Master's thesis. University of Florida, Gainesville.

Hayes, D. B., J. R. Bence, T. J. Kwak, and B. E. Thompson. 2007. Abundance, biomass, and production. Pages 327–374 in C. S. Guy and M. L. Brown, editors. Analysis and interpretation of freshwater fisheries data. American Fisheries Society, Bethesda, Maryland.

Hightower, J. E., J. R. Jackson, and K. H. Pollock. 2001. Use of telemetry methods to estimate natural and fishing mortality of striped bass in Lake Gaston, North Carolina. Transactions of the American Fisheries Society 130:557–567.

Hilborn, R., and C. J. Walters. 1992. Quantitative fisheries stock assessment: choice, dynamics, and uncertainty. Chapman and Hall, New York.

Hilborn, R., C. J. Walters, R. M. Peterman, and M. J. Staley. 1984. Models and fisheries: a case study in implementation. North American Journal of Fisheries Management 4:9–14.

Hoenig, J. M., N. J. Barrowman, W. S. Hearn, and K. H. Pollock. 1998. Multiyear tagging studies incorporating fishing effort data. Canadian Journal of Fisheries and Aquatic Sciences 55:1466–1476.

Isely, J. J., and T. B. Grabowski. 2007. Age and growth. Pages 187–228 in C. S. Guy and M. L. Brown, editors. Analysis and interpretation of freshwater fisheries data. American Fisheries Society, Bethesda, Maryland.

Isermann, D. A., W. L. McKibbin, and D. W. Willis. 2002. An analysis of methods for quantifying crappie recruitment variability. North American Journal of Fisheries Management 22:1124–1135.

Jiang, H., K. H. Pollock, C. Brownie, J. E. Hightower, J. M. Hoenig, and W. S. Hearn. 2007. Age dependent tag return models for estimating fishing mortality, natural mortality and selectivity. Journal of Agricultural, Biological, and Environmental Statistics 12:177–194.

Johnson, B. L. 1995. Applying computer simulation models as learning tools in fishery management. North American Journal of Fisheries Management 15:736–747.

Leaman, B. M., and R. J. Beamish. 1984. Ecological and management implications of longevity in some northeast Pacific groundfishes. International North Pacific Fisheries Commission 42:85–97.

Maceina, M. J. 1997. Simple application of using residuals from catch-curve regressions to assess year-class strength in fish. Fisheries Research 32:115–121.

Maceina, M. J., and D. L. Pereira. 2007. Recruitment. Pages 121–186 in C. S. Guy and M. L. Brown, editors. Analysis and interpretation of freshwater fisheries data. American Fisheries Society, Bethesda, Maryland.

Miranda, L. E., and P. W. Bettoli. 2007. Mortality. Pages 228–277 *in* C. S. Guy and M. L. Brown, editors. Analysis and interpretation of freshwater fisheries data. American Fisheries Society, Bethesda, Maryland.

Myers, R. A., K. G. Bowen, and N. J. Barrowman. 1999. Maximum reproductive rates of fish at low population sizes. Canadian Journal of Fisheries and Aquatic Sciences 56:2402–2419.

Paragamian, V. L., R. C. P. Beamesderfer, and S. C. Ireland. 2005. Status, population dynamics, and future prospects of the endangered Kootenai River white sturgeon population with and without hatchery intervention. Transactions of the American Fisheries Society 134:518–532.

Pine, W. E., III., K. H. Pollock, J. E. Hightower, T. J. Kwak, and J. A. Rice. 2003. A review of tagging methods for estimating fish population size and components of mortality. Fisheries 28(10):10–23.

Ploskey, G. R. 1986. Effects of water level changes on reservoir ecosystems, with implications to management. Pages 86–97 *in* G. E. Hall and M. J. Van Den Avyle, editors. Reservoir fisheries management: strategies for the 80's. American Fisheries Society, Bethesda, Maryland.

Pollock, K. H., J. M. Hoenig, W. S. Hearn, and B. Calingaert. 2002. Tag reporting rate estimation: 2. use of high-reward tagging and observers in multiple-component fisheries. North American Journal of Fisheries Management 22:727–736.

Quinn, T. J., II, and R. B. Deriso. 1999. Quantitative fish dynamics. Oxford University Press, New York.

Quist, M. C. 2007. An evaluation of techniques used to index recruitment variation and year-class strength. North American Journal of Fisheries Management 27:30–42.

Quist, M. C., M. A. Pegg, and D. R. DeVries. In press. Age and growth. *In* A. V. Zale, D. L. Parrish, and T. M. Sutton, editors. Fisheries techniques, 3rd edition. American Fisheries Society, Bethesda, Maryland.

Ricker, W. E. 1975. Computation and interpretation of biological statistics of fish populations. Fisheries Research Board of Canada Bulletin 191.

Rogers, M. W., and M. S. Allen. 2008. Hurricane impacts to Lake Okeechobee: altered hydrology creates difficult management trade offs. Fisheries 33(1):11–17.

Rogers, M. W., M. J. Hansen, and T. D. Beard. 2003. Catchability of walleyes to fyke netting and electrofishing in northern Wisconsin lakes. North American Journal of Fisheries Management 23:1193–1206.

Schnute, J. 1981. A versatile growth model with statistically stable parameters. Canadian Journal of Fisheries and Aquatic Sciences 38:1128–1140.

Slipke, J. W., and M. J. Maceina. 2001. Fishery Analysis and Simulation Tools (FAST), users guide. Auburn University, Auburn, Alabama.

Smith, B. D., and L. W. Botsford. 1998. Interpretation of growth, mortality, and recruitment patterns in size-at-age, growth, increment, and size frequency data. Pages 125–139 *in* G. S. Jamieson and A. Campbell, editors. Proceedings of the North Pacific symposium on invertebrate stock assessment and management. Canadian Special Publication Fisheries and Aquatic Sciences 125.

Smith, B. D., and G. A. McFarlane. 1990. Growth analysis of Strait of Georgia lingcod by use of length-frequency and length-increment data in combination. Transactions of the American Fisheries Society 119:802–812.

Smith, W. E., F. S. Scharf, and J. E. Hightower. 2009. Fishing mortality in North Carolina's southern flounder fishery: direct estimates of instantaneous fishing mortality from a tag return experiment. Marine and Coastal Fisheries: Dynamics, Management, and Ecosystem Science 1:283–299.

Thompson, J. S., D. S. Waters, J. A. Rice, and J. E. Hightower. 2007. Seasonal natural and fishing mortality of striped bass in a southeastern reservoir. North American Journal of Fisheries Management 27:681–694.

Walters, C. J., and S. J. D. Martell. 2004. Fisheries ecology and management. Princeton, University Press, Princeton, New Jersey.

Waters, D. S., R. L. Noble, and J. E. Hightower. 2005. Fishing and natural mortality of adult largemouth bass in a tropical reservoir. Transactions of the American Fisheries Society 134:563–571.

Chapter 3

Scale and Fisheries Management

JAMES T. PETERSON AND JASON DUNHAM

3.1 INTRODUCTION

Scale is an issue of central importance to fisheries managers. This chapter provides an overview of what "scale" is and why its consideration is essential for effective fisheries management. The overview is followed with illustrations of approaches for identifying different scales at which ecological processes may operate and how these approaches relate to incorporating scale into management practices. In contrast to the current trend of providing standard methods for sampling fishes or describing habitat in inland waters (Bain and Stevenson 1999; Bonar et al. 2009), the point of this chapter is that scaling is unique to the question and management situation at hand. With this view in mind, it becomes clear there is no single common protocol or approach that applies to inland fisheries management. Rather it is the question and the scale or scales at which questions are addressed that drive the approach. In a sense, scale is both the question and the answer in fisheries biology and management applications. A primary goal of this chapter is to explore this notion and motivate readers who are relatively new to ideas about scaling in natural systems to appreciate what many view as one of the most daunting challenges in both basic and applied biology. The range of issues and examples covered herein are far from comprehensive, but hopefully the point is made that scale fundamentally controls how fisheries managers see and understand the challenges they confront.

3.2 DEFINING SCALE AND ITS IMPORTANCE

The literature on scale can be confusing and is littered with what King (1997) referred to as "conceptual clutter." This lack of a clear and consistent articulation of basic concepts and terms related to scaling hampers understanding (e.g., Morrison and Hall 2002). This is partly because scale is inherently difficult to define and partly because of disparity on how researchers and practitioners in and among disciplines treat the issue of scaling. We address what we view to be a critical subset of terms, definitions, and considerations for scaling applications in fisheries management.

In the simplest terms, scale is defined by the "grain" and "extent" of observation (Figure 3.1). Grain refers to the finest spatial or temporal resolution possible in a given data set, usually a sample unit. For example in terms of space, grain size may represent the minimum resolution at which a fish length was measured (e.g., millimeter) or the pixel size in geographic data (e.g., square meter). In terms of time, grain size may represent the temporal resolution of

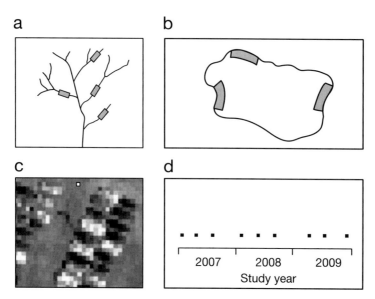

Figure 3.1. Common examples of grain and extent used in fisheries management applications where (a) the spatial extent is a stream network within a watershed and grain is the length of stream sampled (gray rectangles); (b) the spatial extent is a lake and the grain is the area of the individual shoreline electrofishing sites (gray areas); (c) the spatial extent is a geographical information system (GIS) coverage and the grain is represented by three two week sampling periods (squares); and (d) the temporal extent is a 3-year study and the grain is a 2-week sampling period represented by black squares.

a measurement of water temperature (e.g., measured at intervals of 30 min) or time intervals at which populations of fish are monitored (e.g., daily, weekly, or annually). Extent refers to the spatial dimensions or period of time over which observations are made. Common uses include the area of a particular study (e.g., watershed, lake, or in a park boundary) or the "period of record" for stream discharges or lake levels.

The issue of grain and extent of observation determine a fishery manager's ability to observe a phenomenon of interest. Consider, for example, the seemingly simple task of measuring stream temperatures with a digital thermograph. The scale of observation is determined by how frequently the thermograph is set to record measurements (grain) and the temporal interval over which measurements are made (extent). If measurements are made at a relatively coarse grain (e.g., greater than 2-h intervals), daily maximum temperatures may be underestimated by more than 2°C (Dunham et al. 2005). This is obviously more likely in cases in which temperatures can change very quickly (less than 2-h intervals). If the temporal extent of measurement does not include the warmest time of year, daily maximums may similarly be missed in the sample. The spatial grain at which temperatures are measured may also constrain a fishery manager's ability to detect small thermal anomalies that may be important to fish. For example, in warmer streams salmonids may use relatively small patches of cold water ($\leq 10^1$-m grain) in warmer reaches (e.g., extents of 10^1–10^2 m; Torgersen et al. 1999; Ebersole et al. 2001) or larger patches (10^3-m or greater grain) of cold water in the headwaters of river networks by moving over large extents (extents of 10^3 m or greater; Dunham et al. 2002). Smaller patches of cold water may serve as important thermal refugia for short-term survival of individuals or as stopover habitats during migration, whereas larger patches may be more important for persistence of populations. Consequently, different spatial and temporal scales may provide strongly-contrasting views of habitat use and implications for individuals and populations.

The preceding example involving water temperatures indicates that the ratio of grain to extent influences perception of how habitat and other factors influence fish. Within-grain heterogeneity (variability) tends to increase as relative grain size increases (Figure 3.2). For example, the depth, velocity, and substrate composition of a 1-m-long section of a small stream are likely to be homogeneous relative to a 100-m-long section of stream, which is more likely to contain a wide variety of habitat types. As larger grains are considered, differences among them typically decrease owing to increasing variability within grains. For instance, in temperate zones, differences in average annual temperature (large grain) are much less variable from year to year compared with variation from month to month in average monthly temperatures (smaller grain). The ability to detect relationships and patterns in data are weakest when the within-grain heterogeneity is high and between-grain differences are low (Fuller 1987). This means that fisheries managers could arrive at very different conclusions about the relationship between a factor (e.g., habitat characteristics) and fish population response (e.g., abundance) depending on the choice of grain size alone. A good example is coexistence of native and nonnative fish species where both may coexist when considering large grains, but local segregation (e.g., in portions of a lake or stream) occurs and is not detected by coarse-grained analysis (Melbourne et al. 2007). Another common scale dependency is described by "scale–area" curves that examine the effect of grain size on estimates of the area of occupied habitat for different species (Box 3.1). The consequence is that the fishery manager's view of the area of habitat occupied by different species could actually reverse between species for reasons

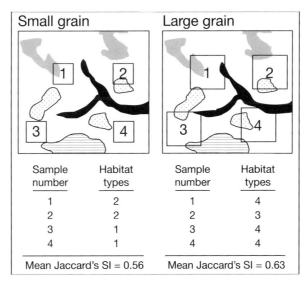

Figure 3.2. An example of a fixed study site (extent) with different habitats shown as shapes (with varying patterns) and two different grain sizes (heavy boxes). Each sample collected with the small grain size (a) contains only one to two habitat types, whereas each sample collected with the large grain size (b) contains three to four habitat types. As grain size increases relative to the extent, within-grain heterogeneity in habitats increases and between-sample differences in habitat types decrease. Jaccard's similarity index (SI) is a measure of similarity among samples.

Box 3.1. Scale–Area Curves

Kunin (1998) demonstrated that the grain of observation can influence the view of how species are distributed across landscapes. Consider the example below that illustrates three sampling scenarios for a species with a patchy distribution (occurrences indicated by dots). In (a) a fine-grained sampling grid indicates occupancy of 39% (percent area of white or occupied grid cells) of the area within the sampling frame (all cells, overall extent). In (b) a doubling of the cell size within the grid indicates an occupied area of 75%, whereas in (c), with only four large cells, the occupied area is 100%. This phenomenon leads to a scale–area curve that defines how the grain (scale) of observation influences the estimate of the area of occupancy (d). Given the patterns of rarity we observe among inland fishes (Box 3.2), this effect should be a cause for concern. The importance of scale–area curves and relevance for assessing rarity and risk was explored by Fagan et al. (2005) for desert fishes in the southwestern USA.

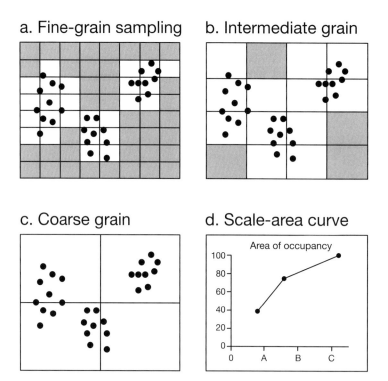

Figure. Illustration of scale–area curve, which defines how the grain (scale) of observation influences the estimate of the area of occupancy.

none other than simply changing the grain size (i.e., sample unit size) of observation (Kunin 1998).

Another issue related to scale is that many processes operate on different temporal or spatial scales. Stream characteristics can influence fish growth, survival, or reproduction over relatively short time spans compared with decades and centuries over which geomorphic processes often operate. For example, interannual or seasonal variation in temperatures of streams in the Missouri Ozarks influences annual variation in fish growth. However, growth in these same streams is currently affected by excessive sediment due to land use changes that occurred more than 150 years ago (Jacobson and Gran 1999). Additionally, variation in both productivity and nutrient availability among these streams has occurred over millions of years due to differences in underlying geologic features and local weathering rates.

A single process may operate on multiple scales. For example, consider the process of gene flow due to dispersal of individuals among local populations of fishes. In some cases, gene flow may be realized through a relatively-constant rate of movement of individuals among populations. A common pattern of gene flow is isolation by distance, by which gene flow is more likely among neighboring populations (Neville et al. 2006). In other cases, infrequent pulses of dispersal or isolation may occur over a wide range of distances in response to episodic disturbances (e.g., extreme floods or droughts). When averaged over multigenerational time scales, patterns of gene flow may appear similar between these two different scenarios. For example, if 10 individuals disperse among locations each year for 10 years or 100 individuals in 1 of 10 years, the average level of dispersal is 10 per year, but the pattern of dispersal is very different between these two scenarios. If gene flow is analyzed at a finer temporal grain, the signatures of these distinctive scenarios may be observable. Over longer time frames (e.g., postglacial colonization), the legacy of gene flow from influences dating back thousands of years may be evident in evolutionary relationships among populations observed today (Avise 1994). The implication is that a clear understanding of fish dispersal requires careful consideration of the processes that influence dispersal at different temporal and spatial scales, as well as the consequences (e.g., gene flow at different scales).

To complicate matters further, it is clear that both pattern and process interact. For example, fish populations that are closer together in space or time may be more likely to interact via dispersal and are more likely to be influenced by common environmental factors (e.g., local climate). The spatial extent occupied by populations also may be important. For instance, those fish populations occupying a greater spatial extent may experience a broader range of environmental conditions that can stabilize populations through time (Bisson et al. 2009). In fact, the view of what processes are important may change with scale. When viewing the distribution of fishes at a broad scale, the influences of climatic variability on temperature may be more evident, whereas observing individual fish at a local scale may highlight the importance of biotic interactions (Fausch et al. 1994). In general, processes at large spatial extents operate over longer time frames and processes at smaller extents operate over shorter time frames. For example, the distribution of freshwater fishes in North America is largely the result of glaciation and fish dispersal operating over thousands of years (Hocutt and Wiley 1986), whereas the distribution of fish in a stream reach is often the result of diel and seasonal habitat use modified by species interactions (Matthews and Heins 1987). This general relationship between spatial and temporal extents indicates that short-term studies are likely to conclude that small-scale processes have greater effects on fishes than do processes operating over larger scales. Thus, local fishery management efforts, such a changing harvest regulations, are likely to fail if the primary processes responsible for modifying the fishery are occurring at larger scales and are not adequately addressed (Lewis et al. 1996; Maceina and Bayne 2001).

Scale determines how fisheries managers perceive patterns and processes believed to be important (Figure 3.3). Scale may not refer to a constant spatial or temporal dimension but may vary according to the process under consideration (Wiens 1989). Because both fisheries and freshwater ecosystems are highly variable in space and time, there is no standardized sampling approach that can provide a consistent frame of observation. This should be a major cause for concern in fisheries management, but the good news is that considerable progress has been made in the acknowledgment of scale as a central issue, along with concurrent developments in methods for understanding scale. Practical approaches to working with scale in fisheries management are discussed in following sections.

3.3 WORKING WITH SCALE

Incorporating scale into practice is not an easy task, but fisheries managers can benefit greatly by adopting a multiscale perspective (Lewis et al. 1996; Fausch et al. 2002). A multiscale perspective means that managers consider the effects of physical and biological factors operating at large to small spatial scales and short- to long-term temporal scales. Then by means of data and (or) theory, the processes and associated scales that have the greatest influence on a fishery and the scales at which management is likely to be most successful are identified. The complexity of ecological systems and the potentially large number of interacting factors operating over multiple spatial and temporal scales makes this quite challenging. There are, however, a few practical steps that fisheries managers can take to facilitate the incorporation or consideration of scale into fisheries management strategies.

Assuming that management objectives have been clearly and explicitly identified (a crucial first step!), the next step to working with scale is to create a conceptual model of system dynamics. By conceptual model, we are referring primarily to ideas, notions, or hypotheses about how the system works rather than mathematical expressions, although the latter can be extremely useful. The best approach is to create a conceptual model with all of the important relationships and processes that influence the management objective. The structure of the conceptual model should be based on local observations, expert opinion, and other salient information and guided by contemporary theories of system dynamics. Ideally, the conceptual model should be in a graphical form called an influence diagram with arrows between components representing causal relationships (Figure 3.3). Often this is not as straightforward as it may seem as most initial model-building attempts result in conceptual models that are very large, unwieldy, and difficult to interpret. However, the conceptual model is crucial for facilitating communication among managers as well as among team members when used in an interdisciplinary effort that involves decision makers and the general public. Accordingly, managers should attempt to make the conceptual model as simple as possible with only key processes, relationships, and outcomes. In practice, it usually takes several iterations to get to the final conceptual model (Box 3.2). Once the conceptual model is complete, the next step is to use the conceptual model as a guide to identify the most important factors and scales influencing the system of interest. Below are some useful approaches for identifying these scales.

3.4 IDENTIFYING THE APPROPRIATE SCALES

The various approaches to identify the most important scales can be roughly classified as theoretical and empirical. Theoretical approaches rely primarily on ecological theory and published studies to identify the most important factors and associated scales influencing the fishery. In contrast, empirical approaches use existing data to identify the most important fac-

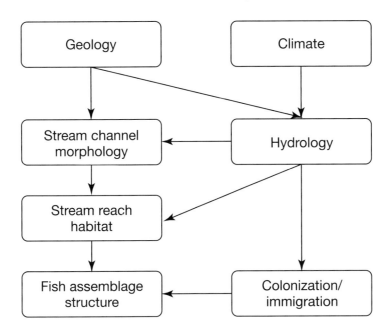

Figure 3.3. A simple conceptual model of the factors affecting the structure of stream fish assemblages in a stream reach. Here the structure of local fish assemblages is influenced by (small scale) stream habitat characteristics, such as the structure and availability of physical habitats, and (larger scale) colonization and immigration dynamics of fishes from connected water bodies. Arrows imply directions of hypothesized causal relationships.

tors and scales. In general, an empirical approach, guided by theoretical considerations, may be more defensible, but when data or available resources for analyses are limited, theoretical approaches may be the only practical means available for identifying appropriate management scales.

3.4.1 Theoretical Approaches

Theory on the role and importance of scale in ecology is large and rapidly expanding (e.g., Allen and Starr 1982; Peterson and Parker 1998; Holyoak et al. 2005). Of the many contemporary theories, hierarchy theory is among the most useful and widely used in fisheries and aquatic ecology (e.g., Frissell et al. 1986; Durance et al. 2006; Cheruvelil et al. 2008). In hierarchy theory, structure and function of biotic communities are viewed as a response to a hierarchical system of constraints in which processes operating at upper levels constrain those operating at lower levels. These levels may correspond roughly to a range of different spatial or temporal dimensions, but in a strict sense the idea of biotic or physical organization is not equivalent to a fixed scale (King 1997). In an oversimplified sense, organization can be thought of as structured interactions among processes, whereas scale in the narrowest sense refers to spatiotemporal dimensions of a phenomenon.

The dual influences of organization at different levels and relationships with scale can be illustrated with examples. Consider that the organization of native stream fish assemblages in local reaches is typically the result of constraints on the fish species pool occurring at larger spatial and temporal scales (e.g., ecoregions; Figure 3.4). In a strict hierarchy, upper-level

Box 3.2. Conceptual Model Development

The first step in evaluating the influence of scale and the response of fisheries to management actions is to create a conceptual model of the system dynamics. The process usually begins with a very complicated and detailed diagram, which is then refined in an iterative process. Below is the initial conceptual model of Rieman et al. (2001) that was used to evaluate the response of native salmonids to land management in the interior Columbia River basin. The initial model contained 45 components, called nodes, with four management action inputs (boxes) and one predicted response, fish population trend.

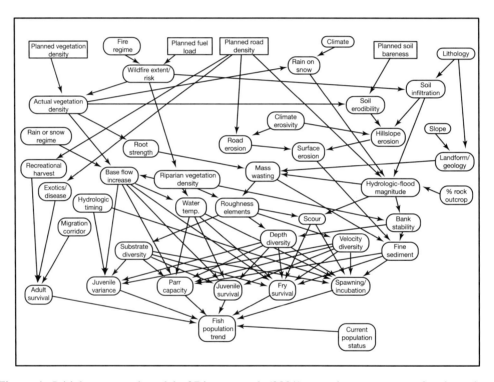

Figure A. Initial conceptual model of Rieman et al. (2001) to evaluate response of native salmonids to land management.

This initial conceptual model was subsequently modified by an interdisciplinary team of scientists and managers over a 3-month period resulting in a much simpler 23-node model (shown below) with four management inputs and one predicted response, future population status. Such substantial changes from initial model to final conceptual model are typical of multiscale evaluations.

(Box continues)

Box 3.2. Continued

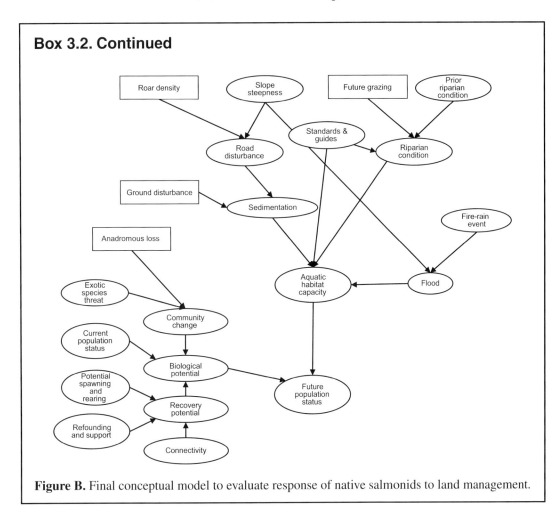

Figure B. Final conceptual model to evaluate response of native salmonids to land management.

factors correspond to larger spatial and longer temporal scales, whereas lower-level factors correspond to smaller spatial scales and shorter temporal scales. Thus, processes operate over shorter spatial extents and with greater frequency at lower levels compared with upper levels. By identifying the hierarchical levels of organization and understanding the relationships among levels (and spatiotemporal dimensions or scales), fisheries managers may identify the primary constraint acting on a fishery and evaluate the feasibility of management actions (Box 3.3). For example, the growth and condition of fishes and population size at a local scale are directly related to the productivity of the larger water body (Waters et al. 1993; Kwak and Waters 1997). Processes influencing water chemistry and productivity of a water body are, in turn, largely controlled by processes tied to watershed geologic features and climate at larger scales (Fetter 2001). In regions where geologic influences constrain productivity, local attempts to increase fish population size, perhaps by installing artificial habitat structures or changing harvest regulations, are likely to fail because population size is constrained by processes operating at a regional scale (Maceina and Bayne 2001). Managing these higher-level processes often requires greater effort or may be impossible (Figure 3.3). For example, consider a situation in which a stream fishery on locally-managed lands is negatively affected

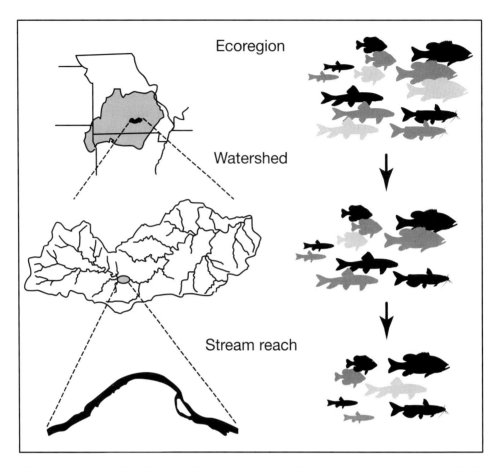

Figure 3.4. An example of the factors influencing fish assemblage structure at three hierarchical levels. At the largest spatial scale of an ecoregion, the fish species pool is the result of geologic and evolutionary processes operating over long time scales. Within the ecoregion, watershed characteristics determine the amount and type of discharge and sediment delivery patterns. At this scale, the watershed places constraints on the assemblage structure by determining the types and amounts and sizes of habitats available. At the smallest scale shown, longitudinal position of a stream reach influences the disturbance regime (i.e., frequency and intensity), which places a constraint on the fish assemblage by limiting the assemblage to the pool of available species that can survive or cope with the disturbance regime.

by habitat degradation. Riparian and stream habitat restoration at the stream reach (local) level, such as replanting riparian vegetation or adding artificial instream structures, are likely to be successful in improving the fishery if the habitat was degraded due to the loss of adjacent riparian vegetation at the local level. However, this scale of restoration is likely to be ineffective if the local habitat is degraded due to larger-scale land uses in the watershed or broad climatic patterns (Bond and Lake 2003).

In addition to these theoretical considerations, the most appropriate scales for management often differ with species-specific characteristics. For example, behavior and life history activities of a species can influence how a species interacts with its environment. Fish species with migratory life histories generally use discrete and often distant areas in a stream system

Box 3.3. An Example of Scale-Dependent Levels of Organization in Fisheries Management

Fisheries management objectives and activities may be scale dependent. The example of threatened bull trout illustrates possible relationships between conceptual scales at which natural populations of salmonid fishes are structured (Dunham et al. 2002) and units of conservation as defined in practice (see table below based on the draft bull trout recovery plan, U.S. Fish and Wildlife Service 2002; table modified from Fausch et al. 2006). Both spatial and temporal scales are implied here: as larger spatial extents are considered, longer time frames become more relevant. For example, local populations in patches can fluctuate substantially either seasonally or from year to year, but the overall range occupied by a species or distinct population segment may remain relatively constant for thousands of years under natural conditions. Specific management objectives may also vary according to unique characteristics that can be identified at each scale. For example, individual habitat patches may be the focus in managing local populations, but individual patches may be less important in considering the species or distinct population segment as a whole. Potential management activities also vary accordingly, with shorter-term actions at smaller scales (e.g., local habitat improvements or removal of nonnative species) and longer-term actions at larger scales (e.g., reintroductions to increase extent of historical habitat occupied or removal of major barriers to restore network connectivity).

(Box continues)

or beyond (e.g., marine migrations; Gross et al. 1988) to complete various life history requirements, such as spawning or juvenile rearing (Schlosser 1991). To identify the most appropriate scales for management, therefore, fisheries managers must also account for the ecological attributes of the species of interest to identify critical elements of the species' biology. The spatial extent over which these activities occur should also be considered when defining the most appropriate spatial and temporal scales for management (Box 3.4).

3.4.2 Empirical Approaches

There are a variety of empirical approaches to identifying relevant scales associated with a fishery, and those most appropriate scales for management are roughly categorized as qualitative or quantitative. Qualitative approaches use classification and ordination techniques to detect patterns in physical and biological data and infer the relative influence of small- and large-scale factors on fishes (Figure 3.5). Quantitative approaches use more detailed measures of fish abundance or community structure (e.g., species richness) to quantify relationships within and among scales and identify those having the greatest influence on fishes. Quantitative approaches are generally superior to qualitative approaches because they can estimate the magnitude of differences and strengths of relationships among scales (see Kwak and Peterson 2007).

Many of the quantitative approaches to dealing with scale are based on the idea that places (e.g., lakes or study sites) close to one another are more similar than are places farther apart. Similarly in the temporal dimension, two observations (e.g., samples) that are made at

Box 3.3. Continued

Table. Possible relationships between conceptual scales at which natural populations of salmonid fishes are structured (Dunham et al. 2002) and units of conservation as defined in practice (based on the draft bull trout recovery plan, U.S. Fish and Wildlife Service 2002; table modified from Fausch et al. 2006).

Conceptual scale	Unit of conservation	Description	Measurable characteristics
Patch	Local population	A discrete unit of suitable habitat, which may or may not be occupied; occupied patches approximate local populations; local populations are characterized by frequent (daily to seasonal) interactions among individuals within them	Patch occupancy, local population size, habitat size (watershed area or stream network length), quality (e.g., habitat conditions within a patch, presence of nonnative species, or barriers), and connectivity as related to fish movement or transport of materials and energy (e.g., water, sediment, or nutrients)
Patch network or metapopulation	Core area	Local aggregation of patches or local populations characterized by less frequent (annual to decadal) interactions	Total number or collective size of patches, rates of occupancy, numbers of individuals summed across patches, overall patterns of connectivity, habitat quality, and variability in the distribution of conditions among patches or local populations
Subbasin	Recovery unit	Naturally discrete aggregations of patch networks or core areas within larger drainage basins that interact potentially over long time frames (hundreds to thousands of years)	Distribution of characteristics among patch networks, network connectivity, and overall habitat conditions (e.g., climate, landform, and geology)
Region	Distinct population segment	Major biogeographic units that characterize distinct evolutionary lineages	Contemporary and historical location and geographic extent of species distributions, suitable habitat conditions, connectivity, and disposal

Box 3.4. Patterns of Rarity among Inland Fishes

The importance of the pattern of rarity for inland fishes was recognized by Minckley and Deacon (1968) for the case of desert fishes in the southwestern USA. Minckley and Deacon (1968) classified these fishes into four categories: (1) species that are widely distributed and positively influenced by human alteration of aquatic systems; (2) species that have not been influenced by human activities and remain widespread; (3) species that require large, special habitats; and (4) species that occupy small and unique habitats or occur as relicts or isolated endemics. Clearly fishes in these different categories have distinctive geographic and biological characteristics and present different management issues. Rabinowitz (1981) expanded these concepts into the "seven forms of rarity" (Rabinowitz et al. 1986) that are possible by considering patterns in the geographic distribution, habitat specificity, and abundance. Rey Benayas et al. (1999) expanded this framework to include habitat occupancy as an additional criterion to yield 10 different potential forms of rarity, and one of "commonness," as follows in the table below.

The implications of these concepts for inland fishes are multifold. First, as indicated earlier (e.g., Box 3.1), patterns of occupancy, and therefore rarity, are strongly scale dependent, and determination of occupancy should be approached with due caution. Second, if there is confidence in the assessments of the criteria recognized by Rabinowitz (1981) or Rey Benayas et al. (1999), it is clear that spatial scaling is critical in terms of both the grain (habitat occupancy) and extent (geographic range) of a species distribution and the inferred pattern of rarity. Finally, whereas the fundamental ideas about scale and pattern of rarity have been known for inland fishes for more than 40 years (Minckley and Deacon 1968), they have rarely been applied (Gaston and Lawton 1990; Fagan and Stephens 2006).

(Box continues)

a single location closer together in time are generally more similar than those made at longer time intervals. These similarities can be due to several factors, such as shared history, climate, and geologic features, intra- and interspecific interactions, and movements of fish. For example, fish assemblages inhabiting two proximate streams often have similar geomorphologic features; hence, there are similar habitats and the fish experience the same weather events, such as floods and droughts. Given that these factors often influence stream fish assemblage structure (Larimore et al. 1959; Matthews 1986; Bayley and Osborne 1993), fish assemblages in these two streams are likely to be similar compared with assemblages in more distant streams. This similarity or dependence among observations located closely in space or time has been defined by statisticians as spatial or temporal autocorrelation, respectively (Sokal and Rohlf 1995). Many statistical techniques have been developed to examine and describe the degree of spatial or temporal autocorrelation and patterns of dependence that may be used to identify characteristic scales of relevance to fish populations (Fortin and Dale 2005; Wagner and Fortin 2005). Of these, hierarchical models are discussed because they are related to linear regression and analysis of variance (ANOVA), two statistical techniques commonly used by fisheries biologists.

Box 3.4. Continued

Table. Framework for patterns of rarity based on Rey Benayas et al. (1999).

Criterion								
Geographic range	Wide				Narrow			
Habitat specificity	Broad		Restricted		Broad		Restricted	
Abundance	Large	Small	Large	Small	Large	Small	Large	Small
Habitat occupancy high	Common	Widespread	Specialist[a]	Locally endangered	Locally common	Nonexistent	Endemic specialist[a]	
Habitat occupancy low	Highly dispersed	Sparse	Locally endangered		Potentially endangered		Endangered	

[a] Rey Benayas et al. (1999) used the term "indicator" (retained here for accuracy), but the term "specialist" is used here to avoid potentially misleading implications or interpretations of what an "indicator" means (Carignan and Villard 2002).

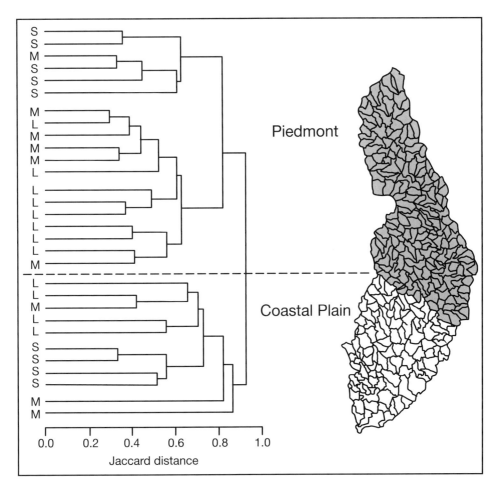

Figure 3.5. A hierarchical cluster analysis of fish assemblages in small (S), medium (M), and large (L) streams in the Flint River basin, Georgia. The cluster analysis shows that assemblages initially cluster most strongly based on physiographic province, with the streams located in the Piedmont and Coastal Plain clustered above and below the broken line, and secondarily by stream size. This suggests that factors at the physiographic province scale have the greatest influence on stream fish assemblage structure, and stream size influences assemblage structure within physiographic province. Jaccard's distance measures the dissimilarity between assemblages.

Hierarchical models are techniques that have been developed to analyze hierarchically-structured data. They may encompass any number of nested levels (i.e., given sufficient data), with each level corresponding to a spatial or temporal scale. In hierarchical model terminology, lower-level units are nested within upper-level units (e.g., streams nested within watersheds) and lower levels correspond to smaller scales and upper levels to larger scales. Hierarchical models include both linear and nonlinear forms, such as logistic and Poisson regression. Here, the description is restricted to a linear form that provides an example of a simple two-level model. Readers interested in additional details on hierarchical models should consult Snijders and Bosker (1999), Bryk and Raudenbush (2002), or Royle and Dorazio (2008).

To describe the hierarchical linear model, we begin with an ordinary linear regression model with one predictor (independent) variable:

$$Y_i = \beta_0 + \beta_1 X_{1i} + r_i, \qquad (3.1)$$

where Y_i is the response (e.g., fish density), and X_{1i} is the predictor variable (e.g., stream width or elevation) for observation i, β_0 is the intercept, β_1 is the regression coefficient (i.e., slope), and r is the residual. Collectively, the residuals are assumed to be normally distributed with a mean of 0 and variance (Bryk and Raudenbush 2002). This model is appropriate for examining the relationship between the response and predictor variable when all observations (i) are independent. However, when observations are collected from within different groups (e.g., fish collected from multiple streams in different watersheds), the observations from a group may be more similar to one another simply due to their close proximity (in time or space) or due to similar environments within each group. As discussed above, this is known as dependence or autocorrelation of observations. Hierarchical models can represent this dependence by modeling the variability in the response within and among groups. To illustrate, suppose that J groups (e.g., watersheds) were randomly selected, and sample units within each group (e.g., streams within watersheds) were randomly selected and sampled. A separate regression model can be fit for each of the groups. Mathematically, this can be represented in a single equation with subscript j as

$$Y_{ij} = \beta_{0j} + \beta_{1j} X_{1ij} + r_{ij}, \qquad (3.2)$$

where the variables are defined above. In hierarchical modeling, this is a level-1 model where observations (i) and groups (j) are defined as level-1 and level-2 units, respectively. The model coefficients can be treated as fixed, that is, their value is assumed equal across level-2 units (e.g., $\beta_{01} = \beta_{02} = \beta_{03}$), or, alternatively, they can be treated as randomly varying, that is, their values differ among level-2 units (e.g., $\beta_{01} \neq \beta_{02} \neq \beta_{03}$). Thus, models with randomly varying coefficients (i.e., intercepts and slopes) differ from single-level models (equation 3.1) in that each level-2 unit (group) can have unique coefficients.

Simpler and more familiar forms of hierarchical models can be obtained when selected terms in equation (3.2) are replaced by 0s. For example, when there is 0 variability among coefficients, equation (3.2) is equivalent to ordinary linear regression (equation 3.1). One of the hierarchical model forms that is widely used by fisheries biologists is a random effects ANOVA, which is a hierarchical model without predictors (i.e., slopes, or β_1, β_2, and so on):

$$Y_{ij} = \gamma_{00} + u_{0j} + r_{ij}, \qquad (3.3)$$

where γ_{00} is the grand mean across groups and u_{0j} is an estimate of how much the response of the jth group differs from the grand mean. Collectively, the values of u_{0j} have a mean of 0 and variance τ_{00}, which is an estimate of variability among groups (i.e., the level-2 variance), whereas the residual variance, σ^2, is an estimate of the variability within groups (i.e., the level-1 variance). The total variance of Y_{ij} is the sum of the estimated variance at levels 1 and 2 ($\hat{\tau}_{00} + \hat{\sigma}^2$) and is used to estimate the intraclass correlation coefficient as

$$\rho = \frac{\hat{\tau}_{00}}{\hat{\tau}_{00} + \hat{\sigma}^2}. \qquad (3.4)$$

The intraclass correlation coefficient is a measure of the proportion of variance that is accounted for by the group level (Bryk and Raudenbush 2002). In the context of hierarchical multiscale models, the intraclass correlation coefficient is used to estimate the amount of variance in the response variable that is due to (unknown) factors at the level-2 spatial scale. An example of fitting and interpreting hierarchical models can be found in Box 3.5.

3.5 INCORPORATING SCALES INTO MANAGEMENT

In this section, two approaches that fisheries managers have taken to incorporate scaling considerations into fisheries management are described—expert judgment and process modeling. As discussed above, delineation of spatial or temporal dimensions of a management problem is possible via a variety of means, but these scales will often lack sufficient details about processes that are ultimately of concern in fisheries management. Lacking these process details, it may not be clear why certain outcomes occur after a management action is implemented. Whenever possible, therefore, managers should formulate competing hypotheses about how management may influence processes at different scales and measure management

---

Box 3.5. Hierarchical Modeling of Stream Fish Density

The use of hierarchical models is illustrated by identifying and quantifying scales of influence with empirical fish abundance and habitat data collected from 236 streams in 23 watersheds (6–16 streams per watershed) in central Idaho (Rieman et al. 2006). Here the level-1 units are streams nested within level-2 units, watersheds, and the response of interest is the density (number/100 m) of Westslope cutthroat trout. The analysis was begun by partitioning the variance in trout density within and among watersheds by means of a random effects ANOVA. The variance among watersheds, $\hat{\tau}_{00}$, was 2,972.5 and within watersheds, σ^2, was 1,706.2, which means that $\frac{2972.5}{2972.5+1706.2} = 0.635$ or 63.5% of the variation in Westslope cutthroat trout density was due to factors at the watershed scale. Previous studies reported that salmonid density is related to the geology of the watershed, particularly the amounts of productive lithology (Thompson and Lee 2000). To quantify the influence of a large-scale factor, watershed geology, a level-2 model was fit with a single predictor, percent mafic (productive) lithology. After fitting the model, the level-2 variance, $\hat{\tau}_{00}$, was 1,227.9, which means that $\frac{2972.5-1227.9}{2972.5} = 0.587$, or 58.7% of the variability in Westslope cutthroat trout density among watersheds, and $\frac{2972.5-1227.9}{2972.5+1706.2} = 0.373$, or 37.3% of the total variability, was accounted for by differences in lithology. Previous studies also have suggested that the density of stream-dwelling salmonids is influenced by the characteristics of stream sample reaches. Thus, a complete two-level hierarchical model was fitted by including stream gradient in the mafic lithology model, a (smaller-scale) level-1 predictor. After fitting the model the level-1 variance, σ^2, was 1,305.3, indicating that gradient accounted for $\frac{1706.2-1305.3}{1706.2} = 0.235$, or 23.5% of the variation among streams within a watershed, and $\frac{1706.2-1305.3}{2972.5+1706.2} = 0.086$, or 8.6% of the total variability in Westslope cutthroat trout density. The fact that most variability in density was related to large-scale factors was fairly strong evidence that the greatest influences in Westslope cutthroat populations are large-scale, watershed-level factors.

outcomes against predictions from these hypotheses. It is recognized that not every management action can be elevated to the level of a research project, but it can be stressed that a clear articulation of processes and testable hypotheses (e.g., Box 3.2) is essential for accurate identification, refinement, and functional understanding of management scales. In that spirit, some practical advice and methods for incorporating scale into inland fisheries management are offered.

3.5.1 Expert Judgment Approaches

Fisheries managers are sometimes reluctant to use models because they (1) distrust models and only believe "data" or (2) believe that creating explicit models is too difficult when existing data and knowledge of system dynamics are incomplete. In response to the first concern, it can be argued that data are only numbers until a human brain (using a model) interprets them. Thus, data and model are in essence equivalent if data are to have meaning. The second concern is more challenging to address. One means of coping with the chronic lack of information and uncertainty that characterizes fisheries management is through the use of modeling approaches based on expert judgment, which have a long history in economic applications (Clemen 1996), but are only recently beginning to be used in inland fisheries management. Expert models are based on how people think ecological systems work and require an explicit definition of relationships or key processes and uncertainties. Usually, these models are represented graphically as influence diagrams (Figure 3.3). The models need not be excessively large and should represent only those key processes believed to influence the fishery of interest. The most common expert models take the form of probabilistic networks (Haas 1991). Probabilistic networks, also known as Bayesian belief networks (BBNs), consist of model components defined as nodes and causal links between components represented by directional arcs. Each node consists of a set of mutually-exclusive states, and the relationships between components (nodes) are modeled using probabilistic (conditional) dependencies. To illustrate, consider a simple three-node model for which pool habitat availability in a stream is modeled as a function of stream channel morphology and streamflow (Figure 3.6). Here habitat availability is represented by three states: low, moderate, and high; channel morphology by two states: unconfined and confined channel; and streamflow by three states: low, normal, and high. In practice, each node state would be delineated by a value or range of values that are mutually exclusive, such as less than 4.9 pools/km (low pool habitat availability), 5.0–14.9 pools/km (moderate pool availability), and greater than 15 pools/km (high pool availability). In the pool habitat BBN, the probability that pool habitat availability is low, moderate, or high is conditional (dependent) on the stream channel morphology and streamflow (Figure 3.6). When the channel morphology is unconfined and streamflow is low, the BBN estimates that the probabilities of low, moderate, or high pool habitat availability are 80, 15, and 5%, respectively (Figure 3.6a). In contrast, when the stream channel is confined the probability that pool habitat availability is low, moderate, or high is 50, 40, and 10%, respectively (Figure 3.6b).

Although the conditional probabilities that produced these estimates are hypothetical, in real-world applications conditional probabilities are based on judgments of one or more experts in the field (Clemen 1996; Marcot et al. 2001). The drawbacks of using expert judgment approaches are that a substantial burden of proof is placed on the decision maker (Morgan and Henrion 1990) and model development can be a time consuming process where modifications are made as assumptions are evaluated and changed in an iterative process (Clemen 1996).

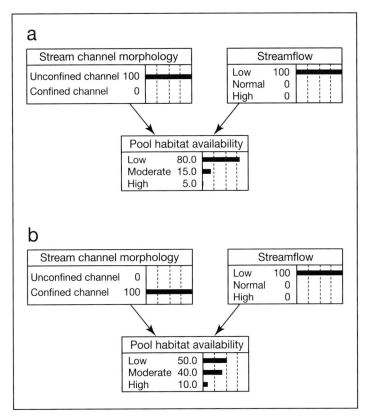

Figure 3.6. Bayesian belief network for pool habitat availability in stream reach as a function channel morphology and streamflow for two combinations of probabilities: (a) unconfined stream channel 100% and low streamflow 100%, and (b) confined stream channel 100% and low streamflow 100%. Numbers in the boxes are probabilities of a particular state expressed as a percentage.

The benefits of expert models are that they convey how managers believe a system functions, how key assumptions may be tested, and where uncertainties are greatest. In practice, uncertainties about a system may play a greater role in decision making. Thus, identifying key areas of uncertainty can be critical. Greater detail about BBNs and expert models can be found in Clemen (1996), Marcot et al. (2001), and Martin et al. (2005).

A recent example of an expert judgment approach was an assessment of alternative conservation strategies for salmonids in the inland northwestern USA (Rieman et al. 2001). This study evaluated the effects of conservation strategies and management actions at multiple scales ranging from stream reach to watershed to large river basins. The study faced difficulties regarding a lack of data on and knowledge of the effects of multiple physical, biological, and anthropogenic factors operating over multiple spatial and temporal scales. Interdisciplinary teams developed and modified conceptual models of the study system (Box 3.2). These models were modified and refined by external reviewers and parameterized by groups of area-specific experts. For example, conditional dependencies for hydrologic and geomorphic nodes were parameterized by hydrologists and geomorphologists familiar with the respective management areas. Following parameterization, a sensitivity analysis was performed on the models to identify key components (i.e., those with the greatest influence on the estimated change in fish populations) and model assumptions were tested. Investigators then evaluated whether the best management strategy for recovering migratory Pacific salmon populations could be influenced by the removal of large hydroelectric dams on the Snake River in Idaho. The effects of management actions with and without dams were assessed, and the expert model suggested the influence of the dams did not change the relative benefits of conservation actions. Through the use of BBNs, this study communicated how the scientists believed the ecological system functioned, quantified the relative value of conservation actions, predicted the outcome of alternative conservation scenarios, evaluated the sensitivity of their predictions to assumptions, and identified key uncertainties. Identifying these uncertainties is important for prioritizing future monitoring and research efforts aimed at increasing the understanding of processes operating at multiple scales and thereby improving future management.

3.5.2 Process-Based Simulations

A less common approach to incorporating scale into fisheries management uses process-based simulations. These models often relate rates of change in both physical and biological components of aquatic systems to factors such as management actions, land use, and climate change. The relative scarcity of empirical simulation modeling approaches is presumably due to a lack of knowledge of and data on the combined effects of large- and small-scale ecological processes. The most common use of process models is as a guide during development of management strategies (Box 3.3). Direct estimation of fish population (e.g., presence and abundance) and assemblage (e.g., species richness) responses to multiscale processes is relatively uncommon. This is because accurate predictions about complex nonlinear responses across multiple spatial and temporal scales often require process-based models. Given the complexity of aquatic systems, development of multiscale process-based models requires interdisciplinary collaboration and matching of spatial and temporal scales from outputs of physical and ecological models.

An example of a process-based simulation is described by Peterson and Kwak (1999). They simulated the effects of large-scale land use and climate change on riverine smallmouth bass populations based on empirical precipitation runoff models, climate projection models, and an age-structured smallmouth bass population dynamics model parameterized by means of a 13-year data set. Another example is the across trophic level system simulation (ATLSS) project in the Florida Everglades. The ATLSS project is a multidisciplinary effort that includes scientists from a variety of fields, including hydrologists, landscape ecologists, and fisheries managers. One objective is to develop multiscale simulation models for predicting the abundance of fishes in response to restoration activities in the Everglades (Gaff et al. 2004). Simulation models were based, in part, on empirical field observations and laboratory studies combined with expert opinion as to how demographic processes scale up to the entire Everglades. These and other studies highlight the need for greater interdisciplinary collaboration (Vaughan et al. 2007).

3.5.3 Adaptive Approaches to Management

An important theme of this chapter is the influence of scale on how fisheries managers perceive fish populations and communities. Significant uncertainties are associated with even the simplest management decision and managers can never be 100% confident of achieving desired outcomes. This uncertainty stems from three basic sources (following Williams et al. 2002): (1) environmental uncertainty, which is composed of environmental and demographic variation and has both spatial and temporal components; (2) statistical uncertainty, due to the use of sample data to estimate model parameters; and (3) structural (or ecological) uncertainty, due to an inability to determine accurately the processes or models that best represent system dynamics (e.g., the relationship between geomorphology, streamflow, habitat availability, and fish population demographics). Consequently, special attention is devoted to structural uncertainty because it represents the main source of uncertainty associated with managing across multiple spatial and temporal scales. Overviews of several approaches for describing and estimating multiscalar systems have been given, but because structure and function of aquatic systems are so complex, there are often several plausible hypotheses to explain observed ecological patterns and processes. Sorting through these hypotheses to determine those that are most accurate can reduce structural uncertainty.

Several methods are available for reducing structural uncertainty. The gold standard is experimentation, which involves replication, randomization, and treatments. Conducting experiments, however, is labor intensive, which precludes application at large spatial scales and long time frames that are often necessary in fisheries management. In contrast to experiments, observational studies use statistical control to describe patterns in data that may be collected across broad spatial or temporal extents. These types of studies often provide the basis for constructing the empirical, theoretical, or simulation models discussed above. Relationships derived from observed data, however, are often confounded with other factors. Thus, a third, complementary approach is advocated, adaptive resource management (ARM; Walters 1986), which is a technique well-suited to reducing structural uncertainty and improving decision making in management.

In ARM, structural uncertainty is explicitly considered by postulating feasible alternative models, with each model representing a hypothesized relationship among multiscale inputs, system dynamics, and objectives (Williams et al. 2002). Each model is assigned a plausibility or probability of being "true." The best management decision then is selected based on the current system state (e.g., population size) and a prediction of the expected future state taking into account environmental, structural, statistical, and other sources of uncertainty. Model probabilities are updated through time by comparing each of the model-specific predictions to conditions as they are realized. The updated model probabilities are again used to predict future conditions and the "best" management decision made for the following time step. This feedback facilitates adaptive learning and the resolution of competing hypotheses over time. Thus, ARM allows managers to make decisions and enables them to learn about the nature of scale.

3.6 CONCLUSIONS

It should be clear by now that scale is an essential consideration for effective fisheries management. Scale is not just an "academic" issue but one that profoundly influences all stages of management, from the interpretation of patterns in research or monitoring data to the creation and implementation of management strategies. Because the scales at which fisheries managers observe ecological systems directly influence the scales that are perceived to be important, it is sometimes difficult to know whether the proper scales for management have been identified. Lacking this knowledge, managers may incorrectly identify and diagnose the source of problems with a fishery and potentially waste resources and management efforts implementing ineffectual or harmful actions. Potential problems can be avoided if careful thought is given to processes and scales influencing the problem of interest. Uncertainties regarding how physical and ecological processes operate over multiple scales can be specified and incorporated into the process. Experience has proven this is much easier said than done (e.g., Walters 1997), but examples here show that scaling in practice is possible and most definitely desirable in fisheries management. Many of the chapters in this book provide essential background and information on the key scales and processes influencing inland fisheries and can provide guidance regarding multiscale approaches. Finally, fisheries managers are encouraged to consider the importance of scale and incorporate an adaptive approach to management as a means to improve ecological knowledge and future management.

3.7 REFERENCES

Allen, T. F. H., and T. B. Starr. 1982. Hierarchy: perspectives for ecological complexity. University of Chicago Press, Chicago.

Avise, J. C. 1994. Molecular markers, natural history and evolution. Chapman and Hall, New York.

Bain, M. B., and N. J. Stevenson, editors. 1999. Aquatic habitat assessment: common methods. American Fisheries Society, Bethesda, Maryland.

Bayley, P. B., and L. L. Osborne. 1993. Natural rehabilitation of stream fish populations in an Illinois catchment. Freshwater Biology 29:295–300.

Bisson, P. A., J. B. Dunham, and G. H. Reeves. 2009. Freshwater ecosystems and resilience of Pacific salmon: habitat management based on natural variability. Ecology and Society 14(1):45 Available: www.ecologyandsociety.org/vol14/iss1/art45. (March 2010).

Bonar, S. A., D. W. Willis, and W. A. Hubert, editors. 2009. Standard sampling methods for North American freshwater fishes. American Fisheries Society, Bethesda, Maryland.

Bond, N. R., and P. S. Lake. 2003. Local habitat restoration in streams: constraints on the effectiveness of restoration for stream biota. Ecological Management and Restoration 4:193–198.

Bryk, A. S., and S. W. Raudenbush. 2002. Hierarchical linear models: applications and data analysis methods, second edition. Sage, Newbury Park, California.

Carignan, V., and Villard, M.-A. 2002. Selecting indicator species to monitor ecological integrity: a review. Environmental Monitoring and Assessment 78:45–61.

Cheruvelil, K. S., P. A. Soranno, M. T. Bremigan, T. Wagner, and S. L. Martin. 2008. Grouping lakes for water quality assessment and monitoring: the roles of regionalization and spatial scale. Environmental Management 41:425–440.

Clemen, R. T. 1996. Making hard decisions, second edition. Duxbury, Belmont, California.

Dunham, J., G. Chandler, B. Rieman, and D. Martin. 2005. Measuring stream temperature with digital data loggers: a user's guide. U.S. Department of Agriculture, Forest Service, Rocky Mountain Research Station, General Technical Report, RMRSGTR-150WWW, Fort Collins, Colorado.

Dunham, J. B., B. E. Rieman, and J. T. Peterson. 2002. Patch-based models of species occurrence: lessons from salmonid fishes in streams. Pages 327–334 in J. M. Scott, P. J. Heglund, M. Morrison, M. Raphael, J. Haufler, and B. Wall, editors. Predicting species occurrences: issues of scale and accuracy. Island Press, Covelo, California.

Durance, I., C. Lepichon, and S. J. Ormerod. 2006. Recognizing the importance of scale in the ecology and management of riverine fish. River Research and Applications 22:1143–1152.

Ebersole, J. L., W. J. Liss, and C. A. Frissell. 2001. Relationship between stream temperature, thermal refugia and rainbow trout *Oncorhynchus mykiss* abundance in arid-land streams in the northwestern United States. Ecology of Freshwater Fish 10:1–10.

Fagan, W. F., C. Aumann, C. M. Kennedy, and P. J. Unmack. 2005. Rarity, fragmentation, and the scale dependence of extinction risk in desert fishes. Ecology 86:34–41.

Fagan, W. F., and A. J. Stephens. 2006. How local extinction changes rarity: an example with Sonoran Desert fishes. Ecography 29:845–852.

Fausch, C., B. Rieman, J. Dunham, and M. Young. 2006. Strategies for conserving native salmonid populations at risk from nonnative invasions: tradeoffs in using barriers to upstream movement. U.S. Department of Agriculture Forest Service, Rocky Mountain Research Station, GTR-RMRS-174, Fort Collins, Colorado.

Fausch, K. D., S. Nakano, and K. Ishigaki. 1994. Distribution of two congeneric charrs in streams of Hokkaido Island, Japan: considering multiple factors across scales. Oecologia 100:1–12.

Fausch, K. D., C. E. Torgersen, C. V. Baxter, and H. W. Li. 2002. Landscapes to riverscapes: bridging the gap between research and conservation of stream fishes. BioScience 52:1–16.

Fetter, C. W. 2001. Applied hydrogeology, fourth edition. Prentice Hall, Upper Saddle River, New Jersey.

Fortin, M. J., and M. R. T. Dale. 2005. Spatial analysis: a guide for ecologists. Cambridge University Press, Cambridge, UK.

Frissell, C. A., W. J. Liss, C. E. Warren, and M. D. Hurley. 1986. A hierarchical framework for stream habitat classification: viewing streams in a watershed context. Environmental Management 10:199–214.

Fuller, W. A. 1987. Measurement error models. Wiley, New York.

Gaff, H., J. Chick, J. Trexler, D. DeAngelis, L. Gross, and R. Salinas. 2004. Evaluation of and insights from ALFISH: a spatially explicit, landscape-level simulation of fish populations in the Everglades. Hydrobiologia 520:73–87.

Gaston, K. J., and J. H. Lawton. 1990. The population ecology of rare species. Journal of Fish Biology 37:97–104.

Gross, M. R., R. M. Coleman, and R. M. McDowall. 1988. Aquatic productivity and the evolution of diadromous fish migration. Science 239:1291–1293.

Haas, T. C. 1991. A Bayesian belief network advisory system for aspen regeneration. Forest Science 37:627–654.

Hocutt, C. H., and E. O. Wiley, editors. 1986. The zoogeography of North American freshwater fishes. Wiley, New York.

Holyoak, M., M. A. Leibold, and R. D. Holt, editors. 2005. Metacommunities: spatial dynamics and ecological communities. University of Chicago Press, Chicago.

Jacobson, R. B., and K. B. Gran. 1999. Gravel routing from widespread, low-intensity landscape disturbance, Current River basin, Missouri. Earth Surface Processes and Landforms 24: 897–917.

King, A. W. 1997. Hierarchy theory: a guide to system structure for wildlife biologists. Pages 185–214 *in* J. A. Bissonette, editor. Wildlife and landscape ecology: effects of pattern and scale. Springer-Verlag, New York.

Kunin, W. E. 1998. Extrapolating species abundance across spatial scales. Science 281:1513–1515.

Kwak, T. J., and J. T. Peterson. 2007. Community indices, parameters, and comparisons. Pages 667–763 *in* M. Brown and C. Guy, editors. Analysis and interpretation of freshwater fisheries data. American Fisheries Society, Bethesda, Maryland.

Kwak, T. J., and T. F. Waters. 1997. Trout production dynamics and water quality in Minnesota streams. Transactions of the American Fisheries Society 126:35–48.

Larimore, R. W., W. F. Childers, and C. Heckrotte. 1959. Destruction and re-establishment of stream fish and invertebrates affected by drought. Transactions of the American Fisheries Society 88:261–285.

Lewis, C. A., N. P. Lester, A. D. Bradshaw, J. E. Fitzgibbon, K. Fuller, L. Hakanson, and C. Richards.

1996. Considerations of scale in habitat conservation and restoration. Canadian Journal of Fisheries and Aquatic Sciences 53 (Supplement 1):440–445.

Maceina, M. J., and D. R. Bayne. 2001. Changes in the black bass community and fishery with oligotrophication in West Point Reservoir, Georgia. North American Journal of Fisheries Management 21:745–755.

Marcot, B., R. Holthausen, M. Raphael, M. Rowland, and M. Wisdom. 2001. Using Bayesian belief networks to evaluate fish and wildlife population viability under land management alternatives from an environmental impact statement. Forest Ecology and Management 153:29–42.

Martin, T. G., P. M. Kuhnert, K. Mengersen, and H. P. Possingham. 2005. The power of expert opinion in ecological models using Bayesian methods: impacts of grazing on birds. Ecological Applications 15:266–280.

Matthews, W. J. 1986. Fish faunal structure in an Ozark stream: stability, persistence and a catastrophic flood. Copeia 1986:388–397.

Matthews, W. S., and D. C. Heins, editors. 1987. Community and evolutionary ecology of North American stream fishes. Oklahoma University Press, Norman.

Melbourne, B. A., H. V. Cornell, K. F. Davies, C. J. Dugaw, S. Elmendorf, A. L. Freestone, R. J. Hall, S. Harrison, A. Hastings, M. Holland, M. Holyoak, J. Lambrinos, K. Moore, and H. Yokomizo. 2007. Invasion in a heterogeneous world: resistance, coexistence or hostile takeover? Ecology Letters 10:77–94.

Minckley, W. L., and J. E. Deacon. 1968. Southwestern fishes and the enigma of "endangered species." Science 159:1424–1432.

Morgan, M. G., and M. Henrion. 1990 Uncertainty: a guide to dealing with uncertainty in quantitative risk and policy analysis. Cambridge University Press, Cambridge, UK.

Morrison, M. L., and L. S. Hall. 2002. Standard terminology: toward a common language to advance ecological understanding and application. Pages 43–52 *in* J. M. Scott, P. J. Heglund, M. Morrison, M. Raphael, J. Haufler, and B. Wall, editors. Predicting species occurrences: issues of accuracy and scale. Island Press, Covelo, California.

Neville, H., J. B. Dunham, and M. Peacock. 2006. Assessing connectivity in salmonid fishes with DNA microsatellite markers. Pages 318–342 *in* K. R. Crooks and M. Sanjayan, editors. Connectivity conservation. Conservation Biology Series 14. Cambridge University Press, Cambridge, UK.

Peterson, D. L., and V. T. Parker, editors. 1998. Ecological scale: theory and applications. Columbia University Press, New York.

Peterson, J. T., and T. J. Kwak. 1999. Modeling the effects of land use and climate change on riverine smallmouth bass. Ecological Applications 9:1391–1404.

Rabinowitz, D. 1981. Seven forms of rarity. Pages 205–217 *in* H. Synge, editor. The biological aspects of rare plant conservation. John Wiley, Chichester, UK.

Rabinowitz, D., S. Cairns, and T. Dillon. 1986. Seven forms of rarity and their frequency in the flora of the British Isles. Pages 182–204 *in* M. E. Soulé, editor. Conservation biology: the science of scarcity and diversity. Sinauer, Sunderland, Massachusetts.

Rey Benayas, J. M., S. M. Scheiner, M. García Sánchez-Colomer, and C. Levassor. 1999. Commonness and rarity: theory and application of a new model to Mediterranean montane grasslands. Conservation Ecology 3(1):5.

Rieman, B. E., J. T. Peterson, J. Clayton, W. Thompson, R. F. Thurow, P. Howell, and D. C. Lee. 2001. Evaluation of the potential effects of federal land management alternatives on the trends of salmonids and their habitats in the interior Columbia River basin. Journal of Forest Ecology and Management 153:43–62.

Rieman, B. E., J. T. Peterson, and D. L. Myers. 2006. Have brook trout displaced bull trout in streams of central Idaho? An empirical analysis of distributions along elevation and thermal gradients. Canadian Journal of Fisheries and Aquatic Sciences 63:63–78.

Royle, J. A., and R. M. Dorazio. 2008. Hierarchical modeling and inference in ecology: the analysis of data from populations, metapopulations and communities. Academic Press, London.

Schlosser, I. J. 1991. Stream fish ecology: a landscape perspective. BioScience 41:704–712.

Snijders, T. A. B., and R. J. Bosker. 1999. Multilevel analysis: an introduction to basic and advanced multilevel modeling. Sage, Thousand Oaks, California.

Sokal, R. R., and F. J. Rohlf. 1995. Biometry: the principles and practice of statistics in biological research. Freeman, New York.

Thompson, W. L., and D. C. Lee. 2000. Modeling relationships between landscape-level attributes and snorkel counts of Chinook salmon and steelhead parr in Idaho. Canadian Journal of Fisheries and Aquatic Sciences 57:1834–1842.

Torgersen, C. E., D. M. Price, H. W. Li, and B. A. McIntosh. 1999. Multiscale thermal refugia and stream habitat associations of Chinook salmon in northeastern Oregon. Ecological Applications 9:301–319.

U.S. Fish and Wildlife Service. 2002. Bull trout (*Salvelinus confluentus*) draft recovery plan. U.S. Fish and Wildlife Service, Portland, Oregon.

Vaughan, I. P., M. Diamond, A. M. Gurnell, K. A. Hall, A. Jenkins, N. J. Milner, L. A. Naylor, D. A. Sear, G. Woodward, and S. J. Ormerod. 2007. Integrating ecology with hydromorphology: a priority for river science and management. Aquatic Conservation: Marine and Freshwater Ecosystems 19:113–125.

Wagner, H. H., and M. J. Fortin. 2005. Spatial analysis of landscapes: concepts and statistics. Ecology 86:1975–1987.

Walters, C. 1986. Adaptive management of renewable resources. MacMillan, New York.

Walters, C. 1997. Challenges in adaptive management of riparian and coastal ecosystems. Conservation Ecology (online) 1(2):1. Available: www.consecol.org/vol1/iss2/art1. (March 2010).

Waters, T. F., J. P. Kaehler, J. T. Polomis, and T. J. Kwak. 1993. Production dynamics of smallmouth bass in a small Minnesota stream. Transactions of the American Fisheries Society 122:588–598.

Wiens, J. A. 1989. Spatial scaling in ecology. Functional Ecology 3:385–397.

Williams, B. K., J. D. Nichols, and M. J. Conroy. 2002. Analysis and management of animal populations. Academic Press, San Diego, California.

Chapter 4

The Legal Process and Fisheries Management

JEFFERY A. BALLWEBER AND HAROLD L. SCHRAMM, JR.

4.1 INTRODUCTION

Ownership and management responsibility for fish and wildlife resources and the land and water on which they are dependent have been serious legal considerations as far back as the Roman era. As different forms of government evolved and became increasingly more complex, management responsibilities for fish, wildlife, and other natural resources were shared among different branches (i.e., executive, legislative, and judicial) and levels (e.g., national, state or provincial, county, and municipal) of government.

This evolution has created a labyrinth of seemingly overlapping and conflicting governmental authorities and agency goals. To chart a path through the maze, the initial issue to resolve is who "owns" or has management responsibility for a particular fishery. This question may be answered explicitly by national constitutions or implied from other governmental powers. Wild fish and wildlife are public resources that the government manages to ensure the resources' persistence for future generations. Depending on the nation and, in some cases, the geographic area (e.g., international water bodies such as the Great Lakes) or specific fishery (e.g., salmon in the Pacific Northwest), management authority is shared among different levels of government (national, sub-national, and local). In addition, different branches of government have different roles in fisheries management.

Within this framework, fisheries professionals manage fisheries resources and are also charged to be fisheries advocates in other water and land management and development decisions. This chapter provides information that will enable fisheries managers to function more effectively in the legal realm of fisheries management and to enhance their effectiveness in representing and advocating for fisheries in issues of land and water management. The chapter begins with an overview of North American governmental organization and then provides some background on the historical basis for governmental management of natural resources. These two topics are then integrated to show the interrelationship of different levels of government in fisheries management. Opportunities to interject fisheries management concerns within broader watershed and ecosystem management efforts are also discussed. The specific information presented, except as noted, pertains to the United States (USA); in most cases Canada and the United Mexican States (Mexico) have somewhat similar concepts and principles with different terminology.

4.2 OVERVIEW OF NORTH AMERICAN GOVERNMENTS

National and sub-national governments have a pervasive role in fish and wildlife management and conservation both as management entities and by enacting a plethora of rules and regulations controlling private activities related to fish and wildlife, their habitats, and their uses. North America's three largest national governments all have constitutions that specify the details of both national and sub-national governments' structures. Despite some significant differences in procedures and terminology, Canada, Mexico, and the USA share three important characteristics: (1) a strong democratic foundation with elected representatives; (2) national and sub-national constitutions that allocate authority between executive (the Crown in Canada), legislative, and judicial branches of government; and (3) a system of "dual sovereignty," or shared powers, among the national (also known as federal or central) government and various sub-national (e.g., states, provinces, territories, and protectorates) and local (e.g., county and municipal) governments. In analyzing the political and legal aspects of inland fisheries management it is vital to understand (1) the roles of the different branches of government and the system of horizontal (at the same government level) checks and balances between branches of government and (2) the concept of dual sovereignty or vertically-shared powers between the different levels of government.

In each of the three largest North American nations, the national constitutional authority is the highest level of authority for all governmental action in a country. At the national level, treaties and land claim agreements between the national government and indigenous peoples can provide an additional level of legal authority and may grant indigenous people governmental powers that are different from that of states or provinces. In general, regardless of the level of government, legal authority flows from the constitution, to legislative laws, and then to agency regulations. As such, regulations are invalid if not authorized by a law, and laws are invalid if they conflict with the constitution. For example, a law creates an agency and specifies its purpose; the agency then promulgates regulations to carry out its legislated purpose. All legal authority can be changed to reflect new circumstances. Though difficult, constitutions can be amended or replaced. Similarly, legislatures regularly pass new laws and amend existing laws as part of their oversight of executive branch agencies. Finally, executive branch agencies routinely promulgate or amend regulations to implement their statutory authority. As a rule, it is more difficult to change national legal authority than sub-national authority. Likewise, it is more difficult to modify constitutions than laws (statutes), and laws are more difficult to enact or modify than regulations. As a simple rule, the legislative branch makes laws, the executive branch executes those laws, and the judicial branch interprets the laws.

4.2.1 The Legislative Branch

The legislative process is designed to allow broad agency and public participation in making laws and allocating funding to reflect policy priorities (Figure 4.1). Laws are made for the overall good of the people. In the case of fisheries resources, the "good of the people" usually means the conservation of the resource for the present and the future, but it may also mean to achieve benefits from the resource. In the USA, the Library of Congress has a complete and regularly updated web site to track pending legislation. The following discussion focuses on the national legislative process, but all legislatures follow a similar format.

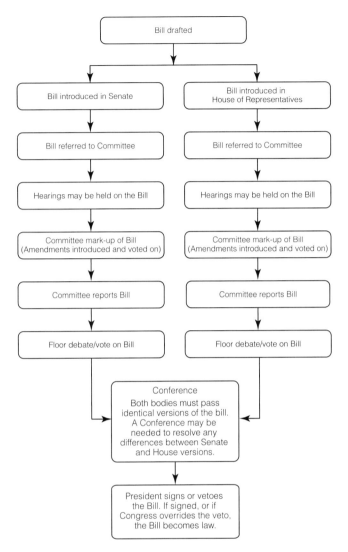

Figure 4.1. From idea to law: how federal laws are established in the USA. A similar process occurs for state laws.

Both Canada and the USA have bicameral legislatures. Laws must pass both chambers, and the executive branch has to assent to the legislation for it to become law. Laws are collected and published in the national code, called the U.S. Code (USC) in the USA. The USC is organized into 50 titles that cover the entire range of Congress' authority, and each title is further organized into numerous chapters. Conservation is addressed in Title 16, but many other titles also have a significant bearing on fisheries management. The Conservation Title has in excess of 87 chapters. The U.S. states and Canadian provinces use a similar structure for their laws. Tribal governments in the USA follow a similar structure, whereas Canada's First Nations and Inuit do not. The constitutions, laws, and regulations of most states, provinces, and tribes are available on-line and are regularly updated.

The USA uses a basic, two-part legislative process that distinguishes between (1) laws that establish federal agencies (organic acts) or programs (substantive acts) and (2) laws that fund them (appropriation acts). Organic acts create agencies and governmental entities and prescribe their organizational structure and duties. The legislature can create additional departments (e.g., U.S. Department of the Interior or U.S. Department of Agriculture) and cabinet positions (the appointed head [secretary] of each department is a member of the cabinet), new agencies or bureaus under existing cabinet positions (e.g., the National Park Service Organic Act of August 15, 1916 created the National Park Service, and the Fish and Wildlife Act of 1956 created the U.S. Fish and Wildlife Service [USFWS] within the Department of the Interior), or new independent agencies (e.g., the Environmental Protection Agency [EPA]). Much like the U.S. legislature created the USFWS, sub-national-level legislatures create state, provincial, or tribal game and fish departments whose missions include conserving and managing the state's (or province's or tribe's) fisheries resources for current and future generations.

Substantive laws create new programs to be carried out by one or more existing agencies. The Endangered Species Act (ESA, Box 4.1) and the Clean Water Act (CWA) represent two of the many substantive laws that allocated new authority to the USFWS and the EPA, respectively. Frequently, substantive laws specify laws of authorized funding or appropriations for the program for a given number of years. When a program's authorized appropriation expires, traditionally every 5 years, the relevant congressional committees reevaluate the program and how it is administered and reauthorize or amend the substantive law. States often attach "sunset clauses" to legislation to accomplish this purpose. Legislatures periodically amend organic laws and amend or enact additional substantive laws granting agencies additional responsibilities. Alternatively, appropriations are required annually to fund agencies' programs.

4.2.2 The Executive Branch

Constitutions lay out a structure for the executive branch that identifies a chief executive (e.g., President or Governor General) as well as specifying certain minimal agencies or ministries. Chief executives commonly have some authority independent of the legislature to issue executive orders to implement or interpret the constitution, a treaty, or a law.

Executive branch agencies draw their authority from statutes and then promulgate rules to interpret or implement their organic or substantive statutory authority (Ballweber and Jackson 1996; Nylander 2006). The legislatures depend on agency rule making to refine laws based on more detailed scientific or economic information. For example, a statute authorizes a fisheries agency to manage inland fisheries and then the agency would uses rule making to set seasons, creel limits, and other fisheries management rules and regulations (see section 4.4.2.1.). In Canada, regulations are made by the Cabinet through an Order of the Council. Much as legislatures codify and organize statutes, federal agencies also publish information about their actions in the *Canada Gazette* or the *Federal Register* (USA) and codify and organize their rules in the Statutory Instruments Act and Regulations in Canada or the Code of Federal Regulations in the USA. Most states, provinces, and tribes follow a similar pattern that can be found on their particular e-government site.

4.2.2.1 Promulgation of Regulations

Recognizing that an agency's organic or substantive laws may contain some procedural requirements, there is also a body of administrative laws that does not grant any specific new authority but specifies how agencies use their authority. As the name implies, the National Ad-

Box 4.1. The Endangered Species Act

Paul Hartfield[1]

The purpose of the U.S. Endangered Species Act (ESA) is to conserve endangered and threatened species and the ecosystems upon which they depend. Conserve is defined in the ESA as "all methods and procedures which are necessary to bring any endangered species or threatened species to the point at which the measures [provided under the ESA] … are no longer necessary." Such measures include research, census, law enforcement, habitat acquisition and maintenance, propagation, live trapping, transportation, and any other activity associated with scientific resources management. Conserve, or conservation of species listed under the ESA, therefore, includes both protection and recovery to the point at which the species no longer requires the protection of the ESA.

As of 2008, 74 fish species were listed as endangered and 65 fish species were listed as threatened in the USA and its territories. Endangered species are those that are likely to become extinct in the foreseeable future, and threatened species are those likely to become endangered. Plans outlining recovery objectives and tasks have been prepared for 101 of these fishes. All endangered and threatened fish species require high levels of research and management to be conserved, and the involvement of professional fisheries biologists is essential.

The ESA provides several avenues to achieve the conservation of species after they have been listed as endangered or threatened, and these avenues can present opportunities for involvement by research-oriented fisheries biologists. Section 4 of the ESA requires the U.S. Fish and Wildlife Service (USFWS) to develop and implement recovery plans for species listed under the ESA and specifically authorizes the USFWS to seek the assistance and services of appropriate public and private agencies and institutions and other qualified persons. Recovery plans contain objective and measurable criteria for recovery of the species and descriptions of site-specific management actions necessary for conservation. Development of a recovery plan for an endangered or threatened fish requires knowledge of the species' taxonomy, distribution, demographics, and life history, as well as threats to the species. Implementing recovery tasks may involve application of survey methods and techniques, statistics, genetics, and hatchery or habitat management.

Section 6 of the ESA requires the USFWS to cooperate with the states in achieving the conservation of listed species. This includes helping states establish conservation programs, acquiring land or aquatic habitat, and providing financial assistance to implement recovery actions. All states currently have cooperative agreements with the USFWS to conserve listed animal species. Under these "Section 6 agreements" the states are provided an annual budget based on the number of listed species that reside in their state. The states, in consultation with the appropriate USFWS field office and regional office, may allocate Section 6 funds to survey, monitor, implement specific research, or conduct other tasks related to the recovery of individual fish or other listed species. Proposals can usually be submitted through the nongame or endangered species branch of a state resource agency or to a state USFWS field office.

[1]U.S. Fish and Wildlife Service, Jackson, Mississippi.

(Box continues)

> **Box 4.1. Continued.**
>
> Another avenue to recovery is through the provisions of Section 7 of the ESA. Section 7(a)(1) requires all federal agencies to use their authorities to carry out programs for the conservation (i.e., recovery) of endangered and threatened species. Section 7(a)(2) requires federal agencies to avoid jeopardizing the continued existence of any listed species by any action they may conduct, fund, or permit. Section 7 places responsibility on federal agencies whose actions affect listed species not only to avoid harming listed species but also to contribute to their recovery. The greater the impact of agency activities and programs on an endangered or threatened species, the greater their role in conservation of that species will be. Complying with the various components of Section 7 (e.g., surveys, biological assessments, avoidance, mitigation, and monitoring) requires knowledge of the distribution, demographics, life history, or contaminant sensitivity of listed species. This expertise is often lacking in federal agencies, and fisheries professionals can play an important role in filling these information gaps. Federal agencies highly involved with fish species listed under the ESA include the U.S. Army Corps of Engineers, Environmental Protection Agency, Federal Highway Agency, U.S. Department of Agriculture Forest Service, Natural Resources Conservation Service, and Bureau of Land Management.
>
> Becoming familiar with regional fish species listed under the ESA, as well as with state and local federal agency activities and their relationship to those species, may help fisheries professionals identify research and management opportunities. Links to species lists, recovery plans, *Federal Register* publications, and the Threatened and Endangered Species Database System can be found at http://www.fws.gov/endangered/.

ministrative Procedure Act in the USA is a template (1) to require agencies to keep the public informed of their organizations, procedures, and rules; (2) to provide for public participation in informal rulemaking processes; and (3) to prescribe uniform standards for the conduct of formal rulemaking and agency adjudicative proceedings (Nylander 2006). State fisheries agencies generally follow the same three-step process to promulgate or change regulations.

There is a myriad of other administrative acts, but two are particularly relevant to fisheries management. The National Environmental Policy Act (NEPA) and the Fish and Wildlife Coordination Act (FWCA) in the USA ensure that all federal agencies explicitly consider the impacts of their proposed activities on natural resources, including fisheries. Similarly, Canada's Fisheries Act requires anyone who would destroy fish habitat or kill fish to have the Minister of Fisheries and Oceans' permission to do so. The Fisheries Act applies to government agencies and provides direction on how to go about "informing" the Minister to determine if a person needs "permission." These acts guarantee national and sub-national fisheries management agencies an opportunity to review and comment on actions proposed by others agencies (Ballweber and Jackson 1996). These procedural consultation and cooperation requirements provide formal mechanisms to interject fisheries management concerns into emerging interagency and federal–state watershed and ecosystem management efforts.

State fish and game agencies usually are overseen by a commission (or commissioner), so while the agency is responsible for the technical aspects of rule making, such as why a regulation (a rule) is necessary and what it should be, the commission usually has the final authority to approve or disapprove any proposed rule. The rule-making process in most states is similar to the process described for Arizona in Box 4.2 and Figure 4.2. Additional information on rule making is provided in section 4.4.2.1, and Chapter 7 discusses the fisheries regulation process in more depth.

4.2.2.2 Fisheries Management Funding Mechanisms

National and sub-national game and fish agencies are fairly unique among governmental agencies in that they often have a variety of different funding mechanisms and sources. Most, but not all, states receive at least some legislative appropriations from the general revenue generated by state taxes. All state fisheries agencies are supported by license sales, entry fees at public fishing areas, and specially-designated funds generated from national excise taxes on the sale of fishing-related equipment (Box 4.3). Use of funds provided by anglers for conservation of nongame fishes and other aquatic biota has historically been contested. Special funds are now available for conservation of nongame species (Box 4.4).

4.2.3 The Judicial Branch

Unlike the other two branches of government, the judicial branch is largely immune from public influence and political pressures. Furthermore, at the national level, the judicial branch's direct authority is, except for criminal trials, largely limited to resolving disputes between the other two branches of government, national and sub-national levels of government, and individuals and the government. The amorphous and changing relationships between different branches of governments and levels of governments related to fisheries and other natural resources are increasingly litigated in federal courts. In general the judicial branch is the final arbiter of disputes between the legislative and executive branches of government, between national and sub-national governments, and between different sub-national governments.

4.3 HISTORICAL BASIS OF GOVERNMENTS' INLAND FISHERIES MANAGEMENT AUTHORITY

Many of North America's laws regarding wild animals, including fishes, and use of riverbanks and the edge of the sea (riparian and littoral law) are inherited from the ancient Roman era. The USA and Canada largely follow those tenets as they were modified by early English "common law." Common law is not based on any express legislative enactment but is composed of prerevolutionary, or preindependence, statutory and English case law (judicial rulings) applicable to the protection of people and property from the government. Alternatively, the Mexican government is based more directly on the ancient Roman system of codified law known as the "civil law" system. Even today, courts cite ancient Roman legal treatises as a precedent and basis for their rulings (Adams 1993).

The ancient Roman legal tradition of the "law of things" (*res*) is the foundation of fisheries management authority in North American legal systems. The Romans recognized two categories of things: private property (*res in patrimonio*) and public property (*res extra patrimonium*). Public property had several additional categories, such as highways, rivers, and

Box 4.2. The Legal Process of the Arizona Game and Fish Commission

We provide a single state's fisheries management framework to help better understand how the legal system and process works. Not all states are the same, but this example is offered as a representative example. Arizona's constitution does not specifically address fisheries management, but the legislature claimed Arizona's fish and wildlife as state property. Fish and wildlife are managed by the five-member Game and Fish Commission (AGFC). Commissioners are appointed by the Governor, subject to approval by Arizona's senate, to serve staggered five-year terms. The AGFC directs the activities of the Arizona Game and Fish Department and hires the department's director. The Game and Fish Act requires the AGFC to undertake certain activities that include the following.

1. Make rules it deems necessary to carry out the Game and Fish Act.
2. Establish broad policies and long-range programs to manage, preserve, and harvest fish and wildlife.
3. Establish fishing rules and prescribe the manner and methods that may be used to take wildlife.
4. Enforce wildlife protection laws.
5. Publish and distribute public information on wildlife and the department's activities.
6. Prescribe rules for the expenditure of all funds arising from appropriations, licenses, gifts or other sources.
7. Exercise powers and duties necessary to carry out fully the act and in general exercise powers and duties related to adopting and carrying out the department's policies and control of its financial affairs.
8. Cooperate with the Arizona–Mexico Commission in the Governor's office and with researchers at universities in this state to collect data and conduct projects in the USA and Mexico on issues within the scope of the department's duties that relate to quality of life, trade, and economic development in Arizona.

On fisheries issues outside the AGFC's direct control, the legislature requires the AGFC to confer and coordinate with the director of Arizona Water Resources on (1) restoration projects where water development and use are involved, (2) the abatement of pollution injurious to wildlife, and (3) the development of fish and wildlife aspects of the director of Arizona Water Resources' plans. Furthermore, the AGFC has jurisdiction over fish and wildlife resources and activities on projects constructed under or pursuant to the director of Arizona Water Resources' jurisdiction.

In addition to the mandatory responsibilities described above the act also gives the AGFC discretionary authority to undertake the following.

1. Conduct investigations, inquiries, or hearings.
2. Establish game management units or refuges to preserve or manage wildlife.
3. Construct and operate fish hatcheries, fishing lakes, or other facilities relating to fish and wildlife preservation or propagation.

(Box continues)

Box 4.2. Continued.

4. Remove or permit to be removed from public or private waters fish that hinder or prevent propagation of sport or food fish.
5. Purchase, sell, or barter wildlife to stock public or private lands and waters and take wildlife for research, propagation and restocking purposes or for use at a fish hatchery and declare wildlife salable when in the public interest or the interest of conservation.
6. Enter into agreements with the federal government, other states or political subdivisions of the state, and private organizations to construct and operate facilities; to produce management studies, measures, or procedures for or relating to wildlife preservation and propagation; and to expend funds for carrying out such agreements.
7. Prescribe rules for the sale, trade, importation, exportation, or possession of wildlife.
8. Consider the adverse and beneficial short-term and long-term economic impacts on resource-dependent communities, small businesses, and the state of Arizona of policies and programs for wildlife management, preservation, and harvest by holding a public hearing to receive and consider written comments and public testimony from interested persons.

The AGFC may also enter into agreements with a multi-county water conservation district and other parties to participate in the lower Colorado River multispecies conservation program, including the collection and payment of any monies authorized by law for the program. With the Governor's approval the AGFC may acquire land or water for fish hatcheries, game farms, firing ranges, reservoir sites, or access to fishing waters. Reflecting the system of checks and balances between branches of government, the AGFC must obtain prior legislative approval before using eminent domain to acquire more than 65 ha (160 acres) of land for these purposes. In addition, any money derived from the sale or lease of departmental property is deposited in the game and fish fund.

Furthermore, statutory authority reflects intergovernmental relations between the AGFC and the U.S. Fish and Wildlife Service (USFWS). Specifically, with the AGFC's approval, the USFWS can conduct fish hatching, fish culture, and related operations, including acquiring land. However, Arizona's legislature also clearly asserts its sovereignty by ensuring that this cooperation does not give the USFWS any right to interfere with the department's activities or facilities, nor does this cooperation contravene any Arizona law relating to public health or water rights.

Arizona's Administrative Procedures Act defines the roles and responsibilities for both the AGFC and the department and mandates the process for implementing the Game and Fish Act's substantive authority. In addition, the Governor's Regulatory Rules Commission requires an impact assessment for proposed AGFC rules. Similar to the U.S. Code of Federal Regulations, Arizona rules are organized and codified in the Arizona Administrative Code (AAC). Under Arizona's notice and comment rule-making process, the AGFC issues the rules. Procedures to set season types (such as catch and release, artificial fly, and lure

(Box continues)

> **Box 4.2. Continued.**
>
> only) and special methods of take (such as archery) are all established in the AAC. The AGFC rules codified in the AAC include:
>
> lawful methods of taking aquatic wildlife;
> possession of live fish;
> possession, transportation, or importation of live baitfish, crayfish, or waterdogs;
> seasons for lawfully taking fish, mollusks, crustaceans, amphibians, and aquatic reptiles;
> aquatic wildlife stocking permit;
> live bait dealer's license; and
> white amur (grass carp) stocking and holding licenses.
>
> The AGFC must use a less rigorous procedure to promulgate AGFC orders that open and close seasons and set bag and possession limits on an annual or biennial basis.

harbors (*res publicae*), and theaters, universities, and other public institutions (*res institutiones*). The Romans also recognized that certain property, such as air and water, was owned by everyone and open to all (*res nullius*). Similarly, fish and nondomesticated wildlife were classified as wild (*ferae naturae*) and owned by no one. The central government held some types of public property such as highways and public buildings much like a private owner and could sell those assets. However, the government held resources such as seashores and navigable rivers (*jus publicum*) in trust for the public good and these resources could not be transferred to private ownership. Fish and wildlife seem to fall under the public trust (Etling 1973; Adams 1993). These traditions were established by the Emperor Justinian and his successors in a body of work commonly cited as the *Corpus Juris Civilis* and are the foundation of civil law legal systems.

The amorphous boundary between land and water has always been a complicated interweaving of private and public interests. Access to rivers, especially those that could be used for navigation and commerce, were held in trust (*res communes*) as a public right (*jus publicum*) and could not be transferred to private ownership. In general, the shore (*littus*) extended inland to the point reached by a river's highest floods and were *res communes*, generally open to all and not available for private ownership (Adams 1993). However, private structures could be built on the shore in the floodplain if they were in the public interest and had proper governmental approval. Many of these types of restrictions are still being refined and debated today regarding wetlands regulations and building in the 100-year floodplain. Another increasingly controversial issue is the concept of public waters, which may allow the general public access to private property when that land is temporarily inundated by water from a nearby river or stream or when a body of water otherwise separated by land from navigable waters becomes connected to the navigable water during a high-water event.

From the decline of the Roman Empire until the advent of the Magna Carta in 1215, the Roman traditions were transformed across Europe into a sovereign's right. Wild animals were no longer *res nullius* but became the property of the landowner who had properly received title to the land from the sovereign. As sovereigns were forced to cede political power to

The Legal Process and Fisheries Management

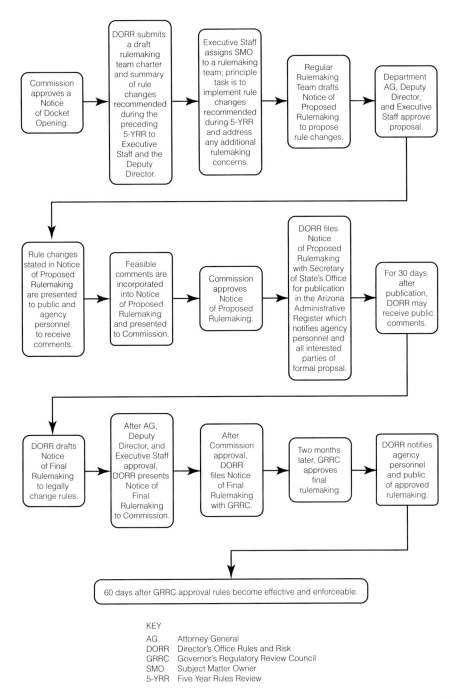

Figure 4.2. The rule-making process used by the Arizona Game and Fish Department. Similar processes are used in other states.

Box 4.3. Sport Fish Restoration

Revenues from fishing license sales provide a substantial portion of the funding for fisheries management by the states in the USA, but money provided by the Federal Aid in Sport Fish Restoration Program (SFRP) significantly augments state funds. The SFRP is a good example of a user pay–user benefit tax and creates a strong partnership among anglers, fisheries management, and the sportfishing industry.

Modeled after a similar program directed at wildlife management, the SFRP was created in 1950 through the Federal Aid in Sport Fish Restoration Act, also known as the Dingell–Johnson Act after the congressman and senator who championed the legislation. The original act imposed a 10% excise tax on rods, reels, creels, lures, and related fishing tackle. The revenue was deposited in a dedicated account and apportioned to the states by a formula based on number of anglers and land and water area. Very importantly, the act had a provision to ensure that no state fishing license revenues were diverted to other, non-fisheries uses.

The act has been modified four times since its initial enactment. In 1984, the Wallop–Breaux amendment created the Aquatic Resources Trust Fund. The fund contains two accounts, the Boat Safety Account and the Sport Fish Restoration Account. The amendment increased revenues into the fund by expanding the original excise tax to include (1) nearly all items of fishing tackle and equipment; (2) a portion of the federal fuel taxes paid on fuel used in motorboats; and (3) import duties on fishing tackle and boats. The SFRP funds available for fisheries jumped from US$38 million in 1985 to $122 million in 1986, the first year after the Wallop–Breaux amendment. In addition to the obvious benefit of more funds to state fisheries management efforts, the amendment required funding for boating access facilities, allowed funding for aquatic resources education, and required equitable division of funds between freshwater and saltwater fisheries management. With trust fund status, the two accounts also accrue substantial income.

Amendments in 1990 transferred federal fuel taxes on small gasoline engines (e.g., lawn mowers, snow blowers, and string trimmers) from the Highway Trust Fund to the Sport Fish Restoration Account. Legislation dedicated funds from the Sport Fish Restoration Account to the National Wetlands Program; the amount of funds to the National Wetlands Program was expected to approximate the amount of funds received from the small-engine tax.

An amendment in 1992 created a more equitable distribution of SFRP funds, provided funds for marine pump-out facilities to handle sewage from boats equipped with marine sanitation devices, increased funding for boating access and facilities from 10% to 12.5% of available funds, and added the word "outreach" to the aquatic education component of the SFRP.

Amendments in 1998 increased the monies received from the motorboat and small engines fuel tax, and new funds were allocated for outreach and communications and for boating facilities for non-trailerable recreational vessels.

In 2008, almost $400 million were available to assist freshwater and coastal marine fisheries management. These funds are distributed to states and territories based on land and water area (including coastal and Great Lakes waters) in proportion to total land

(Box continues)

> **Box 4.3. Sport Fish Restoration (Continued)**
>
> and water area of the USA (40% of total funds) and the number of paid fishing-license holders (60% of total funds). No state may receive more than 5% or less than 1% of the total apportionment. Further, Puerto Rico receives 1%, and the District of Columbia and U.S. territories (American Samoa, Guam, the Northern Mariana Islands, and the U.S. Virgin Islands) each receive one-third of 1%. These funds can be used to pay for up to 75% of fisheries management and other allowed activities but must be matched with 25% of funds, including "in kind" contributions, from nonfederal sources. There is a substantial volume of rules for the proper expenditure of these funds.
>
> The SFRP forms the funding backbone for fisheries management in the USA. The SFRP has grown in fiscal magnitude, in supporting partners, and in agencies and interests that receive funds. This growth helps ensure longevity of the funds.

governments, the public trust or common ownership reemerged in the common law to protect general rights of access for fishing, trading, and other uses claimed and used by all subjects (Sax 1970).

These ancient precepts came to North America as a "public trust" that applies to certain resources (fish, fowl, and game) in certain locations (the transitory boundary between water bodies and land). The New Jersey Supreme Court gave perhaps the most succinct statement of the public trust in a case decided in 1821; "the ports, the bays, the coasts of the sea, including both the water and the land under the water, for purposes of [access], navigation, fishing, fowling, sustenance and all the other uses of the water and its products according to their pleasure, subject only to the laws which regulate that use" (*Arnold v. Mundy*, 6 N.J.L. 1 [1821]). That use is subject to the government's regulation or management of those resources for current and future generations.

In brief, the public trust has two elements: (1) a geographic limit associated with oceans and navigable rivers and their beds (for coastal shores, areas subject to the ebb and flow of the tide; for upland rivers, up to the ordinary high water line); and (2) uses (commerce, fishing, and fowling were the traditional uses). Over time, courts and legislatures have significantly expanded the geographic reach and permissible uses of the public trust (Sax 1970; Lazarus 1986). In addition to the public trust, the concept of a public right of navigation on international and interstate rivers and lakes (*jus publicum*) has had tremendous repercussions in the U.S. legal system. The national government has a preeminent "navigation servitude" to regulate the waters in those rivers and lakes to promote and protect navigation for commerce. Alternatively, states have sovereign ownership of those same water bodies' submerged beds (Lazarus 1986). Much of the longstanding wetlands controversy stems from efforts to define the geographic reach of waters of the USA that are *jus publicum*. In Canada, the public right to fish exists only in tidal waters; otherwise the right is vested with the owner of the bed of the water. In most cases, the bed of the water is owned by the Crown (usually a province), and most provincial legislatures have passed acts that treat Crown lands as if they were held in the public trust.

Box 4.4. State Wildlife Grants

WALTER HUBBARD[1]

A long-standing concern in fisheries management has been the sometimes competing demands of conservation and management. This does not seem like much of an issue if you define fisheries management as "the wise use of fisheries resources as compatible with conservation of the species," but conservation versus management becomes significant in terms of fiscal accountability. Although the issue can be stated from different perspectives, fisheries administrators can be hard-pressed to justify expenditure of funds generated by recreational fishing (see Box 4.3) to conserve sensitive or imperiled species or their habitats.

To address the need for funding for fish and wildlife conservation issues, the U.S. Congress created the State Wildlife Grants Program (SWGP) in 2002 and mandated each state to develop a conservation strategy for wildlife and fish species having the greatest conservation need. Approved state strategies, known as wildlife action plans, were required by 2005 for states to participate in SWGP funding. Only animals are eligible for SWGP funds; however, conservation efforts on key habitats of these animals are also eligible for funding.

Wildlife action plans, by design in the SWGP, required many partners and perspectives for a broad conservation vision. Specific guidance provided to the states directed that their plans provide (1) information on species distribution and abundance; (2) descriptions of key habitats; (3) priority research and survey needs; (4) implementation priorities for conservation actions; (5) evaluations for the effectiveness of conservation actions; (6) periodic plan reviews; (7) coordination among various agencies and organizations; and (8) broad public participation. These planning guidelines, based across broad geographies and various professional disciplines, were a desirable approach for aquatic species because threats and impacts affecting them are often at the watershed or basin scale.

Partnerships also have helped funding requirements of SWGP grants. Grants for developing wildlife management plans required a nonfederal match of 25%. Much of the real effort of state wildlife plans was intended to be implementation, and implementation has required a 50% nonfederal match. Sharing labor forces and financial resources between more organizations has helped facilitate creative cost sharing as well as more meaningful, productive, and effective projects.

[1]National Audubon Society, Holly Springs, Mississippi.

4.4 PRIMARY INLAND FISHERIES MANAGEMENT STRUCTURES IN NORTH AMERICA

From a legal perspective, the fishery manager's job is primarily to manipulate human beings, their institutions, and aspects of habitat to conserve and enhance fish populations and assemblages (Coggins and Ward 1981). Having discussed the basic structures of North American governments and the historical basis for governmental responsibilities for fisheries management, this section analyzes how fisheries management is coordinated and allocated among different levels of government.

National constitutions are the supreme law of the land, but that does not mean that the national government always has supremacy. Regardless of the apparent hierarchy between national and sub-national governments, national constitutions grant each level of government some degree of primacy over different issues related to managing fisheries or fish habitats. Accordingly, under constitutional systems in place in North America, natural resources and environmental authority can be broadly categorized as exclusive national authority, exclusive sub-national authority, or shared national and sub-national authority.

4.4.1. National Inland Fisheries Management Authority

Both Mexico's 1927 Constitution (Article 27) and Canada's Constitution Act, 1867 (Section 91 § 12) directly address the allocation of fisheries management authority between the national and sub-national governments. Conversely, the U.S. Constitution does not mention fish or wildlife. Notwithstanding this omission, the USA has considerable "implied" fisheries management authority under some of U.S. Congress' express constitutional powers. Specifically, Congress has asserted national fisheries management authority under its express rights (1) to regulate international and interstate commerce; (2) to make treaties with other nations and aboriginal tribes; and (3) to manage and protect property belonging to the USA.

International relations are exclusively a national government function. Fish and fish migrations frequently occur in interjurisdictional waters and are often the subject of international treaties and interstate management compacts. Canada, Mexico, and the USA have a history of formally cooperating on various issues of mutual concern including fisheries management. This is reflected in the many bilateral and multi-lateral agreements North America's nations are parties to, including those listed below:

1. Canada and the USA ratified the bilateral Great Lakes Fisheries Convention in 1954 to establish a Great Lakes Fishery Commission and determine the need for and establish appropriate measures to control sea lamprey in the Great Lakes. This treaty subsequently has been used to coordinate other state and provincial fisheries management activities.

2. Mexico and the USA entered into the Treaty Relating to the Utilization of Waters of the Colorado and Tijuana Rivers and of the Rio Grande on February 3, 1944 (also known as the 1944 Water Utilization Treaty).

3. Canada, Mexico, and the USA have also ratified the Convention on Nature Protection and Wildlife Preservation in the Western Hemisphere (1940), also known as the Western Hemisphere Convention.

4. Canada, Mexico, and the USA entered into the North American Free Trade Agreement (NAFTA) on December 17, 1992. The North American Agreement on Environmental Cooperation (NAAEC or the NAFTA Environmental Supplemental Agreement), September 13, 1993, established a mechanism to encourage and monitor environmental enforcement in the three NAFTA countries and established a tripartite Commission for Environmental Cooperation.

Again, although some national and many state constitutions specifically address fish and wildlife, such is not the case with the U.S. Constitution; thus, any federal role in managing fish and wildlife or their habitats is implied or derived from one of the expressed constitutional powers (Coggins and Ward 1981; Adams 1993). For example, the commerce clause was the justification for the Lacy Act of 1900, which prohibits the import, export, sale, or purchase of fish, wildlife, or plants taken, possessed, transported, or sold in violation of national, tribal, or state law. The Lacy Act remains an important criminal statute to enforce international, national, tribal, and state fish and game laws. Canada meets this international obligation through the Wild Animal and Plant Protection and Regulation of International and Interprovincial Trade Act. The U.S. Congress has used its property power to authorize the acquisition of lands for fish and wildlife habitat (e.g., National Wildlife Refuge System Administration Act and Migratory Bird Conservation Act). Similarly in Canada, the Migratory Birds Convention Act of 1917 allows the Canadian government to pass and enforce regulations to protect migratory birds included in the convention. Finally, the U.S. Congress has established various dedicated revenue streams that can be shared with states to fund the acquisition of fish and wildlife habitat (e.g., Land and Water Conservation Fund). Knowledge of the lengthy list of laws that affect fish and fisheries management is why all fisheries management agencies have attorneys.

4.4.2. Sub-National Inland Fisheries Management Authority

The seemingly clear legal divisions of fisheries management authority between national and sub-national governments' roles and responsibilities often become blurred in practice. The bulk of fisheries management and regulations take place at the province, state, or tribe level. All provinces and states and many tribal governments have comprehensive Web sites that provide ready access to statutes and agency rules.

Under the Canadian Constitution, federal and provincial governments share authority to protect and manage fisheries. The national government has authority over the sea coast and inland fisheries, while provincial governments have authority over property and civil rights and the management of public lands in the province. The judiciary's interpretation of the federal government's responsibilities does not extend to dealing with the ownership of fishing rights but charges the national government to manage or control how the rights can be exercised for the conservation and the general benefit of Canadians. Fisheries and Oceans Canada (the national fisheries management agency) determines the allowable harvest and the provinces are responsible for deciding who can fish and how the allowable harvest is allocated. The provincial Crowns also have an ownership or stewardship interest in the fish resource. In practice, governments delegate some responsibilities to the other government levels.

Fisheries and Oceans Canada regulates the harvest of fishes and other ocean-dwelling species of wildlife. Provincial ministries of environment and wildlife have authority over all other wildlife, including endangered species, and provincial wildlife acts often include the power to designate wildlife areas to preserve important wildlife populations. The federal government, through the Canadian Wildlife Service of Environment Canada, manages migratory birds and other migratory species, as well as threatened species of national significance.

In the USA, the states traditionally have had broad authority to regulate and manage their fish and wildlife resources. All states and most tribes in the USA have departments of fish and game or some analogous agency responsible for regulating hunting and fishing and managing state wildlife conservation areas within their borders. Although fisheries management remains largely a state and tribal activity, the formal balance of power between national and sub-national governments is flexible and depends on specific issues and geographic areas (Coggins 1980). For example, state efforts to stock fish in wilderness areas for anglers may be in conflict with ecological or social values in federal wilderness areas. The federal agency may strongly prefer only native fish be stocked in such areas, whereas a state may wish to stock desirable, nonnative sport fish. Federal agency regulations recognize state authority for fish stocking, but judicial opinions allow federal agencies authority of direct involvement pertaining to fish stocking in wilderness areas (Landres et al. 2001).

4.4.2.1. State and Provincial Authority

The U.S. Supreme Court has ruled that states have the authority to manage inland fisheries subject to some national constitutional limitations (*Manchester v. Massachusetts*, 139 U.S. 240 [1890]). Much as at the national level, state laws create an agency in the executive branch of government to manage fish and wildlife resources for the use of its citizens now and into the future. To accomplish this goal, fisheries agencies are empowered to promulgate rules or regulations in the context of state statutes (laws). For example, legislation states a fisher may "take" fish; an agency will promulgate rules to prescribe how an angler may take them. The scope of regulatory authority allocated to agencies varies considerably among states, but many fisheries agencies can set license requirements and harvest restrictions (e.g., bag limits, length limits, and seasons). Legislative bodies retain oversight of agency decisions and in extreme cases may enact legislation to override agency management decisions (e.g., remove or add size limits, or revise methods of take [legal fishing methods]). Fisheries managers must become familiar with the authority the legislature delegates to the fishery agency in the state where they are working, which in turn determines the legal process of fisheries management. An example of the legal process in one state is provided in Box 4.2.

In the USA, state fisheries agencies operate under the guidance of a commission. Commissions vary among states in their size and the duration of commissioners' tenure. Commissioners are appointed from a cross section of citizens with a general interest in the outdoors to terms defined by statute, except in Missouri where the commissioners are elected. Commissioners are responsible for setting the management agency's policy, assisting the agency in implementing that policy, and acting as the intermediary between the agency and the political decision makers. The commissioners interact with fishery agency administrators, and occasionally biologists, at regularly scheduled meetings and serve to guide the fishery agency's actions.

4.4.2.2. Tribal and Indigenous Peoples' Authority

The interactions of tribes and indigenous peoples with state, provincial, and national governments are increasingly complicated. Both Canada and the USA realize a trust relationship with tribal governments. Fish and wildlife management responsibilities are often contentious as they may be of significant cultural and religious importance to indigenous peoples, and fulfilling those cultural or religious needs may conflict with state fish and wildlife management regulations.

In the USA, the Constitution and numerous treaties grant Native Americans significant rights of self-government. Tribal governments usually have powers very similar to states. The U.S. Department of the Interior Bureau of Indian Affairs is the primary federal agency responsible for carrying out the USA's trust responsibility to Native American tribes. This trust includes the protection and enhancement of Native American lands and the conservation and development of natural resources including fish and wildlife, outdoor recreation, and water, range, and forest resources. Native American tribes usually are exempt from state law except under limited circumstances (*Cabazon Band of Mission Indians v. California*, 480 U.S.202 [1987]).

In Canada, the Crown, whether federal or provincial, has a duty to uphold and protect aboriginal rights. In essence, aboriginal people are accorded first access to fish and wildlife resources. Governments can only restrict aboriginal uses for conservation reasons. Further, many of the provincial laws and permits do not apply on First Nations' lands because the reserves are federal property. As of 2008, Canada was developing modern treaties or land claims with aboriginal groups to create fish and wildlife management authorities.

In Mexico, some municipalities are mainly populated by indigenous peoples with distinct laws, religions, languages, and customs. Mexican law grants these indigenous groups special protection as minorities, but generally they are subject to all applicable federal and state laws and provisions. In August 2001, the Mexican Constitution underwent an "indigenous reform" in which some articles were amended to include special provisions for indigenous groups. One of the key reforms recognizes Mexico's pluri-cultural makeup and acknowledges and guarantees the rights and autonomy of indigenous peoples and communities to decide the form of their social, economic, political, and cultural organization.

4.5. INLAND FISHERIES MANAGEMENT WITHIN A WATERSHED OR ECOSYSTEM MANAGEMENT FRAMEWORK

In Canada, the Fisheries Act, which requires ministerial permission or authorization to alter fish habitat or kill fish, creates an opportunity for cooperation among agencies to conserve fisheries resources. Ecosystem-based management is considerably more challenging in the USA. Fisheries managers have direct and preeminent authority to manage aquatic and terrestrial fisheries habitat within the boundaries of the lands that they manage (e.g., National Wildlife Refuges, federal lands surrounding U.S. Army Corps of Engineers or U.S. Bureau of Reclamation reservoirs, and state lands managed by game and fish agencies). Outside those geographic boundaries, fisheries managers and management agencies must build formal or informal partnerships with other federal, state, and local agencies and private landowners and strongly represent the fisheries resource in other agencies' decision-making processes. In short, outside the boundaries of the public land they directly manage, fisheries managers and management agencies are limited to indirect authority such as NEPA and the FWCA to influence other agencies' regulatory and management decisions that impact fisheries (Ballweber and Jackson 1996). As discussed in Chapter 3, there is a growing recognition of the need to integrate water, land, and living-resource management along natural watershed or ecosystem boundaries. While these efforts are not, and should not be, fishery-centric, they do offer fisheries managers an opportunity to interject fisheries needs into the process to influence the management of critical fishery habitat outside the boundaries of lands owned by state and federal governments for fisheries benefits.

4.6 WATER

Water quality and quantity are essential to fish and valuable commercial and recreational fisheries. Impacts on fisheries are one of many considerations relevant to water quality and quantity decisions, but they seldom are a seminal or paramount factor. In Canada, the Fisheries Act grants control of water quality and quantity to the federal fisheries agency. In the USA, authority over water quality and quantity is allocated to different levels of government and different agencies at the federal, tribal, and state levels. Subject to some notable exceptions, such as the ESA discussed in Box 4.1, fisheries agencies have no legal authority to set water quality or quantity criteria. Nonetheless, fisheries managers should be prepared to provide guidance and, when possible, be strong advocates to ensure that fisheries impacts of proposed actions that influence water quality or quantity are fully documented and fairly presented to other decision-makers in federal or state agencies or courts.

4.6.1. Water Quality

A spectrum of potential water quality impairments ranging from contaminants (e.g., toxins, silt, pathogens, and nutrients) to thermal enrichment (discharge of heated but otherwise clean or safe water) challenge fisheries management. Water quality pollutants commonly are categorized by their source. Point-source pollutants, as the name implies, are those that enter a water body from a clearly identifiable source such as a pipe or ditch that can be traced back to a responsible party. Nonpoint sources, on the other hand, are essentially all pollutants that come from anywhere else. Point-source pollution is relatively easy to monitor and regulate, but nonpoint pollution is more difficult to ascribe to sources and is difficult to regulate.

In Canada, the Fisheries Act can be used to guard against discharge of pollutants that injure fisheries, and the provinces have similar laws to prevent pollution. In the USA, the CWA was enacted in 1972 to restore and maintain the chemical, physical, and biological integrity of the nation's surface waters. Simply put, the CWA was intended to establish water quality standards that would result in "drinkable, fishable, and swimmable" waters. To accomplish this, the CWA prohibits the discharge of any pollutant into the "waters of the United States" without a permit. The two most prevalent permits are (1) EPA's National Pollutant Discharge Elimination System (NPDES), which is generally delegated to state agencies to regulate point-source discharges and (2) the U.S. Army Corps of Engineer's wetlands permit program to control the discharge of dredge and fill materials into "waters of the United States."

The CWA has made significant progress in restoring water quality in the USA by controlling point-source pollution through the NPDES program. Unfortunately, many water quality issues that adversely impact productive fisheries, such as sedimentation and cultural eutrophication, are from nonpoint sources and, therefore, outside the CWA's regulatory structure.

Continued water quality improvements will require vigorous efforts to address these non-point sources of water pollution. The most effective mechanisms to address nonpoint sources of pollution are land-use best management practices, planning, zoning, and building codes. However, these activities are usually implemented and enforced by county or municipal governments or agencies. Whether due to point-source or nonpoint-source pollutants, water quality is a watershed issue and often beyond the sole authority of a single agency to regulate holistically. Civil and criminal penalties can help ensure compliance with individual water quality permits, but institutionally fisheries managers are expected to represent fish and fisheries issues aggressively in inter-agency consultations related to fish water quality and habitat concerns. In Canada, natural resource agencies can actually override water quality and quantity decisions when necessary to address fisheries issues.

4.6.2 Water Quantity

The quantity of water is the other part of the foundation for healthy fisheries. Increasingly, fish and fisheries habitat are competing with agricultural, municipal, and industrial interests, so competition among users can be expected. Predicted changes in precipitation patterns resulting from global climate change are likely to increase competition for water resources. Allocation of water is not a new issue. Water allocation is largely a matter of state, provincial, or tribal law; in the case of interstate waters in the USA, multi-state compacts are negotiated between the various states and then, in the USA, validated by Congress. As competition for water increases, water rights in the USA have become increasingly contentious and complicated, with new and often competing demands being pursued by federal mandates, interstate water compacts, and within tribal and state water law regimes (Tarlock et al. 2002). For example, because water rights in the Oregon portion of the Klamath Basin were not adjudicated, the U.S. Bureau of Reclamation's Klamath Project could not legally prevent junior water rights holders from exercising their right to divert water for out-of-stream beneficial uses. As a result, when the U.S. Bureau of Reclamation needed to meet ESA requirements and provide a minimum instream flow and lake elevation for federally-listed threatened and endangered species in California and Oregon, the Bureau had to obtain water through groundwater pumping and land idling to provide instream flow and meet ESA requirements. The Federal Energy Regulatory Commission has also been important in ensuring consideration of water needs for fisheries when hydroelectric facilities affect aquatic resources (Box 4.5).

Water law in the USA has evolved from one of two different foundations that diverge roughly in the middle of the country along the 98th meridian. Historically, states east of that line have enjoyed a fairly abundant water supply, whereas states to the west of that line have frequently endured a scarcity of water. Recognizing these drastically different climates and, hence, water-availability conditions, two distinct types of state water law were established to govern private water rights.

Riparian, or eastern, water law is largely common law doctrine that connects the right to use surface water with ownership of the contiguous land. Riparian water rights cannot be sold or transferred separately from the land. The allocation of water between riparian owners is governed by one of two legal approaches: (1) natural flow, which prohibits any riparian owner from using water that would diminish the natural flow downstream to other riparian lands, and (2) reasonable use, which gives riparian owners the right to alter the flow if the use is deemed reasonable when balanced against the rights of downstream riparian owners.

Box 4.5. Water Rights and the Federal Energy Regulatory Commission

CINDY WILLIAMS[1]

The Federal Energy Regulatory Commission is the interstate regulatory authority for electric power, natural gas, oil pipelines, and the hydroelectric industry in the USA. The Office of Energy Projects (formerly the Office of Hydropower Licensing) administers the production and operation of the non-federal hydropower program. The Federal Water Power Act of 1920 provided the initial legislation and authority from Congress for the Federal Power Commission, which was placed under the direction of the Secretaries of War, Agriculture, and Interior. In 1930, the Federal Power Commission was reorganized into an independent commission with five appointed commissioners, and the Department of Energy Organization Act of 1977 created the Federal Energy Regulatory Commission (herein the Commission).

The Commission plays a significant role in inland fisheries management in its decisions to approve development and operation of hydroelectric facilities. In licensing a hydroelectric facility, the Commission is required to give "equal consideration" to power and development; energy conservation; protection of, mitigation of damage to, and enhancement of fish and wildlife (including spawning grounds and habitat); protection of recreational opportunities; and preservation of other aspects of environmental quality. Each license includes conditions to protect, mitigate, and enhance fish and wildlife affected by the project. These conditions are to be based on recommendations received pursuant to the Fish and Wildlife Coordination Act from the USFWS, the National Marine Fisheries Service, and state fish and wildlife agencies. The Commission is empowered to resolve any instances in which such recommendations are viewed as inconsistent while according "due weight to the recommendations, expertise, and statutory responsibilities" of the resource agencies. The Commission is also required to mandate the construction, maintenance, and operation of fish passage facilities as prescribed by the Secretary of Commerce or the Secretary of the Interior.

Hydropower licenses are issued by the Commission to private parties and municipalities for a period of 30 to 50 years based on the license application. The Commission conducts an independent analysis of the license application and the resources the project will affect through the National Environmental Policy Act (NEPA) process. Through NEPA, the Commission must ensure the project minimizes environmental impacts and is in compliance with applicable state and federal laws, such as the Clean Water Act, Endangered Species Act, National Historic Preservation Act, Coastal Zone Management Act, and the Wild and Scenic Rivers Act, while it produces an economically-feasible hydroelectric generation of power. Fisheries and water quality and quantity are generally the resources most affected by hydropower construction and operation. The Commission depends on the initial review and comments from state and federal fish, wildlife, and water management agencies throughout the licensing process. The license applicant's response to these comments and concerns influences licensing decisions. The Commission determines whether or not the applicant has provided sufficient information in the license

[1]U.S. Fish and Wildlife Service, Atlanta, Georgia.

(Box continues)

> **Box 4.5. Water Rights and the Federal Energy Regulatory Commission (Continued)**
>
> application for the Commission to conduct its analysis and produce a NEPA document, which recommends whether or not the Commission will issue a license. The applicant must also request and obtain a CWA § 401 Water Quality Certification from the responsible state agency. The state has 12 months from the date of the request to issue the certificate. If the state does not act within this time frame, the Commission considers the water quality certificate waived and can proceed with licensing. In addition, hydroelectric projects require state water rights to divert and store water. Without a water right, the hydroelectric project has no protection against subsequent appropriators. Existing water rights do not excuse compliance with water quality laws as part of relicensing. It is not uncommon for the Commission staff to impose licensing requirements based on their independent NEPA analysis that were not part of the license application or comments from concerned agencies to address operation issues and impacts to natural resources.
>
> The amount of water flow needed to support the natural aquatic system is an essential but unresolved issue. In 1995, many federal and state fish and game agencies in the USA participated in the National Instream Flow Program Assessment Project funded by the U.S. Fish and Wildlife Service. This assessment compared state instream flow provisions and evaluated existing and emerging instream flow criteria. In 1998, this project resulted in the creation of the Instream Flow Council (IFC), a nonprofit organization that includes state and provincial fish and wildlife agencies. The IFC's mission is to improve the effectiveness of current instream flow policies and programs to conserve aquatic resources (IFC 2002).

Prior appropriation, or western, water law developed in the western USA to separate water rights from land ownership so water could be "claimed" from a source to be used for a beneficial use somewhere possibly far removed from that water source. Under the prior appropriation doctrine, the first person (regardless of land ownership) to divert water from its source and put it to a *beneficial use* (the "senior" appropriator) has a superior right over all subsequent "junior" appropriators. This is commonly referred to as "first in time, first in right." Under this system, water rights can be sold, leased, or contracted to other parties. The system may be administered by special water courts or an administrative agency. Also the water must be regularly applied to the use for which it was appropriated or the appropriation may be forfeited, a provision referred to as "use it or lose it." In times of water shortage, available water is allocated in order of priority with no balancing of harm or need between appropriators.

A third system, commonly called "regulated riparianism," blends pure riparian and prior appropriation doctrines by instituting some type of water use permit that follows the reasonable use requirement. This system is also known as the California Doctrine after the California Water Code initially adopted the system and is the system generally used in Canada.

In times of shortage, the riparian system (eastern doctrine) spreads the limitation among all riparian users equitably, whereas the prior-appropriation system shuts down junior approprietors to protect the rights of senior appropriators regardless of the consequences. Both systems have key phrases, such as *reasonable use* and *beneficial use*, that are subjective and regularly reviewed, modified, and amended by legislatures and the courts. Initially, beneficial uses were limited and required diversions to take water out of rivers and lakes for mining, agriculture, manufacturing, and water supply. Over time, state legislatures and courts have expanded beneficial uses and recognized that some such uses (e.g. providing recreation, maintaining in-stream flow, and sustaining fish and wildlife) actually require the water to stay in the river or lake.

Despite the predominance of state water law, the U.S. Supreme Court has found that the Constitution implies that water rights may be owed to the federal government; this is called the doctrine of implied federal reserved water rights. In many western, prior-appropriation states, rivers and streams are increasingly becoming over-appropriated, so that even in years of normal precipitation, insufficient water is available to satisfy the rights of all appropriators. The doctrine of implied federal water rights allows the federal government to file a law suit in federal court to argue for an appropriation of water necessary for a specific type of federal property necessary to satisfy its intended purpose. This concept is perhaps best understood in the context of Native American water rights. Basically, the creation of a reservation by the federal government implicitly reserved a water right to the tribe or tribes occupying the reservation as necessary to carry out the purpose for which the land was set aside (*Winters v. United States*, 207 U.S. 564 [1908]; Royster 1994). Federally-reserved water rights have been pursued for a variety of other federal lands, such as national parks and wilderness areas, to allow the managing agency to achieve specific purposes for which the land was designated. The Devils Hole case is one such example. In 1972, a suit was filed by the U.S. Department of the Interior to keep the level of spring water high enough in the Devils Hole portion of Death Valley National Park, Nevada, to assure the continued existence of the Devils Hole pupfish. In 1976, the U.S. Supreme Court unanimously upheld the federally reserved water rights, thereby facilitating the continued existence of the Devils Hole pupfish (*Cappaert v. United States* [426 U.S. 128 1976]; Minckley and Deacon 1991).

4.7 LAND

Land is important to fisheries because the land directly and indirectly affects the water. Land ownership (public, private) and that land's designated use (e.g., forest, recreation, or residential) or location (urban, suburban, or rural) significantly impacts the quality and quantity of aquatic and riparian habitat as well as limits who has access to any fisheries that may be found in waters on or flowing through that land. The various public land and resource management agencies use their discretionary authority and the numerous administrative procedures for interagency "cooperation and coordination" to implement watershed or ecosystem management on an *ad hoc* basis. A working knowledge of these administrative mechanisms will prepare fisheries managers to be effective advocates for fishery resources in this process. For this discussion, land is broadly categorized as being either (1) public and managed by a particular public agency to achieve some goal specified by law or (2) private property not managed by the government but still subject to varying degrees of public regulation.

4.7.1 Public or Crown Land

A variety of national, sub-national, and local public lands are referred to as Crown land in Canada. The provincial governments "own" the public lands in the 10 provinces, and the federal government owns the public land in the three territories. The national government is the largest landowner in the USA and owns approximately one-third of the nation's lands (Adams 1993; Mansfield 1993). The amount of national public lands varies among states and is greatest in western states. Congress, as the steward of those lands, has allocated management authority or use of that land among several agencies. Although federal lands are often managed for multiple compatible uses, a law usually specifies a primary designated use (e.g., wildlife or national park). This designation usually dictates which agency has management responsibility for the land. In addition specific laws may designate management priorities and goals specific to that property. Many public lands, such as fish and wildlife refuges and national parks, are generally open to the public and actively managed to provide fishing opportunities. Some lands, such as those set aside for national defense, may have restricted public access and limited active fisheries management.

The U.S. Bureau of Land Management (BLM) and its land management practices defined by the Federal Land Policy and Management Act of 1976 (FLPMA) provide a good example of the complexity of achieving multiple use and sustained yield in land management. Under FLPMA, the BLM must inventory all of its lands and develop land-use plans that, among other things (1) reflect multiple-use and sustained-yield principles; (2) take a multidisciplinary approach that includes physical, biological, economical, and other sciences; (3) consider present and potential future uses; and (4) generally conform with state, local, and tribal land use policies. In addition to FLPMA, the Multiple-Use Sustained-Yield Act of 1960 provides another statutory overlay on the management activities of both the BLM and the U.S. Forest Service. Generally the practices of the BLM are consistent with provincial and federal Crown land management in Canada.

Multiple-use management tries to balance uses of the different surface resources available on public lands, including outdoor recreation, grazing, mining, logging, watershed protection, and fish and wildlife conservation. Multiple use does not necessarily give priority to the combination of uses that will give the greatest economic return or the greatest unit output. However, sustained yield is achieving and maintaining a high level of annual or regular output of various renewable resources on public lands consistent with perpetual multiple use. The challenge is for multiple federal and state agencies, as well as the public, to function within these statutory labyrinths to reach a consensus on how to implement new interdisciplinary management regimes that can exceed single-resource management expectations.

Sub-national governments have patterns of public lands that may include open space and other classifications. Every state has a system of protected areas, which can provide a diversity of conservation benefits and recreational opportunities. In addition, local and county parks and playgrounds often protect small natural areas or open spaces. Still, despite the fragmentation of public land management across myriad agencies, the basic decision-making processes are much the same.

4.7.2 Private Land

Private property is not totally immune from government regulation. As a general rule, local and municipal governments have the most direct authority to "manage" activities on private property through land use planning, accomplished by enforceable zoning regulations in Canadian provinces and many U.S. states. Some states have adopted statewide land use plans that include urban-growth boundaries. In more rural areas, state soil and water conservation agencies or their equivalent may also have the authority to adopt and enforce land use plans for soil and water conservation needs. Although hotly contested, the federal government has certain regulatory mechanisms to restrict activities on private property. For example, the ESA prohibits "the taking" (removal) of plants or animals listed as endangered. Thus, a habitat alteration that harms a plant or animal listed under the ESA as threatened or endangered is a violation of the ESA. The USFWS, which enforces the ESA, therefore can influence activities on private land if they threaten the habitat of listed species (*Babbitt v. Sweethome Chapter of Communities for a Greater Oregon*, 515 U.S. 687 [1995]). Similarly, Canada's Fisheries Act and other laws also apply on private property.

4.8 CONCLUSIONS

The governmental side of fisheries management may seem daunting, but laws and regulations are necessary elements of fisheries management. It is impossible to understand fully the fisheries management process solely from reading statutes, rules, and regulations, but it is essential to know the roles different parties (e.g., federal or state legislators, judges, presidents and governors, mayors, agency staff, commissions, and the public) have in the process.

The subject of fisheries management legislation is ripe for discussions and role-playing in which a student or student teams assume different roles (e.g. legislative, agency, user group, and environmental group) in discussing fisheries issues. The press and news media are reporting on national and local resource controversies almost daily and should provide timely and undecided fisheries issues for discussion. Emotions flair and battle lines are drawn.

Current fisheries management processes are a blend of legislative and administrative authorities that have been and are being reviewed and refined by the federal courts; in most cases, long-term solutions to fisheries management concerns will require a cooperative blending of these same parties' authority. Relationships and trust are critical to keep the process running smoothly. Legislators must trust that they are getting the best advice possible from agencies and professional societies. Likewise, the angler must also trust that regulations are in the best interest of the fishery. A key element of this trust is understanding who has what role and authority in the fisheries management process.

4.9 REFERENCES

Adams, D. A. 1993. Renewable resource policy: the legal-institutional foundations. Island Press, Washington, D.C.

Ballweber, J. A., and D. C. Jackson. 1996. Opportunities to emphasize fisheries concerns in federal agency decision-making. Fisheries 21(4):14–19.

Coggins, G. C. 1980. Wildlife and the Constitution: The walls come tumbling down. Washington Law Review 55:295–1980.

Coggins, G. C., and M. E. Ward. 1981. The law of wildlife management on the federal public lands. Oregon Law Review 60:59–155.

Etling, C.D. 1973. Who owns the wildlife? Environmental Law, Spring 1973:23–31.

IFC (Instream Flow Council). 2002. Instream flows for riverine resource stewardship. Instream Flow Council, Cheyenne, Wyoming.

Landres, P., S. Meyer, and S. Matthews. 2001. The Wilderness Act and fish stocking: an overview of legislation, judicial interpretation, and agency implementation. Ecosystems 4:287–295.

Lazarus, R. J. 1986. Changing conceptions of property and sovereignty in natural resources: questioning the public trust doctrine. Iowa Law Review 71:631–715.

Mansfield, M. E. 1993. A primer on public land law. Washington Law Review 68:801–857.

Minckley, W. L., and J. E. Deacon, editors. 1991. Battle against extinction: native fish management in the American West. University of Arizona Press, Tucson.

Nylander, J. 2006. The Administrative Procedure Act, a public policy perspective. Michigan Bar Journal, November 2006:38–41.

Royster. J. V. 1994. A primer on Indian water rights: more questions than answers. Tulsa Law Journal 30:61–104.

Sax, J. L. 1970. The Public Trust Doctrine in natural resource law: effective judicial intervention. Michigan Law Review 68:471–566.

Tarlock, A. D., J. N. Corbridge, Jr., and D. H. Getches. 2002. Water resource management: a casebook in law and public policy. Foundation Press, New York.

Chapter 5

The Process of Fisheries Management

STEVE L. MCMULLIN AND EDMUND PERT

5.1 INTRODUCTION

Fisheries management is a challenging and exciting process of planning and taking actions to manipulate fish populations, fish habitat, and people to achieve specific human objectives (Figure 5.1). The three components of fisheries management could be thought of as the legs of a stool. If any one of the three legs is weak, the stool will not bear the weight it is intended to support. The process is challenging because fisheries managers rarely have all of the information needed to manage fish populations, habitats, and people with maximum effectiveness. Despite the uncertainty created by the lack of information, fisheries managers must make decisions that are important to a wide variety of stakeholders (i.e., anyone who has an interest in the issue or who may be affected by an issue, either positively or negatively; Decker and Enck 1996), including anglers, conservation organizations, government officials at federal, state, and local levels, farmers, ranchers, industries, and many others. Fisheries management is exciting for many reasons, but one of the more exciting aspects is that it is the nexus of science and policy. Fisheries managers serve as central points of communication, translating scientific principles and data into terms their stakeholders can understand and receiving public input regarding management of the resource that is transmitted to policymakers.

It is difficult to overstate the importance of managing people in fisheries management. If you ask fisheries managers what is the most challenging aspect of the job, they will almost certainly tell you it is managing people. Natural resource managers from forestry to wildlife to fisheries have been saying that for a century. Gifford Pinchot, the first chief of the U.S. Forest Service, once said, "To start with, I had to know something about the people, the country and the trees. And of the three, the first was the most important" (Pinchot 1947:32). Aldo Leopold, widely recognized as the father of wildlife management, said, "The real problem ... is not how we shall handle the deer... The real problem is one of human management. Wildlife management is comparatively easy; human management difficult" (Meine 1988:444). Peter Larkin (1988) described the difficult job of a fisheries manager as balancing the desires of recreational anglers (making people content), artisanal fishermen (keeping people employed), and commercial fishermen (making money). Larkin suggested that, "because it is not possible to optimize for several kinds of things simultaneously, it is necessary to find a common currency for contentment, employment, and economic performance" (Larkin 1988:8). Note that all aspects of this difficult balancing act focus on people.

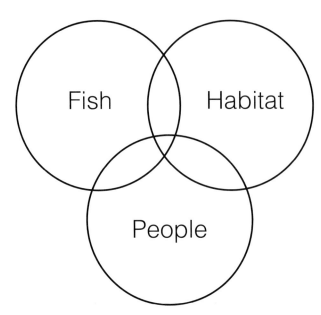

Figure 5.1. Fisheries management consists of the interrelated processes of planning and taking actions to manipulate fish populations, fish habitat and people to achieve specific human objectives.

One of the reasons that the people management aspect of fisheries management is so challenging is that frequently managers have less scientifically sound information about people than they do about fish populations or habitat. Most fisheries managers would be uncomfortable making important decisions about management of a fishery without sound data on fish populations and the status of fish habitat. However, until the emergence of human dimensions as a subdiscipline of fisheries science in the late 20th century, fisheries managers frequently made decisions without the benefit of good data on people and their preferences and opinions regarding management of the resource.

The graphical representation of fisheries management depicted in Figure 5.1 shows the circles surrounding populations, habitat, and people overlapping because, in reality, most issues require that fisheries managers simultaneously deal with two or all three of those areas. For example, fisheries managers throughout North America face the issue of water withdrawal from rivers and lakes. Humans use water for domestic, agricultural, and industrial purposes. While reducing streamflows or lake levels may seem to be simply a habitat issue, it also is a fish population issue because lower flows often result in greater fish mortality, thus reducing overall abundance and possibly species diversity as well. Water withdrawal also becomes a people issue when streamflows or lake levels are reduced to the point of affecting adversely the ability of anglers and boaters to use a body of water. Water provides esthetic and economic as well as utilitarian benefits for people. For example, property with waterfront footage usually has higher economic value than does other property because people value the esthetic qualities associated with the sights and sounds of water near their homes.

The issue of intentional introduction of a new species into a body of water also includes aspects of all three areas (see Box 5.1). Species introductions were once a commonly-used tool of fisheries management and introduced species provide a large portion of the recreational angling in North America. Today, fisheries managers make few intentional species

Box 5.1. Management of Native Species in California's High Mountain Lakes

Fisheries management in high-elevation lakes in California's Sierra Nevada has transitioned from the early days of stocking nonnative trout for fishing to managing fisheries with a strong sensitivity for native species. This change in direction was met with public outcry from stakeholders. Fisheries managers needed to find creative ways to balance fisheries, native species, and stakeholders.

Sierra Nevada fisheries management was based on the goal of improving fishing. This goal was widely supported by the public but did not meet public trust responsibilities placed on the California Department of Fish and Game (CDFG) by the state legislature. Initially, trout were moved by early settlers above natural fish barriers or across watershed divides, thereby extending the distributions of rainbow trout, cutthroat trout, and golden trout. Eventually high-country fish stocking was coordinated by game wardens, hatchery managers, and fisheries managers. The result is that 89% of Sierra Nevada lakes larger than 2.5 ha and deeper than 3 m have extant trout populations where there once were none. Thousands of additional smaller lakes and streams also support self-sustaining populations of trout. Historically, little effort was applied toward evaluating fishery performance and management efficacy. Even less was known about the landscape-scale impact of introduced trout on native animals, especially amphibians and invertebrates.

Almost all high mountain lakes in the Sierra Nevada were fishless, but the term "barren" is hardly applicable. Native amphibians, aquatic invertebrates, and their terrestrial predators were plentiful. For example, the two species of mountain yellow-legged frogs were once thought to be the most abundant aquatic vertebrates in numerous Sierra Nevada basins. These frogs are now candidate species for listing under the U.S. Endangered Species Act (ESA).

Results of early stockings of trout into high elevation lakes were impressive, producing legacy fisheries. Not many anglers that experienced these fisheries are around today; however, there remains a strong expectation that fisheries managers can achieve similar results through increased trout stocking. Some anglers, and many stakeholders that depend on tourism, continue to judge the "quality" of fisheries management by the numbers of trout or lakes stocked rather than on actual performance of fisheries.

Given an increasing awareness of the value of native species, high-elevation fisheries management has evolved toward a more ecosystem-conscious approach that incorporates recreational interests and conservation and protection of native fauna. A rift between those who believed introduced fisheries were benign and those who believed introduced trout exert an impact on native species could only be closed through completing comprehensive assessments of high elevation lakes and incorporating results into basin-scale management plans. The CDFG increased funding for assessments and management in 2001 leading to over 90% (approximately 10,000 waters) of Sierra Nevada Wilderness lakes and ponds being surveyed by 2008, of which approximately one-third were included in management plans.

Most resource managers and scientists now agree that introduced trout can cause local extinctions of amphibians, such as the mountain yellow-legged frog. Other amphibians with shorter larval stages (e.g., Pacific treefrog) or anti-predator toxic skin glands (e.g., Yosemite and western toads) are less affected by nonnative trout.

(Box continues)

Box 5.1. Continued.

In addition to resource assessments, an analysis of public use of fisheries was included in basin management plans. This approach facilitated the development and protection of networked habitats for native species, and the maintenance and improvement of important fisheries. The goal was to implement balanced management between native fauna and historic recreation, stated as follows: manage high mountain lakes and streams in a manner that maintains or restores native biodiversity and habitat quality, supports viable populations of native species, and provides for recreational opportunities, considering historical and future use patterns.

Critical to success was the involvement of stakeholders, whose perspectives and values varied widely. Some anglers questioned the value of native species, whereas the wilderness conservation community opposed changes to the pristine nature of the environment. Initial engagement of stakeholders indicated that the CDFG might be sued by environmental advocates to force compliance with the California Environmental Quality Act. At the same time, some anglers threatened that alterations in angling opportunity would be met with a "bucket stocking" response. Fisheries managers feared the potential movement and further proliferation of brook trout, a species introduced from the eastern USA that is the most common fish in Sierra Nevada lakes but, with few exceptions, has not been stocked for decades. Unfortunately, the majority of these brook trout fisheries consist of stunted individuals. Opposition to fish stocking intensified when both the mountain yellow-legged frog and the Yosemite toad were petitioned for listing under the ESA. To implement basin management plans successfully, it was necessary to involve and educate both anglers who did not want change and environmental advocates who were opposed to maintaining fisheries for nonnative trout.

Fisheries managers with the CDFG presented draft management concepts that integrated comments from stakeholders, including county supervisors and commissioners, chambers of commerce, angling and environmental advocacy groups, professional societies, the Declining Amphibian Task Force conferences, the popular media, and stakeholder workshops. The public perception that all trout should be killed and that no effort was being made to protect native amphibians remained strong. The most directly affected stakeholders were the "pack stock operators," people whose livelihood depended on transporting anglers to mountain lakes. Pack stock operators were historically active in stocking many high mountain lakes and in some areas were still involved. From this relatively small, but intensely controversial, segment of California's fish stocking program, opposition grew that questioned stocking impacts in nearly all waters of the state. The CDFG began working with the U. S. Fish and Wildlife Service to disclose the impacts of its fish stocking program and establish mitigation measures to offset or reduce impacts.

Though changing the management approach for high mountain lakes by the CDFG has been difficult, it has been a success in many ways. Some key factors include:

- a strong resource assessment program that generates pertinent data that are used by resource managers for adaptive decision-making and are shared with stakeholders;

(Box continues)

> **Box 5.1. Continued.**
>
> - development of basin-scale management plans that include detailed objectives for both sport fisheries and protection and recovery of native species;
>
> - early involvement of stakeholders to help refine basin plans to minimize conflict between recreation and native species recovery; and
>
> - successful implementation of plans with improvements in fisheries, in cases which that is the management objective, and recovery of native species, especially mountain yellow-legged frogs.
>
> Through more intensive resource assessment and management, CDFG has improved fisheries where appropriate and the stewardship of native species throughout the Sierra Nevada.

introductions; however, well meaning but poorly informed anglers introduce species into new waters with alarming frequency. Some introduced species alter the habitat. For example, common carp can reduce water clarity and overall primary production in small impoundments and lakes when they stir up bottom sediments as they feed. When an introduced species does well in the new body of water, it may alter the composition of the entire ecological community. For example, predation by illegally-introduced lake trout in Yellowstone Lake significantly reduced the population of native cutthroat trout, which in turn, reduced a seasonal food source for grizzly bears (Ruzycki et al. 2003). An altered aquatic community, in addition to the ecological effects, may have major impacts on human use of the resource. For example, the intentional introduction of opossum shrimp by fisheries managers into lakes throughout the western United States to improve forage for kokanee actually caused crashes in many kokanee populations as the two species competed for food and opossum shrimp provided excellent forage for the juvenile life stages of kokanee predators. In lakes where kokanee populations crashed, popular fisheries were severely affected (Martinez and Bergersen 1989). Fisheries managers today have become more sensitive to both positive and negative ecosystem-wide effects of their management practices than in the past. In fact, fisheries managers can implement activities that purposefully affect not only fisheries but nongame species as well to balance both needs (Box 5.1).

5.2 ROLES OF FISHERIES PROFESSIONALS, STAKEHOLDERS, AND POLICY MAKERS

Fisheries management should be an adaptive process that involves fisheries managers, stakeholders, and policymakers when decisions are being made that combine values and technical choices. Each of the groups involved in the management process has a unique role. When the groups fulfill their roles well, the process tends to work smoothly. When one or more of the groups either does not fulfill its role well or intrudes upon the roles of other groups, the process is more likely to function poorly. Ideally, the role of the fisheries professional should be to inform the public regarding the status of the fishery in question, the factors that affect the status of the fishery, and the implications of various options for managing the fishery. The specialized training that fisheries professionals receive at universities prepares them to make the technical choices associated with fisheries management. Examples of technical choices include assessing and monitoring fish populations, identifying factors that affect fish populations, and analyzing the implications of management options. Effective fisheries managers also are able to translate highly technical fisheries science and communicate the implications of management options in a way that the average citizen can understand (Figure 5.2).

Fisheries professionals working for government agencies also have a public trust responsibility. The North American model of resource management states that natural resources are public resources held in trust for all citizens to enjoy (Geist et al. 2001; Chapter 4, this volume). Public agency managers are charged with the responsibility of managing those resources with the long-term interest of all the citizens in mind. However, fisheries managers do not determine what is in the public interest by themselves; the public interest is usually determined through policy processes in the political arena. Fisheries managers normally implement programs that reflect policies designed to protect the public interest.

The role of members of the public as stakeholders should be to assist in defining the public interest (i.e., the benefits that the resource should provide), usually indirectly through elected or appointed officials. In the past, many proposed fisheries regulations were developed by fisheries professionals, sometimes with little or no public input. However, because changes in management of a fishery can have profound effects on stakeholders, it is important for fisheries managers to solicit input from the public. Since much of "the public" has little interest in a specific fisheries management issue, fisheries managers usually focus on engag-

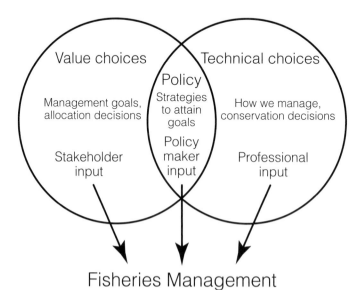

Figure 5.2. Decisions about fisheries management include value choices that should primarily reflect stakeholder values, technical choices that should primarily reflect the expertise of fisheries professionals, and policy choices which overlap both value and technical choices and are primarily the responsibility of policy makers.

ing stakeholders in defining the public interest. Over time, the list of stakeholders interested in fisheries issues has increased dramatically from what was once primarily anglers to what now includes, but is not limited to, anglers, environmental organizations, the business community, and local citizens. Fisheries managers often facilitate collection of information on stakeholder opinions to be used by elected or appointed officials when making policy decisions. Defining desired benefits of the resource is the values aspect of fisheries management.

Policymakers are those elected or appointed officials who represent stakeholders and who make decisions incorporating both values and technical aspects of fisheries management. A common example of appointed policy makers are the state commissioners (sometimes referred to as board members) who are usually appointed by state governors to set policies for state fish and wildlife agencies (Chapter 4). Their qualifications for the position usually include a strong interest in fisheries and wildlife management and a history of support for the political campaigns of the governors who appointed them. Although the great majority of commissioners take their position and its duties seriously, they also face steep learning curves as they become educated about how to integrate fisheries biology and policy, that is, the balancing of technical choices and values choices. The history of fisheries management includes many examples of the struggle to find the appropriate balance between values aspects and technical aspects of fisheries management.

5.3 A CONTRAST OF MARINE AND INLAND FISHERIES MANAGEMENT

Marine and inland fisheries management have taken contrasting but parallel paths to arrive at similar destinations in the search for balancing values and technical aspects of fisheries management. Marine fisheries management frequently has placed greater emphasis on allocation of resources (value choices) than on conservation of resources (technical choices). Marine fisheries management systems in the USA and Canada have traditionally given substantial weight to the role of stakeholders in making technical as well as value choices, leading some experts in the field to question whether foxes were being asked to guard the henhouse (McCay 1996). Conversely, inland fisheries management has often relied more on the input of fisheries managers than stakeholders to make both values choices and technical choices. Studies of effective fish and wildlife management agencies have shown that highly effective agencies maintain a solid biological basis for management decisions (technical choices) while also involving stakeholders in making values choices (McMullin 1993).

5.3.1 Marine Fisheries Management

Marine fisheries management is a complex process of balancing the interests of commercial fishing, recreational fishing, and conservation of fishes. In the USA, lead authority for managing marine fisheries is determined by distance from the shore. Most coastal states have lead authority from the shore to a distance of 3 miles (4.8 km) at sea. Interstate commissions for the Atlantic, Gulf of Mexico, and Pacific states also play a role in marine fisheries management because most marine fisheries overlap multiple state jurisdictions. From 3 to 200 miles (321.8 km) from shore, the federal government has lead authority in what is called the Exclusive Economic Zone (EEZ). Beyond 200 miles from shore are international waters, where fisheries management is determined by multinational treaties. In Canada, the federal government regulates all marine fisheries from shore to the 200-mile limit, while the provinces and territories regulate on-shore fish processing. Similarly, marine fisheries in the United Mexican States (Mexico) are regulated by the federal government.

The Magnuson Fishery Conservation and Management Act, first passed by Congress in 1976 and modified in 1996 and 2006 (now called the Magnuson–Stevens Fishery Conservation and Management Act and hereafter referred to as the Magnuson–Stevens Act), created regional fisheries management councils that gave commercial and recreational fishing stakeholders nearly equal standing with fisheries managers when making decisions about management of marine fisheries. The councils were charged with developing management plans for all commercially- and recreationally-important fish stocks in the EEZ. Thirty years after creation of the councils, 24% of the commercially important fish stocks in U.S. waters were overfished because the regional fisheries management councils frequently made management decisions that weighed value choices (keeping commercial fishers fishing) more heavily than technical information, which suggested the harvest of many fish stocks should be reduced (National Marine Fisheries Service 2008).

Comprehensive reviews of the marine fisheries management system resulted in recommendations to redefine the decision-making authority of regional councils and create greater separation of value choices and technical choices (Eagle et al. 2003). The 2006 modifications of the Magnuson–Stevens Act specified that committees composed of primarily fisheries scientists would set annual catch limits (i.e., the amount of fish that can be removed from a fish stock on a sustainable basis), a technical decision. The councils' roles were better defined to focus their decisions on allocation of opportunities to catch fish subject to the annual catch limit (value choices). Prior to the 2006 changes, the councils could override the specified catch limits if they felt such actions were in the best interest of fishing communities, and they often did. The 2006 revisions to the Magnuson–Stevens Act mandated that overfishing (i.e., harvesting fish at a rate greater than they can be replaced on a sustainable basis) must end by specified dates for stocks that are overfished (i.e., stocks in which the biomass is less than the biomass that will sustain maximum sustainable yield). The revisions have created a better balance between value (allocation) and technical (conservation) choices.

Although Canadian and Mexican managers of marine fisheries involve stakeholders in the development of management plans, they do so without the framework of the regional councils established in the USA. Fisheries and Oceans Canada has responsibility for managing marine fisheries in that country, whereas in Mexico, fisheries are regulated by the Ministry for Agriculture, Livestock, Rural Development, Fisheries, and Food.

5.3.2 Inland Fisheries Management

Fisheries management decisions in inland waters of North America have traditionally relied more on fisheries managers than on stakeholders to make both technical choices and values choices. Fisheries professionals have frequently made values choices, such as deciding which species to stock in reservoirs. Therefore, they defined much of the benefits the fishery would produce with little or no input from stakeholders. In addition, stakeholders have rarely been given the authority to make technical decisions regarding fisheries management in inland systems (although stakeholders frequently make suggestions of a technical nature).

The importance of stakeholder involvement in making fisheries management decisions in inland waters began to grow in the latter third of the 20th century, when increased interest in the environment led to passage of several landmark environmental laws (e.g., the Endangered Species Act, the National Environmental Policy Act, and the Clean Water Act). In the USA the National Environmental Policy Act (NEPA), passed into law in 1970, had a particularly important effect on stakeholder participation in public-arena decision-making processes, as it mandated public involvement during the scoping (identification) of issues and selection of management alternatives by federal resource management agencies. The elevated importance of stakeholders in decision-making processes created by NEPA contributed to a proliferation of nongovernmental special interest groups, many of which were formed to influence federal resource management policies (Figure 5.3).

Although stakeholder input into management decisions in inland waters has grown in importance, it has rarely attained the formal authority or achieved the level of importance seen in marine fisheries management. The major difference between the systems is that far fewer commercial fisheries exist in inland waters than exist in marine waters. Most inland waters support predominately recreational fisheries with more emphasis on catching fish than on harvesting fish. Although harvesting fish is an important component of many inland fisheries, far fewer stakeholders (compared with marine systems) make a living directly from harvesting fish.

Most inland waters in the USA are managed by the individual states within which the waters are located. Federal involvement in inland fisheries management varies considerably. Inland waters located in national parks and national wildlife refuges generally are managed by federal agencies (the National Park Service and the U.S. Fish and Wildlife Service, respectively). The U.S. Forest Service and the U.S. Bureau of Land Management focus most of their efforts in fisheries management to habitat protection and improvement while leaving management of fish populations in waters located on the lands they control to the individual states. When inland waters border more than one state, the states involved usually collaborate to determine how the water will be managed, but each state normally has its own regulations.

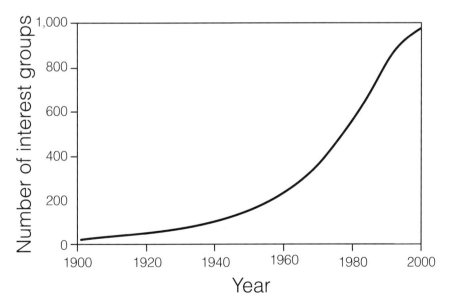

Figure 5.3. A growth curve of nongovernmental organizations with a primary interest in natural resources, 1900–2000 (data from Street 2003).

The Great Lakes Fishery Commission coordinates fisheries research and facilitates cooperative fisheries management among the states and Canadian provinces that border the Great Lakes. Native American tribes that have received federal recognition generally have authority to manage waters located on their reservation lands.

Another important distinction between fisheries management in inland systems and marine systems is the increasing emphasis placed on conservation and restoration of nongame fishes and other species in inland systems (See Box 5.1). Prior to passage of the Endangered Species Act (ESA) in 1972 (and its precursors, the Endangered Species Preservation Act of 1966 and the Endangered Species Conservation Act of 1969), nongame fishes were frequently regarded only as food for sport fish, objects for scientific curiosity, or even pests to be controlled in favor of sport fish species. The ESA represented an important values choice that elevated the importance of nongame species. However, unlike management of sport fish species, and despite the controversy that sometimes accompanies decisions to list species as threatened or endangered, conservation of nongame species has been accomplished mostly by fisheries managers making technical choices with little input from stakeholders. However, see Box 5.1 for a recent example where stakeholder input was very important in the management of nongame species.

It is important to note that even when the biological basis for fisheries management decisions is sound, decisions may be difficult or impossible to implement if stakeholders feel that important values they hold are not being adequately addressed (Churchill et al. 2002). The case of Norris Reservoir, Tennessee provides an excellent example of how difficult it can be to balance public values and good biology (see Box 5.2).

5.4 ADAPTIVE FISHERIES MANAGEMENT

Adaptive fisheries management is an iterative process of describing the current status of the fishery (Where are we?), identifying goals and objectives (Where do we want to go?), designing and implementing strategies to attain those goals and objectives (How will we get there?), evaluating the effects of strategies, and using the evaluation information to learn more about the system and to modify goals and objectives (Did we make it? and, What can we do differently?; Figure 5.4). In-depth discussions of adaptive resource management are available in Walters (1986) and Lee (1993).

5.4.1 Describing the Current Status of a Fishery

Good fisheries management is supported by scientifically-sound information on fish populations, fish habitat, and the stakeholders. One of the primary tasks of fisheries management agencies is to collect data on the fisheries they manage. Although most agencies have databases of information about fish populations (e.g., species abundance, age, and growth) and habitats (e.g., river flows and water quality), agencies rarely are satisfied that they have enough data for every fishery they manage. The resources available to an agency to inventory fish populations and habitats are never sufficient to generate reliable data for every fishery it manages. Additionally, the human dimensions of fisheries management (e.g., angler effort and angler opinions) have received less attention because this field is young compared with other aspects of fisheries management. Because fisheries managers always need additional infor-

Box 5.2. Value Choices and Technical Choices in the Norris Reservoir Fishery, Tennessee

The Tennessee Valley Authority (TVA) constructed Norris Reservoir in 1936 to provide flood control and hydroelectric power. As with many reservoirs, Norris Reservoir supported an initially productive fishery that began to wane as the reservoir aged. Thirty years after impoundment, anglers began expressing concern about declining catch and average size of walleye and sauger. As these fisheries declined, the Tennessee Game and Fish Commission (later to become the Tennessee Wildlife Resources Agency, TWRA) created a new fishery in the reservoir by stocking striped bass. The striped bass grew rapidly and established a popular trophy fishery. Despite the success of the striped bass fishery, many anglers, particularly those interested in fishing for walleye, sauger, crappies, and black basses, were unhappy with management of the reservoir.

At a meeting between anglers and the TWRA in 1988, anglers suggested that striped bass were responsible for poor fishing for other species due to predation and competition for prey fish. Many of the anglers at the meeting demanded that TWRA cease its stocking of striped bass. Although TWRA fisheries managers assured the anglers that predation by striped bass on other sport fishes was inconsequential, they agreed to increase their emphasis on management of native sport fishes in the reservoir. Attacks by the percid and black bass anglers on the striped bass management program led striped bass anglers to get involved in defending their fishery, and subsequently anglers became polarized into two camps (Churchill et al. 2002).

As the controversy escalated, the TWRA formed the Norris Reservoir Task Force and charged it with researching the issues related to fisheries management and developing recommendations for improving fishing in Norris Reservoir (i.e., making both values choices and technical choices). The task force included representatives from universities, TVA, TWRA, anglers, and a boat dock owner. Fisheries managers on the task force drove much of the decision making that resulted in the Norris Reservoir Adaptive Fisheries Management Plan. The plan proposed multiple field studies over a 5-year period designed to determine if changes in fisheries management strategies in Norris Reservoir were warranted. The plan called for a reduction in stocking of striped bass but did not call for a cessation of stocking altogether, as some anti–striped bass interests demanded. Although fisheries professionals, agency administrators, politicians, and some anglers approved of the plan, another group of anglers became increasingly alienated and vocal. The latter group of anglers formed the Tennessee Sportsman's Association (TSA) with the goal of increasing their influence on decisions regarding fisheries management in Tennessee.

The TSA convinced state legislators to introduce bills in the 1995 legislative session that would have prohibited stocking of striped bass in Norris Reservoir and removed harvest regulations for striped bass. These bills and some other attempts to introduce legislation regarding fisheries management were defeated, but legislators ordered the TWRA to fund a new research project to study competition and predation by striped bass. The TWRA solicited proposals to conduct the study from only out-of-state scientists in an attempt to avoid perceived bias by in-state scientists who had worked with TWRA on previous efforts. The independent study eventually concluded that striped bass predation

(Box continues)

Box 5.2. Continued.

on other game fish was negligible but that potential existed for competition between striped bass and other predators during periods of low prey abundance (Raborn et al. 2002, 2003, 2007). The TSA criticized the independent study and remained opposed to TWRA's management strategies for Norris Reservoir.

Following completion of the independent study, the TWRA initiated a second effort to develop a management plan for Norris Reservoir with a different set of stakeholders participating in the Norris Lake Fishery Advisory Committee. Unlike its predecessor, the advisory committee was comprised entirely of stakeholders, and TWRA fisheries managers served only as advisors. Members of the advisory committee included representatives of the antagonists (TSA and striped bass anglers) as well as nonaffiliated anglers, business interests, and county governments. Independent facilitators conducted meetings of the advisory committee. The advisory committee also differed from the task force in that it was tasked with devising management goals for the major sport fisheries in Norris Reservoir (making value choices), while development of alternative management strategies (technical choices) was left to the TWRA. The advisory committee chose which management strategies would become part of the new management plan.

The strategies approved in the new management plan did not differ greatly from those of the previous plan, calling for reduced stocking rates of striped bass and increased emphasis on management of native sport fishes. However, stakeholder acceptance of the management strategies was substantially greater. Churchill et al. (2002) concluded that the improved definition of stakeholder and professional roles in the separation of value choices and technical choices, along with further reductions in stocking rates of striped bass, led to substantial reduction of criticism from stakeholder groups.

Among the most important lessons learned from the Norris Reservoir case study was that dissatisfied anglers cannot be ignored in development of management plans and that research in the human dimensions of fisheries management is just as important as biological research as an underpinning for management. The TWRA had excellent information regarding fish populations and fish habitat during the Norris Reservoir controversy and gained even more information from the studies done as the controversy progressed. Although minor controversies continue to emerge regarding management of the Norris Reservoir fishery, the level of controversy dropped substantially when TWRA fisheries managers were able to focus stakeholders on making value choices and fisheries professionals on making technical choices. For more detail see Churchill et al. (2002).

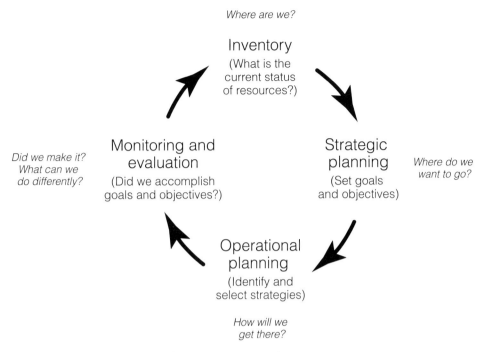

Figure 5.4. The adaptive fisheries management cycle.

mation, and they have limited resources to obtain this information, they must prioritize what additional data are most important to collect. Adaptive fisheries management is a process of setting priorities that enables managers to strive for a management goal, testing the effects of strategies designed to reach that goal by monitoring implemented strategies, and then adjusting future strategies based on what is learned from the monitoring.

5.4.2 Identifying Goals and Objectives

Goals are broad statements of benefits desired from resources. Goals tend to be qualitative and value-laden expressions of the public interest. For example, a goal for management of a reservoir in the southeastern USA could be "to maintain a high-quality fishery for largemouth bass, including the opportunity to catch trophy-sized fish." Note the qualitative nature of the goal (i.e., high quality is not defined) and the values component (i.e., the emphasis on trophy-sized fish reflects just one of many reasons why people may be interested in fishing the reservoir). Due to the emphasis on the values component of goals, it is important for stakeholders to play an important role in determining them. Policymakers frequently identify tentative goals, after which fisheries managers facilitate public input on their desirability, which helps the policymakers to make decisions regarding final goals (Box 5.3). Since goals rarely have a significant technical choice component, the role of fisheries managers usually is limited to facilitating public input into their development and transmitting that input to policymakers.

Objectives are *specific*, *measurable*, *achievable*, *realistic*, and *time-bound* (SMART) statements of what must be achieved in order to attain goals. Often, multiple objectives are needed to attain a goal. For example, one objective related to the example goal listed above could

Box 5.3. The Process of Setting Fishing Regulations

Fishing regulations represent one of the primary management tools of fisheries managers. In the USA, inland fishing regulations are typically set by individual states through state fish and game regulatory processes. Fish and game commissions are generally separate entities from fish and game agencies, although they work together in various capacities. Fish and game commissioners are typically appointed by state governors, and they are generally appointed because they have a strong interest in some aspect of fisheries and wildlife. In some cases, regulations are established by state legislatures passing bills that are signed into law by the governor. State fish and game commissions use strict public processes to vet issues and receive public input. Fisheries managers must understand how their state commissions function to maximize the chances that fishing regulations are based on an appropriate balance of science and social factors. Because fisheries regulations require blending scientific information and social factors, fish and game commissioners must balance the needs of fisheries resources and the public interest.

Establishing functional, durable, and flexible fishing regulations requires much work and patience by fisheries professionals. Identification of a fisheries problem that can be improved through a regulation change is the first step toward development of new fish regulation. A fish and game commission can become aware of a problem through state fisheries managers, a commission member, members of the public, nongovernmental organizations, elected officials, or other stakeholders. If the commission feels a problem requires a regulatory solution, they will direct the state fisheries agency to bring forth to them a range of recommendations for their consideration. The process of preparing a packet for consideration by fish and game commissions can be a very simple exercise that can be completed in few months or can be extremely complex and last many years.

The four main factors to consider when developing a range of alternatives for a fishing regulation are (1) technical, (2) legal, (3) enforcement, and (4) social. Technical factors include information about the species of interest such as habitat requirements, life history characteristics, or harvest rates. In general, fisheries managers are comfortable in collecting and compiling technical information, though often uncertainties and data gaps exist that open up a decision to one based outside pure science. A legal review of proposed regulations should always occur to ensure that they comply with state and federal legal mandates and requirements. Enforcement staff should also review proposed regulations to ensure that they can be enforced. Establishing regulations that cannot be enforced or do not comply with legal requirements represents not only a waste of time in producing a regulatory package but can also cause problems between the agency and the public by creating confusion or conflict. Hence, it is important to consult early in the process with legal and enforcement staff. Accounting for social factors in developing fishing regulations is typically the most difficult step for fisheries professionals. Rarely do all segments of the fishing public agree on a regulatory solution. For example, there are often differing opinions among constituents regarding how protective a regulation should be (e.g., should a fishery be closed?). Those affected by a regulation because they live near a fishery may have different perspectives than those who travel far to fish there. Different user groups often feel differently about whether a fishery should be managed to allow for the use of bait or only single, barbless hooks.

(Box continues)

Box 5.3. Continued.

A critical component of preparing a range of alternatives for commission consideration is understanding the range of values associated with the proposed regulation. To understand the range of values, outreach should be conducted to include members of the fish and game agency itself, other agencies that could be affected by the regulation, user groups, local land owners, industry leaders, and local fishing businesses that may be affected. In addition to better understanding the issues associated with the proposed regulation, it is less likely the fisheries management agency will be surprised when proposing the regulation to the commission at the public hearing, something neither the management agency nor the commission tends to react to positively. Public outreach can be difficult and can range from contacting organized angling groups to posting information in tackle shops and on appropriate internet sites regarding proposed regulations and how interested parties may offer input about the proposal. This type of outreach can make a world of difference in developing balanced regulations that are palatable to most constituents.

Under the simplest of scenarios, when the technical, legal, enforcement, and social factors are as well understood as possible, fisheries managers develop a regulatory package that generally includes a problem statement with background information, current data on the issue, any pertinent legal and enforcement ramifications, and impacts to anglers and other constituents. This packet should be reviewed through the management agency's chain of command and then forwarded to the fish and game commission. The commission will publicly publish notice of the proposed regulation as required by law. There may be a number of commission hearings on the matter before a vote is taken to ensure adequate public notice. It is typical that fisheries agency staff present the information to the commission and answer any questions. It is important to prepare a clear and concise presentation. Preparation should include a review by agency management as well as anticipating questions that may arise. When presenting material to a fish and game commission it important to understand the roles of the agency and of the commission. Specifically, the agency prepares and presents materials, makes recommendations, and answers questions of the commission. The commission considers information presented, including public testimony, and makes the decision to promulgate a regulation or not.

If the commission adopts a regulation, the regulatory language is sent to the state's office of administrative law (or equivalent) to ensure the language meets legal requirements for clarity and other legal standards. In the event that a regulation is adopted the commission typically has a mechanism to disseminate the new regulations to interested parties.

Any regulation should include a monitoring component to determine if the regulation has the desired effect. Agency staff should plan to update the commission on the effect of the regulation to the fishery and stakeholders. This communication allows the commission to work toward an adaptive management approach by considering adjusting regulations as needed.

Details of the process of establishing fisheries regulations varies somewhat from state to state; however, the basic framework is consistent. Understanding the process for those who engage in regulation changes is critically important. Equally important is a clear grasp and appreciation of the role of the fishery agency, the fish and game commission, and the stakeholders. When all of these parties play their role appropriately, fisheries regulations with the right balance of science and social values should emerge.

be "to increase the average length of largemouth bass caught by anglers to 350 mm within 5 years." Note that the SMART objective is specific and easily measurable (i.e., it states a measurable feature that constitutes success, and it can be measured relatively easily through creel surveys or monitoring of fishing tournaments). It also is time-bound (i.e., within 5 years). We cannot tell from this hypothetical example how realistic or achievable the objective may be; judging how realistic and achievable the objective may be is the job of the fisheries manager. If objectives are SMART, they contribute to the credibility of fisheries management by making it easy to determine if objectives have been attained. Fisheries professionals play an important role in development of objectives because much technical knowledge is required to translate broad, general goals into SMART objectives. Policymakers and stakeholders have to agree that selected objectives are appropriate stepping stones to attaining goals, but they generally play a lesser role in developing the objectives.

5.4.3 Designing and Implementing Strategies

Strategies are the methods used to attain goals and objectives. While goals and objectives focus on what we (fisheries managers, stakeholders and policymakers) want to achieve, strategies focus on how we will achieve goals and objectives. Fisheries managers play a key role in developing strategies and in evaluating the likelihood that strategies will attain an objective. Fisheries managers also should evaluate the effects that proposed strategies are likely to have on the resource and the stakeholders.

Balancing the roles of fisheries managers and stakeholders in development and evaluation of strategies is complex. Stakeholders frequently suggest strategies to fisheries managers (e.g., requesting a change in fishing regulations on a specific body of water) without giving much thought to the objective they want to attain. For example, when I (S. L. McMullin) worked as a fisheries manager in Montana, a local chapter of Trout Unlimited requested that the regulations on a portion of a popular trout river be changed from a 5-fish-per-day limit to catch-and-release angling. Queries as to why the group wanted a catch-and-release regulation identified the organization's goal as the desire to have fish of larger average size in the catch. When information was presented to the group that indicated the fish population in that portion of the river was increasing and that average size was decreasing as the population increased, the group agreed that harvesting some fish was a desirable strategy.

Stakeholders share with fisheries managers the key role of evaluating strategies because choosing how to attain objectives often creates winners and losers. For example, when restrictive regulations like the catch-and-release proposal discussed above are implemented, they usually are accompanied by restrictions on the use of live bait because mortality of fish caught and released on live bait is typically higher than is mortality of fish caught and released on artificial flies and lures (Taylor and White 1992; Muoencke and Childress 1994). The losers in this case are the anglers who prefer to use live bait because they are excluded from fishing the area under the restrictive regulation. The winners may be the anglers who use flies and artificial lures because they now have fewer anglers using the area, unless the restrictive regulation attracts more fly and lure anglers to the area.

Fisheries managers, stakeholders, and policymakers all play important roles in selecting strategies to pursue. Depending on the type of strategy, either fisheries managers or policymakers will probably make the final decision, but stakeholders should have input into most of those decisions. For example, fisheries managers usually make decisions about strategies related to management of fish habitat, whereas policymakers (commissions or boards) usually make final decisions about strategies related to fishing regulations at the state level (see Box 5.3).

5.4.4 Monitoring and Evaluation

Monitoring and evaluation provide the feedback loop that allows fisheries managers to determine if selected strategies are having intended effects. This critical step in the adaptive management process should be based on scientifically-sound sampling to determine if goals and objectives are being attained. The sampling should be designed to answer questions such as the following.

1. Are fish populations responding as desired or predicted?

2. Is the quality of fish habitat improving, declining, or staying about the same?

3. Are stakeholders satisfied with the quality of the fishery or their fishing experiences?

Fisheries professionals play key roles when designing and implementing monitoring programs, analyzing the data, and reporting to stakeholders and policymakers on the progress toward attaining goals and objectives. The judgment of fisheries managers is important in suggesting why strategies may not be working and what alternative strategies might be more effective. Stakeholder input is important in determining satisfaction or dissatisfaction with progress made toward attaining goals and in modifying goals if that is deemed necessary. Policymakers, as elected or appointed representatives of stakeholders, have oversight responsibility to ensure that fisheries professionals are doing all that they can to attain management goals.

The information gained from monitoring and evaluation should feed back into the adaptive management loop to help determine the status of fish populations, habitat, and people (e.g., stakeholder satisfaction) after strategies have been implemented. The important point is that all participants in the adaptive management process learn more about the system as a result of the monitoring and evaluation activities. When monitoring and evaluation information is shared, fisheries managers, stakeholders, and policymakers may decide that new goals and objectives are needed or that the current goals and objectives are appropriate. In either case, the process should continue through successive iterations to continue learning about the system and to improve management of the fishery. Box 5.4 discusses two contrasting cases of fisheries management on Montana rivers and Box 5.5 discusses a controversial fisheries management project in California that illustrates the importance of gaining public acceptance for management.

5.5 CONCLUSIONS

Fisheries management is a challenging process of understanding fish populations, fish habitat, and stakeholders interested in a fishery and taking actions to manipulate those components of the fishery to achieve goals and specific objectives. The process of fisheries management involves fisheries managers, stakeholders, and policymakers when making values

Box 5.4. Adaptive Fisheries Management in Montana

The Bighorn River is a world-famous trout fishery in south–central Montana. Prior to construction of Yellowtail Dam (1960s) near the Wyoming–Montana border, the Montana portion of the river was virtually ignored by anglers because it held few fish of interest to them. However, the dam created cold, clear water ideal for trout and soon after impoundment, the river was producing many large brown trout and rainbow trout. The river's reputation for producing good trout fishing led to a rapid increase in angler use, but that ended in 1975 when the Crow Tribe of Native Americans closed access to the river to all but members of the tribe. A lengthy legal battle between the Crow Tribe and the State of Montana ensued until 1981 when the U.S. Supreme Court decided that the State of Montana had responsibility for managing the Bighorn River fishery. The court decision opened the river to public access and fishing pressure increased rapidly.

Starting in July 1981, fisheries personnel from the Montana Department of Fish, Wildlife and Parks (MDFWP) began monitoring population abundance, average size, and growth rates of brown trout and rainbow trout in particular as well as angler use. Monitoring indicated that river flows significantly affected fish abundance. Low average flows resulted in higher mortality and lower population abundance, whereas higher average flows produced higher population abundance. Despite high angler use, mortality due to fishing appeared to be insignificant, probably because most anglers voluntarily practiced catch-and-release fishing.

Despite the evidence that habitat factors were more important than was fishing pressure in determining fish population abundance, a series of events led stakeholders to exert pressure on the MDFWP to implement more restrictive regulations. A highly publicized fish kill during the summer of 1984 was followed by extremely low flows and cold temperatures (which reduced growth rates of trout) during the summer of 1985, which led many stakeholders to believe that the fishery, especially the trophy fish component, was in trouble. As a result, MDFWP fishery managers decided to develop a management plan for the Bighorn River based on an adaptive fisheries management process.

The process began with public meetings designed to involve stakeholders in identifying desired goals for management of the fishery (focus on making value choices). Stakeholders identified the opportunity to catch large, wild trout as the highest-priority management goal. With the goal identified, MDFWP fisheries managers developed specific objectives and strategies designed to achieve the goal (technical choices). The goal, objectives, and strategies were published in a draft management plan that included detailed descriptions of trends in fish populations and the factors affecting fish populations. Strategies recommended by the MDFWP focused on improving management of flows in the river with only minor changes in fishing regulations.

Results of a survey distributed with the draft management plan indicated that an overwhelming majority of people who read the plan agreed with the goal, objectives, and recommended strategies. Ninety-three percent of the survey respondents agreed that reading the plan improved their understanding of the Bighorn River fishery. Furthermore, 50% of the survey respondents agreed that they had changed their minds about how the fishery should be managed as a result of reading the plan. The latter result was especially

(Box continues)

Box 5.4. Continued.

gratifying to MDFWP fishery managers because a high percentage of the people who received copies of the draft management plan were stakeholders who had previously contacted the agency with critical comments about management of the fishery.

When the Montana Fish, Wildlife and Parks Commission (the policymakers) met to consider adoption of the management plan, one stakeholder group that disagreed with the proposed strategies presented a competing proposal that would have implemented more restrictive fishing regulations than were called for in the plan. After hearing both the agency proposal and the competing proposal, the commission approved the management plan as proposed. The commissioners cited the adaptive management process used by MDFWP fisheries managers, including its focus on stakeholder involvement in making value choices and professional involvement in making technical choices, as a key factor in their decision to adopt the management plan rather than the competing proposal.

At the same meeting in which the commissioners adopted the Bighorn River management plan, they also denied a request by MDFWP fishery managers in another region of the state to implement new regulations on a portion of the Missouri River. The Missouri River fishery managers had fish population monitoring information that was as good as that available on the Bighorn River. They also faced a competing proposal from stakeholders who opposed the change in regulations proposed by the MDFWP. The major difference between the two fisheries was the adaptive fisheries management process utilized by managers of the Bighorn River fishery. The commissioners expressed their concern that stakeholders on the Missouri River did not have the same level of involvement in the management process as stakeholders on the Bighorn River. The side-by-side contrast of the Bighorn and Missouri river cases led the commission to instruct MDFWP fisheries administrators to develop management plans for the 10 most highly-used fisheries in Montana by means of the adaptive fisheries management process.

It is important to note that the adaptive management process did not eliminate controversy over how the Bighorn River should be managed, as demonstrated by the competing proposal presented at the commission meeting. However, it did defuse most controversies. In addition, it provided opportunity for stakeholders to define the management goal (value choice) and for MDFWP fishery managers to educate stakeholders regarding the rationale for their recommended strategies (technical choices).

Box 5.5. Lake Davis Case Study

A controversial fisheries management project involving California's Lake Davis highlights the importance of the human dimension in fisheries management. Lake Davis is a 1,600-ha reservoir in California's Sierra Nevada. Proposals to use rotenone to eradicate illegally-introduced northern pike from the reservoir seemed straightforward from a technical perspective, but the human dimension added substantial complexity to the process of deciding the course of action and implementation of the eradication project.

A highly controversial eradication effort in 1997 and a generally well-accepted eradication project 10 years later highlight the importance of a broad and integrated fisheries management approach that recognizes and addresses human values as an essential element of a project. In particular, the values of clean drinking water and public health were explicitly integrated into the project as an objective. This was key as managers worked to eradicate an invasive species.

In the 1980s, reproducing northern pike were discovered in California's 1,011-ha Frenchman Reservoir. Soon after, they were discovered in nearby streams. The species was eradicated using rotenone but then were found several miles away in Lake Davis.

The establishment of northern pike in Lake Davis adversely affected the local trout fishery. In addition, the growing potential for their escape or illegal transport from Lake Davis to other parts of California posed significant risk to a number of sensitive species should they become established elsewhere. Consequently, California fisheries managers chose to eradicate the population.

Lake Davis, however, is more than a premier trout fishery. Prior to 1997, the reservoir was used as the drinking water supply for the community of Portola and other local residents. In preparing for the first Lake Davis eradication project in 1997, fisheries managers proposed to take the drinking water off line (alternative groundwater wells were constructed) and treat the reservoir with rotenone. The proposal was greeted with numerous concerns from members of the local community about the safety of chemicals proposed for use. The California Department of Fish and Game (CDFG) fisheries managers repeatedly assured the community that the chemicals were safe. However, because the chemical formulation was considered proprietary by the rotenone formulation manufacturers, the CDFG was unable to provide legally to the public the list of constituents in the rotenone formulation that was proposed for use. As a result, the fisheries management proposal was perceived by the local community as being suspect and harmful to public health and safety. The community responded with protests and legal action, which were reportedly widely by the media.

On the morning of the treatment, about 300 state and local law-enforcement personnel had arrived and were stationed around the reservoir, in part due to death threats to CDFG personnel. To the community, however, the large enforcement presence appeared heavy handed and fueled more anger. Despite the controversy, the CDFG proceeded with the project.

Although the treatment initially appeared successful, a year and a half later in 1999 northern pike were found again in Lake Davis. It has not been determined if these northern pike were illegally reintroduced or survived the 1997 treatment. The vehemence and emotion surrounding the treatment were such that after the rediscovery, the CDFG did not propose a retreatment. Instead the agency's director met with a group of local elected

(Box continues)

Box 5.5. Continued.

officials and community members who formed a steering committee to address the issue. The CDFG and the steering committee prepared a plan to deal with the invasive northern pike that focused on education, enforcement, and intensive manual removal. Rotenone was removed as an option for dealing with the problem. The CDFG established a small office in the local community, and the CDFG fisheries managers became both ambassadors to the community and northern pike removal specialists as they proceeded with removal efforts.

Despite the removal of tens of thousands of northern pike over the next few years by means of backpack and boat electrofishers, gill, trap, and fyke nets, and a purse seine, the population grew; the control and containment efforts were ineffective. The northern pike caused the trout fishery to decline, impacting the local economy, and the risk of escape of northern pike downstream or their transport by humans was increasing. By 2003, the steering committee requested that the CDFG evaluate eradication but only via methods that were completely safe to the public and that considered economic effects to the community. The value of clean drinking water and public health and safety needed to be the primary considerations in fisheries management objectives.

The CDFG considered many options, but rotenone came to the forefront as the safest and most reliable means of eradication. However, if rotenone was to be approved for use in Lake Davis, all of the water quality concerns raised by the public would need to be addressed in an open and public forum. The fact that the formulation had been evaluated by the state and federal regulators was not sufficient to allay concerns of the local citizens. All of the technical questions and concerns needed to be addressed by knowledgeable specialists in a community forum that included all of the relevant public health agencies.

The U.S. Forest Service (USFS), which owned the land around Lake Davis, became a full partner in the project. In September 2006 the CDFG and the USFS issued a notice of preparation of an Environmental Impact Report and Environmental Impact Statement (EIR–EIS), a joint environmental document for full public disclosure under the California Environmental Quality Act and the National Environmental Policy Act. Initial public scoping meetings revealed skepticism by many locals regarding the safety of rotenone to the environment. The CDFG committed to a series of public meetings to disclose fully any information about the chemicals and to address other community concerns. Many concerns regarding the project were also raised by a variety of local, state, and federal agencies. A draft EIR–EIS was produced that examined an array of alternatives for northern pike eradication. After review and response to the public comment, the CDFG and the USFS certified the final environmental document, and a project involving the use of rotenone was approved.

Public outreach continued throughout the planning process; openness and transparency continued to be a top priority. This helped to demystify the project and demonstrate to the public why the project was necessary and that success was likely.

Under a structured leadership system, eradication took place over 1 month and included treatment of the reservoir and all the tributaries to Lake Davis. More than 500 participating staff were housed and fed locally to benefit the local economy. Subsequent to

(Box continues)

> **Box 5.5. Continued.**
>
> the eradication, extensive monitoring and reporting were carried out. The reservoir was restocked with rainbow trout and included a highly publicized event aimed at promoting fishing and tourism in the area.
>
> In 2007 much of the community supported the project because the CDFG and the USFS engaged the community, openly provided information about the project, and answered all questions from the public. The turnaround in the community response to the CDFG resulted from recognition that fisheries managers must address the human elements as well as technical issues to complete the project.

choices (defining desired benefits) and technical choices (how those benefits can be achieved). The process is most effective when stakeholder involvement focuses on making values choices and professional involvement focuses on making technical choices. Policymakers must bridge the gap between values choices and technical choices. Adaptive fisheries management is an iterative process of describing the current status of the fishery (where are we?), identifying goals and objectives (where do we want to go?), designing and implementing strategies to attain those goals and objectives (how will we get there?), evaluating the effects of strategies, and using the evaluation information to learn more about the system and to modify goals and objectives (did we make it and what can we do differently?).

5.6 REFERENCES

Churchill, T. N., P. W. Bettoli, D. C. Peterson, W. C. Reeves and B. Hodge. 2002. Angler conflicts in fisheries management: a case study of the striped bass controversy at Norris Reservoir, Tennessee. Fisheries 27(2):10–19.

Decker, D. J., and J. W. Enck. 1996. Human dimensions of wildlife management: knowledge for agency survival in the 21st century. Human Dimensions of Wildlife 1:60–71.

Eagle, J., S. Newkirk, and B. H. Thompson, Jr. 2003. Taking stock of the regional fishery management councils. Island Press, Washington, D.C.

Geist, V., S. P. Mahoney, and J. F. Organ. 2001. Why hunting has defined the North American model of wildlife conservation. Transactions of the North American Wildlife and Natural Resources Conference 66:175–185.

Larkin, P. A. 1988. The future of fisheries management: managing the fisherman. Fisheries 13(1):3–9.

Lee, K. N. 1993. Compass and gyroscope: integrating science and politics for the environment. Island Press, Washington, D.C.

Martinez, P. J., and E. P. Bergersen. 1989. Proposed biological management of *Mysis relicta* in Colorado lakes and reservoirs. North American Journal of Fisheries Management 9:1–11.

McCay, B. 1996. Foxes and others in the henhouse? Environmentalists and the fishing industry in the U. S. regional council system. Pages 380–390 *in* R. M. Meyer, C. Zhang, M. L. Windsor, B. McCay, L. Hushak, and R. Muth, editors. Fisheries resource utilization and policy; proceedings of the World Fisheries Congress, Theme 2. Oxford & IBH Publishing Company Pvt. Inc. New Delhi, India

McMullin, S. L. 1993. Characteristics and strategies of effective state fish and wildlife agencies. Transactions of the North American Wildlife and Natural Resources Conference 58:206–210.

Meine, C. 1988. Aldo Leopold: his life and work. The University of Wisconsin Press, Madison, Wisconsin.

Muonecke, M. I., and W. M. Childress. 1994. Hooking mortality: a review for recreational fisheries. Reviews in Fisheries Science 2(2):123–156.

National Marine Fisheries Service. 2008. 2007 Status of U.S. fisheries. U.S. Department of Commerce. National Oceanic and Atmospheric Administration, National Marine Fisheries Service. Silver Spring, Maryland.

Pinchot, G. 1947. Breaking new ground. Harcourt, Brace and Company, New York.

Raborn, S. W., L. E. Miranda, and M. T. Driscoll. 2002. Effects of simulated removal of striped bass from a southeastern reservoir. North American Journal of Fisheries Management 22:406–417.

Raborn, S. W., L. E. Miranda, and M. T. Driscoll. 2003. Modeling predation as a source of mortality for piscivorous fishes in a southeastern U.S. reservoir. Transactions of the American Fisheries Society 132:560–575.

Raborn, S. W., L. E. Miranda, and M. T. Driscoll. 2007. Prey supply and predator demand in a reservoir of the southeastern United States. Transactions of the American Fisheries Society 136:12–23.

Ruzycki, J. R., D. A. Beauchamp, and D. L. Yule. 2003. Effects of introduced lake trout on native cutthroat trout in Yellowstone Lake. Ecological Applications 13:23–37.

Street, B. 2003. 2003 conservation directory: the guide to worldwide environmental organizations, 48th edition. Island Press, Washington, D.C.

Taylor, M. J., and K. R. White. 1992. A meta-analysis of hooking mortality of nonanadromous trout. North American Journal of Fisheries Management 12:760–767.

Walters, C. 1986. Adaptive management of renewable resources. MacMillan Publishing, New York.

Chapter 6

Communication Techniques for Fisheries Scientists

SCOTT A. BONAR AND MICHAEL E. FRAIDENBURG

6.1 INTRODUCTION

The lone figure shuffled to the podium in the dark and crowded auditorium. He silently reached into the breast pocket of his coat for some handwritten notes, snapped on the podium light, and placed the notes in front of him. The first slide, a boring, two-color affair crowded with text, was indecipherable to anyone sitting farther than the second row. The next slide was more of the same. He marched on and on through his slides and droned and droned in a hollow monotone reading from his notes, never lifting his head. The audience found paying attention increasingly difficult. That his head stayed down was probably good, because he did not see what happened next.

Halfway through the talk there was a soft "pop" as the projector bulb blew and the screen went dark. Behind the security of his notes and bathed in the soft glow of the podium light the speaker never noticed. The audience kept their collective mouth shut as he continued to drone, his voice moving forward alone. Except for one or two titters most of the group sat in shocked silence. Then suddenly he finished. He lifted his head. There was a smattering of clapping. He left the podium as unobtrusively as he came. Nobody in the audience could quite remember what his talk was about. Because of his poor performance he wasted hours of his time preparing the talk and missed the opportunity to inform or influence the audience.

All of us have seen more of these poor presentations than we care to admit. Each time we promise ourselves to practice our talks and work hard to keep the audience engaged. As a professor, I (S. A. Bonar) do everything I can to help my students approach the podium thoroughly prepared. As a consultant, I (M. E. Fraidenburg) always present to a client after doing an audience analysis to ensure I meet the needs of the audience.

If you are reading this book, you are probably preparing for an exciting and exacting career in fisheries management. You will get to work on interesting, relevant issues and hopefully work in appealing locations. You will have the satisfaction of making the lives of future generations better by protecting our environment today. Do not miss your opportunity to have an effect and make substantive contributions because of poor communication skills.

Although knowing the technical aspects of fisheries science is the first step, you can make a difference only by communicating the science in ways that others can understand and by building a community that is willing to take action on your advice. When we think of the great conservationists of history, most have been talented communicators and motivators as well as scientists. Where would our profession be if Louis Agassiz, Rachel Carson, David Starr Jordan, Aldo Leopold, or Spencer Fullerton Baird could not have communicated their ideas and led us to new accomplishments?

Why is communication and community building important for the fisheries professional? Your job responsibilities will require you to protect fishes and their habitats and provide people with opportunities to enjoy or harvest fish. You will have to know fisheries science so that you can recommend what needs to be done to achieve objectives and convince people to do them. That is where communication becomes important. The fisheries manager who does not add these skills to his or her toolkit risks having his or her advice ignored. This chapter is about methods that will help you work with people so your ideas can influence decision makers and help affect management decisions. Skills used by successful conservationists, politicians, managers, and psychologists are discussed. If you use some of these techniques, you will find that your effectiveness and your relationships with others are likely to improve markedly. First—how do you influence someone?

6.2 INFLUENCE TECHNIQUES

Suppose the following: (1) you want to influence anglers to reduce poaching and otherwise protect the limited number of largemouth bass in a lake; (2) you want to ensure that a rancher keeps his cattle out of a creek that has a sensitive population of cutthroat trout; or (3) you want a rich donor to fund the new research wing of your fisheries laboratory. How would you persuade these people that your ideas are worthwhile? Fortunately, we can borrow techniques from advertising, psychology, and other fields to convince people to take action (Cialdini 2001; Bonar 2007). In fisheries management or conservation biology, advocacy is often discouraged because some feel it damages the credibility of those who take a stand on an issue (see Brouha 1993; Noss 1999; Roberts 2004; Walters and Martell 2004 for discussion of pros and cons of advocacy). However, the reality is that even a professional that does nothing is, in effect, taking an advocacy position: usually implicit support for the status quo. People influence others continually. Doing so in an appropriate manner is up to you. The choice is not that you try to influence (you will), but rather the subjects on which you try to exert influence and the manner in which you try to influence people; these decisions must be appropriate for your particular position. A fisheries manager may try to influence commercial harvesters to reduce bycatch. A fisheries scientist may try to influence a state agency to fund her project. A fisheries advocacy group may try to influence the public to vote for a politician. Each of these can be appropriate, given the context of each position.

Use of influence techniques can be incredibly effective. If they were not so effective, advertisers could not sell us their soap, cars, or toys—or pet rocks, bell bottom jeans, or fake tattoos! However, if you use influence techniques in fisheries management, it is always necessary to use them in an ethical manner to protect your credibility and implement the mission of your organization. The goal is to influence people in a way that gets your point across while not manipulating them.

How do you determine if you are being ethical? Ask yourself, "Will this person be happy I influenced him about this subject 6 months or 10 years from now or will he feel that I took him for a ride?" and "Can this person agree that I stated my needs (i.e., have been strong about content) while being fair and evenhanded (i.e., did not manipulate them)?" Also, give your influencing behavior a "gut check." Deep down you know if you are being underhanded. Always strive to be ethical. Assuming that you will use these techniques in an ethical manner, let us discuss how to influence people.

6.2.1 Maslow's Hierarchy of Needs

Abraham Maslow (1908–1970) was a humanistic psychologist who developed a "hierarchy of needs" (Maslow 1970). Essentially Maslow stated that people have needs that must be met, and they must meet basic needs before higher needs. The most basic are physiological needs such as breathing, eating, shelter, and water. Once physiological needs have been met, safety needs must be met followed by social and esteem needs. For example, how can you become popular with your friends and increase your self-esteem? At the height of the hierarchy of needs is "self-actualization." This is the point at which most of your needs have been met and you have a feeling of having "arrived." According to Maslow, only about 2% of people reach the self-actualization level.

How do you use Maslow's hierarchy of needs to influence someone? You begin by ascertaining where your audience is within the hierarchy. For example, if you limit the amount of freshwater mussels a commercial fisher can take and he is worried about feeding his family (safety and physiological needs), arguments designed to meet higher needs, such as protecting the diversity of mussel species because it is the right thing to do (esteem and social needs) will probably fall on deaf ears. Your arguments are more effective if they are designed to meet his basic needs first. For example, argue that "if we continue to harvest mussels at the current rate they will disappear, and you will completely lose your business. If we cut back on the harvest rate, you will probably be able to save some of your mussel harvesting business." Alternatively, if you are trying to get a rich philanthropist to buy land surrounding a section of creek to protect the mussels, you would probably design your argument to appeal to their social or esteem needs. For example, "If you buy this property and use it to protect this mussel population, we will give you a conservation award or name a streamside park after you" (esteem and social needs met).

6.2.2 Cialdini's Influence Strategies

Robert Cialdini is a social psychologist who has developed an understanding of how advertisers and sellers influence people to buy things. He found that people are influenced in six basic ways (Cialdini 2001), and because we process a large amount of information each day, most of us respond to these influence strategies automatically.

6.2.2.1 Liking and similarity

People will be influenced more by those they know and like and by those who are most similar to themselves rather than those who are disliked or are dissimilar. For example, I (Bonar) like my staff at the Arizona Cooperative Fish and Wildlife Research Unit. Therefore, I am strongly influenced by their opinions and commonly do the things they ask. However, there are some people I either dislike or do not trust, and I am reluctant to accept their advice. How might you use the liking and similarity principle in fisheries management? Place a premium on getting along with the people you are trying to influence whether they are members of another agency, an angling group, or a conservation organization. Emphasize your similarities with these people, not your differences. This does not mean becoming insincere, but it does mean giving people as many reasons as you can to get them to like and trust you despite your current differences.

6.2.2.2 Reciprocation

If you give somebody something, that person feels compelled to return a favor. For example, Cialdini (2001) cites a study in which a person asks another if he would be willing to buy some raffle tickets. The number of positive responses was low. However, if the person gave the other person a soft drink before asking him to purchase raffle tickets, the number of positive responses increased. People felt compelled to reciprocate by buying raffle tickets after they were given the gift of a soft drink. At the University of Arizona, I (Bonar) always tell students to volunteer on agency projects to establish reciprocation and, thereby, increase their chance of being offered a future job with the agency. In addition, an impasse at many negotiations can be broken if one side gives a small concession and the other side feels obligated to respond in kind.

6.2.2.3 Commitment and consistency

People feel a strong desire to be consistent with previous commitments they have made. For example, when most people say they are going to do something, they will try hard to do it. This principle is one reason it is hard to change directions in an agency or start a new program. For instance, in some agencies there is strong pressure to keep fishing regulations simple even though separate water bodies are better managed individually. This is often because "we have always done it that way." It is also difficult to get biologists to update fish sampling programs or use standard sampling procedures because of this principle. "I have always sampled fish in this manner, and I will continue to do so," is a commonly heard refrain—even if other sampling programs would improve data collection and decisions.

You can use this principle to your advantage in fisheries management. If you know someone has a history of conservation behavior, you can remind that person how his future actions will be consistent with his past actions. Assume you want to convince a bass angling club to follow new regulations needed to protect length structure of bass populations. Club members are more likely to be influenced if you remind them of their strong history of protecting largemouth bass populations and that following new regulations will continue their past commitment to conservation.

6.2.2.4 Scarcity

People feel that things that are scarce are more valuable than things that are common. For example, large common carp are prized by anglers in the UK because they are rare. In contrast, common carp are considered an abundant pest and are not generally valued in North America.

This principle can help protect resources such as threatened and endangered fishes, unique water bodies, and rare angling opportunities by emphasizing what will be lost if an action is not taken. People are more sensitive about losing things than they are about gaining things (Bazerman 2006). Therefore, during agency budget exercises, you will often see experienced fisheries administrators state what will be lost if a particular program is eliminated rather than what will be gained if the program is left in place.

6.2.2.5 Social proof

We tend to like what others like. Remember in high school when one person was very popular and seemed to get all the dates and have a lot of friends? This is social proof in action. If some people liked that person, he or she seemed to get even more popular. This principle works in fisheries management as well. If many people think fly fishing for trout is a fun activity, others will be influenced as well. Similarly, if many people like noodling for catfishes, others will want to do it too. To use social proof in fisheries management, ensure that you have as much support as possible for your proposals or ideas. For instance, if you are trying to influence a bass angling club to follow new regulations, having evidence that other clubs support your regulations would help influence the club you are addressing.

6.2.2.6 Authority

People are influenced by others they perceive as authority figures. Famous conservationists, noted anglers, and celebrities are commonly used to sell products or promote points of view. Actor Marlon Brando fished with Native Americans to promote their treaty rights for salmon fishing in the Pacific Northwest. Brando's participation in these protests was, in part, credited for calling attention to the rights of the tribes to fish in-kind with Anglo fishers (American Friends Service Committee 1970). You, too, can use the authority principle by getting respected people to support your point of view or program. For example, if you need to approach an angler group about new agency procedures, get the support of the club president or senior members of the organization or bring an authority figure the group respects to your meeting so that person can influence others.

You cannot avoid the advocacy question as both action and inaction are taking a stand on an issue. Instead, manage advocacy by being clear what you stand for, being ethical in your approach, and using proven techniques like Maslow's hierarchy of needs or Cialdini's six influence strategies. See Bonar (2007) for more information on influencing people regarding natural resources; Cialdini (2001) for general information on how to influence; and the example of how the American Fisheries Society used Cialdini's influence strategies in developing standard fish sampling protocols (Box 6.1).

So, perhaps during your attempts to influence, conflict has arisen. How do you deal with conflict effectively and still maintain a good working relationship with the other person?

6.3 VERBAL JUDO AND CONFLICT RESOLUTION

In communication, conflict resolution is analogous to first aid—you often need to de-escalate conflict immediately so you can then use other types of communication skills. There are times that you will be approached by an angry angler, landowner, or even someone who is part of your own agency. How do you deflect criticism and calm someone down so you can use other methods of communication? Try "verbal judo."

"Government workers are lazy and just mooch off the rest of us who pay taxes." If you work for an agency, you may hear this statement from a constituent, a commissioner, or even

Box 6.1. An Example of How the American Fisheries Society Used Influence Techniques

Data collection in many professions, such as medicine, water quality monitoring, meteorology, and geology, has been standardized for years, which has led to great advances and improved communication. However, routine inland fisheries monitoring, which often has similar goals among regions, has not been standardized in North America. This lack of standardization was mainly due to stiff social resistance, not shortfalls in the available science (Bonar and Hubert 2002). The American Fisheries Society (AFS), in collaboration with 10 federal, state, and private organizations, used Cialdini's influence strategies to help overcome resistance when developing standard sampling methods for North American freshwater fishes. The goals of the project were to develop a book containing standard sampling methods for major fish indices (length structure, body condition, catch per unit effort, and growth) in various North American freshwater body types (see Bonar et al. 2009) and have these methods accepted widely.

Influence strategies were used by the AFS and included the following.

- Authority—A group of top sampling experts was asked to author each book chapter.
- Liking and similarity—Authors represented a wide range of managers, academics, and biologists from across Canada, Mexico, and the USA who were similar to the intended users.
- Social proof—Most agencies in the USA, Canada, and Mexico were asked to provide input to the book, and many provided funding.
- Scarcity—The AFS does not intend to force these methods on anyone. However, biologists will lose their ability to compare their data to standardized data collected by others if they do not use these methods.
- Reciprocation—To fund the book, potential contributors were told of the funds (seed money) given by other agencies. Users of the book are provided regional and rangewide averages of catch per unit effort, body condition, growth, and length structure for common fish species collected by means of these standard techniques.
- Commitment and consistency—Numerous funders, agencies, and individuals have previously committed to using standardized procedures, and this book is a continuation of those prior commitments.

The book was published in 2009. The results included the following.

- The project was funded by 11 federal and state agencies and organizations.
- Two hundred and eighty-four biologists and managers from 107 agencies, universities, and organizations volunteered their time to contribute to the volume;
- The volume was reviewed three times by different groups of scientists and managers.
- Chapter authors were invited to present their methods at several regional, national, and international symposiums.

(Box continues)

> **Box 6.1. Continued.**
>
> - Comparison data for the chapter containing index averages were obtained from over 4,000 fish surveys from 43 states and provinces (68% response rate of a mail survey).
> - Other countries have expressed interest as well. The lead editor was invited to give a keynote talk at a European Union fish sampling conference, and biologists from six countries reviewed and commented on the methods.
>
> Will these procedures be widely accepted? Time will tell, but reviews of the work to date have generally been quite positive, and several fisheries agencies have started adopting the methods. Cialdini's influence strategies are proving critical to the AFS in meeting its objectives for this project.

your own friends and family. How do you answer such a difficult question? You have three possible choices. You could respond in anger with a statement like "What are you talking about! I work hard to keep people like you from ripping off our natural resources so you can make a quick buck!" You could respond by ignoring the statement or remark sadly "Yeah, I'm really sorry we in government just can't seem to get it together." Or you could comment in an assertive but upbeat manner, "You are definitely right—there are lazy government workers! I'll tell you that I am not one, and I know that those who work with me certainly work hard to make good things happen. I hope you'll understand that our work is about protecting natural resources so your kids will be able to enjoy these in the future as well as meet your needs today!"

As you might guess, the first response would probably alienate your critic and escalate the conflict. The second response would accomplish little except for a loss of your self-esteem. The last response is verbal judo and is a way to use the power of the other person's argument to respond to negative comments and get that person on your side. Although there are times when you will want to respond forcefully to a person's attack or ignore a comment because the battle is not one you want to fight, verbal judo represents a positive way to get your point across while preserving or improving your relationship with the other person.

Verbal judo is about sticking up for yourself without putting down the other person. It is about enforcing the ground rule that honorable people can honorably disagree. Verbal judo uses three steps: (1) understanding the other person's comment, (2) responding with agreement, empathy, or strokes, and (3) diplomatically stating your point of view.

Step 1—Understand the other person's comment. Before you respond to the other person, ask questions to enhance your understanding of his or her position, which enables you to address the expressed concerns.

To understand what led the person to think that government workers are lazy, you could ask some open-ended questions. Open-ended questions are ones that do not have a "yes" or "no" answer. In our example this could be a question like, "Wow, sounds like you have had a bad time with some government workers! Can you tell me what happened?" He might answer, "Yeah, I saw some folks goofing off at one of the city offices downtown—they wouldn't help me." Now you know specifically what the problem was. Other open-ended questions can also draw out a person.

- "Why do you feel that way?"
- "Could you tell me a bit more? What happened?"
- "Hmm, why would you say that?"

Step 2—Verbal judo. When you have drawn the person out and believe that you know what is really upsetting the individual, you can address the problem using agreement, empathy, or strokes. With agreement you agree in some way with what your critic says. The key is that your response has to be true and sincere or your critic will not believe you and the method will not work. Below are examples of how you might agree with the "lazy" statement above.

- "You're right! Some of them are lazy!"
- "Most definitely! Sometimes government workers don't look like they work that hard."
- "Absolutely, sounds like those folks down at the city office were not taking care of business when you were down there."

The good news is that if you are creative, you can agree with almost anything your critic says without lying and still being true to yourself. When someone criticizes you, the last thing they expect is for you to agree. Agreement is likely to turn them away from attacking you and put both of you into a problem-solving mode.

Another way to respond to the criticism is through empathy. Empathy is putting yourself "in someone's shoes" and trying to understand the criticism through that person's eyes. For the criticism above you might give a response like one of the following.

- "Wow, I know exactly what you mean. Visiting a government office and not getting your questions answered is frustrating."
- "I know how you feel. I went to the county assessor's office the other day and couldn't get my questions answered either."

Empathy is a powerful technique that shows the critic that you can sympathize with his or her problem, even if you might not agree with his or her overall statement.

Strokes are compliments and another way to start positive communication. People do good things all the time and often are not recognized for them. Your job is to recognize their talents and contributions honestly. For our critic who thinks government workers are lazy, examples of strokes include the following.

- "You are a really good business owner, so I'm sure seeing employees goofing off must frustrate you."
- "You seem very efficient to me, so I'm sure seeing something like that makes you mad."

Should we compliment people about things that are not true? Everyone does something well. If you keep your compliment honest and about things the person is truly good at you keep your credibility intact and avoid manipulating people.

Step 3—Diplomatically state your point of view. Only after you believe that you understand the person's concern and you have deflected criticism by using verbal judo is the other person ready to hear your point of view. At this point you can diplomatically state your point of view in a positive, assertive manner. State your opinion about the problem, not about the other person. Again, to answer the above example, you could use any of the following statements.

- "Most of the government workers I know bust their butts for very little salary."
- "I really haven't seen that many government workers goof around in my office."
- "There are a few goof-offs in fish and wildlife conservation; however, most of the people I know got into this field because they love their job and generally work many more hours than they are paid."

Step 4—Put it all together. We provide an example from start to finish illustrating an exchange between an angler and one of the authors (Bonar) when he was conducting a creel survey. You will probably get a similar criticism at some time if you deal with resource users. An old man was trout fishing on the bank of a lake.

Bonar: "Sir, can I ask you some questions about your fishing today?"

Angler: "You aren't one of those damned ecologists are you?" (As you can see, things got off to a good start!)

Bonar: "Why do you ask?" **(Inquiry)**

Angler: "Because those ecologists are filled with more c_ _p than a Christmas goose."

Bonar: (upbeat with a smile) "You're right sir, some of them are!" **(Agreement)** "Why do you say that?" **(Inquiry)**

Angler: "Because you guys lowered the catch limit from five trout per day last year to three this year." **(Now I understand his specific problem)**

Bonar: "Yeah, it's a bummer not to be able to catch as many fish this year." **(Empathy)** "This lake just has a lot more people fishing on it, and there are only so many fish to go around. Therefore, so everybody gets a chance to catch some fish, we had to lower the limit." **(Diplomatically stating point of view)**

Angler: "Well if you guys managed the lake right, this wouldn't have to happen."

Bonar: "I know it is tough fishing here for several years under the old limit and now you are not allowed to catch as many fish." **(Empathy)** "I wish our hatcheries could produce more fish for you to catch, but they are producing as much as they can right now." **(Diplomatically stating your point of view)** "Is there any way I could ask you about your fishing? If you are upset, I will be able to take your comments back to the agency. I can't promise changes, but I do promise to take your views back to the right people. I would really appreciate it if you took some time to talk with me!"

Angler: "Well, I suppose so."

Although verbal judo is not always successful, it can increase your odds of turning a negative encounter into a positive one. There are several forms of verbal judo. The above procedure incorporates some of the tactics. To learn more about verbal judo see Burns (1980), Horn (1996), Thompson and Jenkins (2004), or Bonar (2007).

Now you have used verbal judo to tame the conflict and get people talking to you in a productive way. How do you negotiate to get what you want?

6.4 PRINCIPLES OF NEGOTIATION

All of us have negotiated for something at one time or another, even if we are not professional negotiators. You may have negotiated salary with your boss or summer employees, a memorandum of understanding with another agency, or the amount of staff time that will be contributed by a project sponsor. Typically, people think that traditional negotiation (taking a firm position on an issue) is the best approach. You see something you want, you offer an exceptionally low price for it, the seller sets an exceptionally high price, and then you go back and forth until you meet somewhere in between.

Interest-based negotiation is a type of negotiation that is considered to be more efficient than the traditional, position-based negotiation (Fisher et al. 1991; Susskind et al. 2000). In interest-based negotiation, the parties examine their underlying interests and try to find creative solutions that meet the interests of each so that all can walk away from the table with an agreement that satisfies them to some degree.

In traditional bargaining, the parties take a position. For instance, a fisheries manager might take the position, "I will not allow your cattle in the creek." The position of the rancher might be, "I am going to let my cattle walk in the creek. You just try to stop me." They are at an impasse. However, if you look at their underlying interests, or why they hold these opposed positions, you may come up with creative solutions that will meet the interests of both parties. Why does the fisheries manager want to keep the cattle from going into the creek? He wants to protect the spawning habitat of a sucker species that lives there. Why does the rancher want his cattle to walk in the creek? He wants it to be easy for his cattle to drink. By knowing the interests of each party, it may be possible to discuss and negotiate solutions to the disagreements instead of getting gridlocked into only a "cows in–cows out" argument. For example, perhaps they could decide to have a pipe from the creek that goes to a watering trough on shore so the cattle do not have to walk in the creek. A pipe in this instance would not take much of the flow from the stream. Or perhaps the fisheries manager could help the rancher search for grants that allow him to bring water for the cattle from another water source. Interest-based negotiation is based on research conducted by the Harvard Negotiations Project (Fisher et al. 1991). Project participants examined a huge number of successful negotiations and determined why these negotiations succeeded when others did not. They recommend the following procedure.

Step 1—Know and develop your best alternative to a negotiated agreement. Cast doubt on theirs. "Best alternative to a negotiated agreement" (BATNA) is what you have when negotiations either do not occur or fail. Your BATNA might be "if we cannot negotiate a settlement, I will take you to court," or "if we fail to work out an agreement, I will ask my boss to decide." It might also be, "I might not like how things are going, but I will live with the current circumstances."

It is important to have a strong BATNA, or fallback position, if negotiations fail. Throughout the negotiation you are striving to build up your BATNA and casting doubt on the BATNA of the opposing party. For example, to cast doubt on the BATNA of the rancher in the cattle example above, the biologist might ask disturbing questions such as "Don't you think it is really in your best interest to work out an agreement regarding your cattle's use of the stream? Aren't you concerned you could face a hefty fine from some of the game and fish agents and bad publicity if we can't figure something out here?"

Step 2—Focus on underlying interests, not positions. The position of a homeowners' group that wants to kill nuisance algae in a lake is that they want to treat it with copper sulfate. The position of the state water quality agency is that it does not want to treat the lake with copper sulfate. However, if one looks at the underlying interests, there are many similarities:

- both want to control the overabundance of algae;
- both want clean, unpolluted water; and
- both want to avoid a lot of criticism from the broader community living around the lake.

By identifying each of the party's interests, we see both have shared interests with which they may work to arrive at a solution. The key to identifying interests is to ask why the other group is taking a particular position. Although you might not want to share your bottom-line position with the other party, you always want to share your interests.

Step 3—Invent options for mutual gain. Brainstorm as many options as you can that will meet as many of the interests of both parties as possible. The key to coming up with options is that you do not have to divide up an outcome and have everyone take a loss as is the situation with traditional (position-based) bargaining. You can think of your options not as a divided pie, but as an expanding pie designed to meet as many interests as you can. During brainstorming, list as many options as possible but do not comment as to whether they are good or bad at this stage. A group is not held to any option it suggests. Options for the algae example could include

- use copper sulfate this one time and next time treat the lake with something else;
- do nothing and live with the algae;
- plant a buffer of vegetation along the shoreline to prevent nutrients from entering the water;
- find where the nutrients are being produced and clean up those sources;
- add water to the lake to flush the system more rapidly; and
- treat the lake with alum to bind the phosphorus to the sediments.

Step 4—Use objective criteria to argue for "the package" you favor. Here is where you and the other side narrow all the available options to arrive at a particular option or set of options that best meet the interests of both sides. Using objective criteria can help narrow the choices. For example, the water quality agency cannot allow the lake owners to use copper sulfate because the chemical is not currently approved for algae control in lakes in this area, the lake already has an excess of copper in the sediments, and this solution will not meet the interest for clean water for either the homeowners or downstream users. Doing nothing and living with the algae certainly does not cost much, but is unsuitable for either group.

This leaves the last four options listed above. Flushing the lake is a proven technique for reducing algae; however, in this case there is not a ready source of water to use. The lake owners and the state finally agree to a package of treating the lake with alum, finding where the nutrients are being produced and cleaning up those sources, and planting a buffer of vegetation along the shoreline. Doing this meets the interests of both parties—cleaning the lake, keeping pollution down, and satisfying the homeowners.

Step 5—Negotiate as if relationships matter. Finally, to get the best results from negotiation, treat the people on the other side as you would like to be treated. Separate the people from the problem and fight hard to solve the problem, but do what you can at all times to treat the people on the other side with respect. In natural resource negotiations, placing a premium on the relationship with the other party is very important. You will find that across North America the conservation community is small, and even if you move to another area, the people with whom you work have a way of cropping up throughout your life. If you are always "hard" on the problem but "soft" on the people, you stand the best chance of succeeding with negotiations throughout your career.

Go to Fisher et al. (1991), Susskind et al. (2000), or Bonar (2007) for more information on negotiation techniques. Perhaps you have negotiated a solution and need to implement the preferred solution package. How do you get a group to work together to execute the project?

6.5 MANAGEMENT OF A GROUP PROJECT

Congratulations! By getting up and out of the house this morning you successfully completed a project. The goal was to engage in the opportunities of your life outside your home. The objectives were to get ready to leave the house, dress appropriately for the day's activities, and safely commute. You implemented a set of discrete, time-dependent tasks including setting your alarm the night before, making the bed (well, some of you may have), taking a shower, eating breakfast, packing a lunch, and planning your route. And you evaluated whether or not your project succeeded based on several criteria: you arrived on time for your first appointment, you did not crash into anything during your commute, and you are dressed professionally. So, is that all there is to project management? Defining a goal, setting objectives, and accomplishing tasks? Completing a set of steps like this is project management, but it can and should be more, especially when you need to get a group to work together. Project management in a group context becomes a tool for conflict resolution and for creating change.

"Let's form a committee. . ." is often a response to implementing fisheries management. Creating successful group projects from committees requires agreement about the goal of a project; the specific objectives that you will achieve; how money, staff, and equipment will be allocated; what success will mean; and, importantly, accountability, or who is going to do what. Getting agreements means you are also resolving operational questions to get your group to do the actual work.

Project management should also be about creating change, but change is often difficult to achieve. If there is a need to break away from the past and try something new, project management, especially goal and objective setting, can take a frustration about the status quo and turn it into a search for something new. Fisheries managers with good project-planning skills are in a position to exert considerable leadership when those about them are puzzled about the way forward. Below are basic steps of project management you can use to exert leadership. As an example, we will assume you have to prepare a report from your inland fish management division to the state fish and wildlife commission.

Step 1—Paint the big picture. The Organization of Wildlife Planners (www.owpweb.org) has a four-step model that is a useful scheme for thinking about how to conduct a project (see

Figure 5.4, page 145). Their model is a version of management by objectives and stresses the importance of projects going through a repeating cycle to ensure they remain relevant and effective. Consider writing a statement or task list to answer the questions in each of these four steps. For your report to the commission example above, the response to the step "where do we want to go?" on Figure 5.4 may be "to get the commissioners to approve a new set of regulations."

Step 2—Analyze stakeholders. Create a list of people or organizations that care about your project. Consider both the people who will work on the project and the stakeholders who will be affected by the project. Be especially careful to think about who may object to what you are proposing and prepare to deal with these objections. Also, list the role in your project of those who may object so you can include them in the planning. For example, they may be responsible for the work, they may participate in it, they may approve it, they may need to be consulted, or they may simply just need to be informed about your project. For your report to the commission, your stakeholder analysis might include the agency program leads who will be responsible for writing different sections, your boss who will need to approve your report before it goes to the commission, the key commissioners who are most likely to veto your idea, and any stakeholder group that will be affected by your recommendations.

Step 3—Work out "project logic." There are goals, purposes, outcomes, and tasks for projects, all linked by inherent logic. A goal is the highest-level aspiration for all the work (e.g., fisheries conservation). The purpose is the result desired from the specific project being undertaken (e.g., save species *X* and species *Y* from extinction in management areas 3 and 4). An outcome is the specific effect needed to achieve the purpose (e.g., restore the hydrologic regime of the system from river mile 12 to 16). A task is a specific, measurable activity (e.g., replant the six most common species of native plants in riparian zone).

Use if–then statements to clarify project logic and link project activities with the overall goal. Think of it this way: if you do the right tasks, then you achieve the right outcomes; if you achieve the right outcomes, then you achieve the right purpose; and if you achieve the right purpose, then you achieve the right goal. As you move down this logical path from goal to tasks, you are answering the questions about *how* the goal will be achieved. As you move up this logic path from tasks to goal, you are answering the questions of *why* you are doing these particular activities. If your answer is that an activity is not necessary to achieve the next higher level in this chart, delete that activity. An example of logic progression for your commission report shows how project activities are linked (Figure 6.1).

Step 4—Create a project time and activities chart. Gantt charting (Morris 1994) is useful for displaying an overall picture of the project tasks. This is done by listing tasks on the left of a matrix and drawing the dates that work begins and ends for each task on the right (Figure 6.2). The tasks are sequenced to ensure that things get done in the proper order and that there is enough time scheduled to complete each task. This charting activity is a good way to brainstorm all the tasks that need to get done in a project.

The task list can get large. If this happens, managers should evaluate the importance and urgency of each task and separate the "nice to have" from the "critical to have" tasks. The criteria for making this decision include (1) how firmly the task is linked to other tasks; (2) the potential for a task to solve multiple implementation problems; or (3) cost versus benefit. The objective of this exercise is to identify which tasks can be omitted if funding or time is short. For example, having reviewed the if–then logic for the commission presentation (Figure 6.1), it might be nice to have an attractive multimedia presentation, but it is not necessary for success. Therefore, this step may be carried out only if there is time.

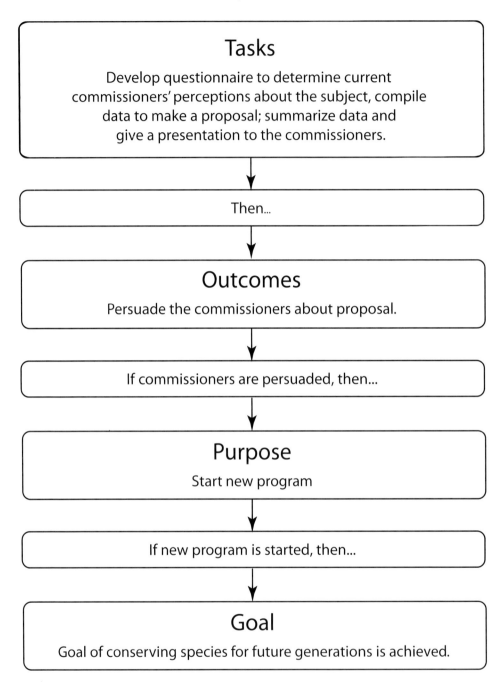

Figure 6.1. Project logic progression for a commission report. This example shows how project activities are linked to the goal using if–then statements.

Step 5—Analyze your assumptions to manage risk. Committing to do a project task also commits a fisheries manager to one or more assumptions that the task can get done as planned. Common in fisheries is the assumption that work will be infrequently disrupted by bad weather. Perhaps the project plan calls for installing sampling equipment on June 1st. What could be done to maintain the integrity of the overall project if a flood prevents instal-

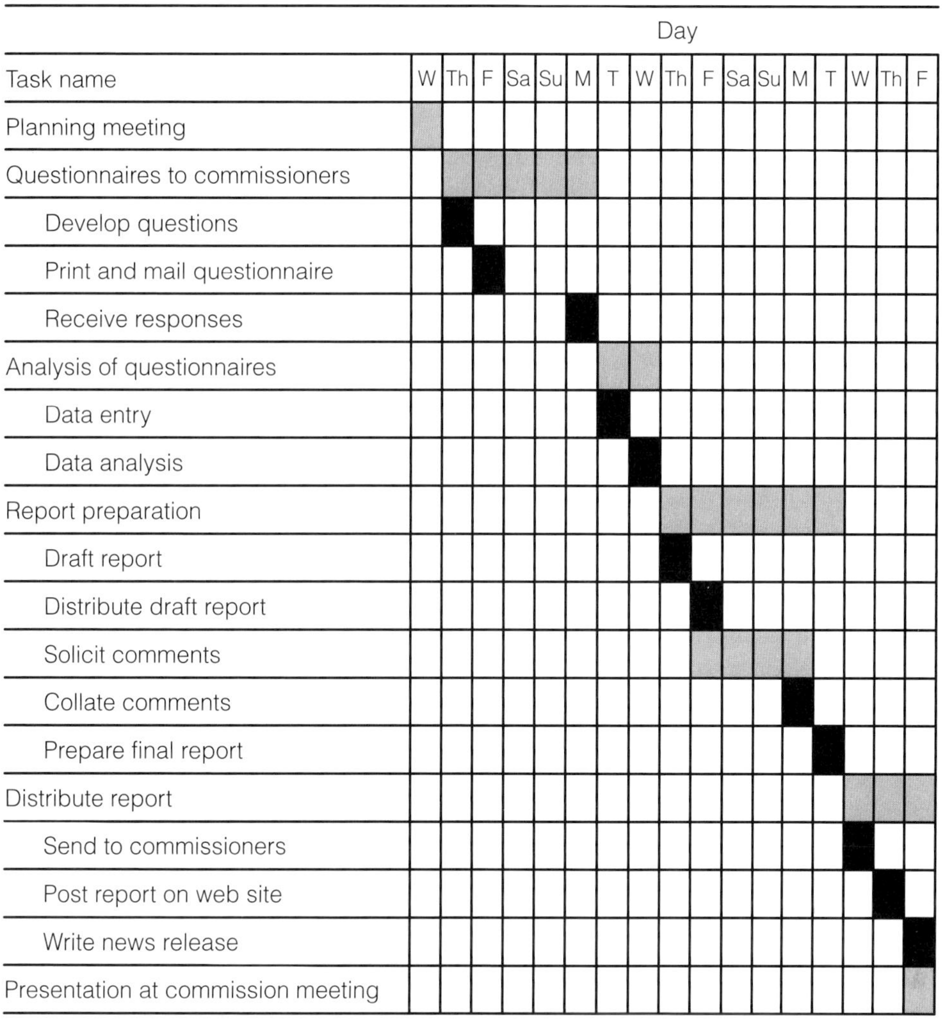

Task name	Day																	
	W	Th	F	Sa	Su	M	T	W	Th	F	Sa	Su	M	T	W	Th	F
Planning meeting	■																
Questionnaires to commissioners		▨	▨	▨	▨												
Develop questions		■															
Print and mail questionnaire			■														
Receive responses						■											
Analysis of questionnaires							▨	▨									
Data entry							■										
Data analysis								■									
Report preparation									▨	▨	▨	▨	▨				
Draft report									■								
Distribute draft report										■							
Solicit comments											▨	▨	▨				
Collate comments													■				
Prepare final report														■			
Distribute report															▨	▨	▨
Send to commissioners															■		
Post report on web site																■	
Write news release																	■
Presentation at commission meeting																	▨

Figure 6.2. An example of a Gantt chart (resource and time allocation matrix) for preparing a hypothetical presentation to a fish and wildlife commission.

lation until June 15th? Think about the inherent assumptions associated with the project and anticipate how to manage things if the assumptions prove wrong. This is risk management. In our example of making a presentation to the commission, it may be assumed that the commissioners will understand the meaning of a specific chart or data you present, but it is known that two of the seven commissioners did not complete high school. Therefore, the assumption that they will understand the data to be presented may not be true. Therefore, risk could be managed by calling those two commissioners before the meeting to discuss the data and answer any questions they might have.

Project management in contemporary fisheries management is not just a task list. It is increasingly about getting groups to cooperate and leading them to implement new or more effective ways of conserving fisheries resources. A person with the desire to achieve positive outcomes for fisheries resources can use project planning skills to exert considerable leadership (Box 6.2).

So, now you have a project plan. How will you present it to your commission?

Box 6.2. A Conversation with a Project Planning Expert

Terry Schmidt is an internationally-recognized expert in planning. A former engineer, he is the founder of ManagementPro (www.ManagementPro.com), part of the Centre for Strategic Management, a global alliance of experts in applying systems thinking to strategic management. Terry's fifth book, *Strategic Action Planning*, is about connecting day-to-day work to strategic intent. The follow are excerpts from Mike Fraidenburg's conversation with him.

Fraidenberg: Imagine this scenario. I am a fish biologist sitting next to you on an airplane. After I learn about your work I do a bit of complaining that I am attending ever more "coordination" meetings that do not coordinate much of anything. As a matter of fact, when I get off the plane I have to lead a meeting of diverse folks who are supposed to plan a fish restoration, but they are not working well together. What would you say if I asked for advice about how to manage this meeting?

Schmidt: First, stop the negative self-talk. Before you can manage others you must manage yourself. Otherwise you create a self-fulfilling prophecy that you can't create positive change. Then set up the usual process for having a good meeting: agenda, shared expectations, clear roles, active decision making, and plans for a "what comes next" discussion. Then guide your group to answer four questions by the time the meeting is over.

1. What are we trying to accomplish and why? Get agreement about "why" you are doing the project, such as "to restore the native fish fauna between river miles 3 and 12 and make it sustainable." The answer to why is consensus building about the project purpose (i.e., what is the goal we share). The answer to "what?" provides the list of objectives that need to be achieved (i.e., what are the major deliverables that will ensure a sustainable population).
2. How will we know if we are successful? Get agreement on what success looks like and the measures you will use. What quantity, quality, time, and cost measures might apply? Is adding one fish to the population within 10 years sufficient? Probably not. Then what are the measures? The answer provides a clear and objective way to track project success.
3. What other conditions must exist? The answer to this question illuminates assumptions we are making. Assumptions influence success but are often hidden or ignored. Assumptions are the other things that must come true for our plan to work. Every project carries assumptions, and these have different risks. When a risk is large or an assumption is uncertain, incorporate contingency steps in your plan to manage these.
4. How will we get there? This is the schedule of tasks and list of materials, people, timing, money, and consultations that you put together to get the actual work done in the field. People tend to jump to action steps first. Do not let your group do that. Answer questions 1–3 first. These four questions form the basis of the logical framework, a systems-thinking tool that ensures successful project management. Readers can go to my Web site for a free article on this topic.

(Box continues)

> **Box 6.2. Continued.**
>
> One hundred years from now, when people look back at us from the condition of the environment we've left them, will they ask, "Why did you let this happen" or will they say, "Thanks for your good work"? That is why project planning has a place in every manager's toolbox. Project management is all about planning to change, executing that change, planning more change, executing again, and planning to change yet again.

6.6 THE SUCCESSFUL PRESENTATION

No matter who you are—a student, biologist, or administrator—you will be asked to give a talk. Unfortunately, a poorly designed talk can bore an audience and hurt your ability to communicate and influence people. Remember the speaker described at the beginning of this chapter. You do not want to be that kind of presenter. Instead, consider these tips when developing talks in order to get your point across and keep the audience engaged (Fraidenburg 2005).

Most talks by fisheries managers are now supported with computerized presentation programs such as Microsoft Office PowerPoint®. A few speakers use overhead transparencies or slides in carousels. Regardless of what technology is used, most successful talks have the same characteristics.

Step 1—Remove anything that distracts the audience or departs from your message. Keep the talk simple! Slides and other graphics should contain only necessary information. If you are not sure about something, leave it out. Simplify the information on the slides. Just include the key points and orally fill in the details for the audience. If people are reading a lot of extra information on a slide, they are not listening to the talk. When presenting a computerized slide show such as PowerPoint, resist the urge to add numerous flashy transitions, color explosions, and "flying pages." The audience might be impressed by your computer wizardry, but they are likely to miss your message.

Step 2—Manage overload. Most speakers are able to use the commonly allotted 20 min for a talk at a scientific meeting. In fact, it is easy to fill even hour-long lectures. More often, the task is to manage overload (i.e., not to cram too many words into a short presentation). If you are struggling to figure out how to prepare a presentation, analyze precisely what the audience needs to know and how to use the limited time. The audience relies on the speaker to conduct high-quality, relevant work and present the most important points.

Second, manage the amount of information put on each slide. The audience is depending on the speaker to manage its limited ability to assimilate. Ask yourself, "What do I want the audience to remember?" Most of us remember about three things from a presentation (Arredondo 1990). Do not try to have too much information on one slide and if the slide presents only text, consider replacing it with a picture or drawing that conveys the message in a different way. That way the audience can simultaneously "hear" the message and "see" it in a different way.

Step 3—Deliver your information so it is easily assimilated by the audience. An audience can silently read a slide to themselves much faster than the presenter can read it out loud. Therefore, if you simply read line after line of text from bullet slides, people's minds will drift and they will loose track of what you are saying. This was illustrated by Mayer (2001) who discovered that information transfer diminishes when a narrator's words appear on a slide at the same time the words are being spoken. Ask yourself if the audience really needs you to interpret what it can see (e.g., like a tour guide) or if you are simply repeating information it can easily read. Keep text slides to a minimum and when used highlight only the most important points.

People learn in different ways, so when possible vary your delivery style. Some people learn best from hearing information, others need to see it, and still others need to feel it (Druckman and Swets 1988). Advice from people who study how adults learn is that a speaker should vary communication modes at intervals of 8 min or less. Optimally a 20-min presentation should include three or four distinctly different communication modes to maximize impact. Different communication modes include talking, showing some slides, creating a small group exercise, interviewing some audience members, walking into the audience and telling a story, using humor, using a prop, presenting a demonstration, asking a question, or doing something altogether different. Of course, given the constraints of giving a presentation at a scientific conference or other type of formal meeting, using all of these modes is not possible. However, adding variety is possible, even at these venues, if the presenter is creative. Variety will keep a talk interesting and relevant for a larger number of people.

Manage the room where the talk is given. Do not completely darken the room. If there are concerns about how well images will project, fix the images to project well in a lighter room instead of putting the audience in the dark. When giving a talk in a dark room, images should be the focus of the presentation, not the presenter. Also, when in a dark room the speaker looses eye contact and the ability to respond to the audience's nonverbal behavior—such as members of the audience falling asleep! If the speaker plans to engage the audience in something besides listening, the speaker should have the meeting organizers arrange the room and seating configuration to meet the communication needs. And finally, if problems with the room arise (e.g., excess noise from the kitchen) it is usually better to deal with the problems than to try to ignore them.

Designing effective slides and graphics can dramatically improve how your information is delivered. Essentially, bigger and simpler are better. Slides should be designed to be understood in 6 seconds or less, contain one point per slide, and feature consistent transitions. A few design elements will improve a majority of presentations (Table 6.1), and using these and some of the tips discussed in Box 6.3 will put you in the best position to influence others.

So, now you have presented well to your audience to meet its needs and get members to engage in your topic. How would you write an effective report or publication to get your point across to a larger audience?

6.7 THE SUCCESSFUL PUBLICATION

Publications are the primary means to transmit information in the scientific professions. A good publication record is necessary for advancement in many positions, especially for those involved in research. Therefore, knowing how to write quickly, clearly, and accurately is important for any fisheries professional.

Writing a publication can be a daunting task for the beginning fisheries professional. Many scientists do not like to write. Charles Darwin said that "a naturalist's life would be a happy one if he had only to observe and never to write" (Trelease 1958), but fortunately for us he conquered his dislike of writing and provided us with *On the Origin of Species* and other

Table 6.1. Design elements for effective slides (from Fraidenburg 2005).

Design feature	Description
Headers	Use no more than five words.
Bullet list	Use no more than five items.
Structure for phrases	Use parallel structure (e.g., begin every bulleted phrase with a noun or verb).
Font size	Make titles > 30–36 points; subtitles > 28 points; text > 24 points
Font design	Use sans serif fonts such as Arial, Tahoma, Franklin Gothic, or Avante garde; avoid serif fonts such as Times and Palatino.
Font colors	Use white to yellow for dark backgrounds and black to dark blue for light backgrounds; avoid red and green to ensure readability to color-blind audience members.
Contrast	Use either most common—dark background, light-colored text—or less common—dark text, light background, which is sometimes useful in rooms with considerable ambient light.
Images	Scan with a resolution of 100–150 pixels per inch; an image size of 640 × 480 pixels is nearly ideal; use JPEG and TIFF image formats for the majority of computerized presentations.

work! An experiment or study is not concluded unless it is written up in a report or publication, and learning a few tips makes writing much easier.

The style of scientific publications is much different than are other types of writing. Most fisheries research articles are written in the IMRAD (introduction, methods, results, and discussion) format (Day and Gastel 2006). This format makes it easier for both the reader and the author because there is a clear template in which to present the information. The object of scientific writing is brevity and clarity: present information as clearly and succinctly as possible so others have enough information to understand your work, repeat the experiment if necessary, and judge scientific quality.

Each element of the IMRAD format can be improved by careful adherence to details. Day and Gastel (2006) recommend that introductions state the nature and scope of the problem and review the relevant literature to orient the reader. Furthermore, they recommend that the introduction state the principle methods, results, and conclusion of the study, but others leave these statements to their respective sections further in the paper. Introductions are typically written in the present tense.

Other sections describe the work itself in more detail. The methods section describes procedures used to conduct the study, which should be repeatable by other competent fisheries professionals. Results are the core findings of the work, presented in crystal clarity. Day and Gastel (2006) recommend that an overall picture of the study be presented in the results as well as representative—but not all—data. Methods and results are usually written in past tense. The discussion section conveys what the results mean. Day and Gastel (2006) state that good discussions: (1) discuss but do not recapitulate the results; (2) point out any exceptions or lack of correlation and define any unsettled points; (3) show how results and interpretations agree or contrast with previous published work; (4) discuss the theoretical and practical impli-

Box 6.3. A Conversation with a Presentation Expert

Vanna Novak is an expert in designing and delivering presentations that motivate audiences to take action. As owner of Speak to Persuade™ (www.speaktopersuade.com) and a board member of the National Speakers Association, Vanna has spent the past 20 years training people to deliver presentations that influence decision making. The following is an excerpt from Mike Fraidenburg's conversation with her.

Fraidenburg: Why are "'persuasive" communication skills important for fish managers these days?

Novak: In our system, most natural resource management is done by government employees. There was a time when a government authority could stand behind a lectern and be automatically perceived as credible. We no longer instinctively trust these "talking heads." Today it is difficult to build trust and capture attention when there is a perceived barrier between you and the audience—including simple things like standing behind a lectern. Instead, use a strategy to connect with the audience, a style that communicates, "'I have nothing to hide . . . I'm accessible . . . I'm not afraid to have a conversation with you." That is where trust begins.

Think about presenting to persuade versus to inform. An informative speaker has no investment in what the audience does. The moment you care about what the audience does, your presentation becomes persuasive. At that point one enters the realm of pathos not just logos—that is, delivering an appropriate mix of emotional content along with appeals to logic. This approach can be something as simple as using fewer charts and more pictures or as simple as talking less about data and more about mission and why that sense of purpose is good for society. So, persuasive communication is really about connecting with people at some level of authenticity.

In my workshops I talk about three principles for being more persuasive. First, be appropriately unpredictable. That means craft both a delivery and structure for your speech that is meant to keep the audience tuned into what you are saying. If you only prepare the content and then assume people will listen, you risk lulling the audience into indifference. Be stimulating. Be different from the other speakers. Second, balance rational appeal with emotional appeal. People understand how to take action from the facts you deliver. They are motivated to act by the emotion they feel for your topic. Use some techniques like explicitly describing benefits, telling stories, or giving real-world examples. These are appeals to the pathos or emotional side of their decision making. And third, know what your audience wants and needs more of or less of in their lives. Then speak to these. People are reluctant to act unless they perceive a gap. The gap also needs to be theirs, not mine. My gap is not important to them until they feel the same way I do about an issue.

(Box continues)

Box 6.3. Continued.

Fraidenburg: Give me a flavor for what you mean by persuasive skills. Is it selling or something else?

Novak: I do not teach selling. Selling is more about technique and less about the personal importance you place on the message. Becoming a more persuasive communicator is about passion; having a genuine belief in what you are talking about combined with a desire for that message to get through because it is important for your audience to take action. From what you've told me about your readers, they clearly seem to have a deep passion for their mission. That is the kind of distinction I am talking about. Persuasive speaking is communicating that important mission, not just selling an idea.

Fraidenburg: How does one become a more persuasive communicator?

Novak: First, have courage to experiment. Just because you've always communicated a particular way does not make it the best way. Being an effective communicator does not have much to do with what you feel comfortable doing. Clients ask me if I want them to be somebody they are not. The answer is, of course, no. But, realizing that the behaviors that make up who we are today are a collection of patterns and habits we put together over time is important. Just because these routine behaviors come naturally—feel normal—does not mean they are effective. Stay the way you are now if you are getting the results you want. If not, think about what changes are needed to get the effect you want. Assess the negative stereotypes your audience holds about you, your profession, or your current communication model. Give them the unexpected—a reason to sit up and pay greater attention. Get feedback from three to five people. Look for consistencies among their evaluations and try some changes to deal with these. Do more of what they thought worked in your presentations and change the things that turned them off. More than anything, a speaker needs to have an experience with him or herself. By that I mean have someone video tape you and then analyze what worked and what did not. I find watching a tape twice—with and without sound—to be particularly revealing. When being informative is not enough, speak to persuade. It is a way to improve your personal effectiveness.

---

cations of the work; (5) clearly state conclusions; and (6) summarize evidence for *each* conclusion. Discussions are usually written in a combination of present and past tense. Published work is cited in present tense, while the author's work is cited in past tense. Many discussions are too verbose—here is a good opportunity to eliminate wordiness. References such as Day and Gastel (2006) and the latest edition of *Scientific Style and Format* (Council of Science Editors) include lists of words that can replace phases to reduce wordiness and jargon.

Graphs and tables can enhance manuscripts but are much more costly to produce than is text. If you must present repetitive data, use a table. If the data show clear trends or produce an interesting picture, use a figure. If there is a small amount of data, present them in the text.

Writing quickly and accurately is a skill that improves with practice. Do not agonize over the grammar of each sentence when writing the first draft. One option is to write as quickly as possible to get basic thoughts on paper and then rework the manuscript until grammar, punctuation, and spelling are suitable. Write your introduction and methods when you are still working on the experiment. They will be fresh in your mind when you are conducting the experiment and you will not be swamped with work when the experiment is over.

Use active voice over passive voice wherever possible. "We found" is much clearer and less wordy than "it was found." Passive voice was used in the past because some thought it conveyed more humility than does active voice. However, modern scientific writers usually do not adhere to this convention.

References in most fisheries journals are included in a literature cited section at the end of the manuscript. Reference format varies by journal, and most journals have instructions online that are readily accessible to authors. Proficiency using reference software (e.g., Endnote®, Carlsbad, California) can speed entry and formatting of references because this software automatically formats references in the style of the particular journal. Make sure all references in the text are cited in the literature cited section and vice versa. Missing entries indicate sloppy work to editors and reflect poorly on the accuracy and care with which the study was conducted.

Several references give detailed information on how to write a scientific article. See Day and Gastel (2006) for techniques on scientific articles in general and Hunter (1990) or the American Fisheries Society publications Web site (www.fisheries.org) for information on writing articles for fisheries journals. The latest editions of *Scientific Style and Format*, *The Chicago Manual of Style* (University of Chicago Press, Chicago), and Strunk and White's *The Elements of Style* (4th edition, Allyn and Bacon, Needham Heights, Massachusetts) can provide valuable tips.

Once you have written the article, it is ready to submit for publication. Become familiar with the submission procedures of the publisher. Have colleagues examine and edit the article before you submit it. They can often catch mistakes that you cannot when you have been working with a manuscript for a long time. If you are submitting to a journal, hope for acceptance and prepare for rejection. Rejection is common and a necessary part of scientific writing. If rejected, use the comments to improve your article and submit it somewhere else. At least 35 articles that would eventually earn a Nobel Prize were rejected outright during the initial inspection by reviewers (Campanario 2002), so do not give up with the first rejection. Now that you are able to communicate your work orally and in writing, knowing how to use other collaboration and communication techniques to implement your findings is important.

6.8 EFFECTIVE MEETINGS

People are often reluctant to attend meetings because we all have been to many bad ones. Even so, people still communicate best in person despite the many electronic tools now available. This is especially true when there are differences of opinion. Therefore, if managed correctly, meetings can be important for exchanging information.

You will need to determine if the meeting will be a conference, where presentation skills are the focus and most of the attendees are listening; if the meeting will be a conference call or small board meeting, where there is free exchange among the participants; or if the meeting will be somewhere in between. In the following section we discuss tips for managing smaller, interagency or public meetings, where there is considerable interplay among participants. Presentation skills provided in section 6.6 will help with the larger, more presentation-oriented meetings.

In meetings, most important is the quality of work generated, not the time spent. If you ensure people in meetings efficiently share information or make solid decisions, meetings become catalysts for change. Begin by focusing everyone's efforts on the desired outcome of the meeting.

Step 1—Have a purpose for the meeting. Planning effective meetings begins with defining the specific deliverables that must be produced. Pretend the meeting has just ended and it was wildly successful. What did successes look like? What essential topics were covered? What essential problems were resolved? What decisions were made? What products resulted that others were waiting for? In short ask, "What deliverables are needed that make this meeting necessary?" Once the purpose of the meeting is clearly understood, you can focus the activities of the meeting participants.

Step 2—Write an agenda. An agenda is needed for meetings. The most effective show logistical details (e.g., time and place), state clear meeting objectives and outcomes, are ruthless in jettisoning or relegating to last all unimportant topics, clearly notify participants of any premeeting preparation required (name names if it creates greater accountability), provide a logical sequence of topics, and estimate an appropriate amount of time to complete each part of the meeting (Table 6.2).

Above all, a good agenda does not confuse content and process: *content* is the task-related aspect of a meeting—it describes what participants will discuss. *Process* is the method the group uses to tackle the task—it describes how participants will proceed. Both must be managed carefully; otherwise expect poor-quality meetings.

Step 3—Assign tasks. You would probably not implement a large research project by simply calling people together and asking them to start working on the project. Everyone would be assigned jobs. Similarly, at the best meetings, roles for leadership, support, and participation are assigned. Typical meeting roles include the chair, who may not run the meeting but serves as the meeting "sponsor" and an active participant. The chair sets constraints for the meeting and often approves the agenda but does not use his or her power to trump others. The recorder writes down the ideas and decisions of the group in a form that can be easily used after the meeting and does not insert his or her ideas without approval of the group. The facilitator manages the meeting process, keeping the conversation focused and everyone on task. This person does not stray into content but suggests methods for managing the meeting and coordinates with the

Table 6.2. Example of tasks listed on a meeting agenda.

Time	Topic	Person	Process	Product
1:00–1:15	Meeting roles	Mike	Presentation	Assigned responsibilities
1:15–1:45	Meeting purpose and objectives	Scott	Facilitated discussion	Agreement on specific deliverables from this meeting
1:45–3:00	Deploying water sampling equipment	Group	Brainstorming, categorizing, group discussion	Map showing location and scheduled installment dates

recorder and chair to ensure relevance of the meeting outcome. Balancing power among the people in the meeting and helping plan meeting logistics are other duties of the facilitator. Meeting participants keep the facilitator and the recorder in their neutral roles. Participants should ensure quality control of the content and completeness and argue actively for their own point of view. Participants contribute in good faith, owning responsibility for achieving the meeting's purpose. The supervisor may or may not attend, but when present should not let the group give answers designed to satisfy him or her. The supervisor establishes the larger-order purpose for the meeting (e.g., to implement the strategic plan) without steering the method that will be used. Additionally, the supervisor may make the final decision, especially if the group is unable to come to a final decision, or may approve the final decision made by the group.

Step 4—Follow up the meeting. There will be few results from a meeting if there is not follow-up or accountability. Make clear decisions about what happens after the meeting by creating an assignment list that uses the formula, 4W + F + H = accountability. This formula includes the following components: *what* (be specific in defining the task), *who* (identify people assigned to the job and the lead), *when* (specific deadlines for tasks), *where* (frame any sideboards that bound the assignment), *format* (specify format and method of delivering result), and *how* (usually left to the people in charge of task but, if necessary, specify how work will be done). Last, evaluate how well the meeting worked so you can improve future meetings. Brainstorm with the participants a list of all those things that worked well in the meeting and those that did not.

Effective meetings generate a sense of involvement among the people who have devoted time and creativity; create a stronger team; and improve communication, quality and quantity of decisions made, and information shared. Create meetings that are catalysts for change. Now you have been exposed to skills to present your information and work within small groups. How can you create change in an entire community?

6.9 COMMUNITY CHANGE THROUGH COMMUNITY PARTICIPATION

Sometimes a fisheries manager has to go beyond convincing one person or a small group of people to change and must change the behavior of an entire community, city, or state. Changing an attitude of an entire community on large, controversial issues can be facilitated through community participation (Box 6.4; see also Box 5.5).

How would you go about setting up a process to have an entire state write a plan for managing its contribution to global warming? Tom Peterson, who founded the Center for Climate Strategies, helps state governments do just that. Below are seven principles of stakeholder self-determination used by Peterson that can work for fisheries managers. This material is from an interview with Peterson that appears in Fraidenburg (2007).

Peterson suggests that managers find opportunities by meshing a natural resource need with people who accept that the present situation will not work in the future and who are ready to accept that they need help. The fishery manager can become the "go-to" person by positioning, which essentially means being in the right place doing the right thing. Once the fishery manager is so positioned, principles to implement change can be used.

Peterson advises the first principle of implementing change is to decouple people from their own history. One major barrier to change is that people just accept hand-me-down beliefs without much examination. Only a relatively small percentage of people are insightful about science issues in general and natural science in particular. As a result, most processes initiating change need to begin with fact finding to broaden everyone's understanding that the

Box 6.4. Successful Change in Community Opinion Facilitated by California Fish and Game Department

In 1997, things did not look good for the California Fish and Game Department (CFGD) in Plumas County. Members of the community vehemently protested their planned use of rotenone to kill illegally introduced northern pike in Lake Davis. Community members marched on the state capitol, threatened CFGD fisheries managers with arrest, chained themselves to buoys in the lake prior to treatment of the lake, and pelted CFGD employees with Halloween candy during the treatment (Bonar 2007). When chemical residues of the rotenone treatment persisted in the lake longer than predicted, CFGD settled on a payout of US$9.1 million to the community. However, when another treatment was required in 2007 to kill northern pike that reappeared, there was widespread community support, even from some of those officials who had chained themselves to buoys during the 1997 treatment. Why did public opinion change so drastically?

The CFGD learned from past mistakes. Within 2 years of the 1997 rotenone treatment, northern pike were again seen in the lake. The CFGD staff moved to an office in town to be closer to the constituents. They held dozens of meetings, working diligently to educate the public and include them in their decision making (Fimrite 2007; McKinley 2007; Ritter 2007). The CFGD discussed options at these meetings, listened to public input, and tried to address the public's concerns. Doing nothing about the newly found northern pike was considered but quickly dismissed—town businesses were impacted severely by their presence, which curtailed the lucrative trout fishery. In addition, there was fear that northern pike could escape downstream to areas supporting imperiled fishes and wildlife susceptible to northern pike predation. Other methods besides rotenone, including electrofishing, nets, and explosives, were first tried but were unsuccessful in ridding the lake of northern pike. Although 65,000 northern pike were removed, many still remained in the lake. By this time, the public realized another rotenone treatment might be one of the few solutions left. However, many things were done this time to make a rotenone treatment more palatable. Previously, Lake Davis supplied drinking water to the community. The CFGD had drilled water wells so that groundwater could supply Portola with water. Residents were concerned about the synergists used with the liquid rotenone in 1997. For the 2007 treatment, a different, less-controversial formulation of rotenone was used. The Plumas County Public Health Agency and the California Department of Health Services determined that the treatment plan would not adversely affect the public. The U.S. Forest Service, which owned the land surrounding the lake, was asked to support the treatment and close the area during treatment.

Bill Powers, one of the local officials who previously chained himself to a buoy, stated, "In '97 there were secrets; there were unknowns. The more the local government, people like myself, asked questions, the more we were stonewalled. This time, every question we asked has been answered." Furthermore, Powers stated, "I've seen the evidence now that there are no public safety issues and that it is a necessary evil. I wish we didn't have to do it, but we do. In 1997, it all came down to no communication."

For the second treatment, the CFGD implemented a classic example of how to create community change through community participation. The result was a treatment of Lake Davis with community support, not condemnation.

problems and solutions are potentially far greater than first impressions suggested. Adopt the ground rule that points of view be stated as objectively as possible followed by an explanation, in some factual form, why that view is accurate. If the manager thinks something is too expensive, is that based on a study? If so, what study? Why is this approach a better way to look at the question or problem?

Providing a degree of self-determination is a safety net for people—that is the second principle. Peterson explains that most people are afraid of change and of rearranging the power dynamic. Democracy does that. It says that if you have a little fiefdom and you join a democratic group, you may no longer be the big fish in the little pond. People are more willing to rearrange the power dynamic when choices are voluntary. Do not avoid hard issues but simultaneously confer two responsibilities—problem definition and solution finding—in a process in which people can decide for themselves.

Peterson explains that getting diverse people to a new, shared place is where real change happens. This is more than compromise; it consists of finding a new way, or a constructive alternative, that the participants think is better than leaving things alone or "splitting the difference." Be prepared for people who say they want collaboration when, in reality, they are just pursuing their own agenda. In the end, though, creating a good faith process is the third principle.

Then build confidence that there is a way to move forward. Psychologically, if people hear there is a problem but they do not hear solutions, they will tune you out. In climate change work, the message about global warming and the science behind it are frightening. People say, "Great. You just laid a really big bummer on us. Now we feel hopeless so we give up." The conversation is different when you can follow the problem statement with concrete, workable ideas on how to make progress. So the fourth principle is to anticipate that complaints about insufficient information will emerge as a reason not to deal with tough issues. It is the manager's job to be there with the information laying out problem, solution, and process in the same breath. Managers who do not build that kind of capacity for technical support risk failure.

The fifth principle is leadership. Convening a large, controversial process requires somebody in a position of great authority to say, "Managing this issue is important. I'm not satisfied with where it stands right now. I want to get folks around a table to help me make the right decision." That kind of commitment tells everybody that business as usual and hugging old turf are not alternatives, and things are going to change no matter what. Find someone in a leadership position who can champion fisheries work if it is large and controversial.

With all of the above in place, the best available alternative now becomes figuring out how to work effectively with the new crowd. Of course, the first thing participants want to do is steer the process in their own direction. One should expect this behavior at the beginning. And, of course, the participants try to manipulate each other at some point. So the sixth principle is to be prepared for steering behaviors. One thing to do in response, which is typical in processes like these, is to establish some criteria for how your group is going to make decisions. For issues like climate change this typically involves costs and benefits. Establishing decision criteria amplifies the importance of being able to do expert technical analysis. Managers that have technical staff available can separate out all the different, intertwined issues. Getting specific is the key to finding potential solutions. Expect to dissect general criticisms constantly to find the specific cause of the heartburn. Only then can you intervene effectively. Negotiating decision criteria up front and making people be very specific about their concerns are pretty successful counters to steering behaviors.

The seventh principle is to maintain full transparency and full inclusion so everybody's voice gets heard and everything is up for consideration. Be careful to protect this credibility of your process. Decisions should not be made secretly. If that principle is violated, the discussion explodes. A common way things go wrong is succumbing to pressure to create private side conversations to cut deals. Instead, get a written agreement on this point to serve much like a contract. Then, if a party wants the advantage of private deal cutting, you have a ground rule upon which to fall back.

Community participation increases understanding of not only issues and trends that cause problems but also the dynamics of how positive change occurs and becomes permanent. Mobilize communities and increase public confidence in fisheries management by involving them in decision making.

6.10 CONCLUSIONS

A personal point of satisfaction for the authors is seeing people we know grow as communicators. We see these colleagues gain confidence and show conservation leadership that changes the future in positive ways by being verbal judo black belts who transform difficult conversations into positive dialogs; by knowing how to use successful techniques ethically to influence the thinking and behavior of others; by being apt negotiators who get their needs met while doing what they can to meet the needs of others and, at the same time, build positive relationships; by possessing the ability to conceive and execute projects that involve others in productive work; by delivering informative and engaging presentations; and by working with their communities and causing them to change in ways that are more conservation minded about valuable fisheries resources. These people get things done.

We all know people who seem to be naturally good at these skills, but everyone gets better with practice. No one is an expert the first time they put on ice skates. Success takes some falls and practice. The same is true of communication skills. Realize you will get better the more you practice these skills, and do not berate yourself if you suffer the occasional setback. If you are interested in learning more about these skills, look in the communication, psychology, and business sections of your bookstore or library. These disciplines have many good references on these topics and many of those we have cited in the above sections.

Working with people is an exciting area of conservation! Knowing the science of our profession is necessary, but by itself, insufficient. Adding communication skills allows you to be more successful, makes your job more fulfilling and helps you make a difference in fisheries management!

6.11 REFERENCES

American Friends Service Committee. 1970. Uncommon controversy, fishing rights of the Muckleshoot, Puyallup, and Nisqually Indians. University of Washington Press, Seattle.
Arredondo, L. 1990. How to present like a pro: getting people to see things your way. McGraw-Hill, New York.
Bazerman, M. 2006. Judgment in managerial decision making. Wiley, New York.
Bonar, S. A. 2007. The conservation professional's guide to working with people. Island Press, Washington, D.C.
Bonar, S. A., and W. A. Hubert. 2002. Standard sampling of inland fish: benefits, challenges, and a call for action. Fisheries 27(3):10–16.

Bonar, S. A., W. A. Hubert, and D. W. Willis, editors. 2009. Standard methods for sampling North American freshwater fishes. American Fisheries Society, Bethesda, Maryland.

Brouha, P. 1993. The emerging science-based advocacy role of the American Fisheries Society. Journal of the North American Benthological Society 12:215–218.

Burns, D. D. 1980. Feeling good: the new mood therapy. Signet, New York.

Cialdini, R. B. 2001. Influence: science and practice, 4th edition. Allyn and Bacon, Boston.

Campanario, J. M. 2002. A new approach to making scientific journals actively compete for good manuscripts. European Science Editing 28:78–79.

Day, R. H., and B. Gastel. 2006. How to write and publish a scientific paper, 6th edition. Greenwood Press, Westport, Connecticut.

Druckman, D., and J. A. Swets, editors. 1988. Enhancing human performance: issues, theories, and techniques. National Academy Press, Washington, D.C.

Fimrite, P. 2007. Thousands of fish go belly up as poisoning of Lake Davis starts. San Francisco Chronicle, September 26, 2007.

Fisher, R., W. Ury, and B. Patton. 1991. Getting to yes: negotiating agreement without giving in., 2nd edition. Penguin Books, New York.

Fraidenburg, M. E. 2005. Snooze alarm! Avoiding PowerPoint perils. Fisheries 30(5):34–38.

Fraidenburg, M. E. 2007. Intelligent courage—natural resource careers that make a difference. Krieger Publishing, Malabar, Florida.

Horn, S. 1996. Tongue fu! How to deflect, disarm, and defuse any verbal conflict. St. Martin's Griffin, New York.

Hunter, J. 1990. Writing for fishery journals. American Fisheries Society, Bethesda, Maryland.

Maslow, A. H. 1970. Motivation and personality, 2nd edition. Harper and Row, New York.

Mayer, R. E. 2001. Multimedia learning. Cambridge University Press, Cambridge, UK.

McKinley, J. 2007. California officials tackle a toothy lake predator. New York Times, September 12, 2007.

Morris, P. W. G. 1994. The management of projects. Thomas Telford, London.

Noss, R. 1999. Is there a special conservation biology? Ecography 22:113–122.

Ritter, J. 2007. Calif. hopes to hook lake's pike problem. USA TODAY, August 23, 2007.

Roberts, C. M. 2004. Advocating against advocacy in fisheries management. Trends in Ecology and Evolution 19:462–463.

Susskind, L., P. F. Levy, and J. Thomas-Larmer. 2000. Negotiating environmental agreements. Island Press, Washington, D.C.

Thompson, G. J., and J. B. Jenkins. 2004. Verbal judo: the gentle art of persuasion, 2nd edition. Quill, New York.

Trelease, S. 1958. How to write scientific and technical papers. Massachusetts Institute of Technology Press, Cambridge.

Walters, C. J., and S. J. D. Martell. 2004. Fisheries ecology and management. Princeton University Press, Princeton, New Jersey.

Chapter 7

Regulating Harvest

Daniel A. Isermann and Craig P. Paukert

7.1 INTRODUCTION

Humans harvest fish from inland waters for a variety of reasons including consumption, display as trophies, use as bait, feed for cultured fish or livestock, and fertilizer. Harvest is an integral component of many inland fisheries and can have significant impacts on population viability, community interactions, and fishery quality. Consequently, regulating harvest is one of the most common management practices used by natural resource agencies. It is difficult to pinpoint the exact origins of harvest regulations in North America, but they were in use well before the turn of the 20th century. Season closures for some saltwater fisheries were implemented in the 1600s (Redmond 1986), and by the time of the American Revolution numerous statutes were in place regulating the harvest of fishes (see Chapter 1).

7.1.1 History of Harvest Regulations in Inland Fisheries

Temporal trends in the use of harvest regulations have varied among fisheries (e.g., Redmond 1986; Paukert et al. 2001; Lester et al. 2003). Redmond (1986) suggested there have been three periods in the evolution of harvest regulations for inland fisheries in North America. From 1630 to 1940, regulations progressively changed from very liberal, based on the assumption that fisheries resources were inexhaustible, to more restrictive, in recognition that fisheries were finite. The expanding number of harvest regulations during this era reflected adherence to the concept of maximum sustainable yield (MSY), the prevailing fisheries management paradigm of the early 19th century (see Chapter 1). The primary objective of this management philosophy was to maximize harvest in terms of yield or numbers. In the late 1800s, resource agencies began to recognize that harvest of small fish did not always achieve this objective, and many regulations were implemented to delay harvest until fish had time to reach larger sizes. Furthermore, many agencies began enacting regulations that were designed to allow fish to spawn at least once before being harvested.

From 1940 to 1960, resource agencies recognized that based on population dynamics and the purpose of regulations, more restrictive regulations were not always necessary. Furthermore, many agencies and anglers began to recognize that implementing harvest regulations did not always ensure better fishing. By 1959, 34 states had removed previously established length restrictions on black bass, and by 1967 only 14 states continued to enforce minimum-length restrictions on sport fishes (Redmond 1986).

Since the 1960s there has been a general trend toward more restrictive harvest regulations. Reductions in the number of fish that can be legally harvested and the use of restrictive length limits have increased over the past several decades, and application of water- and species-specific regulations has become common practice (Quinn 1996; Radomski et al. 2001; Lester et al. 2003). By the 1970s, aspects other than harvest were given more consideration when formulating harvest policies. Fisheries were being managed under the paradigm of optimum sustainable yield (OSY), which accounted for harvest but suggested that management should be based on biological, sociological, economic, and political justifications. For example, Paukert et al. (2007) concluded that many black bass regulations have been implemented primarily for social rather than biological reasons. Similarly, catch-and-release regulations have become more popular in some fisheries (Quinn 1996). From the mid-1980s to the mid-1990s, 86% of U.S. states increased catch-and-release regulations for black bass and 88% of U.S. states increased catch-and-release regulations for salmonids. Conversely, during most of the 20th century, harvest regulations remained relatively liberal for some inland species (e.g., bluegill, yellow perch, and channel catfish). In many cases, harvest of these species was not regulated, with the exception that fishers may have been required to purchase a license or permit to engage in hook-and-line fishing. Anglers often were allowed to harvest relatively large numbers of fish on a daily basis (e.g., >50 fish/d). In recent decades, this pattern has changed, and more restrictive harvest regulations exist for most sport fisheries. For instance, recreational anglers on Lake Winnibigoshish, Minnesota, were allowed to harvest 100 yellow perch per day prior to 2000; in that year harvest was limited to only 20 yellow perch per day (Radomski 2003). Additionally, the complexity of harvest regulations has increased over time. Creel limits and season closures were often the only regulations used by resource agencies at the turn of the 20th century, whereas now slot length limits and length-based creel limits are used widely to restrict harvest.

7.1.2 Types of Fisheries

While most inland fisheries occurring in North America are recreational in nature, a diversity of fisheries exists across the continent, including commercial and subsistence fisheries. Fisheries that include combinations of recreational, commercial, or subsistence user groups are often referred to as "mixed" fisheries. Conventional rod-and-reel angling comprises the majority of recreational fishing, but other techniques are used, including snagging, various forms of set lines, hand grabbing, and spearing. Harvests in recreational fisheries typically occur for the purposes of consumption or for the selective removal of large fish that are considered trophies. Rarely is it legal for fish captured by recreational fishers to be sold for profit.

Over the past several decades, competitive or tournament angling has increased as a form of recreational fishing for many popular sport fish species. While many competitive fishing events are generally catch-and-release ventures, the frequency and timing of these events are often regulated by state and provincial resource agencies (Kerr and Kamke 2003).

Commercial fisheries are far less common in inland waters than they are in marine systems and are often relegated to species that are not pursued as popular sport fish; however, important commercial fisheries for popular sport fish (e.g., walleye and yellow perch) still exist on portions of the Great Lakes and elsewhere. Commercial fishers harvest fish that are largely sold for human consumption, but some fish are harvested and sold for other purposes such as bait, fertilizer, or animal feed.

Native American tribes use harvest restrictions to manage fisheries on reservation or ceded territory waters. Many fisheries managed by Native American tribes involve fish solely used by tribal members, primarily for food. However, tribal fisheries can include a commercial component, in which fish are sold for profit or as sportfishing opportunities.

7.2 REGULATORY AUTHORITY, IMPLEMENTATION, AND ENFORCEMENT

In North America, regulating harvest occurring on public waters is typically the responsibility of state and provincial natural resource agencies. Depending on the state or province, fisheries occurring on private property may be subject to harvest regulations. The role of the federal government in regulating harvest is typically restricted to management of endangered or anadromous species, fisheries occurring on federal property, or international fisheries. In the Great Lakes, management of certain species is conducted under the auspices of the Great Lakes Fishery Commission, an international consortium that guides and coordinates the implementation of harvest regulations among bordering states and provinces. Harvest regulations on waters that border multiple states or provinces are often conducted jointly and, in some cases where waters cross state or provincial boundaries, agencies may work together in managing particular species. Where tribal fisheries exist, state or provincial agencies interact with tribal and federal agencies in setting harvest policies (see Chapter 4).

7.2.1 Enforcement

Because most harvest regulations on inland fish populations are established by state and provincial natural resource agencies, these agencies are also responsible for enforcing regulations. Enforcement of harvest regulations is usually accomplished by conservation officers who contact fishers and inspect their catch. Many tribes also employ enforcement officers and in some cases state and tribal officers share enforcement responsibilities. Failure to adhere to established harvest regulations can result in a variety of penalties, ranging from monetary fines or restitution fees to jail time or seizure of personal property. Enforcement of regulations is often difficult due to logistic and budgetary constraints. Often only a few enforcement officers cover large geographical areas and are required to monitor a wide variety of outdoor activities (e.g., hunting, fishing, boating, and trapping) that are regulated by state or provincial statutes. However, enforcement of harvest regulations is aided in many cases by concerned citizens who provide information regarding potential infractions that may have gone undetected by law enforcement personnel. This often occurs through agency tip lines that are established for reporting violations.

7.2.2 The Regulatory Process

The process of implementing harvest regulations varies widely among states and provinces and can be complicated by joint jurisdictions (e.g., border waters and the Great Lakes) and by the need to incorporate federal mandates regarding some waters and fisheries. A conceptual framework that represents the regulatory process of many jurisdictions is provided in Figure 7.1. In most states and provinces, proposals for new regulations typically come from biologists working in the associated natural resource agency or from the public. Proposals

Figure 7.1. A conceptual framework depicting the generalized process required to implement new harvest regulations in many states and provinces.

for new regulations are often first reviewed at the local level by fisheries managers directly responsible for management on the waters where the regulation is proposed. This level of review typically includes an assessment of available biological and fishery data to determine whether the new regulation has initial merit. In some cases, opinion surveys or public meetings may be conducted to gauge public opinion regarding proposed regulations. Based on the perceived merit and public support of proposed regulations, a fisheries manager may then seek review from higher supervisory levels (e.g., district or regional fisheries supervisors) to pursue implementation of the regulation. The degree of review will vary with the internal hierarchy of the agency in question. Following this process, the public is typically allowed to review and comment on proposed regulations. Public review and comment often involve public meetings at which agency biologists present information regarding the regulations and address questions from the public. Following this process, agency personnel again scrutinize the regulation proposal in light of comments provided by the public. Following this step, fisheries administrators in a particular agency typically must approve the proposed regulations. Fisheries managers often supply these administrators with information detailing the previous steps in the process, including pertinent biological observations and a summary of public comments. This packet may include a recommendation on whether the regulations should be enacted. Following approval by fisheries administrators, top-level administrators in the natural resource agency, such as directors or fish and game commissions, must then approve regulations. Following this approval, regulations may require approval by the state or provincial legislature, but legislative approval is not always required. In certain cases, emergency actions may be taken to enact regulations, and portions of the approval process may be circumvented to act quickly. See Chapter 5 for a more extensive review of the process of implementing regulations.

7.2.3 Justification for Regulations

Harvest regulations are implemented for a variety of reasons, not all of which are biological. Most often, harvest regulations are enacted to improve fishing quality or to maintain population viability, but regulations may also be implemented to alter community dynamics or to remove undesirable fishes. Requiring individuals to purchase licenses or permits to participate in fishing may serve some biological purpose, but it also provides a source of revenue for natural resource agencies, contact information for survey purposes, and demographic data that may be used to assess the use of fisheries. In many situations, harvest regulations may be enacted for social purposes, possibly to attach some value to a particular fishery or to accommodate individual user groups. Consequently, understanding the sociological aspects associated with a particular fishery represents a critical component when selecting harvest regulations, as individual desires often vary within a group using the same fishery (Gigliotti and Peyton 1993; Allen and Miranda 1996; Edison et al. 2006). In some cases, harvest regulations or general restrictions on fishing are enacted for public safety reasons, such as consumption concerns due to contaminants or safety concerns regarding dangerous areas near dams. Usually both sociological and biological factors are considered when implementing or altering harvest policy in inland fisheries.

7.3 TYPES OF HARVEST REGULATIONS

A diverse array of regulations is used to regulate harvest in inland waters, and regulations vary across jurisdictions and fishery types. The most commonly used harvest restrictions and specific reasons for their implementation are described in the following sections.

7.3.1 Licenses and Permits

In most cases, an individual must obtain a license to participate legally in various forms of fishing. Licenses are issued by state, provincial, and tribal agencies and allow individuals to participate in fishing during a specified time interval (e.g., annual, 24 h, or 3 d) and to harvest the numbers of fish allowed by law. Certain portions of the public (e.g., children, senior citizens, and active military personnel) may not be required to purchase a license to engage in certain types of fishing. Sale of fishing licenses typically occurs on an open-entry basis, meaning fishing privileges are available to all individuals willing to purchase the proper license. However, some commercial fisheries are limited entry; that is, only a fixed number of licenses are made available. Limited-entry systems have also been used to regulate some recreational fisheries (Box 7.1), often on a species- or water-specific basis. Limited-entry systems may be used for a variety of reasons. Harvest occurring in open-entry fisheries may not be biologically sustainable (Johnson and Stein 1986; Lester et al. 2003; Sullivan 2003) or may result in poor or suboptimal fishing quality (Cox and Walters 2002a; Lester et al. 2003; Sullivan 2003). Limited-entry may offer regulatory agencies more control in monitoring and regulating harvest, increasing profitability in commercial fisheries or reducing angler conflicts in high-use recreational fisheries. Occasionally, recreational fishing effort is managed by means of a daily permit system, ensuring that only a specified number of anglers use a particular fishery on a given day.

Laws may also require that additional permits or licenses be acquired to participate in specific types of fishing. In addition to buying a standard recreational fishing license, anglers in many jurisdictions are required to purchase an additional stamp to fish for salmonid species. Revenue from stamp sales is often used for salmonid propagation and management activities. Fisheries may also be regulated by means of a harvest-tag system, by which anglers must obtain a special tag to participate in a specified fishery and can only harvest fish in accordance with the number of tags they possess (Figure 7.2; Scarnecchia and Stewart 1997). Often these tags are issued using a lottery-based approach, representing a system similar to those used to manage big game hunting. For example, anglers desiring to participate in the paddlefish-snagging fishery below Gavins Point Dam in South Dakota must acquire a special tag prior to snagging (Box 7.1).

7.3.2 Catch-and-Release Regulations

Catch-and-release or "no-kill" regulations, which require individuals to release all of their catch, have become increasingly popular for managing recreational fisheries. Catch-and-release regulations are sometimes enacted to improve fishing quality (Barnhart 1989; Perry et al. 1995) under the premise that fishing mortality is negatively influencing population density and size structure. Harvest of fish may be prohibited due to contaminant concentrations that render the fish harmful for human ingestion (Orciari and Leonard 1990; Carline et al. 1991) or to protect an imperiled species or stock considered highly susceptible to overexploitation (Barnhart 1989; Sullivan 2003). Catch-and-release regulations may be seasonal, designed to protect fish from harvest during certain periods of the year when fish are highly concentrated and more vulnerable to capture or to protect mature fish before and during spawning. A common example is regulation of black bass fisheries with catch-and-release restrictions during the spring spawning period. Furthermore, catch-and-release regulations may be enacted to protect broodfish in waters where sexually-mature adults are used for propagation programs. Last, catch-and-release regulations may be implemented on one species to improve population size structure of another species (e.g., Schneider and Lockwood 2002; Shroyer et al. 2003). In Minnesota, catch-and-release regulations have been instituted for largemouth bass to increase size structure of bluegills in some lakes (Shroyer et al. 2003). The premise of these regulations is to increase largemouth bass density to promote increased predation on bluegills, resulting in reduced bluegill density and enhanced growth (Shroyer et al. 2003; also see Chapter 16).

Box 7.1. Regulating Paddlefish Harvest below Gavins Point Dam— The Progression Towards Limited Entry

Paddlefish are a large, long-lived species with a life history that makes them particularly sensitive to exploitation (i.e., late maturation). Following the 1955 closure of Gavins Point Dam on the Missouri River in South Dakota, a substantial paddlefish-snagging fishery developed in the tailwater between Nebraska and South Dakota. The South Dakota Department of Game, Fish, and Parks and the Nebraska Game and Parks Commission jointly manage this fishery. Effectively regulating paddlefish harvest downstream from Gavins Point Dam has proven difficult due to the open-entry nature of the fishery (Mestl and Sorenson 2009). The first paddlefish harvest regulations were implemented in 1957 as a daily bag limit of two and a possession limit of four paddlefish. From 1970 to 1988, harvest regulations were altered several times due to increased fishing pressure and concerns regarding overharvest. During this time, snagging seasons were instituted, and in 1987 the snagging season lasted for 30 d and daily bag and possession limits were 2 paddlefish per angler. However, concerns regarding overharvest persisted. A harvest quota system was adopted by which the fishery would close when 1,600 paddlefish had been harvested from the tailwater or when the 30-d snagging season (October 15– November 15) ended. Check stations were established and anglers were required to report all paddlefish harvested from the tailwater.

In 1989, the season lasted only 4 d during which time 1,364 paddlefish were harvested. The decision to close the season was made on the second day of the season when 1,006 paddlefish had been harvested, but due to a law requiring a 48-h notice of closure the season remained open for two additional days. Anglers, aware of the quota, turned out in large numbers on opening day: the number of anglers in the tailwater on opening day increased 48% from the previous year. The 1990 snagging season closed after 4 d with an estimated total harvest exceeding 2,000 paddlefish.

It was apparent that the paddlefish harvest quota was inadequate for regulating harvest, and it was not effective in protecting sexually-mature fish. Furthermore, fishing pressure in the Gavins Point Dam tailwater was creating safety issues and decreasing the quality of the angling experience. The tailwater creel census resulted in long delays for anglers and exhausted check station crews. A better means of managing the fishery was needed. Proposed regulation changes were needed to control harvest, limit the harvest of larger paddlefish to promote reproduction, and decrease fishing pressure in the Gavins Point Dam tailwater. Many possible regulation changes had been considered including closed areas, restricted areas, shorter seasons, limited entry through a free permit system, and elimination of catch-and-release except for large fish.

A closed area in the Gavins Point Dam tailwater was adopted for the 1991 season, and catch and release of paddlefish was allowed. A protected slot of 89–114 cm measured from the front of the eye to the fork of the tail and a maximum hook size of 13 mm (point to shaft) was adopted for the 1992 season, along with changing the opening day of the season.

Conditions appeared to improve in 1991 and 1992, when the season ran for the full 30 d; however, the season was closed after just 3 d in 1995. The total harvest quota

(Box continues)

Box 7.1. Continued.

(1,600 paddlefish) was exceeded by over 600 paddlefish and fishing pressure was once again an issue.

During March 1996, a public meeting was held to discuss the status of the fishery. Concerns included the fact that the season had become a "race." Check station clerks noted that anglers were "double dipping," coming out in the morning and harvesting a fish and then returning later in the day. Anglers were also returning each day, indicating that it was likely they were ignoring the possession limit. The regulations and notification requirements continued to result in the quota being exceeded. Season-closure notices came on opening day due to the harvest quota being met or exceeded. Overcrowding was getting worse and the quality of the angling experience was declining. Anglers were not supportive of the protective slot limit because the slot included the traditional lengths of fish that were targeted. Crowds were becoming so large that there were concerns for emergency vehicle access if needed. Compounding all of this was that the snagging season was becoming a tourist attraction, further exacerbating crowding problems.

The decision to close the 1996 season was made the afternoon of the first day. The season was closed after 3 d, as per the requirements of the 48-h closure notice, and the harvest quota (1,600 paddlefish) for the entire river was exceeded by 150 fish in the tailwater alone. The open-entry approach to management had once again failed to regulate harvest effectively.

In 1997, a limited-entry system was introduced to regulate paddlefish harvest in the Gavins Point fishery. A free tag would be required for any person attempting to harvest paddlefish during the snagging season and each of the two states would issue 1,200 tags. A paddlefish must be tagged in the dorsal fin immediately following harvest. There would be a limit of two tags per person during the season for each state issuing tags (i.e., maximum number of tags per angler = 4). The season would continue to run for 30 d (October 1–October 30). In 2000, the number of permits that each state issued was increased to 1,400.

Since 1997, the harvest tag system has done an adequate job of controlling paddlefish harvest while eliminating crowding in the tailwater. Anglers have been very receptive to the changes and appreciate the fact that they have 30 d to snag fish with virtually no crowds to compete with in the process. This evolution of harvest regulations represented a concerted effort to integrate fishing effort dynamics, biological information, and angler opinions to develop a sustainable fishery.

7.3.3 Creel and Possession Limits

Creel limits (also called bag or catch limits) are the most common form of regulation used to restrict harvest. Creel limits are typically defined as the number of fish that can be legally harvested by an individual during a single day. Possession limits represent the number of fish that can be possessed by a person at any one time. Possession limits typically include those fish that an individual harvested on a current fishing trip plus fish that were harvested previ-

Figure 7.2. Harvest tags or permits are used in some cases to regulate and monitor harvest in some inland fisheries, a system very similar to those used to regulate and monitor harvest of big game animals such as elk and deer.

ously and remain in the individual's possession (e.g., in a home freezer). Most states and provinces prohibit the act of "culling" or "high grading" fish, meaning that once a fish has been reduced to possession (e.g., placed in a live well or on a stringer) it cannot legally be returned to the water in exchange for another fish that is captured later. Creel and possession limits may be established for individual species or for a few closely-related species in aggregate (e.g., all trout species occurring in a particular stream). Daily creel limits in North American waters often vary among species. Larger piscivores such as northern pike and black bass tend to be regulated with more restrictive daily creel limits than are smaller species that are lower in the food chain, such as bluegill and yellow perch (Radomski et al. 2001). Daily creel limits have generally become more restrictive over time, but fisheries for many species (e.g., bluegills and yellow perch) are still regulated with relatively liberal creel limits (i.e., ≥ 25 fish/d).

Creel limits are typically enacted in an effort to reduce harvests and improve quality of fisheries, although in some instances creel limits may be liberalized to encourage removal of fish from certain populations (Goeman et al. 1993). One rationale for implementing creel limits is to distribute harvest over a larger number of participants (Fox 1975); however, as noted by Radomski et al. (2001), no study has effectively demonstrated that this occurs. It has also been suggested that creel limits may serve a purpose by reminding anglers that fisheries resources are finite (Radomski et al. 2001). Additionally, some anglers may use daily creel limits as a goal or benchmark while fishing (Fox 1975; Snow 1982) and achieving a daily limit or a certain proportion of a daily limit may be important to angler satisfaction (Snow 1982; Cook et al. 2001).

7.3.4 Length Limits

Another common means of regulating harvest is the use of length-based restrictions defining the lengths of fish that may or may not be harvested. Length limits are typically expressed in terms of total length, but in some cases other length designations are used. For example, length limits for paddlefish have been defined in terms of body length, length of eye to fork of tail, or lower jaw fork length (excludes the rostrum). Length limits typically apply for the duration of a fishing season; however, length limits may be in effect for only certain portions of a season to accommodate angler desires to harvest fish (Hurley and Jackson 2002).

The most common length limit is the minimum length limit, which dictates that an individual can harvest only fish that have attained or exceeded a specified length (Figure 7.3). For instance, with a 381-mm minimum length limit all fish less than 381 mm in length must be immediately returned to the water. The second most common form of length-based harvest restriction is a length-based creel limit that restricts the number of fish in a certain length

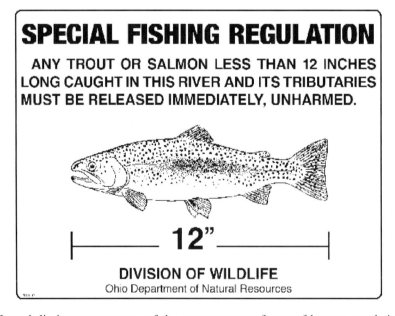

Figure 7.3. Length limits represent one of the most common forms of harvest regulations in inland fisheries.

range that can be harvested. Most often these regulations are used to limit the harvest of larger fish. For example, the number of catfishes that can be harvested by an individual angler is not restricted in Tennessee; however, only one catfish over 86 cm (34 in) can be harvested daily. Another common form of length-based regulation is referred to as a slot length limit. Slot limits involve a range of fish lengths. A protected slot length limit prohibits the harvest of fish within a defined range of lengths. For example, a protected slot length limit of 381 mm to 457 mm indicates that only fish less than 381 mm or greater than 475 mm can be harvested (Figure 7.4). A harvest slot length limit represents the opposite scenario. A harvest slot length limit of 381 mm to 457 mm means that only fish between 381 mm and 457 mm can be harvested. Harvest slot limits are used less frequently than are protected slot limits. Although less common, maximum length limits are also used to restrict harvest. Maximum length limits prohibit harvest of fish equal to or exceeding a specified length; a maximum length limit of 508 mm means that all fish greater than or equal to 508 mm in length must be released.

Length limits are enacted for a wide variety of reasons but have been largely implemented to prevent overexploitation (Maceina et al. 1998; Fayram et al. 2001; Stone and Lott 2002). Length limits have also been used to increase population density and size structure and to increase catch rate and the size structure of fish caught by anglers (Novinger 1987; Colvin 1991; Isermann 2007). In some instances, minimum length limits have been enacted to protect spawning stocks (Scarnecchia et al. 1989; Munger et al. 1994; Paukert et al. 2001) or stocks deemed highly susceptible to overexploitation (Post et al. 2003; Sullivan 2003). Length limits may be enacted to provide a trophy fishery for large fish, such as muskellunge (Wingate 1986). Length limits may also be enacted for one species in an effort to improve a fishery for

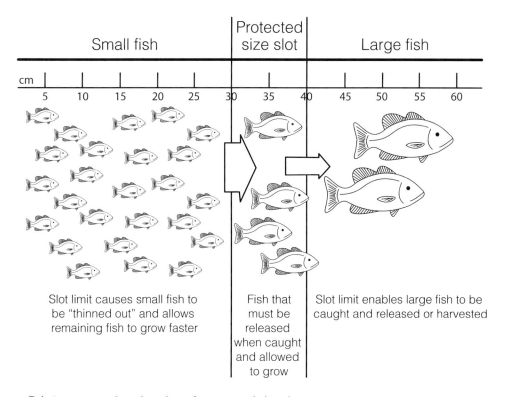

Figure 7.4. A conceptual explanation of a protected size slot.

another species. For example, high-density bluegill populations existing in small impoundments frequently exhibit slow growth and small size structure. Consequently, largemouth bass length limits may be enacted to increase largemouth bass density and subsequent predation on bluegills in an effort to reduce bluegill density and improve bluegill growth and size structure (Rasmussen and Michaelson 1974).

The rationale for applying protected slot length limits differs from that of minimum length limits. In some cases, high densities of small fish that are undesirable to anglers can result in poor growth and size structure, a situation that can be created or exacerbated by the implementation of a minimum length limit (Eder 1984; Novinger 1990). Protected slot length limits may improve growth and size structure by allowing harvest of smaller fish (thereby reducing intraspecific competition) while protecting medium-to-larger fish that have reached lengths more desirable to anglers (Eder 1984; Novinger 1990; Buynak and Mitchell 2002). However, this response has not always been observed, largely because anglers do not harvest sufficient numbers of small fish (Gabelhouse 1987; Martin 1995). Harvest slot length limits are often used to focus harvest on a certain segment of a population. For instance, Scarnecchia et al. (1989) proposed a harvest slot of 57–86 cm for paddlefish in the upper Mississippi River to prevent harvest of smaller, fast-growing fish and large females. Maximum length limits have seen comparatively little use in regulating harvest, but they have been enacted for black bass in Minnesota (e.g., Carlson and Isermann 2010) with the goal of protecting larger fish while allowing harvest of smaller individuals and circumventing problems with density-dependent growth.

7.3.5 Seasons

Agencies routinely establish seasons to prohibit the harvest of certain species during defined time periods. These regulations differ from catch-and-release regulations in that fishers are not allowed to fish for a specified species during a designated time period. Closed seasons are often designed to protect fish during or prior to spawning (Schramm et al. 1995; Philipp et al. 1997) or when fish are concentrated in certain areas. Closed seasons may also be instituted to prevent conflicts among individuals participating in different fisheries (Walker et al. 1993) or other water-based recreation activities. Closed seasons may also be considered beneficial from a socioeconomic perspective as the concept of a "fishing opener" promotes excitement among anglers and may encourage participation.

7.3.6 Closed Areas or Sanctuaries

Sanctuaries, or selected areas that are closed to fishing, have been used on a relatively limited basis in regulating harvest in inland fisheries. In some cases, restrictions are enacted merely as a safety precaution to prevent fishers from using dangerous areas, such as areas around dams. Closed areas may also be used in mixed fisheries to prevent conflicts among different users of a particular resource. For instance, Quinn (1988) reported that restricting commercial netting to only designated areas in a Georgia reservoir reduced bycatch of popular sport fish, which represented an important concern for recreational anglers using the same water body. Sanctuaries have been implemented for biological reasons, typically with the goal of protecting certain portions of a population from harvest (Sztramko 1985; Madenjian and DeSorcie 1999; Bettoli et al. 2007). Sanctuaries may be seasonal, especially if the goal of the sanctuary is to protect spawning individuals (e.g., Sztramko 1985). Sanctuaries have been used to protect black bass spawning areas and portions of some streams or rivers where various species make spawning runs.

7.3.7 Fishery closures

The most extreme form of harvest regulation occurs when fishing for a specified species or stock is completely prohibited. This largely occurs when a stock has collapsed and the fishery is closed as a rehabilitation measure (e.g., Francis et al. 1996; Olney and Hoenig 2001; Wilberg et al. 2003). In other cases, fishing on certain waters may be prohibited to protect broodfish used in propagation efforts or to protect humans due to contaminant levels in fish.

7.3.8 Gear restrictions

The type of gear used by fishers may also be regulated. Gear restrictions represent an indirect form of harvest management that prohibits anglers from using certain capture methods. A wide array of gear restrictions exists. In most jurisdictions, individuals are not allowed to use explosives, electricity, or fish toxins to capture fish (i.e., unless specially permitted), but regulation of other fishing techniques varies widely across jurisdictions.

In conventional hook-and-line fisheries, agencies often restrict the number of lines that can be used by an individual or define how many and what type of hooks may be present on an individual line. For instance, anglers participating in certain fisheries may be required to use only a single hook or hooks that are barbless. Use of live bait is not permitted in some recreational fisheries, and in some cases only certain types of artificial baits can be used. In certain instances, specific techniques for using baits or lures may be prohibited. For example, on some Wisconsin waters forward trolling with a motorized boat is prohibited.

In inland commercial fisheries or in recreational fisheries that do not use conventional hook-and-line techniques, restrictions on the number or design of nets, lines, hooks, or traps used by individuals may be implemented to regulate harvest. In South Carolina, fishers using lines suspended from jugs to capture fish are restricted to a maximum of 50 jug lines, and hooks on trotlines in Illinois must be spaced at least 61 cm apart. Due to the size-selective nature of most net types, mesh-size restrictions are commonly used to regulate harvest in net-based fisheries (Sullivan 2003).

In many cases, gear restrictions are not enacted to reduce harvest explicitly. In many recreational fisheries, the incidence of catch-and-release is high, either due to regulations or prevailing angler attitudes. In these fisheries, gear restrictions may promote the survival of released fish by eliminating fishing techniques that may induce high mortality. Gear restrictions may also be enacted for social reasons. Some fishing techniques are prohibited due to public opinion related to the effectiveness of the gear and the perceived impacts on fish populations. In some instances gear restrictions are enacted purely for social reasons, as is the case with fly-fishing only regulations, which are in place for some salmonid fisheries.

7.3.9 Quotas

Although quotas are used to regulate harvest in a variety of fisheries, they are most frequently associated with commercial or mixed fisheries. The use of quotas is relatively rare in areas where strictly recreational fisheries exist. A quota represents the number or weight of fish that can be legally harvested during a specified time interval (see Box 7.2) and is often referred to as the safe harvest level or total allowable catch. Quotas are typically set on an annual basis or in relation to a defined fishing season and may be specific to certain areas within a water body based on surface area, fishing effort, or fish abundance. Quota-based harvest regulations are most frequently used to prevent overexploitation and to promote allocation of fishery resources across multiple user groups. The rationale for harvest quotas is that once the quota has been met, the fishery is closed to harvest. In some cases, bycatch quotas are used in commercial fisheries to regulate unintentional harvest of certain species that are caught in gears designed to target other species. Lake whitefish and cisco fisheries in Alberta lakes have been regulated with bycatch quotas for walleyes (Sullivan 2003), meaning that once a certain number of walleyes have been taken the fishery for coregonids must cease.

7.3.10 Regulation of Competitive Fishing

The prevalence of competitive or tournament fishing events has increased dramatically over the past few decades and a large segment of anglers that routinely fish for certain species (e.g., black bass) may participate in competitive events. Many states regulate the frequency of competitive events held on individual waters via permit systems; however, permits are often required for only events of a certain size, typically expressed in terms of the number of boats or participants involved. Large numbers of smaller competitive events often occur on water bodies and are not required to obtain permits (Quinn and Paukert, 2009). Competitive events are often promoted as catch-and-release events, but frequently anglers must transport fish to a predetermined weigh-in site prior to release and therefore must comply with state and provincial regulations governing the harvest of fish. If harvest regulations are considered too stringent for competitive events, event organizers may petition regulatory agencies for exemptions (Guy et al. 1999; Edwards et al. 2004). Resource agencies may not allow catch-and-release tournaments during certain times of the year when survival of fish following weigh-in procedures is expected to be low. This is true of walleye tournaments conducted during summer months (e.g., late June–August) in several Midwestern states. When tournaments occur during sensitive time periods, tournaments are considered kill events and anglers may be allowed to keep their catch or the fish may be used by local charity organizations.

7.4 SELECTION OF HARVEST REGULATIONS

Selecting the proper harvest regulation to meet a defined objective is a difficult endeavor, largely due to the inherent variability among fish populations and the complex nuances of public opinion and human behavior. Several papers have offered conceptual and biological guidance in selecting harvest regulations (e.g., Allen and Miranda 1995; Johnson and Martinez 1995). Obviously, understanding the dynamics of a fish population and fishery is an important component in selecting harvest regulations. Reality dictates that fisheries managers must often make regulatory decisions with limited biological information. In many jurisdictions, a large number of fishable waters dictates that pertinent biological information will be lacking for many populations due to budgetary and logistical constraints (Shuter et al. 1998; Cox and Walters 2002b). In particular, useful information on fishing mortality rates is rarely available for most exploited populations in inland waters (e.g., Cox and

Box 7.2. Regulating Harvest in Mixed Fisheries—Walleyes in the Ceded Territory of Northern Wisconsin

In 1983, a federal appellate court decision affirmed the rights of six Wisconsin Ojibwe bands to fish off-reservation waters in lands the bands ceded to the USA in treaties of 1837 and 1842 and allowed the use of traditional methods (e.g., spearing) during the spawning season. More than 800 lakes containing walleyes were opened to these new tribal fisheries. Recreational fishing for walleyes represents an important component of the thriving tourism economy in the northern portion of the Wisconsin.

The State of Wisconsin was charged with equally-distributing walleye harvest in these lakes between recreational anglers and the Chippewa people. This situation posed a new and complex management challenge for fisheries managers.

The management objective for these walleye fisheries is to prevent annual exploitation rates of adult fish from exceeding 35%. A risk criterion is used to allow for uncertainty in meeting this objective and mandates that exploitation of adult walleyes in these fisheries should not exceed 35% in more than 1 out of 40 lakes. To accomplish this objective, state and tribal biologists use an innovative approach incorporating walleye population estimates, harvest quotas, and a sliding bag limit system to regulate harvest by recreational anglers (see figure below).

The initial step of this process is to estimate adult walleye abundance in each lake for each year. In some cases, this is accomplished with mark–recapture studies conducted during early spring (Hansen et al. 2000). By means of predetermined adjustment factors (Hansen et al. 1991), these population estimates are used for 2 years after the mark–recapture process occurred. However, given the large number of lakes involved, recent

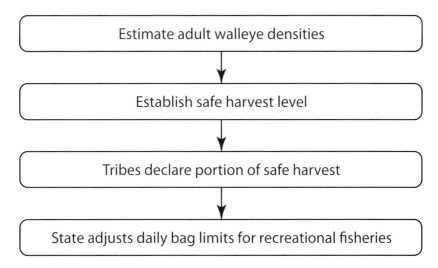

Figure. A conceptual framework of the system used to manage mixed walleye fisheries in northern Wisconsin.

(Box continues)

> **Box 7.2. Continued.**
>
> population estimates are not available for each lake. In lakes where a population estimate has not been made in the last 2 years, adult walleye abundance is predicted from lake area and the primary source of walleye recruitment (i.e., natural reproduction or stocking) in the lake by means of a log-linear model (Nate et al. 2001). Based on these estimates of adult walleye abundance, safe harvest levels (i.e., harvest quotas) are established for each lake based on the 35% exploitation rate and the 1-out-of-40 risk criterion (Staggs et al. 1990).
>
> Following this process, tribes declare a percentage of the safe harvest that will be primarily targeted by spearing fisheries. The State of Wisconsin must then establish recreational harvest regulations for each lake. This is accomplished using a sliding bag limit system, by which daily bag limits for recreational anglers can be set between zero and five walleyes per angler, depending on the percentage of safe harvest declared by the tribes. The effectiveness of this management system is evaluated by estimating exploitation rates in both spearing and angling fisheries.

Walters 2002b). Consequently, decisions regarding harvest regulations are often made under the premise that fishing mortality has an important effect on population status.

In some instances, various biological indicators (e.g., mortality rates, yield per recruit, spawning stock biomass, and maturity rates) are used to describe the relative "health" and future viability of a particular fish stock and to guide the selection of harvest policies (Slipke et al. 2002; Gangl and Pereira 2003). Biological reference points (BRPs) or thresholds are sometimes established to denote points at which reductions in harvest may be necessary (Miranda 2002; Quist et al. 2004). A BRP is a limit or boundary that should not be exceeded and may represent a target or desired range for a particular biological indicator. For instance, when walleye abundance in Lake Erie is estimated to be less than 15 million fish, a point at which the fishery is considered to be in a crisis state, attempts are made to limit fishing mortality by reducing annual harvest quotas for sport and commercial fisheries (Lake Erie Walleye Task Group 2005).

Defining the objectives for and exploring potential consequences of reducing harvest also represent important initial steps in selecting harvest regulations. In general, the impetus for implementing new regulations stems either from agencies that perceive that overfishing is occurring or from complaints from fishers who are dissatisfied with their catch and believe that fishing mortality is a primary cause. In many cases, defining the primary biological objectives associated with regulating harvest is relatively simple. For instance, anglers in recreational fisheries may be dissatisfied with the lack of large fish they catch and perceive this problem to be a result of fishing mortality. Consequently, harvest regulations may be enacted with the objective of increasing abundance and angler catch rates of larger fish. In other instances where recruitment overfishing (see Chapter 2) may be occurring, the objective might be to increase the number of mature fish in the population or to ensure that fish are given more opportunities to spawn prior to being harvested (Pitlo 1997; Scholten and Bettoli 2005). However, achieving certain objectives may affect other facets of a population and other components of an aquatic community. Growth of fish in certain segments of a population may decline after a regulation has been implemented as a result of density-dependent interactions (see section 7.6.6). Increasing the abundance of one species through harvest reductions may alter food web dynamics and affect populations of other aquatic organisms. Enacting a harvest regulation on one water body may merely displace effort to other nearby waters, confounding ability to discern the true effectiveness of a regulation and possibly transferring harvest-related concerns from one population to another. Thus, many potential biological outcomes and attitudes and behavior of fishers must be carefully considered when selecting regulations.

Predictive modeling based on existing population and fishery data has seen increased use as a means of evaluating potential population responses to harvest regulations (Quinn et al. 1994; Allen and Miranda 1995; Scholten and Bettoli 2005; see Chapter 2). Modeling attempts vary from relatively simplistic approaches incorporating a limited amount of data to rather complex models that integrate many forms of data from multiple sources. Regardless of model complexity, reliance on models to make decisions regarding harvest policies has risks due to uncertainty in correct model structure and input parameters (Radomski and Goeman 1996). Consequently, addressing model uncertainty has become an important step in evaluating harvest strategies for many fisheries.

Decisions on harvest regulations are not made in a biological vacuum. Political and public pressures strongly influence the decision-making process and can sometimes override biological information when regulation choices are made. Opinions regarding harvest regulations may vary widely among anglers using a particular fishery. In an increasing number of recreational fisheries, harvest regulations are implemented at the request of the public, sometimes in cases where no clear biological evidence suggests that regulations are necessary (e.g., Boxrucker 2002). Conversely, fishers and other interested parties (e.g., fishing guides and resort owners) may oppose regulations when reductions in harvest are deemed too severe, regardless of the biological impacts. Whether a specific regulation is deemed necessary or successful is often relegated to the realm of public perception. Gathering and assessing public opinions on the desired status of a fishery or the implementation of potential regulations (see Chapter 14) remains an important facet in selecting harvest regulations.

7.5 EVALUATION OF HARVEST REGULATIONS

Evaluating impacts of harvest regulations is an important practice for fisheries managers; however, effective evaluation can be difficult due to population and environmental variability, limited agency resources, and difficulty in sampling fish populations and estimating fish harvests adequately. Effective evaluation starts before a regulation has been implemented. Meaningful and measurable objectives must be established to gauge the effects of regulations and their success in meeting those objectives. Fisheries biologists can disagree as to whether a harvest regulation has been successful or not (Radomski et al. 2009); therefore, criteria used for determining success should be clearly defined prior to evaluation. Efforts to assess the effects of regulations should include both population and fishery metrics, as fishery trends may not accurately reflect population status and vice versa (Shuter et al. 1998; Hansen et al. 2005). In cases where fishing efficiency may be high or anglers exaggerate catch rates, catch rates may remain relatively high despite the fact that population abundance has declined (i.e., hyperstability; Sullivan 2003; Hansen et al. 2005). In other instances, harvest rates may decline faster than does abundance (i.e., hyperdepletion; Hilborn and Walters 1992), possibly because certain portions of a stock may be highly vulnerable and rapidly depleted, while other portions are less vulnerable to capture and remain abundant.

Many previous evaluations of harvest regulations have been short in duration, often occurring over 5 years or less. Additionally, population and fishery data collected prior to regulations are necessary to assess the effects of a harvest regulation; evaluations often include limited information from preregulation periods. Short evaluation periods and a lack of preregulation data undermine ability to assess the effects of a harvest regulation. Short evaluation periods may fail to capture the true effects of harvest regulations due to variation in population dynamics. Harvest regulations designed to improve population size structure may take many years to achieve measurable changes because of the amount of time required for fish to reach large sizes (e.g., Margenau and AveLallemant 2000; Pierce et al. 2003).

Researchers occasionally use reference populations or stocks to examine effects of harvest regulations (e.g., Lyons et al. 1996; Isermann 2007). Harvest in reference populations is typically restricted to a lesser degree than where the "treatment" or the harvest regulation has been applied. Reference populations can serve as a form of "control" designed to compare trends occurring in a population deemed similar to one that has been subject to a more stringent form of regulation. Use of reference populations can prove useful in disentangling effects that prevailing environmental conditions might exert on important metrics (e.g., recruitment and growth) and are especially helpful when preregulation data are lacking for populations where new harvest regulations are to be evaluated.

7.6 FACTORS CONFOUNDING EVALUATION AND EFFECTIVENESS OF HARVEST REGULATIONS

Numerous factors can confound evaluation and effectiveness of harvest regulations. Six factors that typically have serious implications with regards to evaluation and effectiveness of harvest regulation are highlighted below.

7.6.1 Noncompliance

For a harvest regulation to be effective, users of a fishery must comply with a regulation. Levels of noncompliance with harvest regulations in inland fisheries have varied widely, from very low levels that might be deemed innocuous to levels sufficient to negate the effectiveness of regulations. Several studies have demonstrated that relatively low levels of noncompliance can negate the expected benefits of some harvest regulations (Gigliotti and Taylor 1990; Sullivan 2002; Post et al. 2003).

Many factors can influence noncompliance rates including lack of regulation awareness, regulation complexity, catch rates, failure to identify species correctly, and angler measurement error (Page et al. 2004; Page and Radomski 2006). In addition, lack of enforcement could also promote noncompliance if fishers believe that the probability of being caught with illegally-harvested fish is low (Walker et al. 2007).

Biologists need to ensure that the public is aware of harvest regulations and carefully consider the potential for noncompliance when selecting harvest regulations, as certain fisheries may be prone to high levels of illegal harvest. Biologists should also attempt to monitor the extent of noncompliance within regulated fisheries. Although this has typically been accomplished by examining harvested fish during creel surveys, this approach may not reflect the true rate of illegal harvest occurring in a fishery because anglers can hide fish or curtail illegal activities if they are aware a creel clerk is present.

7.6.2 Mortality following Release

The implementation of harvest regulations often means that fishers must release fish following capture. Mortality of released fish can affect the effectiveness of a harvest regulation. In cases where mortality may be high among released fish, the public may perceive harvest regulations as wasteful. Besides direct mortality associated with capture and release of fish, which can often be delayed, released fish can exhibit increased nonlethal stress that may increase predation rates (Coggins et al. 2007) or reduce fitness of fish (e.g., immunological suppression or reduced feeding; Lupes et al. 2006; Cooke and Schramm 2007). The topic of release (i.e., hooking) mortality has been reviewed extensively (Muoneke and Childress 1994; Bartholomew and Bohnsack 2005; Cooke and Schramm 2007). The effects of mortality following release on the effectiveness of harvest regulations varies based on the extent of this mortality, natural mortality rates of the fish population, and the harvest regulation in place. Coggins et al. (2007) determined that effects of mortality following release were related to life history traits of the fish (i.e., short lived and high productivity versus long lived and low productivity), and mortality rates of released fish as low as 5–20% were sufficient to render length limits ineffective. Release mortality may occur at levels threatening the sustainability of a particular fishery and may require that more drastic harvest restrictions be implemented (e.g., limited entry). In some fisheries, the estimated number of fish lost to mortality following release is included in total kill estimates that must remain under a mandated quota. Consequently, mortality following release can strongly influence decisions regarding harvest restrictions for some fisheries.

7.6.3 Natural Mortality

High natural mortality may reduce or negate the expected benefits of harvest regulations, as large numbers of fish may die prior to being harvested. For instance, several studies have indicated that natural mortality rates in crappie populations may be too high to warrant intensive harvest regulations (Larson et al. 1991; Reed and Davies 1991; Hale et al. 1999). Similar natural mortality conditions may exist for a variety of other fishes (Bronte et al. 1993; Quinn et al. 1994; Paukert et al. 2002).

Furthermore, there is evidence that natural mortality may compensate for decreases in fishing mortality, such that reductions in fishing mortality achieved by a harvest regulation may be negated to some extent by corresponding increases in natural mortality (Kempinger and Carline 1978; Allen et al. 1998; Boxrucker 2002). Understanding the magnitude of natural mortality occurring in a population would seem an important aspect in selecting and evaluating harvest regulations, but in practice this information is often lacking due to difficulties associated with estimating natural and fishing mortality rates. Many harvest restrictions are enacted under the premise that exploitation represents an important source of mortality in a population to be regulated, but the failure of this assumption may explain some of the unsuccessful attempts at improving fisheries through harvest restrictions.

7.6.4 Recruitment Variation

Most species sought by fishers exhibit variable recruitment, which can have a profound effect on the status of a fishery. Emergence of strong year-classes can result in periods of "good" fishing, while periods of poor recruitment can result in "poor" fishing. Consequently, recruitment dynamics can influence angler effort and exploitation rates in individual fisheries. Moreover, recruitment variability can have a pronounced effect on a manager's ability to assess impacts of harvest regulations. Specifically, emergence of strong or weak year-classes can greatly alter metrics of abundance and length structure typically used to monitor the effects of harvest regulations (Miranda and Allen 2000; Isermann 2007). This variation can make it difficult to detect potential changes in populations resulting from implementation of harvest restrictions (Allen and Pine 2000). In theory, reducing harvest via regulations should mitigate impacts of recruitment variation on fishery status by stabilizing adult abundance. While this may occur in some instances, regulations may not be sufficient to overcome population and fishery fluctuations that arise due to inconsistent recruitment (Miranda and Allen 2000; Isermann 2007).

Understanding the extent of recruitment variation in a population is a critical component when evaluating harvest regulations. This often proves difficult because recruitment patterns are rarely cyclic or predictable, and recruitment is frequently linked to unpredictable changes in environmental conditions. Often data required to describe recruitment patterns are lacking. When recruitment data are available, biologists must carefully consider the influence of recruitment variation when selecting and evaluating harvest regulations, especially when choosing the duration of these evaluations.

7.6.5 Fishing Effort Dynamics

Monitoring fishery trends is a critical component when evaluating impacts of harvest regulations. Unfortunately, ability to monitor fisheries is often limited by budgetary and logistic constraints. Creel surveys, which are typically used to estimate angler effort and catch in recreational fisheries, are time consuming and expensive. Therefore, assessing angler effort is often ignored when assessing impacts of harvest regulations. Changes in regulations may deter or attract anglers, depending on angler attitudes and the degree of change. Harvest-oriented anglers may choose to fish elsewhere if harvest restrictions become too restrictive, whereas more stringent regulations or regulations that prove initially successful in improving fishing quality may attract anglers (Johnson and Carpenter 1994; Cox and Walters 2002a). Consequently, changes in effort can influence the outcome of harvest regulations and may mask the true effects of the regulations. Harvest regulations may appear to succeed in meeting certain management objectives; however, an important question is whether those objectives were met by reducing fishing mortality (e.g., fishing effort remained relatively stable, regulation reduced harvest) or by influencing angler use of a particular fishery (e.g., reduced effort led to reduced harvest). In other instances, harvest regulation may appear to fail in meeting specified objectives. Apparent failure could merely be a result of increased effort and subsequent harvest manifested by the initial or perceived success of the regulation.

7.6.6 Density Dependence

If a harvest regulation effectively alters mortality rates, population density may change. Population density can influence a variety of population parameters and density-dependent responses can influence effectiveness of harvest regulations. Density-dependent declines in growth have been noted in several evaluations of minimum-length regulations for largemouth bass (Dean and Wright 1992), where abundance of largemouth bass below a minimum length limit increased (often referred to as "stockpiling"). Declines in growth can be problematic as more fish are lost to natural mortality before reaching harvestable length or fewer fish attain lengths desirable to anglers. In many instances, density-dependent growth has led to the application of protected slot length limits, which allow for some harvest of smaller fish in an attempt to reduce density and increase growth. Increased density achieved under harvest regulations may also result in compensatory increases in natural mortality rates. Harvest regulations are often enacted on important predator and prey species; altering the densities of these species can reverberate through the food web and the aquatic ecosystem as a whole. Consequently, biologists must carefully consider the holistic effects that may occur due to changes in harvest policy.

7.7 CONCLUSIONS

The success of various harvest regulations in meeting management objectives has been mixed. Harvest regulations have served to protect populations from overexploitation and have improved fishing quality (e.g., Colvin 1991; Newman and Hoff 2000; Linton et al. 2007). Conversely, harvest regulations have had little apparent impact in some fisheries. Creel limits have been deemed largely ineffective for controlling harvest in recreational fisheries, largely because limits are set too high; the majority of anglers harvest relatively few fish and rarely harvest the allowable creel limit (Baccante 1995; Cook et al. 2001; Isermann et al. 2007). Furthermore, enacting daily creel limits that are low enough to be effective may be socially unacceptable in fisheries that are typically harvest oriented. For several types of harvest regulations (e.g., closed seasons and sanctuaries) little is known regarding their effects on inland fisheries. For example, it has been suggested that closed seasons during black bass spawning periods should benefit some fisheries through enhanced recruitment (Gross and Kapuscinski 1997; Ridgway and Shuter 1997). However, definitive studies evaluating this assumption are lacking (Philipp et al. 1997) and other research indicates that closed seasons may provide only limited benefits (Chance et al. 1975; Fox 1975; Gillooly et al. 2000). When applied in a proper situation, harvest regulations can have positive impacts on fish populations and associated fisheries, but enacting these regulations is no guarantee that fishing will improve.

Harvest regulations will continue to remain an important tool for fisheries managers and the proliferation of fishery-specific regulations (e.g., a 381-mm minimum length for largemouth bass in Lake X) will likely continue. The first length-based harvest regulation for an individual walleye population in Minnesota went into effect in 1986; in 2005 more than 30 of these population-specific regulations were in place across the state (Isermann 2007). Similarly, Lester et al. (2003) reported that the summary of fishing regulations for the province of Ontario expanded from a 2-page pamphlet in the mid-1970s to a 100-page booklet in 2001. Population variability and differences in angler behavior may dictate that numerous regulations are necessary to manage fisheries effectively on a large scale; however, there are several limitations to this approach. Pertinent information for individual fisheries is often lacking. Consequently, a broadscale approach might offer a better means of applying harvest regulations in some regions. Furthermore, there exists a poorly-defined balance between participation and the complexity or severity of harvest regulations. Overregulation or increasingly complex regulations may deter fishers from participating in certain fisheries, and while this may be a necessary step in some instances, in other cases it may merely mean a loss of revenue and political support for natural resource agencies.

There will continue to be a need for innovative and meaningful studies exploring the effects of harvest regulations on populations and angler behavior so that managers have a better understanding of when and how to regulate harvest in certain fisheries. Active, adaptive management programs that incorporate learning opportunities into the management process will be necessary to apply harvest regulations more effectively (see Chapter 5). It must also be recognized that typical harvest regulations like bag and length limits may not be enough to maintain quality fishing opportunities, and broader control of fishing effort may become necessary to meet public demands for better fishing.

7.8 REFERENCES

Allen, M. S., and L. E. Miranda. 1995. An evaluation of the value of harvest restrictions in managing crappie fisheries. North American Journal of Fisheries Management 15:766–772.

Allen, M. S., and L. E. Miranda. 1996. A qualitative evaluation of specialization among crappie anglers. Pages 145–151 in R. E. Miranda and D. R. DeVries, editors. Multidimensional approaches to reservoir fisheries management: proceedings of the third national reservoir fisheries symposium. American Fisheries Society, Symposium 16, Bethedsa, Maryland.

Allen, M. S., L. E. Miranda, and R. E. Brock. 1998. Implications of compensatory and additive mortality to the management of selected sportfish populations. Lakes and Reservoirs: Research and Management 3:67–79.

Allen, M. S., and W. E. Pine III. 2000. Detecting fish population responses to a minimum length limit: effects of variable recruitment and duration of evaluation. North American Journal of Fisheries Management 20:672–682.

Baccante, D. 1995. Assessing catch inequality in walleye angling fisheries. North American Journal of Fisheries Management 15:661–665.

Barnhart, R. A. 1989. Symposium review: catch-and-release fishing, a decade of experience. North American Journal of Fisheries Management 9:74–80.

Bartholomew, A., and B. A. Bohnsack. 2005. A review of catch and release angling mortality with implications for no-take reserves. Reviews in Fish Biology and Fisheries 15:129–154.

Bettoli, P. W., G. D. Scholten, and W. C. Reeves. 2007. Protecting paddlefish from overfishing: a case history of the research and regulatory process. Fisheries 32(8):390–397.

Boxrucker, J. 2002. Rescinding a 254-mm minimum length limit on white crappies at Ft. Supply Reservoir, Oklahoma: the influence of variable recruitment, compensatory mortality, and angler dissatisfaction. North American Journal of Fisheries Management 22:1340–1348.

Bronte, C. R., J. H. Selgeby, and D. V. Swedberg. 1993. Dynamics of a yellow perch population in western Lake Superior. North American Journal of Fisheries Management 13:511–523.

Buynak, G. L., and B. Mitchell. 2002. Response of smallmouth bass to regulatory and environmental changes in Elkhorn Creek, Kentucky. North American Journal of Fisheries Management 22:500–508.

Carline, R. F., T. Beard Jr., and B. A. Hollender. 1991. Response of wild brown trout to elimination of stocking and no-harvest regulations. North American Journal of Fisheries Management 11:253–266.

Carlson, A. J., and D. A. Isermann. 2010. Mandatory catch and release and maximum length limits for largemouth bass in Minnesota: is exploitation still a relevant concern. North American Journal of Fisheries Management 30:209–220

Chance, C. J., A. O. Smith, J. A. Holbrook II, and R. B. Fitz. 1975. Norris reservoir: a case history in fish management. Pages 399–407 in H. Clepper, editor. Black bass biology and management. Sport Fishing Institute, Washington, D.C.

Coggins, L. G., Jr., M. J. Catalano, M. S. Allen, W. E. Pine III, and C. J. Walters. 2007. Effects of cryptic mortality and the hidden costs of using length limits in fishery management. Fish and Fisheries 8:196–210.

Colvin, M. A. 1991. Evaluation of minimum-size limits and reduced daily limits on the crappie populations and fisheries in five large Missouri reservoirs. North American Journal of Fisheries Management 11:585–597.

Cook, M. F., T. J. Goeman, P. J. Radomski, J. A. Younk, and P. C. Jacobson. 2001. Creel limits in Minnesota: a proposal for change. Fisheries 26(5):19–26.

Cooke, S. J., and H. L. Schramm. 2007. Catch and release science and its application to conservation and management of recreational fisheries. Fisheries Management and Ecology 14:73–79.

Cox, S. P., and C. J. Walters. 2002a. Maintaining quality in recreational fisheries: how success breeds failure in the management of open-access sport fisheries. Pages 107–119 *in* T. J. Pitcher and C. Hollingworth, editors. Recreational fisheries: ecological, economic, and social evaluation. Blackwell Science, Oxford, UK.

Cox, S. C., and C. J. Walters. 2002b. Modeling exploitation in recreational fisheries and implications for effort management on British Columbia rainbow trout lakes. North American Journal of Fisheries Management 22:21–34.

Dean, J., and G. Wright. 1992. Black bass length limits by design: a graphic approach. North American Journal of Fisheries Management 12:538–547.

Eder, S. 1984. Effectiveness of an imposed slot length limit of 12.0–14.9 in on largemouth bass. North American Journal of Fisheries Management 4:469–478.

Edison, T. W., D. H. Wahl, M. J. Diana, D. P. Philipp, and D. J. Austen. 2006. Angler opinion of bluegill regulations on Illinois lakes: effects of angler demographics and bluegill population size structure. North American Journal of Fisheries Management 26:800–811.

Edwards, G. P., Jr., R. M. Neumann, R. P. Jacobs, and E. B. O'Donnell. 2004. Impacts of small club tournaments on black bass populations in Connecticut and the effects of regulation exemptions. North American Journal of Fisheries Management 24:811–821.

Fayram, A. H., S. W. Hewett, S. J. Gilbert, S. D. Plaster, and T. D. Beard Jr. 2001. Evaluation of a 15-in minimum length limit for walleye angling in Wisconsin. North American Journal of Fisheries Management 21:816–824.

Fox, A. C. 1975. Effects of traditional harvest regulations on bass populations and fishing. Pages 392–398 *in* R. H. Stroud and H. Clepper, editors. Black bass biology and management. Sport Fishing Institute, Washington, D.C.

Francis, J. T., S. R. Robillard, and J. E. Marsden. 1996. Yellow perch management in Lake Michigan: a multijurisdictional challenge. Fisheries 21(2):18–20.

Gabelhouse, D. W, Jr. 1987. Responses of largemouth bass and bluegills to removal of surplus largemouth bass from a Kansas pond. North American Journal of Fisheries Management 7:81–90.

Gangl, R. S., and D. L. Pereira. 2003. Biological performance indicators for evaluating exploitation of Minnesota's large-lake walleye fisheries. North American Journal of Fisheries Management 23:1303–1311.

Gigliotti, L. M., and R. B. Peyton. 1993. Values and behaviors of trout anglers, and their attitudes toward fishery management, relative to membership in fishing organizations: a Michigan case study. North American Journal of Fisheries Management 13:492–501.

Gigliotti, L. M., and W. W. Taylor. 1990. The effect of illegal harvest on recreational fisheries. North American Journal of Fisheries Management 10:106–110.

Gillooly, J. F., T. C. O'Keefe, S. P. Newman, and J. R. Baylis. 2000. A long-term view of density-dependent recruitment in smallmouth bass in Nebish Lake, Wisconsin. Journal of Fish Biology 56:542–551.

Goeman, T. J., P. D. Spencer, and R. B. Pierce. 1993. Effectiveness of liberalized bag limits as management tools for altering northern pike population structure. North American Journal of Fisheries Management 13:621–624.

Gross, M. L., and A. R. Kapuscinski. 1997. Reproductive success of smallmouth bass estimated and evaluated from family-specific DNA fingerprints. Ecology 78:1424–1430.

Guy, C. S., M. N. Burlingame, T. D. Mosher, and D. D. Nygren. 1999. Exemption of bass tournaments from fishing regulations: an opinion survey. North American Journal of Fisheries Management 19:188–191.

Hale, R. S., M. E. Lundquist, R. L. Miller, and R. W. Petering. 1999. Evaluation of a 254-mm limit on crappies in Delaware Reservoir, Ohio. North American Journal of Fisheries Management 19:804–814.

Hansen, M. J., T. D. Beard Jr., and S. W. Hewett. 2000. Catch rates and catchability of walleyes in angling and spearing fisheries in northern Wisconsin lakes. North American Journal of Fisheries Management 20:109–118.

Hansen, M. J., T. D. Beard Jr., and S. W. Hewett. 2005. Effect of measurement error on tests of density dependence of catchability for walleyes in northern Wisconsin angling and spearing fisheries. North American Journal of Fisheries Management 25:1010–1015.

Hansen, M. J., M. D. Staggs, and M. H. Hoff. 1991. Derivation of safety factors for setting harvest quotas on adult walleyes from past estimates of abundance. Transactions of the American Fisheries Society 120:620–628.

Hilborn, R. and C. J. Walters. 1992. Quantitative fisheries stock assessment. Chapman and Hall, London.

Hurley, K. L., and J. J. Jackson. 2002. Evaluation of 254-mm minimum length limit for crappies in two southeast Nebraska reservoirs. North American Journal of Fisheries Management 22:1369–1375.

Isermann, D. A. 2007. Evaluating walleye length limits in the face of population variability: case histories from western Minnesota. North American Journal of Fisheries Management 27:551–558.

Isermann, D. A., D. W. Willis, B. G. Blackwell, and D. O. Lucchesi. 2007. Yellow perch in South Dakota: population variability and predicted effects of creel limit reductions and minimum length limits. North American Journal of Fisheries Management 27:918–931.

Johnson, B. L., and R. A. Stein. 1986. Competition for open-access resources: a class exercise that demonstrates the tragedy of the commons. Fisheries 11(3):2–6.

Johnson, B. M., and S. R. Carpenter. 1994. Functional and numerical responses: a framework for fish-angler interactions. Ecological Applications 4:808–821.

Johnson, B. M., and P. J. Martinez. 1995. Selecting harvest regulations for recreational fisheries: opportunities for research/management cooperation. Fisheries 20(10):22–29.

Kempinger, J. J., and R. F. Carline. 1978. Dynamics of the northern pike population and changes that occurred with a minimum size limit in Escanaba Lake, Wisconsin. Pages 382–389 *in* R. L. Kendall, editor. Selected coolwater fishes of North America. American Fisheries Society, Special Publication 11, Bethesda, Maryland.

Kerr, S. J., and K. K. Kamke. 2003. Competitive fishing in freshwaters of North America: a survey of Canadian and U.S. jurisdictions. Fisheries 28(3):26–31.

Lake Erie Walleye Task Group. 2005. Report for 2004 by the Lake Erie Walleye Task Group. Lake Erie Committee, Great Lakes Fishery Commission, Ann Arbor, Michigan.

Larson, S. C., B. Saul, and S. Schleiger. 1991. Exploitation and survival of black crappies in three Georgia reservoirs. North American Journal of Fisheries Management 11:604–613.

Lester, N. P., T. R. Marshall, K. Armstrong, W. I. Dunlop, and B. Ritchie. 2003. A broad-scale approach to management of Ontario's recreational fisheries. North American Journal of Fisheries Management 23:1312–1328.

Linton, B. C., M. J. Hansen, S. T. Schram, and S. P. Sitar. 2007. Dynamics of a recovering lake trout population in eastern Wisconsin waters of Lake Superior. North American Journal of Fisheries Management 27:940–954.

Lupes, S. C., M. W. Davis, B. L. Olla, and C. B. Schreck. 2006. Capture related stressors impair immune system function in sablefish. Transactions of the American Fisheries Society 135:129–138.

Lyons, J., P. D. Kanehl, and D. M. Day. 1996. Evaluation of a 356-mm minimum length limit for smallmouth bass in Wisconsin streams. North American Journal of Fisheries Management 16:952–957.

Maceina, M. J., P. W. Bettoli, S. D. Finley, and V. J. DiCenzo. 1998. Analyses of the sauger fishery with simulated effects of a minimum size limit in the Tennessee River of Alabama. North American Journal of Fisheries Management 18:66–75.

Madenjian, C. P., and T. J. DeSorcie. 1999. Status of lake trout rehabilitation in the Northern Refuge of Lake Michigan. North American Journal of Fisheries Management 19:658–669.

Margenau, T. L., and S. P. AveLallemant. 2000. Effects of a 40-in minimum length limit on muskellunge in Wisconsin. North American Journal of Fisheries Management 20:986–993.

Martin, C. C. 1995. Evaluation of slot length limits for largemouth bass in two Delaware ponds. North American Journal of Fisheries Management 15:713–719.

Mestl, G., and J. Sorenson. 2009. Joint management of an interjurisdictional paddlefish snag fishery in the Missouri River below Gavins Point Dam, South Dakota and Nebraska. Pages 235–259 in C. P. Paukert and G. D. Scholten, editors. Paddlefish management, propagation, and conservation in the 21st century: building from 20 years of research and management. American Fisheries Society, Symposium 66, Bethesda, Maryland.

Miranda, L. E. 2002. Establishing size-based mortality caps. North American Journal of Fisheries Management 22:433–440.

Miranda, L. E., and M. S. Allen. 2000. Use of length limits to reduce variability in crappie fisheries. North American Journal of Fisheries Management 20:752–758.

Munger, C. R., G. R. Wilde, and B. J. Follis. 1994. Flathead catfish age and size at maturation in Texas. North American Journal of Fisheries Management 14:403–408.

Muoneke, M. I., and W. M. Childress. 1994. Hooking mortality: a review for recreations fisheries. Reviews in Fisheries Science 2:123–156.

Nate, N. A., M. A. Bozek, M. J. Hansen, and S. W. Hewett. 2001. Variation in adult walleye abundance in relation to recruitment and limnological variables in northern Wisconsin lakes. North American Journal of Fisheries Management 21:441–447.

Newman, S. P., and M. H. Hoff. 2000. Evaluation of a 16-in minimum length limit for smallmouth bass in Pallette Lake, Wisconsin. North American Journal of Fisheries Management 20:90–99.

Novinger, G. D. 1987. Evaluation of a 15.0-in minimum length limit on largemouth and spotted bass catches at Table Rock Lake, Missouri. North American Journal of Fisheries Management 7:260–272.

Novinger, G. D. 1990. Slot length limits for largemouth bass in small private impoundments. North American Journal of Fisheries Management 10:330–337.

Olney, J. E., and J. M. Hoenig. 2001. Managing a fishery under a moratorium: assessment opportunities for Virginia's stocks of American shad. Fisheries 26(2):6–12.

Orciari, R. D., and G. H. Leonard. 1990. Catch-and-release management of a trout stream contaminated with PCBs. North American Journal of Fisheries Management 10:315–329.

Page, K. S., G. C. Grant, P. Radomski, T. S. Jones, and R. E. Bruesewitz. 2004. Fish total length measurement error from recreational anglers: causes and contribution to noncompliance for the Mille Lacs walleye fishery. North American Journal of Fisheries Management 24:939–951.

Page, K. S., and P. Radomski. 2006. Compliance with sport fishery regulations in Minnesota as related to regulation awareness. Fisheries 31(4):166–178.

Paukert, C. P., J. A. Klammer, R. B. Pierce, and T. D. Simonson. 2001. An overview of pike regulations in North America. Fisheries 26(6):6–13.

Paukert, C. P., M. McInerny, and R. Schultz. 2007. Historical trends in creel limits, length-based limits, and season restrictions for black basses in the United States and Canada. Fisheries 32(2):62–72.

Paukert, C. P., D. W. Willis, and D. W. Gabelhouse Jr. 2002. Effect and acceptance of bluegill length limits in Nebraska natural lakes. North American Journal of Fisheries Management 22:1306–1313.

Perry, W. B., W. A. Janowsky, and F. J. Margraf. 1995. A bioenergetics simulation of the potential effects of angler harvest on growth of largemouth bass in a catch-and-release fishery. North American Journal of Fisheries Management 15:705–712.

Philipp, D. P., A. Toline, M. F. Kubacki, D. B. F. Philipp, and F. J. S. Phelan. 1997. The impact of catch-and-release angling on the reproductive success of smallmouth and largemouth bass. North American Journal of Fisheries Management 17:557–567.

Pierce, R. B., C. M. Tomcko, and M. T. Drake. 2003. Population dynamics, trophic interactions, and production of northern pike in a shallow bog lake and their effects on simulated regulation strategies. North American Journal of Fisheries Management 23:323–330.

Pitlo, J., Jr. 1997. Response of upper Mississippi River channel catfish populations to changes in commercial harvest regulations. North American Journal of Fisheries Management 17:848–859.

Post, J. R., C. Mushens, A. Paul, and M. Sullivan. 2003. Assessment of alternative harvest regulations for sustaining recreational fisheries: model development and application to bull trout. North American Journal of Fisheries Management 23:22–34.

Quinn, N. W. S., R. M. Korver, F. J. Hicks, B. P. Monroe, and R. R. Hawkins. 1994. An empirical model for lentic brook trout. North American Journal of Fisheries Management 14:692–709.

Quinn, S. P. 1988. Effectiveness of restricted areas in reducing incidental catches of game fish in a gill-net fishery. North American Journal of Fisheries Management 8:224–230.

Quinn, S. P. 1996. Trends in regulatory and voluntary catch and release fishing. Pages 152–162 in L. E. Miranda and D. R. DeVries, editors. Multidimensional approaches to reservoir fisheries management: proceedings of the third national reservoir fisheries symposium. American Fisheries Society, Symposium 16, Bethesda, Maryland.

Quinn, S. P., and C. P. Paukert. 2009. Centrarchid fisheries. Pages 312–339 in S. J. Cooke and D. P. Phillip, editors. Centrarchid fishes: diversity, biology and conservation. Blackwell Publishing, West Sussex, UK.

Quist, M. C., J. L. Stephen, C. S. Guy, and R. D. Schultz. 2004. Age structure and mortality of walleyes in Kansas reservoirs: use of mortality caps to establish realistic management objectives. North American Journal of Fisheries Management 24:990–1002.

Radomski, P. J. 2003. Initial attempts to actively manage recreational fishery harvest in Minnesota. North American Journal of Fisheries Management 23:1329–1342.

Radomski, P. J., C. S. Anderson, and K. S. Page. 2009. Evaluation of largemouth bass length limits and catch-and-release regulations, with emphasis on the incorporation of biologists' perceptions of largemouth bass frequency distributions. North American Journal of Fisheries Management 29:614–625.

Radomski, P. J., and T. J. Goeman. 1996. Decision making and modeling in freshwater sport-fisheries management. Fisheries 21(12):14–21.

Radomski, P. J., G. C. Grant, P. C. Jacobson, and M. F. Cook. 2001. Visions for recreational fishing regulations. Fisheries 26(5):7–18.

Rasmussen, J. L., and S. M. Michaelson. 1974. Attempts to prevent largemouth bass overharvest in three northwest Missouri lakes. Pages 69–83 in J. L. Funk, editor. Symposium on overharvest of largemouth bass in small impoundments. American Fisheries Society, North Central Division, Special Publication 3, Bethesda, Maryland.

Redmond, L. C. 1986. The history and development of warmwater fish harvest regulations. Pages 186–195 in G. E. Hall and M. J. Van Den Avyle, editors. Reservoir fisheries management: strategies for the 80's. American Fisheries Society, Southern Division, Reservoir Committee, Bethesda, Maryland.

Reed, J. R., and W. D. Davies. 1991. Population dynamics of black and white crappies in Weiss Reservoir, Alabama: implications for the implementation of harvest restrictions. North American Journal of Fisheries Management 11:598–603.

Ridgway, M. S., and B. J. Shuter. 1997. Predicting the effects of angling for nesting male smallmouth bass on production of age-0 fish with an individual-based model. North American Journal of Fisheries Management 17:568–580.

Scarnecchia, D. L., T. W. Gengerke, and C. T. Moen. 1989. Rationale for a harvest slot for paddlefish in the upper Mississippi River. North American Journal of Fisheries Management 9:477–487.

Scarnecchia, D. L., and P. A. Stewart. 1997. Implementation and evaluation of a catch-and-release fishery for paddlefish. North American Journal of Fisheries Management 17:795–799.

Schneider, J. C., and R. N. Lockwood. 2002. Use of walleye stocking, antimycin treatments, and catch-and-release angling regulations to increase growth and length of stunted bluegills populations in Michigan lakes. North American Journal of Fisheries Management 22:1041–1052.

Scholten, G. D., and P. W. Bettoli. 2005. Population characteristics and assessment of overfishing for an exploited paddlefish population in the lower Tennessee River. North American Journal of Fisheries Management 134:1285–1298.

Schramm, H. L., Jr., P. E. McKeown, and D. M. Green. 1995. Managing black bass in northern waters: summary of the workshop. North American Journal of Fisheries Management 15:671–679.

Shroyer, S. M., F. L. Brandow, and D. E. Logsdon. 2003. Effects of prohibiting harvest of largemouth bass on the largemouth bass and bluegill fisheries in two Minnesota lakes. Minnesota Department of Natural Resources, Division of Fisheries, Investigational Report 506, St. Paul.

Shuter, B. J., M. L. Jones, R. M. Korver, and N. P. Lester. 1998. A general life history based model for regional management of fish stocks: the inland lake trout (*Salvelinus namaycush*) fisheries of Ontario. Canadian Journal of Fisheries and Aquatic Sciences 55:2161–2177.

Slipke, J. W., A. D. Martin, J. Pitlo Jr., and M. J. Maceina. 2002. Use of the spawning potential ratio for the upper Mississippi River channel catfish fishery. North American Journal of Fisheries Management 22:1295–1300.

Snow, H. E. 1982. Hypothetical effects of fishing regulations in Murphy Flowage, Wisconsin. Wisconsin Department of Natural Resources, Technical Bulletin 131, Madison.

Staggs, M. D., R. C. Moody, M. J. Hansen, and M. H. Hoff. 1990. Spearing and sport angling for walleye in Wisconsin's ceded territory. Wisconsin Department of Natural Resources, Fisheries Management, Administrative Report 31, Madison.

Stone, C., and J. Lott. 2002. Use of a minimum length limit to manage walleyes in Lake Francis Case, South Dakota. North American Journal of Fisheries Management 22:975–684.

Sullivan, M. G. 2002. Illegal angling harvest of walleyes protected by length limits in Alberta. North American Journal of Fisheries Management 22:1053–1063.

Sullivan, M. G. 2003. Active management of walleye fisheries in Alberta: dilemmas of managing recovering fisheries. North American Journal of Fisheries Management 23:1343–1358.

Sztramko, L. K. 1985. Effects of a sanctuary on the smallmouth bass fishery of Long Point Bay, Lake Erie. North American Journal of Fisheries Management 5:233–241.

Walker, J. R., L. Foote, and M. G. Sullivan. 2007. Effectiveness of enforcement to deter illegal angling harvest of northern pike in Alberta. North American Journal of Fisheries Management 27:1369–1377.

Walker, S. H., M. W. Prout, W. M. Taylor, and S. R. Winterstein. 1993. Population dynamics and management of lake whitefish stocks in Grand Traverse Bay, Lake Michigan. North American Journal of Fisheries Management 13:73–85.

Wilberg, M. J., M. J. Hansen, and C. R. Bronte. 2003. Historic and modern abundance of wild lean lake trout in Michigan waters of Lake Superior: implications for restoration goals. North American Journal of Fisheries Management 23:100–108.

Wingate, P. J. 1986. Philosophy of muskellunge management. Pages 199–202 *in* G. E. Hall, editor. Managing muskies: a treatise on the biology and propagation of muskellunge in North America. American Fisheries Society, Special Publication 15, Bethesda, Maryland.

Chapter 8

Managing Undesired and Invading Fishes

CINDY S. KOLAR, WALTER R. COURTENAY, JR., AND LEO G. NICO

8.1 INTRODUCTION

Throughout much of history, humans have directly or indirectly facilitated the introduction of fishes and other aquatic organisms into areas where they had not previously existed, places outside their natural geographic distributions. These introductions have dramatically changed biological communities throughout the world. North America is no exception and the continent is now home to multiple species native to other parts of the world. In addition, many aquatic animals native to one or a few drainages and regions in North America have been transported by humans to other drainages and regions within the continent (Fuller et al. 1999). Through time, the many foreign introductions and intra-continental transplants, in conjunction with loss of native (often endemic) species, have resulted in aquatic faunas across North America that are increasingly homogenized and biologically less distinctive.

The motives behind aquatic organism introductions, the means by which they are introduced, and the ultimate outcomes of these introductions are many. A large proportion of introductions have been deliberate, usually a result of authorized stocking by governments or other institutions. However, there have also been a large number of illegal or otherwise unauthorized introductions. In addition, a wide variety of introductions have occurred that are considered accidental or unintended, a by-product of human activities (e.g., construction of a canal that allows dispersal of fish into new areas). In terms of motives, many fishes and certain other aquatic organisms have been introduced to establish food sources, create new fisheries, and restore depleted stocks (Fuller et al. 1999; Wydoski and Wiley 1999). In addition, a diverse array of nonnative and native fishes has been stocked for biological control of unwanted plants, invertebrates, and other fishes, as well as for conservation purposes. Introductions have also occurred as a result of unauthorized liberation of small fishes used as bait (i.e., bucket releases), releases of aquarium and water garden plants and pets, and escapes from aquaculture facilities. Aquatic species have also invaded new environments by way of water craft, usually by attaching to vessel hulls or by being carried in ship ballast water; others have invaded adjacent drainages by way of excavated canals or other artificial water channels (Courtenay 1993; Fuller et al. 1999).

Although introductions do not always result in reproducing populations, introductions have occurred over centuries and have involved large numbers and a wide diversity of organisms. There are now hundreds of nonnative aquatic species with established or permanent populations in North America (Fuller et al. 1999). Depending on a variety of factors (e.g., the type of species or genetic variant), individuals of a nonnative species may—at least over the short term—remain few in number and not disperse far beyond where initially introduced. Often, locally-established populations cause little or no detectable ecological or economic harm (Courtenay 1993). However, a proportion of introduced aquatic organisms become abundant and widespread, and many of these either cause or have great potential to cause substantial ecological and economic damage. These organisms are termed invasive.

The term "invasive species" has been defined a number of ways. In the USA, Executive Order 13112, signed by President Clinton in 1999, defines an invasive species as "an alien species whose introduction does or is likely to cause economic or environmental harm or harm to human health." In the "Executive Summary" of the National Invasive Species Management Plan (NISC 2001), an invasive species is characterized as "a species that is nonnative to the ecosystem under consideration and whose introduction causes or is likely to cause economic or environmental harm or harm to human health." For purposes of this chapter, we use the phrase "nuisance species" to describe invasive, nonnative aquatic species as well as certain native species whose populations have grown to such an extent that they are deemed undesirable.

Because nuisance fishes—whether a nonnative invasive or an undesirable native population—can be very difficult and expensive to eradicate, it is prudent to consider carefully potential unintended consequences prior to introducing nonnative fish deliberately and to look for ways to reduce the likelihood of introductions from by-products of human activities. In some situations, a nuisance aquatic species may be present in one or a few isolated ponds or lakes or other confined water body. In these instances, eradication is sometimes relatively easily achieved. However, in those cases in which eradication is implausible or perhaps too costly, emphasis is placed on developing a management and control program. The ultimate goal of any particular fish eradication or control project depends on multiple factors, such as the type, size, and complexity of the aquatic system, as well as the type, abundance, and geographic distribution of organism or organisms targeted, including their potential for harm. Some methods for management and control have been used for decades whereas other approaches are relatively new or are currently under development. There have also been marked shifts in fisheries management paradigms or basic practices, including acceptance of integrated pest management (IPM) and adaptive management practices. Moreover, it is increasingly recognized that management of nuisance species needs to take into account other environmental stressors, such as global climate change.

In this chapter the history of fish introductions in North America is reviewed and the many different motives and ways in which introductions have occurred and changed through time are described. The challenges faced by natural resource managers, who are forced to remediate complex situations involving multiple vectors of introduction, old and new, and the continual and sometimes unanticipated introduction of new and potentially harmful species are also addressed. Various aspects of prevention, management, and eradication of nuisance fishes, focusing heavily on advances in the field and new ways of thinking about invasive species problems are then discussed.

8.2 HISTORY OF FISH INTRODUCTIONS

Fishes and other aquatic organisms introduced to the inland waters of North America have come from all over the globe (Courtenay 1993, 1995; Fuller et al. 1999). Many of these introduced species have caused little or no observable change in their novel habitats. Some introductions have lead to beneficial uses, such as the numerous valuable recreational and commercial fisheries created throughout the continent. Other introductions, however, have resulted in biotic disruptions, sometimes threatening native habitats and native species resulting in ecological harm and economic costs to society (Fuller et al. 1999). Understanding how and why nonnative nuisance fish were introduced and have become established, as well as how society's perception of introductions has changed through time, helps to inform and enhance current management practices. To highlight the situation and improve understanding, we recognize and describe three distinct periods of fish introductions in North America.

8.2.1 Early Introductions (1800–1950)

Introductions of nonnative fishes into North America began with colonization, population growth, and expansion of European immigrants across the continent. Welcomme (1981) suggested that this period of fish introductions began around the middle to latter part of the 1800s with the development of international trade. However, DeKay (1842) reported the first release of goldfish into the Hudson River, New York, in the late 1600s; thus, the first individuals of this species must have arrived from Europe in the era of sailing ships. The goldfish was the first nonnative fish to become established in the USA and later in Canada and the United Mexican States (Mexico) (Courtenay and Stauffer 1984).

Following the U.S. Civil War (1861–1865), native wild fish stocks were being overexploited and depleted. This led to a growing interest among U.S. government officials, partly influenced by immigrants of European ancestry, to revitalize the commercial fish industry and human food resources by importing and culturing food and sport fishes native to European waters. Many of these fishes were already familiar to immigrants and most of these fishes were already being cultured in Europe. To address this task, the U.S. Fish Commission was created in 1871. The newly formed agency was placed under the direction of the renowned scientist, Spencer F. Baird. Baird immediately arranged for the import of several species of fish from Europe, and, once received, these fishes were cultured and then distributed throughout the USA and its territories (Baird 1893). The transport of live fish into the USA and across the continent was made possible by recent advances in transportation, such as steam-powered ships and transcontinental rail.

One of Baird's early assignments was to send ichthyologist Barton Warren Evermann to Yellowstone National Park to explore its native fish resources and to recommend nonnative fishes to introduce to create sport fisheries in the region (Jordan 1891). Yellowstone National Park, the world's first national park, was established in 1872. Creation of the park signaled the beginning of U.S. conservation efforts, although the push to introduce nonnative fishes into Yellowstone waters was not yet considered to be in conflict with the newly evolved conservation philosophy of the period (Courtenay 1993). In addition, politicians and local governments of many states and U.S. territories also requested fishes of foreign origin for stocking into their waters. In particular, they wanted what was then termed a "wonder fish" from Europe, common carp (Courtenay 1993). Also on the wish list was Loch Leven or German trout, which we know as brown trout, and stocks of this fish were imported from both Scotland and Germany and initially cultured in Michigan (Laycock 1966). Largely because of those early introductions, both common carp and brown trout became established in the wild and remain widely distributed throughout most of the USA (Fuller et al. 1999).

Two other fishes from Europe, ide and tench, were also imported by the U.S. Fish Commission. These species along with goldfish and common carp were being cultured and displayed by the commission in ponds near the banks of the Potomac River in Washington, D.C., when a flood in 1889 reached the ponds and most of the fish escaped into the river (Baird 1893). It is believed that the escaped fishes persisted in the Potomac River for a time but eventually disappeared and are not part of the fauna now (Courtenay 1993; Jenkins and Burkhead 1994). However, ide and tench were subsequently introduced into other U.S. waters and reproducing populations exist, but these remain localized (Fuller et al. 1999).

Beginning in the late 1800s (and continuing until the late 1940s), fertilized eggs, fry, and juveniles of fishes of European origin (and eastern U.S. native fishes including brook trout, American shad, and striped bass) were moved from hatcheries into containers for transport behind railroad steam engines hauling specially designed "fish cars." These rail cars were equipped with cooling (initially with ice) and aeration (initially by hand); the methods improved over time because of innovations in cooling and aeration technologies (Leonard 1979). Transport of fishes from the eastern to western USA by rail began in 1873 with the intent of establishing new fish populations in targeted novel waters. On return trips from the western USA, fish cars carried western native fishes such as rainbow trout for introduction into midwestern and eastern waters. By 1923, over 72 billion fish had been moved by railroad fish cars. The last such rail fish car was retired in 1947 (Leonard 1979).

In addition to planned and deliberate fish introductions, transport of fishes by rail car also resulted in a number of unauthorized fish introductions. For example, in 1873 a bridge collapsed and the train's fish cars, carrying some 300,000 live fish (mostly American eels and American shad and some yellow perch), plunged into the Elkhorn River near Omaha, Nebraska, releasing their live cargo into the wild (Fuller et al. 1999). Information in early U.S. Fish Commission reports also indicates that, on occasion, fish cars were "parked" on or near a bridge over a river, and train personnel released a portion of their containers of live fish into the rivers below. Some of these releases were authorized, but it is presumed that others were not. Unfortunately, few official records of these releases exist except in notes of personnel traveling in the fish cars, so information on localities, dates, or species released is not available or is hidden within the massive U.S. Fish Commission reports.

8.2.2 Second Period of Introductions (1950 to cerca 1975)

Except for a few foreign commercial fishes, mostly of European origin, very few nonnative fishes of foreign origin were introduced to North American waters during the early decades of the 1900s (Courtenay et al. 1984, 1986; Fuller et al. 1999). However, after World War II (post-1945) and the advent of intercontinental jet cargo aircraft in the early 1950s, live fish could be rapidly transported from one continent to another, often in a few hours. Ornamental fish importers pioneered the use of plastic bags for carrying and shipping live fishes by air and land. The plastic bags are sometimes injected with pure oxygen to ensure fish survival and further protected by placement in Styrofoam™ containers. This traffic increased dramatically, with most imported fishes and plants destined for the aquarium fish industry and hobbyists. Other species were later imported similarly for potential use as biological control organisms and in aquaculture. This practice continued and Ramsey (1985) estimated that over 100 million fish were imported by air annually during the early 1980s.

During this same period, the number of facilities designed for culturing fishes to supply a growing aquarium fish trade and hobby in North America increased, especially in subtropical areas of Florida. Boozer (1973) estimated that 80% of all aquarium fishes sold in North America were being cultured in Florida. Based on estimates of imports by Ramsey (1985), 20% of aquarium fish imports were fishes not cultured in North America at that time.

There have been escapes of fishes from a number of culture facilities into open waters (Courtenay and Stauffer 1990; Courtenay and Williams 1992). In addition, as the aquarium fish hobby grew, the number of species released by hobbyists into North American waters increased (Fuller et al. 1999); additional species and introductions continue to be observed in open waters from this vector. In warmer waters such as those in southern Florida, Texas, and California, in Hawaii, and in thermal spring outflows in the American West (as far north as Montana and Alberta), a variety of released aquarium fishes have become established as reproducing populations. Admittedly, it is not always possible to ascertain the true source of introduced populations—whether a pet release or escape from a culture facility. Regardless of origin and motive of introduction, some of these introductions have had dire consequences to native fishes, particularly those with populations endemic to the American Southwest and northern Mexico (Courtenay et al. 1985; Deacon and Minckley 1991; Jelks et al. 2008).

During this period, nonnative fishes also were commonly introduced because of their potential as biological control agents. A wide variety of small and large fishes were involved in this endeavor, and, depending on species, their intended uses included control of rooted aquatic plants, algae, and mosquitoes (even though most fishes are opportunistic feeders) (Courtenay and Meffe 1989; Courtenay 1993; Fuller et al. 1999). Some managers, focused on the beneficial uses of these species and perhaps under the impression that correcting unintended consequences of using nonnative fishes for biological control would be relatively easy, gave little attention to assessing the potential risks associated with these introductions.

Some introductions of fishes as biological control agents resulted in achievement of management goals. Stocking the Asian cyprinid grass carp has led to control of macrophytes in many waters across the USA (Cassani 1996); stocking Pacific salmonids in the Great Lakes led to declines in invasive alewives (Madenjian et al. 2002); and stocking striped bass has resulted in reductions of native nuisance threadfin and gizzard shad populations in various reservoirs around the country (Axon and Whitehurst 1985). There are nuances to each of these successes, however: population size of grass carp is important to success—overstocking or subsequent reproduction can lead to loss of all submerged vegetation to the detriment of desired species (Hanlon et al. 2000); reliance on alewives as prey can lead to thiamine deficiency in native lake trout as well as in introduced Pacific salmonids, resulting in early mortality syndrome in their progeny (Honeyfield et al. 2005); and the circumstances under which control of shad species can be achieved depends on the size structuring of the prey base (Dettmers et al. 1998).

Stocking of grass carp deserves special attention due to the numerous success stories of its effective control of aquatic vegetation, its continued use, its widespread availability, and the unintended consequences of using grass carp for biological control. Grass carp was identified in the 1960s as a potential biological control for nuisance aquatic macrophytes owing to its preference for consuming such vegetation (Courtenay 1993). However, those promoting introduction and use of grass carp for plant control assumed that this Asian species was so specific in its reproductive habits and habitat requirements that it could not establish reproductive populations in North American waters (Courtenay 1993). However, not all agreed with this assessment. For instance, Lachner et al. (1970) expressed concern that natural reproduction of grass carp in North American waters was probable. Their fear was proven correct by Conner et al. (1980), who confirmed that grass carp were reproducing in the wild in the Mississippi River basin as early as about 1975. Since that time grass carp has become widespread and it is now presumed established in 18 states in the USA (Nico et al. 2010). An additional concern associated with grass carp is that it carries a nonnative parasite, the Asian carp tapeworm, which spread to baitfishes in polyculture with grass carp. Infected baitfishes, cultured in midwestern states and sold in western states, passed this parasite to endangered native fishes (Kolar et al. 2007). Additional species of Asian carps (black carp, bighead carp, and silver carp) introduced in the USA after grass carp are also known to be carriers of this particular parasite, as well as other parasitic species.

Although wild populations of grass carp spread and reproduce, there remained a strong desire for continued use of the species for control of nuisance aquatic plants. To reduce the risk of additional reproducing populations and widespread establishment, substantial effort was invested in the research and development of sterile or nonreproductive grass carp. In 1985 an apparently reliable technique, one using pressure to shock fertilized eggs, was found to produce near 100% triploid (sterile) fry (Cassani and Caton 1985). This new and relatively inexpensive technology expanded interest in grass carp for biological control, but uncertainty regarding ploidy of fry lingered. However, shortly thereafter an economical method for determining ploidy of grass carp fry by means of a Coulter counter was made available (Wattendorf 1986). The U.S. Fish and Wildlife Service (USFWS) developed a standard operating procedure using Wattendorf's method to validate that shipments of grass carp were 100% triploid (Mitchell and Kelly 2006) and began offering this service to states. States responded very favorably to the program, with over 30 states participating through time (Mitchell and Kelly 2006).

8.2.3 Third Era of Introductions (Post-1975 and Continuing)

Recently, three additional species of Asian carps have been imported into North America, and all three have escaped or been released into inland waters. Two species, the bighead carp and silver carp, are well established, very abundant, and alarmingly invasive. Rationales for importation of these two species were biological control of nuisance phytoplankton in sewage treatment ponds, enhancement of water quality in aquaculture ponds, and potential as food fishes (Kolar et al. 2007). Because these carps feed primarily on plankton, the base of the food chain for all larval and some juvenile and adult native fishes, there is concern that they could have major negative effects on native fish populations. A third species, the black carp, used by the aquaculture industry to control snails in culture ponds, is possibly established in the lower Mississippi River basin (Nico et al. 2005). Because black carp feed almost exclusively on mollusks and many native freshwater mussels are in decline and are imperiled, there is concern that black carp will further threaten their survival.

The aquaculture industry in North America experienced rapid growth during the 1960s and 1970s (Courtenay and Stauffer 1990). To lower production costs and provide convenient access to water, culture facilities were initially often sited in lowland areas, typically near canals or flowing waters, thereby increasing the risk that any escaped individuals could find suitable waters to colonize downstream of the fish farm. Many states have since enacted more stringent legislation and regulations pertaining to the construction of aquaculture ponds and location of aquaculture facilities. Escape from aquaculture facilities (of all types—baitfishes, food fishes, and aquarium species) has been reported as a means of introduction into the wild for over 90 fish species in the USA (http://nas.er.usgs.gov). For some of these species, aquaculture was only one of several vectors by which the species was introduced (e.g., common carp, goldfish, Asian carps, and mosquitofishes). For others, however, releases from culture facilities are almost certainly the primary vector of introduction (e.g., a nonnative species found downstream of a fish farm producing that species). This phenomenon spawned a series of articles in a leading aquarium hobbyist magazine documenting the numerous locations in Florida where nonnative tropical fishes can be easily collected "without leaving the country" (Ganley and Bock 1998).

Escapes from the aquarium fish culture industry, in addition to releases by aquarium hobbyists, have resulted in the introductions and establishment of suckermouth armored catfishes of the genus *Pterygoplichthys*, commonly known as sailfin catfishes, in the USA and Mexico (CEC 2009). In aquaria, these sailfin catfishes are used to control algae. However, members of this genus are large fish that grow rapidly, often becoming too large for their aquaria. Release of adult sailfin catfishes by aquarists is presumed to be the main reason why these fishes have become established in many streams and lakes in Florida, Hawaii, Texas, and Mexico (Nico and Martin 2001; Wakida-Kusunoki et al. 2007; Nico et al. 2009). Once established, sailfin catfishes excavate large burrows in the banks of lakes and streams. These burrows are used as spawning and nesting sites and also contribute to bank erosion (Nico et al. 2009). In some waters of Mexico, such as Infiernillo Reservoir, sailfin catfishes have become dominant, replacing native species and even some nonnative tilapias that used to dominate economically-important fisheries there and causing collapse of local commercial fisheries (Mendoza et al. 2007).

Stocking of fishes by state and provincial agencies to enhance angling opportunities, a practice that began with the U.S. Fish Commission in the late 1800s, has remained an important vector by which nonnative fishes are introduced into North American waters (Courtenay 1993, 1995; Fuller et al. 1999). In the USA these introductions have been aided by annual federal funding. Anglers have been typically pleased with the results of stocking, and a large number of economically-valuable fisheries rely on stocking. However, this practice is controversial because of irreversible changes to systems' natural ecology and aquatic biota. Nonnative foreign and transplanted fishes used in stocking are typically cultured in federal, state, or provincial hatcheries. In some regions, the arguments against stocking center on documented problems created by the culture and release of various salmonids popular for sportfishing. Rainbow trout, native to waters west of the Continental Divide, has been widely introduced. In western states it has hybridized with some native trout to their detriment (Fausch 2008). Brook trout, introduced to western U.S. waters from its native distribution in eastern North America has displaced cutthroat trout from portions of its native distribution. Ironically, introduced rainbow trout has displaced brook trout in portions of its native distribution (Fausch 2008). Brown trout, introduced to the northeastern USA and eastern Canada in the late 1800s to early 1900s, has hybridized with native Atlantic salmon to the detriment of the native species (McGowan and Davidson 1992).

Transoceanic shipping has been the vector of introduction for many nonnative aquatic species to both coasts of North America and the Laurentian Great Lakes, particularly in recent decades. Although there are other vectors by which ships can release organisms (e.g., from hull fouling), clearly the most important in recent decades has been the emptying of ballast tanks while cargo is being loaded. Filling ballast tanks at one port for release at another is necessary to ensure stability of the ship during overseas voyages if the ship is not loaded with cargo. However, the water from foreign ports contains aquatic species, usually invertebrates but sometimes small fishes. Since the completion of the St. Lawrence Seaway in 1959, at least 28 nonnative species have become established in the Great Lakes from ballast water releases (Grigorovich et al. 2003). Some of these species have become invasive, most notably zebra and quagga mussels. Since their initial introduction in the 1980s, these mussels have spread throughout the Great Lakes, down the Mississippi River, and throughout parts of that basin; since 2007 quagga mussels have also been spreading in waterways of the western USA, including lakes Mead, Mohave, and Havasu (Benson et al. 2010). Both mussel species have had a history of clogging intake pipes to power plants, dam operating structures, and boat engines. Controlling invasive mussels cost electric power generating facilities on the Great Lakes alone an estimated US$10–30 million annually between 1989 and 2004 (Connelly et al. 2007). Since their initial introduction, spread of zebra and quagga mussels has been facilitated unintentionally by activities of recreational boaters (in live wells, bilge water, or on the hulls of boats).

The introduction of invasive mussels into the Great Lakes may have "paved the way" for the successful establishment of other species also introduced by ballast water from the Ponto-Caspian region, particularly the round goby, a fish whose diet includes zebra mussels in its native distribution. Both round goby and tubenose goby were first discovered in the St. Clair River, Michigan, in 1990 (Crossman 1991; Jude et al. 1995). The round goby, in particular, has since become widely distributed throughout the Great Lakes. During the 1980s, another European fish, the ruffe, entered North America via ballast water and became established in the Great Lakes region (Ricciardi and Rasmussen 1998).

Akin to the situation in the Great Lakes, California currently has four species of introduced gobies from the western Pacific, whose introduction originated from ballast water released from transoceanic ships that entered California harbors. Two of the gobies venture into and can become established in inland waters. One of these, the shimofuri goby, an aggressive species with high reproductive potential, has become invasive in freshwater areas. Discovered in California waters in 1985, its introduction was undoubtedly via ship ballast water and its invasion partly explained by the presence of some of its preferred foods, Asian invertebrates that also arrived through ballast water release (Moyle 2002).

Another vector by which nonnative fishes have been introduced into North America has been as human food, beginning with introductions of European fishes by European immigrants and, more recently, introductions of Asian fishes by Asian immigrants. There is a wide tradition in many Asian cultures of providing live fish at food markets. The purchase and occasional release of live fish from the live food fish industry in North America is an increasingly common vector of introduction. Among fishes imported to satisfy this trade were snakeheads. Snakeheads are top predators, mostly native to subtropical and tropical regions of Asia and Africa; thus, parts of North America with similar climates have a higher risk for establishment of these species. The northern snakehead is an exception to this rule and can survive in waters that freeze (Courtenay and Williams 2004).

Courtenay and Williams (2004) reported introductions of other snakehead species and reviewed the history of establishment of the northern snakehead and its subsequent eradication from a pond in Maryland. After the eradication, additional established populations of this cold-tolerant snakehead were found in the Potomac River of Maryland and Virginia and later in Arkansas, Pennsylvania, and New York. Other species of snakehead, also probably introduced via the live fish food markets, are also present. For example, in 2000, the bullseye snakehead was found established in southeastern Florida, the first record of this species in North American waters (Shafland et al. 2008).

8.3 PREVENTION OF UNINTENDED INTRODUCTIONS AND REDUCTION OF RISKS OF DELIBERATE INTRODUCTIONS

In part because of a lag time between introduction, detection, and identification by management agencies of introduced species, managing nuisance fish populations often begins only after the species has already become established and achieved nuisance levels. Management options at this point in the invasion sequence are generally fewer and more difficult than if the species had been detected earlier, when less abundant and widespread (Figure 8.1). Not

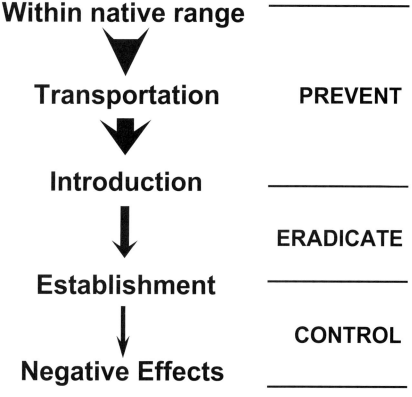

Figure 8.1. In order to persist in a novel ecosystem, nonnative species must survive a series of transitions that occur in a predictable sequence. Individuals must survive transportation from their native ecosystem and introduction into the novel ecosystem in sufficient numbers; their ecological requirements must be met to an extent that allows reproduction and establishment; at this point, some species go on to become a nuisance, either economically, ecologically, or both (modified from Lodge et al. 2006). Usually, a small percentage of species introduced successfully make these transitions (as indicated by the thinning arrows from one transition to the next). As species become more entrenched in the invaded ecosystem, management options become more limited and move from prevention to eradication or, more often, to control of populations. Control, hopefully below nuisance levels, requires sustained and costly effort.

only are more management options available early in the invasion sequence, but economic analyses have shown that an ounce of prevention can, indeed, be worth a pound of cure when it comes to reducing impacts of invasive species (Keller et al. 2009). For these reasons and because eradicating nuisance fishes is not always a viable option, preventing unintended introductions and taking precautions to lower the risks associated with deliberate introductions is prudent. Therefore, we consider preventing the introduction of nuisance fishes relevant to discussing their management.

In this section we (1) discuss the role of government in the prevention and reduction of nonnative aquatic species; (2) present information on legislation and regulations, best management practices, and education and outreach to prevent unintended introductions of nonnative fishes; and (3) describe measures that fisheries management entities can take to reduce unintended consequences of deliberate introductions (Table 8.1).

8.3.1 Government Agencies and Legislative Authorities

A variety of national and state or provincial agencies and legislative authorities play a role in the management of invasive and other undesirable aquatic organisms across North America. For instance, in the USA issues regarding invasive species are handled by more than 20 federal agencies (U.S. Congress 1993) and a wide variety of state agencies; in Canada, primary responsibility and authority regarding aquatic invasive species rests with two government departments, Fisheries and Oceans Canada (DFO) and Environment Canada; and in Mexico, prevention and control of aquatic invasive species is split between several governmental agencies.

In the USA, the Nonindigenous Aquatic Nuisance Prevention and Control Act of 1990 established an interagency committee to develop and implement a program to prevent the introduction and dispersal of aquatic nuisance species in U.S. waters; to monitor, control, and research these species; and to disseminate information. Two regional panels were created to identify priorities, coordinate nonnative species program activities, and advise public and private interests on control efforts. These and additional regional panels have proven useful and effective in coordinating issues regarding aquatic invasive species (e.g., developing national management plans for invasive species of particular concern, disseminating information of regional scope, and developing rapid response plans).

Federal efforts in the USA dealing with invasive species are coordinated by the National Invasive Species Council (NISC), an interagency group established in 1999. In 2001, NISC released its first National Invasive Species Management Plan, which it revised in 2008 (NISC 2008). The 2008 plan calls for federal agencies to use relevant programs and authorities to prevent introductions of invasive species; find and eliminate or reduce new invasive species through early detection and rapid response; stop their spread and minimize impacts through control and management; restore native species and habitats; rehabilitate high-value ecosystems and processes; and maximize effectiveness on invasive species issues through organizational collaboration. However, most fisheries management of inland waters in the USA falls to state agencies. At least 39 states currently have invasive species councils or committees (R. Westbrooks, U.S. Geological Survey, personal communication). These groups are typically composed of people with diverse backgrounds but share an interest in the control and management of invasive species within the state; their goal is to facilitate prevention and control efforts. Some groups deal exclusively with invasive plants, whereas others are inclusive of aquatic species.

Table 8.1. Common vectors by which nuisance fishes are introduced into inland waters and methods used to minimize their associated risks (in the case of authorized stocking by governmental agencies) or to minimize introduction events (remaining vectors). Capitalized jurisdiction represents the more common level of regulation.

Vector of introduction	Prevention measures		
	Regulation	Best management practices	Education and outreach
Ballast water	FEDERAL, state	X	
Aquaculture industry	Federal, STATE	X	
Live food fish industry	Federal, STATE	X	X
Stocking by government agencies	Federal, STATE	X	X
Water garden and aquarium pets	Federal, STATE		X
Unauthorized stocking	Federal, STATE		X
Bait bucket releases	State		X
Recreational activities	State	X	X
Research activities	State	X	X
Diffusion from neighboring waters			X

Both Canada and Mexico responded to the adoption of the 1992 United Nations Convention on Biodiversity by producing national biodiversity strategies (Minister of Supply and Services Canada 1995; CONABIO 2000; Muñoz et al. 2009). Each of these strategies recognizes invasive species as a threat to national biodiversity, and both countries followed these national biodiversity strategies with developing national plans addressing the prevention and control of invasive species (CCFAM 2004; SMARN 2009; Muñoz et al. 2009).

Because aquatic and other invasive species are introduced through specific human-mediated vectors, Canada addresses vectors separately to reduce the threat. For example, a set of guidelines addresses authorized introductions, including aquaculture and fish stocking (DFO 2003); ballast water is regulated by the Canada Shipping Act; and other aquatic vectors of introduction are addressed by the national plan addressing aquatic invasive species threats (CCFAM 2004). In effort to help control, eradicate, and prevent introductions that threaten ecosystems, the DFO formed the Centre of Expertise for Aquatic Risk Assessment (CEARA). This center has developed scientifically defensible risk assessment standards and tools to identify key points in the invasion pathway and maximize efficient use of limited resources to the greatest effect.

In Mexico, prevention and control of pest species, especially those associated with agriculture and aquaculture, are under the purview of the federal agency Servicio Nacional de Sanidad, Inocuidad y Calidad Agroalimentaria (SENASICA) (Muñoz et al. 2009). Although a draft national plan on invasive species has been prepared (SMARN 2009), and Mexico is party to several international cooperatives addressing invasive species issues, the country does not yet have a national policy addressing invasive species in natural areas (Muñoz et al. 2009). Authorities and mandates regarding prevention and control of aquatic invasive species are split among governmental agencies in Mexico. Ballast water is addressed by the Secretariat of Communication and Transports through the Main Directorate of Merchant Marine; the National Secretariat of Environment and Resources, by means of the Office of the Judge

Advocate General, Federal Protection to the Atmosphere and the Main Directorate of Inspection of Wildlife, inspects ports, international airports, and border points; and the National Environmental Policy for the Sustainable Development of the Oceans and Coasts proposes specific strategic targets and tactics to control aquatic invasive species.

In addition to efforts of individual countries, there is an international effort to address the importation of potentially invasive species into North American through the Council for Environmental Cooperation (CEC) via the North American Free Trade Agreement. That cooperation recently resulted in the release of tri-national risk assessment guidelines (CEC 2009).

8.3.2 Prevention of Unintended and Unauthorized Introductions

Consequences of some unintended and unauthorized fish introductions have been costly economically and ecologically (e.g., sea lamprey in the Great Lakes) (Fuller et al. 1999). Because eradication is not always achievable and control efforts are costly, these introductions can have long-term negative effects. Therefore, the most cost effective means of dealing with invasive species is preventing their initial introduction (NISC 2001). Leung et al. (2002), for example, determined that society could benefit by spending up to $324,000 per year to prevent invasion of zebra mussels into a single lake with a power plant (which Leung et al. 2002 pointed out is more than a third of what the USFWS spent in 2001 to manage all aquatic invaders in all U.S. lakes). Below we review the three major avenues that can be pursued to reduce the likelihood of unintended or unauthorized introduction of fishes as the by-products of human activities.

8.3.2.1 Legislation and Regulation

One method of preventing introductions of nonnative aquatic species is to regulate their initial importation and to define or regulate what is allowable with regard to the species after importation. In general, comprehensive legislation addressing potentially invasive plants and animals is lacking in the USA, Canada, and Mexico. In each country, a patchwork of federal and, in some countries, state or provincial, agencies oversee the different facets of the import process while often focusing on only particular types of plants or animals.

Most relevant to the importation of potentially invasive fishes in the USA are duties falling to the USFWS (Stanley et al. 1991), which may prohibit importation and interstate transport of injurious terrestrial and aquatic animals (including mammals, birds, fish, amphibians, reptiles, mollusks, and crustaceans) under the injurious wildlife provisions of the Lacey Act. However, the current list of injurious taxa is short (Table 8.2), the process to list a species as injurious is lengthy, and there are no provisions for emergency listing (Short et al. 2004). The other main federal agency charged with regulating the entry of fishes into the country is the Animal and Plant Health Inspection Service (APHIS) within the Department of Agriculture. The authority of APHIS arises from laws such as the Plant Protection Act and a number of statues collectively referred to as the animal quarantine laws. By means of these laws APHIS can prohibit, inspect, treat, quarantine, or require mitigation measures prior to allowing entry of plant species, their pests, biological control organisms, animals, animal products and by-products, or their host commodities or conveyances (NISC 2001). Currently APHIS has a number of domestic quarantines in place to prevent invasive species from moving within the country (such as the emergency order now in place restricting the movement of certain fishes

Table 8.2. Aquatic animals currently listed under the injurious wildlife provisions of the Lacey Act.

Common name	Scientific name
Salmon[1]	Family Salmonidae
Walking catfish	Family Clariidae
Mitten crabs	Genus *Eriocheir*
Zebra mussels	Genus *Dreissena*
Snakeheads	Genera *Channa* and *Parachanna*
Silver carp (and hybrids)	*Hypophthalmichthys molitrix*
Largescale silver carp (and hybrids)	*Hypophthalmichthys harmandi*
Black carp (and hybrids)	*Mylopharyngodon piceus*

[1] except those accompanied by proper health certification

in the USA to prevent the further spread of the fish pathogen viral hemorrhagic septicemia and special importation restrictions placed on several fishes that are known carriers of spring viremia of carp virus). Lodge et al. (2006) evaluated current U.S. national policies and practices on biological invasions in light of current scientific knowledge and provide a series of recommendations to improve protection from invasive species.

In the USA, states have developed diverse legislation to prevent or regulate the introduction of fishes deemed nuisances (reviewed in Filbey et al. 2002). A variety of actions are variously regulated, including importation, possession, transportation, and introduction. Sometimes paralleling federal regulatory control of aquatic invasive species, state responsibility for prevention and control is often divided, sometimes overlapping, among a variety of state agencies and typically involves departments of natural resources and agriculture as well as a department of environment (Reeves 1999). In some regions, neighboring states regulate the import or introduction of nonnative aquatic species very differently. The state-by-state patchwork of invasive species legislation complicates and can frequently hinder effective prevention and management (Nico et al. 2005; Kolar et al. 2007). In some cases, different states located in the same river basin may clash, for example, with one promoting introduction of a commercial fish while that species may be strictly prohibited by another. Because escaped aquatic organisms do not recognize political boundaries, a misjudgment by one state that results in introduction into the wild of an invasive aquatic species may ultimate cause ecological and economic harm to other states within the same basin. An example is the situation with Asian carps in the Mississippi River basin, whereby some states permitted wide distribution of some Asian carps whereas other states strictly prohibited these species.

Primary legal authority over importation of potentially invasive fishes in Canada lies within the Department of Fisheries and Oceans (DFO) and Environment Canada; in Mexico this authority resides largely with the Secretary of the Environment and Natural Resources (i.e., Secretaría de Medio Ambiente y Recursos Naturales). The umbrella legislation governing introductions of fishes to the provinces and territories of Canada is the Fisheries Act of Canada (Leach and Lewis 1991). According to Canada's plan to address the threat of aquatic invasive species (CCFAM 2004), efforts to coordinate laws and regulations with a bearing on aquatic invasive species are in their early stages across Canada. Although broad regulatory mechanisms already address the introduction of aquatic invasive species in many cases, enforcement responsibility and funding must yet be addressed. The National Commission for the Knowledge and Use of Biodiversity has created a system of information on invading species in Mexico, including biological inventories, places of origin, vectors of introduction, and negative impacts of the species (Muñoz et al. 2009). With this information in hand, in combination with the risk assessment guidelines developed through the CEC, Muñoz et al. (2009) indicate that it is important to establish policies and rules now for preventing and controlling the introduction of invasive species in Mexico.

8.3.2.2 Best Management Practices

A second strategy for preventing unintentional introductions of invasive species is the development and implementation of best management practices (BMPs). Sets of policies, practices, and procedures, BMPs are designed to reduce unintended consequences from on-the-ground activities. Typically developed by agencies or industries for use by personnel in the field, BMPs can be designed to reduce the unintended introduction or spread of invasive aquatic species. They can also be developed for use by members of the public for activities such as proper cleaning of recreational boats and equipment before leaving a water body infested with nuisance species. There are many examples of BMPs in use to minimize the risk of introducing aquatic invasive species (e.g., Arkansas Bait and Ornamental Fish Growers Association 2002; U.S. Environmental Protection Agency 2008).

Hazard and critical control points (HACCP) planning is a rigorous type of BMP borrowed and modified from the food industry and applied to natural resource work. An international standard (ASTM 2008), HACCP planning is a framework for reducing or eliminating the spread of unwanted species during specific processes or practices or in materials or products. Through HACCP planning specific actions of personnel in the field that incur risk of spreading nonnative species can be identified and then addressed to reduce risk at each of these critical control points. The USFWS encourages the use of HACCP planning and provides training and planning tools on the internet (http://www.haccp-nrm.org).

8.3.2.3 Education and Outreach

A last method employed to reduce the risk of introductions is providing information about aquatic invasive species and nuisance fishes to the public and various user groups through education and outreach. In general, information is provided on the various types of vectors by which species are transported from place to place; steps individuals can take to ensure that they do not accidentally spread species; and the environmental and ecological risks associated with releasing pets, emptying bait buckets, and illegally stocking fishes. There are several active, widespread education and outreach programs targeting particular user groups in the USA to reduce risks of introducing and spreading aquatic invasive species. An important aspect of conducting an education and outreach campaign is crafting a unified message reaching target audiences.

One campaign developed in the USA, sponsored by the U.S. Fish and Wildlife Service and the U.S. Coast Guard, is the program "Stop Aquatic Hitchhikers!". This program targets recreational boaters in an attempt to educate them on the importance of cleaning recreational boats after leaving an infested water body. Program information is disseminated through a wide variety of outlets, including highway radio messages, billboards, and television, radio, and newspaper advertisements; displays at rest areas; kiosks at retail and other outlets; ads on gas pumps; lawn banners; advertisements in regulations booklets; signs at water accesses; windshield flyers;

displays at airports; brochures; stickers; and other media (Jensen 2008). Another program in the USA, known as "Habitattitude™," is a national initiative developed by the Aquatic Nuisance Species Task Force and its partner organizations, including, among others, the pet industry (see http://www.habitattitude.net/). This program targets aquarium and water garden owners to convey the message that it is not wise to release plants or animals from home aquaria into the wild. The group provides alternative solutions for those who might be contemplating releasing an unwanted pet. This campaign includes in-store signage; brochures; advertisements in newsletters, television, and magazines; bookmarks; pet care sheets; a website; and other media (Jensen 2008). Both Stop Aquatic Hitchhikers! and Habitattitude programs represent broad collaborations among federal and state management agencies and the private sector.

8.3.3 Reduction of Risks Associated with Deliberate Introductions

Stocking of nonnative fishes, primarily for recreational angling, remains a common practice in fisheries management across North America. As of 1995, all continental U.S. states included nonnative fisheries in their recreational fisheries programs, and, primarily due to stocking, approximately one-third of states were home to more nonnative sport fish species than sport fish native to the state (Horak 1995). Widespread stocking is a concern because each stocking event carries risk of unintended consequences (e.g., spread beyond the intended area, stock contamination, or genetic or other detriment to native species). Occasionally, the consequences are substantial. The numbers and distribution of native fishes have been reduced owing to presence of nonnative competitors and predators, and some species have been locally extirpated (Clarkson et al. 2005). Also, because eradication or population control of an introduced species cannot be guaranteed, negative consequences of a deliberate stocking have the potential to be felt for many years to come. Because of the possibility of unintended consequences, it is prudent to consider carefully the ecological and economic risks prior to stocking nonnative species. Over recent decades, most fisheries management agencies have increasingly recognized the risks involved. Multiple symposia and workshops, many sponsored by the American Fisheries Society (AFS), have been published that also address the issue (Stroud 1986; Schramm and Piper 1995; Nickum et al. 2004). In addition, various fisheries journals have devoted entire issues to the subject of fish introductions, impacts, and regulations (e.g., *Fisheries* 1986, volume 11, issue 2; *Canadian Journal of Fisheries and Aquatic Sciences* 1991, volume 48, supplement 1).

Jackson et al. (2004) reported on results of a recent survey administered to U.S. and Canadian fisheries managers containing questions pertaining to stocking practices. Respondents did not indicate a change in emphasis placed on stocking within their jurisdiction from the 1980s through the 1990s compared with the time period since the 1990s. However, 91% of those responding indicated increased justification required to stock fish since the 1990s. Survey results also indicated that trends in agency policies regarding stocking and biodiversity issues changed over recent decades. Only 8 of the 42 (13%) respondents stated that their agencies made decisions not to stock due to biodiversity concerns before 1980. In contrast, between 1980 and 1990, such decisions were made by 37% of respondents, and since 1990, 71% of responding agencies have decided not to stock in order to conserve biodiversity of receiving waters (Jackson et al. 2004). In recent decades, concern over the effect of stocked fish on native fish assemblages and on the genetic integrity of wild fish populations has also grown (Jackson et al. 2004).

Many guidelines exist to aid fisheries management agencies in making stocking decisions (e.g., see articles in *Fisheries* 1986, volume 11, issue 2). Canada has instituted a national set of guidelines pertaining to the movement of aquatic organisms from one water body to another, including, but not limited to, aquatic organisms from fish-rearing facilities (DFO 2003). These guidelines also provide jurisdictions in Canada with a consistent process for assessing potential impacts that may result from deliberate introductions and transfers of aquatic organisms (DFO 2003). The AFS has endorsed a policy statement on introductions of aquatic species (AFS Policy Statement 15; Box 8.1). Some U.S. states follow these guidelines as part of their decision process when considering whether to stock and, if stocking is warranted, how to best proceed. Participants at a recent workshop on the uses of propagated fishes in resource management suggested the following be considered in stocking decisions: ensure stocking activity is part of a comprehensive fish management plan; assess the biological and environmental feasibility of the introduction; complete a risk and benefits analysis and an economic evaluation; seek input from the public; and cooperate with other agencies (Mudrak and Carmichael 2005).

There are also specific actions that can be taken by fisheries managers to reduce the risk that deliberately introduced fishes might adversely affect native species. One such tool is the stocking of sterile fish. Many states require that grass carp stocked for biological control be certified as being sterile (Mitchell and Kelly 2006). State guidelines for appropriate water bodies to stock with nonnative fishes is another means of limiting potential negative impacts of deliberate introductions. It is the policy of the Arizona Game and Fish Department, for example, to allow stocking of nonnative fishes only in waters where native fishes are absent (Rinne et al. 2004). Although potential unintended consequences of stockings are increasingly recognized by fisheries management agencies, the fact that 26% of respondents on a recent survey of fisheries management agencies indicated their agency does not have formal stocking criteria or conditions that must be met prior to stocking cultured fishes (Jackson et al. 2004) indicates that additional improvements are possible.

8.4 ERADICATION AND POPULATION CONTROL

Eradication of invasive species may be a laudable goal but is usually difficult or, in many situations, impossible to achieve. Success depends on a variety of factors, including the type, abundance, and geographic distribution of the targeted species, as well as the physical and biological composition, size, complexity, and sensitivity of the invaded environment. Invasive and other undesirable fish species are abundant and widespread throughout much of North America (Fuller et al. 1999). Few entire populations of an invasive or unwanted fish species have been targeted for eradication, and, among those targeted, relatively few have been eliminated from all locations. This is true in North America as well as in other parts of the world. Nevertheless, because of the increased recognition that invasive species pose a substantial ecological as well as economic threat, eradication remains an important management option. In those cases in which eradication is considered infeasible, management emphasis shifts to controlling or containing the introduced population.

Most successful eradications of nuisance fishes have been in small, easily-accessible, closed aquatic systems (e.g., ponds or small lakes) that are shallow and sparsely vegetated (e.g., Courtenay and Williams 2004; Lozano-Vilano et al. 2006). In more open or complex systems such as large streams or extensive marsh habitats, eradication may be impossible or, at best, difficult

> **Box 8.1. American Fisheries Society Position on Introductions of Aquatic Species**
>
> Below are recommended evaluations to be made prior to introducing nonnative fishes into an ecosystem (taken from American Fisheries Society Policy Statement 15).
>
> *Rationale.* Reasons for seeking an import should be clearly stated and demonstrated. It should be clearly noted what qualities are sought that would make the import more desirable than native forms.
>
> *Search.* Within the qualifications set forth under rationale, a search of possible contenders should be made, with a list prepared of those that appear most likely to succeed, and the favorable and unfavorable aspects of each species noted.
>
> *Preliminary assessment of the impact.* This should go beyond the area of rationale to consider impact on target aquatic ecosystems and general effect on game and food fishes or waterfowl, aquatic plants, and public health. The published information on the species should be reviewed, and the species should be studied in preliminary fashion in its biotope.
>
> *Publicity and review.* The subject should be entirely open, and expert advice should be sought. It is at this point that thoroughness is in order. No importation is so urgent that it should not be subject to careful evaluation.
>
> *Experimental research.* If a prospective import passes the first four steps, a research program should be initiated by an appropriate agency or organization to test the import in confined waters (e.g., experimental ponds).
>
> *Evaluation or recommendation.* Again, publicity is in order and complete reports should be circulated amongst interested scientists and presented for publication in the Transactions of the American Fisheries Society.
>
> *Introduction.* With favorable evaluation, the release should be effected and monitored, with results published or circulated.

and expensive—and therefore is often not considered. Whether eradication is a viable option depends on many factors including whether reliable plans and methods exist or can be developed, sufficient funding is available, and human resources, including expert leaders and trained crews, are available in addition to other necessary resources (e.g., Donlan and Wilcox 2007).

Eradication and control become more problematic and costly once the targeted population becomes widespread. Consequently, eradication is best attempted almost immediately upon discovery of the new invader population (Simberloff 2009). Unfortunately, many waterways or drainages are not monitored or sampled adequately on a regular basis. For this reason, many nonnative populations are already large and widely dispersed by the time biologists become aware of the invasion. Such was the case with Asian swamp eels, now established in Florida.

Field surveys conducted shortly after their discovery revealed that the geographic range of each of the three known Florida populations was extensive. For instance, Asian swamp eels were first discovered in the Miami area in 1997 (Collins et al. 2002). During the following few years, U.S. Geological Survey biologists sampled connected waterways and found that the invader occupied over 50 km of southeastern Florida canals and concluded, because of its wide distribution and other factors, that the swamp eel was likely present a decade or more prior to their discovery (L. Nico, unpublished data).

8.4.1 Recognition of a New Invader and Determination of Appropriate Action

Successful eradication and control of an undesired or invasive fish species requires basic knowledge of the species and the invaded environment. A critical first step in dealing with a potential introduction is positive identification of the species to confirm the species is nonnative (Fuller et al. 1999). Unfortunately, many fisheries biologists and managers have inadequate knowledge of fish systematics and ichthyology to identify unfamiliar fishes accurately (Courtenay 2007). This phenomenon is not unique to the field of fisheries; the widespread loss of taxonomic expertise has been widely acknowledged (e.g., Agnarsson and Kuntner 2007) and was the impetus for the creation of Partnerships for Enhancing Expertise in Taxonomy, a program that the National Science Foundation continues to fund. Positive identification of foreign organisms may be especially problematic because many are poorly known (Fuller et al. 1999). Following confirmation of a new introduction, rapid but comprehensive field surveys using appropriate gear are needed to determine geographic extent. Armed with the resulting information, appropriate agencies can decide whether eradication is warranted and feasible. If eradication is viable, it is essential to gather basic information on the species rapidly, typically by a combination of literature review and original research. Although a fair amount of data can be gathered during preliminary field surveys, pre-eradication or control efforts may require detailed information on abundance and reproductive status, life history, environmental tolerances, and population dynamics.

Time and effort needed to procure basic biological information depends on the characteristics of the targeted species, characteristics of the invaded habitat, risk that the population will rapidly or easily spread, and potential undesirable effects of the species. Simberloff (2009) argued that successful eradication calls for quick action—in some situations a scorched-earth approach—with minimal time spent conducting research. Nevertheless, Simberloff also recognized that some cases require sophisticated scientific research prior to action. Obviously, a basic understanding of the biology of the targeted fish is necessary in all cases so that the eradication method chosen is appropriate and offers the greatest chance of success.

Each eradication campaign is unique, and problems vary considerably, but those that have been successful share several key elements (Simberloff 2009): (1) early detection of an invasion and quick action to eradicate invader; (2) sufficient resources allocated to the project from start to finish (including posteradication surveys and follow-up, if necessary); (3) a person or agency with the authority to enforce cooperation; (4) sufficient study of the targeted species to suggest vulnerabilities (often basic natural history suffices); and (5) optimistic, persistent, and resilient project leaders.

8.4.2 Methods of Eradication and Control

Methods for eradication and control of invasive organisms can be divided into three categories: chemical, physical, or biological. Some techniques have a long history whereas others are largely experimental or untested. Although there are a wide range of strategies and methods available for controlling unwanted fishes, relatively few have proven useful for eradication. In many instances, an integrated approach is chosen, using multiple methods in combination (e.g., Lee 2001; USFWS 2002; Diggle et al. 2004). A crucial question for eradication is whether the targeted species can be removed at a greater rate than the population grows. Many invasive fishes have high reproductive potentials, and the survival and successful spawning of even one adult pair can potentially lead to thousands of offspring. For this reason, spawning grounds are often a primary target of eradication and control efforts (Diggle et al. 2004).

8.4.2.1 Chemical Methods

Use of fish toxicants (i.e., ichthyocides, piscicides, or fish poisons) is a primary method for eradication or control of undesired and invasive fishes (Marking 1992; Bettoli and Maceina 1996; Wydoski and Wiley 1999; Moore et al. 2008). Some now refer to these chemicals as "biocides." According to Cailteux et al. (2001), as many as 30 different fish toxicants have been used for fisheries management in the USA and Canada, but only three are currently registered for use in both countries. These include rotenone, a widely-used general piscicides (i.e., used for complete fish kills), and TFM (3-triflouromethyl-4-nitrophenol) and Bayluscide (5, 2'-dichloro-4'-nitrosalicylanilide), two chemicals effective against sea lamprey (Smith and Tibbles 1980; Finlayson et al. 2000; Cailteux et al. 2001). Antimycin A (Fintrol®) is another general-use piscicide registered for use in the USA and previously registered in Canada. In the USA, these four chemicals are designated as "restricted use pesticides," and only certified applicators may purchase or supervise their application; however, individual states may have additional requirements regarding use of piscicides (Finlayson et al. 2005). In Mexico the legal status of rotenone and other fish toxicants is less clear. For example, neither Brian Finlayson of California Department of Fish and Game (CDFG) nor the well-known ichthyologist Dr. Salvador Contreras-Balderas (formerly of the Universidad Autonoma de Neuvo Leon, Monterrey, Mexico, personal communication) were aware of instances in Mexico in which rotenone was used to control nuisance fish populations.

According to Rinne and Turner (1991), the scientific literature is "woefully lacking" in documenting the extent, results, and techniques used in many operations using fish toxicants. Even today, most data on the outcome of these types of projects are available only in the gray literature, although many unpublished reports can now be accessed via the internet. In addition, most published papers deal only with projects with successful outcomes (e.g., Table 8.3). Fortunately, updated guidelines for the effective, legal, and safe planning and execution of projects using toxicants are now widely available (e.g., Finlayson et al. 2000; Moore et al. 2008). Of particular value over recent years have been periodic training courses offered by the AFS and other entities for those interested in learning how to plan and execute rotenone and antimycin projects.

Rotenone is a naturally-occurring compound found in the roots of certain plants of the family Leguminosae. It remains the most widely used fish toxicant in control and eradication campaigns in much of the world (Wydoski and Wiley 1999; Cailteux et al. 2001; Clearwater et al. 2003). In North America, rotenone has been used by fish biologists and managers as a

Table 8.3. Summary of selected campaigns conducted in North America to eradicate nonnative fishes, listed in chronological order.

Location	Site description	Targeted taxa	Method	Years	Outcome	Remarks	References
Streams in Craven, Pitt, and Jones counties, North Carolina	14 large coastal streams	Longnose gar	Dynamite	1957	Partial, temporary success	In one day removed 12,707 longnose gars; all or most streams were reinvaded	Johnston (1961)
Leon Creek system, Texas	8 km of small stream and adjacent marshes	Sheepshead minnow and its hybrids	Rotenone and antimycin	1976	Unsuccessful	Subsequent intensive and repeated seining reportedly removed survivors, resulting in success	Hubbs (1980); Minckley and Deacon (1991)
Sinkhole pool, Dade County, Florida	10 × 15 m pool	Black piranha	Rotenone	1977	Successful		Shafland and Foote (1979)
Streams in Great Smoky Mountains National Park, USA	Portions of 5 small streams	Rainbow trout	Backpack electrofishing	1977–1981	Unsuccessful	Target fish were greatly reduced but not eradicated	Moore et al. (1986)
Bylas Springs, Arizona	3 small spring brooks (each 0.2–1 m wide and 2–20 cm deep)	Western mosquitofish	Antimycin	1982	Unsuccessful		Meffe (1983)
Arnica Creek, Yellowstone National Park, Wyoming	Stream drainage, including a 23.6-ha lagoon	Brook trout	Antimycin	1985–1986	Successful		Gresswell (1991)

Table 8.3. Continued.

Location	Site description	Targeted taxa	Method	Years	Outcome	Remarks	References
Susquehanna River, Pennsylvania	Thermal effluent area of electric power plant	Blue tilapia	Temperature manipulation	1987	Partial success	Local population appeared eliminated, but blue tilapia persisted in drainage	Stauffer et al. (1988)
Knife Lake and Knife River, Minnesota	512-ha lake and 113 km of river	Common carp	Rotenone	1989	Successful		Brastrup (2001)
Strawberry Valley, Utah	3,327-ha reservoir, 259 km of stream, and multiple springs	Utah chub, Utah sucker, and other nonnative species	Rotenone	1990	Successful (?)	Targeted fishes reappeared in reservoir, but numbers reportedly remain low; reservoir previously treated with rotenone in 1961	Lentsch et al. (2001)
Frenchman Lake, California	639-ha lake (maximum depth 31m) and tributaries	Northern pike	Rotenone	1991	Successful		Lee (2001)
Streams in Bridger National Forest, Wyoming	3 small streams (1.2–1.6-m mean width)	Brook trout	Backpack electrofishing	1992–1993	Unsuccessful	Population reduced but not eradicated	Thompson and Rahel (1996)

Table 8.3. Continued.

Location	Site description	Targeted taxa	Method	Years	Outcome	Remarks	References
Streams in Great Smoky Mountains National Park, Tennessee	Reaches of 2 small streams	Rainbow trout	Backpack electrofishing	1996–1997	Partially successful	Treatment area consisted of 858 m of stream	Kulp and Moore (2006)
Bighorn Lake, Banff National Park, Canada	2.1-ha (maximum depth 9.2 m) alpine lake	Brook trout	Gill nets	1997–2000	Successful	Naturally fishless lake first stocked in 1965; evidence of reproduction beginning cerca 1980	Parker et al. (2001); Schindler and Parker (2002)
Little Moose Lake, Adirondack Park, New York	271 ha (maximum depth 44 m)—only littoral zone targeted	Smallmouth bass	Boat electrofishing	1998–2005	Successful (temporarily)	Study was conducted to evaluate littoral fish community response to removal of nonnative predator; purpose was not to eradicate entire bass population	Weidel et al. (2007)
Pozo San Jose del Anteojo, Mexico	<0.1-ha pool	Spotted jewelfish	Traps	2000–2002	Successful		Lazano-Vilano et al. (2006)

Table 8.3. Continued.

Location	Site description	Targeted taxa	Method	Years	Outcome	Remarks	References
Green Pond, Alachua County, Florida	0.2-ha sinkhole	Convict cichlid	Rotenone	2001	Successful	Over 1,000 convict cichlids removed	Hill and Cichra (2005)
Crofton Pond, Maryland	1.6-ha pond (average depth 1.4 m)	Northern snakehead	Rotenone	2002	Successful	8 adult and 834 juvenile northern snakeheads were recovered during rotenone treatment	Lazur et al. (2006)
Diamond Lake, Oregon	1,200-ha lake (maximum depth 16m)	Tui chub	Rotenone	2006	Successful	Previous rotenone treatment occurred in 1954	Truemper (2008)
Lake Davis, California	1,619-ha reservoir (maximum depth 33 m) and tributaries	Northern pike	Rotenone	2007	Successful	Previous rotenone treatment occurred in 1997	Lee (2001); B. Finlayson (personal communication)
Piney Creek drainage, Arkansas	63 km of main stream channel, 660 km of ditches and laterals, and additional shallow standing-water sites	Northern snakehead	Rotenone	2009	Unknown, currently being assessed	Northern snakeheads discovered in drainage in 2008; as of mid-2009 follow-up surveys to assess results not yet completed	M. L. Armstrong (Arkansas Game and Fish Commission, personal communication)

piscicide since the 1930s. Eradication campaigns have targeted a wide diversity of fish species inhabiting a broad range of aquatic habitats, including still and flowing waters (Rinne and Turner 1991; McClay 2005). The first use of rotenone as a fish management tool with intent to eradicate nonnative fishes occurred in Michigan in 1934, an operation involving the removal of common carp and goldfish from two small ponds (Krumholz 1948). Because of the chemical's long history and wide use in fisheries management, there is substantial fisheries literature on rotenone (Wydoski and Wiley 1999; Cailteux et al. 2001; McClay 2005). Over the past several years, the AFS Fish Management Chemicals Subcommittee, rotenone registrants, and the U.S. Environmental Protection Agency have worked on the re-registration process for rotenone. A re-registration eligibility decision was issued March 2007, allowing for continued use of rotenone as a piscicide in the USA. The decision required changes in the label (i.e., written directions for proper use) and preparation of a rotenone standard operating procedure manual, which was recently released (Finlayson et al. 2010). Because information on the environmental effects of rotenone in marine waters is minimal, the new federal regulations will likely permit rotenone use in only freshwater systems unless special approval is obtained.

Antimycin A (antimycin) is a fungal antibiotic whose potential for use as a general fish toxicant was recognized in the early 1960s (Wydoski and Wiley 1999; Finlayson et al. 2002; Moore et al. 2008). The only formulation available for use in North America is Fintrol (Clearwater et al. 2003). It is generally more toxic to fish than is rotenone, and therefore less chemical is needed to achieve similar results (Finlayson et al. 2002). Moreover, at piscicidal concentrations, antimycin does not elicit an avoidance response in fishes, an advantage over rotenone (Dawson et al. 1998; Finlayson et al. 2002). Scaled fish and some rotenone-resistant species tend to be highly susceptible to antimycin, and it has been the piscicide of choice for some native salmonid restoration projects in streams of the western states (Burress and Luhning 1969a, 1969b; Finlayson et al. 2002). Because efficacy depends somewhat on water and habitat characteristics (e.g., pH, water flow, and amount of leaf litter), some fish biologists prefer antimycin for use in small systems and rotenone for larger systems (Finlayson et al. 2002).

Both rotenone and antimycin are general piscicides, but, depending on the water body to be treated and the fish species to be controlled, they have sometimes been used selectively (Lowman 1959; Willis and Ling 2000; Moore et al. 2008). Application of the chemicals typically involves release of diluted liquid solutions directly into the water, although use of powdered rotenone is also common. Some resource managers report success using an ingestible, rotenone-laced feed pellet known as fish management bait to remove grass carp (Mallison et al. 1995). The main advantage cited for antimycin is its effectiveness at lower concentrations and its nondetectability by fish, whereas advantages associated with rotenone are its broad range of toxicity to all species of fish and its effectiveness under a broader range of chemical conditions (i.e., pH) (Finlayson et al. 2002). Rotenone also is presently much less expensive than is antimycin. Both rotenone and antimycin have the benefit of relatively rapid degradation into harmless compounds and both are neutralized by potassium permanganate (Moore et al. 2008). Depending on water temperature and sunlight exposure, degradation may occur within days or weeks with rotenone or within hours or days with antimycin (Dinger and Marks 2007). Depending on concentration, both chemicals can have deleterious effects on aquatic invertebrates. However, because of a shortage of studies, much less is known about the effects of antimycin on these other organisms (Finlayson et al. 2002; Dinger and Marks 2007).

In North America, TFM and the molluscicide Bayluscide have been used extensively since the late 1950s and early 1960s to control sea lamprey in the Laurentian Great Lakes (Heinrich et al. 2003). Both chemicals are nonpersistent in the environment and are nontoxic to other fishes at the low levels used for sea lamprey control. Several different formulations of both chemicals have been developed to treat different habitats of tributary streams and rivers where sea lampreys spawn (Boogaard 2003; Heinrich et al. 2003). In addition, TFM may have the potential to control other nuisance fishes. Boogaard et al. (1996) demonstrated that ruffe were three to six times more sensitive than were native fish species to TFM, but the chemical is currently not approved for control of ruffe.

More than 40 other chemicals have been used as fish toxicants worldwide but have not been fully developed and tested or have not received government approval in North America for use in fish management (Marking 1992; Clearwater et al. 2003; Dawson 2003). One of the more interesting is a diverse group of plant-derived compounds referred to as saponins or triterpene glycosides. Two of the more frequently-cited piscicidal saponins are tea seed cake and mahua oil cake (Clearwater et al. 2003). Dawson (2003) evaluated chemicals known to be used as piscicides and identified Squoxin (1,1'-methylenedi-2-naphthol), a piscicide selective against northern pikeminnows, as a candidate chemical for further development. He also suggested several other chemicals that show promise based on selectivity, ease of application, low toxicity to nontarget organisms, safety to humans, persistence in the environment, low tendency to bioaccumulate, and low cost. Although there is need and continued interest in developing additional piscicides, particularly taxonomically selective piscicides, costs and time associated with research and registration may be prohibitive or, at minimum, preclude their availability in the foreseeable future.

A major drawback with most fish toxicants is their nonspecificity, causing death or harm to nonnative and native fishes as well as to aquatic invertebrates. In some cases, targeted species may have a higher sensitivity to the fish toxicant than do co-occurring native species, thereby allowing some degree of selectivity. However, in many situations the targeted nonnative is either more tolerant (e.g., Asian swamp eels; Schofield and Nico 2007) or has similar sensitivity (e.g., round gobies; Schreier et al. 2008) to many of the nontargeted native fishes present. Consequently, if fish toxicants are used, restocking of native species is typically necessary.

Presence of imperiled aquatic species in an invaded habitat creates special problems. Although a nonnative species may further threaten the imperiled species, use of fish toxicants or other measures to control the invader may also be harmful. Perhaps the greatest reported misuse of a fish toxicant with disastrous results occurred in 1962 when 715 km of the Green River and its tributaries in southwestern Wyoming and northeastern Utah were treated with rotenone as part of a massive fish renovation project (Holden 1991). Because of an inadequate supply of the detoxicant, potassium permanganate, some of the rotenone remained in the river and continued killing fish as it flowed downstream, resulting in heavy losses of native fishes including some imperiled species. It was also suggested that repeated use of rotenone to rid the Little Colorado River of invasive common carp contributed to the disappearance of the Little Colorado spinedace, an endemic minnow (Miller 1963).

Recognizing the need to balance reasonable environmental safeguards with prudent use of rotenone and other fish toxicants, the AFS, in addition to their training course (mentioned above), recently developed and implemented a rotenone stewardship program. Information on rotenone, along with important links, is available via the AFS Web site (www.fisheries.org/units/rotenone/).

8.4.2.2 Physical Methods

A wide array of physical or mechanical methods has been used to control invasive fish populations but have limited potential in eradication (Roberts and Tilzey 1996; Mueller 2005; CDFG 2007). Physical methods include removal of fish by use of nets, traps, gigs and spears, electrofishing gear, explosives, and management of water levels and flows as well as other methods (Smith and Tibbles 1980; Roberts and Tilzey 1996; Wydoski and Wiley 1999).

In general, nets and traps may be useful for controlling nonnative fishes, but eradication is rare and limited to small, isolated water bodies or in portions of drainages. For instance, an intensive seining program conducted during 1976–1978 apparently was successful in completely removing nonnative sheepshead minnow and its hybrids from a small stream system in Texas (Minckley and Deacon 1991). More recently, gill netting was used to eradicate nonnative rainbow and brook trout in high mountain lakes in the Sierra Nevada of California (Knapp and Matthews 1998; Vredenburg 2004) and Banff National Park, Canada (Parker et al. 2001). Small traps were successfully used to eradicate a nonnative fish from a small isolated pool in Mexico (Lozano-Vilano et al. 2006). On the Baja Peninsula of Mexico, removing nonnative fishes from Pozo Largo by means of a variety of nets and traps corresponded with a temporary recovery of an endemic native fish, although it was uncertain if recovery was related to the removals (Ruiz-Camposa et al. 2006). Nonetheless, use of nets to remove unwanted nonnative fishes has also failed to achieve elimination of targeted populations (Neilson et al. 2004).

Researchers in North America have tested backpack electrofishing gear for removal or reduction of nonnative salmonid populations in small upland streams with mixed results (Moore et al. 1986; Thompson and Rahel 1996). A study conducted 1996–1997 in a small southern Appalachian stream showed that four separate removal efforts successfully halted nonnative rainbow trout reproduction and five removal efforts eliminated the species in one reach (Kulp and Moore 2000). In larger or more complex water bodies, electrofishing is less effective. For example, since 2001 biologists have used boat-mounted electrofishing gear several times each year in an attempt to control a large Asian swamp eel population inhabiting a canal system (>15 km long) adjacent to Everglades National Park. In the first year 1,400 Asian swamp eels were captured, but the removal seemed to have little effect on the overall population size or length structure (L. Nico, unpublished data). This removal effort with electrofishing boats has continued and the population appears to have declined. However, as of late 2008, Asian swamp eels remained common in the canal system.

Potential for underwater explosives to kill or injure fishes has been well documented (Teleki and Chamberlain 1978; Keevin 1998). Permit requirements for underwater blasting in the USA and Canada vary considerably among the different states and provinces. A series of experiments determining the efficacy of explosives (dynamite) in removing longnose gar from large coastal streams in North Carolina had limited success and the fish typically recolonized the treated areas (Johnston 1957). Recently, in an attempt to control and eradicate nonnative northern pike in Lake Davis, California, detonation cord was tried but found to be labor-intensive, expensive, and minimally effective (CDFG 2007). There may be considerable variation in blast effects depending on the charge type (e.g., low-velocity versus high-velocity detonation; linear versus point source), charge weight, blast design (e.g., detonation depth), and habitat characteristics (e.g., depth and bottom configuration) (Keevin 1998). Mortality rate and severity of injury may vary between fish species. Fish with swim bladders suffer great harm whereas those that lack swim bladders often survive underwater explosions

(Goertner et al. 1994), indicating some selectivity if explosives are used against certain nonnative fishes.

Fish species exhibit differences in their thermal tolerance limits, but there are relatively few instances in which use of water temperature manipulation to eradicate or control nonnative fish is feasible. In a unique situation, Stauffer et al. (1988) evaluated ways to eradicate a population of blue tilapia in the Susquehanna River in Pennsylvania. Laboratory tests indicated that the lower lethal temperature of the resident blue tilapia population was about 5°C. Because the local wild population in the Susquehanna overwintered in the thermal effluent of the Brunner Island electric power plant, Stauffer and colleagues recommended that the facility temporarily lower the water temperature. In February 1987, electrical output from the power station was purposefully reduced, causing water in the discharge canal to drop below 5°C for at least 25 h. Blue tilapia in the canal were unable to maintain position in the current and were carried into the river. In spring and summer 1987 the canal and adjacent river were sampled, and no blue tilapias were collected. Although the local population appeared eliminated, it was recognized that blue tilapia populations likely persisted in the drainage because of other thermal discharges along the river.

In some instances, water level of a lake or reservoir can be reduced in conjunction with other methods, especially fish toxicants (CDFG 2007). Partial draining is typically performed to reduce the amount of toxicant needed, to increase freeboard above dam so chemical neutralization is not needed, and to contain the targeted fish within a smaller and more exposed area, thereby increasing the chances of eradication. Complete dewatering of a water body to eradicate nonnative fish populations has been proposed for some large reservoirs (CDFG 2007), but in North America complete draining has largely been limited to small water bodies, usually aquaculture ponds (Alvarez et al. 2003; Mueller 2005).

Restoring natural flows in regulated systems has the potential to affect the distribution of nonnative fishes. Impounded systems in California, for instance, tend to be dominated by nonnative fishes, whereas unregulated systems tend to be dominated by native fishes because would-be colonists are regularly flushed downstream (Moyle and Light 1996). In a Mojave Desert oasis, researchers restored natural flow in a small stream thereby promoting recolonization by native fishes while simultaneously deterring invasion and proliferation of nonnative sailfin molly and western mosquitofish, species that prefer lentic habitats (Scoppettone et al. 2005). It was concluded, however, from recent analysis of 19 years of fish assemblage and water flow data from the Gila River basin (New Mexico) that natural flow restoration alone would be unlikely to ensure persistence of native fish assemblages in that system (Propst et al. 2008).

Various barriers have been constructed or proposed for preventing dispersal of nonnative fishes (Hunn and Youngs 1980; Carpenter and Terrell 2005). As with other physical methods, barriers are often used in conjunction with other control or eradication methods. During initial attempts in the 1950s and 1960s to prevent sea lamprey migration, mechanical and screen-type weirs were installed in tributary streams of the Great Lakes (Hunn and Youngs 1980). These included some electromechanical barriers. The barriers provided partial control and an opportunity for design evaluation, but most proved expensive to build and maintain (Smith and Tibbles 1980). Based on attempts to protect a population of Gila topminnows from nonnative fishes, Meffe (1983) concluded that the only sure method of eradication would involve repeated massive applications of antimycin combined with construction of downstream barriers.

Other barriers in use include electricity, bubble curtains, and acoustics, either alone or in combination. The largest electrical barrier project currently operational in the USA is in the artificial Chicago Sanitary and Ship Canal (which connects Lake Michigan to the Mississippi River basin). These barriers have received much attention as a means of preventing Asian carps, present in large numbers in the Mississippi and Illinois rivers, from entering the Great Lakes (Egan 2009). An initial barrier with an expected service life of 3–5 years was activated near Romeoville, Illinois, in 2002 (Conover et al. 2007). A more permanent and more powerful barrier is currently being constructed. A portion of the improved barrier was activated in 2009, amid further safety testing. Barriers have the associated disadvantages of high initial construction costs, continued costs for maintenance, environmental impacts, and the potential to impede native fish migration and movements (Hubert and Dawson 2003).

Increased harvest is often suggested, especially by some members of the public, as a means of controlling invasive or unwanted fishes. This alternative may take a variety of forms, for example, modification of regulations to promote angling and commercial harvesting of targeted species. Increased exploitation may be encouraged by incorporating derbies or by offering bounties (Lee 2001). Unfortunately, because fishes vary in their susceptibility to fishing gears, fishing methods used by anglers and commercial fishers are generally size and species selective. Few small fish are captured by these methods, and many adult fish and some species often evade capture. Consequently, the likelihood of removing an entire population by these methods is extremely low (Thresher 1996). Even if substantial numbers of fish are removed, promotion of angling for or creating commercial interest in an invasive fish may also increase the risk that humans will illegally transport the invasive fish to other water bodies (Fuller et al. 1999).

During the 1960s the blue tilapia was introduced into Florida waters and spread rapidly. An early attempt to eradicate the blue tilapia was unsuccessful (Buntz and Manooch 1969; Hale et al. 1995). Over recent years a major commercial fishery for blue tilapia developed in Florida (Hale et al. 1995), but these nonnative cichlids remain abundant and widespread.

8.4.2.3 Biological Methods

A variety of biological methods have been attempted or proposed to eradicate or control unwanted fishes. Release of predatory species to prey on undesirable or invasive species has a long history although it is not particularly common in the fight against invasive fishes. One of the more unusual attempts occurred in 1891 when authorities released 19 sea lions into Lake Merced, California, to prey on and eliminate the reservoir's large common carp population (Smith 1896). In south Florida, predatory South American butterfly peacock bass were stocked into canals to control spotted tilapia and other nonnative cichlids (Shafland 1995). This was judged to be a biological success by some (Shafland 1995; Thresher 2008), but introduced cichlids remain abundant and several new nonnative fishes (e.g., Asian swamp eels and bullseye snakeheads) have colonized canals already occupied by butterfly peacock bass (Collins et al. 2002). Release of nonnative and native predators to control other nonnative fishes also may result in unintended consequences (Moyle et al. 1986), such as stressing native prey fish species. This was the outcome of stocking additional predators to control ruffe in Lake Superior (Mayo et al. 1998).

There is a potential for use of contagious diseases (e.g., koi [common carp] herpes virus, or KHV) to control invasive fishes, but this alternative is highly controversial because of the potential to harm related desirable species (Gilligan and Rayner 2007) and difficulty in correcting any unintended consequences from the introduction. Moreover, it is likely that surviving fish would have immunity to the disease, rendering the method useless after one application. Still, the use of contagious disease could be helpful in combination with other methods.

More technologically advanced biological methods have promise but, with the exception of the release of sterile males to help control sea lamprey in the Great Lakes, these methods remain largely untested in controlling nuisance fish populations. The two main genetic manipulation techniques proposed include (1) chromosome set manipulations involving production and release of triploid sterile nonnative fish with the intent of reducing the population size of targeted naturalized individuals; and (2) recombinant DNA methods involving transgenic techniques designed to produce sterile fish or spread deleterious transgenes (i.e., "Trojan horse" genes) to a target nonnative species (Kapuscinski and Patronski 2005; Gilligan and Rayner 2007; Thresher 2008). In North America, the only field application using a genetic strategy to control an invasive fish involves the annual release of chemically sterilized (by injection with bisazir [P,P-bis(1-azirindinyl)-N-methylphosphinothioic amide]) male sea lamprey into the Great Lakes (Heinrich et al. 2003; Bergstedt and Twohey 2007). This method has been shown to contribute to the sustained reduction of sea lamprey in that ecosystem (Twohey et al. 2003). Researchers in Australia have been interested in use of "daughterless genetic technology" to combat introduced fishes, especially common carp. This genetic technique involves creating a heritable gene that suppresses the production of female offspring thereby causing a reduction in the nuisance population over successive generations (Gilligan and Rayner 2007). Recently U.S. Geological Survey scientists began investigating use of daughterless and under-dominance inheritance as a possible method to control Asian silver and bighead carp populations in North America (King 2009).

Production of genetically-modified food fish within the aquaculture industry has become increasingly common (Howard et al. 2004). In contrast to the use of this technology for biocontrol purposes, commercially-desirable species are genetically manipulated to increase their fitness and, therefore, many scientists are concerned that the resulting engineered fish, if released to the wild, will cause substantial ecological harm (Howard et al. 2004). Numerous concerns exist regarding field application of genetic strategies proposed for control of invasive fishes, and decisions will likely require wide government, scientific, and public approval.

One of the more promising and possibly most benign biological control methods in development is use of pheromones, natural chemicals secreted by many fish and important in influencing their behavior (Sorensen and Stacey 2004; Fine et al. 2006). To date, most laboratory and field application of pheromone techniques have been directed at control of sea lamprey in the Laurentian Great Lakes (e.g., Teeter 1980; Li 2005). Wagner et al. (2006) conducted field tests and demonstrated that pheromone signals were highly effective as attractants in trapping migrating sea lampreys. Even chemically-sterilized males release a sex pheromone that attracts ovulating females (Siefkes et al. 2003). Use of pheromones in sea lamprey control would have the added benefit of reducing reliance on the effective but costly fish toxicants such as TFM.

8.4.3 Past Eradication Campaigns: Successes and Failures

Attempts to eradicate nonnative fish populations in North America have had widely mixed results (Minckley and Deacon 1991; Rinne and Turner 1991; Cailteux et al. 2001). Generally, invasive species are more difficult to eradicate from larger water bodies than from smaller ones (but see Meronek et al. 1996). Most documented successful eradication projects have relied entirely or heavily on the use of rotenone. Nevertheless, a substantial number of eradication projects involving application of rotenone fell short of eradication. Some of these failures are likely explained by poor planning or project implementation rather than rotenone being a poor choice. Rinne and Turner (1991) reviewed past efforts using fish toxicants to remove or eradicate various unwanted fishes from western streams of the USA. Among 26 such projects for which results were provided, 9 (35%) were considered successful, 15 (58%) were listed as either unsuccessful or failures, and 2 others were judged as being a short-term success or of variable success. Among 51 projects reportedly designed to eliminate one or more target fish, Meronek et al. (1996) judged 32 to be successful. However, the Meronek et al. definition of success did not necessarily mean eradication.

Table 8.3 provides a summary of a few recent campaigns in North America to eradicate nonnative or other undesirable fishes. Examples include a range of targeted species and different small and large bodies of water. Following is a brief review of two projects, one an example of a small project in Mexico using traps and the other a large project in California.

In 1996, the African spotted jewelfish was discovered in a natural clear-water pool known as Pozo San Jose del Anteojo in the Cuatro Ciénegas valley of northern Mexico (Lozano-Vilano et al. 2006). The region is home to a number of endemic native fishes and there was evidence that native species in the pool had greatly declined following appearance of the invasive fish. To eradicate the invasive fish, minnow traps were deployed, resulting in removal of over 19,000 spotted jewelfish over the course of 3 years (2000–2002). The invaded water body was quite small (28-m wide, 0.8-m deep), and follow-up trapping indicated the eradication effort was probably successful, although researchers noted continued monitoring of the site.

One of the more recent large-scale attempts to eradicate a nonnative fish population occurred in California (Lee 2001; CDFG 2007). In 1994 northern pike was discovered in Lake Davis, a large artificial reservoir covering 1,600 ha with a maximum depth of 33 m at full pool (see Box 5.5). After substantial planning and various attempts to reduce the northern pike populations by means of nets, electrofishing gear, and other fish capture devices, a decision was made to apply rotenone to the lake and its tributary streams. In the early 1990s, the CDFG had already had success using rotenone to eradicate northern pike from nearby Frenchman Lake, another large and deep lake. The Lake Davis project faced considerable public opposition, especially because of fears that the domestic water supply would be harmed. The CDFG received the necessary permits and cleared the legal challenges in late 1997, and rotenone treatment of the lake began in mid-October of that year (Lee 2001). Despite delays and a few technical difficulties, the 1997 Lake Davis operation initially appeared to be a success, with toxic concentrations of rotenone throughout the lake for several weeks (Siepmann and Finlayson 1999). However, 2 years later northern pike were rediscovered in the lake. It remained unclear whether the 1999 fish were survivors of the 1997 treatment or had been illegally reintroduced (Lee 2001; CDFG 2007). Following rediscovery of northern pike, several years were spent attempting to control and contain the fish without use of piscicides. This included use of explosives (i.e., primacord), a variety of nets, and electrofishing boats. Although the strategy removed large numbers of northern pike, the population continued to increase in size. By that time the public better understood the various threats posed by presence of northern pike. With the added benefit of public support, CDFG developed plans for executing a second rotenone application. Rotenone treatment of the lake and tributary streams occurred in September 2007 and, based on the presence of toxic rotenone concentrations throughout the lake (McMillin and Finlayson 2008) and the absence of northern pike in subsequent fish samples (into 2009), the second rotenone treatment has been considered a success. According to Brian Finlayson (personal communication), state biologists and resource managers learned much about how best to prepare for and conduct rotenone projects as a result of problems that arose during the first Lake Davis treatment.

8.4.4 Research Needs

Throughout North America there is a need for improved methods to eradicate or control nuisance fishes. The need is especially great in situations where introduced fishes are causing the decline of endemic or imperiled native species (Ruiz-Camposa et al. 2006). Future research will likely be devoted to the control or eradication of a few of the more notorious invasive and undesirable species. However, many of the techniques and strategies developed in combating one species can be applied to other undesirable taxa. Future needs include reexamination and adjustment of certain methodologies, such as using currently registered piscicides to greater efficacy. There is also the need to develop and test other chemicals that may be more selective and less harmful to nontarget species. Use of newer biological techniques, including Trojan genes and pheromones, may permit future biologists and resource managers to be highly selective in targeting fish and other aquatic species for removal. Unfortunately, it is likely that many of these advances will be costly to develop, and field applications are decades away.

8.5 INTEGRATED PEST MANAGEMENT

No "silver bullet" exists to prevent or control the introduction or spread of nuisance fishes. Management takes many forms and may, depending on the goal, be aimed at preventing the initial introduction, controlling populations already present, or eradicating established populations. Appropriate management actions and their likelihood of success depend on a variety of factors, including the number and types of introduction vectors, characteristics of the undesired species (e.g., population size, life history, and habitat needs), characteristics of the ecosystem (e.g., habitat available, size of the area invaded, accessibility for application of management measures, and physiochemical characteristics), and a variety of human influences (e.g., willingness to regulate or to apply control measures and economic feasibility). These factors vary not only among species considered for prevention and control measures but also within one species across a variety of human and wild landscapes. Likewise, effective management alternatives appropriate for a given situation depend on whether the species has already been introduced or is newly established, spreading, or already widespread. Successful prevention and control of nuisance fishes often use an approach that identifies and combines all viable management alternatives into an integrated management framework.

8.5.1 Integrated Pest Management: What is it?

Integrated pest management was originally developed out of the need to control agricultural pests and became formalized into a cohesive strategy around the late 1950s (Forney

1999). The beginnings of IPM, however, can be traced to the late 1800s when ecology was identified as the scientific foundation for plant protection (Kogan 1998). Although founded in agriculture, IPM has been applied widely. For example, IPM has been used in shrubs and trees used as landscape plants (Raupp et al. 1992) and to control reed canarygrass (Kilbride and Paveglio 1999), white-tailed deer (Coffey and Johnston 1997), and sea lice on farmed salmonids (Mordue and Pike 2002).

The approach taken by IPM is to incorporate the best management options and control tools available into an overall management plan with the goal of restricting, reducing, and maintaining the target species at levels of insignificant impact while minimizing danger to the environment, human health, and the economy (Hart et al. 2000). Traditionally, therefore, the goal of IPM has to maintain pest populations below nuisance levels (Smith and Reynolds 1966; Dent 1995); however, applying IPM to eradicate targeted pest populations is intuitive, and some management plans with eradication as a goal have proposed IPM (e.g., Asian carps in the Mississippi River drainage; Conover et al. 2007).

8.5.2 Development of an Integrated Pest Management Strategy: Fitting the Pieces Together

A focused IPM strategy begins with determining the type and form of IPM system needed to achieve the desired level of control (Dent 1995). According to Hubert and Dawson (2003), the process is guided by answering a series of questions: (1) who will be using the IPM control techniques? (2) on what scale will the program be implemented? (3) what control measures will be used? (4) how will the control measures be applied? (5) what are the perceived benefits? and (6) over what temporal scale will benefits be realized?

Substantial information must be collected prior to developing an IPM strategy, beginning with detailed information on site conditions and complexity of the system. As thorough an understanding as possible of the current and potential distribution and the life history and ecology of the targeted species, as well as the biology and distribution of sensitive or important nontarget species, is also needed to determine which control measures should be included in the IPM strategy and to identify life stages most appropriate to target for individual control methods. Managers must also consider the economic resources available to tackle the problem, understand the likelihood and severity of undesirable effects of no action as well as understand the likelihood and severity of unanticipated negative effects of the proposed management actions, and work with relevant management authorities to develop common goals and determine responsibilities. It is also important to include all management authorities in developing and implementing an IPM strategy, with input from all stakeholders regarding the managed resources. Greater involvement by outside agencies and entities better ensures that management goals remain relevant and encompass all appropriate resource uses. Integrated pest management requires a more multidisciplined approach than does traditional natural resource management, and logistics of implementing IPM strategies are often not trivial and require specialized personnel training.

Because IPM is typically seen as a longer-term commitment rather than a quick fix, it should not be seen as a stagnant process. Developing and implementing the optimal IPM strategy functions best when practiced iteratively—when the outcomes of management actions are monitored, the results are evaluated, and management actions are adjusted on the basis of what has been learned from previous management efforts. Thus, IPM is improved by adaptive management, a structured, iterative process used to make optimal decisions in the face of uncertainty (See Chapter 5; Holling 1978; Walters 1986).

Another important component of IPM is educating the public not only about the negative effects of nonnative and nuisance fishes but also about the types of control being implemented and potential effects of control efforts. In some cases, public involvement in controlling nuisance fishes has been incorporated into management plans in the form of bounties and encouragement to increase harvest. Many educational materials that target a variety of audiences are available from federal, state, and nongovernmental organizations for this purpose.

With the realization that nonnative and nuisance species do not adhere to jurisdictional boundaries the most effective IPM strategies are developed at a scale sufficient to address the problem. In many applications, this approach may encompass at least the watershed to be managed, but the strategies could be adjusted to regional or larger scales if necessary to achieve greatest effect. Overall, IPM is a holistic approach that seeks to improve management by promoting the exchange of all types of information regarding problems associated with the nuisance fish population and the implemented control actions; allows for increased sharing of data generated from monitoring and control efforts; and aids in developing and communicating a common message across agencies and partners from the public and other entities. One of the best examples of a large-scale program to control an invasive animal that uses IPM as a foundation for success, outside an agricultural context, is the sustained control of sea lamprey in the Great Lakes (Christie and Goddard 2003). This control program exemplifies some of the benefits of IPM: it has a basis in research to discover life stages most vulnerable to control; it continually seeks to develop new and improve existing control techniques; it looks to minimize effects of control on native and desired species; it efforts to minimize environmental impacts of control measures; and it continues to develop efficiencies in research, monitoring, control, and information sharing by extensive use of partnerships. See Box 8.2 for a detailed description of the sea lamprey control program in the Great Lakes.

8.6 CONCLUSIONS

In general, it is society, sometimes in the form of natural resource management goals, other times in the form of irate anglers and members of other user groups, that decides which fish populations are a "nuisance." Moreover, society's perceptions can change considerably over time. Management goals and the methods used to achieve them, therefore, typically do not remain static. Decisions made a century ago to develop or enhance recreational fisheries by liberal stocking of nonnative fishes are judged differently today. Management agencies are more aware of uncertainties associated with deliberate introductions, although stocking fish for recreational and other purposes is still widely practiced throughout North America. Still, it is apparent that today's resource managers have learned from earlier introductions gone awry, and there is a discernible trend toward greater consideration being put into recent stocking decisions. Moreover, continued interest among fisheries managers to reduce risks associated with deliberate introductions indicates that this trend is likely to continue. Likewise, federal, state, tribal, and other management entities, as well as the public, have greater awareness of the unintended consequences that can result from introductions (whether accidental or deliberate). Development of new tools to identify potentially invasive species and risky vectors of introduction demonstrate this greater awareness. New legislation and

Box 8.2. Sea lamprey in the Great Lakes

The campaign to control sea lamprey in the Laurentian Great Lakes has been long, intensive, groundbreaking, successful, and expensive. More than five decades of research and management have been devoted to combating this invasive fish species. Research continues today to fine-tune control techniques in use for decades; to make better decisions about how, when, and where to apply control measures; and to explore new methods using genetic manipulation and pheromones to exploit weaknesses in sea lamprey biology further.

The sea lamprey is a primitive and jawless fish that as an adult is a parasitic predator on larger fish. It is an anadromous species, native to the Atlantic Ocean and spawning in streams along the Atlantic coasts of Europe and the USA. Although native to Lake Ontario (Bryan et al. 2005), modifications to the Welland Canal in 1919 allowed passage of sea lampreys to Lake Erie from Lake Ontario (Christie 1974). By the 1940s, sea lamprey flourished in the upper lakes, drastically reducing the abundance of lake trout, lake whitefish, and cisco populations (Christie 1974). Commercial catches plummeted, and lake trout were eliminated from all of the Great Lakes except for Lake Superior (Elrod et al. 1995). In response to the collapse of commercial fisheries, the governments of the USA and Canada created the Great Lakes Fishery Commission (GLFC) in 1954 with a goal of controlling sea lamprey in the Great Lakes (Kolar et al. 2003). Ashworth (1987) provides a good review of the invasion of sea lamprey into the Great Lakes.

The GLFC began searching for a selective piscicide that could be used to target sea lamprey at the larval stage, when larvae burrow in sediments of small streams for between 5 and 7 years (Kolar et al. 2003). Over 6,000 chemicals were screened, and sensitivity to nitrophenols was discovered (Applegate et al. 1958). Eventually TFM (3-trifluoromethyl-4-nitrophenol) was identified as being selective for sea lamprey (Applegate et al. 1961). A control program evolved based on the application of TFM to streams with larval sea lamprey. Wounding rates (fresh wounds and scars from previous sea lamprey attachment) of stocked lake trout began to decline and abundance of lake trout increased (Kolar et al. 2003). Chemical treatments have been applied to natal streams of sea lamprey in the Great Lakes basin every year since. Even though suppression of sea lamprey populations to 10% of peak abundance through time was achieved, it became apparent that eradication of the species was unlikely and a longer-term control program was developed. As concerns about release of chemicals into the environment grew, the GLFC began to search for other means by which to control sea lamprey. As early as 1980, an integrated pest management approach was suggested for sea lamprey (Sawyer 1980). This concept provided the framework that the GLFC used to develop its integrated management of sea lamprey policy (Davis et al. 1982).

The integrated management program developed to include several different types of control, though chemical application remains important today. An excellent synopsis of the program can be found in Christie and Goddard (2003). In addition, over 800 pages of research articles dedicated to the biology and control of sea lamprey can be found in a 2003 special issue of the *Journal of Great Lakes Research* (volume 29, supplement 1). Current sea lamprey control techniques include applying chemicals (TFM formulations,

(Box continues)

> **Box 8.2. (Continued)**
>
> as well as several formulations of niclosamide [2′,5-dichloro-4′-nitrosalicylanilide], another chemical that selectively kills larval sea lamprey), trapping and removing adults that return to spawn, releasing chemically sterilized males (sterilized by injection with bisazir) to compete with fertile males for spawning females, and deploying physical barriers of various types (mechanical and electrical) to interfere with migration to spawning habitats. Promising new control techniques using pheromones are under evaluation. These techniques affect migration and spawning behaviors of sea lampreys to increase capture efficiency and reduce spawning success (Christie and Goddard 2003; Li et al. 2003; Sorensen and Vrieze 2003). There is also hope that genetic and molecular techniques may be added to the tool chest to manage sea lamprey in the Great Lakes (C. Goddard, GLFC, personal communication).
>
> Sustained control of sea lamprey in the Great Lakes has allowed for revitalization of the commercial and recreational fishery in the Great Lakes, estimated to be worth over US $7 billion annually (Anonymous 2008). The program is successful because of a multitude of factors: realization that successful management of the species would require participation by all fisheries management authorities among the Great Lakes, resulting in formation of the GLFC; study of the biology of sea lamprey and identification of a vulnerable life stage for control; identification and development of a selective chemical agent; use of an integrated management approach that develops and adds new control techniques as they become available; continual funding of research on the biology and control of sea lamprey; management and decision-making infrastructure provided by the GLFC; and sustained social and political will to continue control. At an annual cost of approximately $20 million (M. Gaden, GLFC, personal communication), the sea lamprey control program offers a success story as well as a warning: even for those few species for which specific control measures can be developed, prolonged control comes at substantial economic cost. Therefore, preventing the introduction of nuisance species into additional areas is usually the most cost-effective means of control.

regulations addressing this concern, development of new BMPs, monitoring and reporting networks, and national management plans for particular species of concern further indicate that we, as a society, are more informed and looking to reduce unintended consequences of our actions. Perhaps these concerted efforts portend a brighter future, one with fewer nuisance fish populations and, as a result, less of a need to spend valuable time and resources on eradication and control efforts.

The situation with nonnative fishes is dynamic. Many nonnative fish populations already present will persist, and some will become more abundant, invade new areas, and cause ecological and economic harm. Almost certainly, new introductions will occur, some involving species not currently in the pathway and presenting unique problems. Some nonnative fishes will be viewed as benign and others as nuisance. Certain native fish populations may also expand, impinge on meeting management goals, and thereby also earn the tag of nuisance species. Given this scenario, it is almost certain that nuisance fishes and other aquatic organisms will continue to pose challenges for natural resource managers far into the future. Aquatic ecosystems, because of a limited number of approved control methods and ecosystem complexity and connectivity, offer unique obstacles to implementation of effective management. Global changes such as changes in land and water use and human population growth will affect the distribution, habitat, and management of undesired fishes in North America, presumably often in unforeseen and complex ways. Most introduced species currently established in inland waters of North America exist mostly in what have been considered warm-to-cool temperate areas. With climate change, currently established introduced species can be predicted to expand their ranges in coming decades. Climate change was not a factor considered when past introductions were made, so the full effects of those earlier introductions may not yet have been realized. Consideration of the implications of climate change as well as other global change drivers should be incorporated into current and future fisheries management plans.

To deal with these challenges better, fisheries managers should work toward (1) managing at an appropriate scale, encompassing at least the entire watershed area inhabited by the nuisance species whenever possible; (2) developing IPM strategies after collecting baseline information and carefully considering all control and eradication options and enlisting participation from all relevant partners and management agencies; (3) managing adaptively through time to incorporate lessons learned from past efforts; (4) keeping abreast of and implementing new developments in monitoring, early detection, control and management, and information sharing as appropriate; (5) making concerted efforts to educate and inform individuals, groups, industries, and other entities about risks of introducing aquatic species associated with human activities; and (6) being cognizant of global changes that may be affecting the management area.

Additional research in several key areas might give fisheries managers much needed tools to better combat nuisance fishes. Often nuisance fishes are not discovered until they have already become numerous, have spread to a large area, or both. New methods to detect newly introduced or established nonnative fishes would increase opportunities for successful eradication. Once a species is detected, deciding the appropriate course of action would be improved if more were known about which introductions carry the highest risk to the environment and the economy. Further research in screening and risk assessment techniques and tool development would aid the field of fisheries management. Once it is decided to proceed toward eradication or control, there are many methods from which a manager has to choose. However, there is ample room for improving these existing techniques, and the need to develop additional tools for controlling aquatic species remains great. Lastly, managing nuisance fishes would be better informed by research in the social sciences. Understanding human behavior as it relates to risks of spreading aquatic species unintentionally, factors affecting the decision to liberate animals into inland waters, and natural resource use patterns to assess vectors of introduction better and identifying the most effective means of educating user groups about potential risks would all improve management efforts.

8.7 REFERENCES

Agnarsson, I., and M. Kuntner. 2007. Taxonomy in a changing world: seeking solutions for a science in crisis. Systematic Biology 56:531–539.

Alvarez, J. A., C. Dunn, and A. Zuur. 2003. Response of California red-legged frogs to removal of nonnative fish. Transactions of the Western Section of the Wildlife Society 38/39:9–12.

Anonymous. 2008. Today's angler 2008. American Sportfishing Association, Alexandria, Virginia.

Applegate, V. C., J. H. Howell, J. W. Moffett, B. G. H. Johnson, and M. A. Smith. 1961. Use of 3-trifluoromethyl-4-nitrophenol as a selective sea lamprey larvicide. Great Lakes Fishery Commission Technical Report 1, Ann Arbor, Michigan.

Applegate, V. C., J. H. Howell, and M. A. Smith. 1958. Use of mononitrophenols containing halogens as selective sea lamprey larvicides. Science 127:336–338.

Arkansas Bait and Ornamental Fish Growers Association. 2002. Best management practices for bait and ornamental fish farms. Arkansas Bait and Ornamental Fish Growers Association, Lonoke, Arkansas.

Ashworth, W. 1987. The late Great Lakes, an environmental history. Wayne State University Press, Detroit, Michigan.

ASTM (American Society for Testing and Materials) International. 2008. Standard guide for conducting hazard analysis-critical control point (HACCP) Evaluations. ASTM E2590-08, West Conshohocken, Pennsylvania.

Axon, J. R., and D. K. Whitehurst. 1985. Striped bass management in lakes with emphasis on management problems. Transactions of the American Fisheries Society 114:8–11.

Baird, S. F. 1893. Report of the Commissioner for 1889–91. U.S. Commission of Fish and Fisheries, Government Printing Office, Washington, D.C.

Benson, A. J., M. M. Richerson, and E. Maynard. 2010. *Dreissena rostriformis bugensis*. USGS (U.S. Geological Survey) nonindigenous aquatic species database, Gainesville, Florida. Available: http://nas3.er.usgs.gov/queries/FactSheet.aspx?SpeciesID=95. (March 2010).

Bergstedt, R. A., and M. B. Twohey. 2007. Research to support sterile-male-release and genetic alteration techniques for sea lamprey control. Journal of Great Lakes Research 33 (Special Issue 2):48–69.

Bettoli, P. W., and M. J. Maceina. 1996. Sampling with toxicants. Pages 303–333 *in* B. R. Murphy and D. W. Willis, editors. Fisheries techniques, 2nd edition. American Fisheries Society, Bethesda, Maryland.

Boogaard, M.A. 2003. Delivery systems of piscicides. Pages 39–50 *in* V. K. Dawson and C. S. Kolar, editors. Integrated management techniques to control nonnative fishes. U.S. Geological Survey, Upper Midwest Environmental Sciences Center, La Crosse, Wisconsin. .

Boogaard, M. A., T. D. Bills, J. H. Selgeby, and D. A. Johnson. 1996. Evaluation of piscicides for control of ruffe. North American Journal of Fisheries Management 16:600–607.

Boozer, D. 1973. Tropical fish farming. American Fish Farmer 4(8):4–5.

Brastrup, T. J. 2001. Knife Lake and Knife River rehabilitation project. Pages 9–28 *in* R. L. Cailteux, L. DeMong, B. J. Finlayson, W. Horton, W. McClay, R. A. Schnick, and C. Thompson, editors. Rotenone in fisheries: are the rewards worth the risks? American Fisheries Society, Bethesda, Maryland.

Bryan, M. B., D. Zalinski, K. B. Filcek, S. Libants, W. Li, and K. T. Scribner. 2005. Patterns of invasion and colonization of the sea lamprey (*Petromyzon marinus*) in North America as revealed by microsatellite genotypes. Molecular Ecology 14:3757–3773.

Buntz, J., and C. S. Manooch. 1969. *Tilapia aurea* (Steindachner), a rapidly spreading exotic in south-central Florida. Proceedings of the Annual Conference Southeastern Association of Game and Fish Commissioners 22(1968):495–501.

Burress, R. M., and C. W. Luhning. 1969a. Field trials of antimycin as a selective toxicant in channel catfish ponds. U.S. Fish and Wildlife Service, Bureau of Sport Fisheries and Wildlife, Investigations in Fish Control 25, Washington, D.C.

Burress, R. M. and C. W. Luhning. 1969b. Use of antimycin for selective thinning of sunfish populations in ponds. U.S. Fish and Wildlife Service, Bureau of Sport Fisheries and Wildlife, Investigations in Fish Control 28, Washington, D.C.

Cailteux, R. L., L. DeMong, B. J. Finlayson, W. Horton, W. McClay, R. A. Schnick, and C. Thompson, editors. 2001. Rotenone in fisheries: are the rewards worth the risks? American Fisheries Society, Bethesda, Maryland.

Carpenter, J., and J. W. Terrell. 2005. Effectiveness of fish barriers and renovations for maintaining and enhancing populations of native southwestern fishes. U.S. Geological Survey, Final Report to U.S. Fish and Wildlife Service, Fort Collins, Colorado. Available: http://www.usbr.gov/pmts/fish/Reports/CarpenterBarrierEvaluationReport.pdf. (March 2010).

Cassani, J. R., editor. 1996. Managing aquatic vegetation with grass carp: a guide for water resource managers. American Fisheries Society, Introduced Fish Section, Bethesda, Maryland.

Cassani, J. R., and W. E. Caton. 1985. Induced triploidy in grass carp, *Ctenopharyngodon idella* (Val.). Aquaculture 46:37–44.

CCFAM (Canadian Council of Fisheries and Aquaculture Ministers). 2004. A Canadian action plan to address the threat of aquatic invasive species. Canadian Council of Fisheries and Aquaculture Ministers, Aquatic Invasive Species Task Group. Available: http://www.dfo-mpo.gc.ca/science/enviro/ais-eae/plan/plan-eng.pdf. (November 2009).

CDFG (California Department of Fish and Game). 2007. Lake Davis pike eradication project: final environmental impact report/environmental impact statement (EIR/EIS). California Department of Fish and Game and U.S. Forest Service, Plumas National Forest, Portola. Available: http://www.dfg.ca.gov/lakedavis/EIR-EIS/. (November 2008)

CEC (Commission for Environmental Cooperation). 2009. Trinational risk assessment guidelines for aquatic alien invasive species: test cases for the snakeheads (Channidae) and armored catfishes (Loricariidae) in North American inland waters. Commission for Environmental Cooperation, Montreal.

Christie, G. C., and C. I. Goddard. 2003. Sea lamprey international symposium (SLISS II): advances in the integrated management of sea lamprey in the Great Lakes. Journal of Great Lakes Research 29 (Supplement 1):1–14.

Christie, W. J. 1974. Changes in the fish species composition of the Great Lakes. Journal of the Fisheries Research Board of Canada 31:827–854.

Clarkson, R.W., P. C. Marsh, S. E. Stefferud, and J. A. Stefferud. 2005. Conflicts between native fish and nonnative sport fish management in the southwestern United States. Fisheries 30(9):20–27.

Clearwater, S. J., C. W. Hickey, and M. I. Martin. 2003. Overview of potential piscicides and molluscicides for controlling aquatic pest species in New Zealand. Science for Conservation Series 283, New Zealand Department of Conservation, Wellington.

Coffey, M. A., and G. H. Johnston. 1997. A planning process for managing white-tailed deer in protected areas: integrated pest management. Wildlife Society 25:433–439.

Collins, T. M., J. C. Trexler, L. G. Nico, and T. A. Rawlings. 2002. Genetic diversity in a morphologically conservative invasive taxon: swamp eel introductions in the southeastern United States. Conservation Biology 16:1024–1035.

CONABIO (Comisión Nacional para el Conocimiento y Uso de la Biodiversidad). 2000. Estrategia nacional sobre biodiversidad de México. Available: http://www.inafed.gob.mx/wb/ELOCAL/ELOC_Estrategia_Nacional_sobre_Biodiversidad_de_Me. (November 2009).

Connelly, N. A., C. R. O'Neill Jr., B. A. Knuth, and T. L. Brown. 2007. Economic impacts of zebra mussels on drinking water treatment and electrical power generation facilities. Environmental Management 40:105–112.

Conner, J. V., R. P. Gallagher, and M. F. Chatry. 1980. Larval evidence for natural reproduction of the grass carp (*Ctenopharyngodon idella*) in the lower Mississippi River. Pages 1–19 *in* L. A. Fuiman, editor. Proceedings of the fourth annual larval fish conference. U.S. Fish and Wildlife Service FWS/OBS-80/43.

Conover, G., R. Simmonds, and M. Whalen, editors. 2007. Management and control plan for bighead, black, grass, and silver carps in the United States. Aquatic Nuisance Species Task Force, Asian Carp Working Group, Washington, D.C. Available: http://www.anstaskforce.gov/Documents/Carps_Management_Plan.pdf. (March 2010).

Courtenay, W. R., Jr. 1993. Biological pollution through fish introductions. Pages 35–61 *in* B. N. McKnight, editor. Biological pollution: the control and impact of invasive exotic species. Indiana Academy of Sciences, Indianapolis.

Courtenay, W. R., Jr. 1995. The case for caution with fish introductions. Pages 413–424 *in* J. H. L. Schramm, and R. G. Piper, editors. Uses and effects of cultured fishes in aquatic ecosystems. American Fisheries Society, Symposium 15, Bethesda, Maryland.

Courtenay, W. R., Jr. 2007. Introduced species: what species do you have and how do you know? Transactions of the American Fisheries Society 136:1160–1164.

Courtenay, W. R., Jr., J. E. Deacon, D. W. Sada, R. C. Allan, and G. L. Vinyard. 1985. Comparative status of fishes along the course of the pluvial White River, Nevada. Southwestern Naturalist 30:503–524.

Courtenay, W. R., Jr., D. A. Hensley, J. N. Taylor, and J. A. McCann. 1984. Distribution of exotic fishes in the continental United States. Pages 41–77 *in* W. R. Courtenay Jr. and J. R. Stauffer Jr., editors. Distribution, biology, and management of exotic fishes. Johns Hopkins University Press, Baltimore, Maryland.

Courtenay, W. R., Jr., D. A. Hensley, J. N. Taylor, and J. A. McCann. 1986. Distribution of exotic fishes in North America. Pages 675–698 *in* C. H. Hocutt and E. O. Wiley, editors. Zoogeography of North American freshwater fishes. John Wiley & Sons, New York.

Courtenay, W. R., Jr., and G. K. Meffe. 1989. Small fishes in strange places: a review of introduced poeciliids. Pages 319–331 *in* G. K. Meffe and F. F. Snelson, editors. Ecology and evolution of livebearing fishes (Poeciliidae). Prentice Hall, Englewood Cliffs, New Jersey.

Courtenay, W. R., Jr., and J. R. Stauffer, Jr. editors. 1984. Distribution, biology, and management of exotic fishes. Johns Hopkins University Press, Baltimore, Maryland.

Courtenay, W. R., Jr., and J. R. Stauffer, Jr. 1990. The introduced fish problem and the aquarium fish industry. Journal of the World Aquaculture Society 21:145–159.

Courtenay, W. R., Jr., and J. D. Williams. 1992. Dispersal of exotic species from aquaculture with emphasis on freshwater fishes. Pages 49–81 *in* A. Rosenfield and R. Mann, editors. Dispersal of living organisms into aquatic environments. University of Maryland, Maryland Sea Grant Program, College Park.

Courtenay, W. R., Jr., and J. D. Williams. 2004. Snakeheads (Pisces, Channidae)—a biological synopsis and risk assessment. U.S. Geological Survey Circular 1251.

Courtenay, W. R., Jr., J. D. Williams, R. Britz, M. N. Yamamoto, and P. V. Loiselle. 2004. Identity of introduced snakeheads (Pisces, Channidae) in Hawai'i and Madagascar, with comments on ecological concerns. Bishop Museum Occasional Papers 77:1–13.

Crossman, E. J. 1991. Introduced freshwater fishes: a review of the North American perspective with emphasis on Canada. Canadian Journal of Fisheries and Aquatic Sciences 48 (Supplement 1):46–57.

Davis, J., P. Manion, L. Hudson, B. G. H. Johnson, A. K. Lamsa, W. McCallum, H. Moore, and W. Pearce. 1982. A strategic plan for integrated management of sea lamprey in the Great Lakes. Great Lakes Fishery Commission, Ann Arbor, Michigan.

Dawson, V. K. 2003. Successes and failures of using piscicides. Pages 33–38 *in* V. K. Dawson and C. S. Kolar, editors. Integrated management techniques to control nonnative fishes. U.S. Geological Survey, Upper Midwest Environmental Sciences Center, La Crosse, Wisconsin.

Dawson, V. K., T. D. Bills, and M. A. Boogaard. 1998. Avoidance behavior of ruffe exposed to selected formulations of piscicides. Journal of Great Lakes Research 24:343–350.

Deacon, J. E. and W. L. Minckley. 1991. Western fishes and the real world: the enigma of "endangered species" revisited. Pages 405–413 in W. L. Minckley and J. E. Deacon, editors. Battle against extinction. University of Arizona Press, Tucson.

DeKay, J. E. 1842. Zoology of New-York, or the New-York fauna. Part IV. Fishes. W. and A. White and J. Visscher, Albany, New York.

Dent, D. R. 1995. Defining the problem. Pages 86–104 in D. R. Dent, editor. Integrated pest management. Chapman and Hall, London.

Dettmers, J. N., R. A. Stein, and E. M. Lewis. 1998. Potential regulation of age-0 gizzard shad by hybrid striped bass in Ohio reservoirs. Transactions of the American Fisheries Society 127:84–94.

DFO (Department of Fisheries and Oceans Canada). 2003. National code on introductions and transfers of aquatic organisms. Available: http://www.dfo-mpo.gc.ca/Science/enviro/ais-eae/code/Code2003-eng.pdf. (November 2009).

Diggle, J., J. Day, and N. Bax. 2004. Eradicating European carp from Tasmania and implications for national European carp eradication. Inland Fisheries Service, Hobart, Tasmania.

Dinger, E. C., and J. C. Marks. 2007. Effects of high levels of antimycin A on aquatic invertebrates in a warmwater Arizona stream. North American Journal of Fisheries Management 27:1243–1256.

Donlan, C. J., and C. Wilcox. 2007. Complexities of costing eradications. Animal Conservation 10:154–156.

Egan, D. 2009. New carp barrier to be activated. River Crossings 18(1):1–3.

Elrod, J. H., R. O'Gorman, C. P. Schneider, T. H. Eckert, T. Schaner, J. N. Bowlby, and L. P. Schleen. 1995. Lake trout rehabilitation in Lake Ontario. Journal of Great Lakes Research 21 (Supplement 1):83–107.

Fausch, K. D. 2008. A paradox of trout invasions in North America. Biological Invasions 10:685–701.

Filbey, M., C. Kennedy, J. Wilkinson, and J. Balch. 2002. Halting the invasion: state tools for invasive species management. Environmental Law Institute, Washington, D.C.

Fine, J. M., S. P. Sisler, L. A. Vrieze, W. D. Swink, and P. W. Sorensen. 2006. A practical method for obtaining useful quantities of pheromones from sea lamprey and other fishes for identification and control. Journal of Great Lakes Research 32:832–838.

Finlayson, B. J., R. A. Schnick, R. L. Cailteux, L. DeMong, W. D. Horton, W. McClay, and C. W. Thompson. 2002. Assessment of antimycin A use in fisheries and its potential for reregistration. Fisheries 27(6):10–19.

Finlayson, B. J., R. A. Schnick, R. L. Cailteux, L. DeMong, W. D. Horton, W. McClay, C. W. Thompson, and G. J. Tichacek. 2000. Rotenone use in fisheries management: administrative and technical guidelines manual. American Fisheries Society, Bethesda, Maryland.

Finlayson, B., R. Schnick, D. Skaar, J. Anderson, L. Demong, D. Duffield, W. Horton, and J. Steinkjer. 2010. Planning and standard operating procedures for the use of rotenone in fish management—rotenone SOP manual. American Fisheries Society, Bethesda, Maryland.

Finlayson, B., W. Somer, D. Duffield, D. Propst, C. Mellison, T. Pettengill, H. Sexauer, T. Nesler, S. Gurtin, J. Elliot, F. Partridge, and D. Skaar. 2005. Native inland trout restoration on national forests in the western United States: time for improvement? Fisheries 30(5):10–19.

Forney, D. R. 1999. Importance of pesticides in integrated pest management. Pages 174–197 in N. N. Ragsdale and J. N. Seiber, editors. Pesticides: managing risks and optimizing benefits. American Chemical Society, Washington, D.C.

Fuller, P. L., L. G. Nico, and J. D. Williams. 1999. Nonindigenous fishes introduced into the inland waters of the United States. American Fisheries Society, Special Publication 27, Bethesda, Maryland.

Ganley, T., and R. Bock. 1998. Fish collecting in Florida: collecting tropical fish without leaving the country. Aquarium Fish Magazine (November):31–41.

Gilligan, D., and T. Rayner. 2007. The distribution, spread, ecological impacts and potential control of carp in upper Murray River. New South Wales Department of Primary Industries—Fisheries Research Report Series 14, Cronulla, Australia.

Goertner, J. F., M. L. Wiley, G. A. Young, and W. W. McDonald. 1994. Effects of underwater explosions on fish without swim bladders. Naval Surface Warfare Center, Dahlgren Division, White Oak Detachment, Technical Report NSWC TR 88–114, Silver Spring, Maryland.

Gresswell, R. E. 1991. Use of antimycin for removal of brook trout from a tributary of Yellowstone Lake. North American Journal of Fisheries Management 11:83–90.

Grigorovich, I. A., R. I. Coautti, and H. J. MacIsaac. 2003. Ballast mediated animal introductions in the Laurentian Great Lakes: retrospective and prospective analysis. Canadian Journal of Fisheries and Aquatic Sciences 60:740–756.

Hale, M. M., J. E. Crumpton, and J. R. J. Schuler. 1995. From sportfishing bust to commercial fishing boon: a history of the blue tilapia in Florida. Pages 425–430 *in* J. H. L. Schramm and R. G. Piper, editors. Uses and effects of cultured fishes in aquatic ecosystems, American Fisheries Society, Symposium 15, Bethesda, Maryland.

Hanlon, S. G., M. V. Hoyer, C. E. Cichra, and D. E. Canfield Jr. 2000. Evaluation of macrophyte control in 38 Florida lakes using triploid grass carp. Journal of Aquatic Plant Management 38:48–54.

Hart, S., M. Klepinger, H. Wandell, D. Garling, and L. Wolfson. 2000. Integrated pest management for nuisance exotics in Michigan inland lakes. Michigan State University Extension, Water Quality Series: WQ-56, East Lansing.

Heinrich, J. W., K. M. Mullett, M. J. Hansen, J. V. Adams, G. T. Klar, D. A. Johnson, G. C. Christie, and R. J. Young. 2003. Sea lamprey abundance and management in Lake Superior, 1957 to 1999. Journal of Great Lakes Research 29 (Supplement 1):566–583.

Holden, P. B. 1991. Ghosts of the Green River: impacts of Green River poisoning on management of native fishes. Pages 43–54 *in* W. L. Minckley and J. E. Deacon, editors. Battle against extinction: native fish management in the American West. University of Arizona Press, Tucson.

Holling, C. S., editor. 1978. Adaptive environmental assessment and management. John Wiley & Sons, New York.

Honeyfield, D. C., S. B. Brown, J. D. Fitzsimons, and D. E. Tillitt. 2005. Early mortality syndrome in Great Lakes salmonines. Journal of Aquatic Animal Health 17:1–3.

Horak, D. 1995. Native and nonnative fish species used in state fisheries management programs in the United States. Pages 61–67 *in* H. L. Schramm and R. G. Piper, editors. Use and effects of cultured fishes in aquatic ecosystems. American Fisheries Society, Symposium 15, Bethesda, Maryland.

Howard, R. D., J. A. DeWoody, and W. M. Muir. 2004. Transgenic male mating advantage provides opportunity for Trojan gene effect in a fish. Proceedings of the National Academy of Sciences 101:2934–2938.

Hubbs, C. 1980. The solution to the Cyprinodon bovinus problem: eradication of a pupfish genome. Proceedings of the Desert Fishes Council 10:9–18.

Hubert, T. D., and V. K. Dawson. 2003. Developing an integrated pest management strategy. Pages 81–86 *in* V. K. Dawson and C. S. Kolar, editors. Integrated management techniques to control nonnative fishes. U.S. Geological Survey, Upper Midwest Environmental Sciences Center, La Crosse, Wisconsin.

Hunn, J. B., and W. D. Youngs. 1980. Role of physical barriers in the control of sea lamprey (*Petromyzon marinus*). Canadian Journal of Fisheries and Aquatic Science 37:2118–2122.

Jackson, J. R., J. C. Boxrucker, and D. W. Willis. 2004. Trends in agency use of propagated fishes as a management tool in inland fisheries. Pages 121–138 *in* M. J. Nickum, P. M. Mazik, J. G. Nickum, and D. D. MacKinlay, editors. Propagated fish in resource management. American Fisheries Society, Symposium 44, Bethesda, Maryland.

Jelks, H. L., S. J. Walsh, N. M. Burkhead, S. Contreras-Balderas, E. Díaz-Pardo, D. A. Hendrickson, J. Lyons, N. E. Mandrak, F. McCormick, J. S. Nelson, S. P. Platania, B. A. Porter, C. B. Renaud, J. J. Schmitter-Soto, E. B. Taylor, and M. L. Warren Jr. 2008. Conservation status of imperiled North American freshwater and diadromous fishes. Fisheries 33(8):372–407.

Jenkins, R. E., and N. M. Burkhead. 1994. Freshwater fishes of Virginia. American Fisheries Society, Bethesda, Maryland.

Jensen, D. A. 2008. Successful education efforts to meet the aquatic invasive species challenge. Michigan's call to action on AIS, East Lansing, Michigan. Available: http://www.michigan.gov/documents/deq/Minn-AIS-Ed-efforts_230392_7.pdf. (November 2009).

Johnston, K. H. 1957. Removal of longnose gar from rivers and streams with the use of dynamite. North Carolina Wildlife Resources Commission, Raleigh.

Johnston, K. H. 1961. Removal of longnose gar from rivers and streams with the use of dynamite. Proceedings of the Annual Conference Southeastern Association of Game and Fish Commissioners 15(1961):205–207.

Jordan, D. S. 1891. A reconnaissance of the streams and lakes of Yellowstone National Park, Wyoming, for the purposes of the U.S. Fish Commission. U.S. Fish Commission Bulletin 9:41–63.

Jude, D. J., J. Janssen, and G. Crawford. 1995. Ecology, distribution, and impact of the newly introduced round and tubenose gobies on the biota of the St. Clair and Detroit rivers. Pages 447–460 *in* M. Munawar, T. Edsall, and J. Leach, editors. The Lake Huron ecosystem: ecology, fisheries and management. SPB Academic Publishing, Amsterdam.

Kapuscinski, A. R., and T. J. Patronski. 2005. Genetic methods for biological control of nonnative fish in the Gila River basin. Final Report to the U.S. Fish and Wildlife Service, Minnesota Sea Grant Publication F20, St. Paul. Available: http://www.seagrant.umn.edu/downloads/f20.pdf. (March 2010).

Keevin, T. M. 1998. A review of natural resource agency recommendations for mitigating the impacts of underwater blasting. Reviews in Fisheries Science 6:281–313.

Keller, R. P., D. M. Lodge, M. A. Lewis, and J. F. Shogren. 2009. Bioeconomics of invasive species: integrating ecology, economics, policy, and management. Oxford University Press, New York.

Kilbride, K. M., and F. L. Paveglio. 1999. Integrated pest management to control reed canarygrass in seasonal wetlands of southwestern Washington. Wildlife Society Bulletin 27:292–297.

King, T. L. 2009. Microsatellite DNA markers for assessing phylogeographic and population structure in three invasive Asian carp species: silver carp (*Hypophthalmichthys molitrix*), big head carp (*Hypophthalmichthys nobilis*), and grass carp (*Ctenopharyngodon idella*). Available: http://biology.usgs.gov/genetics_genomics/geos.html. (March 2010).

Knapp, R. A., and K. R. Matthews. 1998. Eradication of nonnative fish by gill netting from a small mountain lake in California. Restoration Ecology 6:207–213.

Kogan, M. 1998. Integrated pest management: historical perspectives and contemporary developments. Annual Review of Entomology 43:243–270.

Kolar, C. S., M. A. Boogaard, and T. D. Hubert. 2003. Case study of integrated pest management control of sea lamprey in the Great Lakes. Pages 93–104 *in* V. K. Dawson and C. S. Kolar, editors. Integrated management techniques to control nonnative fishes. U.S. Geological Survey, Upper Midwest Environmental Sciences Center, La Crosse, Wisconsin.

Kolar, C. S., D. C. Chapman, W. R. Courtenay Jr., C. M. Housel, J. D. Williams, and D. P. Jennings. 2007. Bigheaded carps; a biological synopsis and environmental fish assessment. American Fisheries Society, Special Publication 33, Bethesda, Maryland.

Krumholz, L. A. 1948. The use of rotenone in fisheries research. Journal of Wildlife Management 12:305–317.

Kulp, R. A., and S. E. Moore. 2000. Multiple electrofishing removals for eliminating rainbow trout in a small southern Appalachian stream. North American Journal of Fisheries Management 20:259–266.

Lachner, E. A., C. R. Robins, and W. R. Courtenay Jr. 1970. Exotic fishes and other aquatic organisms introduced into North America. Smithsonian Contributions to Zoology 59:1–29.

Laycock, G. 1966. The alien animals. Natural History Press, Garden City, New York.

Lazur, A., S. Early, and J. M. Jacobs. 2006. Acute toxicity of 5% rotenone to northern snakeheads. North American Journal of Fisheries Management 26:628–630.

Leach, J. H., and C. A. Lewis. 1991. Fish introductions in Canada: provincial views and regulations. Canadian Journal of Fisheries and Aquatic Sciences 48 (Supplement 1):156–161.

Lee, D. P. 2001. Northern pike control at Lake Davis, California. Pages 55–61 *in* R. L. Cailteux, L. DeMong, B. J. Finlayson, W. Horton, W. McClay, R. A. Schnick, and C. Thompson, editors. Rotenone in fisheries: are the rewards worth the risks? American Fisheries Society, Bethesda, Maryland.

Lentsch, L. D., C. W. Thompson, and R. L. Spateholts. 2001. Overview of a large-scale chemical treatment success story: Strawberry Valley, Utah. Pages 63–79 *in* R. L. Cailteux, L. DeMong, B. J. Finlayson, W. Horton, W. McClay, R. A. Schnick, and C. Thompson, editors. Rotenone in fisheries: are the rewards worth the risks? American Fisheries Society, Bethesda, Maryland.

Leonard, J. R. 1979. The fish car era. U.S. Government Printing Office, Washington, D.C.

Leung, B., D. M Lodge, D. Finnoff, J. F. Shogren, M. A. Lewis, and G. Lamberti. 2002. An ounce of prevention or a pound of cure: bioeconomic risk analysis of invasive species. Proceedings of the Royal Society B 269(1508):2407–2413.

Li, W. 2005. Potential multiple functions of a male sea lamprey pheromone. Chemical Senses 30 (Supplement 1):i307–i308.

Li, W., M. J. Siefkes, A. P.Scott, and J. H. Teeter. 2003. Sex pheromone communication in the sea lamprey: implications for integrated management. Journal of Great Lakes Research 29 (Supplement 1):85–94.

Lodge, D. M., S. Williams, H. J. MacIsaac, K. R. Hayes, B. Leung, S. Reichard, R. N. Mack, P. B. Moyle, M. Smith, D. A. Andow, J. T. Carlton, and A. McMichael. 2006. Biological invasions: recommendations for U.S. policy and management. Ecological Applications 16:2035–2054.

Lowman, F. G. 1959. Experimental selective rotenone killing of undesirable fish species in flowing streams. Texas Game and Fish Commission, Federal Aid in Fish Restoration, Project F-9-R-6, Job E-3, Segment Completion Report, Austin.

Lozano-Vilano, M. L., A. J. Contreras-Balderas, and M. E. Garcia-Ramirez. 2006. Eradication of spotted jewelfish, *Hemichromis guttatus*, from Poza San Jose del Anteojo, Cuatro Cienegas Bolson, Coahuila, Mexico. Southwestern Naturalist 51:553–555.

Madenjian, C. P., G. L. Fahnenstiel, T. H. Johengen, T. F. Nalepa, H. A. Vanderploeg, G. W. Fleisher, P. J.Schneeberger, D. M. Benjamin, E. B. Smith, J. R. Bence, E. S. Rutherford, D. S. Lavis, D. M. Robertson, D. J. Jude, and M. P. Ebener. 2002. Dynamics of the Lake Michigan food web, 1970–2000. Canadian Journal of Fisheries and Aquatic Sciences 59:36–53.

Mallison, C. T., R. S. Hestand, and B. Z. Thompson. 1995. Removal of triploid grass carp with an oral rotenone bait in two central Florida lakes. Lake and Reservoir Management 11:337–342.

Marking, L. L. 1992. Evaluation of toxicants for the control of carp and other nuisance fishes. Fisheries 17(6):6–12.

Mayo, K. R., J. H. Selgeby, and M. E. McDonald. 1998. A bioenergetics modeling evaluation of top-down control of ruffe in the St. Louis River, western Lake Superior. Journal of Great Lakes Research 24:329–342.

McClay, W. 2005. Rotenone use in North America (1988–2002). Fisheries 30(4):29–31.

McGowan, C., and W. S. Davidson. 1992. Unidirectional natural hybridization between brown trout (*Salmo trutta*) and Atlantic salmon (*S. salar*) in Newfoundland. Canadian Journal of Fisheries and Aquatic Sciences 49:1953–1958.

McMillin, S., and B. Finlayson. 2008. Chemical residues in water and sediment following rotenone application to Lake Davis, California 2007. California Department of Fish and Game, Office of Spill Prevention and Response, Administrative Report 08–01, Rancho Cordova.

Meffe, G. K. 1983. Attempted chemical renovation of an Arizona springbrook for management of the endangered Sonoran topminnow. North American Journal of Fisheries Management 3:315–321.

Mendoza, R., S. Contreras-Balderas, C. Ramirez, P. Alverez, and V. Aguilar. 2007. Los peces diablo: especies invasoras de alto impacto. Biodiversitas 68:2–5.

Meronek, T. G., P. M. Bouchard, E. R. Buckner, T. M. Burri, K. K. Demmerly, D. C. Hatleli, R. A. Klumb, S. H. Schmidt, and D. W. Coble. 1996. A review of fish control projects. North American Journal of Fisheries Management 16:63–74.

Miller, R. R. 1963. Distribution, variation, and ecology of *Lepidomeda vittata*, a rare cyprinid fish endemic to eastern Arizona. Copeia 1963:1–5.

Minckley, W. L., and J. E. Deacon, editors. 1991. Battle against extinction: native fish management in the American West. University of Arizona Press, Tucson.

Minister of Supply and Services Canada. 1995. Canadian biodiversity strategy: Canada's response to the Convention on Biological Diversity. Available: http://www.cbin.ec.gc.ca/documents/national_reports/cbs_e.pdf. (November 2009).

Mitchell, A. J. and A. M. Kelly. 2006. The public sector role in the establishment of grass carp in the United States. Fisheries 31(3):113–121.

Moore, S. E., M. Kulp, B. Rosenlund, J. Brooks, and D. Propst. 2008. A field manual for the use of antimycin A for restoration of native fish populations. U.S. Department of the Interior, National Park Service, National Resource Report NPS/NRPC/NRR-2008/001, Fort Collins, Colorado.

Moore, S. E., G. L. Larson, and B. Ridley. 1986. Population control of exotic rainbow trout in streams of a natural area park. Environmental Management 10:215–219.

Mordue, L. A. J., and A. W. Pike. 2002. Salmon farming: towards an integrated pest management strategy for sea lice. Pest Management Science 58:513–514.

Moyle, P. B. 2002. Inland fishes of California. University of California Press, Berkeley.

Moyle, P. B., H. W. Li, and B. A. Barton. 1986. The Frankenstein effect: impact of introduced fishes on native fishes in North America. Pages 415–425 *in* R. H. Stroud, editor. Fish culture in fisheries management. American Fisheries Society, Fish Culture and Fisheries Management Section, Bethesda, Maryland.

Moyle, P. B., and T. Light. 1996. Fish invasions in California: do abiotic factors determine success? Ecology 77:1666–1670.

Mudrak, V. A., and G. J. Carmichael. 2005. Considerations for the use of propagated fishes in resource management. American Fisheries Society, Bethesda, Maryland.

Mueller, G. A. 2005. Predatory fish removal and native fish recovery in the Colorado River main stem: what have we learned? Fisheries 30(9):10–19.

Muñoz, A. A., and R.A. Mendoza. 2009. Especies exoticas invasoras: impactos sobre las poblaciones de flora y fauna, los procesos ecologicos y la economia, Paginas 277–318 *en* Capital natural de México, vol. II: estado de conservación y tendencias de cambio. CONABIO (Comisión Nacional para el Conocimiento y Uso de la Biodiversidad), Tlalpan, Mexico.

Neilson, K., R. Kelleher, G. Barnes, D. A. Speirs, and J. Kelly. 2004. Use of fine-mesh monofilament gill nets for the removal of rudd (*Scardinius erythropthalmus*) from a small lake complex in Waikato, New Zealand. New Zealand Journal of Marine and Freshwater Research 38:525–539.

Nickum, M. J., P. M. Mazik, J. G. Nickum, and D. D. MacKinlay, editors. 2004. Propagated fish in resource management. American Fisheries Society, Symposium 44, Bethesda, Maryland.

Nico, L. G., P. L. Fuller, and P. J. Schofield. 2010. *Ctenopharyngodon idella*. USGS (U.S. Geological Survey) nonindigenous aquatic species database, Gainesville, Florida. Available: http://nas3.er.usgs.gov/queries/FactSheet.aspx?SpeciesID=514. (February 2010).

Nico, L. G., H. L. Jelks, and T. Tuten. 2009. Nonnative suckermouth armored catfishes in Florida: description of nest burrows and burrow colonies with assessment of shoreline conditions. Aquatic Nuisance Species Research Program Bulletin 9:1–30.

Nico, L. G., and T. R. Martin. 2001. The South American suckermouth armored catfish, *Ptergoplichthys anisitsi* (Pisces: Loricariidae), in Texas, with comments on foreign fish introductions in the American Southwest. Southwestern Naturalist 46:98–104.

Nico, L. G., J. D. Williams, and H. L. Jelks. 2005. Black carp: biological synopsis and risk assessment of an introduced fish. American Fisheries Society, Special Publication 32, Bethesda, Maryland.

NISC (National Invasive Species Council). 2001. Meeting the invasive species challenge: national invasive species management plan. National Invasive Species Council, Washington, D.C. Available: http://www.invasivespeciesinfo.gov/docs/council/mpfinal.pdf. (March 2010).

NISC (National Invasive Species Council). 2008. 2008–2012 National invasive species management plan. National Invasive Species Council, Washington, D.C. Available: http://www.invasivespeciesinfo.gov/council/mp2008.pdf. (June 2009).

Parker, B. R., D. W. Schindler, D. B. Donald, and R. S. Anderson. 2001. The effects of stocking and removal of a nonnative salmonid on the plankton of an alpine lake. Ecosystems 4(4):334–345.

Propst, D. L., K. B. Gido, and J. A. Stefferud. 2008. Natural flow regimes, nonnative fishes, and native fish persistence in arid-land river systems. Ecological Applications 18:1236–1252.

Ramsey, J. S. 1985. Sampling aquarium fishes imported by the United States. Journal of the Alabama Academy of Science 56:220–245.

Raupp, M. J., C. S. Koehler, and J. A. Davidson. 1992. Advances in implementing integrated pest management for woody landscape plants. Annual Review of Entomology 37:361–385.

Reeves, E. 1999. Analysis of laws and policies concerning exotic invasions of the Great Lakes. Office of the Great Lakes Michigan Department of Environmental Quality, Lansing.

Ricciardi, A., and J. B. Rasmussen. 1998. Predicting the identity and impact of future biological invaders: a priority for aquatic resource management. Canadian Journal of Fisheries and Aquatic Sciences 55:1759–1765.

Rinne, J. N., L. Riley, R. Bettaso, R. Sorenson, and K. Young. 2004. Managing southwestern native and nonnative fishes: can we mix oil and water and expect a favorable solution? Pages 445–466 *in* M. J. Nickum, P. M. Mazik, J. G. Nickum, and D. D. MacKinlay, editors. Propagated fish in resource management. American Fisheries Society, Symposium 44, Bethesda, Maryland.

Rinne, J. N., and P. R. Turner. 1991. Reclamation and alteration as management techniques, and a review of methodology in stream renovation. Pages 219–244 *in* W. L. Minckley and J. E. Deacon, editors. Battle against extinction: native fish management in the American West. University of Arizona Press, Tucson.

Roberts, J., and R. Tilzey, editors. 1996. Controlling carp: exploring the options for Australia. CSIRO (Commonwealth Scientific and Industrial Research Organisation) Land and Water, Canberra, Australia. Available: http://www.clw.csiro.au/publications/controlling_carp.pdf. (December 2008).

Ruiz-Camposa, G., F. Camarena-Rosalesa, C. A. Reyes-Valdeza, J. de la Cruz-Agüeroc, and E. Torres-Balcazar. 2006. Distribution and abundance of the endangered killifish *Fundulus lima*, and its interaction with exotic fishes in oases of Central Baja California, Mexico. Southwestern Naturalist 51:502–509.

Sawyer, A. J. 1980. Prospects for integrated pest management of the sea lamprey (*Petromyzon marinus*). Canadian Journal of Fisheries and Aquatic Resources 37:2081–2092.

Schindler, D. W., and B. R. Parker. 2002. Biological pollutants: alien fishes in mountain lakes. Water Air and Soil Pollution 2:379–397.

Schofield, P. J., and L. G. Nico. 2007. Toxicity of 5% rotenone to nonindigenous Asian swamp eels. North American Journal of Fisheries Management 27:453–459.

Schramm, H. L., Jr. and R. G. Piper, editors. 1995. Uses and effects of cultured fishes in aquatic ecosystems. American Fisheries Society, Symposium 15, Bethesda, Maryland.

Schreier, T. M., V. K. Dawson, and W. Larson. 2008. Effectiveness of piscicides for controlling round gobies (*Neogobius melanostomus*). Journal of Great Lakes Research 34:253–264.

Scoppettone, G. G., P. H. Rissler, C. Gourley, and C. Martinez. 2005. Habitat restoration as a means of controlling nonnative fish in a Mojave Desert oasis. Restoration Ecology 13:247–256.

Shafland, P. L. 1995. Introduction and establishment of a successful butterfly peacock fishery in southeast Florida canals. Pages 443–451 *in* H. L. Schramm Jr. and R. G. Piper, editors. Uses and effects of cultured fishes in aquatic ecosystems, American Fisheries Society, Symposium 15, Bethesda, Maryland.

Shafland, P. L., and K. J. Foote. 1979. A reproducing population of *Serrasalmus humeralis* Valenciennes in southern Florida. Florida Scientist 42:206–214.

Shafland, P. L., K. B. Gestring, and M. S. Stanford. 2008. Florida's exotic freshwater fishes—2007. Florida Scientist 71:220–245.

Short, C. I., S. K. Gross, and D. Wilkinson. 2004. Preventing, controlling, and managing alien species introduction for the health of aquatic and marine ecosystems. Pages 109–125 *in* E. E. Knudsen, D. D. MacDonald, and Y. K. Muirhea, editors. Sustainable management of North American fisheries. American Fisheries Society, Symposium 43, Bethesda, Maryland.

Siefkes, M. J., R. A. Bergstedt, M. B. Twohey, and W. Li. 2003. Chemosterilization of male sea lampreys (*Petromyzon marinus*) does not affect sex pheromone release. Canadian Journal of Fisheries and Aquatic Sciences 60:23–31.

Siepmann, S., and B. Finlayson. 1999. Chemical residues in water and sediment following rotenone application to Lake Davis, California. California Department of Fish and Game, Office of Spill Prevention and Response, Administrative Report 99–2, Sacramento.

Simberloff, D. 2009. We can eliminate invasions or live with them: successful management projects. Biological Invasions 11:149–157.

SMARN (Secretaría de Medio Ambiente y Recursos Naturales). 2009. Estrategia nacional sobre especies invasoras en México: prevención, control y erradicación. CONABIO (Comisión Nacional para el Conocimiento y Uso de la Biodiversidad), Tlalpan, Mexico.

Smith, B. R., and J. J. Tibbles. 1980. Sea lamprey (*Petromyzon marinus*) in lakes Huron, Michigan, and Superior: history of invasion and control, 1936–78. Canadian Journal of Fisheries and Aquatic Science 37:1780–1801.

Smith, H. M. 1896. A review of the history and results of the attempts to acclimatize fish and other water animals in the Pacific states. U.S. Fish Commission Bulletin 15(1895):379–472.

Smith, R. F., and H. T. Reynolds. 1966. Principles, definitions and scope of integrated pest control. Pages 11–17 *in* Proceedings of the FAO Symposium on Integrated Pest Control. Food and Agriculture Organization of the United Nations, Rome.

Sorensen, P. W., and N. E. Stacey. 2004. Brief review of fish pheromones and discussion of their possible uses in the control of nonindigenous teleost fishes. New Zealand Journal of Marine and Freshwater Research 38:399–417.

Sorensen, P. W., and L. A. Vrieze. 2003. The chemical ecology and potential application of the migratory pheromone in the sea lamprey. Journal of Great Lakes Research 29 (Supplement 1):66–84.

Stanley, J. G., R. A. Peoples, and J. A. McCann. 1991. U.S. Federal policies, legislation, and responsibilities related to importation of exotic fishes and other aquatic organisms. Canadian Journal of Fisheries and Aquatic Sciences 48 (Supplement 1):162–166.

Stauffer, J. R., S. E. Boltz, and J. M. Boltz. 1988. Cold shock susceptibility of blue tilapia from the Susquehanna River, Pennsylvania. North American Journal of Fisheries Management 8:329–332.

Stroud, R. H., editor. 1986. Fish culture in fisheries management. American Fisheries Society, Fish Culture and Fisheries Management Section, Bethesda, Maryland.

Teeter, J. 1980. Pheromone communication in sea lampreys (*Petromyzon marinus*): implications for population control. Canadian Journal of Fisheries and Aquatic Science 37:2123–2132.

Teleki, G. C., and A. J. Chamberlain. 1978. Acute effects of underwater construction blasting on fishes in Long Point Bay, Lake Erie. Journal of the Fisheries Research Board of Canada 35:1191–1198.

Thompson, P. D., and F. J. Rahel. 1996. Evaluation of depletion-removal electrofishing of brook trout in small Rocky Mountain streams. North American Journal of Fisheries Management 16:332–339.

Thresher, R. E. 1996. Physical removal as an option for the control of feral carp populations. Pages 58–72 *in* J. Roberts and R. Tilzey, editors. Controlling carp: exploring the options for Australia. CSIRO (Commonwealth Scientific and Industrial Research Organisation) Land and Water, Canberra, Australia. Available: http://www.clw.csiro.au/publications/controlling_carp.pdf. (December 2008).

Thresher, R. E. 2008. Autocidal technology for the control of invasive fish. Fisheries 33(3):114–121.

Truemper, H. 2008. Diamond Lake 2006–2007: tui chub presence/absence report. Oregon Department of Fish and Wildlife, Roseburg, Oregon. Available: http://www.dfw.state.or.us/fish/diamond_lake/docs/Lake_Condition_Index_Report_2006–2007.pdf. (June 2009).

Twohey, M. B., J. W. Heinrich, J. G. Seelye, K. T. Fredricks, R. A. Bergstedt, C. A. Kaye, R. J. Scholefield, R. B. McDonald, and G. C. Christie. 2003. The sterile-male-release technique in Great Lakes sea lamprey management. Journal of Great Lakes Research (Supplement 1):410–423.

U.S. Congress. 1993. Harmful non-indigenous species in the United States. U.S. Congress, Office of Technology Assessment, OTA-F-565, U.S. Government Printing Office, Washington, D.C.

U.S. Environmental Protection Agency. 2008. Aquaculture operations—best management practices. Available: http://www.epa.gov/oecaagct/anaqubmp.html. (March 2010).

USFWS (U.S. Fish and Wildlife Service). 2002. Final finding of no significant impact: tilapia removal program on the Virgin River, Clark County, Nevada, and Mohave County, Arizona. U.S. Fish and Wildlife Service, Ecological Services, Southern Nevada Field Office, Las Vegas, Nevada. Available: www.fws.gov/Nevada/protected_species/fish/documents/vr/virgin_river_fonsi.pdf. (December 2008).

Vredenburg, A. T. 2004. Reversing introduced species effects: experimental removal of introduced fish leads to rapid recovery of a declining frog. Proceedings of the National Academy of Sciences 101:7646–7650.

Wagner, C. M., M. L. Jones, M. B. Twohey, and P. W. Sorensen. 2006. A field test verifies that pheromones can be useful for sea lamprey (*Petromyzon marinus*) control in the Great Lakes. Canadian Journal of Fisheries and Aquatic Sciences 63:475–479.

Wakida-Kusunoki, A., R. Ruiz-Carus, and E. Amador-del-Angel. 2007. Amazon sailfin catfish, *Pterygoplichthys pardalis* (Castelnau, 1855) (Loricariidae), another exotic species established in southeastern Mexico. Southwestern Naturalist 52:141–144.

Walters, C. 1986. Adaptive management of renewable resources. Macmillan, New York.

Wattendorf, R. J. 1986. Rapid identification of triploid grass carp with a Coulter counter and channelyzer. Progressive Fish-Culturist 48:125–132.

Weidel, B. C., D. C. Josephson, and C. E. Kraft. 2007. Littoral fish community response to smallmouth bass removal from an Adirondack lake. Transactions of the American Fisheries Society 136:778–789.

Welcomme, R. L. 1981. Register of international transfers of inland fish species. FAO (Food and Agriculture Organization of the United Nations) Fisheries Technical Paper 213, Rome.

Willis, K. and N. Ling. 2000. Sensitivities of mosquitofish and black mudfish to a piscicide: could rotenone be useful to control mosquitofish in New Zealand wetlands? New Zealand Journal of Zoology 27:85–91.

Wydoski, R. S., and R. W. Wiley. 1999. Management of undesirable fish species. Pages 403–430 *in* C. C. Kohler and W. A. Hubert, editors. Inland fisheries management in North America, 2nd edition. American Fisheries Society, Bethesda, Maryland.

Chapter 9

Use of Hatchery Fish for Conservation, Restoration, and Enhancement of Fisheries

Jesse Trushenski, Thomas Flagg, and Christopher Kohler

9.1 INTRODUCTION

Aquaculture is the propagation of aquatic organisms under circumstances that facilitate greater productivity than would be observed in a natural setting. In terms of tangible resources and labor, culture methods and inputs vary from extensive (little effort or resources expended, minimal confinement of animals) to semi-intensive (pond production, limited input such as provision of supplemental feed or pond fertilization to enhance zooplankton productivity) to intensive (indoor production in tanks or raceways, provision of complete formulated feeds). Culture of finfishes, herein referred to as fish culture, is conducted for differing purposes, but most fish are raised for direct consumption as food fish or for stocking into natural habitats. Ornamental fishes are also cultured for the pet and aquarium trades. The approach taken by fish culturists differs among these scenarios with respect to production goals (rapid growth and food conversion efficiency versus genetic diversity and reproductive success in the wild) and specific culture methods (intensive production, high densities, and high performance feeds versus lower densities, reduced exposure to habituating elements, seminatural habitats, and predator avoidance and foraging training).

The approach undertaken to produce hatchery fish varies by management strategy. Stocking programs are implemented when increasing the number of fish in a population is desired, but the underlying reasons for increasing population size, and thus the preferred characteristics of the fish, will differ from one situation to the next. Selection of broodstock and day-to-day husbandry techniques can influence population genetics, individual behavior, and the ultimate success of propagated fishes in the wild. Misconceptions about the practice of fish culture and mismatch between the means (hatchery operation and culture techniques) and the ends (management objectives) of using propagated fishes in fisheries management has fostered some criticism of fish culture and hatchery fish. However, increased communications among culturists, geneticists, fisheries managers, and other stakeholders have supported development of best management practices for fish culture and an age of hatchery reform. In this chapter, a general description of fish culture practices is provided, how propagated fishes can be used to meet fisheries management objectives is described, and recommendations to improve the success of stocking programs under various management scenarios are given.

9.1.1 Agency Goals and Public Pressure

Fish stocking has long been an important tool used by natural resource agencies to manage a variety of fisheries. Accordingly, agencies have established fish production programs relying on state, provincial, or federal hatchery facilities to meet their stocking needs (Heidinger 1999; Hartman and Preston 2001; Halverson 2008). The extent to which recreational fisheries rely on stocking programs varies among states and provinces and with the type of water body. In the state of Michigan, for example, 40% of all recreational fishing depends on stocked fish, with at least 70% of the Great Lakes' trout and salmon fishery resulting from stockings. Agencies are mandated to manage waters in their purview for the betterment of the resource and to meet the needs and demands of the public. Management plans are developed by fisheries management professionals and are guided by information gleaned from population assessments and other surveys, as well as pressures exerted by the fishing public. Recreational fishing is often an overriding factor in this regard and, in many cases, stocking programs become essential, if only from a public relations standpoint. Thus, stocking programs have biological, ecological, and political underpinnings. When political considerations have an overriding influence, the ends and means of the stocking program may not be based on the best available science. Public hatcheries operate to meet the needs of the agency's stocking plans, and hatchery professionals often have little input into why, where, or how the fish they raise are stocked. However, hatchery reform and increased communications among fisheries biologists, managers, culturists, and their respective oversight bodies are refining the process of fish culture to support stocking programs that better suit stakeholder needs and management objectives.

9.1.2 Roles of Individuals and the Private Sector

A majority of hatcheries and stocking programs are operated in the public sector; however, it is important to recognize that private individuals and organizations, including commercial and nonprofit groups, may also be involved. Propagation programs are often initiated by government agencies to support stocking efforts for public benefit (e.g., to improve commercial or recreational fishing opportunities or to restore imperiled species). Once the programs are well developed, individuals and the private sector may also become engaged in the production of fish for the public good. In these cases, the public sector will often retain responsibility for technical aspects of propagation and rely on the private sector for practical matters (Lorenzen et al. 2001). For example, Alaska's "ocean ranching" efforts are largely supported by private nonprofit hatcheries operating with public agency oversight (Heard 2003). Canada's Salmon Enhancement Program (SEP) includes provisions for Public Involvement Project (PIP) hatcheries operated largely by community volunteers: of the approximately 174 million fish released under the SEP umbrella in 2002, roughly 10% were produced in PIP hatcheries (MacKinlay et al. 2005). Alternatively, natural resource agencies may simply purchase fish for stocking from private producers or public hatcheries may produce fry and fingerlings and then contract with private aquaculture operations to grow them out to preferred stocking sizes. Increasingly, fish culture, even for the purpose of public benefit, is represented by individuals and groups from the private and public sectors.

9.1.3 A Word regarding Terminology

The term "introduced fish" takes on a number of meanings because it has both geopolitical and ecological connotations (see Chapter 8). In fisheries management, it is often used whenever a species is stocked into a system where it did not previously exist. It could be a transplant (i.e., moved within its native distribution) or an exotic (i.e., moved from outside its native distribution). However, some species have been stocked so widely across North America (e.g., largemouth bass, rainbow trout, and striped bass) that "native distribution" has little meaning beyond a historic context. In all practicality, the term "exotic" is most often used to refer to a species originating from another continent (e.g., Asian carps, zander, and tilapias).

9.2 STOCKING PHILOSOPHY

There are numerous reasons for stocking fish as part of a comprehensive program to manage public waters (Noble 1986). New or newly-renovated waters usually require an introductory stocking of appropriate fish species. For example, a new reservoir might be inhabited by riverine species existing in the drainage prior to impoundment, but these species are often poorly suited to the newly created, lacustrine environment. Likewise, farm ponds and other small impoundments must initially be stocked with appropriate assemblages (see Chapter 16). In some states, such as Illinois, state hatcheries will provide largemouth bass, sunfishes, and channel catfish for the initial stocking of newly constructed or renovated farm ponds.

Many stockings are conducted as "value-added" fishery augmentations to increase or diversify recreational fishing opportunities. Stockings often serve the purpose of filling voids. This might include stocking a fish species such as striped bass to establish a pelagic fishery or flathead catfish to create or augment the benthic fishery. Another management goal of value-added stocking may be to establish trophy fisheries for popular sport fish such as muskellunge.

Although some fish are stocked in the hope of increasing recruitment, other stockings take place with no expectation of establishing a self-sustaining fishery. For instance, many reservoirs in warmer climates support "two-story fisheries" for rainbow trout, wherein cool, deep waters provide refuge when temperatures above the thermocline are too warm. Rainbow trout grow well in many reservoirs but are unlikely to find sufficient suitable habitat in reservoirs to spawn and create a self-sustaining population. Harvest and natural mortality in these fisheries are compensated through routine supplemental stockings. Another example of a stocking-dependent, "put-and-take" fishery is stocking catchable-size coldwater species such as rainbow trout into warmwater streams in the late fall. In this case, the goal is to create an intermittent fishery in which nearly all stocked fish are returned to the creel before rising water temperatures cause mortality in the spring.

Anadromous species (e.g., salmonids or striped bass) and interspecies hybrids (e.g., saugeye, tiger muskellunge, and hybrid striped basses) are often stocked with little expectation of establishing self-sustaining populations, though some have occurred. For example, several species of anadromous salmonids have become established in Japan (coho salmon), Patagonia (brown trout, rainbow trout, and Chinook salmon), New Zealand (sockeye salmon and Chinook salmon) and in various locations outside their normal distribution in North America (pink salmon in Maine; pink salmon, coho salmon, and Chinook salmon in the Great Lakes); however, in some instances, the species became established by developing land-locked life histories (Pascual and Ciancio 2007). Striped bass, introduced to California in 1879, became established in the San Joaquin River estuary system and once supported large sport and commercial fisheries (Stevens et al. 1985). In the case of interspecies hybrids, such as hybrid striped basses (various crosses of *Morone* species, e.g., white bass × striped bass), natural reproduction is observed but contributes little to recruitment (Avise and Van Den Avyle 1984).

Even when natural reproduction occurs, the size of the breeding population or recruitment may still be insufficient to support a self-sustaining population. Supplemental stockings are often necessary to overcome habitat modifications or limitations, intense harvest, or a combination of anthropogenic effects. In circumstances where habitat or environmental quality is unlikely to be restored, routine supplemental stocking may be required. In these situations, poor recruitment is compensated by stocking juvenile fish with the expectation they will grow to a size to be caught by anglers or commercial fishers. These are called "put-grow-take" fisheries. For example, species such as walleye and northern pike may be able to spawn in reservoirs, but limited nursery habitat often results in poor year-classes in the absence of supplementation. The salmonid stocking programs of the Pacific Northwest, sometimes referred to as ocean ranching, are another example of supplemental stocking used to compensate for high fishing mortality and restricted access to spawning grounds. Fish stockings may also be necessary following natural or, more likely, human-induced fish kills. It is also not uncommon for electric utility companies to establish hatchery facilities to stock fish routinely to mitigate losses resulting from their operations (e.g., intake impingements or thermal pollution from discharges). Supplemental stocking of prey species (e.g., threadfin shad or mysid shrimp) is another means to augment established fish populations that may be underperforming due to inadequate prey availability. It should come as no surprise that the concept of predator–prey balance is often blurred in systems receiving the dual anthropogenic forces of stocking and high fishing mortality.

Fish may also be stocked as biological controls of undesired organisms. Examples include stocking western mosquitofish or fathead minnows for mosquito control and grass carp for control of aquatic nuisance plants. Large piscivores, such as muskellunge, may be introduced to control large-bodied prey species. Piscivores may also be stocked as part of "biomanipulation" strategies to enhance water quality (Lathrop et al. 2002; Mehner et al. 2002) or to enhance fishing opportunities (Neal et al. 1999).

Propagation and stocking programs also play an important role in enabling the recovery of rare or endangered fishes (Johnson and Jensen 1991). In most cases, federal hatcheries are responsible for undertaking these efforts. These hatcheries serve as refugia, sites to conduct controlled research, and as sources for re-introductions or supplemental stockings of imperiled fishes (see section 9.3.3).

9.3 HATCHERIES AND APPROACHES TO FISH CULTURE

Hatcheries can be generally categorized according to operational strategies geared to various production goals and stocking philosophies discussed above. Production, supplementation, and conservation hatcheries have distinct directives that shape how they function and influence the physical, genetic, and behavioral characteristics of fish they produce. However, it is important to recognize that many modern hatcheries function as categorical hybrids and may conduct all three types of propagation programs in a single location. Commercial food fish and ornamental culture are beyond the scope of this chapter; however, goals of these operations are most similar to those of production hatcheries.

9.3.1 Production Hatcheries

The primary focus of production hatcheries is to produce large numbers of fish to increase recreational or commercial harvest opportunities or as mitigation to maintain fisheries affected by anthropogenic activities. These strategies attempt to increase demographic abundance, and success is typically measured in numbers of fish raised and stocked. Production hatcheries are typically medium to large facilities producing hundreds of thousands to millions of juveniles per year. Production hatcheries most commonly use industrialized rearing techniques that are focused on efficiency of juvenile fish production (see Piper et al. 1982 and Pennell and Barton 1996 for fish culture history and techniques). Fish are often reared outside in large raceways or ponds and are released in large numbers into receiving waters. Modern production hatcheries are instrumental in supplying fish to public waters. However, this industrialized approach to fish production has been criticized as contributing to the overall decline of wild populations through negative ecological interactions between hatchery and wild fish, genetic "swamping" of natural populations with inferior alleles selected for in the hatchery environment (artificial selection, inadvertent or otherwise), and fostering continued harvest of highly exploited populations (see Naish et al. 2007 for review). However, production-oriented culture methods are commonly used, particularly in support of intermittent or other put-grow-take fisheries.

9.3.2 Supplementation Hatcheries

Supplementation programs are designed to produce fish that, once reintroduced into the natural environment, will become naturally spawning wild fish. Supplementation projects generally use production hatchery rearing facilities (section 9.3.1). However, they utilize wild-caught broodstock or gametes collected from feral fish and may employ sophisticated breeding programs to ensure minimal genetic drift or artificial selection pressure (see section 9.4). Supplementation has potential benefits of reducing short-term risk of extinction, speeding recovery, recolonizing vacant habitat, and increasing harvest opportunity. Supplementation hatcheries, as opposed to production hatcheries, are a relatively recent development and one that has fueled controversy and uncertainty. The key question for supplementation programs is whether or not the contributions of wild-spawning, hatchery-origin fish are beneficial. To date, little information is available regarding the performance of supplementarily stocked fish and their progeny in the natural environment. However, the documented risks of hatchery rearing and propagation techniques (see section 9.5) should be considered prior to implementation of a supplementation program to help gauge whether supplementation will be beneficial. When supplementation is used, it should be regarded as experimental and carried out in an adaptive management framework (see section 9.6; Chapter 5).

9.3.3 Conservation Hatcheries

The goals, operational approaches, and measures of success for conservation hatcheries differ considerably from those of production or supplementation hatcheries. The mission of a modern conservation hatchery is two-fold: preservation of the gene pool and recovery of wild populations. Intensive monitoring and oversight of breeding programs are provided to ensure that sourcing, rearing, and mating protocols protect genetic integrity. Conservation hatcheries should function in ways that reflect the latest scientific information and conservation practices to maintain genetic diversity and natural behavior and to reduce the short-term risk of extinction. A conservation hatchery approach requires application and integration of a number of rearing protocols that are known to affect the inherent fitness of the fish to survive and breed in its natural environment. A conservation hatchery approach for salmonids, for example, requires a specialized rearing facility to breed and propagate a stock of fish genetically equivalent to the native stock with the full ability to return to reproduce naturally in the native habitat. A conservation hatchery must be equipped with a full complement of culture strategies to produce very specific stocks of fish with specific attributes. Fish husbandry in a conservation hatchery must be conducted in a manner that (1) mimics the natural life history patterns, (2) improves the quality and survival of hatchery-reared juveniles, and (3) lessens the genetic and behavioral influences of propagation techniques on hatchery fish and, in turn, the genetic and ecological impacts of hatchery releases on wild stocks. Operational guidelines have been described for conservation hatcheries rearing of Pacific salmon (Flagg et al. 2005; Table 9.1); however, many of the recommendations would apply to any conservation-based propagation program.

Although conservation hatchery concepts have not been in operation long enough to be

Table 9.1. Principles for hatchery management and systemwide recommendations developed by the Hatchery Scientific Review Group (modified from Mobrand et al. 2005).

Well-defined goals:

- Set goals for all stocks and manage hatchery programs on a regional scale.
- Measure success in terms of contribution to harvest, conservation, and other goals.
- Have clear goals for educational programs.

Scientific defensibility:
- Operate hatchery programs within the context of their ecosystems.
- Operate hatchery programs as either genetically integrated or segregated relative to naturally spawning populations.
- Size hatchery programs consistent with stock goals.
- Consider both freshwater and marine carrying capacity in sizing hatchery programs.
- Ensure productive habitat for hatchery programs.
- Use in-basin rearing and locally adapted broodstocks.
- Spawn adults randomly throughout the natural period of adult return.
- Use genetically benign spawning protocols that maximize effective population size and minimize potential artificial or domestication selection under hatchery conditions.
- Emphasize quality, not quantity, in fish releases.
- Reduce risks associated with outplanting (releasing hatchery fish to rear or spawn in streams).

Informed decision making:
- Adaptively manage hatchery programs.
- Incorporate flexibility into hatchery design and operation.
- Evaluate hatchery programs regularly to ensure accountability for success.

fully developed or tested, initial information indicates that rearing fish under conservation strategies may reduce aberrant behavioral and ecological interactions and increase survival (Maynard et al. 2005; Flagg et al. 2005; Hebdon et al. 2005; also see section 9.6.2). Salmon restoration in the Pacific Northwest has focused on the use of conservation hatchery strategies to aid restoration of spawning runs and rebuilding of depleted natural spawning runs (Anders 1998; Flagg and Nash 1999; Flagg et al. 2005). By means of a conservation hatchery approach, the potential for conservation and enhancement of Pacific salmon based on artificial propagation appears well grounded in other vertebrate species recovery actions worldwide (DeBlieu 1993; Olney et al. 1994; Bryant 2003).

9.4 GENETIC CONSIDERATIONS

Although there are no unambiguous, empirical studies demonstrating adverse genetic consequences of hatchery fish on wild fish populations (Campton 1995; Williamson 2001), it is still incumbent on fisheries professionals to use all practical means to limit any such effects.

9.4.1 Inbreeding

Special care must be taken in hatcheries to avoid crossing closely-related broodstock. When offspring are produced from parents sharing one or more recent ancestors they may be subject to inbreeding depression. Inbreeding depression occurs due to higher incidences of recessive (often deleterious) traits being expressed in a homozygous (identical alleles at a given locus) state. Inbred individuals may suffer from reduction in fitness due to physical abnormalities, metabolic deficiencies, or developmental anomalies (Busack and Currens 1995; Williamson 2001). Even when recessive traits are not overly prevalent among offspring, loss of fitness can occur because of an overall loss of heterozygosity. Depending on the relatedness between individuals (e.g., mating between full siblings versus half-siblings), identifiable losses of heterozygosity can happen within a few generations. Although phenotypic changes have been documented in association with the loss of heterozygosity, reductions in overall fitness are very difficult to measure because of confounding environmental effects (effects masked by ideal environmental conditions may become problematic when water quality, habitat, or prey abundance declines), imperfect correlation between measured phenotypes and absolute fitness (survival to adulthood encompasses part, but not all, of absolute fitness), and ploidy of the species (the tetraploid genome of salmonids is more resistant to loss of heterozygosity compared with the diploid genome of other fishes; Wang et al. 2002). Nonetheless, inbreeding should be avoided in propagation programs, particularly when hatchery fish are likely to interbreed with wild fish and receiving populations are at risk for loss of genetic variation.

9.4.2 Genetic drift

Whether collected recently from the wild or maintained in the hatchery for many years, hatchery broodstocks are, by definition, finite populations. Further, these populations typically represent a small subset of the wild breeding population. Small populations are more vulnerable to the actions of genetic drift, or changes in allele frequencies within a population arising from random, stochastic events rather than from selective pressures. In some respects, genetic drift can be thought of as sampling error. If only a small number of individuals are collected to establish the broodstock population, the odds are against rare alleles being represented. Assuming less-common alleles are represented in the hatchery broodstock, those alleles are vulnerable. In a small group, rare alleles are likely to be represented by a single individual; if this individual is lost, so is the allele. Genetic drift due to small effective population sizes was established as the primary explanation for divergence in allele frequencies between hatchery and wild populations of Chinook salmon (Waples and Teel 1990). As with inbreeding depression, the consequences of genetic drift may not be evident in all circumstances (i.e., populations may undergo genetic drift without an identifiable loss of fitness). Nonetheless, loss of genetic diversity can reduce the ability of populations to cope with environmental change (less "raw material" to support the process of adaptation and natural selection). Accordingly, to avoid the consequences of genetic drift in the brood and receiving populations, hatcheries should either maintain large captive populations of broodstock or consistently re-introduce new individuals from wild breeding populations.

9.4.3 Effective Population Size

To avoid inbreeding and genetic drift, hatcheries must strive to maintain an adequate number of broodstock. The field of population genetics provides a useful relationship, called effective population size (N_e), as follows:

$$N_e = 4N_m N_f / (N_m + N_f),$$

where N_m is the number of mature males and N_f is the number of mature females.

Both small numbers of spawners and unequal sex ratios will reduce the effective population size, which can be defined as the size of an ideal population of broodstock having the same rate of genetic drift as the wild population serving as the broodstock source or wild population being supplemented by the stocking. Tave (1986) recommended an N_e of 424–685 individuals. The higher number assures virtually no alleles will be lost. However, Tave recognized these numbers may not always be possible and suggested taking all steps to keep N_e as high as practical.

9.4.4 Domestication

Domestication results from the selective forces of the artificial hatchery environment or husbandry practices. From a genetics standpoint, domestication is change in the quantity, variety, or combination of alleles within a captive population or derivative broodstock in comparison with the source or donor population (Williamson 2001). Individual fish in the brood that are better suited to the hatchery setting will undergo positive selection and will survive and contribute to subsequent generations more than do their less-tolerant counterparts. The most serious form of domestication occurs when subsequent generations of captive broodfish are spawned, even when maintaining an "adequate" effective population size. Though not domestication per se, in a practical sense selection for fish with superior hatchery performance will also occur among nonbrood animals. The longer fish are held in captivity, the more they become behaviorally accustomed to their artificial surroundings and feeding protocols (see Berejikian 1995), and habituated individuals will undergo positive selection in terms of survival and growth while in the hatchery. While beneficial in the hatchery setting, acceptance of prepared feeds, reduced aversion to predators, increased aggression, and other learned behaviors are not generally advantageous in natural environments. Logistically and economically, hatcheries are limited: only so many broodfish can be maintained in a single facility; often, offspring must be held for extended periods to meet management objectives and ensure survival of the stocked fish; and, in many cases, the natural environment cannot be adequately mimicked to prevent habituation completely. Accordingly, some domestication and habituation is inevitable.

Concerns with respect to inbreeding, genetic drift, effective population sizes, and domestication are exacerbated when dealing with rare or endangered species (Rinne et al. 1986; Kohler 1995; Williamson 2001). In these specialized cases, it is crucial for hatchery personnel to take steps to preserve genetic integrity and wild-like behavior in the hatchery fish. When possible, a high N_e is vital when dealing with imperiled species, as is a conservation hatchery approach.

9.5 DISTINCTIONS BETWEEN WILD AND HATCHERY FISH

In the early days of fish culture, little thought was given to the characteristics of the fish produced, so long as they survived to be stocked into the receiving waters. However, as wild populations continued to decline despite supplementation efforts, additional consideration was given to the nature of fish produced in hatcheries. As early as the mid-1930s, researchers began questioning whether fish reared in hatcheries were somehow inferior to fish produced in the wild (Davis 1936).

Research has identified morphological, behavioral, physiological, and genetic differentiation between wild- and hatchery-reared fish. Traditional hatchery broodstock management often selected for individuals that spawned outside the normal spawning period (Flagg et al. 1995; Ford et al. 2006), resulting in early or late spawning runs among returning hatchery-origin fish. Hatchery culture practices can also alter juvenile growth and life history events such as size and age of out-migration by anadromous salmonids (Beckman et al. 1998, 1999; Larsen et al. 2001). Such life history alterations can have dramatic results, including increased male precocity and skewed temporal spawning distributions. The protective nature of hatchery rearing (i.e., reduced pressure of natural selective processes) can also increase spawned egg-to-smolt survival of hatchery-reared compared with wild salmonids (70–90% hatchery compared with only a few percent for wild; Leitritz and Lewis 1976; Piper et al. 1982; Pennell and Barton 1996). The postrelease survival and reproductive success of cultured fish are often considerably lower than that of wild-reared fish (Nickelson et al. 1986; Berejikian and Ford 2004; Naish et al. 2007), though factors unrelated to the fish themselves (e.g., stocking methods and timing of release) can also greatly influence poststocking survival (see section 9.6.3). Because of these direct or indirect changes in selective pressures, reproductive fitness of hatchery populations and fitness of their wild-spawned progeny may be reduced compared with wild populations (Berejikian and Ford 2004; Kostow 2004; Araki et al. 2007).

It is likely that the most immediate effect of traditional fish-rearing practices is disruption of innate behaviors (see section 9.6.2). In a hatchery setting, fish experience current velocities that are normally lower and more uniform than they are in nature; are not typically provided structure in which to seek refuge from predators or larger fish of the same species; are held at high, stress-inducing densities; are surface fed prepared diets; and are conditioned to approach large, moving objects at the surface (Maynard et al. 1995; Olla et al. 1998; Maynard et al. 2005). Resultant behaviors among released animals have been cited as contributing to failure in re-establishing wild populations (Johnson and Jensen 1991; DeBlieu 1993; Olney et al. 1994). Studies suggest that traditional hatchery rearing environments can profoundly influence social behavior of Pacific salmon (Maynard et al. 1995; Berejikian and Ford 2004; Naish et al. 2007). Differences in behavior may appear quite early in development, as noted by Berejikian et al. (1999), who observed greater dominance and tolerance of resource-limiting conditions among newly hatched coho salmon fry of captive-bred parentage compared with the offspring of wild fish. Social divergence of cultured fish may begin as early as the incubation stage. Lack of incubation substrate and exposure to light in the hatchery incubation environments can induce higher activity levels, resulting in reduced energetic efficiency, size, and survival (Poon 1977; Murray and Beacham 1986; Fuss and Johnson 1988). Food availability and fish rearing densities in hatcheries far exceed those found in natural streams and may contribute to differences in agonistic behavior between hatchery- and wild-reared fish (Berejikian 1995; Berejikian et al. 2001; Olla et al. 1998). Compared with wild fish, hatchery-reared brown trout have been described as inefficient foragers, expending more energy but feeding less frequently (Bachman 1984). Deverill et al. (1999) observed cultured brown trout to be similarly inefficient and thus poor growing, though most of the activity observed in these fish was associated with heightened aggressive behavior.

Reproductive success and contributions of hatchery fish to wild populations have become a double-edged sword for culturists and managers (see Box 9.1). Numerous studies have demonstrated reduced reproductive success of hatchery fish due to intentional or unintentional selection for spawning time (Chilcote et al. 1986). In these situations, hatcheries are rebuked for not producing fish that contribute maximally to subsequent generations in the wild. Conversely, hatcheries have also been admonished for producing fish that dominate wild gene pools either through behavioral dominance or simply by numbers. Fish culturists need to incorporate techniques to minimize the differences between hatchery and wild fish and to help mitigate the effects of artificial rearing and any negative influences of hatchery fish once stocked (see section 9.6.2).

9.6 BEST MANAGEMENT PRACTICES FOR PROPAGATION AND STOCKING PROGRAMS

9.6.1 Selection of Propagation and Stocking Strategies to Achieve Management Objectives

Hatchery conditions and operational procedures must reflect potential differences between wild and hatchery fish and whether these differences (see section 9.5) will pose considerable risk to the wild population. A key factor in determining the type of stocking or management option a hatchery will undertake is the biological significance of the stock. Biological significance is a function of the stock's origin, inherent genetic diversity, biological attributes and uniqueness, local adaptation, and genetic structure relative to other conspecific populations. McElhany et al. (2000) described four key population parameters that can be used to assess

Box 9.1. Hatcheries and the Endangered Species Act

From the 1950s to the 1960s, natural resource industries (i.e., timber and capture fisheries) were key components of the Pacific Northwest economy. This period also heralded the industrial phase of Pacific salmon hatchery operation to cope with increasing harvest pressures, regional development (e.g., expansion of transportation systems, construction of impoundments, and clearing of forested watersheds) and the attendant loss of freshwater habitat. Production peaked in the 1980s, with more than 420 million fry, fingerlings, and smolts released annually along the west coast of North America (Mahnken et al. 1998).

The 1980s and early 1990s saw an increased understanding of impacts of natural resource exploitation in the Pacific Northwest (e.g., the spotted owl issue and the effects of hatcheries on wild coho salmon populations in the Columbia River [Flagg et al. 1995]). In addition, Pacific Northwest economies were growing, diversifying, and shifting away from natural resource exploitation. A major change in salmon management philosophy occurred in the early 1990s with the listing of Columbia River salmon populations under the U.S. Endangered Species Act (ESA). The ESA is the cornerstone of U.S. legislation to prevent the loss of biodiversity and includes among its provisions a prohibition against harassing, harming, killing, capturing, or collecting a listed species (defined by the statute as a "take"; 16 U.S.C. §1538[a]) and the requirement that all federal agencies ensure their actions are "not likely to jeopardize the continued existence of any [listed species] or result in the destruction or adverse modification of habitat of such species" (16 U.S.C. §1536[a][2]). In 2007, 27 stocks of anadromous salmon and steelhead on the Pacific coast (states of Washington, Idaho, Oregon, and California) were listed by the National Oceanic and Atmospheric Administration National Marine Fisheries Service as threatened or endangered under the ESA. Unlike other listed species, Pacific salmonids are unique in that they co-exist with large hatchery-supplied, ocean-ranching populations. Current hatchery practices and harvest methods have been considered contributing factors leading to the overall decline of wild populations (Waples 1991; Lichatowich 1999; Levin et al. 2001), and thus the traditional, production-oriented strategies of fish propagation have come into conflict with the mandates of the ESA. The need to preserve wild fish biodiversity and meet the demands of the ESA has led biologists, managers, and culturists to rethink propagation of Pacific salmon and steelhead on the west coast of the USA.

Where hatchery operations conflict with recovery of ESA-listed stocks, the options appear to be (1) manage hatchery production as a reproductively distinct population (i.e., genetically segregated from naturally spawning populations) or (2) manage hatchery production as a genetically integrated component of a natural population by means of conservation-oriented approaches (see section 9.3.3; Flagg and Nash 1999; Flagg et al. 2005; Mobrand et al. 2005 for details of conservation hatchery operation). Mobrand et al. (2005) described two genetic management options to complement integrated versus segregated strategies, and each leads to a different set of operational guidelines (detailed information on integrated versus segregated approaches can be found on the Hatchery Scientific Review Group [HSRG] Website: www.hatcheryreform.us).

(Box continues)

Box 9.1. Continued.

In a segregated hatchery program, the goal is to produce a distinct hatchery-supported population that is reproductively isolated from wild populations (HSRG 2004a). A segregated program creates a new, hatchery-adapted population intended to meet goals for harvest or other purposes (e.g., research and education) while allowing for imperiled stocks to recover. In a segregated program, broodstocks are sourced from returning hatchery fish and, thus, little to no gene flow occurs between natural-origin spawners (NOS) and hatchery-origin spawners (HOS). Over time, a genetically distinct, hatchery-adapted population develops because of founder effects, genetic drift, and domestication selection in the hatchery environment (Mobrand et al. 2005). To prevent undesired transfer of hatchery-adapted characteristics to wild populations, the HSRG (2004a) recommends the percent of HOS should be less than 5% of the number of NOS on the spawning grounds. The degree to which segregated hatchery programs are successful depends significantly on the degree to which genetic and ecological risks to natural populations can be minimized. A critical aspect of this strategy is complete, or near complete, harvest of returning HOS to prevent genetic or other interactions with NOS.

In an integrated hatchery program, the goal is to minimize the genetic effects of domestication by allowing selection pressures in the natural environment to drive the genetic constitution and mean fitness of wild- and hatchery-origin fish (HSRG 2004b). The intent of an integrated program is to increase the abundance of a natural population demographically while at the same time minimizing the genetic effects of hatchery propagation. Genetically integrated broodstocks must include a prescribed proportion of wild fish in the broodstock each year to maintain genetic integration with a natural population (Mobrand et al. 2005). For any fixed proportion of NOS incorporated into the hatchery broodstock (pNOB), the smaller the proportion of HOS on the spawning grounds (pHOS), the stronger the opportunity for the natural environment to drive adaptation (HSRG 2004b). Thus, the HSRG (2004b) recommends the pNOB exceed the pHOS for an integrated program and that the pNOB should be a minimum of 10% to avoid divergence of the hatchery population from the natural component, even when pHOS is 0. Further, for stocks of moderate or high biological significance and viability (or to maintain or improve the current biological significance and viability of the stock), the HSRG (2004b) recommends the "realized spawning composition" (pNOB / [pHOS + pNOB]) be greater than 0.7. A successful integration program thus requires sufficient returns of NOS to supplement the broodstock each year and natural habitat capable of sustaining this natural population. Therefore, the size of composite populations generated by integrated programs will be limited by habitat availability and the ability to restrict natural spawning by hatchery-origin adults.

Implementation of either a genetically segregated or a genetically integrated strategy requires the ability to distinguish hatchery- and natural-origin adults, both in the hatchery and on the natural spawning grounds, to assess the genetic risks and gene flow rates. Both strategies require that a majority (or preferably all) of the fish carry discernable distinguishing

(Box continues)

> **Box 9.1. Continued.**
>
> marks (e.g., tags or fin clips). Both types of programs require methods to remove hatchery-origin fish prior to reaching spawning grounds to control hatchery-to-wild fish ratios adequately on the spawning grounds. Often, achievement of these goals will require a combination of directed selective fisheries and control structures such as weirs to remove adequate numbers of fish prior to arrival on spawning grounds.

the viability and biological importance of salmon populations: (1) abundance, (2) growth rate, (3) spatial structure, and (4) diversity. The viability and biological importance of the supplemented population will determine, in part, the propagation strategy—more sensitive or unique wild populations will demand more conservative propagation strategies.

Several generalized guides have been published regarding the use of propagation for stocks of Pacific salmon, including those at high risk of extinction (Hard et al. 1992; Flagg and Nash 1999; Flagg et al. 2005). However, these guides are designed to be broadly applicable and do not offer specific recommendations to ensure success of the stocking program. Essentially, the guides suggest the stocking and management strategy should depend on the particular stock of fish, its level of depletion, the physical and management limitations of each individual hatchery action, and the biodiversity of the ecosystem. Reviewing the characteristics of the target population, receiving ecosystem, and available propagation strategies is critically important to tailoring the guidelines to achieve specific management objectives. We focus on the particulars of propagation and stocking strategies in the following sections. However, effective use of hatcheries and cultured fishes must also include an assessment of the relative risks and benefits of management and supplementation strategies.

In light of possible effects of stocked fish, either introduced or supplemental, on wild populations and the high cost of producing fish for stocking, not stocking should always be considered as an option. For example, in cases where habitat restoration would also elicit the desired outcome, directing resources to habitat improvements instead of stocking efforts may ultimately be more cost-effective and ecologically sustainable. Increasing diversity may not always be achievable because the stocked species may flourish at the expense of existing species already popular with the fishing public. Genetic alterations of populations being supplemented with hatchery fish can also be problematic. In light of these and other considerations, risk assessment approaches may be useful in determining the most economically- and ecologically-appropriate course of action. Risk assessments assist decision makers by assessing a proposed activity in terms of the probability and consequences of a negative outcome. Risk assessments also attempt to describe, if not measure, uncertainty associated with the activity and its effects. Applied to stocking programs, risk assessments can be used to summarize genetic and other interactions between hatchery and wild fish, effects on other species in the ecosystem, and the uncertainty associated with ecological responses to stocking (Pearsons and Hopley 1999). While these assessments can be very useful in terms of describing risk, whether the level of risk is acceptable is a separate question that should be addressed by means of the best available science, professional experience, and the opinion of stakeholders and the general public. If the decision to use propagated fish is made, the following guidelines may be used to improve culture and stocking methods.

9.6.2 Husbandry Techniques and Hatchery Operation

Mobrand et al. (2005) described three foundational principles for best management practices for operation of hatcheries (Table 9.2).

Principle 1—Every hatchery stock must have well-defined goals in terms of desired benefits and purpose. Well-defined goals for operation of hatcheries provide both explicit targets and measures for success. Stocking goals must reflect the purpose and desired benefits of the program (e.g., harvest, conservation, research, or education). An integrated hatchery program should include short-term and long-term goals for production and outcomes, as well as monitoring plans in place to track progress. Hatcheries should delineate specific objectives to meet these goals. Goals and objectives should be explicit and include (1) the intended number of fish to be harvested each year, (2) the number of fish returning to a hatchery or spawning naturally in a watershed (i.e., escapement), (3) the expected results of any associated scientific research, and (4) the benefits to be derived from education and outreach components.

Principle 2—Hatchery programs must be scientifically defensible. The stated goals of stocking programs and the day-to-day operations of hatcheries must be scientifically defensible. They must represent a logical approach to achieve the management goals and should be based on knowledge of the target ecosystem and the best scientific information available. Once the goals for a program are established, the scientific rationale for the design and operation of the program must be explicitly described so that the scientific basis of operation and day-to-day activities may be understood by all personnel and, ideally, the general public. In line with principle 1, a written, comprehensive management plan for every hatchery program, covering broodstock management, fish rearing, and release and harvest management components, is imperative to program acceptance and successful implementation. These guidelines should include a decision-making procedure to use in developing the initial project goals and a course of action to achieve these goals. The decision-making guide can also be very useful in justifying hatchery operation or dealing with contingencies after the plan has been implemented. Further, scientific oversight and peer review should be integral components of every hatchery program.

Principle 3—Hatchery programs must respond adaptively to new information. Scientific monitoring and evaluation of the hatchery and stocked fish are necessary to ensure that hatcheries are achieving their goals. Evaluation should include assessment of juvenile-to-adult survival and, where applicable, returns to spawning grounds; contributions of hatchery-origin adults to harvest and natural reproduction; and rates of hatchery fish migration or straying to nontarget waters. Where possible, evaluation should include assessments of genetic and ecological interactions (e.g., interbreeding, competition, and predation) between hatchery- and natural-origin fish. Results should be evaluated annually to allow timely programmatic adjustments, and hatcheries should always be managed adaptively to respond to new goals, new scientific information, and changes in the status of natural stocks and habitat.

To summarize these principles, there is a need for increased monitoring and evaluation, scientific oversight, and accountability of hatchery operations. Hatcheries need to operate in scientifically defensible modes with well-defined goals and substantially increased data collection and evaluation. Hatcheries also need to be flexible and adaptable; that is, they need to operate and be evaluated in the context of both the ecosystem in which the hatcheries occur and the ecological processes on which hatchery-origin fish depend. Scientific uncertainties associated with hatchery operations are numerous. The science to manage these risks is still inadequate and some of the risks are poorly understood (see Currens and Busack 2005). It

Table 9.2. Operational comparisons between production and conservation hatchery strategies for rearing of Pacific salmon (modified from Flagg et al. 2005).

Parameter and factor	Production hatchery		Conservation hatchery	
	Action	Objective	Action	Objective
Egg collection				
Spawn timing	Directed (e.g., early or late component)	Synchronize adult return or harvest opportunities	Synchronized to wild; representative numbers collected over range of run	Maintain wild timing
Number	Directed (probably large number of eggs taken)	Maximize output	Directed (relatively small number of eggs needed)	Stage production to habitat carrying capacity
Egg fertilization				
Mating strategy	Directed (for characteristics)	Selected desired attributes (e.g., return size and age)	Directed (to maintain genetic diversity	Maintain diversity
Egg incubation				
Incubator type	Use accepted guidelines for species	Maximize output	Include substrate	Approximate wild conditions and maximize hatch size
Temperature	Surface or well	Time hatch to production needs	Controlled to ambient for stock	Synchronize hatch with wild timing

Table 9.2. Continued.

Parameter and factor	Production hatchery		Conservation hatchery	
	Action	Objective	Action	Objective
Fish rearing				
Vessel type	Standard (typically smooth with no internal structure)	Maximize output	Altered to include enriched (seminatural) habitats with cover, structure, substrate, or other	Reduce domestic conditioning
Temperature	Surface or well	Time rearing to production needs	Controlled to ambient for stock	Synchronize rearing with wild stock
Culture techniques	Standard (designed to maximize fish output)	Maximize output	Innovative (designed to maximize fish quality)	Reduce domestic conditioning and improve fitness
Pond timing	Variable	Maximize culture opportunity	Synchronized to wild	Approximate wild rearing scenario
Photoperiod	Natural	Provide ambient conditions	Natural	Provide ambient conditions
Density	Up to maximum safe levels	Maximize space use	Use low rearing density	Minimize behavioral and health concerns

is clear that maintaining healthy habitat is critical not only for viable, self-sustaining natural populations but also to control risks of hatchery programs adequately and realize the benefits of hatcheries to recover populations and sustain healthy harvests in increasingly populated environments (Mobrand et al. 2005). Hatcheries cannot be regarded as surrogates or substitutes for lost habitat, declining environmental quality, or adequate regulation and management of capture fisheries; rather, hatcheries should be viewed as a complementary component of broader natural resource management and restoration activities (see Box 9.2).

In a recent review, Brown and Day (2002) highlighted the similarities between the goals of fisheries management and conservation biology.

Lessons from conservation biology. Although conservation biologists may be concerned with preserving the unique attributes of populations, whereas managers may be more interested in maintaining commercial or recreational fisheries, in either case the aim is to establish (or re-establish) self-sustaining populations. For hatchery fish to contribute to this common goal, at a minimum, they must survive long enough to reach a harvestable size and sexual maturity. Brown and Day (2002) outlined a series of changes in rearing methods that might enhance postrelease success of hatchery fish. Although these suggestions are from the conservation biology perspective and may be most applicable or feasible in a conservation hatchery setting, it is important to recognize that any hatchery or stocking program would benefit from enhanced postrelease survival.

Brown and Day (2002) acknowledge the importance of broodstock selection and maintenance in order to avoid problems such as inbreeding and domestication (see section 9.4) but focus on morphological differences and learned behaviors (see section 9.5) as the primary contributors to postrelease mortality of hatchery fish. To some extent, morphological alterations (e.g., coloration, fin morphology, growth rate and size, and tissue composition) can be corrected by providing natural or seminatural foods, reducing production densities, or using seminatural lighting (Maynard et al. 1995). However, use of natural or seminatural rearing environments appears to result in the most comprehensive restoration of wild morphology and behavior. The authors noted several behaviors that are critical to the success of stocked fishes: predator recognition and avoidance, recognition and acquisition of food, appropriate social interactions with conspecifics, identification or construction of needed habitats (e.g., nests), and navigation and locomotion in complex environments. Exposure to some level of structural complexity prior to stocking (seminatural streambeds, submerged structure, overhead cover, or use of earthen ponds instead of tanks or raceways), particularly in conjunction with the opportunity for natural or seminatural foraging, appears to provide some sort of behavioral "life skills training" and increase postsurvival stocking of hatchery fishes (Brown and Day 2002). Although these techniques are certainly more labor- and resource-intensive than are traditional propagation protocols, if greater poststocking survival is achieved, they may be more cost-effective in the long term.

9.6.3 Stocking Techniques

9.6.3.1 Transportation of fish

Regardless of where fish originate, some amount of travel must occur to transport fish to the receiving system. Most agencies possess fish hauling vehicles that consist of vehicle-mounted tanks and some type of aeration device (e.g., agitators, blowers, or pressurized

Box 9.2. Hatchery Fish and Restoration of Striped Bass in the Chesapeake Bay

Historically, striped bass populations along the U.S. Atlantic coast supported lucrative and popular commercial and recreational fisheries. Annual harvests peaked in 1973 at 6,700 metric tons but rapidly declined over the next 10 years under the pressures of zealous harvest, habitat modification, and reduced water quality (Richards and Rago 1999). By 1983, the striped bass catch had dwindled to approximately 15% of peak harvest. In response, the Atlantic States Marine Fisheries Commission developed an Interstate Fisheries Management Plan for the striped bass and these recommendations were later vested with regulatory authority by the 1984 Atlantic Striped Bass Conservation Act (Public Law 98–613). Recognizing striped bass recruitment on the Atlantic coast is largely supported by the Chesapeake Bay spawning and nursery grounds, biologists targeted the bay for intensive restoration efforts. Although striped bass had been stocked along the coast since the late 1800s (Rulifson and Laney 1999), intensive stocking of the Chesapeake Bay began in earnest in 1985. By 1993, 7.5 million fingerling striped bass had been released into the bay (Richards and Rago 1999). Given the strong influence of temporal and stochastic effects on larval striped bass survival, fry were also released into the Chesapeake Bay system to compensate for high mortality among natural spawns (Secor and Houde 1998).

The Chesapeake Bay striped bass fishery was declared recovered in 1995 and today all migratory stocks of the Atlantic striped bass are considered restored to historic levels (Rulifson and Laney 1999). Some have argued the rebound of striped bass in eastern U.S. waters is due largely to restricted harvest pressure and stocking efforts contributed little to the restoration of Chesapeake Bay striped bass (Richards and Rago 1999). However, hatchery-origin fish are represented in the growing subadult and reproductively-mature year-classes, particularly in the Patuxent River (Rulifson and Laney 1999). Reduced fishing pressure combined with habitat improvement certainly enhances the likelihood of wild- and hatchery-origin fish contributing to natural recruitment.

It has been suggested that hatchery supplementation alone cannot restore overexploited populations (Lorenzen 2008). Rather, successful fishery enhancements are generally characterized by a "major transformation of the fisheries system," including propagation and release of cultured fish, implementation of more restrictive harvest limits, increased population monitoring and regulatory oversight, and greater involvement of stakeholders to speed acceptance and compliance with new regulation (Lorenzen 2008). In the case of the Chesapeake Bay striped bass, it is unlikely that stocking alone would have resulted in an equally rapid and robust resurrection. However, recovery of the Atlantic striped bass populations is an excellent example of how hatchery fish can be implemented as part of a multifaceted approach to fishery restoration and enhancement.

---

cylinders of gaseous or liquid oxygen). Proper care during transport is critical to the success of a stocking program because of the high potential for stress-induced mortality due to handling, crowding, disease exposure, osmotic shock, or temperature shock (see Carmichael et al. 2001). Feed is typically withheld for 24–48 h prior to transport to reduce subsequent egestion-related fouling in the hauling tank. Fish may be given a prophylactic treatment with an antibiotic or other therapeutant prior to or during transport; although prophylactic treatment may reduce the incidence of disease, it is not routinely used because of withdrawal times (typically 7–21 d) required for many approved drugs. In the event drugs are used, whether in the hatchery or during transport, only approved products should be applied in strict accordance with guidelines for their use in aquaculture (up-to-date information on drugs approved for use in U.S. aquaculture can be obtained from the U.S. Fish and Wildlife Service Aquatic Animal Drug Approval Partnership Program, www.fws.gov/fisheries/aadap/home.htm).

Hauling tanks may be disinfected with calcium hypochlorite and rinsed thoroughly prior to filling with culture water. When necessary, the water can be cooled to desirable hauling temperatures (15–20°C for cool- and warmwater fishes; 10–15°C for coldwater fishes), but care must be taken to acclimate the fish slowly to the new temperature. A good rule of thumb is to temper fish for 30 min for each 1°C change in excess of a 2°C difference in temperature. Assuming other differences in water chemistry between hatchery and receiving waters are minor, acclimation is generally unnecessary if the temperature difference (cooler or warmer) is less than 2°C. Hauling waters should be supersaturated with oxygen at the outset and regulated to maintain dissolved oxygen content in excess of 5 mg/L for the entire haul. Depending on species, salt (NaCl) is often added to hauling water to attain a salinity level of 3–7 ‰ to assist fish in maintaining their osmotic balance. Water changes are sometimes necessary during long hauls; all the preceding precautionary steps should be taken when exchanging water during a haul. Loading rates by weight will vary depending upon distance to the stocking site, species of fish, size (smaller fish = smaller loading weight), temperature, water hardness (lower levels of divalent cations = lower loading rates), and aeration efficiency. An example of loads and distances are shown in Table 9.3.

It is critical to ensure transport and stocking activities do not inadvertently contribute to the spread of aquatic animal diseases or invasive species. This includes compliance with recommended procedures to prevent transfer of aquatic nuisance species (e.g., removal of organisms and debris from vehicles and vessels and disinfection of equipment prior to traveling to another location) and transportation restrictions or prohibitions (e.g., federal restrictions on

Table 9.3. Pounds of catfish that can be transported per gallon of 18.3°C (65°F) water (from Piper et al. 1982).

Number of fish per pound	Transit period in hours		
	8	12	16
1	6.30	5.55	4.80
2	5.90	4.80	3.45
4	5.00	4.10	2.95
50	3.45	2.50	2.05
125	2.95	2.20	1.80
250	2.20	1.75	1.50
500	1.75	1.65	1.25
1,000	1.25	1.00	0.70
10,000	0.20	0.20	0.20

interstate transport of susceptible species to control the spread of viral hemorrhagic septicemia in the Great Lakes region).

9.6.3.2 Choice of taxa stocked

Considering all the rationales for stocking, it should come as no surprise that numerous taxa, encompassing nearly all trophic levels and biological characteristics, have been stocked in North America (Heidinger 1999). In 2004, government agencies stocked 104 taxa (species, subspecies, or hybrids) in U.S. waters (Halverson 2008). Of the 1.75×10^9 individual fish stocked in 2004, a majority (82%) were stocked for sportfishing or as forage to benefit sport fisheries. Stockings of imperiled or rare species were minor in terms of numbers and biomass stocked; however, these species represent roughly half of the species propagated in 2004. Increasing production of ESA-listed or otherwise at-risk fishes reflects the growing importance of "conservation aquaculture" and imperiled species restoration in aquatic ecosystem management (see section 9.8). Table 9.4 provides some generalized information on representative species and pertinent characteristics that influence decisions to stock.

9.6.3.3 Sizes and numbers stocked

Depending on the species and stocking goal, fish can be stocked as fry (larvae), fingerlings, advanced fingerlings, or adults (herein, catchables). Large numbers of fry might be stocked with the assumption that some will survive, or, alternatively, a smaller number of larger fish might be stocked. In general, larger fish will have greater survival rates because they are more tolerant of stressors associated with transport and stocking (Pitman and Gutreuter 1993) and are less vulnerable to predation. However, larger fish are more difficult and costly to produce. Decisions to rear and stock fish at various life stages and sizes may be based on biology, ease of culture, management goals, economics, politics, or any combination thereof (Hartman and Preston 2001). For example, walleye are routinely stocked at smaller sizes because of issues associated with rearing large numbers of advanced fingerlings in captivity (i.e., difficulty of feed training, cannibalism, and economics). In 2004, walleye represented nearly 60% of the total number of fish stocked by state and federal agencies in the USA but less than 1% of the total biomass stocked (Halverson 2008). Conversely, species such as rainbow trout are highly tolerant of culture procedures and are often stocked at catchable sizes. Rainbow trout represented about 5% of the number of fish stocked in the USA in 2004 but approximately 50% of the biomass (Halverson 2008). Stocking catchable-size fish, such as for a put-and-take trout fishery, can result in an instant public relations boon because of the speed at which the hatchery-produced fish find their way to the fisher's creel. It is not uncommon for greater than 90% of hatchery trout to be caught within weeks of a put-and-take stocking (see section 9.6.3.4). Urban fishing programs also rely on stockings of catchable-size fish, with catfishes, sunfishes, and common carp being quite popular.

The decision on size and number to stock is often based on what is most practical as opposed to a detailed cost-effectiveness analysis. However, cost-effectiveness analyses can be very helpful in determining the most efficient means (releasing fry, fingerlings, or catchables) of meeting a management goal (e.g., increased recruitment or greater returns to the creel; Leber et al. 2005) and may be increasingly used by natural resource agencies facing budget cuts. Ideally, stocking rates should also account for density-dependent effects on survival—while low stocking rates may not achieve the desired outcome, very high stocking rates may also result in poor

Table 9.4. List of selected taxa that are stocked and some of the pertinent biological characteristics to consider when choosing a fish for stocking (modified from Heidinger 1999).

Taxon	Characteristics
Atherinidae	
Inland silverside	Forage fish; winterkills but can tolerate colder temperatures than can threadfin shad; young of the year reproduce in Midwest
Centrarchidae	
Black crappie	Easier to handle and transport than is white crappie; tends to predominate over white crappie in northern and southern portions of USA; does not readily accept a prepared diet
Bluegill	Becomes stunted in small ponds and can limit largemouth bass recruitment; readily accepts prepared diets
Green sunfish	Very vulnerable to largemouth bass predation
Hybrid sunfishes	Grows faster than parental species; certain F_1s are predominately males; F_1s tend to be fertile
Largemouth bass	Sport fish; Florida largemouth bass cannot survive in cold water as well as can northern largemouth bass
Redear sunfish	Harder to catch than are bluegill; capable of eating mollusks; does not readily accept a prepared diet
Smallmouth bass	Grows well on insects and crayfishes as forage; grows well at warm temperatures but does not recruit in southern states in ponds with largemouth bass and sunfishes present; in the southern part of its range it recruits in streams
White crappie	Tends to overpopulate in small ponds and lakes or does not recruit; tends to dominate over black crappie in turbid water; does not readily accept a prepared diet
Clupeidae	
Gizzard shad	Very fecund forage species; not desired in small pond or lake if managing for sunfishes; spawns at 2 years
Threadfin shad	Very fecund forage species; young of the year spawn; winterkills at temperatures below 8°C
Cyprinidae	
Common carp	Commercial species; capable of eating infauna; highly fecund; long lived; wide temperature tolerance
Fathead minnow	Forage fish; so vulnerable that species tends to be eliminated by largemouth bass
Golden shiner	Has been stocked in small lakes and ponds as forage for largemouth bass; tends to be more successful in northern part of largemouth bass range; in Midwest may overpopulate and limit largemouth bass recruitment
Grass carp	Used as biological control of vegetation; stocked at 5 to 15 fish per hectare; triploids are available; not approved in all states; commonly reaches 14 kg; very vulnerable to largemouth bass predation below 20 cm

Table 9.4. Continued.

Taxon	Characteristics
Esocidae	
Muskellunge	Trophy sport fish; fry are very vulnerable to fish predation
Tiger muskellunge	Accepted as trophy sport fish by most muskellunge anglers; easier to raise to advanced fingerling stage on prepared diet than are parental species; sterile
Ictaluridae	
Black and yellow bullheads	Used in urban fisheries; tend to reproduce at small size (15 cm); dense populations capable of keeping a pond muddy in areas of colloidal clay
Channel catfish	Requires cavity in which to spawn; very vulnerable to largemouth bass predation below 15–20 cm; may not recruit in small ponds; will readily accept prepared diets
Moronidae	
Hybrid striped bass	Cross using female striped bass and male white bass (palmetto bass) grows larger than reciprocal (sunshine bass); easier to train to take prepared diet than parental species; will backcross
Striped bass	Pelagic sport fish capable of eating large forage fish not vulnerable to other piscivores; floating eggs require large headwater stream to recruit; some populations are maintained by stocking fry

survival and recruitment due to increased intraspecific competition for resources (Fayram et al. 2005). The presence of predators, abundance of prey, and carrying capacity of the receiving system should all be considered when determining stocking rates and size at release.

9.6.3.4 Timing and stocking site

Most of the North American fish fauna spawn in spring, so fry will be stocked at that time, fingerlings in late spring to early summer, and advanced fingerlings in fall. It is advantageous to time fry and small fingerling stockings to peaks in zooplankton populations, but in practice this occurs more serendipitously than as a result of planning (Heidinger 1999; Hartman and Preston 2001). As temperatures rise, so does the risk of stress and disease outbreaks after stocking, so, excluding northern regions, fish are not routinely stocked in summertime. Winter also presents a set of problems in that fish handled at cold temperatures (particularly warmwater and coolwater species) are more prone to fungal infection. Ice coverage of receiving systems can also be a deterrent to stocking in winter.

Larval fish can be harmed when subjected to bright sunlight, therefore, evening or early morning stocking is recommended. Fish species or life stages that are pelagic in nature should be stocked in open waters as opposed to near a convenient boat-launching ramp. Care should also be taken to stock littoral species or life stages along shorelines containing structural habitat such as aquatic vegetation or woody debris. In the case of anadromous fishes, juvenile releases normally occur at acclimation sites to encourage returns to target watersheds. In all cases, prior to releasing fish, water quality should be measured at prospective stocking sites to avoid unsuitable areas (e.g., low dissolved oxygen) and to determine the amount of acclimation needed to compensate for differences in water chemistry between the hauling and receiving waters (Pitman and Gutreuter 1993).

Risk of immediate postrelease mortality can also be minimized by stocking at multiple locations in a receiving water body and spreading stocking efforts over a period of days or weeks. Stocking at different times and locations minimizes the risk of complete failure but is not routinely done because of the need for additional personnel and other logistical problems associated with multiple releases. Similarly, so-called "soft releases" can improve postrelease survival but are often impractical. Whereas a "hard release" will involve little more than tempering prior to stocking, a soft release includes an extended acclimation period prior to release and, in some cases, may involve release into a protected confinement (i.e., a cage or net pen) prior to full release (Brown and Day 2002). Although routine in reintroduction of terrestrial species, soft releases are relatively uncommon in fisheries enhancement and restoration.

Put-and-take stockings are different than those described above in that the management goal is often to maximize return to the angler's creel. Accordingly, fish are often stocked in open view at locations easily accessible to the hauling vehicle and the fishing public. Upcoming stocking events might also be covered by local media, drawing even more anglers to the stocking site. This process often results in "truck following," whereby anglers are casting lines even before the truck pulls away.

9.6.3.5 Evaluation of stocking programs

A critical part of any management plan involving a stocking program should include an assessment of its successes, failures, or unintended consequences (Murphy and Kelso 1986; Wahl et al. 1995). Management plans should clearly state the rationale(s) for the stocking and the intended outcome(s). An introduction stocking can readily be assessed by sampling for the presence of offspring from stocked fish and, ultimately, their progeny if a self-sustaining population develops. Likewise, supplemental stockings can be evaluated on the extent they strengthen a given year-class, but some sort of marking program (genetics, chemical markers, or tags; see Guy et al. 1996) is necessary to confirm the relative contributions of stocked and wild fish. Creel surveys are a simple and effective method to evaluate whether a stocking has improved fishing from a quantity aspect. On the other hand, creating a trophy fishery via stocking can improve the quality of the fishing experience, but more detailed interviews of the fishing public are necessary to gather this sort of information (see Knuth and McMullin 1996). Unintended consequences of stockings (e.g., genetic pollution, interspecific competition, or habitat disruption) are more difficult to evaluate in that they are often not readily apparent or do not occur for many years. Unless stocking has previously been undertaken over an extended period, a management plan should include a research component to evaluate ecological considerations in addition to the more immediate impact on the fishing public.

9.7 CRITICISMS OF FISH CULTURE AND HATCHERY FISH

Many fisheries professionals and lay people share an outdated view of fish culture and its use in fisheries management and restoration. Many inaccurately equate fish culture with juggernaut factory farms and dismiss the role of hatcheries in fisheries management because of an assumed "quantity-not-quality" driven approach. For example, Helfman (2007) stated,

> Aquaculture as currently practiced may create additional pressure on wild stocks because of competition for (1) larvae and other fishes that are fed to cultured stocks, (2) coastal ecosystems and their services, and (3) world markets where products are sold. Also, capture fisheries and marine ecosystems will likely suffer due to problems of waste production, chemical pollution, exotic species invasions, and pathogen transmission. Supplementation programs, therefore, have an effect opposite of their purported goals, reducing wild fish abundance when wild fish can least afford additional insults and populations are at historic lows. Hatchery activities and population supplementation cannot be justified on the grounds of conserving wild populations, now or in the future.

Helfman (2007) concluded, "Hatcheries accelerate extinction." This indictment of fish culture is not supported by scientific information and does not reflect the opinion of most fisheries professionals. The vast majority of fisheries managers have concluded that fish culture is integral to fisheries conservation and restoration, and management strategies cannot be divorced from the culture practices upon which they rely. It is correct that fish culture alone will not compensate for the effects of overharvest, habitat degradation, or other stressors; however, many fisheries exist only because of the efforts of fish culturists. Globally, stock enhancement and culture-based fisheries activities yield approximately 20% of freshwater and diadromous capture fisheries (Lorenzen et al. 2001). In the Pacific Northwest, it is estimated that 70–80% of some coastal fisheries are based on hatchery releases (Mahnken et al. 1998; Naish et al. 2007). Further, the availability of commercially-cultured seafood reduces harvest pressure on wild populations. Arguments to the contrary are fueled largely by cultural and socioeconomic ties to traditional capture fisheries.

With respect to hatchery and stocking programs, it is important to recognize that traditional hatchery operating procedures were not established with modern goals of supplementation or restoration in mind. Hatcheries have and continue to operate at the behest of public interest. When stocking rates were the sole measure of success, hatchery managers concerned themselves with production volumes. As maintenance of genetic diversity and the role of local adaptation became prominent paradigms, hatchery managers changed their focus from numbers to genotypic and phenotypic characteristics—modern hatcheries aim for quantity and quality, a central theme of American Fisheries Society symposia addressing the use of propagated fishes (Stroud 1986; Schramm and Piper 1995; Nickum et al. 2005) and a topic increasingly emphasized during revision of this chapter for this volume (see Kohler and Hubert 1993, 1999). Scientific uncertainty and unanticipated effects of hatchery rearing and stocking may have limited the positive impacts of hatcheries in the past. However, the negative consequences of traditional stock enhancement are now being used in an adaptive management strategy to inform the process of hatchery reform.

Some exotic species introductions can be tied to accidental releases from aquaculture facilities, but a large number of introductions blamed on fish culturists were purposeful stockings conducted under the direction of natural resource agency initiatives (Mitchell and Kelly 2006). While some introductions and stocking programs have had unexpected, negative consequences (e.g., habitat degradation and competition with native species), fish culture and stocking cannot be dismissed as a management tool because some strategies have proven ill advised. It is a popular assumption that all nonnative species are overwhelmingly destructive, and most reports have focused on cases for which negative consequences have been observed (Gozlan 2008). In fact, many introductions have had positive economic effects and have enhanced biodiversity without negative ecological impacts. In a review of the economic and ecological costs of nonnative species, Pimentel et al. (2005) generally lamented the negative effects of exotic fishes on endemic populations in the USA but also conceded that introductions of nonnative fishes have yielded considerable economic benefits in the form of sport fishery enhancement. Further, in some instances exotic species (e.g., grass carp and western mosquitofish) have proven tremendously useful as biological controls for enhancing environmental quality and restoring ecological function. For the majority of freshwater species that have been introduced to systems outside their native distribution, the risk of negative ecological impact following introduction is only 10% (Gozlan 2008). Of course, the relative ecological risk varies among species and can be minimized by implementing additional preventative measures to avoid accidental introductions of higher-risk species. For further discussion of the positive and negative impacts of introduced fishes, we refer the reader to Chapter 8.

Pathogen transmission from culture facilities to wild populations is a hotly-debated issue that continues to be fueled by conflicting data. Because of the nature of fish and their pathogens, it is essentially impossible to prevent pathogen transfer during movement of fish from one locale to another. When transferred to "naïve" fish populations by relocation or introduction, introduced pathogens can become problematic. In addition to the usual concerns associated with exotic species introduction, transfer of nonindigenous cultured fish can be a vector for disease or infestation of wild populations, as evidenced by transfer-related outbreaks of whirling disease among rainbow trout in the Pacific Northwest and sea lice infestations of Atlantic salmon in Norway (Waples 1999). However, when propagation and stocking efforts are restricted to regionally-sourced, indigenous species, pathogen transfer between hatcheries and wild fish is less likely. In any event, cases of pathogen transmission from hatcheries to wild populations are largely unsubstantiated in the USA and current measures to control pathogen releases appear to be effective (LaPatra 2003).

Many of the other criticisms leveled at fish culture can be traced to public perception of commercial operations, particularly those rearing marine, carnivorous fishes. Aspects of culturing these species remain contentious, but the long-term success of the industry relies upon increasing sustainability and reducing impacts. The question of competition for larvae, that is, capturing wild-spawned larvae for grow-out in captivity, is largely restricted to commercial culture of marine food fishes, such as tunas, because spawning or larval husbandry techniques are as yet lacking for these species. Use of "trash" fishes as food for large carnivorous animals is primarily restricted to commercial operations, where it is dwindling as suitable formulated feeds are being developed.

Fish feeds are also controversial, owing to the use of reduction fishery products, that is, fish meals (FM) and oils (FO), in their formulation. Concern regarding FM- and FO-based feeds is a widespread issue, affecting both private and public fish culture operations. Driven by concerns of feed cost, product availability, and, more recently, transfer of persistent organic contaminants, the Food and Agriculture Organization of the United Nations (FAO) has described the development of animal feeds free of FM and FO as "a major international research priority" (FAO 2005). Many argue against the transformation of small pelagic fishes into FM and FO, contending these fishes could be directly consumed by humans (Naylor et al. 2000). In the Asian-Pacific region, most of the "trash fishery" landings are consumed directly by humans; only 25% of landings are incorporated into aquaculture feeds (FAO 2005). However, most Western consumers are unwilling to accept these trash fish as foods, and landings of wild food-grade fishes are simply insufficient to keep pace with demand (FAO 2009). Strides have been made in reducing or eliminating reduction fishery products in aquafeeds (New and Wijkström 2002). In the case of salmon feeds, FM inclusion rates have decreased from 60% in 1985 to a current average of approximately 35% (Tacon 2005). Modern grow-out feeds formulated for herbivorous or omnivorous species typically contain 2–15% FM, whereas feeds for carnivorous species contain 20–50% FM (Tacon 2005). Nutritionists continue to refine formulations and have been increasingly successful at partially or completely replacing FM and FO without affecting production performance (reviews by Hardy and Tacon 2002; Sargent et al. 2002; Trushenski et al. 2006; Gatlin et al. 2007). Formulations are also continually modified to enhance digestibility and nutrient retention, which reduces production cost and waste production (Cho and Bureau 2001). In time, these experimental formulations will be adopted by commercial feed manufacturers and, in turn, fish culture operations.

9.8 THE FUTURE OF FISH CULTURE AS A MANAGEMENT TOOL

The effective use of cultured fishes and stocking as a management tool is guided by science but is also subject to political, social, and economic forces. The decision whether to use cultured fishes is complex, and it is impossible to gauge the future of fish propagation and stocking programs with certainty. By examining recent trends in the use of cultured fishes, however, we can gain some insight into the ways in which fish culture and stocking practices might evolve. To assess current stocking practices in terms of the historic record, Halverson (2008) reviewed stocking activities conducted by U.S. federal and state agencies from 1931 to 2004. The review revealed a number of trends that we can reasonably expect to continue into the future: (1) decreasing involvement of federal agencies in stocking programs; (2) larger individuals and greater total biomass being stocked; and (3) greater diversity of taxa being stocked, particularly rare or imperiled fishes.

Decreasing federal involvement in fish stocking programs reflects a broader trend in natural resource management, specifically decentralization and the transfer of federal responsibilities to states or communities. The transfer of responsibility to regional or local governments has been touted as a means to provide greater flexibility and efficiency and greater incentives for program execution, compliance, and success because the regulatory power is put in the hands of the stakeholders (Andersson et al. 2004). Many federal hatcheries in the USA and Canada have been transferred to the states or provinces, and a majority of those that remain federally operated have transitioned to propagation of native or imperiled species used in restoration or mitigation programs (Edwards and Nickum 1993; Jackson et al. 2005). In terms of individuals produced, federal contributions to fish stocked in U.S. waters have declined from roughly 70% in the 1930s to less than 8% in 2004 (a conservative estimate based on the 33 states for which data were available; Halverson 2008). In turn, state agencies are increasingly involved in fish culture and stocking efforts: nationwide, approximately one-third of full-time state fisheries personnel are involved in fish production and distribution (Gabelhouse 2005) and one-third of state fisheries expenditures are for the purposes of hatchery and stocking programs (Ross and Loomis 1999).

Historically, propagation and stocking techniques involved little more than seeding systems with fertilized eggs. As the limitations of these strategies became evident, culturists increasingly focused on production of larger individuals that would have a greater likelihood of survival in the wild. In the 1940s, large fish (>15.2 cm) represented roughly 20% of all fish stocked in U.S. waters; in recent years, the contribution of large individuals has grown to more than 50%. Although production of advanced fingerlings and catchables required greater inputs, it was assumed these costs would be outweighed by the benefits accrued to the target population and fishing public. Stocking of larger individuals has proven successful in terms of increasing creel returns and continues to be the standardized approach for many species. During this same period, inland commercial fisheries were becoming less relevant to the U.S. food supply and economy, and management priorities were shifting from commercial fisheries enhancement to sport fisheries and recreational fishing opportunities. In many cases, this transition meant rigorous efforts to expand the range of sport fishes, often with little expectation of establishing self-sustaining populations (see section 9.2). Current stocking efforts are dominated by sport fishes, by number (72% of total) and by biomass (82% of total). Given the demands of the sportfishing public, recreational fisheries enhancement is unlikely to wane considerably in the near future. However, widespread recognition of the importance of ecosystems instead of individual species has placed a premium on ecosystem-based approaches to aquatic resource management. Fisheries agencies in the USA and Canada have begun to view stocking in the context of broader management strategies, and management plans are less likely to rely solely on propagated fishes (Jackson et al. 2005). Stocking programs will be increasingly paired with habitat rehabilitation, pollutant and stressor mitigation, harvest restrictions, and other methods in a more holistic approach to restoring aquatic ecosystems and inland fisheries. Further, as philosophical and statutory imperatives to protect imperiled species become more prominent among fisheries professionals and the lay public, propagation of threatened and endangered fishes will grow.

9.9 CONCLUSIONS

The use of propagated fishes in inland fisheries management has a long, though controversial, history in North America. Early efforts were hampered by incomplete knowledge of husbandry and stocking techniques; later efforts became limited by their own success, as the differences between hatchery and wild fishes and the impacts of stocking became evident. Nonetheless, the usefulness of stocking and the importance of cultured fishes to achieving management programs cannot be denied. Hatchery reform, adoption of risk management and decision-making tools, and increased emphasis on conservation aquaculture and ecosystem-based approaches will ensure the continued relevance of cultured fishes to adaptive management of aquatic resources.

9.10 REFERENCES

Anders, P. J. 1998. Conservation aquaculture and endangered species. Fisheries 23(11):28–31.
Andersson, K. P., C. C. Gibson, and F. Lehoucq. 2004. The politics of decentralized natural resource governance. PS: Political Science & Politics 37:421–426.
Araki, H., B. Cooper, and M. S. Blouin. 2007. Genetic effects of captive breeding cause a rapid, cumulative fitness decline in the wild. Science 318:100–103.

Avise, J. C., and M. J. Van Den Avyle. 1984. Genetic analysis of reproduction of hybrid white bass × striped bass in the Savannah River. Transactions of the American Fisheries Society 113:563–570.

Bachman, R. A. 1984. Foraging behavior of free-ranging wild and hatchery brown trout in a stream. Transactions of the American Fisheries Society 113:1–32.

Beckman, B. R., D. A. Larsen, B. Lee-Pawlak, and W. W. Dickhoff. 1998. The relationship of fish size and growth to migratory tendencies of spring Chinook salmon (*Oncorhynchus tshawytscha*) smolts. North American Journal of Fisheries Management 18:537–546.

Beckman, B. R., W. W. Dickhoff, W. S. Zaugg, C. Sharpe, S. Hirtzel, R. Schrock, D. A. Larsen, R. D. Ewing, A. Palmisano, C. B. Schreck, and C. V. W. Mahnken. 1999. Growth, smoltification, and smolt-to-adult return of spring Chinook salmon (*Oncorhynchus tshawytscha*) from hatcheries on the Deschutes River, Oregon. Transactions of the American Fisheries Society 128:1125–1150.

Berejikian, B. A. 1995. The effects of hatchery and wild ancestry and experience on the relative ability of steelhead trout fry (*Oncorhynchus mykiss*) to avoid a benthic predator. Canadian Journal of Fisheries and Aquatic Sciences 52:2076–2082.

Berejikian, B. A., and M. J. Ford. 2004. Review of relative fitness of hatchery and natural salmon. NOAA (National Oceanic and Atmospheric Administration) Technical Memorandum NMFS (National Marine Fisheries Service) NWFSC-61, Northwest Fisheries Science Center, Seattle. Available: www.nwfsc.noaa.gov/publications. (December 2009).

Berejikian, B. A., E. P. Tezak, S. Riley, and A. LaRae. 2001. Social behavior and competitive ability of juvenile steelhead (*Oncorhynchus mykiss*) reared in enriched and conventional hatchery tanks and a stream environment. Journal of Fish Biology 59:1600–1613.

Berejikian, B. A., E. P. Tezak, S. L. Schroder, T. A. Flagg, and C. M. Knudsen. 1999. Competitive differences between newly emerged offspring of captive-reared and wild coho salmon. Transactions of the American Fisheries Society 128:832–839.

Brown, C., and R. L. Day. 2002. The future of stock enhancements: lessons for hatchery practice from conservation biology. Fish and Fisheries 3:79–94.

Bryant, P. J. 2003. Captive breeding and reintroduction. In Biodiversity and conservation, a hypertext book. Available: http://darwin.bio.uci.edu/~sustain/bio65/Titlpage.htm.

Busack, C. A., and K. P. Currens. 1995. Genetic risks and hazards in hatchery operations: fundamental concepts and issues. Pages 71–80 in H. L. Schramm and R. G. Piper, editors. Uses and effects of cultured fishes in aquatic ecosystems. American Fisheries Society, Symposium 15, Bethesda, Maryland.

Campton, D. E. 1995. What do we really know? Pages 337–353 in H. L. Schramm and R. G. Piper, editors. Uses and effects of cultured fishes in aquatic ecosystems. American Fisheries Society, Symposium 15, Bethesda, Maryland.

Carmichael, G. J., J. R. Tomasso, and T. E. Schwedler. 2001. Fish transportation. Pages 641–660 in G. A. Wedemeyer, editor. Fish hatchery management, 2nd edition. American Fisheries Society, Bethesda, Maryland.

Chilcote, M. W., S. A. Leider, and J. J. Loch. 1986. Differential reproductive success of hatchery and wild-run steelhead under natural conditions. Transactions of the American Fisheries Society 115:726–735.

Cho, C. Y., and D. P. Bureau. 2001. A review of diet formulation strategies and feeding systems to reduce excretory and feed wastes in aquaculture. Aquaculture Research 32:349–360.

Currens, K. P., and C. A. Busack. 2005. Practical approaches for assessing risks of hatchery programs. Pages 277–290 in M. J. Nickum, P. M. Mazik, J. G. Nickum, and D. D. MacKinlay, editors. Propagated fish in resource management. American Fisheries Society, Symposium 44, Bethesda, Maryland.

Davis, H. S. 1936. Hatchery trout versus wild trout. The Progressive Fish-Culturist 3:31–35.

DeBlieu, J. 1993. Meant to be wild: the struggle to save endangered species through captive breeding. Fulcrum Publishing, Golden, Colorado.

Deverill, J. I., C. E. Adams, and C. W. Bean.. 1999. Prior residence, aggression and territory acquisition in hatchery-reared and wild brown trout. Journal of Fish Biology 55:868–875.

Edwards, G. B., and J. G. Nickum. 1993. Use of propagated fishes in Fish and Wildlife Service programs. Pages 41–44 in M. R. Collie and J. P. McVey, editors. Interactions between cultured species and naturally occurring species in the environment. Alaska Sea Grant UJNR (U.S.–Japan Cooperative Program in Natural Resources) Technical Report 22, Fairbanks.

FAO (Food and Agriculture Organization of the United Nations). 2005. FAO Aquaculture Newsletter 34. Available: www.fao.org/docrep/009/a0435e/a0435e00.htm. (December 2009).

FAO (Food and Agriculture Organization of the United Nations). 2009. State of world fisheries and aquaculture 2008. FAO Fisheries and Aquaculture Department, Rome.

Fayram, A. H., M. J. Hansen, and N. A. Nate. 2005. Determining optimal stocking rates using a stock–recruitment model: an example using walleye in northern Wisconsin. North American Journal of Fisheries Management 25:1215–1225.

Flagg, T. A., F. W. Waknitz, D. J. Maynard, G. B. Milner, and C. V. W. Mahnken. 1995. The effects of hatcheries on native coho salmon populations in the lower Columbia River. Pages 366–375 in H. L. Schramm and R. G. Piper, editors. Uses and effects of cultured fishes in aquatic ecosystems. American Fisheries Society, Symposium 15, Bethesda, Maryland.

Flagg, T., and C. Nash. 1999. A conceptual framework for conservation hatchery strategies for Pacific salmonids. NOAA (National Oceanic and Atmospheric Administration) Technical Memorandum 38. Available: www.nwfsc.noaa.gov/publications. (December 2009).

Flagg, T., C. Mahnken, and R. Iwamoto. 2005. Conservation hatchery protocols for Pacific salmon. Pages 603–620 in M. J. Nickum, P. M. Mazik, J. G. Nickum, and D. D. MacKinlay, editors. Propagated fish in resource management. American Fisheries Society, Symposium 44, Bethesda, Maryland.

Ford, M. J., H. Fuss, B. Boelts, E. LaHood, J. Hard, and J. Miller. 2006. Changes in run timing and natural smolt production in a naturally spawning coho salmon (*Oncorhynchus kisutch*) population after 60 years of intensive hatchery supplementation. Canadian Journal of Fisheries and Aquatic Sciences 63:2343–2355.

Fuss, H. J., and C. Johnson. 1988. Effects of artificial substrate and covering on growth and survival of hatchery-reared coho salmon. Progressive Fish Culturist 50:232–237.

Gablehouse, D. W. 2005. Staffing, spending, and funding of state inland fisheries programs. Fisheries 30(2):10–17.

Gatlin, D. M., F. T. Barrows, P. Brown, K. Dabrowski, T. G. Gaylord, R. W. Hardy, E. Herman, G. Hu, Å. Krogdahl, R. Nelson, K. Overturf, M. Rust, W. Sealey, D. Skonberg, E. J. Souza, D. Stone, R. Wilson, and E. Wurtele. 2007. Expanding the utilization of sustainable plant products in aquafeeds: a review. Aquaculture Research 38:551–579.

Gozlan, R.E. 2008. Introduction of nonnative freshwater fish: is it all bad? Fish and Fisheries 9:106–115.

Guy, C. S., H. L. Blankenship, and L. A. Nielsen. 1996. Tagging and marking. Pages 353–383 in B. R. Murphy and D. W. Willis, editors. Fisheries techniques, 2nd edition. American Fisheries Society, Bethesda, Maryland.

Halverson, M. A. 2008. Stocking trends: a quantitative review of governmental fish stocking in the United States, 1931 to 2004. Fisheries 33(2):69–75.

Hard, J. J., R. P. Jones Jr., M. R. Delarm, and R. S. Waples. 1992. Pacific salmon and artificial propagation under the Endangered Species Act. NOAA (National Oceanic and Atmospheric Administration) Technical Memorandum NMFS (National Marine Fisheries Service) NWFSC-2, Northwest Fisheries Science Center, Seattle.

Hardy, R. W., and A. G. J. Tacon. 2002. Fish meal: historical uses, production trends and future outlook for sustainable supplies. Pages 311–325 in R. R. Stickney and J. P. McVey, editors. Responsible marine aquaculture. CABI, Wallingford, UK.

Hartman, K. J., and B. Preston. 2001. Stocking. Pages 661–686 *in* G. A. Wedemeyer, editor. Fish hatchery management, 2nd edition. American Fisheries Society, Bethesda, Maryland.

Heard, W. R. 2003. Alaska salmon enhancement: a successful program for hatchery and wild stocks. Pages 149–169 *in* Y. Nakamura, J. P. McVey, K. Leber, C. Neidig, S. Fox, and K. Churchill, editors. Ecology of aquaculture species and enhancement of stocks: proceedings of the thirtieth U.S.–Japan meeting on aquaculture, Sarasota, Florida, Dec. 3-4. UJNR (U.S.–Japan Cooperative Program in Natural Resources) Technical Report 30, Mote Marine Laboratory, Sarasota, Florida.)

Hebdon, J. L., P. A. Kline., D. Taki, and T. A. Flagg. 2005. Evaluating reintroduction strategies for Redfish Lake sockeye salmon captive broodstock progeny. Pages 401–413 *in* M. J. Nickum, P. M. Mazik, J. G. Nickum, and D. D. MacKinlay, editors. Propagated fish in resource management. American Fisheries Society, Symposium 44, Bethesda, Maryland.

Heidinger, R. C. 1999. Stocking for sport fisheries enhancement. Pages 375–401 *in* C. C. Kohler and W. A. Hubert, editors. Inland fisheries management in North America, 2nd edition. American Fisheries Society, Bethesda, Maryland.

Helfman, G. S. 2007. Fish conservation—a guide to understanding and restoring global aquatic biodiversity and fishery resources. Island Press, Washington, D.C.

HSRG (Hatchery Scientific Review Group). 2004a. Segregated Hatchery Programs, June 21, 2004. Hatchery Scientific Review Group, Washington Department of Fish and Wildlife, and the Northwest Indian Fisheries Commission Technical Discussion Paper 2. Available: http://hatcheryreform.us.

HSRG (Hatchery Scientific Review Group). 2004b. Integrated Hatchery Programs, June 21, 2004. Hatchery Scientific Review Group, Washington Department of Fish and Wildlife, and the Northwest Indian Fisheries Commission Technical Discussion Paper 1. Available: http://hatcheryreform.us.

Jackson, J. R., J. C. Boxrucker, and D. W. Willis. 2005. Trends in agency use of propagated fishes as a management tool in inland fisheries. Pages 121–138 *in* M. J. Nickum, P. M. Mazik, J. G. Nickum, and D. D. Mackinlay, editors. Propagated fish in resource management. American Fisheries Society, Symposium 44, Bethesda, Maryland.

Johnson, J. E., and B. L. Jensen. 1991. Hatcheries for endangered freshwater fishes. Pages 199–217 *in* W. L. Minckley and J. E. Deacon, editors. Battle against extinction—native fish management in the American West. University of Arizona Press, Tucson.

Knuth, B. A., and S. L. McMullin. 1996. Measuring the human dimensions of recreational fisheries. Pages 651–684 *in* B. R. Murphy and D. W. Willis, editors. Fisheries techniques, 2nd edition. American Fisheries Society, Bethesda, Maryland.

Kohler, C. C. 1995. Captive conservation of endangered fishes. Pages 77–85 *in* E. F. Gibbons Jr., B.S. Durrant, and J. Demarest, editors. Conservation of endangered species in captivity. State University of New York Press, Albany.

Kohler, C. C., and W. A. Hubert, editors. 1993. Inland fisheries management in North America. American Fisheries Society, Bethesda, Maryland.

Kohler, C. C., and W. A. Hubert, editors. 1999. Inland fisheries management in North America, 2nd edition. American Fisheries Society, Bethesda, Maryland.

Kostow, K. 2004. Differences in juvenile phenotypes and survival between hatchery stocks and a natural population provide evidence for modified selection due to captive breeding. Canadian Journal of Fisheries and Aquatic Sciences 61:577–589.

LaPatra, S. E. 2003. The lack of scientific evidence to support the development of effluent limitations guidelines for aquatic animal pathogens. Aquaculture 226:191–199.

Larsen, D. A., B. R. Beckman, and W. W. Dickhoff. 2001. The effect of low temperature and fasting during the winter on growth and smoltification of coho salmon. North American Journal of Aquaculture 63:1–10.

Lathrop, R. C., B. M. Johnson, T. B. Johnson, M. T. Vogelsang, S. R. Carpenter, T. R. Hrabik, J. F. Kitchell, J. J. Magnuson, L. G. Rudstam, and R. S. Stewart. 2002. Stocking piscivores to improve fishing and water clarity: a synthesis of the Lake Mendota biomanipulation project. Freshwater Biology 47:2410–2424.

Leber, K. M., R. N. Cantrell, and P. Leung. 2005. Optimizing cost-effectiveness of size at release in stock enhancement programs. North American Journal of Fisheries Management 25:1596–1608.

Leitritz, E., and R. C. Lewis. 1976. Trout and salmon culture. California Fish and Game Bulletin 164.

Levin, P. S., R. W. Zabel, and J. G. Williams. 2001. The road to extinction is paved with good intentions: negative association of fish hatcheries with threatened salmon. Proceedings of the Royal Society of London B 268:1153–1158.

Lichatowich, J. 1999. Salmon without rivers: a history of the Pacific salmon crisis. Island Press, Washington, D.C.

Lorenzen, K. 2008. Understanding and managing enhancement fisheries systems. Reviews in Fisheries Science 16:10–23.

Lorenzen, K., U. S. Amarasinghe, D. M. Bartley, J. D. Bell, M. Bilio, S. S. de Silva, C. J. Garaway, W. D. Hartmann, J. M Kapetsky, P. Laleye, J. Moreau, V. V. Sugunan, and D. B. Swar. 2001. Strategic review of enhancements and culture-based fisheries. Pages 221–237 in R. P. Subasinghe, P. B. Bueno, M. J. Phillips, C. Hough, S. E. McGladdery, and J. R. Arthur, editors. Report of the conference on aquaculture in the third millennium. FAO (Food and Agriculture Organization of the United Nation) Fisheries Report 661, Rome.

MacKinlay, D., S. Lehmann, J. Bateman, and R. Cook. 2005. Pacific salmon hatcheries in British Columbia. Pages 57–75 in M. J. Nickum, P. M. Mazik, J. G. Nickum, and D. D. MacKinlay, editors. Propagated fish in resource management. American Fisheries Society, Symposium 44, Bethesda, Maryland.

Mahnken, C., G. Ruggerone, W. Waknitz, and T. Flagg. 1998. A historical perspective on salmonid production from Pacific rim hatcheries. North Pacific Anadromous Fish Commission Bulletin 1:38–53.

Maynard, D. J., T. A. Flagg, and C. V. W. Mahnken. 1995. A review of innovative culture strategies for enhancing the postrelease survival of anadromous salmonids. Pages 307–314 in H. L. Schramm and R. G. Piper, editors. Uses and effects of cultured fishes in aquatic ecosystems. American Fisheries Society, Symposium 15, Bethesda, Maryland.

Maynard, D. J., T. A. Flagg, R. N. Iwamoto, and C. V. W. Mahnken. 2005. A review of recent studies investigating seminatural rearing strategies as a tool for increasing Pacific salmon postrelease survival. Pages 573–584 in M. J. Nickum, P. M. Mazik, J. G. Nickum, and D. D. MacKinlay, editors. Propagated fish in resource management. American Fisheries Society, Symposium 44, Bethesda, Maryland.

McElhany, P., M. H. Ruckelshaus, M. J. Ford, T. C. Wainwright, and E. P. Bjorkstedt. 2000.Viable salmon populations and the recovery of evolutionarily significant units. NOAA (National Oceanic and Atmospheric Administration) Technical Memorandum NMFS (National Marine Fisheries Service) NWFSC-42, Northwest Fisheries Science Center, Seattle.

Mehner, T., J. Benndorf, P. Kasprzak, and R. Koschel. 2002. Biomanipulation of lake ecosystems: successful applications and expanding complexity in the underlying science. Freshwater Biology 47:2453–2465.

Mitchell, A. J., and A. M. Kelly. 2006. The public sector role in the establishment of grass carp in the United States. Fisheries 31(3):113–121.

Mobrand, L., J. Barr, D. Campton, T. Evelyn, T. Flagg, C. Mahnken, L. Seeb, P. Seidel, and W. Smoker. 2005. Hatchery reform in Washington State: principles and emerging issues. Fisheries 30(6):11–23.

Murphy, B. R., and W. E. Kelso. 1986. Strategies for evaluating freshwater stocking programs: past practices and future needs. Pages 306–313 in R. H. Stroud, editor. Fish culture in fisheries management. American Fisheries Society, Bethesda, Maryland.

Murray, C. B., and T. D. Beacham. 1986. Effect of incubation density and substrate on the development of chum salmon eggs and alevins. Progressive Fish Culturist 48:242–249.

Naish, K. A., J. E. Taylor III, P. S. Levin, T. P. Quinn, J. R. Winton. D. Huppert, and R. Hilborn. 2007. An evaluation of the effects of conservation and fishery enhancement hatcheries on wild populations of salmon. Advances in Marine Biology 53:61–194.

Naylor, R. L, R. J. Goldburg, J. H. Primavera, N. Kautsky, M. C. M. Beveridge, J. Clay, C. Folke, J. Lubchenco, H. Mooney, and M. Troell. 2000. Effect of aquaculture on world fish supplies. Nature 405:1017–1024.

Neal, J. W., R. L. Noble, and J. A. Rice. 1999. Fish community response to hybrid striped bass introduction in small warmwater impoundments. North American Journal of Fisheries Management 19:1044–1053.

New, M. B., and U. N. Wijkström. 2002. Use of fish meal and fish oil in aquafeeds: further thoughts on the fish meal trap. FAO (Food and Agriculture Organization of the United Nations) Fisheries Circular 975, Rome.

Nickelson, T. E., M. F. Solazzi, and S. L. Johnson. 1986. Use of hatchery coho salmon (*Oncorhynchus kisutch*) presmolts to rebuild wild populations in Oregon coastal streams. Canadian Journal of Fisheries and Aquatic Sciences 43:2443–2449.

Nickum, M. J., P. M. Mazik, J. G. Nickum, and D. D. MacKinlay, editors. 2005. Propagated fish in resource management. American Fisheries Society, Symposium 44, Bethesda, Maryland.

Noble, R. L. 1986. Stocking criteria and goals for restoration and enhancement of warmwater and coolwater fisheries. Pages 139–159 *in* R. H. Stroud, editor. Fish culture in fisheries management. American Fisheries Society, Bethesda, Maryland.

Olla, B. L., M. W. Davis, and C. H. Ryer. 1998. Understanding how the hatchery environment represses or promotes the development of behavioral survival skills. Bulletin of Marine Science 62(2):531–550.

Olney, P. J. S., G. M. Mace, and A. T. C. Feistner. 1994. Creative conservation: interactive management of wild and captive animals. Chapman and Hall, London.

Pascual, M. A., and J. A. Ciancio. 2007. Introduced anadromous salmonids in Patagonia: risks, uses, and a conservation paradox. Pages 333–353 *in* T. Bert, editor. Ecological and genetic implications of aquaculture activities. Kluwer Academic Publishers, Netherlands.

Pearsons, T. N., and C. W. Hopley. 1999. A practical approach for assessing ecological risks associated with fish stocking programs. Fisheries 24(9):16–23.

Pennell W., and B. A. Barton, editors. 1996. Principles of salmonid culture. Elsevier, Amsterdam.

Pimentel, D., R. Zuniga, and D. Morrison. 2005. Update on the environmental and economic costs associated with alien-invasive species in the United States. Ecological Economics 52:273–288.

Piper, R. G., I. B. McElwain, L. E. Orme, J. P. McCraren, L. G. Fowler, and J. R. Leonard. 1982. Fish hatchery management. U.S. Fish and Wildlife Service, Washington, D.C.

Pitman, V. M., and S. Gutreuter. 1993. Initial poststocking survival of hatchery-reared fishes. North American Journal of Fisheries Management 13:151–159.

Poon, D. C. 1977. Quality of salmon fry from gravel incubators. Doctoral dissertation, Oregon State University, Corvallis.

Richards, R. A., and P. J. Rago. 1999. A case history of effective fishery management: Chesapeake Bay striped bass. North American Journal of Fisheries Management 19:356–375.

Rinne, J. N., J. E. Johnson, B. L. Jensen, A. W. Ruger, and R. Soreson. 1986. The role of hatcheries in the management and recovery of threatened and endangered fishes. Pages 271–285 *in* R. H. Stroud, editor. Fish culture in fisheries management. American Fisheries Society, Bethesda, Maryland.

Ross, M. R., and D. K. Loomis. 1999. State management of freshwater fisheries resources: its organizational structure, funding, and programmatic emphases. Fisheries 24(7):8–14.

Rulifson, R. A., and R. W. Laney. 1999. Striped bass stocking programs in the United States: ecological and resource management issues. Department of Fisheries and Oceans Canadian Stock Assessment Secretariat Resource Document 99-07, Ottawa.

Sargent, J. R.. D. R. Tocher, and J. G. Bell. 2002. The lipids. Pages 182–246 *in* J. E. Halver and R. W. Hardy, editors. Fish Nutrition, 3rd edition. Academic Press, San Diego, California.

Schramm, H. L., and R. G. Piper, editors. 1995. Uses and effects of cultured fishes in aquatic ecosystems. American Fisheries Society, Symposium 15, Bethesda, Maryland.

Secor, D. H., and E. D. Houde. 1998. Use of larval stocking in restoration of Chesapeake Bay striped bass. ICES Journal of Marine Science 55:228–239.

Stevens, D. E., D. W. Kohlhorst, and L. W. Miller. 1985. The decline of striped bass in the Sacramento–San Joaquin Estuary, California. Transactions of the American Fisheries Society 114:12–30.

Stroud, R. H., editor. 1986. Fish culture in fisheries management. American Fisheries Society, Bethesda, Maryland.

Tacon, A. G. J. 2005. State of information on salmon aquaculture feed and the environment. World Wildlife Fund Salmon Aquaculture Dialogue, Washington, D.C. Available: www.worldwildlife.org/cci/pubs/Feed_final_resavedwithdate.pdf. (December 2009).

Tave, D. 1986. Genetics for fish hatchery managers. AVI Publishing, Westport, Connecticut.

Trushenski, J. T., C. S. Kasper, and C. C. Kohler. 2006. Challenges and opportunities in finfish nutrition. North American Journal of Aquaculture 68:122–140.

Wahl, D. H., R. A. Stein, and D. R. DeVries. 1995. An ecological framework for evaluating the success and effects of stocked fishes. Pages 176–189 *in* H. L. Schramm and R. G. Piper, editors. Uses and effects of cultured fishes in aquatic ecosystems. American Fisheries Society, Symposium 15, Bethesda, Maryland.

Wang, S., J. J. Hard, and F. Utter. 2002. Salmonid inbreeding: a review. Reviews in Fish Biology and Fisheries 11:301–319.

Waples, R. S. 1991. Genetic interactions between hatchery and wild salmonids: lessons from the Pacific Northwest. Canadian Journal of Fisheries and Aquatic Sciences 48:124–133.

Waples, R. S. 1999. Dispelling some myths about hatcheries. Fisheries 24(2)12–21.

Waples, R. S., and D. J. Teel. 1990. Conservation genetics of Pacific salmon, I. Temporal changes in allele frequency. Conservation Biology 4:144–156.

Williamson, J. H. 2001. Broodstock management for imperiled and other fishes. Pages 397–482 *in* G. A. Wedemeyer, editor. Fish hatchery management, 2nd edition. American Fisheries Society, Bethesda, Maryland.

Chapter 10

Habitat Improvement in Altered Systems

MARK A. PEGG AND JOHN H. CHICK

10.1 INTRODUCTION

Most freshwater ecosystems have been altered by human activities such as channelization, increased nutrient inputs, and shoreline development. Human activities can change fish habitat so that a system may not be able to achieve fishery management objectives, thus providing impetus to improve habitat conditions. Changes in habitat quantity and quality have been identified as primary reasons for declines in fish populations in North America (Ricciardi and Rasmussen 1999; Venter et al. 2006). There has been a recent, marked increase in support from resource managers and the public to initiate activities to improve habitat for fishes. For example, the number of habitat improvement projects reported in the National River Restoration Science Synthesis database has increased from a few dozen projects in 1985 to nearly 6,000 projects in 2005 (Bernhardt et al. 2005). A standardized approach to coordinate, plan, and implement habitat improvement projects is difficult to apply given the differences among ecosystems and the causes of habitat loss throughout North America. However, a systematic and logical management approach to mitigate or restore habitats that meets fisheries management objectives can streamline the process and is needed.

The concept of habitat improvement is complex and dynamic because it occurs across multiple spatial and temporal scales (Bohn and Kershner 2002; Feist et al. 2003). For example, habitat improvements can occur at a local scale, such as placement of habitat-forming structures in a stream segment during a single year, or at a watershed scale, with habitat improvements made throughout a system over many years. Habitat improvement at small spatial scales is generally conducted at a single lake, stream segment, or reservoir; requires little coordination by fisheries managers with other entities; and is performed with relatively limited personnel and resources. Habitat improvements at larger scales, such as whole streams or entire watersheds, often require considerable planning and coordination because large systems have diverse management and interest groups and are subjected to many different alterations.

The process for improving habitat in aquatic systems that have been physically, chemically, and (or) biologically altered with subsequent impairment of fisheries is the topic of this chapter. Specifically, the term "habitat" is defined, the process used in approaching habitat improvement is described, and agency roles in habitat improvement are addressed. In addition, considerations for habitat improvements in rivers, streams, natural lakes, and reservoirs are discussed. More information focused on natural lakes (Chapter 15), small impoundments

10.2 HABITAT CONCEPTS

The concept of habitat is simultaneously simple and difficult to grasp. Definitions of habitat can be found in numerous references, but it is generally defined as the area where an organism, population, or community occurs in the environment (Ricklefs 1973). The habitat of a given species should not be confused with the ecological niche (sensu Hutchinson 1957) of the species, which encompasses habitat needs, physical and chemical tolerances, and the role of a species in its ecosystem. Disentangling what is habitat from what is an ecosystem can also be confusing at times. Ecosystems comprise many habitats, but the definitions of habitat and ecosystem must be scaled to particular organisms and processes of interest (Allen and Hoekstra 1992). The distinction between habitat and ecosystem is difficult when addressing habitat improvement issues for some fishes, such as anadromous Pacific salmon. Pacific salmon require specific streams with specific habitats to reproduce successfully, although adults reside in streams for a relatively short period of their adult lives.

The following scenario highlights the many complexities of describing habitat in altered systems. One could argue that the only time suitable stream habitat for Pacific salmon must be present is during and after spawning or until the offspring move downstream. Another argument could be that dams have impeded migrations of Pacific salmon, meaning they cannot reach their natural spawning grounds, so there is no need for habitat improvement. Conversely, the stream and its substrates are a composite of geologic time that likely requires continuous natural flow to maintain suitability for spawning by Pacific salmon. Is the stream still Pacific salmon habitat if they can no longer migrate to the stream? The answer is yes, no, or maybe!

A confusing aspect to understanding habitat and habitat improvement can be the jargon that is used. Many terms such as restoration, rehabilitation, mitigation, and enhancement are used to describe efforts to improve habitat in degraded systems. However, there is considerable disagreement about the exact definition of these terms. Most of the disagreement comes from differences in the degree of improvement applied. Habitat improvements can range from something close to a full ecosystem restoration, where habitat is returned to a prior state (e.g., pre-European settlement), or something less extensive, such as habitat improvements to attract sport fish. Therefore, using a standard set of terms is critical to maintain clarity (Table 10.1). In this chapter, we use the term "habitat improvement" to encompass all activities outlined in Table 10.1.

The need and extent of habitat improvement largely depends on the goals and objectives identified to address specific management needs. Ultimately, anthropogenic habitat alterations put fisheries managers in positions in which they are faced with making timely decisions about not only how much and what kinds of habitats are available but also how much and what kinds are needed to fulfill fisheries management goals.

10.3 ASSESSMENT OF HABITAT FOR FISHES

Reasons for improving fish habitats include (1) fulfilling public mandates such as the Clean Water Act of 1972 and the Water Resources Development Acts; (2) undertaking remediation of degraded systems such as natural lakes and streams acidified by coal mine leachates (Ger-

Table 10.1. Definition of terms used to address aquatic habitat improvements in altered systems (terms are modified from Roni 2005).

Term	Definition
Restoration	Return of an ecosystem to its original, undisturbed state (e.g., pre-European settlement, pre-impoundment, or prior to some other major disturbance event).
Rehabilitation	Repair and (or) improve certain components or functions of an ecosystem, but the ecosystem is not returned to an undisturbed state.
Reclamation	Return an area to a habitat state prior to disturbance; function is not always fully returned in reclamation efforts (e.g., sediment removal from a lake).
Mitigation	Alleviate habitat problems negatively influenced by human activities (e.g., create new stream habitat to replace that lost through road development).
Enhancement	Improve habitat through direct manipulation (e.g., placement of fish attractors or fertilization).

emias et al. 2003); (3) improving human living conditions (Golet et al. 2006); (4) enhancing or maintaining aquatic biodiversity; (5) protecting endangered species (Moser 2000); or (6) providing high-quality fisheries. All of these reasons can involve attempts to repair alterations to aquatic ecosystems that are no longer functioning as managers believe they should. Therefore, the underlying theme to improving habitat is reinstatement of "goods and services" that a given ecosystem no longer provides (Dobson et al. 2006; Kumar and Kumar 2008). Goods and services, from a fisheries management perspective, largely center on providing a fishery that meets constituent needs and may include sport fishes, native fishes, or endangered species.

Alteration, degradation, or loss of habitat can be classified into three general categories: (1) loss of habitat quantity, (2) loss of habitat quality, or (3) loss of processes that once sustained habitat for fishes. All of these categories play an integral role in where fish can exist, grow, and reproduce. Habitat improvement measures to replace or repair losses are often implemented when aquatic systems have been altered by human activities.

Natural resource management agencies and other stakeholders often agree that habitat improvement should be implemented, but unanimous agreement on what should be implemented and how is rarely achieved. Risk assessment techniques, adapted from human health assessments, are emerging as part of aquatic habitat improvement plans and are being used to facilitate identification of risks associated with various habitat improvement techniques (Mattson and Angermeier 2007).

Risk in implementing improvement plans is unwanted, and risk assessments help identify sources of problems during habitat improvement planning (Wissmar and Bisson 2003). Risk assessment can also help managers visualize concerns associated with implementing various options for a habitat improvement project (Wissmar and Bisson 2003; Mattson and Angermeier 2007). For example, removing dams on rivers is gaining momentum as a habitat improvement alternative because their removal returns longitudinal connectivity to river systems (Bednarek 2001; Stanley et al. 2007). However, dam removal studies have not shown a universal pattern in fish responses (Catalano et al. 2007; Stanley et al. 2007). A risk commonly associated with dam removal is rapid release of sediments stored above dams, which may alter productivity and contaminate downstream reaches (Frissell and Ralph 1998). An additional risk is movement of invasive species upstream to areas previously protected by the dam. Understanding these risks can facilitate arguments for or against dam removal and contribute to making the best decision on a case-by-case basis. In many situations, improving habitat for one species may eliminate or reduce habitat for other species. Fortunately, risk assessment techniques that provide assistance in planning habitat improvements are available to fisheries managers.

10.4 HABITAT IMPROVEMENT PROCESS

From a fisheries management perspective, the desired outcome of habitat improvement is the reinstatement of ecological goods and services that are no longer provided by an ecosystem; principally, a fishery that satisfies constituent needs. The reasons that an ecosystem no longer provides desired goods and services are usually the result of human alterations having impaired ecosystem structure and function. Therefore, the primary goal of habitat improvement in altered systems is to re-establish structure and function of ecosystems (Cairns 1988; Downes et al. 2002). Conceptually, structure pertains to biotic and abiotic diversity, whereas function pertains to processes that drive the ecosystem, such as sedimentation rates, nutrient transport, or nutrient loading. Fish habitat improvement efforts are therefore often nested in larger ecosystem restoration or improvement projects with the goals of identifying and redirecting the trajectory of altered systems toward more desirable conditions. Desirable conditions for habitat improvement are often identified as historic conditions (e.g., preimpoundment or pre-European settlement), but they can also be identified as another state in systems for which full restoration is not the goal. For example, a habitat improvement plan may be implemented in a reservoir to enhance sport fish habitat and improve fishing. The reservoir scenario has no realistic historical condition but demonstrates a reasonable management goal for structuring habitat improvements.

A generalized fisheries management process is discussed in Chapter 5. Habitat improvement is a major component of that process, and improving habitat in aquatic systems targets all three of the major fisheries management components (i.e., fish populations, habitat, and people). Habitat improvement processes should be thorough, relatively long term, and committed to using new information as it becomes available through adaptive management processes (Williams et al. 1997). Regardless of whether a habitat improvement project is nested within a larger ecosystem project or is a stand-alone fisheries management project, the same framework of project development, implementation, and assessment should be used (Figure 10.1). Note that habitat improvement is an iterative process, just like the fisheries management process, requiring measurements of success and reassessment of objectives.

The first step in the habitat improvement process is to understand the mechanisms that led to habitat alteration (Figure 10.1). This understanding facilitates an informed decision that helps to ensure goals are properly defined (Smith and Jones 2007). Identifying reasonable goals is imperative. A simplistic example of an unrealistic goal would be to create habitat for rainbow trout, a coldwater species, in a warmwater stream. Unfortunately, most habitat improvement goals are not that simple but should still adhere to the idea of setting realistic goals based on knowledge from similar habitat improvement projects and historical information.

Setting goals and implementing a habitat improvement plan generally requires building partnerships and collaborative funding for projects. Fisheries managers employed by agencies are involved in habitat improvement projects that take place in publically-owned water bodies where there are multiple users and stakeholders. Representatives from all user and stakeholder groups are needed to ensure that all pertinent concerns are integrated into the process. For

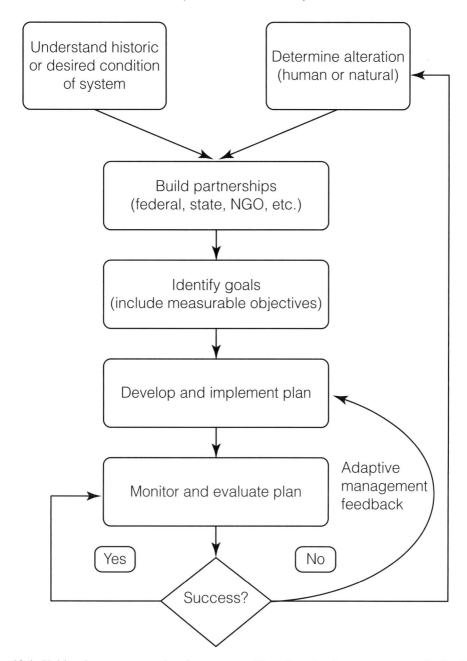

Figure 10.1. Habitat improvement planning process. Note the adaptive management feedback loop can be used when success is not achieved to adjust the plan as new information is gathered or when objectives change. Nongovernmental organizations (NGOs) may include fishing clubs, environmental groups, or civic clubs.

example, when planning to improve habitat for a fish species in a reservoir there is a need to consider the effects on other reservoir uses besides fisheries, such as flood control, irrigation, municipal water supply, and various recreational activities. Most habitat improvement projects are costly, and securing funding from a single agency is not likely. Ensuring all stakeholders know what to anticipate and why will reduce public relations issues and potential conflicts. Partnerships often involve federal, state, provincial, and local government agencies, tribal governments, nongovernment organizations (NGOs), corporations, and private interests. There are usually one or two lead organizations with administrative and logistical responsibilities, but several other stakeholders are likely to collaborate. Commitments can vary from financial support for large portions of the habitat improvement work by federal, state, or provincial agencies to providing staff for coordination of volunteer time or participating in planning efforts. All of these activities are important to ensure the habitat improvement process is implemented.

Goals provide a conceptual framework for a habitat improvement project and direction for future steps in the process, such as identifying objectives that are measureable and meaningful. Measuring changes in response to habitat improvement actions is an obvious extension of the process. Therefore, clearly identifying the objectives of a habitat improvement project is critical for success. Examples of measureable objectives of habitat improvement efforts might be to increase numbers or biomass of targeted fish species sampled during standardized monitoring efforts or increase harvest rates of targeted species by anglers. Regardless of the metric, it is important to identify measurable objectives. Therefore, identifying methods to measure the success or failure of habitat improvement efforts should be identified prior to implementation of habitat improvement activities. Subsequent monitoring efforts are critical to measuring success (see section 10.7). A contemporary example of the habitat improvement process is provided in Box 10.1.

10.5 AGENCY ROLES

Following legislated or executive mandates, federal, state, and provincial natural resource agencies are charged with managing public resources (see Chapter 4). These public mandates often put agencies in positions of leadership and fiscal responsibility for habitat improvement projects on properties entrusted to the public. Other groups such as NGOs, corporations, and private citizens may have ownership of land being considered for habitat improvement as well. Regardless of the agency or group identified as the leader, the leader must build a nexus among both public and private partners to achieve the goal(s) and objective(s) of the habitat improvement project.

10.5.1 Leadership

Agencies charged with managing aquatic habitat should play a leadership role in improving habitat. For instance, agency fisheries managers should be leaders in situations that require flowing water to pass beneath roadways to facilitate upstream fish passage. The design and construction of culverts, bridges, or fords are critical in allowing passage by fish (Warren and Pardew 1998; MacDonald and Davies 2007). Therefore, fisheries managers must be actively involved in the planning process to ensure that the needs of fish are met.

Leadership from agencies happens in many forms and ranges in scale from local projects to projects involving large watersheds, numerous agencies, and activities extending across state, provincial, or international borders. National habitat initiatives can provide broad leadership important in fisheries habitat management, but national initiatives provide less direction at regional or local levels. A good example of a national program is the U.S. National Fish Habitat Action Plan developed to facilitate the creation of partnerships that address habitat

Box 10.1. Habitat Improvement Process

The Great Lakes Regional Collaboration (GLRC) is a group tasked with coordinating habitat improvement and restoration in the North American Great Lakes ecosystem. The GLRC takes advantage of ongoing efforts in the Great Lakes by incorporating existing species-specific and ecosystem improvements with new initiatives where needed. The habitat improvement plan has not yet reached an end point, but actions to date highlight the overall process outlined in Figure 10.1. Additional details on the GLRC structure and habitat improvement process are available from the GLRC (www.glrc.us).

Since the early 1800s, the Great Lakes have been subjected to many ecosystem stressors, including pollution, habitat degradation, competition for water resources, altered hydrology, fish overharvest, and invasive species introductions, leaving a legacy of ecosystems with altered function compared with pre-1800 conditions. Many state, provincial, and federal agencies, Native American and First Nation people, nongovernmental organizations (NGOs), and other private organizations have implemented piecemeal habitat improvement projects in relation to the entire Great Lakes ecosystem. Restoration activities date back as early as the International Waterways Commission in 1905, which then formed a joint committee between the USA and Canada to advise on water management issues. Several other initiatives have worked toward ecosystem improvement, but none had an entirely holistic perspective. By the late 20th century, it was clear that a more unified approach was needed to make significant inroads to sustaining an estimated US$4 billion per year fishery and other critical elements of ecosystem function.

In 2004 a presidential executive order in the USA called for the creation of a cabinet-level interagency task force that ultimately led to the creation of the GLRC. This collaboration is supported by over 1,500 participants representing the Canadian and U.S. federal governments, Native American and First Nation people, eight states and two provinces surrounding the Great Lakes, commissions, cities, universities, and associations with interests in the Great Lakes. Participants work in eight focal areas: (1) habitat and species issues, (2) aquatic invasive species, (3) coastal health, (4) sediments, (5) nonpoint source pollution, (6) toxic pollutants, (7) indicators and information, and (8) sustainable development. These focal areas are the basis for strategy teams charged with recovery planning under the GRLC.

Each strategy team developed a problem statement and identified goals to address its respective issues. For example, the Habitat Strategy Team developed several long-term goals (>10-year timeline) and associated activities that could be implemented in the short term to facilitate recovery for open and nearshore waters, wetlands, riverine habitats, riparian areas, and coastal and upland habitats. Goals of habitat improvement activities include projects that facilitate self-sustaining fish populations, ensure wetland function, and help the entire Great Lakes ecosystem provide all the desired goods and services outlined in the plan.

The goals outlined by each strategy team also provided guidance on immediate actions that should be taken. The Habitat Strategy Team recommended actions such as re-

(Box continues)

> **Box 10.1. Continued.**
>
> storing or protecting nearly 225,000 ha of wetland habitat, developing initiatives to reintroduce lake sturgeon and coregonid fishes to areas where they have been extirpated, and identifying and prioritizing coastal wetlands for protection or improvement.
>
> The GLRC also created a strategy team to identify the biological, physical, chemical, and social information needs to ensure that resource managers, stakeholders, policymakers, and the general public get the best information possible. Making monitoring and assessment an equal partner in any habitat improvement plan is critical to weigh changes in fish populations or other desired variables.

management needs (Box 10.2). Similarly, Canada has a national fish habitat program authorized by the Fisheries Act (R.S.C. 1985, c. F-14). The Fisheries Act formalized habitat improvement for fishes and encourages Department of Fisheries and Oceans staff to participate in habitat improvement activities.

Coordination for multijurisdictional, large-scale habitat improvement projects requires substantial effort. Such projects are typically attempting to address several causes of habitat degradation, such as increased sedimentation, poor water quality, or urban development. The leadership structure usually develops a committee providing programmatic oversight for several subcommittees and working groups. Oversight committees vary but generally include major funding organizations and agencies, researchers, and other groups or individuals with a stake in a project (Figure 10.2). Depending on the goals, one or several subcommittees may be created to address specific issues or address needs of specific taxonomic groups. The overall organization is complex, but the committee structure provides a means of conducting large habitat improvement projects by parsing responsibilities into manageable units.

Project funding is always a major concern when planning habitat improvements because they are expensive, labor intensive, and time consuming. Long time spans may also be needed to detect meaningful changes in fish populations. Hence, proper financial commitment is necessary to implement habitat improvement plans and determine if objectives are attained following the project. Many agencies use innovative approaches to fund habitat improvement programs, such as a percentage of sales tax revenue to fund agency activities (i.e., Missouri, Arkansas, and Minnesota), required purchase of stamps or licenses to fish for highly-valued fishes (e.g., trout stamps), required purchase of habitat stamps (Nebraska; see Box 10.3), or donations made through income tax returns. The funds raised by state agencies are often matched with federal, NGO, corporate, or private funds to implement individual habitat improvement projects. Successful partnerships can provide further assistance through collaborative efforts to secure funding and apply for grants.

10.5.2 Public Acceptance

Critical to any habitat improvement plan is general acceptance of the proposed activities by the public and stakeholders influenced by the plan. Without a general level of acceptance, the plan will likely not be supported. This highlights the need to provide avenues for public input such as public meetings and Internet sites where written comments can be made. Pro-

Box 10.2. The National Fish Habitat Action Plan: Leading the Way to Restore Fish Habitat across the USA

CRAIG P. PAUKERT[1]

About 20% of the aquatic species in the USA are critically imperiled (Heinz 2002). Human activities have led to destruction of aquatic habitats through channelization, creation of impoundments, urbanization, agriculture, pollution, and other factors. Miller et al. (1989) indicated that physical habitat alteration was the most common cause of the extinction of freshwater fishes. Although the destruction and loss of aquatic habitat has been well established, there have been limited large-scale efforts to assess fish habitat. The National Fish Habitat Action Plan (action plan) was developed to help restore fish habitat in the USA to healthy and sustainable levels. The mission of the action plan is to protect, restore, and enhance the nation's fish and aquatic communities through partnerships that foster fish habitat conservation and improve the quality of life for the American people. The action plan is nonregulatory and voluntary and will leverage locally-based partner support to implement on-the-ground projects to help fish habitat. In 2008, there were over 450 partners from state and federal agencies, NGOs, Native American tribes, foundations, and corporations.

The goals of the action plan are to (1) protect and maintain intact and healthy aquatic ecosystems, (2) prevent further degradation of fish habitats that have been adversely affected, (3) reverse declines in the quality and quantity of aquatic habitat to improve the overall health of fish and other aquatic organisms, and (4) increase quality and quantity of fish habitats that support a broad natural diversity of fish and other aquatic species.

The action plan is being implemented by supporting fish habitat partnerships and fostering new partnerships. These partnerships are on-the-ground efforts to restore and enhance fish habitats. First, the action plan helps partners identify priority habitats and provide science-based tools to measure the success of various projects. Second, the action plan helps develop and build support for projects so the public is aware of the importance of the efforts. Third, the action plan helps refine metrics that measure the status of the nation's fish habitat. The action plan also includes an objective of developing a report on the state of the nation's fish habitats, as well as providing input for partnerships to develop baseline conditions that can be used to measure the success of fish habitat projects. Finally, the action plan provides national leadership to coordinate partnerships and on-the-ground efforts.

A large effort such as the National Fish Habitat Action Plan needs governance in place to lobby for support of the program, establish national partners, prioritize and deliver funds to projects, establish measures of success of the partnerships, and report to the U.S. Congress and other partners on the status of the nation's fish habitat. A national board was developed with 20 members representing the Association of Fish and Wildlife

(Box continues)

[1] Missouri Cooperative Fish and Wildlife Research Unit, University of Missouri, Columbia.

Box 10.2. Continued.

Agencies; federal agencies (Departments of Agriculture, Commerce, Interior, and Defense and the Environmental Protection Agency); conservation, science, and academic representatives; and at-large representatives including tribes, interstate management agencies, industry, and elected officials.

Action plan partnerships will be the key to the success of the action plan. These partnerships comprise various state and federal agencies, tribal governments, NGOs, and other interested stakeholders and are usually focused on a region or a species (see Table below). Partnerships must be approved by the national board. The goal was to have at least an additional 12 partnerships by 2010. These partnerships are implemented to engage stakeholders at a local or regional level, work with conservation groups for on-the-ground conservation and restoration efforts, and leverage local funding with action plan funding.

The National Fish Habitat Action Plan is a vision to protect and restore aquatic habitats throughout the USA. One challenge of this effort is to find ways to leverage limited nationwide funds for local and regional projects. In addition, it is important to note that the action plan is intended to enhance all suitable aquatic habitats. The goal of the action plan is not only to improve fish habitat in streams and rivers but also to improve habitat in reservoirs. Therefore, this nationwide effort will provide a framework for evaluation of future aquatic habitat protection and restoration. For more information, please see www.fishhabitat.org.

Table. Partnerships approved by the National Fish Habitat Initiative, December 2008.

Partnership	Web site
Eastern Brook Trout Joint Venture	www.easternbrooktrout.org
Southeast Aquatic Resources Partnership	www.sarpaquatic.org
Western Native Trout Initiative	www.WesternNativeTrout.org
Midwest Driftless Area Restoration Effort	www.fws.gov/Midwest/LaCrosseFisheries
Matanuska–Susitna Basin Salmon Conservation Partnership	www.nature.org/wherewework/northamerica/states/alaska/preserves/art18561.html
Southwest Alaska Salmon Habitat Partnership	www.swakcc.org

viding opportunities for public input is time consuming and not easy when dealing with contentious issues. Furthermore, gaining unanimous support for a habitat management plan is unlikely. However, providing opportunities for input can reduce conflicts and provide a sense of public ownership of a project. See Chapter 6 for additional information on this topic.

10.6 UNIQUE SYSTEMS AND ISSUES

Watersheds are defined as the entire land area contributing surface and groundwater to a particular stream, river, or lake (Williams et al. 1997). The nature of watersheds is that small watersheds are nested within larger watersheds, making standardized references to watershed

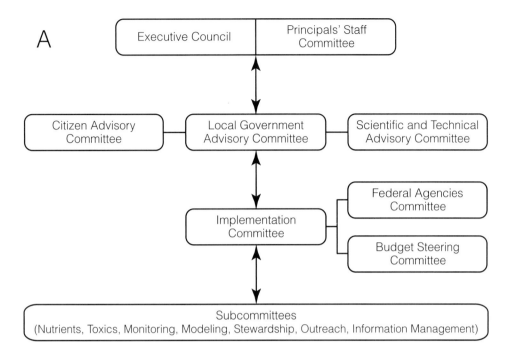

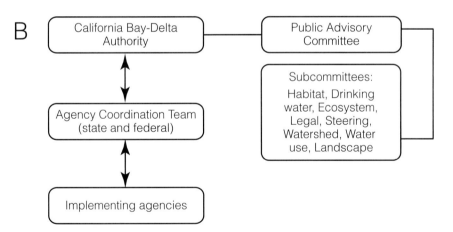

Figure 10.2. Administrative and work group structure for the Chesapeake Bay (Chesapeake Bay Program, www.chesapeakebay.net) (A) and California Bay (B) (CALFED Bay–Delta Program, www.calwater.ca.gov) habitat improvement plans. Flow charts are modified from each project.

size difficult. All aquatic ecosystems fit into one watershed, or more likely many, as spatial scale increases (Figure 10.3), so it is fitting that much of the habitat improvement literature is organized at the watershed level. A watershed approach is also appealing because it encompasses climate, geology, water quality, and biotic conditions in a relatively easily defined area for assessment. However, identifying one or a small set of causes that have altered a

Box 10.3. New Life for Aging Waters: Nebraska's Aquatic Habitat Plan

Resource management agencies are implementing habitat improvement projects across North America in an attempt to conserve, sustain, improve, and create new fisheries for the public. Funding for these projects comes in many forms, such as federal cost shares with nonfederal funds, nongovernment organization (NGO) support, donations, and portions of angler license fees, among many other options. Unfortunately, the needs for habitat improvement projects seem nearly boundless, whereas financial support and personnel to implement plans are extremely limited, forcing agencies to seek alternative funding approaches. The Nebraska Aquatic Habitat Plan is one such plan that has had marked success since its inception in 1996.

About 80% of all fishing trips in Nebraska occur in reservoirs, but Nebraska's reservoirs are rapidly aging. Many anglers had observed changes in fish assemblages from predominantly sport fish, such as largemouth bass, crappies, and walleye, to less desirable fish such as common carp. Other observed changes included loss of aquatic vegetation and frequent algal blooms that can impede angling. Many of Nebraska's anglers and the Nebraska Game and Parks Commission knew that the future of sportfishing would require improving deteriorating conditions associated with the reservoir aging process. The result of this concern was state legislation that specifically dedicated funds to aquatic habitat projects across the state. The plan, as developed in coordination with anglers and state officials, requires anglers, when purchasing a fishing license, to purchase an aquatic habitat stamp in support of habitat improvement. The compelling aspect of this program is that it is a collaborative effort with private citizens and state agencies.

Most habitat improvement plans developed from the habitat stamp have dealt with sediment removal, prevention of sediment and nutrients input, shoreline stabilization, proper water level management, aeration, and alum treatments to sequester nutrients as well as removal and blocking movements of undesirable fish species in reservoirs. River and stream projects typically focus on bank stabilization, riparian buffer zone improvement, increased instream habitat diversity, and construction of fish bypass structures around small dams. Each habitat improvement project also contains an element of evaluation.

The first 10 years of the program, 1996–2006, generated US$9.5 million from the purchase of over 1.9 million stamps. Funds raised by habitat stamp purchases have been leveraged against funds from over 70 other agencies and organizations to generate a total of US$26 million for aquatic habitat improvements in Nebraska. The end result has been the implementation and completion of 43 habitat improvement plans. Nebraska's Aquatic Habitat Stamp program is continuing to lead implementation of new habitat plans throughout the state. The program has been well accepted and will continue for the foreseeable future to repair fish habitat around the state.

---

watershed or its subcomponents can be difficult due to a complex array of human activities that may affect a watershed. This complexity can lead managers to simplify their approach by developing broad plans encompassing a large watershed (Mattson and Angermeier 2007) rather than nesting individual habitat improvement projects to address particular elements of

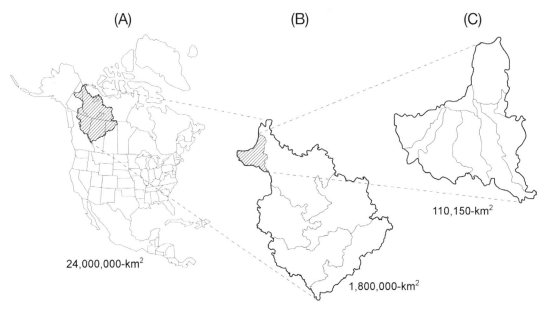

Figure 10.3. Example of how spatial scales can be nested from largest watershed (left) to smallest (right). Each scale represents (A) the Mackenzie River watershed within North America; (B) watersheds within the Mackenzie River watershed; and (C) watersheds within the Peel River watershed as part of the Mackenzie River system.

a watershed (e.g., a specific lake, stream, or reservoir). The intent is to address habitat issues at smaller, more manageable spatial scales through addressing overall watershed needs. Of course, some habitat issues such as an altered hydrograph or nonpoint source pollution are broad in their effect and must be addressed throughout a watershed. Consequently, managers must have the flexibility and insight to identify when and where, and at what scale, projects should be applied. Therefore, habitat alteration is scale dependent, and the scale at which habitat improvements may occur is based on the goals of the habitat improvement plan.

Habitat loss or degradation caused by human activities among the predominant freshwater systems in North America (streams, rivers, lakes, and reservoirs) is discussed in the sections that follow. Many human activities occur simultaneously to create a diverse array of issues that are difficult to disentangle within systems. Co-occurring habitat alterations add complexity to habitat improvement plans. Consequently, fisheries managers should recognize the need to determine all potential mechanisms and sources of habitat degradation and their effects on the system.

10.6.1 Lotic Ecosystems

Rivers and streams are an integral part of every major ecotone in North America, and there is a long and varied history of their alteration. Many changes are a direct result of management practices such as flood control, power generation, navigation, irrigation, or recreation. Dominant river or stream management practices have involved altering streamflow and habitats through impoundments, channelization, levees, and water diversions (Søndergaard and Jeppesen 2007). These practices often fragmented or isolated fish populations and reduced longitudinal and lateral connectivity within systems, having far-ranging influences on physical and biological processes that define lotic ecosystems. For example, about 14% of the world's total annual runoff is stored in reservoirs. The result has been changes to both the biotic and abiotic characteristics of lotic systems because impounded segments have been converted to function more like lentic systems (Downes et al. 2002). Biotic alterations range from changes in fish assemblage structure to extirpations of species and (or) changes in fundamental ecological processes such as nutrient cycling.

The science of habitat improvement in rivers and streams is relatively young, but methodologies are being developed and attempts to improve damaged lotic systems are being made (e.g., Roni et al. 2002; Palmer et al. 2005; Rosgen 2006). Common techniques used to address damages to lotic systems include ensuring water availability through instream-flow protection, improving water quality by eliminating sources of pollution, removing dams, reconnecting the floodplain and the channel, remediating flow, and increasing sinuosity. Most habitat improvement efforts in lotic systems are aimed at restoring small rivers and streams or localized reaches of large rivers (Lake 2001). However, a handful of projects are taking a more holistic approach to riverine habitat improvement. For example, the Kissimmee River rehabilitation effort in Florida has been the impetus of habitat improvement activities since the early 1970s, with the goals of repairing the river's flow regime, improving water quality, and increasing habitat diversity (Toth et al. 1997). Similarly, the Upper Mississippi River Environmental Management Plan has goals of maintaining and improving river structure and function (see Box 10.4). Even though spatial and temporal scales and the specific objectives vary among projects, the overriding goal of these projects has been to improve habitat for fishes and other aquatic organisms.

10.6.2 Lentic Ecosystems

Eutrophication has been identified as a primary source of fish habitat loss in lentic systems (Cooke et al. 2005; Schindler 2006). However, shoreline development (Søndergaard and Jeppesen 2007), increased sedimentation rates due to human activities in the watershed (Marburg et al. 2006), alteration or loss of habitats (Sass et al. 2006), and removal of water (Havens and Gawlik 2005) have all played roles in fish habitat loss in natural lakes.

Efforts to slow eutrophication have focused on reducing nutrient input, primarily phosphorous (Schindler 2006). Techniques such as protecting the watershed, treating inflowing water, flushing the lake, oxygenating deep water, diverting nutrients, and manipulating the food web have all been attempted to reduce nutrient inputs (Cooke et al. 2005) and have met varying degrees of success. Biotic responses can be substantially delayed in systems where nutrient loading from within the system (i.e., natural and historical anthropogenic loading processes) can sustain an altered state for many years or decades (Jeppesen et al. 2007).

10.6.3 Reservoirs

Reservoirs are by definition altered ecosystems because dams change lotic systems to more lentic systems by impounding water. Reservoirs are prevalent throughout North America (Graf 1999) and have dramatically influenced fisheries management. Reservoirs are constructed to achieve many human needs including public water supplies, flood control, navigation, irrigation, recreation, and power generation (Ligon et al. 1995; Graf 2006). Fish populations and assemblages in lotic systems before impoundment can be eradicated or sig-

Box 10.4. Example of Habitat Improvement in the Upper Mississippi River System: The Swan Lake Habitat Restoration and Enhancement Project

The Upper Mississippi River System (UMRS) has been extensively altered by humans, as have the vast majority of major rivers. Major modifications and environmental stressors include (1) construction of 26 locks and dams to allow commercial navigation, (2) isolation of floodplain habitats through the construction of levees for flood protection, (3) conversion of floodplain habitats to agriculture and municipalities, (4) changes in land use throughout the watershed, converting natural areas to agriculture and municipalities, and (5) introduction of invasive species. These changes have had deleterious effects on fish habitat, including (1) longitudinal fragmentation and disruption of migration patterns; (2) an altered annual hydrograph; (3) reduction in access to, and the loss of, critical floodplain habitats; (4) increased sediment and nutrient loads from the watershed; and (5) associated changes to habitat structure, such as the abundance and composition of aquatic vegetation. There are numerous connections among these alterations and effects. For example, access to floodplain habitats such as backwater lakes has been greatly reduced in some areas. Simultaneously, the habitat quality of backwater lakes has been degraded by loss of depth from sedimentation, and loss of aquatic vegetation that results from altered water clarity, reduced substrate quality, and negative impacts of invasive species such as common carp and grass carp. The Swan Lake Habitat Restoration and Enhancement Project provides a good example of the challenges managers face on the UMRS and the attempts to address these challenges.

Swan Lake is a 1,200-ha Illinois River backwater lake located near the confluence of the Illinois and Mississippi rivers. The lake was created in 1938 following the construction of Lock and Dam 26 (Mel Price Lock and Dam) on the Mississippi River. In the years following its creation, Swan Lake supported abundant populations of aquatic plants and fishes and provided valuable habitat for wildlife (USACE 1993; Theiling et al. 2000). After a few decades, however, habitat quality in Swan Lake began to decline as a result of accumulation of sediment. Much of the sediment input remained unconsolidated due to the lack of a summer dry period, which, in turn, was caused by the maintenance of water levels suitable for navigation on the Illinois and Mississippi rivers. These effects were especially detrimental to aquatic vegetation because unconsolidated sediments increase turbidity, reduce light penetration, and are poor substrates for rooted vegetation. Through time, rooted aquatic vegetation decreased in Swan Lake and was eliminated after a 1993 flood. Ultimately, the value of Swan Lake for sportfishing declined along with the aquatic vegetation.

Key partners in designing, constructing, managing, and assessing the Swan Lake habitat project included two federal agencies, two state agencies, and one university. Swan Lake is part of the U.S. Fish and Wildlife Service's (FWS) Two Rivers National Wildlife Refuge, and the FWS is the principal management agency responsible for Swan Lake. The goals and objectives for the project were led by the U.S. Army Corps of Engi-

(Box continues)

Box 10.4. Continued.

neers (ACE) and FWS in consultation with the Illinois Department of Natural Resources, NGOs, and the general public. Funding, engineering design, and construction for the project were largely provided by ACE. Researchers from the Illinois Natural History Survey and Southern Illinois University–Carbondale conducted ecological assessments of the habitat project and helped refine project goals and objectives.

The Swan Lake Habitat Rehabilitation and Enhancement Project is one of a few large-scale management projects that has attempted to restore backwater lake habitat for waterfowl and other wildlife, as well as fishes and other riverine organisms. The main goals of the project were to reduce sediment inputs and allow water level management that would compact sediments to promote growth of aquatic vegetation, thereby improving habitat for fish and wildlife.

To achieve these goals, Swan Lake was sequestered into three distinct units. These units provided an opportunity to implement experimentation through an adaptive management process (i.e., each unit was managed differently with supporting hypotheses) to identify strategies to meet the sometimes conflicting habitat needs of fish and wildlife. The units were initially sequestered as follows.

Upper unit.—Prior to this project, this unit was separated from the main lake by construction of a levee and was managed with aggressive annual drawdowns to promote the growth of moist-soil vegetation, principally to benefit waterfowl. Monitoring data from this unit were used as a control because changes to this unit predated the habitat improvement project.

Middle unit.—This unit was separated from the lower portion of Swan Lake and the Illinois River by construction of a levee. The unit has been managed to promote the growth of emergent vegetation. A gate is used to allow or restrict water outflow, and a pump allows the unit to be drawn down either partially or fully. Full drawdowns of the middle unit were conducted in 2002 and 2005 and partial drawdowns were conducted in 2003 and 2004.

Lower unit.—This unit was separated from the Illinois River by construction of a levee. A stop-log structure, a type of gate used to control the connection between the Illinois River and lower unit, and a pump were installed to enable water level management. This unit was drawn down in 2002 but remained connected to the river through the stop-log structure from 2003 until the summer of 2006.

The above changes greatly reduced the connectivity of Swan Lake with the Illinois River, thereby reducing sediment inputs from the river and allowing water level management (i.e., drawdowns). Drawdowns mimic natural processes that occurred in most backwater lakes in the UMRS floodplain prior to system alterations, whereby flood waters increase lake depths during the spring and fall and lake depths decrease during the summer dry period, allowing sediments to compact.

(Box continues)

> **Box 10.4. Continued.**
>
> Pre- and postproject monitoring was conducted to assess project success. Turbidity and Secchi disk transparency data suggested the habitat project and management strategies used in the middle unit appeared to have led to increases in water clarity, whereas little improvement was observed in the lower unit. Turbidity in the middle unit was often at levels low enough (e.g., turbidity ≤ 40 nephelometric turbidity units) to allow for growth of submersed aquatic vegetation. By 2008, however, submersed and rooted floating vegetation had not re-established in either the lower or middle units. Furthermore, comparisons of pre- and postproject monitoring data indicated that most of the desired benefits for fishes had not been fully achieved. Further reductions in turbidity and hardening of sediments likely are needed in the lower unit before vegetation can re-establish. However, the introduction of seeds and tubers, possibly combined with protection from herbivores, may be needed to re-establish vegetation. Experiments conducted in the middle unit demonstrated that introduced sago pondweed can grow if protected from invasive fish species (common carp and grass carp) and herbivorous native turtles (red-eared sliders). As a result of these findings, managers with the FWS are refining the water management plans to improve conditions further in the lower and middle units of Swan Lake.

nificantly altered in reservoirs. Maintenance of viable fisheries in reservoirs can require considerable management effort involving supplemental stocking, introduction of new species, harvest regulations, habitat improvements, or a combination of all of these practices. Changes that occur after impoundment that influence fishes and their habitats include changes in water temperature and dissolved oxygen (Wetzel 2001), nutrient inputs and outputs (Matzinger et al. 2007), physical structure (Ligon et al. 1995), and sediment deposition (Hayes et al. 1999). Water management that affects water level variability can also play a role in affecting fish populations in reservoirs (Jones and Rogers 1998).

The reservoir aging process also affects habitat for fishes (Thornton 1990). Reservoirs are highly productive for the first several years following impoundment because there is a release of nutrients from flooded areas. This period of productivity is followed by a slow decline in primary and secondary productivity as well as a loss of habitat diversity. Consequently, the reservoir aging process often creates a need for habitat improvements to meet fisheries management goals.

Targets of habitat improvement measures for reservoirs generally fall into two general categories: watershed management or within-reservoir management. Upstream watershed management largely deals with problems associated with land use and land cover. Soil erosion is a prime example of a watershed management problem. Reservoirs have a limited lifespan for holding water because watersheds supply suspended sediments to reservoirs (Schilling and Wolter 2005). Sedimentation rates and reservoir filling are typically calculated prior to reservoir construction and provide estimates of a reservoir's functional life span (Dendy et al. 1973). However, changes in land use practices within the watershed, such as urbanization and poor agricultural practices, can lead to higher sedimentation rates, shorter functional life spans, and more rapid loss of available fish habitat (Kimmel and Groeger 1986). Fine sediments can also carry nutrients (i.e., nitrogen and phosphorus) or contaminants that can have long-term effects on reservoir ecosystems. Efforts to address sediment delivery from watersheds may include incentives or mandates for erosion control efforts on private or public lands or construction of sediment traps just upstream from the reservoir (Columbo et al. 2006; Luo et al. 2006).

Management of habitat in reservoirs has primarily focused on water level manipulation. Drastic water level fluctuations can reduce habitat availability for spawning and nursery areas (Baldwin and Mitchell 2000). Water level management reducing the magnitude of water level fluctuations and controlling the timing of fluctuations can have a positive effect on spawning and nursery habitat for several fish species (Miranda and Lowery 2007). Additional in-reservoir habitat improvements may include placement of natural or artificial materials in the reservoir to diversify cover for fishes and establishing vegetation tolerant of varying water levels in the littoral zone (Naselli-Flores and Barone 2005).

10.7 EVALUATION OF HABITAT IMPROVEMENTS

The key to evaluating habitat improvements is an ability to disentangle the complex interactions among natural variability, human activity, and responses to habitat improvement efforts (Bryce and Hughes 2003). These issues are magnified because the factors altering ecosystems traverse many gradients and landscapes, and most of these systems are unique with no opportunity for replicated studies. Additionally, responses to habitat improvements can occur at varying time scales that are dependent upon not only the processes driving the system but also the extent of the habitat improvement efforts. These complex interactions create unique challenges when evaluating habitat improvement projects. Issues such as appropriate scales to measure responses, logistical limitations, and financial constraints also pose significant obstructions to evaluations (Box 10.5). Recent advances in technology, such as remote sensing and geographic information systems, have helped overcome some of these obstructions, but approaches are required that assess responses through time and at multiple spatial scales. Consequently, it is critical to establish, a priori, scientifically rigorous and spatially and temporally explicit assessment plans to ensure effective use of time and money. It is equally important to establish baseline, preproject data to assess changes in habitat conditions and fish populations following the project. In rare cases, existing data from other sources may provide a baseline, but targeted collection of preproject data will most likely be required. Preproject data usually demand several years of measurements to account for a range of water levels, water temperatures, and other sources of variability. Preproject assessment is often overlooked or half-heartedly conducted but is critical to assess achievement of goals and objectives.

Habitat improvement plans can take considerable time to develop and implement, but there is often an underlying political urgency to detect short-term responses in a fishery or ecosystem. Some short-term responses may occur, but responses are more likely to be gradual and take many years or decades to detect (Pegg and McClelland 2004), thereby requiring long-term commitments of field staff and funding to provide assessment of a habitat improvement project. Such long-term commitments often conflict with agency and organization timelines because monitoring and assessment efforts do not correspond with funding cycles and changes in priorities. Resource managers and policymakers have often done a poor job of ensuring that adequate monitoring and assessment are coupled with habitat improvement

Box 10.5. Habitat Improvement, Ecosystem Restoration, and What to Measure: The Case of the Everglades

JOHN H. CHICK[1] AND JOEL C. TREXLER[2]

Today, most people consider the Everglades to be a prized natural treasure for the USA. However, the relation between American citizens and the Everglades throughout most of our history could be considered "bipolar." There is a long history of draining and converting Everglades land for human use, dating back to Hamilton Disston's efforts in the late 1800s. Alternatively, efforts to preserve the Everglades as a unique and valuable ecosystem date back at least to Ernest Coe's efforts in the 1920s to create a national park. Throughout most of the 20th century, state and federal managers struggled to maintain a balance between efforts to alter the Everglades to support human population growth with efforts to protect the ecological integrity and natural resources of the Everglades. Societal attitudes toward the Everglades tilted to saving the ecological integrity of the system by the end of the 20th century, leading to attempts at ecosystem restoration of staggering proportions. The restoration efforts include both the Everglades and its watershed, the Kissimmee River and Lake Okeechobee.

There are several human imposed stressors that have negatively affected the ecological integrity of the Everglades (Porter and Porter 2001; Frederick and Ogden 2003), including the following:

1. A reduction in the size of the ecosystem. Originally the Everglades was a wetland encompassing 1.2 million hectares; it has now been reduced to half this area by drainage and conversion to agriculture and urban land uses.
2. Changes in water delivery. A variety of human needs, including flood protection, agricultural uses, and drinking water use, have led to significantly less water being delivered to the Everglades. Diversions of water from Lake Okeechobee to the east and west coasts of Florida, along with extensive construction of levees and canals, have disrupted the sheet flow of water through wetland habitats. In many areas, the annual cycle of flooding, or hydroperiod, has changed due to agriculture and urban water demands and the impoundment and management of water in the water conservation areas in the northern and central Everglades.
3. Increased delivery of nutrients to the Everglades. The conversion of over 275,000 ha of wetlands in the northern Everglades to agricultural uses in what is now the Everglades Agricultural Area has increased the delivery of phosphorus to the remaining Everglades wetland habitats. This eutrophication can lead to dramatic changes in ecosystem structure and function because these habitats were formerly oligotrophic, with primary productivity limited by soil and water phosphorus content.
4. Introduction of invasive species. A seemingly ever-increasing number of nonnative

(Box continues)

---

[1] Illinois Natural History Survey, Alton
[2] Department of Biological Sciences, Florida International University, Miami.

Box 10.5. Continued.

and invasive species of animals and plants have become established in the Everglades to the detriment of native species.

The combined effect of these and other stressors on the Everglades have been pronounced. Changes in hydroperiod and water delivery have greatly homogenized the formerly diverse mosaic of wetland habitats, with the conversion of large areas to uniform stands of saw grass being especially notable. Increased phosphorus inputs and changes to hydroperiod have led to vast, uniform stands of cattails in several areas. Hydroperiod changes and nutrient additions also affect the establishment of invasive species, and even a modest increase in phosphorus content can lead to the loss of characteristic periphyton communities. Perhaps most notable in the eyes of the public are the great reductions in the wading bird abundance, likely linked to a similar reduction in abundance of fishes that are important prey for wading birds (Frederick and Ogden 2003; Trexler et al. 2003).

Details of the restoration plans for the Everglades are available elsewhere (Porter and Porter 2001) and are beyond the scope of this example. In brief, the principal goals are to restore a more natural flow of water through the system, naturalize hydroperiod throughout Everglades wetlands, and reduce phosphorus inputs. Changes in hydroperiod and water delivery will mean both lengthening hydroperiod in some areas, especially in the southern Everglades, and shortening hydroperiod in other areas, especially in the central and northern water conservation areas where water pools behind levees. The overall objective of these efforts is to improve ecological integrity and functioning of the Everglades.

Unlike previous examples in this chapter, the habitat goals of the Everglades restoration are not driven by the desire to improve recreational or commercial fisheries. Rather, they are driven by the overall goal of ecosystem restoration. In this case, the role of fishes in the food web of the Everglades is a primary concern. Therefore, setting restoration and monitoring targets for fishes is somewhat less straightforward. Rather than focusing on improvements to specific habitats (e.g., spawning substrates or nursery habitats) for a specific species or group of fishes, the goal of the Everglades restoration is to return the fish assemblage to a more natural condition. Therefore, a community scale approach was needed to set targets but was complicated by the lack of historic data (Trexler et al. 2003).

The Everglades fish assemblage is frequently divided into two groups, both for ecological and logistical (i.e., sampling methodology) reasons: (1) small fishes (adult lengths usually less than 8 cm standard length [SL]) and (2) large fishes (adult lengths generally greater than 8 cm SL). Small fishes are dominant in Everglades marshes, both in terms of density and biomass, and are critical to the diet of wading birds and other organisms. This group is effectively sampled by 1-m^2 throw traps, which yield density and biomass-per-unit-area estimates. Large fishes, which include several important aquatic predators,

(Box continues)

Box 10.5 Continued.

are monitored by means of electrofishing from an airboat; catch-per-unit-effort data are used as an index of density. These methods yield several measures that can be used to evaluate the fish assemblage, including (1) fish density and biomass, (2) size distribution, (3) relative abundance, and (4) proportions of native and nonnative species.

Given a lack of historic information on Everglades fishes, what are reasonable restoration targets? Although predisturbance data are lacking, recent monitoring data can be combined with empirical analyses and simulation modeling to predict responses associated with restoration efforts, particularly for hydroperiod changes (Trexler et al. 2003). In areas where hydroperiod will be lengthened, abundance (across all species) of both small and large fishes is expected to increase, whereas abundance is expected to decrease, particularly for large fishes, in areas where hydroperiod will be shortened. Similarly, length distributions should increase with increases in hydroperiod, as the proportions of larger fishes and older individuals are expected to increase. Not all species will respond in the same way to hydroperiod manipulations. Some species do better in short hydroperiod marshes, whereas others do better in long hydroperiod habitats. Empirical data analyses suggest there are predictable changes in the relative abundance of species that should occur as restoration efforts alter hydroperiod (Trexler et al. 2001; Chick et al. 2004). By moving to a more natural hydroperiod and nutrient regime, it is reasonable to expect that the abundance of nonnative species should be less likely to increase.

As understanding of fish assemblages in the Everglades wetlands increases, it is likely that the restoration targets will be refined. For example, recent experiments are likely to yield predictions about how reductions in phosphorus inputs will affect fish abundance and assemblage structure. Additionally, it is clear that restoration targets will need to be defined by region and habitat type. Fish assemblage structure varies predictably at both local and regional scales, so restoration targets will need to vary accordingly. As with many of the examples in this chapter, the lack of perfect understanding means that adaptive approaches will be required in the assessment of habitat improvement efforts.

---

projects. Bernhardt et al. (2005) reported that only about 10% of the riverine habitat improvement projects in the USA that they surveyed had an associated monitoring program.

An important aspect of measuring responses to habitat improvement practices is selecting appropriate variables. Collecting data on a wide array of biotic and abiotic variables is not logistically or financially feasible. However, monitoring a few variables can provide substantial information on ecosystem responses (Karr 1992). Useful variables may assess either biological (Table 10.2) or physicochemical (Table 10.3) changes. Tables 10.2 and 10.3 are not comprehensive, but provide general categories of variables from which inference may be made regarding changes associated with habitat management projects. In many cases, the information can be separated into subcategories or other measures of change that may summarize information about a fishery, such as population metrics (Karr 1992). However, it is important to note that within these categories the information collected should be ecologically meaningful, relevant to the spatial and temporal scales being measured, responsive to the habitat improvement practices that were applied, and easily understood by resource managers and policymakers.

Table 10.2. General biological monitoring parameters and information gained from each parameter. Many other parameters contribute to measuring success of habitat improvements and should be fully investigated when appropriate.

Parameter	Monitoring and assessment information
Aquatic plants	Aquatic vegetation is an important component of many aquatic ecosystems because it provides nutrient remediation characteristics, stabilization of sediments, and also habitat and food for many aquatic organisms. Therefore, aquatic vegetation is highly valued and establishing, re-establishing, or maintaining stands of aquatic vegetation have been the crux of many habitat remediation efforts.
Fish	Fish have been used widely in the past to document changes to various systems and are valuable because they provide a cumulative reflection of many trophic levels to environmental changes.
Freshwater mussels	Freshwater mussels are one of the most endangered groups of organisms in North America. Mussels are also likely one of the more sensitive groups of organisms to environmental change because they are dependent upon other aquatic fauna for completing their relatively complex life cycle.
Macroinvertebrates	One of the more important groups that can quickly identify localized changes in habitats is the macroinvertebrates (excluding freshwater mussels). These taxa are important not only because of their rapid response to environmental change, but they also play a significant role in food web dynamics by breaking down organic matter into useable nutrients for themselves and other lower-trophic organisms and also by providing a food source for higher trophic levels such as fish, birds, reptiles, and amphibians.
Plankton	These organisms, including both zooplankton and phytoplankton, are at the base of the food web and are valuable indicators of system productivity. A major drawback in using plankton as a monitoring parameter is a lengthy identification process that can require a relatively high level of training.

10.8 CONCLUSIONS

The framework for the habitat improvement process should be adaptive to incorporate changes as managers learn more about systems in which habitat improvement may be needed. It is important to keep in mind that the habitat improvement process occurs in conjunction with the fishery management process (Chapter 5). Both processes should operate simultaneously to meet management goals.

Fisheries managers must also be aware that activities which appear to improve fish habitat may not result in improvement of a fishery. It is easy to get caught up in addressing a symptom of a fishery's failure rather than a cause of the problem. For example, a physician does not prescribe only aspirin for a patient with a broken leg. Rather, the physician sets the broken limb in a cast so it will heal properly. Similarly, if fish recruitment is low because spawning habitat has been lost through fine sediment delivery in a reservoir, then a one-time effort to

Table 10.3. General abiotic monitoring parameters and information gained from each parameter. Many other parameters can play a role in measuring success of habitat improvements and should be fully investigated where appropriate.

Parameter	Monitoring and assessment information
Geomorphology	Geomorphology of a watershed can help determine which habitat improvement techniques are feasible. Measuring parameters such as channel incision, sedimentation rates, and other processes provide insight on the present and future attributes of a water body.
Hydrology	Measuring surface and groundwater hydrology of lentic and lotic systems is valuable in determining the human effects on such systems. Altered hydrologic regimes that lead to changes in water level in lakes and reservoirs, extreme flows, and many other aspects of how water moves through a watershed are important to understanding the changes and improvements made to the system through time.
Land use	Measuring changes in use of terrestrial habitats for anthropogenic needs (e.g., urbanization and agriculture) can provide information on not only problems but can also measure improvements over the long term.
Water quality	Water quality can be extremely useful in measuring biotic associations and reactions to newly created environmental conditions. Measuring physical attributes of water quality such as turbidity, conductivity, and flow rates as well as variables that can give information on nutrient availability (e.g., nitrogen and phosphorus) are often valuable. Data used to assess general habitat characteristics (e.g., substrate type and amount of underwater structure) at sample sites during the associated water quality monitoring are also useful.

add spawning substrate without reducing sediment deposition is unlikely to solve the problem. The sedimentation problem should be addressed initially to enable long-term availability of spawning habitat and then be evaluated to determine if a desired response is achieved. Box 10.6 provides several guiding principles for identifying and implementing habitat improvement projects to minimize the likelihood that a variety of issues will not jeopardize the success of projects.

Another concern with habitat improvement in fisheries management is that habitat modifications may not provide habitat that will enable targeted fishes to carry out their life cycles. There can still be benefits from habitat improvement projects that do not meet every life history need of a targeted species. For example, if the goal of a habitat improvement project is to attract fish to one or a few locations for easier angler access, then building habitat by placing fish-attracting structures at access points may be pertinent. However, if the goal is to provide habitat for one or several species to establish or maintain a sustainable fishery, then adjustments to the goals and inclusion of habitat improvement efforts beyond fish structure placement may be warranted.

Habitat improvement in altered systems is a difficult and complex process that relies on expertise and knowledge, institutional support, and public acceptance. The upwelling of support from responsible agencies and acceptance by the general public (Bernhardt et al. 2005)

Box 10.6. Guiding Ecological Principles for Habitat Improvement

Charles "Si" Simenstad[1] and Daniel L. Bottom[2]

Interest in fish habitat improvement has greatly increased as part of intensifying efforts to recover at-risk fish stocks and to re-establish natural processes. Unfortunately, few empirical data are available to circumscribe sufficiently the specific habitat needs of many species or to ensure that improvement measures will be ecologically effective. Therefore, efforts to improve aquatic habitats must rely heavily on experience gained in other places, then proceed with caution to minimize risks, and finally learn from the results. In the absence of such a detailed understanding, we suggest that an objective, scientifically based set of guiding ecological principles is needed to guide planning, implementation, and evaluation of habitat improvement activities.

The following guiding ecological principles draw on ecological concepts to identify and implement restoration, rehabilitation, and enhancement projects that will promote fishery and ecosystem recovery. Accordingly, they are biased by their emphasis on (1) an ecosystem perspective, (2) fishery recovery, (3) self-sustaining endpoints, and (4) adaptive management that builds scientific understanding about habitat improvement through rigorous assessment and evaluation. Neither the concept nor many of the specific principles are unique. While drawing on our specific experience and knowledge of the Columbia River estuary and juvenile anadromous salmon ecology, we also used various scientific and technical assessments of fish restoration, enhancement, and creation activities around the world.

1. Protect the Best First

Protection of existing quality habitat is critical. Habitat improvement in the absence of any overlying conservation program that protects good habitats is counterproductive because the ecological integrity of the landscape will continue to erode, jeopardizing the ecological capital upon which habitat improvement depends. All habitat improvement sites should be explicitly incorporated into a broad conservation framework that will ensure both the long-term protection of the recovering site but also seek protection for adjacent habitats. Such a framework should build and protect connectivity among high-quality habitats, which serve as refugia and centers for population re-establishment.

2. Do No Harm

Ensure no net loss of habitat functions and protect unimpeded natural processes upon which they depend. Habitat improvement actions should achieve proposed benefits without degrading other ecological functions of natural habitats or broader ecosystems.

(Box continues)

---

[1] School of Aquatic and Fishery Sciences, University of Washington, Seattle.
[2] National Oceanic and Atmospheric Administration, Northwest Fisheries Science Center, Fish Ecology Division, Seattle.

Box 10.6. Continued.

3. Use Natural Processes to Improve and Maintain Habitat Structure

Habitat improvement measures should re-establish the dynamics of hydrology, sedimentology, geomorphology, and other habitat-forming processes that naturally create and maintain habitat rather than simply implant habitat structures at inappropriate or unsustainable locations. Recognition of and focus on natural ecosystem processes provide for long-term resilience of both fishes and their habitat rather than a short-term, static substitute.

4. Restore Rather than Enhance or Create

Past experience demonstrates that compared with restoration, enhancement (where designed to increase one or more specific functions of a degraded habitat) or creation of habitat is problematic and rarely leads to self-sustaining ecosystems. Anthropogenic stressors and permanent changes (see principle 7) often compromise natural processes. Highly-developed landscapes will challenge the ability to initiate self-sustaining systems and require a greater investment in stewardship, maintenance, and contingency measures.

5. Incorporate Fish Life Histories

The extent of habitat improvement activity must be substantial and sites must be distributed appropriately in the landscape to improve ecosystem health significantly for target fish species. For example, anadromous fish require a continuum of habitats to meet juvenile rearing and migration requirements along the freshwater–estuarine gradient from watershed to ocean. Habitat improvement efforts must account for fish access and the productivity of all habitat elements along this continuum.

6. Develop a Comprehensive Plan Using Landscape Scale Ecological Concepts to Re-Establish Ecosystem Connectivity and Complexity

Ad hoc, opportunistic approaches to habitat improvement will not suffice to restore habitat linkages that fish often require or the natural processes that create and maintain habitat structure. Strategic planning is necessary across multiple scales to set a broad vision, articulate clear goals, and place local habitat improvement activities in an ecologically sound ecosystem context.

7. Use history as a Guide but Recognize Irreversible Change

Understanding the historic landscape structure is essential to comprehending how habitat improvement can be implemented strategically in the modern landscape to achieve natural recovery and maintenance of important habitats. However, most historic recon-

(Box continues)

> **Box 10.6. Continued.**
>
> structions are static representations of dynamic ecosystems, such that variability in habitat structure may be more important than is landscape position. Many systems have been so highly modified that the fundamental processes responsible for historic conditions have been significantly altered, in some cases irrevocably. At the minimum, the modified capacities of natural processes to support habitat improvements under present conditions must be well understood to develop realistic goals.
>
> **8. Establish Performance Criteria Based on Explicit Objectives**
>
> Monitoring and adaptive management are essential components of habitat improvement and management. There are few manuals prescribing everything managers need to know about improving ecosystems and the habitats upon which specific fish depend. The inevitability of unexpected outcomes requires an adaptive management plan that will facilitate effective adjustments. Objectives should be clearly stated, site specific, measurable, and long-term, in many cases greater than 20 years. Performance criteria should derive directly from these objectives, include both functional and structural elements, and be linked to suitable, local reference (target) habitats. Habitat improvement should be evaluated uniformly at individual sites and comprehensively at landscape and ecosystem scales to assess whether the cumulative results of local actions achieve overall recovery goals.
>
> **9. Take Advantage of the Best Interdisciplinary Scientific and Technical Knowledge and Use a Peer Review Process**
>
> All available scientific and technical expertise should be brought to bear on the complex problems of habitat improvement through planning, designing, implementing, and monitoring by interdisciplinary, not just multidisciplinary, groups of experts. Physical (e.g., hydrology, geomorphology, geophysics, and sedimentology), chemical (e.g., sediment geochemistry), mathematical (e.g., biostatistics), and engineering sciences should be represented in addition to the essential biological disciplines (e.g., fish ecology and management). An independent, peer review panel should be established to evaluate scientific assumptions and performance throughout the process and to ensure goals are achieved.
>
> While they were not incorporated here, we urge that a parallel series of social, cultural, and economic principles also need to be defined and discussed to guide implementation of improvement plans based on these ecological principles.

establishes the fact that habitat improvement efforts will be firmly entrenched in present and future fisheries management. Obviously, issues and alterations to aquatic ecosystems found throughout the North American landscape make a straightforward, step-by-step guide to habitat improvement impossible to compile. This constraint requires that fisheries managers focus on problems associated with habitat degradation and loss in a structured manner, yet in a style that allows flexibility to address specific needs. Managers are often faced with questions such as, "Is a species-specific or assemblage approach best?" or "Which technique(s) or tool(s) should be used to improve habitat?" or "What metric(s) should be measured to determine the success or failure of habitat improvements?" All of these questions should be considered, but there is no universally-correct answer to any of them. However, fisheries managers should work within a structured framework such as the one presented in this chapter to ensure accountability in the habitat improvement process. A structured framework also allows managers to adapt to present and future habitat improvement plans, thus ensuring a meaningful approach is implemented.

10.9 REFERENCES

Allen, T. F. H., and T. W. Hoekstra. 1992. Toward a unified ecology. Complexity in ecological systems series. Columbia University Press, New York.

Baldwin, D. S., and A. M. Mitchell. 2000. The effects of drying and re-flooding on the sediment and soil nutrient dynamics of lowland river–floodplain systems: a synthesis. Regulated Rivers: Research and Management 16:457–467.

Bednarek, A. T. 2001. Undamming rivers: a review of the ecological impacts of dam removal. Environmental Management 27:803–814.

Bernhardt, E. S., M. A. Palmer, J. D. Allen, G. Alexander, K. Barnas, S. Brooks, J. Carr, S. Clayton, C. Dahm, J. Follstad-Shah, D. Galat, S. Gloss, P. Goodwin, D. Hart, B. Hassett, R. Jenkinson, S. Katz, G. M. Kondolf, P. S. Lake, R. Lave, J. L. Meyer, T. K. O'Donnell, L. Pagano, B. Powell, and E. Sudduth. 2005. Synthesizing U.S. river restoration efforts. Science 308:636–637.

Bohn, B. A., and J. L. Kershner. 2002. Establishing aquatic restoration priorities using a watershed approach. Journal of Environmental Management 64:355–363.

Bryce, S. A., and R. M. Hughes. 2003. Variable assemblage responses to multiple disturbance gradients: case studies in Oregon and Appalachia, USA. Pages 539–560 *in* T. P. Simon, editor. Biological response signatures: indicator patterns using aquatic communities. CRC Press, Boca Raton, Florida.

Cairns, J., Jr. 1988. Increasing diversity by restoring damaged ecosystems. Pages 333–343 *in* E. O. Wilson, editor. Biodiversity. National Academy Press, Washington, D.C.

Catalano, M. J., M. A. Bozek, and T. D. Pellett. 2007. Effects of dam removal on fish assemblage structure and spatial distributions in the Baraboo River, Wisconsin. North American Journal of Fisheries Management 27:519–530.

Chick, J. H., C. R. Ruetz III, and J. C. Trexler. 2004. Spatial scale and abundance patterns of large fish communities in freshwater marshes of the Florida Everglades. Wetlands 24:652–664.

Columbo, S., J. Calatrava-Requena, and N. Hanley. 2006. Analyzing the social benefits of soil conservation measures using stated preference methods. Ecological Economics 58:850–861.

Cooke, G. D., E. B. Welch, S. A. Petereson, and S. A. Nichols. 2005. Restoration and management of lakes and reservoirs, 3rd edition. CRC Press, Boca Raton, Florida.

Dendy, F. E., W. A. Champion, and R. B. Wilson. 1973. Reservoir sedimentation surveys in the United States. Pages 349–357 *in* W. C. Ackermann, G. F. White, and E. B. Worthington, editors. Man-made lakes: their problems and environmental effects. Geophysical Monograph 7, American Geophysical Union, Washington, D.C.

Dobson, A., D. Lodge, J. Alder, G. S. Cumming, J. Keymer, J. McGlade, H. Mooney, J. A. Rusak, O. Sala, V. Wolters, D. Wall, R. Winfree, and M. A. Xenopoulos. 2006. Habitat loss, trophic collapse, and the decline of ecosystem services. Ecology 87:1915–1924.

Downes, B. J., L. A. Barmuta, P. G. Fairweather, D. P. Faith, M. J. Keough, P. S. Lake, B. D. Mapstone, and G. P. Quinn. 2002. Monitoring ecological impacts: concepts and practice in flowing water. Cambridge University Press, Cambridge, UK.

Feist, B. E., E. A. Steel, G. R. Pess, and R. E. Bilby. 2003. The influence of scale on salmon habitat restoration priorities. Animal Conservation 2003:6:271–282.

Frederick, P., and J. C. Ogden. 2003. Monitoring wetland ecosystems using avian populations: seventy years of surveys in the Everglades. Pages 321–350 *in* D. E. Busch and J. C. Trexler, editors. Monitoring ecosystems: interdisciplinary approaches for evaluating ecoregional initiatives. Island Press, Washington, D.C.

Frissell, C. A., and S. C. Ralph. 1998. Stream and watershed restoration. Pages 599–624 *in* R. J. Naiman and R. E. Bilby, editors. River ecology and management. Springer-Verlag, New York.

Geremias, R., R. C. Pedrosa, J. C. Benassi, V. T. Favere, J. Stolberg, C. T. B. Menezes, and M. C. M. Laranjeira. 2003. Remediation of coal mining wastewaters using chitosan microspheres. Environmental Technology 24:1509–1515.

Golet, G. H., M. D. Roberts, R. A. Luster, G. Werner, E. W. Larsen, R. Unger, and G. G. White. 2006. Assessing societal impacts when planning restoration of large alluvial rivers: a case study of the Sacramento River Project, California. Environmental Management 37:862–879.

Graf, W. L. 1999. Dam nation: a geographic census of American dams and their large-scale hydrologic impacts. Water Resources Research 35:1305–1311.

Graf, W. L. 2006. Downstream hydrologic and geomorphic effects of large dams on American rivers. Geomorphology 79:336–360.

Havens, K. E., and D. E. Gawlik. 2005. Lake Okeechobee conceptual ecological model. Wetlands 25:908–925.

Hayes, D. B., W. W. Taylor, and P. A. Soranno. 1999. Natural lakes and large impoundments. Pages 589–621 *in* C. C. Kohler and W. A. Hubert, editors. Inland fisheries management in North America, 2nd edition. American Fisheries Society, Bethesda, Maryland.

Heinz (H. John Heinz III Center for Science, Economics, and the Environment). 2002. The state of the nation's ecosystems: measuring the lands, waters, and living resources of the United States. Cambridge University Press, New York.

Hutchinson, G. E. 1957. Concluding remarks. Cold Spring Harbor Symposia on Quantitative Biology 22:415–427.

Jeppesen, E., M. Meerhoff, B. A. Jacobsen, R. S. Hansen, M. Søndergaard, J. P. Jensen, T. L. Lauridsen, N. Mazzeo, and C. W. C. Branco. 2007. Restoration of shallow lakes by nutrient control and biomanipulation—the successful strategy varies with lake size and climate. Hydrobiologia 581:269–285.

Jones, M. S., and K. B. Rogers. 1998. Palmetto bass movements and habitat use in a fluctuating Colorado irrigation reservoir. North American Journal of Fisheries Management 18:640–648.

Karr, J. R. 1992. Ecological integrity: protecting earth's life support systems. Pages 223–228 *in* R. Costanza, B. G. Norton, and B. D. Haskell, editors. Ecosystem health: new goals for environmental management. Island Press, Washington, D.C.

Kimmel, B. L., and A. W. Groeger. 1986. Limnological and ecological changes associated with reservoir aging. Pages 103–109 *in* G. E. Hall and M. J. Van Den Avyle, editors. Reservoir fisheries management: strategies for the 80's. American Fisheries Society, Southern Division, Reservoir Committee, Bethesda, Maryland.

Kumar, M., and P. Kumar. 2008. Valuation of the ecosystem services: a psycho-cultural perspective. Ecological Economics 64:808–819.

Lake, P. S. 2001. On the maturing of restoration: linking ecological research and restoration. Ecological Management & Restoration 2:110–115.

Ligon, F. K., W. E. Dietrich, and W. J. Trush. 1995. Downstream ecological effects of dams: a geomorphic perspective. BioScience 45:183–192.

Luo, B., J. B. Li, G. H. Huang, and H. L. Li. 2006. A simulation-based interval stochastic model for agricultural nonpoint source pollution control through land retirement. Science of the Total Environment 36:38–56.

MacDonald, J. I., and P. E. Davies. 2007. Improving the upstream passage of two galaxiid fish species through a pipe culvert. Fisheries Management and Ecology 14:221–230.

Marburg, A. E., M. G. Turner, and T. K. Kratz. 2006. Natural and anthropogenic variation in coarse wood among and within lakes. Journal of Ecology 94:558–568.

Mattson, K. M., and P. L. Angermeier. 2007. Integrating human impacts and ecological integrity into a risk-based protocol for conservation planning. Environmental Management 39:125–138.

Matzinger, A., R. Pieters, K. Ashley, G. A. Lawrence, and A. Wuest. 2007. Effects of impoundment on nutrient availability and productivity in lakes. Limnology and Oceanography 52:2629–2640.

Miller, R. R., J. D. Williams, and J. E. Williams. 1989. Extinctions of North American fishes during the past century. Fisheries 14(6):22–38.

Miranda, L. E., and D. R. Lowery. 2007. Juvenile densities relative to water regime in mainstem reservoirs of the Tennessee River, USA. Lakes and Reservoirs: Research and Management 12:87–96.

Moser, D. E. 2000. Habitat conservation plans under the U.S. Endangered Species Act: the legal perspective. Environmental Management 26:S7–S13.

Naselli-Flores, L., and R. Barone. 2005. Water-level fluctuations in Mediterranean reservoirs: setting a dewatering threshold as a management tool to improve water quality. Hydrobiologia 548:85–99.

Palmer, M. A., E. S. Bernhardt, J. D. Allan, P. S. Lake, G. Alexander, S. Brooks, J. Carr, S. Clayton, C. N. Dahm, J. Follstad Shah, D. L. Galat, S. G. Loss, P. Goodwin, D. D. Hart, B. Hassett, R. Jenkinson, G. M. Kondolf, R. Lave, J. L. Meyer, T. K. O'Donnell, L. Pagano, and E. Sudduth. 2005. Standards for ecologically successful river restoration. Journal of Applied Ecology 42:208–217.

Pegg, M. A., and M. A. McClelland. 2004. Assessment of spatial and temporal fish community patterns in the Illinois River. Ecology of Freshwater Fish 13:125–135.

Porter, J. W., and K. G. Porter. 2001. The Everglades, Florida Bay, and coral reefs of the Florida Keys: an ecosystem sourcebook. CRC Press, Boca Raton, Florida.

Ricciardi, A., and J. B. Rasmussen. 1999. Extinction rates of North American freshwater fauna. Conservation Biology 13:1220–1222.

Ricklefs, R. E. 1973. Ecology. Chiron Press, Portland, Oregon.

Roni, P., editor. 2005. Monitoring stream and watershed restoration. American Fisheries Society, Bethesda, Maryland.

Roni, P., T. J. Beechie, R. E. Bilby, F. E. Leonetti, M. M. Pollock, and G. R. Pess. 2002. A review of stream restoration techniques and a hierarchical strategy for prioritizing restoration in Pacific Northwest watersheds. North American Journal of Fisheries Management 22:1–20.

Rosgen, D. L. 2006. River restoration using a geomorphic approach for natural channel design. Proceedings of the Eighth Federal Interagency Sedimentation Conference 8:394–401.

Sass, G. G., J. F. Kitchell, S. R. Carpenter, T. R. Hrabik, A. E. Marburg, and M. G. Turner. 2006. Fish community and food web responses to a whole-lake removal of coarse woody habitat. Fisheries 31(7):321–330.

Schilling, K. E., and C. F. Wolter. 2005. Estimation of streamflow, base flow, and nitrate-nitrogen loads in Iowa using multiple linear regression models. Journal of the American Water Resources Association 41:1333–1346.

Schindler, D. W. 2006. Recent advances in the understanding and management of eutrophication. Limnology and Oceanography 51:356–363.

Smith, K. L., and M. L. Jones. 2007. When are historical data sufficient for making watershed-level stream fish management and conservation decisions? Environmental Monitoring and Assessment 135:291–311.

Søndergaard, M., and E. Jeppesen. 2007. Anthropogenic impacts on lake and stream ecosystems, and approaches to restoration. Journal of Applied Ecology 44:1089–1094.

Stanley, E. H., M. J. Catalano, N. Mercado-Silva, and C. H. Orr. 2007. Effects of dam removal on brook trout in Wisconsin. River Research and Applications 23:792–798.

Theiling, C. H., R. J. Maher, and J. K. Tucker. 2000. Swan Lake rehabilitation and enhancement project: preproject biological and physical response monitoring. Illinois Natural History Survey Technical Report prepared for U.S. Army Corps of Engineers St. Louis District, Illinois Natural History Survey, Urbana–Champaign.

Thornton, K. W. 1990. Perspectives on reservoir limnology. Pages 1–13 *in* K. W. Thornton, B. L. Kimmel, and F. E. Payne, editors. Reservoir limnology: ecological perspectives. Wiley-Interscience, New York.

Toth, L. A., D. A. Arrington, and G. Begue. 1997. Headwater restoration and reestablishment of natural flow regimes: Kissimmee River of Florida. Pages 425–444 *in* J. E. Williams, C. A. Wood, and M. P. Dombeck, editors. Watershed restoration: principles and practices. American Fisheries Society, Bethesda, Maryland.

Trexler, J. C., W. F. Loftus, and J. H. Chick. 2003. Setting and monitoring restoration goals in the absence of historical data: the case of fishes in the Florida Everglades. Pages 351–376 *in* D. E. Busch and J. C. Trexler, editors. Monitoring ecosystems: interdisciplinary approaches for evaluating ecoregional initiatives. Island Press, Washington, D. C.

Trexler, J. C., W. F. Loftus, F. Jordan, J. H. Chick, K. L. Kandl, T. C. McElroy, and O. Bass. 2001. Ecological scale and its implications for freshwater fishes in the Florida Everglades. Pages 153–181 *in* J. W. Porter and K. G. Porter, editors. The Everglades, Florida Bay, and coral reefs of the Florida Keys: an ecosystem sourcebook. CRC Press, Boca Raton, Florida.

USACE (U.S. Army Corps of Engineers). 1993. Upper Mississippi River System—Environmental Management Program Final Definite Project Report (SL-5) with Integrated Environmental Assessment—Swan Lake Rehabilitation and Enhancement, Main Report, Pool 26, Illinois River, Calhoun County, Illinois. U.S. Army Corps of Engineers, St. Louis District, Missouri.

Venter, O., N. N. Brodeur, L. Nemioff, B. Belland, I. J. Dolinsek, and J. W. A. Grant. 2006. Threats to endangered species in Canada. BioScience 56:903–910.

Warren, M. L., Jr., and M. G. Pardew. 1998. Road crossings as barriers to small-stream fish movement. Transactions of the American Fisheries Society 127:637–644.

Wetzel, R. G. 2001. Limnology: lake and river ecosystems, 3rd edition. Academic Press, San Diego, California.

Williams, J. E., C. A. Wood, and M. P. Dombeck. 1997. Understanding watershed-scale restoration. Pages 1–16 *in* J. E. Williams, C. A. Wood, and M. P. Dombeck, editors. Watershed restoration: principles and practices. American Fisheries Society, Bethesda, Maryland.

Wissmar, R. C., and P. A. Bisson. 2003. Strategies for restoring river systems: sources of variability and uncertainty. Pages 3–7 *in* R. C. Wissmar and P. A. Bisson, editors. Strategies for restoring river ecosystems: sources of variability and uncertainty in natural and managed systems. American Fisheries Society, Bethesda, Maryland.

Chapter 11

Methods for Assessing Fish Populations

KEVIN L. POPE, STEVE E. LOCHMANN, AND MICHAEL K. YOUNG

11.1 INTRODUCTION

Fisheries managers are likely to assess fish populations at some point during the fisheries management process. Managers that follow the fisheries management process (see Chapter 5) might find their knowledge base insufficient during the steps of problem identification or management action and must assess a population before appropriate actions can be taken. Managers will implement some type of assessment during the evaluation step as a means of measuring progress relative to objectives. Choosing how to assess a population is an important decision because managers strive to maximize their knowledge of a population while minimizing the time and money expended to gain that knowledge.

A fish population is defined as a group of individuals of the same species or subspecies that are spatially, genetically, or demographically separated from other groups (Wells and Richmond 1995). A population will have a unique set of dynamics (e.g., recruitment, growth, and mortality) that influence its current and future status. The terms population assessment and stock assessment are used interchangeably by some fisheries managers. In general terms, a fish stock is a portion of a population, or a subpopulation. Stock assessment often refers to that portion of the fish population that is exploitable by a fishery, but we use the more inclusive population assessment throughout this chapter. Distinction is also made between a fish population and a sample of that population. Biologists almost never examine all the fish in a population, but rather base inferences on a sample of individuals from a population. How, where, and when those samples are drawn has a tremendous influence on the quality of data and validity of inferences.

In this chapter, methods for assessing inland fish populations to support management decisions are presented. It is important to consider bias (the unequal probability of sampling members of a population), precision (the degree of reproducibility of results), and the benefits of standardized sampling methods. A variety of population parameters and indices currently used to evaluate fish populations are reviewed, as are their respective strengths and limitations. This chapter will help students understand that proper design, analysis, and interpretation of assessment data are the foundation for appropriate management decisions.

11.2 NEED FOR ASSESSMENT

The best management decisions are based on knowledge that is sufficient to infer cause-and-effect relationships between management actions (e.g., harvest regulations) and a fish population (Radomski and Goeman 1996). Complete knowledge is rare or impossible to obtain, so managers attempt to acquire as much information about a fish population as resources allow. Frequent assessments may be necessary because population size, structure, and distribution fluctuate in response to environmental variation (Lett and Doubleday 1976; McRae and Diana 2005). Natural disturbances, such as floods, droughts, or fires, and anthropogenic changes, such as new fishing technologies, regulation changes, or nonnative fish introductions, can alter fish populations. Thus, status and trends in abundance, size or age structure, maturity schedules, or fecundity of fish in a population are central to informed decision making (Ault and Olson 1996; Post et al. 2003).

Although fisheries managers still spend time attempting to understand the ecology and population dynamics of sport fish species (Francis et al. 2007), the trend toward ecosystem management (Cowx and Gerdeaux 2004) has caused managers to devote more attention to nongame species (Angermeier et al. 1991). Assessments of sport and nongame fish populations are similar but driven by different motivations. For instance, population assessments of sport fish are often influenced by a desire to provide recreation or harvest for anglers, whereas population assessments of nongame fish typically aim at maintaining or enhancing the distribution and abundance of these species. Fisheries professionals must integrate population assessments of both types of fishes to implement ecosystem management properly.

Finally, the tendency of the public to become increasingly involved in resource management decisions (Caddy 1999; Bettoli et al. 2007) has increased the need to understand fish populations. General information about fish populations is widely available (e.g., Froese and Pauly 2008; NatureServe 2008) and sophisticated user groups can gain access to technical data, conduct analyses, and draw independent conclusions about particular fish populations (Beierle 2002). Managers can successfully interact with such groups by providing the results of population assessments with comprehensive analyses and interpretations based on sound scientific practices, including comparison with other findings published in peer-reviewed journals.

11.3 SAMPLING CONSIDERATIONS

11.3.1 Bias and Precision

Choosing how to sample and how to characterize a population are generally accorded the most emphasis in assessment programs. Determining sampling bias and precision are also important because bias or low precision make it difficult to identify the status of a population. For example, using electrofishing and the removal method to estimate the number of trout in a stream reach almost always results in an underestimate of fish abundance because the susceptibility of fish to capture by electrofishing declines as the number of capture attempts conducted over relatively short intervals (e.g., less than 1 h) increases (Riley and Fausch 1992). Moreover, the probability of capture of fish by electrofishing is also related to fish length, habitat complexity, stream size, water depth, water conductivity, species being sampled, and fish density. Adherence to consistent sampling protocols does not correct for bias, but if effort and catchability of fish remain constant among sampling events, the size and direction of the bias tend to remain constant and may permit meaningful population inferences. Still, testing for this constancy is important (Box 11.1). Alternatively, a lack of precision can indicate that sampling efficiency is not constant or that too small a sample has been obtained. More intensive sampling may increase precision and reduce bias (White et al. 1982), but identifying and accounting for the ecological, demographic, or habitat-related factors that affect sampling efficiency will produce the most reliable estimates of fish population parameters.

Understanding bias and precision becomes particularly important when determining whether to estimate population parameters directly or to estimate population parameters indirectly by means of indices. An index is defined as a number or property that is presumably related to a parameter of a fish population. Indices often require less effort or fewer resources than do estimates of population parameters but still provide useful information. For example, obtaining a census of bluegill in a lake is difficult, but counts of fish obtained from nets set overnight are relatively easy to obtain and generally reflect fish abundance. Despite their popularity, indices should be used with caution. Often the form of the relationship between an index and the population parameter of interest may be poorly understood, temporally or spatially variable, or based on untested assumptions (Anderson 2003). Nevertheless, if these relations can be well defined, indices can be a powerful tool for understanding population status and trends (McKelvey and Pearson 2001; Hopkins and Kennedy 2004).

11.3.2 Standardized Sampling

If the bias and precision of sampling gears, especially in variable environments, are unknown, standardized sampling may provide a means to assess trends (Bonar et al. 2009). Standardized sampling is defined as sampling with identical gear during the same season (or set of environmental conditions) in the same manner over time or among fish populations. Doing so does not eliminate bias but theoretically holds the bias constant so that differences in indices computed from samples among years or fish populations can be attributed to relative changes in a population or relative differences among populations. Other benefits of standardized sampling include improved communications among fisheries professionals and production of large-scale data sets beneficial for current and future assessments (Bonar and Hubert 2002).

Failure to adopt standardized sampling approaches can prevent managers from detecting population trends or assessing population status. For example, electrofishing catch rates of smallmouth bass are generally greater at night than during the day (Paragamian 1989). If electrofishing samples are collected during the day in some years and at night in others and the difference in vulnerability to capture is not addressed, a monitoring program may erroneously conclude that the smallmouth bass population is unstable.

Standardized sampling protocols cannot substitute for an understanding of fish biology, population dynamics, and gear selectivity. For instance, fyke nets of a certain mesh size and overall dimensions are regularly used in reservoirs to sample age-0 black crappie during the fall as an index of recruitment, but smaller fish are less likely to be captured than are larger fish (McInerny and Cross 2006). In years with early spawning by adult black crappie, many large age-0 fish may be captured during fall sampling and give the appearance that spawning and early survival were ample. In contrast, delayed spawning might result in fewer fall captures because age-0 black crappie would be smaller, giving the appearance that spawning and early survival were inadequate. A rigorous education in fisheries science that includes sampling theory and fish ecology is a prerequisite for implementing standard fish sampling protocols and analyzing the associated data.

Box 11.1. Removal Model Abundance Estimates: Wrong but Useful?

Amanda E. Rosenberger[1]

All models are wrong, but some models are useful—a truism to live by for fisheries managers. Consider the removal model, which uses standard depletion methods to generate an estimate of fish abundance. A primary assumption of the model is that sampling efficiency, or the proportion of fish removed from a site per capture event, is the same for all depletion capture events. However, fish that remain after the first depletion event are often more difficult to capture during subsequent events because they seek cover that is difficult to sample or continue to evade netters due to their relatively small size. When sampling efficiency declines from depletion event to depletion event, the removal model yields biased results: an underestimation of population size and an overestimation of sampling efficiency (e.g., Riley and Fausch 1992; Peterson et al. 2004).

This was the case for rainbow trout in small, headwater streams in the Boise River basin in Idaho (Rosenberger and Dunham 2005). Rainbow trout were marked and left in 31 sites (approximately 100 m in length) between two block nets to form "known" population sizes (following Peterson et al. 2004). After overnight recovery from initial capture and marking, marked trout were sampled by means of standard backpack electrofishing depletion procedures. The removal model generated rainbow trout abundances from depletion data that nearly always underestimated the number of marked fish actually present, averaging only 75% of marked fish.

The model yielded biased results. But could it still be useful? Managers faced with this kind of bias may assert that, although the estimates are incorrect, removal estimates can still be used as a relative index of fish abundance over space and through time. Methods need only be standardized and consistent, creating a highly precise, though wrong, answer. Further, estimates could be calibrated to known values with a simple correction factor to reflect actual fish numbers. This practice assumes that bias, though present, is consistent and based primary on the methods used. It should not be influenced by variables that will change through space and time.

A study in Idaho unfortunately refutes the assumption of constant sampling efficiency (Rosenberger and Dunham 2005). Not only were the removal estimates of rainbow trout abundance biased, but bias was inconsistent and influenced by stream habitat. Larger streams and streams with more instream structure in the form of dead wood yielded more biased estimates than did smaller streams with less instream cover. These stream features negatively affected electrofishing sampling efficiency, implying that what decreases sampling efficiency can increase the bias of removal estimates (also see Peterson et al. 2004). Common differences among sites over space and through time, including size of habitat, presence of structure, size of fish, water temperature, and the density of fish, can affect the sampling efficiency of electrofishing (e.g., Bayley and Dowling 1993; Dolan and Miranda 2003; Peterson et al. 2004). The Idaho study indicates that thorough validation of the removal model for generating absolute or comparable estimates of fish abundance is needed before use. Therefore, a new motto is suggested: all models are wrong; validate and proceed cautiously.

---

[1] University of Alaska–Fairbanks, School of Fisheries and Ocean Sciences, Fisheries Division.

11.3.3 Probabilistic Sampling

Statisticians frequently separate sampling designs according to whether probability or nonprobability sampling procedures are used (Levy and Lemeshow 1991). Probability sampling occurs when all possible samples are included in the selection process, the probability of selection is known, and the selection process is random (or an approximation thereof). The most basic probability sampling procedure used in fish population sampling is simple random sampling, in which a predetermined number of sampling sites is selected from all possible sampling sites such that every potential site has an equal chance of being selected (Hansen et al. 2007). Estimates of population parameters from probability sampling can enable inferences about the entire population. Furthermore, precision (e.g., standard errors) of estimates can be determined from probability sampling (Wilde and Fisher 1996).

Probability sampling may be impractical in many cases. Examples include small-scale assessments that occur in ponds or specialized habitats of rare organisms or situations in which information needed to design a probability sampling procedure is lacking. Nonprobability sampling may be used to provide information on trends in indices of population parameters (e.g., catch rate or size structure) of interest to managers (Wilde and Fisher 1996). Nonprobability sampling generally involves the nonrandom selection of sample sites, frequently based on judgment or convenience, and limits the scope of inference about fish populations. For example, samples collected from subjectively selected fixed sites, a nonprobability sampling procedure commonly used in fish surveys (King et al. 1981), are applicable only to those individuals or locations actually sampled (Wilde and Fisher 1996)—that is, findings should not be extrapolated to the whole population.

11.3.4 Geographic Boundaries of Fish Populations

Assessing a fish population requires the manager to delineate the extent of the population. In simple aquatic systems (e.g., isolated lakes or headwater streams with movement barriers), the boundaries are obvious. There are few barriers to interbreeding in simple aquatic systems and the population parameters are common to all individuals in a given species. Large, complex aquatic systems, however, make geographic delineation of a population challenging. For example, fish in floodplain lakes may have the opportunity to mate with fish from other floodplain lakes during annual spring floods. Similarly, fish in different tributaries to a large river may not be different populations because of movement of individuals among locations. Alternatively, large lakes or complex riverine networks may host demographically-distinct populations of some species that overlap during some seasons or life history phases (Dunham et al. 2002).

11.4 CHARACTERISTICS, STATISTICS, PARAMETERS, AND INDICES

Assessment is often based on characteristics of individual fish in a population. Typical data include their length and weight (Anderson and Neumann 1996) and sometimes their sex, maturity, gonad weight, or liver weight (Strange 1996). Likewise, hard structures (scales, fin rays, or bone) can be used to age individual fish (DeVries and Frie 1996). The amount and type of food in the stomach can be described (Bowen 1996). Numbers of lesions, parasites, or deformities can be recorded, and blood or tissue samples can be collected for genetic or chemical analyses (Strange 1996).

Data from individual fish are summarized with statistics to estimate parameters of the population from which the sample was taken. Such statistics include the prevalence of fish of different sizes and ages. These statistics can be combined to provide estimates of growth rates (Isely and Grabowski 2007). In addition, length and weight data can be combined to gauge the condition or "plumpness" of fish in a population (Pope and Kruse 2007). Use of data from a single sample to estimate population parameters may be inferior to data from multiple collections, but the practice of generating such data from a single sample is common in fisheries management.

An assessment is also likely to include statistics for a population that are not based on summaries of characteristics from individual fish. For example, recruitment and mortality rates are not averages of individual characteristics of fish in a population. Rather, these rates are generally estimated from trends in abundance across years or age-groups.

Assessment of a fish population may involve comparisons of estimates of population parameters from a current sampling effort to estimates of parameters from other populations or to management objectives. Analysis and interpretation are also likely to include modeling exercises that combine relevant estimates into a yield model (Power 2007). Several computer-based yield (or harvest) models (e.g., GIFSIM [Taylor 1981] and FAST [Slipke and Maceina 2001]) simplify the process for fisheries managers. These models allow the prediction of changes in a population or harvest resulting from management actions to limit fishing mortality (see Chapters 2 and 7). Managers must be aware of and acknowledge the uncertainty inherent in model predictions and in the population parameter estimates because assumptions, bias, and uncertainty are compounded when they are incorporated into yield models.

This chapter contains a presentation (expanded from Gibbons and Munkittrick 1994) of some of the common parameters and indices used by inland fisheries managers (Table 11.1). Each parameter or index has advantages and disadvantages because of inherent assumptions associated with its use, limitations of particular data sets, and preferences of investigators. Thus, it is prudent to use multiple tools in assessments of fish populations (see Box 11.2).

11.4.1 Population Dynamics

Population dynamics are the processes responsible for changes in abundance or biomass of a population through time and are a subset of possible population parameters. Estimates of population dynamics can provide greater insight into fish populations than can indices, which are a static portrayal of the population. Estimates of population dynamics can indicate how a population arrived at its current state and how it might change in the future.

A population assessment might focus on determining whether the size of a population is relatively constant, increasing, or decreasing, for which one would need population abundance data and age data to calculate birth and death rates. Other data, including individual weights, are necessary if population biomass is of interest. For many inland fisheries, such as those in natural lakes and small impoundments, birth and death rates tend to be regarded

Table 11.1. Categorized population characteristics that are frequently assessed or monitored by fisheries scientists and a brief description of the type of data required for quantifying the specific characteristic. Information presented is an expansion of categorizations of population parameters originally outlined by Gibbons and Munkittrick (1994).

Category and parameter or index	Type of data
Population dynamics	
Larval or juvenile abundance	Relative abundance and age
Recruitment	Relative abundance and age
Growth	Age and weight or length data
Mortality	Relative abundance and age
Exploitation	Absolute abundance and harvest (from creel data) or tag-reward data
Genetics	
Genetic composition	Tissue and blood samples
Abundance, density, and distribution	
Absolute abundance	Area subsample, mark–recapture, or depletion
Relative abundance	Catch per unit effort
Density	Population estimate and system size
Distribution	Presence–absence data
Population structure	
Mean length	Length
Proportional size distribution	Length
Mean age	Age
Year classes per sample	Age
Length at age	Age and length
Juvenile : adult ratio	Maturity status
Sex ratio	Sex
Age at maturity	Age, sex, and maturity status
Weight at maturity	Weight, sex, and maturity status
Energy acquisition, storage, and use	
Percent feeding	Stomach status and relative abundance
Relative weight	Length and weight
Hepatosomatic index	Weight and liver weight
Tissue lipid levels	Tissue sample
Gonadosomatic index	Weight and gonad weight
Length- or weight-specific fecundity	Length or weight and fecundity
Contaminants and diseases	
Proportion of population with anomalies	Lesion inspection
Proportion of population with parasites	Parasite inspection
Presence of toxicant, pollutant, or heavy metals	Tissue and blood samples
Viral and bacterial status or load	Tissue and blood samples

as more important than immigration and emigration rates, whereas the influence of movement rates is more widely recognized in migratory fishes. Considerably more effort must be expended to determine immigration and emigration rates, and the task is difficult in systems without barriers to fish movement.

Box 11.2. Pitfalls of Relying Solely on Size Structure Indices and Catch per Unit Effort for Management Decisions

C. Craig Bonds[1] and Brian Van Zee[2]

Size structure indices and catch per unit effort (C/f) are commonly used by fisheries managers to draw inferences about fish population dynamics. Proportional size distribution (PSD) and C/f are numerical descriptors of length frequency and relative abundance, respectively. However, fisheries managers should use caution if basing decisions solely on calculations of one or both of these indices. These indices are best used in conjunction with a suite of diagnostic information, including fish growth, condition (relative weight; W_r), and recruitment, as well as angler creel data. An example illustrating this principle is derived from data collected on largemouth bass in a reservoir in western Texas.

O. H. Ivie Reservoir is a 7,770-ha impoundment on the Colorado and Concho rivers. The reservoir is operated to store and supply municipal water to two cities and numerous smaller communities. Because of its arid location the reservoir is subject to prolonged periods of low water followed by years of partial recovery following occasional floods. Largemouth bass harvest was managed with a 457-mm minimum length limit and five-fish-daily bag limit for the first 11 years that recreational fishing was allowed on the reservoir. Largemouth bass, sunfish, and gizzard shad populations were evaluated in autumn by means of a boat-mounted electrofisher and according to standardized procedures. Surveys were conducted, in total, seven times from 1991 through 2000. However, for the sake of illustrative brevity, this box example focuses on sample years 1999 and 2000. In addition, supplemental sampling in the form of angling was used in 1999 to increase sample size and length distribution of largemouth bass used for age and growth analysis.

During 1999 and 2000, sampling of the largemouth bass population showed, respectively, mean catch rates of 94 and 72 bass per hour of electrofishing (see figure below), PSD (quality length) values of 69 and 64, and PSD-P (preferred length) values of 32 and 30 (see table below). The PSD and PSD-P values fell in the range recommended by Gabelhouse (1984) for "balanced" and "big bass" management strategies (see Chapter 16), and the C/f of largemouth bass indicated their relative abundance was adequate in relation to other area reservoirs. Relying solely on these two indices, fishery managers might infer the largemouth bass population was in desirable condition.

(Box continues)

[1] Texas Parks and Wildlife Department, Tyler.
[2] Texas Parks and Wildlife Department, Waco.

Recruitment, growth, and mortality rates are the primary population dynamics (often termed rate functions) influencing the harvestable segment of a fish population (Brown and Guy 2007). Assessing population dynamics of fish is best achieved using long-term data collected with standardized methods because biotic and abiotic influences on population dynamics typically vary from year to year. Unfortunately, such data sets are rare because they are expensive to obtain. In lieu of long-term data, methods have been developed to estimate population dynamics.

Box 11. 2. Continued

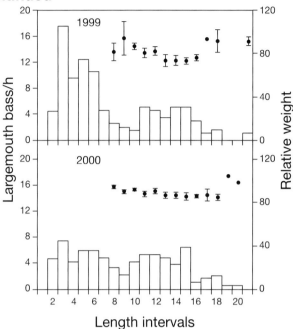

Figure A. Number of largemouth bass caught per hour (*C/f*) and mean relative weights ± SE (W_r; circles) for fall electrofishing surveys, O. H. Ivie Reservoir, Texas, 1999 and 2000. Length-group definitions are in 25.3-mm intervals (e.g., 2 = 50.8–76.1 mm and 3 = 76.2–101.5 mm). In 1999, total *C/f* = 94.5 (16; 189) (the relative standard error [RSE = SE/$\bar{x}$ × 100; *N* = 24 stations] and total number of fish caught are given in parentheses). In 2000, total *C/f* = 72.5 (19; 139).

Table. Proportional size distribution of quality-length (PSD) and preferred-length (PSD-P) largemouth bass collected by electrofishing from O. H. Ivie Reservoir, Texas, 1999 and 2000. Standard errors and sample sizes (number of stock-length largemouth bass) are given in parentheses.

Year	PSD (SE, *N*)	PSD-P (SE, *N*)
1999	69 (5, 71)	32 (7, 71)
2000	64 (7, 77)	30 (6, 77)

(Box continues)

Annual recruitment is typically the most variable factor affecting the dynamics of fish populations but can provide substantial insight into why fish populations may vary in size and structure (Gulland 1982; Allen and Pine 2000; Maceina and Pereira 2007). Larval or juvenile abundance can be an early indication of year-class strength and future recruitment to a fishery (Sammons and Bettoli 1998). Conversely, year-class size may be more closely related to abiotic factors than to larval abundance (Kernehan et al. 1981). Variability in recruitment of fish

Box 11.2. Continued

The relative abundance and size distribution of gizzard shad (see figure below) indicated adequate forage availability, which typically equates to better body condition and growth rates. Gizzard shad electrofishing C/f in 1999 and 2000, respectively, was 242 and 292 per hour, and the index of vulnerability (IOV; DiCenzo et al. 1996) indicated that 75% and 85% of the gizzard shad population were available to most predators those years. The IOV is the percentage of all gizzard shad that are 200 mm or shorter in length and is an index of the proportion of the gizzard shad population that is susceptible to predation by most predators.

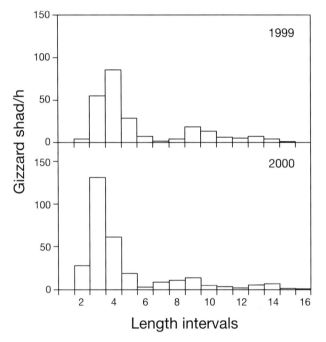

Figure B. Number of gizzard shad caught per hour (C/f) for fall electrofishing surveys, O. H. Ivie Reservoir, Texas, 1999 and 2000. Length-group definitions are in 25.3-mm intervals (e.g., 2 = 50.8–76.1 mm and 3 = 76.2–101.5 mm). In 1999, total C/f = 242.0 (34; 484) (values in parenthesis are as described above). The 1999 index of vulnerability (IOV) = 75.21 (SE = 8.7; N = 484). In 2000 total C/f = 292.2 (24; 560), and IOV = 84.82 (4.6; 560).

(Box continues)

into the harvestable population can be estimated with the recruitment variability index (Guy and Willis 1995), or with the coefficient of determination (r^2) resulting from simple linear regression of $\log_e$(catch at age) on age (Isermann et al. 2002).

Growth rates of fish in a population are intricately linked with mortality and recruitment rates. Growth rate influences survival and age at sexual maturity. Growth of fish is commonly indexed with various coefficients of the von Bertalanffy growth model (see Chapter

Box 11.2. Continued

Examination of additional data, however, revealed a different story. Both the growth rates and conditions of largemouth bass were undesirable. Poor largemouth bass conditions were manifested by mean W_r values less than 90 for many of the sampled groups, especially for fish in length-groups 13–18 (Figure A). In addition, age and growth data indicated a stockpiling of largemouth bass in length-groups 14, 15, 16, and 17, representing bass from 356 to 457 mm in total length, where five to seven different age-groups were represented (see figure below). In an effort to alleviate stockpiling and increase fish growth rates, fishery managers responded by changing the regulation and allowing the harvest of two largemouth bass less than 457 mm per day in 2001. Fishery managers would not have had the insight to modify the harvest regulation in 2001 by relying only on C/f and PSD data as these indices remained consistent concomitant to significant changes in the largemouth bass population.

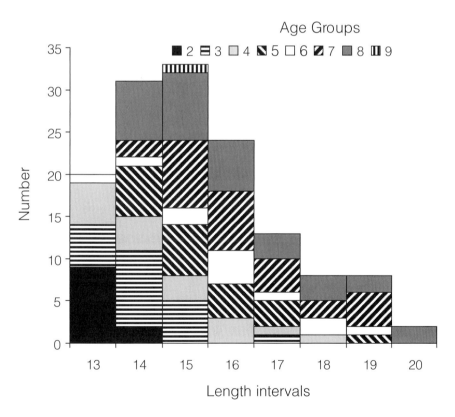

Figure C. Age composition for largemouth bass measuring 330.2–530.8 mm collected from O. H. Ivie Reservoir, Texas, 1999. Length-group definitions are in 25.3-mm intervals (e.g., 13 = 330.2–355.5 mm and 14 = 355.6–380.9 mm).

2), which is widely used to describe the lifetime pattern of somatic growth of organisms, such as fish, with indeterminate growth (Ricker 1975). Other growth models (e.g., Gompertz 1825; Richards 1959) may be more appropriate than the von Bertalanffy model for some situations. In addition, specific growth rate (the change of the logarithm of weight or length per unit time), relative growth rate (the relative change of the weight or length per unit time), and length at age (e.g., mean length at age 3), either measured at time of capture or back-calculated from hard structures such as otoliths, are also used to index growth. Quist et al. (2003) developed a relative growth index (RGI) by which the length at age from a population sample is compared with the age-specific standard length predicted by a von Bertalanffy growth model derived from pooled data for a species throughout its geographic distribution. The RGI is expressed as a percentage of the age-specific standard length achieved by the sampled population, and values greater than 100 indicate that growth is above average. Regardless of how growth is expressed, it is one of the most important rates estimated during a population assessment. Further, it is common for males and females in a population to mature on different schedules (Brown et al. 2006; Coelho and Erzini 2006) or to have different growth rates. Understanding differences in growth rates between sexes can allow managers to design more appropriate management strategies.

A population assessment might include an estimate of total annual mortality (i.e., the proportion of the population that dies in one year). Total annual mortality is related to total instantaneous mortality. Total instantaneous mortality can be estimated using a catch curve. The slope of the regression of $\log_e$(frequency) over age-groups equals the negative instantaneous mortality rate (Ricker 1975; Miranda and Bettoli 2007). Fisheries managers frequently partition total annual mortality of sport or commercial species into two components: (1) natural mortality, attributable to natural processes (e.g., old age, predation, competition, starvation, or disease) or those altered by human activities (e.g., habitat degradation or loss or population isolation), and (2) fishing mortality attributable to harvest or handling by recreational or commercial fishers (see Chapter 2). Agencies attempt to regulate fishing mortality by controlling harvest with gear restrictions, seasons, and length and bag limits (Radomski 2003) and monitor the results by means of creel surveys (Colvin 1991) or tag-reward studies (Reed and Davies 1991).

Exploitation is the portion of the fishing mortality attributed to fishers who harvest what they catch. Exploitation is often considered synonymous with fishing mortality because estimation of other forms of fishing mortality, such as bycatch or postrelease mortality, is difficult. Exploitation can be calculated from an estimate of absolute abundance and a harvest estimate based on creel surveys (Malvestuto 1996). The product of harvest per unit effort multiplied by annual effort provides an estimate of annual harvest. Annual exploitation can be determined by a tag-reward study, by which the ratio of the number of fish tagged and caught in a year divided by the total number of fish tagged the same year is exploitation (Miranda and Bettoli 2007). Exploitation is an important part of an assessment of a fished population because of the influence fishing pressure can have on many other population parameters.

11.4.2 Genetics

The most obvious forms of assessing fish populations involve counting or measuring individual fish, but another suite of characteristics can be very informative—their genes. Genetic assessment can be used to identify cryptic species, determine whether individuals have moved among populations recently or in the distant past, suggest the typical size of a population, and forecast a population's evolutionary future. Assessing these and other properties relies on identifying sets of genetic markers—the proteins expressed by different genes, different-sized fragments of DNA, or variation in the sequence of individual base pairs (White et al. 2005; DeHaan et al. 2006). Markers that are unique to a particular species or population are referred to as diagnostic, and their presence or absence is used to differentiate between groups. Even when no diagnostic markers are present, the relative abundance of shared markers can indicate similarity between populations. The relative abundance of shared markers can also be used to assign individual fish to a particular source population.

For fish species of conservation concern, such as those listed under the U.S. Endangered Species Act, Canada's Species at Risk Act, or the United Mexican States' Norma Oficial Mexicana de Ecología, the genetic characteristics of populations and sometimes of the entire species are relevant to their management. This is in part because the retention of genetic variation is crucial to the short-term avoidance of problems such as inbreeding depression (i.e., expression of deleterious alleles resulting from mating of related individuals; Allendorf and Ryman 2002) and to the long-term potential for continued evolution in changing environments (e.g., climate change). Genetic variation can be expressed in a number of ways. Commonly used parameters include heterozygosity, sometimes measured as the proportion of gene loci for which a population contains more than one allele, and allelic diversity, which is the number of alleles observed in the sample of individuals from a population. A related concept is effective population size, which is the size of an idealized population that would show the same rate of loss of genetic variation as the population in question (Frankham et al. 2002). The retention of genetic variation is directly related to effective population size, which is generally much lower than the number of adult fish present. In addition to diversity, other genetic characteristics of interest may include population genetic structure (Nielsen and Sage 2002), the geographic distribution of genetic variation among and within populations (Wenburg and Bentzen 2001), or historical and current gene flow (Neville et al. 2006a; Box 11.3).

Genetic markers can be used to determine whether exposure of native species to closely related, but typically nonnative, species has led to hybridization between them. This is common between native cutthroat trout subspecies and introduced rainbow trout in the Rocky Mountain region and Guadalupe bass and smallmouth bass in the Edwards Plateau region of Texas. Genetic analyses can determine whether offspring of such crosses are viable and leading to introgression of nonnative genes into the native species' gene pool. Conversely, genetic analyses can determine whether the hybrid crosses are limited to the first generation of offspring. Such information is important because it may inform managers of whether hybrid progeny have low survival rates or are infertile, such as hybrids between bull trout and brook trout (Allendorf et al. 2001). Assessing the prevalence of hybridized populations is sometimes the basis for conservation action (Allendorf et al. 2005).

Until recently, the genetic analysis of fish was often prohibitively expensive and difficult to conduct in a timely fashion, but technological advancements are alleviating these problems. For example, Chinook salmon in the Yukon River may be of U.S. or Canadian origin, and maintaining harvestable stocks in both countries is regulated by treaty (Pacific Salmon Treaty). Genetic monitoring is now used to identify the source of fish as they enter the river, which enables managers to regulate the harvest of the two stocks to ensure that sufficient fish reach their natal streams (Smith et al. 2005). Genetic analyses will play a prominent role in future assessment and monitoring of fishes with commercial or conservation value because genetic analyses provide information not obtainable by other methods and the efficiency and power of such analyses continue to increase (Schwartz et al. 2007).

Box 11.3. Use of Genetic Data to Understand the Biology and Conservation Status of Trout and Salmon Populations

HELEN NEVILLE[1]

In the face of unprecedented environmental change, understanding the ecology, evolution, and conservation status of fish populations is becoming increasingly important. Yet collecting the data needed to assess fish population characteristics—metrics such as effective population size, reproductive success, or dispersal rates—by means of traditional demographic methods (e.g., censuses, mark–recapture, or telemetry) is often logistically difficult and sometimes impossible to achieve. Molecular genetic data offer promising tools for measuring various characteristics of populations and for monitoring changes in them over time. Among other questions in fisheries management, genetic data have been used to monitor the influences of fish from hatcheries or aquaculture on native stocks (e.g., Hansen 2002; Coughlan et al. 2006), assess hybridization between native and nonnative species (e.g., Hitt et al. 2003; Rubidge and Taylor 2005), evaluate the success of captive breeding and translocation or reintroduction efforts (e.g., Dowling et al. 2005; Yamamoto et al. 2006), follow population trends over time (e.g., Nielsen et al. 1999; Hansen et al. 2006), understand the impact of harvest on particular stocks (e.g., Beacham et al. 2004), and contribute to forensic investigations (e.g., Withler et al. 2004). They will also likely be invaluable in monitoring the effects of climate change on fish in the future (Schwartz et al. 2007). Benefits of genetic data include that they are cost-effective to collect, allow evaluation across broad spatial scales (tens to thousands of kilometers) at both ecological (current) and evolutionary (historical) time frames, and require low-impact sampling given the ability to amplify tiny amounts of DNA. Collection of DNA can be achieved with little impact by use of scales or small fin clips from live animals or even from postspawn carcasses and museum specimens.

For example, Neville et al. (2006b) used genetic data to learn about the ecology and conservation needs of Lahontan cutthroat trout (LCT). Historically this subspecies of cutthroat trout had access to many connected headwater and main-stem habitats in the Humboldt River system in northern Nevada, and these large, complex habitats likely sustained resident and migratory life histories of the subspecies. Among inland cutthroat trout, resident fish can complete their entire life cycle in a small headwater reach (tens to hundreds of meters), whereas migratory fish may move long distances (tens to hundreds of kilometers) between headwater spawning areas and main-stem rivers or lakes. However, LCT (and other cutthroat trout subspecies) currently persist mostly as isolated resident populations in small, fragmented headwater streams, and there are few connected watersheds remaining that could potentially allow them to migrate between spawning areas in headwater tributaries and larger habitats in main-stem rivers.

In the study by Neville et al. (2006b), the authors collected small fin clips from over 1,100 individuals and used variability at DNA microsatellite markers to assess population connectivity, size, and stability, and the potential for migratory life history diversity in one

(Box continues)

[1] Trout Unlimited, Boise, Idaho.

Box 11.3. Continued

of these connected watersheds. They observed that several headwater populations were genetically differentiated from samples collected just a couple of kilometers downstream in the same tributary, even where there were no barriers preventing movement between sites. This suggested that fish in these reaches were behaviorally isolated and likely to exhibit resident life histories. Effective sizes in headwater populations were extremely small (N_e varied from 2 to 36), and many populations (particularly those in poor-quality habitats or those isolated by barriers) had undergone genetic bottlenecks, indicating that these LCT populations fluctuate and undergo occasional extreme reductions or even local extinctions. Fish caught in the river main stem, however, were genetically mixed, and their genes "assigned" them as having originated from populations throughout the watershed. This pattern is consistent with the idea that fish from different populations use the main stem as migratory habitat. Overall, genetic data were an efficient and effective way to gain insight about life history diversity and gave important information about LCT conservation needs. The main management implications were that poor habitat quality has negative demographic and genetic impacts on these fish and that habitat connectivity and movement among populations were probably important for the long-term persistence of LCT living in hydrologically variable streams.

In another study, Neville et al. (2006a) used DNA collected from postspawn carcasses to learn about the homing behavior of Chinook salmon. It has long been known that these and other salmon return to the river where they were born after spending several years feeding and growing in the ocean, but how accurately they are able to do this is still not very well understood. From a genetic perspective, if individuals do home to a specific natal site (the site where they were born), they should be surrounded by other individuals born in that area who are their relatives. By looking at patterns of relatedness among individual genotypes throughout the Middle Fork Salmon River in Idaho, it was found that individuals showed patterns of relatedness that indicated accurate homing at scales as small as 2 km. Interestingly, this genetic clustering was found only in females; males showed no patterns of relatedness at any spatial scale within the river basin. Such sex-biased movement makes sense in light of the species' ecology: females, which choose where to deposit eggs, might be under strong selection to return to a known environment (i.e., that where they were born), whereas males, which compete for mates, should benefit more from searching widely for and mating with as many females as they can within their natal river, regardless of exactly where in the river they were born.

11.4.3 Abundance, Density, and Distribution

The absolute abundance of fish in a population is often of fundamental interest to fish managers. However, determining absolute abundance requires extensive data collection, such as a precise estimate of density at sampling sites and a probability-based array of those sites (Hayes et al. 2007; Schwartz et al. 2007). Estimating absolute abundance is costly and time consuming, but may be warranted for highly valued populations (e.g., economically valuable species or species of conservation concern). For many populations, indices of relative abundance are sufficient for assessment. Catch per unit effort (C/f) is a relative abundance index, which is often directly related, though not always in a linear fashion, to absolute abundance (Rose and Kulka 1999; Hubert and Fabrizio 2007). Estimates of C/f are considerably easier to obtain than are estimates of absolute abundance. Like absolute abundance, relative abundance will vary among species and systems. For example, 200 stock-length fish (e.g., largemouth bass ≥ 200 mm or bluegill ≥ 80 mm,; Anderson and Neumann [1996]) per hour of electrofishing in reservoirs is a high catch rate for largemouth bass and only a modest catch rate for bluegill (Brouder et al. 2009). Similarly, 300 stock-length bluegill per hour of electrofishing is a high catch rate for small (<200 ha) reservoirs in the Great Plains ecological region of North America and only a modest catch rate for small reservoirs in the eastern temperate forests ecological region (Brouder et al. 2009). Estimates of fish density (the number or biomass per unit area or volume) are related to C/f, but effort is standardized by the length, area, or volume of water sampled. Precise relative abundance estimates are often collected from one or a few areas thought to be representative of an entire system, but the absence of randomly distributed samples that could account for spatial variation in fish abundance limits the value of this approach (Williams et al. 2004). Nonetheless, this sampling strategy does provide information on trends in indices of population parameters (see nonprobability sampling in section 11.3.3).

The distribution of some species may need to be understood before undertaking relative abundance estimation, particularly for species that are rare or poorly sampled. Assessing the distribution of a population or a species relies only on its presence or absence at particular sites. This relaxes some of the requirements associated with relative abundance estimation but reemphasizes the need for biologists to understand the vulnerability of different fishes to capture. An additional problem is determining where to sample to define a species' distribution. Randomly-distributed samples may lead to robust inferences about fish distributions, but sampling can be made more efficient by focusing on areas that provide potentially suitable habitat (Peterson and Dunham 2003).

11.4.4 Population Structure

Indices of size and age structure rely on estimates of length, weight, and age from a random sample of individual fish (Anderson and Neumann 1996; DeVries and Frie 1996). These indices tend to paint a static picture of a population, although it is possible to draw inferences about population dynamics from a population structure. For example, suppose mean length of fish in standardized samples from a population is used as an index of population structure; shifts to smaller mean lengths of sampled fish may indicate increasing exploitation. One index of population structure that is particularly appropriate for exploited populations is proportional size distribution (Guy et al. 2007), formerly known as proportional stock density (Anderson 1978) and relative stock density (Wege and Anderson 1978). Individual fish are assigned to length categories (many individuals fit in multiple categories), and the index is calculated as the proportion of fish in the stock-length category that also fall into the length category of interest. For example, proportional size distribution of quality-length (PSD) largemouth bass and proportional size distribution of preferred-length (PSD-P) largemouth bass are calculated as the proportions of stock-length largemouth bass (≥200 mm in total length [TL]) that are, respectively, also quality-length largemouth bass (≥300 mm TL) and preferred-length largemouth bass (≥380 mm TL). A value greater than 70 for PSD might result if growth is rapid, if exploitation is low, or if both are true. A value less than 30 for PSD might result if high recruitment leads to "stunting," if size-specific exploitation systematically removes older, larger fish, or if habitat quality is poor. The effectiveness of length-limit regulations are often evaluated relative to changes in proportional size distribution. A PSD-P value that is too low because of size-specific exploitation might be increased by implementation of a minimum length limit (see Chapter 7). The objective of the minimum length limit might be a specified increase in PSD-P over a specified time interval. When anglers are restricted from harvesting older, larger fish, the proportion of those fish in the population should increase, consequently increasing PSD-P. This example further illustrates the challenge of interpreting and using an index for making management decisions. If a PSD-P value is low because of stunting, implementing a minimum length limit would likely reinforce the mechanism(s) that caused the stunting. Care must be used because indices generally do not identify the underlying mechanisms that regulate them.

Estimation of fish population dynamics are often based on age structures (Everhart and Youngs 1981; Isely and Grabowski 2007). To obtain an accurate estimate of age structure, biologists must obtain a random sample of a population. Aging techniques for many important sport fish in North America have been validated (DeVries and Frie 1996), and it is important to use the standard techniques (Beamish and McFarlane 1983). If fish are aged correctly, then estimates of population dynamics will be correct (Marzolf 1955) and should lead to wise management and resource allocation decisions (Isely and Grabowski 2007). Determining age of fish takes considerably more effort than measuring and weighing fish but is usually warranted during population assessments.

The mean age of fish and the number of year-classes in a sample from a population are useful indices because many populations of fish exhibit variable recruitment (i.e., weak year-classes interspersed with strong year-classes). For example, a population of a long-lived species with only young age-groups present in random samples could be experiencing high exploitation or environmental stress. Similarly, another population with several missing age-groups in a random sample could be experiencing poor or failed recruitment (Guy and Willis 1995).

Several indices combine age data with other types of data. Population assessments occasionally rely on a length-at-age index as a proxy for growth rates of fish (Purchase et al. 2005). Aging a subsample of fish within length groups that encompass the possible lengths of fish of a specified age can allow managers to estimate the mean length of fish in that age-group. A large value for a length-at-age index indicates fast growth, whereas a small value for a length-at-age index suggests slow growth. For example, a brook trout that is 160 mm at age 3 is slow growing in a stream, whereas a brook trout in a similar system that is 260 mm at age 3 is fast growing (Brouder et al. 2009). The advantage to this index is that only a few length groups need to be aged. It is important to treat the data as a stratified sample, not a random sample, when size at age is calculated from a subsample of aged fish from several length groups. Calculating the mean length and variance of an age-group from subsampled data can introduce bias unless the subsample is extrapolated to the sample and the statistics are calculated on the entire sample (Bettoli and Miranda 2001).

Maturity status and sex identification allow the calculation of several useful indices. The juvenile-to-adult ratio can indicate important aspects of the dynamics of a fish population

(Reynolds and Babb 1978). A large ratio of juveniles to adults can be an indication of substantial exploitation, optimal spawning conditions in a particular year, or consistently successful recruitment. A population with recruitment difficulties will be characterized by fewer juveniles relative to adults. The ratio of males to females in a population can be important because sex ratio may be altered by size-selective harvest in sexually dimorphic populations. Sex ratio may also provide an indication of anthropogenic influences on the population. Estrogen from municipal sewage and androgens from biotransformation of pulp and paper mill effluents can skew the 1:1 sex ratio common in most healthy fish populations (Larsson et al. 2000). Determining sex and maturity status is relatively easy in fish that are sexually dimorphic but may only be possible during the spawning season in some fish species.

Age at maturity can be determined during a population assessment. Selection pressure (e.g., harvest aimed at larger, older individuals) at the population level could lead to maturity at a younger average age. An abundance of food resources can also lead to maturity at a younger age. Age at maturity is an index that requires age, sex, and sexual maturity data for a year class over several years (i.e., until the year class is fully matured). It can also be indexed with a single assessment, assuming a constant maturity schedule among year classes (Purchase et al. 2005). It is common for males and females of a species to mature on different schedules (Diana 1983), so age at maturity is often calculated separately for the sexes. Changes in weight at maturity can also indicate changes in the population. A decrease in age at maturity typically correlates with a decrease in weight at maturity because a strong relation exists between age and weight. Lifetime reproductive output can be altered when individuals mature at a younger age or smaller size (Quince et al. 2008) because of the strong relation between size and fecundity (Wydoski 2001). Collection of size, age, sex, and maturity information on individual fish can be challenging, but indices give considerable insight into the reproductive pressures acting on a population.

11.4.5 Energy Acquisition, Storage, and Use

Several fish population indices describe the processes of energy acquisition and storage. One simple index is the percentage of fish without food in their stomachs. A high percentage of empty stomachs indicates that acquisition of prey may be problematic for a population, that gut evacuation was rapid since last feeding, or that the sampling method caused regurgitation. Alternatively, a low percentage of empty stomachs indicates a recent availability of prey. Percentage of empty stomachs could also be influenced by time of sampling, especially for non-continuous-feeding fish. If prey availability is not a limiting factor, then growth should be adequate. Determining whether a fish has food in the stomach is fairly simple, and in many instances can be accomplished without sacrificing fish (Kamler and Pope 2001).

A common energy acquisition and storage index is relative weight. Relative weight is the length-specific weight of fish in a sample from a study population relative to a standard for the species. This index is relatively easy to compute. Length and weight data for some fish species have been collected from samples of a large number of populations throughout each species' range and have been used to determine length-specific standards for these species (Blackwell et al. 2000; see Chapter 2 for a description of length–weight equations). If average relative weight for a population or subpopulation exceeds 100, then most individuals in that population or subpopulation are heavier than the standard for the species. This is taken as a positive sign and interpreted as an indication that individuals in a population or subpopulation are doing well. A population or subpopulation with an average relative weight less than 80 is likely prey limited. Care is required when assessing these types of indices because they are highly influenced by season (Pope and Willis 1996).

Another index of energy storage is the hepatosomatic index, which is the liver weight relative to total body weight expressed as a percentage. When fish acquire more energy than necessary to meet basic metabolic and growth requirements, some excess energy is stored in the liver as glycogen. The size of the liver relative to the body is large when a considerable amount of energy is stored in the liver, so a large hepatosomatic index is indicative of a fish that is well fed (Plante et al. 2005). This index is easy to compute but requires sacrificing fish.

Fish also store excess energy as lipids. Lipids are twice as energy dense as proteins or carbohydrates, thus the level of lipid in muscle tissue can be used as an index of energy storage (Kaufman et al. 2007). Large amounts of lipid are likely to be stored in muscle tissue only if a fish is acquiring more energy than is necessary for metabolism and growth. Determining lipid levels in fish tissue is time consuming and requires specialized laboratory equipment.

A different class of population indices expresses or reflects how energy is being allocated by fish populations. Perhaps the simplest index describing energy allocation is the gonadosomatic index (GSI). The GSI is the gonad weight relative to total body weight, usually expressed as a percentage. If the fish population is not nutrition limited, then a large amount of energy will be directed toward gonad development. In this case, GSI values will be relatively large. Large average GSI values usually indicate a favorable status for the fish population.

An important index that indicates energy utilization is length-specific or weight-specific fecundity. On a population level, there is a strong relation between fish size and fecundity. However, there is also variability in fecundity among similar-sized individuals in a population (Taube 1976). Before a fish matures, most energy not required for basic metabolism is used for growth. After a fish matures, excess energy is divided between growth and reproduction. Populations with a variable size-specific fecundity may be manipulating reproductive output to influence growth (Leaman 1991) or responding to specific environmental conditions (Scoppettone et al. 2000). Populations with a low size-specific fecundity may be experiencing stress from lack of adequate prey or from other environmental pressures. This is a moderately complex index to calculate because it requires an estimation of eggs in the ovaries of a large number of individuals.

11.4.6 Contaminants, Diseases, and Parasites

Fish populations are sometimes characterized according to their health status. Inspection of individual fish in a random sample can provide data on the proportion of a population with lesions, parasites, or other anomalies (Wilson et al. 1996). Determining the proportion of a sample with health anomalies can be done rapidly, often with only a visual inspection. A sample with a high proportion of individuals with lesions or parasites is indicative of a population that is likely under stress from either disease or environmental conditions. Stressed individuals may consume less food (Hoffnagle et al. 2006), grow more slowly (Szalai and Dick 1991; Koehle and Adelman 2007), reproduce less successfully (Carter et al. 2005), and suffer a higher mortality rate (Szalai and Dick 1991).

Several other methods of characterizing the overall health of fish in a population are available to managers. Tissue or blood from sampled fish can be screened for toxicants or heavy metals. Techniques for assessing the presence of pollutants are straightforward. Most states or provinces have departments of environmental quality that regularly monitor fish and the environment for contaminants. Such tests are relatively expensive, and particular care must be taken during collection of samples to avoid contamination. Toxicant screening is usually conducted only when a suspicion of pollutant effects on a fish population exists. Fish populations can also be screened for viral or bacterial infections. Mortality, growth, and recruitment rates can be negatively affected by high viral or bacterial loads. Equipment and standard methods for detecting and diagnosing infections are available through fish pathology laboratories. These laboratories often require live specimens for a complete health assessment. If this type of health check is going to be part of a fish population assessment, specific protocols should be acquired from the pathology laboratory prior to fish collection (American Fisheries Society 2007).

11.5 DISTINCTION BETWEEN ASSESSMENT AND MONITORING

Typically, population assessments are one-time evaluations of the status of a fish population. Assessment requires comparisons of population attributes with reference values based on the probable relationships between environmental characteristics and population responses. Examples include comparisons of population density or age structure with regional averages or literature values.

Adaptive resource management (Holling 1978; Walters 1986) requires population monitoring to determine the necessity for changes in management actions based on the outcomes of the original management actions. Managers with a vision for the future of a fish population are more likely to engage in efforts to monitor progress toward management goals. In contrast to population assessment, population monitoring is the continuous or repeated observation, measurement, and evaluation of fish population parameters or indices, ideally using standardized methods for data collection. Comparisons of population parameters to regional averages or literature values are de-emphasized in favor of detecting population trends. Monitoring usually involves the collection of fewer types of data than does an assessment. Often, monitoring allows estimates of population dynamics and their variability. For example, changes in the relative abundance of a year-class over time reflect annual mortality rates, and recruitment variability may be monitored when an index of recruitment is estimated over several years (Quist 2007). It is common for thresholds to be incorporated into a population-monitoring program such that when values of key attributes fall above or below thresholds, a more comprehensive assessment might be undertaken.

11.6 CONCLUSIONS

A variety of parameters, indices, and models are available for assessing and monitoring fish populations. Each provides some insight about a fish population, and the decision of which to include in an assessment should be based on available resources, objectives of the assessment, and management goals. Assessing and monitoring fish populations can be costly and time consuming, but these are not reasons to avoid population assessments. Rather, these costs represent the reality that science-based fisheries management requires a significant investment of resources. It is incumbent on managers to gather the best information possible and base management decisions on sound science. Hence, population assessments will be an important component of fisheries management, and the wise use of appropriate indices will be an important component of fish population assessments.

11.7 REFERENCES

Allen, M. S., and W. E. Pine III. 2000. Detecting fish population responses to a minimum length limit: effects of variable recruitment and duration of evaluation. North American Journal of Fisheries Management 20:672–682.

Allendorf, F. W., R. F. Leary, N. P. Hitt, K. L. Knudsen, M. C. Boyer, and P. Spruell. 2005. Cutthroat trout hybridization and the U.S. Endangered Species Act: one species, two policies. Conservation Biology 19:1326–1328.

Allendorf, F. W., R. F. Leary, P. Spruell, and J. K. Wenburg. 2001. The problems with hybrids: setting conservation guidelines. Trends in Ecology & Evolution 16:613–620.

Allendorf, F. W., and N. Ryman. 2002. The role of genetics in population viability analysis. Pages 50–85 in S. R. Beissinger and D. R. McCullough, editors. Population viability analysis. University of Chicago Press, Chicago.

American Fisheries Society. 2007. Suggested procedures for the detection and identification of certain finfish and shellfish pathogens (blue book edition 2007). American Fisheries Society, Fish Health Section, Bethesda, Maryland.

Anderson, D. R. 2003. Response to Engeman: index values rarely constitute reliable information. Wildlife Society Bulletin 31:288–291.

Anderson, R. O. 1978. New approaches to recreational fishery management. Pages 73–78 in G. D. Novinger and J. G. Dillard, editors. New approaches to the management of small impoundments. American Fisheries Society, North Central Division, Special Publication 5, Bethesda, Maryland.

Anderson, R. O., and R. M. Neumann. 1996. Length, weight, and associated structural indices. Pages 447–482 in B. R. Murphy and D. W. Willis, editors. Fisheries techniques, 2nd edition. American Fisheries Society, Bethesda, Maryland.

Angermeier, P. L., R. J. Neves, and L. A. Nielsen. 1991. Assessing stream values: perspectives of aquatic resource professionals. North American Journal of Fisheries Management 11:1–10.

Ault, J. S., and D. B. Olson. 1996. A multicohort stock production model. Transactions of the American Fisheries Society 125:343–363.

Bayley, P. B., and D. C. Dowling. 1993. The effects of habitat in biasing fish abundance and species richness estimates when using various sampling methods in streams. Polskie Archiwum Hydrobiologii 40:5–14.

Beacham, T. D., M. Lapointe, J. R. Candy, K. M. Miller, and R. E. Withler. 2004. DNA in action: rapid application of DNA variation to sockeye salmon fisheries management. Conservation Genetics 5:411–416.

Beamish, R. J., and G. A. McFarlane. 1983. The forgotten requirements for age validation in fisheries biology. Transactions of the American Fisheries Society 112:735–743.

Beierle, T. C. 2002. The quality of stakeholder-based decisions. Risk Analysis 22:739–749.

Bettoli, P. W., and L. E. Miranda. 2001. Cautionary note about estimating mean length at age with subsampled data. North American Journal of Fisheries Management 21:425–428.

Bettoli, P. W., G. D. Scholten, and W. C. Reeves. 2007. Protecting paddlefish from overfishing: a case history of the research and regulatory process. Fisheries 32(8):390–397.

Blackwell, B. G., M. L. Brown, and D. W. Willis. 2000. Relative weight (W_r) status and current use in fisheries assessment and management. Reviews in Fisheries Science 8:1–44.

Bonar, S. A., and W. A. Hubert. 2002. Standard sampling of inland fish: benefits, challenges, and a call for action. Fisheries 27(3):10–16.

Bonar, S. A., W. A. Hubert, and D. W. Willis, editors. 2009. Standard methods for sampling North American freshwater fishes. American Fisheries Society, Bethesda, Maryland.

Bowen, S. H. 1996. Quantitative description of diet. Pages 513–532 *in* B. R. Murphy and D. W. Willis, editors. Fisheries techniques, 2nd edition. American Fisheries Society, Bethesda, Maryland.

Brouder, M. J., A. C. Iles, and S. A. Bonar. 2009. Length frequency, condition, growth, and catch per effort indices for common North American fishes. Pages 231–282 *in* S. A. Bonar, W. A. Hubert, and D. W. Willis, editors. Standard methods for sampling North American freshwater fishes. American Fisheries Society, Bethesda, Maryland.

Brown, M. L., and C. S. Guy. 2007. Science and statistics in fisheries research. Pages 1–29 *in* C. S. Guy and M. L. Brown, editors. Analysis and interpretation of freshwater fisheries data. American Fisheries Society, Bethesda, Maryland.

Brown, P., K. P. Sivakumaran, D. Stoessel, and A. Giles. 2006. Population biology of carp (*Cyprinus carpio* L.) in the mid-Murray River and Barmah Forest Wetlands, Australia. Marine and Freshwater Research 56:1151–1164.

Caddy, J. F. 1999. Fisheries management in the twenty-first century: will new paradigms apply? Reviews in Fish Biology and Fisheries 9:1–43.

Carter, V., R. Pierce, S. Dufour, C. Arme, and D. Hoole. 2005. The tapeworm *Ligula intestinalus* (Cestoda: Pseudophyllidae) inhibits LH-expression and puberty in its teleost host, *Rutilus rutilus*. Reproduction 130:939–945.

Coelho, R., and K. Erzini. 2006. Reproductive aspects of the undulate ray, *Raja undulata*, from the south coast of Portugal. Fisheries Research 81:80–85.

Colvin, M. A. 1991. Population characteristics and angler harvest of white crappies in four large Missouri reservoirs. North American Journal of Fisheries Management 11:572–584.

Coughlan, J., P. McGinnity, B. O'Farrell, E. Dillane, O. Diserud, E. de Eyto, K. Farrell, K. Whelan, R. J. M. Stet, and T. F. Cross. 2006. Temporal variation in an immune response gene (MHC I) in anadromous *Salmo trutta* in an Irish river before and during aquaculture activities. ICES Journal of Marine Science 63:1248–1255.

Cowx, I. G., and D. Gerdeaux. 2004. The effects of fisheries management practices on freshwater ecosystems. Fisheries Management and Ecology 11:145–151.

DeHaan, P. W., S. V. Libants, R. F. Elliott, and K. T. Scribner. 2006. Genetic population structure of remnant lake sturgeon populations in the upper Great Lakes basin. Transactions of the American Fisheries Society 135:1478–1492.

DeVries, D. R., and R. V. Frie. 1996. Determination of age and growth. Pages 483–512 *in* B. R. Murphy and D. W. Willis, editors. Fisheries techniques, 2nd edition. American Fisheries Society, Bethesda, Maryland.

Diana, J. S. 1983. Growth, maturation, and production of northern pike in three Michigan Lakes. Transactions of the American Fisheries Society 112:38–46.

DiCenzo, V. J., M. J. Maceina, and M. R. Stimpert. 1996. Relations between reservoir trophic state and gizzard shad population characteristics in Alabama reservoirs. North American Journal of Fisheries Management 16:888–895.

Dolan, C. R., and L. E. Miranda. 2003. Immobilization thresholds of electrofishing relative to fish size. Transactions of the American Fisheries Society 132:969–976.

Dowling, T. E., P. C. Marsh, A. T. Kelsen, and C. A. Tibbets. 2005. Genetic monitoring of wild and repatriated populations of endangered razorback sucker (*Xyrauchen texanus*, Catostomidae, Teleostei) in Lake Mohave, Arizona–Nevada. Molecular Ecology 14:123–135.

Dunham, J. B., B. E. Rieman, and J. T. Peterson. 2002. Patch-based models to predict species occurrence: lessons from salmonid fishes in streams. Pages 327–334 *in* J. M. Scott, P. J. Heglund, M. Morrison, M. Raphael, J. Haufler, and B. Wall, editors. Predicting species occurrences: issues of scale and accuracy. Island Press, Covelo, California.

Everhart, W. H., and W. D. Youngs. 1981. Principles of fishery science, 2nd edition. Cornell University Press, Ithaca, New York.

Francis, R. C., M. A. Hixon, M. E. Clarke, S. A. Murawski, and S. Ralston. 2007. Ten commandments for ecosystem-based fisheries scientists. Fisheries 32(5):219–233.

Frankham, R., J. D. Ballou, and D. A. Briscoe. 2002. Introduction to conservation genetics. Cambridge University Press, Cambridge, UK.

Froese, R., and D. Pauly, editors. 2008. FishBase. Available:www.fishbase.org. (April 2008).

Gabelhouse, D. W., Jr. 1984. A length-categorization system to assess fish stocks. North American Journal of Fisheries Management 4:273–285.

Gibbons, W. N., and K. R. Munkittrick. 1994. A sentinel monitoring framework for identifying fish population responses to industrial discharges. Journal of Aquatic Ecology and Health 3:227–237.

Gompertz, B. 1825. On the nature of the function expressive of the law of human mortality, and on a new mode of determining the value of life contingencies. Philosophical Transactions of the Royal Society of London 115:513–585.

Gulland, J. A. 1982. Why do fish numbers vary? Journal of Theoretical Biology 97:69–75.

Guy, C. S., R. M. Neumann, D. W. Willis, and R. O. Anderson. 2007. Proportional size distribution (PSD): a further refinement of population size structure index terminology. Fisheries 32(7):348.

Guy, C. S., and D. W. Willis. 1995. Population characteristics of black crappies in South Dakota waters: a case for ecosystem-specific management. North American Journal of Fisheries Management 15:754–765.

Hansen, M. J., T. D. Beard Jr., and D. B. Hayes. 2007. Sampling and experimental design. Pages 51–120 *in* C. S. Guy and M. L. Brown, editors. Analysis and interpretation of freshwater fisheries data. American Fisheries Society, Bethesda, Maryland.

Hansen, M. M. 2002. Estimating the long-term effects of stocking domesticated trout into wild brown trout (*Salmo trutta*) populations: an approach using microsatellite DNA analysis of historical and contemporary samples. Molecular Ecology 11:1003–1015.

Hansen, M. M., E. E. Nielsen, and K. L. Mensberg. 2006. Underwater but not out of sight: genetic monitoring of effective population size in the endangered North Sea houting (*Coregonus oxyrhynchus*). Canadian Journal of Fisheries and Aquatic Sciences 63:780–787.

Hayes, D. B., J. R. Bence, T. J. Kwak, and B. E. Thompson. 2007. Abundance, biomass, and production. Pages 327–374 *in* C. S. Guy and M. L. Brown, editors. Analysis and interpretation of freshwater fisheries data. American Fisheries Society, Bethesda, Maryland.

Hitt, N. P., A. Frissell, C. C. Muhlfeld, and F. W. Allendorf. 2003. Spread of hybridization between native westslope cutthroat trout, *Oncorhynchus clarki lewisi*, and nonnative rainbow trout, *Oncorhynchus mykiss*. Canadian Journal of Fisheries and Aquatic Sciences 60:1440–1451.

Hoffnagle, T. L., A. Choudhury, and R. A. Cole. 2006. Parasitism and body condition in humpback chub from the Colorado and Little Colorado rivers, Grand Canyon, Arizona. Journal of Aquatic Animal Health 18:184–193.

Holling, C. S. 1978. Adaptive environmental assessment and management. John Wiley and Sons, New York.

Hopkins, H. L., and M. L. Kennedy. 2004. An assessment of indices of relative and absolute abundance for monitoring populations of small mammals. Wildlife Society Bulletin 32:1289–1296.

Hubert, W. A., and M. C. Fabrizio. 2007. Relative abundance and catch per unit effort. Pages 279–325 *in* C. S. Guy and M. L. Brown, editors. Analysis and interpretation of freshwater fisheries data. American Fisheries Society, Bethesda, Maryland.

Isely, J. J., and T. B. Grabowski. 2007. Age and growth. Pages 187–228 *in* C. S. Guy and M. L. Brown, editors. Analysis and interpretation of freshwater fisheries data. American Fisheries Society, Bethesda, Maryland.

Isermann, D. A., W. L. McKibbin, and D. W. Willis. 2002. An analysis of methods for quantifying crappie recruitment variability. North American Journal of Fisheries Management 22:1124–1135.

Kamler, J. F., and K. L. Pope. 2001. Nonlethal methods of examining fish stomach contents. Reviews in Fisheries Science 9:1–11.

Kaufman, S. D., T. A. Johnston, W. C. Leggett, M. D. Moles, J. M. Casselman, and A. I. Schulte-Hostedde. 2007. Relationships between body condition indices and proximate composition in adult walleyes. Transactions of the American Fisheries Society 136:1566–1576.

Kernehan, R. J., M. R. Headrick, and R. E. Smith. 1981. Early life history of striped bass in the Chesapeake and Delaware Canal vicinity. Transactions of the American Fisheries Society 110:137–150.

King, T. A., C. J. Williams, W. D. Davies, and W. J. Shelton. 1981. Fixed versus random sampling of fishes in a large reservoir. Transactions of the American Fisheries Society 110:563–568.

Koehle, J. J., and I. R. Adelman. 2007. The effect of temperature, dissolved oxygen, and Asian tapeworm infection on growth and survival of the Topeka shiner. Transactions of the American Fisheries Society 136:1607–1613.

Larsson, D. G., H. Hällman, and L. Förlin. 2000. More male fish embryos near a pulp mill. Environmental Toxicology and Chemistry 19:2911–2917.

Leaman, B. M. 1991. Reproductive styles and life history variables relative to exploitation and management of Sebastes stocks. Environmental Biology of Fishes 30:253–271.

Lett, P. F., and W. G. Doubleday. 1976. The influence of fluctuations in recruitment on fisheries management strategy. International Commission for the Northwest Atlantic Fisheries Selected Papers 1:171–193.

Levy, P. S., and S. Lemeshow. 1991. Sampling of populations: methods and applications. Wiley, New York.

Maceina, M. J., and D. L. Pereira. 2007. Recruitment. Pages 121–185 *in* C. S. Guy and M. L. Brown, editors. Analysis and interpretation of freshwater fisheries data. American Fisheries Society, Bethesda, Maryland.

Malvestuto, S. P. 1996. Sampling the recreational creel. Pages 591–623 *in* B. R. Murphy and D. W. Willis, editors. Fisheries techniques, 2nd edition. American Fisheries Society, Bethesda, Maryland.

Marzolf, R. C. 1955. Use of pectoral spines and vertebrae for determining age and rate of growth of the channel catfish. Journal of Wildlife Management 19:243–249.

McInerny, M. C., and T. K. Cross. 2006. Factors affecting trap-net catchability of black crappies in natural Minnesota lakes. North American Journal of Fisheries Management 26:652–664.

McKelvey, K. S., and D. E. Pearson. 2001. Population estimation with sparse data: the role of indices versus estimators revisited. Canadian Journal of Zoology 79:1754–1765.

McRae, B. J., and J. S. Diana. 2005. Factors influencing density of age-0 brown trout and brook trout in the Au Sable River, Michigan. Transactions of the American Fisheries Society 134:132–140.

Miranda, L. E., and P. W. Bettoli. 2007. Mortality. Pages 229–277 *in* C. S. Guy and M. L. Brown, editors. Analysis and interpretation of freshwater fisheries data. American Fisheries Society, Bethesda, Maryland.

NatureServe. 2008. NatureServe explorer: an online encyclopedia of life, version 7.0. NatureServe, Arlington, Virginia. Available: www.natureserve.org/explorer. (April 16, 2008).

Neville, H. M., J. B. Dunham, and M. M. Peacock. 2006b. Landscape attributes and life history variability shape genetic structure of trout populations in a stream network. Landscape Ecology 21:901–916.

Neville, H. M., D. J. Isaak, J. B. Dunham, R. F. Thurow, and B. E. Rieman. 2006a. Fine-scale natal homing and localized movement as shaped by sex and spawning habitat in Chinook salmon: insights from spatial autocorrelation analysis of individual genotypes. Molecular Ecology 15:4589–4602.

Nielsen, E. E., M. M. Hansen, and V. Loeschcke. 1999. Genetic variation in time and space: microsatellite analysis of extinct and extant populations of Atlantic salmon. Evolution 53:261–268.

Nielsen, J. L., and G. K. Sage. 2002. Population genetic structure in Lahontan cutthroat trout. Transactions of the American Fisheries Society 131:376–388.

Paragamian, V. L. 1989. A comparison of day and night electrofishing: size structure and catch per unit effort for smallmouth bass. North American Journal of Fisheries Management 9:500–503.

Peterson, J. T., and J. Dunham. 2003. Combining inferences from models of capture efficiency, detectability, and suitable habitat to classify landscapes for conservation of threatened bull trout. Conservation Biology 17:1070–1077.

Peterson, J. T., R. F. Thurow, and J. W. Guzevich. 2004. An evaluation of multipass electrofishing for estimating the abundance of stream-dwelling salmonids. Transactions of the American Fisheries Society 133:462–475.

Plante, S., C. Audet, Y. Lambert, J. de la Noue. 2005. Alternative methods for measuring energy content in winter flounder. North American Journal of Fisheries Management 25:1–6.

Pope, K. L., and C. G. Kruse. 2007. Condition. Pages 423–471 in M. L. Brown and C. S. Guy, editors. Analysis and interpretation of freshwater fisheries data. American Fisheries Society, Bethesda, Maryland.

Pope, K. L., and D. W. Willis. 1996. Seasonal influences on freshwater fisheries sampling data. Reviews in Fisheries Science 4:57–73.

Post, J. R., C. Mushens, A. Paul, and M. Sullivan. 2003. Assessment of alternative harvest regulations for sustaining recreational fisheries: model development and application to bull trout. North American Journal of Fisheries Management 23:22–34.

Power, M. 2007. Fish population bioassessment. Pages 561–624 in C. S. Guy and M. L. Brown, editors. Analysis and interpretation of freshwater fisheries data. American Fisheries Society, Bethesda, Maryland.

Purchase, C. F., N. C. Collins, G. E. Morgan, and B. J. Shuter. 2005. Predicting life history traits of yellow perch from environmental characteristics of lakes. Transactions of the American Fisheries Society 134:1369–1381.

Quince, C., P. A. Abrams, B. J. Shuter, and N. P. Lester. 2008. Biphasic growth in fish II: empirical assessment. Journal of Theoretical Biology 254:207–214.

Quist, M. C. 2007. An evaluation of techniques used to index recruitment variation and year-class strength. North American Journal of Fisheries Management 27:30–42.

Quist, M. C., C. S. Guy, R. D. Schultz, and J. L. Stephen. 2003. Latitudinal comparisons of walleye growth in North America and factors influencing growth of walleyes in Kansas reservoirs. North American Journal of Fisheries Management 23:677–692.

Radomski, P. 2003. Initial attempts to actively manage recreational fishery harvest in Minnesota. North American Journal of Fisheries Management 23:1329–1342.

Radomski, P. J., and T. J. Goeman. 1996. Decision making and modeling in freshwater sport-fisheries management. Fisheries 21(12):14–21.

Reed, J. R., and W. D. Davies. 1991. Population dynamics of black crappies and white crappies in Weiss Reservoir, Alabama: implications for the implementation of harvest restrictions. North American Journal of Fisheries Management 11:598–603

Reynolds, J. B., and L. R. Babb. 1978. Structure and dynamics of largemouth bass populations. Pages 50–61 in G. D. Novinger and J. G. Dillard, editors. New approaches to the management of small impoundments. American Fisheries Society, North Central Division, Special Publication 5, Bethesda, Maryland.

Richards, F. J. 1959. A flexible growth function for empirical use. Journal of Experimental Botany 10:290–300.

Ricker, W. E. 1975. Computation and interpretation of biological statistics of fish populations. Fisheries Research Board of Canada Bulletin 191.

Riley, S. C., and K. D. Fausch. 1992. Underestimation of trout population size by maximum-likelihood removal estimates in small streams. North American Journal of Fisheries Management 12:768–776.

Rose, G. A., and D. W. Kulka. 1999. Hyperaggregation of fish and fisheries: how catch-per-unit-effort increased as the northern cod (*Gadus morhua*) declined. Canadian Journal of Fisheries and Aquatic Sciences (Supplement 1):118–127.

Rosenberger, A. E., and J. B. Dunham. 2005. Validation of abundance estimates from mark–recapture and removal techniques for rainbow trout captured by electrofishing in small streams. North American Journal of Fisheries Management 25:1395–1410.

Rubidge, E. M., and E. B. Taylor. 2005. An analysis of spatial and environmental factors influencing hybridization between native westslope cutthroat trout (*Oncorhynchus clarkii lewisi*) and introduced rainbow trout (*O. mykiss*) in the upper Kootenay River drainage, British Columbia. Conservation Genetics 6:369–384.

Sammons, S. E., and P. W. Bettoli. 1998. Larval sampling as a fisheries management tool: early detection of year-class strength. North American Journal of Fisheries Management 18:137–143.

Schwartz, M. K., G. Luikart, and R. S. Waples. 2007. Genetic monitoring as a promising tool for conservation and management. Trends in Ecology and Evolution 22:25–33.

Scoppettone, G. G., P. H. Rissler, and M. E. Buettner. 2000. Reproductive longevity and fecundity associated with nonannual spawning in cui-ui. Transactions of the American Fisheries Society 129:658–669.

Slipke, J. W., and M. J. Maceina. 2001. Fisheries analysis and simulation tools (FAST). Auburn University, Auburn, Alabama.

Smith, C. T., W. D. Templin, J. E. Seeb, and L. W. Seeb. 2005. Single nucleotide polymorphisms provide rapid and accurate estimates of the proportions of U.S. and Canadian Chinook salmon caught in Yukon River fisheries. North American Journal of Fisheries Management 25:944–953.

Strange, R. J. 1996. Field examination of fishes. Pages 433–446 *in* B. R. Murphy and D. W. Willis, editors. Fisheries techniques, 2nd edition. American Fisheries Society, Bethesda, Maryland.

Szalai, A. J., and T. A. Dick. 1991. Role of predation and parasitism in growth and mortality of yellow perch in Dauphin Lake, Manitoba. Transactions of the American Fisheries Society 120:739–751.

Taube, C. M. 1976. Sexual maturity and fecundity in brown trout of the Platte River, Michigan. Transactions of the American Fisheries Society 105:529–533.

Taylor, C. J. 1981. A generalized inland fishery simulator for management biologists. North American Journal of Fisheries Management 1:60–72.

Walters, C. J. 1986. Adaptive management of renewable resources. McGraw Hill, New York.

Wege, G. J., and R. O. Anderson. 1978. Relative weight (*Wr*): a new index of condition for largemouth bass. Pages 79–91 *in* G. D. Novinger and J. G. Dillard, editors. New approaches to the management of small impoundments. American Fisheries Society, North Central Division, Special Publication 5, Bethesda, Maryland.

Wells, J. V., and M. E. Richmond. 1995. Populations, metapopulations, and species populations: what are they and who should care? Wildlife Society Bulletin 23:458–462.

Wenburg, J. K., and P. Bentzen. 2001. Genetic and behavioral evidence for restricted gene flow among coastal cutthroat trout populations. Transactions of the American Fisheries Society 130:1049–1069.

White, G. C., D. R. Anderson, K. P. Burnham, and D. L. Otis. 1982. Capture–recapture and removal methods for sampling closed populations. Los Alamos National Laboratory, Los Alamos, New Mexico.

White, M. M., T. W. Kassler, D. P. Philipp, and S. A. Schell. 2005. A genetic assessment of Ohio River walleyes. Transactions of the American Fisheries Society 134:661–675.

Wilde, G. R., and W. L. Fisher. 1996. Reservoir fisheries sampling and experimental design. Pages 397–409 *in* L. E. Miranda and D. R. DeVries, editors. Multidimensional approaches to reservoir fisheries management. American Fisheries Society, Symposium 16, Bethesda, Maryland.

Williams, L. R., M. L. Warren Jr., S. B. Adams, J. L. Arvai, and C. M. Taylor. 2004. Basin visual estimation technique (BVET) and representative reach approaches to wadeable stream surveys: methodological limitations and future directions. Fisheries 29(8):12–22.

Wilson, D. S., P. M. Muzzall, and T. J. Ehlinger. 1996. Parasites, morphology and habitat use in a bluegill sunfish (*Lepomis macrochirus*) population. Copeia 1996:348–354.

Withler, R. E., J. R. Candy, T. D. Beacham, and K. M. Miller. 2004. Forensic DNA analysis of Pacific salmonid samples for species and stock identification. Environmental Biology of Fishes 69:275–285.

Wydoski, R. S. 2001. Life history and fecundity of mountain whitefish from Utah streams. Transactions of the American Fisheries Society 130:692–698.

Yamamoto, S., K. Maekawa, T. Tamate, I. Koizumi, K. Hasegawa, and H. Kubota. 2006. Genetic evaluation of translocation in artificially isolated populations of white-spotted char (*Salvelinus leucomaenis*). Fisheries Research 78:352–358.

Chapter 12

Assessment and Management of Ecological Integrity

THOMAS J. KWAK AND MARY C. FREEMAN

12.1 INTRODUCTION

Assessing and understanding the impacts of human activities on aquatic ecosystems has long been a focus of ecologists, water resources managers, and fisheries scientists. While traditional fisheries management focused on single-species approaches to enhance fish stocks, there is a growing emphasis on management approaches at community and ecosystem levels. Of course, as fisheries managers shift their attention from narrow (e.g., populations) to broad organizational scales (e.g., communities or ecosystems), ecological processes and management objectives become more complex. At the community level, fisheries managers may strive for a fish assemblage that is complex, persistent, and resilient to disturbance. Aquatic ecosystem level objectives may focus on management for habitat quality and ecological processes, such as nutrient dynamics, productivity, or trophic interactions, but a long-term goal of ecosystem management may be to maintain ecological integrity. However, human users and social, economic, and political demands of fisheries management often result in a reduction of ecological integrity in managed systems, and this conflict presents a principal challenge for the modern fisheries manager.

The concepts of biotic integrity and ecological integrity are being applied in fisheries science, natural resource management, and environmental legislation, but explicit definitions of these terms are elusive. Biotic integrity of an ecosystem may be defined as the capability of supporting and maintaining an integrated, adaptive community of organisms having a species composition, diversity, and functional organization comparable to that of a natural habitat of the region (Karr and Dudley 1981). Following that, ecological integrity is the summation of chemical, physical, and biological integrity. Thus, the concept of ecological integrity extends beyond fish and represents a holistic approach for ecosystem management that is especially applicable to aquatic systems. The more general term, ecological condition, refers to the state of the physical, chemical, and biological characteristics of the environment and the processes and interactions that connect them. While the concept of ecological integrity may appear unambiguous, its assessment and practice are much less clear.

Ecological integrity made its debut in the USA with the Clean Water Act (CWA) of 1972 (Federal Water Pollution Control Act, as amended through Public Law 107–303, November

27, 2002), which states only one objective, "to restore and maintain the chemical, physical, and biological integrity of the Nation's waters." This legislation compelled resource managers to focus on chemical pollution from point effluent sources, such as industrial and municipal outflows, as well as give attention to diffuse, chronic, and watershed effects on ecological integrity. Further, the CWA allowed pursuit of restoration programs in degraded water bodies and catalyzed the science and practice of restoration ecology.

The term ecosystem health is often raised in discussions of ecological integrity. Perhaps it is natural to anthropomorphize our concern for personal health to ecosystems, so it becomes a useful metaphor for understanding the concept of ecological integrity. However, whether or not an ecosystem should be considered an entity, such as a superorganism, is a debate without end that began with early ecologists and continues today (Clements 1916; Suter 1993; Simon 1999a). Regardless, the ecosystem is indeed a natural unit with a level of organization and properties beyond the collection of those species that occupy it and presents the most appropriate spatial and organizational scale in which to assess and study ecological integrity. Streams and rivers serve as integrators of chemical, physical, and biological conditions across the landscape, and while the theory and practice associated with ecological integrity of aquatic systems is easily applied to flowing waters and is emphasized in this chapter, they are broadly applicable among all aquatic systems.

12.1.1 Factors Influencing Ecological Integrity

The ultimate factor affecting ecological integrity of aquatic ecosystems is human activity, and ecological integrity is inversely related to human impacts on ecosystems (Figure 12.1). Human influence on the earth's ecosystems is substantial and increasing with population and economic growth and technological advances. Humans have transformed over one-third of the earth's land surface, increased atmospheric carbon dioxide by nearly 30%, fixed more nitrogen than all natural terrestrial sources combined, and used more than one-half of all accessible surface freshwater (Vitousek et al. 1997). These global effects have both gradual and catastrophic impacts on biodiversity and ecosystem function at the local level of environments that support aquatic life, fish, and fisheries (Scheffer et al. 2001). Thus, aquatic fauna are declining at a rapid rate globally and locally. In North America, nearly 70% of the approximately 300 freshwater mussel species are extinct or vulnerable to extinction (Williams et al. 1993), 48% of the 363 crayfish species are endangered, threatened, or vulnerable (Taylor et al. 2007), and 39% of the 700 known fish taxa are imperiled (Jelks et al. 2008).

More proximate effects of human activity on ecological integrity of aquatic systems are changes in land use in watersheds, stream channel modifications, point and diffuse chemical and thermal pollution, water impoundment and extraction, species introductions, fishing, and landscape development (Figure 12.1). These activities act at multiple scales to transform the physical and biological components of aquatic systems from their original condition to altered states with reduced native species richness, biological diversity, and ecological function. Factors that influence ecological integrity of aquatic ecosystems are a management priority for fisheries scientists: 15 of the 32 resource policy statements adopted by the American Fisheries Society concern human activities that influence aquatic ecological integrity (Table 12.1), and state and federal agencies are incorporating ecological integrity into their management processes (Box 12.1).

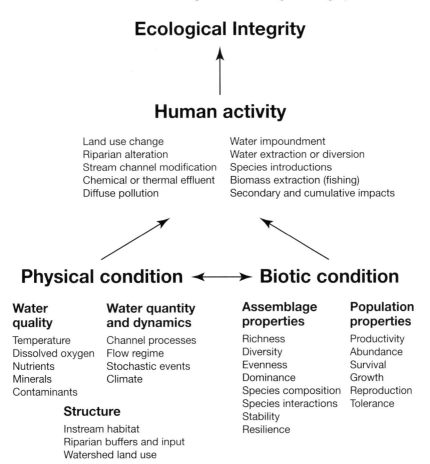

Figure 12.1. Conceptual diagram of physical and biological factors that affect ecological integrity of aquatic systems and how they are altered by the filter of human activity.

12.1.1.1 Land Use

Early environmentalists and modern stream ecologists have long recognized that streams, rivers, and other water bodies are intimately connected with and influenced by their surrounding watersheds (Marsh 1867; Hynes 1975; Vannote et al. 1980; Junk et al. 1989). Water bodies are profoundly affected by changes in the vegetation and land cover of their riparian zones and watersheds. For instance, the effects of agriculture, timber harvest, and urbanization on aquatic systems have been well documented (Allan and Castillo 2007). As native plants are replaced by agricultural crops or livestock, harvested timber and plantations, or urban development, the physical environment in a stream shifts away from its original state and the aquatic biota reflect that change. Mechanisms by which land use changes influence stream ecosystems involve sedimentation, nutrient enrichment, chemical contamination, hydrologic alteration, riparian clearing and canopy destruction, and loss of woody debris (Allan 2004). Although such changes are pervasive, empirical associations and mechanistic details have rarely been elucidated, presumably because of the complex and dynamic processes associated with the land–water interface (Fausch et al. 2002; Allan 2004). While the precise mechanisms

Table 12.1. American Fisheries Society formal policy statements on human activities that influence the ecological integrity of aquatic ecosystems. Full statements and abbreviated summaries may be found at www.fisheries.org/afs/policy_statements.html.

Policy statement number (year approved)	Influence on ecological integrity	Policy statement title
2 (1981)	Human population growth	Man-induced ecological problems: human population growth and technology
3 (1981)	Diffuse water pollution	Nonpoint source pollution
4 (1981)	Sedimentation in aquatic environments	Sedimentation
5 (1981)	Secondary and cumulative impacts	Cumulative effects of small modifications to habitat
6 (1981)	Aquatic contaminants	Effects of toxic substances in surface waters
7 (1981)	Diffuse water pollution	Policies on two issues of environmental concern (acid precipitation)
8 (1981)	Point source water pollution	Coping with point source discharges
9 (1981)	Instream flows and dams	Effects of altered stream flows on fishery resources
12 (1985)	Point source water pollution	Construction and operation of oil and gas pipelines
13 (1985)	Land use alteration	Effects of surface mining on aquatic resources in North America
14 (1986)	Land use alteration	Strategies for stream riparian area management
15 (1986)	Nonnative species	Introductions of aquatic species
23 (1990)	Land use alteration	The effects of livestock grazing on riparian and stream ecosystems
24 (1990)	Nonnative species	Ballast water introduction
25 (1991)	Instream flows and dams	Hydropower development
29 (1996)	Loss of diversity	Biodiversity
32 (2005)	Instream flows and dams	Dam removal

may not be fully understood, the linkage is clear between human actions on the land and the fate of the aquatic environment and its biota.

12.1.1.2 Sources of Water Pollution

Surface waters may receive pollutants in the form of temperature changes, nutrients, or chemicals via effluent point discharges or diffuse terrestrial sources. Thermal pollution may be the result of heated discharge associated with industrial cooling or in the form of cold water released from dams impounding reservoirs. Eutrophication is a widespread problem in surface waters, with nutrients entering aquatic systems from point sources of wastewater

Box 12.1. Indices of Ecological Integrity: a State Agency's Experience

JOHN LYONS[1]

Wisconsin has a rich and diverse array of water and fisheries resources that are critical to its economic, social, and ecological well-being. However, with about 88,000 km of streams and rivers, more than 15,000 lakes, over 150 fish species, hundreds of species of aquatic plants, and almost countless aquatic invertebrate species, the sheer volume and variety of water bodies and biota in Wisconsin presents a major challenge for management. The Wisconsin Department of Natural Resources (WDNR), charged with the conservation of Wisconsin's waters and biota, increasingly relies on indices of ecological integrity based on components of the aquatic biological community to help meet this challenge.

The WDNR has a relatively long history of using indices of ecological integrity. One of the first aquatic macroinvertebrate indices, the Hilsenhoff biotic index (HBI), was developed by William Hilsenhoff of the University of Wisconsin–Madison during the 1970s and 1980s with support from the WDNR (Hilsenhoff 1987). The HBI has become a standard method for WDNR biologists to assess stream water quality, particularly with regard to organic pollution. Advances in fish indices were inspired by the pioneering work of James Karr in the development of the first fish-based index of biotic integrity (IBI; Karr et al. 1986) and incorporation of IBIs into stream monitoring, classification, and regulation by the Ohio and U.S. environmental protection agencies. The first Wisconsin IBI was created for wadeable warmwater streams (Lyons 1992), but it soon became clear that different IBIs would be needed for other aquatic ecosystems. Subsequently, fish-based IBIs were developed for coldwater streams (Lyons et al. 1996), nonwadeable warmwater rivers (Lyons et al. 2001), and intermittent streams (Lyons 2006); one is under development for wadeable coolwater streams. A macroinvertebrate-based IBI has also been developed for wadeable streams (Weigel 2003) and another is in development for nonwadeable rivers.

The WDNR uses indices of ecological integrity for a variety of water resource management applications, including classifying waters, assessing current ecosystem status, monitoring trends in ecosystem health, prioritizing conservation and rehabilitation activities, and evaluating the success of management actions. Within the WDNR, three programs (i.e., watershed management, fisheries management, and fish and habitat research) routinely use indices of ecological integrity. Watershed management is responsible for maintaining and improving water and aquatic habitat quality. Watershed managers use the HBI and IBIs to address water quality issues related to point and nonpoint source pollution and the effects of riparian habitat modifications. The HBI and IBIs are fundamental to development of quantitative biocriteria and implementation of tiered aquatic life use designations in response to mandates of the Clean Water Act. Fisheries man-

(Box continues)

[1] Wisconsin Department of Natural Resources, Madison.

> **Box 12.1. Continued.**
>
> agers are charged with maintaining and improving sport and commercial fisheries. Because it is increasingly apparent that the quality of fisheries is dependent on the condition of entire fish assemblages, fisheries managers use IBIs to assess fish assemblage quality and to gain insight into community level issues that may influence fisheries. Fish and habitat researchers provide technical assistance and evaluation services to both the watershed and fisheries managers. Researchers have helped develop the indices of ecological integrity and have used them to evaluate many management activities.
>
> Even though indices of ecological integrity are widely used by the WDNR, they have critics both in and outside the agency. The main criticisms are that the indices are too imprecise, varying too much over time or among sites to detect real changes, and too nonspecific, identifying general problems but not indicating causes or suggesting solutions. In response, WDNR staff has concentrated on developing better indices, quantifying their inherent variation, and providing guidance on their interpretation. Another concern is whether field personnel have the necessary training and expertise to sample biotic assemblages adequately and to identify collected organisms accurately. To address this issue, WDNR staff provides training and quality assurance to standardize data collection. For example, a comprehensive, photo-based website has been developed (http://wiscfish.org) to aid in fish identification, and a contract has been developed with taxonomic experts at the University of Wisconsin–Stevens Point to identify macroinvertebrates.
>
> Despite initial skepticism, many WDNR biologists and managers now use indices of ecological integrity. The WDNR has conducted or funded numerous studies in which these indices have proven valuable and helped to guide management. For example, fish IBIs have indicated how removal of small low-head dams can improve stream fisheries and how commercial navigation affects ecological integrity of large rivers. Macroinvertebrate and fish indices have suggested effects of various agricultural and urban land uses on stream ecosystems, demonstrated the relative benefits of riparian and watershed land management practices, and contributed to the development of nutrient criteria to protect water quality. Fish IBIs have been used to inventory and assess entire rivers to identify areas worthy of protection or restoration and possible management actions to improve river health and fisheries. Given these diverse and successful applications, indices of ecological integrity are likely to grow in importance in WDNR water resources management.

effluents and nonpoint sources associated with agriculture and urbanization (Carpenter et al. 1998). Organic and inorganic chemicals pose threats to aquatic life in the form of over 100,000 chemicals that are released into the environment, transported through aquatic systems, and transformed with chemical reactions and biological interactions (Shea 2004). While chemical and thermal pollution is widespread and destructive, stream-deposited sediment is considered the greatest single water pollutant in terms of both quantity and economic and ecological impacts (Waters 1995). Recent federal policy and laws in the USA have reduced the discharge of point source pollutants, but nonpoint sources remain relatively unchecked and problematic (Karr et al. 2000; Paulsen et al. 2008). Water pollution can affect ecological integrity of aquatic systems by altering the aqueous environment and habitat for aquatic life and by causing toxicity in organisms, distortion of the food web, and other cascading effects.

12.1.1.3 Instream Habitat and Flow Alteration

The flow dynamics of lotic ecosystems are a primary factor in structuring their habitats and communities. The conceptual basis for understanding ecological processes of streams and rivers is based strongly on longitudinal, lateral, and vertical flow connections as the physical template upon which organic matter processing, energy transfer, and food web dynamics are structured over time (Vannote et al. 1980; Junk et al. 1989). Thus, the dynamics of groundwater, instream, and floodplain flows are integral components affecting ecological integrity of flowing water systems.

Stream and river flow is regulated by a number of natural and anthropogenic factors. Geology, geomorphology, and climate form the physical stage upon which factors such as land use change, water impoundment, and water withdrawal alter flow patterns and modify geomorphic features. Conversion of land cover from grasses, forest, or wetlands to agriculture or urban development affects streamflow by altering patterns of evapotranspiration and runoff. There are many indirect influences on streamflow and instream habitat, but the single most direct impact on flowing waters may be the presence of dams (Collier et al. 1996; Pringle et al. 2000). Dams are abundant and widespread (more than 79,000 dams over 2 m high in the USA; USACE 2005) and provide many important services to society. Dams generate electricity, mitigate flooding, supply water for human uses, enhance navigation, and provide recreation, but they also interrupt ecological functions of rivers by altering the flow regime, water availability, water quality, thermal environments, stream connectivity, and both downstream and upstream habitat. Thus, in many cases where economic and ecological costs of a dam have outweighed its benefits, dams have been removed. Other hydrological alterations, such as small-scale water withdrawal, groundwater extraction, stream channelization, instream mining, and interbasin water transfer, also have negative influences on aquatic ecological integrity.

12.1.1.4 Nonnative Species Introductions

The scale and economic costs of nonnative species introductions in the USA are immense: 50,000 species have been introduced and have an impact exceeding US$120 billion annually (Pimentel et al. 2005). The introduction of nonnative species into aquatic environments is the most direct means to alter biotic integrity and may cause declines or extinctions of native species and have cascading effects on ecosystems and fisheries. At least 536 fish taxa have been introduced to U.S. waters, and 316 taxa native to the USA have been translocated to waters outside their native distributions (Fuller et al. 1999). Eighty-eight species of mollusks have been introduced and established in U.S. waters (Pimentel et al. 2005). While the creation of nonnative fisheries may support local economies in some areas, annual losses due to introduced fishes are estimated at $5.4 billion, and those associated with zebra mussel and Asian clam are about $2 billion (Pimentel et al. 2005). Fewer introductions of exotic and native translocated fishes are known for Canada (18 exotic and 37 translocated species) and the United Mexican States (Mexico; 26 exotic and 29 translocated species), but introductions there are not well documented (Contreras and Escalante 1984; Crossman 1984). Introduced species are frequently cited as the most important threat to native aquatic biodiversity in North America, following habitat degradation and loss (Wilcove et al. 1998; Lodge et al. 2000; Jelks et al. 2008).

Not all fish and aquatic invertebrate introductions result in viable populations—in fact, most introductions fail—but once established, the impacts vary and are unpredictable (see Chapter 8). A general rule, sometimes called the "tens rule," pertaining to biotic invasions is that 10% of introductions become established and 10% of those become pests (Williamson 1996). Aquatic ecologists have studied factors that affect the invasion and establishment of nonnative species on local, regional, and global scales, and several broad observations have emerged (e.g., Moyle and Light 1996; Ruesink 2005). Invasion success depends on interactions of the invader, the invasion site, and the native species present. The biotic resistance hypothesis suggests that ecosystems with more native species and less human disturbance are less vulnerable to invasion by introduced species. This hypothesis appears generally applicable to freshwater fish introductions, but all freshwater systems are highly invasible depending on conditions at the time of introduction, and eradication of established species is rarely feasible (Jeschke and Strayer 2005; Moyle and Marchetti 2006). For example, dams and resulting impoundments enhance mechanisms for nonnative species transport and reduce biotic resistance to invasion, and introduced aquatic species may be up to 300 times more likely to occur in impoundments than in natural lakes in a region (Johnson et al. 2008). This illustrates how the physical and biotic components of ecological integrity are linked.

12.1.1.5 Secondary and Cumulative Impacts

Human alterations of the environment often occur as minor actions that may not produce an immediate measureable effect, but collectively result in serious effects later or at another location. Such effects are termed secondary and cumulative impacts (CEQ 2005). Secondary impacts are caused by human actions and occur later in time or at a distant location but are reasonably foreseeable, and cumulative impacts result from the incremental consequences of an action when added to other past and foreseeable future actions. Secondary and cumulative impacts must be considered in environmental review of proposed federal projects and are common concerns of state and federal transportation and natural resource agencies. For instance, highway effects often occur in three stages: initial construction, presence, and eventual landscape urbanization (Wheeler et al. 2005). Examples of secondary impacts would be those associated with subsequent residential, commercial, or industrial development after a highway is constructed (the primary impact), whereas cumulative impacts would be the sum of impacts across the watershed. In this example, the secondary impact (urbanization) is likely to be more detrimental than the primary impact (construction). Secondary and cumulative impacts are especially important to consider in riparian and watershed assessments. Thus, coarse-scale watershed variables, such as percent impervious surface or road density, may be more descriptive of environmental impacts than are reach scale variables associated with flow or channel form; elucidating trends or thresholds of watershed effects is critical. Secondary and cumulative impacts are certainly the most difficult to quantify in ecological integrity assessment, but probably the most influential.

12.1.2 Ecological Integrity and Fisheries Management

Management of aquatic systems for both ecological integrity and fisheries may be common goals in some systems, but may also present a conflict. Fisheries management incorporates information about not only the biology and ecology (i.e., science) of the resource but also considers economics, esthetics, human attitudes, and the interests of users and the general public (Chapter 5). However, the values of direct fishery users (e.g., catch, harvest and fish quality) may conflict with those of indirect users or nonusers (e.g., esthetics, existence value, or ecological services). Therefore, fisheries management goals, such as enhancing or maximizing fish abundance, population structure, catch, or harvest may be contrary to those to maintain robust communities and natural habitat.

Most natural resource management agencies embrace the concept of ecosystem management. However, fisheries management is often conducted with a single-species objective while management for ecological integrity is more comprehensive. In general, functional aquatic ecosystems with high ecological integrity can support sustainable fisheries, and examples of fisheries that benefit from high ecological integrity exist for anadromous fishes and natural lake and stream-dwelling fishes. However, the productivity of aquatic ecosystems is limited by geology, geography, and water quality, and some aquatic habitats in their undisturbed state do not support fish of any species (e.g., headwater streams and some alpine lakes). Objectives in fisheries management may be to extend these limits of productivity by manipulating the habitat (e.g., fertilization or liming) or by undertaking supplemental stocking or introduction of nonnative species, all of which reduce the ecological integrity of the managed system. As fisheries managers seek to include multiple stakeholders and uses in management, a primary challenge is to incorporate concepts and values of users and nonusers, as well as consumptive and nonconsumptive users, and to consider ecological integrity as they manage the physical and biotic environments of fishes (Chapter 14).

12.2 ECOLOGICAL INTEGRITY ASSESSMENT

There is no standardized approach to assess ecological integrity of aquatic systems, but there are some generalizations and common considerations in planning and conducting such assessments. A typical framework for assessing ecological integrity is to gather physical and biological field data from representative sites in a region or watershed, summarize that information, compare those findings among sites to a set of reference sites, historical data, or data that represent least-disturbed conditions, and interpret the relative distribution of findings in the context of ecological integrity. Such an undertaking can be intensive and costly in terms of data collection, and the resulting data set can be unwieldy to compile, analyze, and interpret. Quantitative approaches to assessment and monitoring provide the highest inference, but are costly. Thus, qualitative techniques that may be less costly and precise have received attention and are generally referred to as rapid assessments (e.g., Barbour et al. 1999). While frameworks for developing assessment tools have been proposed, there is no standard protocol to follow, so many subjective decisions arise with regard to identifying spatial and temporal scale, defining least-disturbed reference conditions, quantifying the suite of parameters measured or estimated, and comparing information among sites to obtain interpretable results and relative conclusions.

12.2.1 Relative Measures and Scale

Just as there is no standard protocol, there are few data standards or threshold values that apply universally to aid in the interpretation of field data or indices relative to ecological integrity. Thus, each investigation into ecological integrity is an independent assessment. It is not ecologically relevant to assess the integrity of a single site in any quantitative way because it is the ecological conditions relative to other sites, in particular reference sites, that yield meaningful conclusions for management. Therefore, the spatial, temporal, and organizational scales of the sampling environment are critical considerations in ecological integrity assessment.

Variation in habitat and biota among regions may be influenced by broad-scale factors, such as geology, river basin boundaries, biogeographic history, glaciation, and evolution (Matthews 1998). Thus, ecological integrity assessment should occur within a defined geographic region. Typical spatial frameworks for delineating regions in this context are generally by ecoregion or drainage basin (Omernik 1995; Omernik and Bailey 1997; Angermeier et al. 2000).

Stream networks are a continuum of physical and biotic processes and conditions that vary from headwater reaches downstream to large rivers in predictable ways along a longitudinal gradient. In most systems, species richness and diversity increase with stream size (Horwitz 1978; Wiley et al. 1990), and other community and population attributes, such as fish and invertebrate species composition, density, biomass, growth, body size, and trophic dynamics, change longitudinally (Vannote et al. 1980; Matthews 1998). Thus, stream size or longitudinal position must be considered when comparing ecological data among sites within and among drainage networks. Site classification may be required to ensure that sites of similar type within a region are being compared to avoid confounding effects of geology, physiography, or other natural influences and to select appropriate references.

Timing of sampling can strongly influence results of field assessments. For example, many freshwater fishes migrate seasonally to fulfill one or more life history requirements (e.g., spawning or juvenile rearing) and to seek refuge during severe environmental conditions (Peterson and Rabeni 1996; Grossman et al. 1998). Thus, the structure of a fish assemblage in a stream or a river is likely to vary among seasons, which can increase sample variance and obscure assessments. To reduce the influence of seasonal effects, scientists may limit comparisons of fish assemblages to similar seasons and, if possible, to seasons with minimal fish movements.

The temporal and spatial design of aquatic macroinvertebrate, plant, physical habitat, or water quality sampling is also an important consideration in ecological assessment. These ecosystem components may be so variable over time that using them to assess ecological integrity is best accomplished through monitoring programs of repeated sampling over time. In the dynamic environment of streams and rivers, aquatic macroinvertebrates may be extremely mobile, relocating by drifting downstream and colonizing new areas by swimming, crawling, or flying and by adult emergence (Smock 1996). Aquatic plant communities follow seasonal growing cycles, and hydrologic patterns resulting from seasonal weather conditions can alter instream and riparian habitat conditions over fine and coarse temporal scales. Flow dynamics and chemical processes also make effective characterization of water quality a challenge. A single site visit or water sample (grab sample) is unlikely to detect features that influence the ecosystem. Approaches that integrate water sampling over time can improve assessment and may serve as alternatives to repeated sampling to characterize water quality dynamics (Heltsley et al. 2005). As with all aquatic sampling, a wide range of macrohabitat types should be sampled to ensure that all important habitat types are represented (e.g., benthos, drift, and woody debris for invertebrates; slack water and flowing water for periphyton and plants; riffles, pools, and runs for instream habitat), and attention should be given to avoid confounding temporal effects.

The resolution to which sampled organisms are identified can confound results and alter ecological conclusions. Identifying fishes to the species level is practical, but aquatic invertebrates can be difficult to identify accurately. If accurate identification varies among samples, then estimates of community indices, such as species richness, dominance, and diversity, will be biased. Thus, it is critical that investigators maintain a constant level of taxonomic precision in comparative analyses among sites and samples. Furthermore, investigators may simplify analyses and reduce variable sampling bias by excluding early life stages or rare species from analyses (Kwak and Peterson 2007). Such an approach is often adopted in assemblage level analyses, but it may reduce estimates of species richness and obscure patterns in occurrence of rare species that could provide insight into community organization and guide conservation strategies.

12.2.2 Definition of Reference Conditions

Reference systems or sites to represent undisturbed or least-disturbed ecological conditions are a critical component of ecological integrity assessment (Hughes 1995; Karr and Chu 1999). Such systems specify the biological potential among ecosystems in a region. They are critical benchmarks for comparison to detect and understand effects of human activities on ecosystems and to identify goals for ecological restoration. Biological criteria are numeric values that describe the reference condition of aquatic communities in waters of a designated aquatic life use (USEPA 1990). Therefore, selection of a reference system or definition of reference conditions is of utmost importance in the assessment of ecological integrity, but no clear criteria exist for such decisions.

Variation over space and time is key to identifying reference systems or conditions. Before indices are established for a region, a system or location in that region with appropriate biotic or physical criteria (e.g., watershed land use or riparian disturbance) should be selected to serve as reference conditions. Similarly, information from the past, recent or historical, may provide qualitative or quantitative descriptions of predisturbance conditions. Museum references, theses and dissertations, and agency reports can be valuable sources of historical data on physical conditions, flora, and fauna for specific regions.

Searches for information on predisturbance conditions may be difficult, but are perhaps the only option in regions exposed to large-scale degradation. For example, only 42 high-quality, free-flowing rivers remain in the conterminous USA and only 2% of the rivers have features sufficient to receive federal protection (Benke 1990). Further telling is Hynes' (1970) conclusion that it would be extremely difficult to find any stream that has not been altered by humans and impossible to find any such river—and that was more than four decades ago. When undisturbed sites are unavailable, "least disturbance" systems or sites are often substituted for undisturbed references. Alternatively, reference conditions from another region may be cautiously applied. Given the lack of options available in some regions, creative approaches may be required. In an ecological integrity assessment among rivers, Radwell and

Kwak (2005) defined least-disturbed conditions as a composite of optimal physical and biotic conditions with theoretically maximum ecological integrity. Environmental filters have also been used to estimate expected species richness for macroinvertebrates and fishes (Chessman and Royal 2004; Chessman 2006). Environmental filters are models that successively exclude a proportion of a regional species pool, leaving a local assemblage to occupy a particular site. Species for a site are filtered according to their evolutionary history, dispersal ability, physiological tolerances, habitat requirements, and interactions with other species. These are approaches that might be useful in an area where all streams have been subjected to human disturbance. When developing approaches to define reference conditions, it is preferable to base them on systematically-gathered data rather than on expert opinion that may be subjective. Hughes (1995) examined approaches for determining regional reference conditions and recommended interpreting a combination of reference sites and historical data by means of linear models and expert judgment.

12.2.3 Measures and Indices

A number of indices that describe physical and biotic conditions of aquatic ecosystems have been developed based on comparisons of existing conditions and knowledge of a variety of relationships between organisms and their environment. An effective ecological index should be (1) socially-relevant, (2) simple and easily understood, (3) scientifically-based, (4) quantitative, and (5) cost-effective (O'Connor and Dewling 1986). Ecological integrity indicators may be regarded within a hierarchical framework from reductionist (e.g., chemical concentrations or presence of species) to holistic (e.g., biodiversity or ecological process rates; Jorgensen et al. 2005). In addition, single indicator organisms or metrics, standard community parameters, or multivariate statistics may be used to reflect environmental conditions and to compare them among sites or ecosystems with regard to reference or least-disturbed conditions as described below.

12.2.3.1 Water Quality and Habitat Indices

The development of physical habitat indices has lagged behind biotic indices, perhaps because of the difficulty in relating habitat features to biologically meaningful consequences or ecological integrity. Individual parameters describing water quality, stream characteristics, or lake attributes can be measured by relatively standard procedures, and indices may be calculated and compared for specific habitat components. A number of rapid, qualitative habitat rating systems have been attempted, but these systems are subject to observer bias and lack quantitative rigor. Unfortunately, there is no single habitat index available to describe general habitat conditions in an aquatic ecosystem.

The U.S. Environmental Protection Agency (EPA) has developed water quality standard recommendations for concentrations of nutrients, contaminants, and other substances in surface waters and has allowed states to develop their own standards from these. These standards are designed to determine appropriate water body uses (e.g., public water supply, recreation, fish and wildlife, agriculture, and industry), to protect aquatic life and human health, and to set water quality goals (USEPA 1994). For example, the EPA recommends a protective minimum dissolved oxygen concentration of 5 mg/L for aquatic life, but many states have adopted 4 mg/L as their standard. Water quality criteria for protection of aquatic life contain two allowable concentrations, one maximum concentration to protect against acute (short-term) effects and another continuous concentration to protect against chronic (long-term) effects. In addition to chemical water quality criteria, the EPA has developed criteria for turbidity and suspended and bedded sediment that relate to physical habitat and aquatic life. As one might imagine, universal water quality criteria are difficult to apply in ecological assessment, as reference or least-disturbed conditions vary among regions. Furthermore, water quality criteria provide threshold concentrations or levels for individual parameters, and no single index combining multiple water quality parameters has been developed to describe ecological integrity.

Protocols have been developed to measure parameters and describe physical habitat characteristics. Instruments and procedures to perform aquatic habitat assessments are available and follow similar approaches to measure and quantify important instream and riparian habitat characteristics (McMahon et al. 1996; Bain and Stevenson 1999). In general, these procedures are designed to describe the bank, riparian zone, and watershed characteristics of streams and rivers and the morphometrics, water quality, hydrodynamics, and trophic state of lakes and reservoirs. Such habitat parameters are typically quantified as the mean, variance, or distribution of measurements, but there are several classifications and indices available that condense multiple measurements into a single index. Examples of indices that quantify specific habitat components are diversity indices to quantify habitat complexity for categories of habitat variables, substrate particle size distributions as indices of spawning-gravel quality in streams (Kondolf 2000), and a shoreline development index that describes the shape of lakes or reservoirs by comparing the shoreline length to the circumference of a circle with the same length (Bain and Stevenson 1999). Several composite habitat indices have been developed by state agencies for specific regions, but none have been applied more broadly. An example of this type of index is the habitat quality index developed for Wyoming trout streams that incorporated stream width, bank condition, substrate, physical cover, water flow, velocity, temperature, and nitrogen concentration into a linear model to predict trout biomass (Binns and Eiserman 1979).

The advent of geographic information systems and associated data sets opened the door to new approaches in aquatic habitat assessment at multiple spatial scales. These approaches are especially applicable to ecological integrity. Habitat may be described at reach, segment, and basin scales, and riparian and watershed attributes can be quantified from existing data layers (Fisher and Rahel 2004). Such landscape level approaches have proven useful for identifying reference conditions and for conducting broad-scale assessments of human disturbance and anthropogenic stress (e.g., Mattson and Angermeier 2007; Wang et al. 2008). Many of the impacts to ecological integrity of aquatic systems act at a broad scale and may be cumulative in nature; thus, the ability to quantify watershed or riparian land cover and associated parameters, such as road density, human population density, and the proportion of a watershed that is forested, urban, or impervious surface, is valuable in assessing ecological condition of aquatic systems.

12.2.3.2 Biotic Indices

The concept of using indicator organisms to assess environmental quality predated the concept of ecological integrity by at least a century. The "Saprobiensystem," developed by Kolkwitz and Marsson (1908) in Europe delineated zones of organic enrichment and classified the animal species that occupied them. That early biotic index was later applied to river systems and modified (Chandler 1970), and this led to development of a variety of biotic indices based on aquatic invertebrates and fishes. Biotic indices have been widely lauded (Davis 1995; Simon 1999a; Karr and Chu 1999), but also criticized (Suter 1993). In general, criticisms include a perceived lack of ecological meaning, predictability, diagnostic power, and application to water resource regulation, but these limitations can be overcome by caution and reason in application of techniques and interpretation of results. In addition to biotic indices, community structure parameters, such as richness or diversity, also serve as informative biotic assessment tools for ecological integrity (Box 12.2).

Biotic indices are developed to describe ecological condition according to known or suspected relationships between indicator organisms and their environment. Indicator organisms or guilds may be selected because they are particularly sensitive or tolerant to environmental degradation, and both sensitive and tolerant organisms may be incorporated into a single biotic index. It is difficult to create effective biotic indices that are universal; as fauna and environmental stresses change regionally, so will suitable indicator organisms. Thus, a biotic index developed for a specific region and environmental stressors may require modification for a different fauna and environmental relationships.

Macroinvertebrate indices. A seemingly endless number of biotic indices for aquatic macroinvertebrates have been developed and applied (Washington 1984; Resh and Jackson 1993; Rosenberg et al. 2008). These indices vary from simple community indices, including a suite of diversity indices, to those based on relative proportions of taxa in a sample or at a site. Taxa richness (the number of taxa in a sample) is the most common index used to describe macroinvertebrate assemblages, and taxa diversity is often applied in biotic assessment (Box 12.2).

Many aquatic macroinvertebrate indices are derived from the relative occurrence of taxa with varying levels of tolerance or sensitivity to environmental degradation or water pollution (Washington 1984; Resh and Jackson 1993; Rosenberg et al. 2008). One index, EPT richness, which is the number of taxa of Ephemeroptera (mayflies), Plecoptera (stoneflies), and Trichoptera (caddisflies), is commonly applied. It is based on the premise that most species in these three orders are sensitive to pollution and it is practical because identification to insect order can be accomplished with minimal training. A related index that includes additional information is the number of EPT individuals expressed as a ratio to the number of Chironomidae (midges) individuals. This represents the ratio of sensitive taxa to a relatively tolerant taxon.

Examining the relative occurrence of macroinvertebrate feeding guilds can be informative with regard to ecological function, rather than structure, of the assemblage. Multimetric combination indices incorporate measures of taxa richness, abundance, tolerance levels, feeding guilds, and other attributes to yield an overall index score. In general, a multimetric index is applicable only within the region in which it was developed.

An approach that integrates multiple aspects of the indices above is to compare observed invertebrate assemblage composition of a site with that expected based on sampling least-disturbed reference sites, which describes taxonomic completeness of a site (i.e., river invertebrate prediction and classification system [RIVPACS]; Hawkins 2006; Yuan 2006). A primary advantage of the taxonomic completeness approach is that it may be consistently applied among regions by comparing a common statistic, the ratio of observed to expected taxa.

Fish indices. Fishes are especially well suited as indicators of environmental quality (Karr et al. 1986; Simon 1999a). They are widely distributed; they can accurately reflect environmental conditions at multiple scales; life history and geographic distribution information are extensive for many species; and effective techniques are available for sampling.

Box 12.2. Community Structure Parameters for Biotic Assessment

Biotic indices developed for aquatic invertebrate or fish assemblages typically include some measure of community structure as well as relationships of indicator organisms with their environment. Once indicator organisms are incorporated into a biotic index, its applicability usually becomes restricted, perhaps to the region where it was developed. In some cases, metrics within an index based on indicator organisms can be calibrated to define expectations relative to reference conditions; in other cases, metrics are substituted when applied in other regions where those metrics may not be sensitive to environmental degradation. However, parameters that describe community structure may be applied more widely and can be useful biotic assessment tools when development or application of a regional biotic index is not practical.

Community ecologists have devoted substantial effort to describing the relative abundance of species or other taxa in a community as a single measure that is intended to reflect the state of the community. The most common of these measures is species diversity, which incorporates the number of species in an assemblage (species richness) and the relative abundance of those species (evenness). While such structural indices are usually applied at the species level, it is equally appropriate to estimate them at any taxonomic or other hierarchical classification level. Diversity is a useful index for describing assemblage structure, but it should not be interpreted as an index of ecological integrity. Higher diversity does not necessarily indicate more integrity, as many ecosystems support few species and low diversity in their natural state (e.g., those in infertile environments or exposed to harsh conditions), and increased species occurrence may indicate anthropogenic impacts in some systems. The relationship between community indices and anthropogenic impacts should be elucidated before indices are applied in ecological integrity assessments.

Species richness

Species richness is the simplest and oldest community structure parameter; it is a count of the number of species represented in an assemblage. In the field, managers rarely conduct a complete count of every species in an assemblage, so the number collected in a sample should be considered a minimal estimate that may vary with sample size, area sampled, or effort. Statistical methods have been developed to estimate the number of species not detected in a sample, but such procedures that derive estimates by extrapolation beyond the boundaries of empirical data should be applied cautiously (Kwak and Peterson 2007).

Species diversity

Diversity indices combine information on the number of species in an assemblage (richness) and their relative abundance (evenness). Unfortunately, there is no "correct" means of assigning proportional weighting between these two components, and thus,

(Box continues)

Box 12.2. Continued.

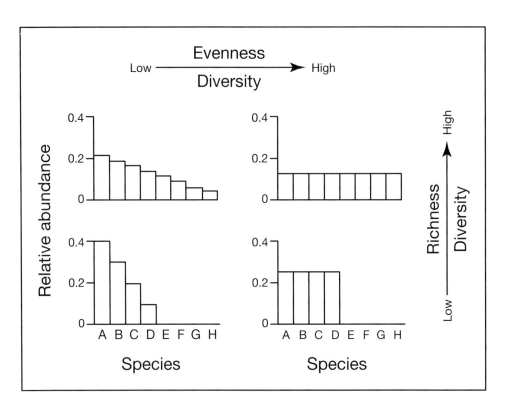

Figure. The concept of species diversity illustrated by variable relative abundance of species (bars) among assemblages (plots). Diversity increases with the number of species (richness) and equitability of distributions among species (evenness).

dozens of diversity indices have been developed and applied (Washington 1984). A diversity index is a parameter describing assemblage structure, but any relationship to ecological function or integrity remains unclear. And while diversity indices have been widely applied, they have also been criticized for lack of biological relevance and should be considered only one of many tools available to describe community structure.

Shannon's index of diversity (H'; Shannon and Weaver 1949) has endured ongoing criticism yet remains the most widely-applied diversity index in aquatic systems. It was independently developed by Shannon and Wiener at about the same time and is often referred to as the Shannon–Wiener index. Shannon's index (H') is based on information theory and is defined as

$$H' = -\sum_{i=1}^{s}(p_i)(\log_e p_i),$$

(Box continues)

Box 12.2. Continued.

where s = number of species and p_i = proportion of the total sample represented by the ith species.

Species evenness

Evenness is a measure of the equitability in relative abundance among species. There are many approaches to quantifying evenness, but the most common is to express it as a proportion of estimated diversity relative to the corresponding maximum diversity for the specific number of species and sample size. For Shannon's diversity index (H'), the corresponding index of evenness (J') is calculated as

$$J' = \frac{H'}{H'_{max}} = \frac{H'}{\log_e s},$$

where $H'_{max} = \log_e s$ = maximum possible value of Shannon's index, and s = number of species.

There is no clear theoretical upper limit for Shannon's index, but in practice, it rarely exceeds 5.0 for biological assemblages.

Species dominance

Another simple index related to evenness is species dominance, which is expressed as the relative abundance of a subset of the most numerous species. For example, the proportion of the assemblage composed of the two or three most abundant species would, in general, be inversely related to evenness. The equation to calculate species dominance for the three most abundant species (D_3) is

$$D_3 = \sum_{i=1}^{3} p_i,$$

where p_i = proportion of the total sample represented by the ith species. Species dominance may be estimated for a variable number of dominant species (usually 2–3).

Richness, diversity, and evenness are among the simplest and most comprehendible community indices and, thus, are widely applied. Example calculations of these indices for fish assemblages may be found in Kwak and Peterson (2007), as well as more complex approaches for comparing aquatic communities that apply to biotic integrity assessments.

Additionally, fishes are more visible, understood, and valued by regulators, politicians, and the general public than are other aquatic organisms.

Indicator fishes may be particularly sensitive or tolerant to environmental degradation. The application is more biologically relevant when indicator guilds are used because guild occurrence may imply ecological function, such as feeding or reproduction, whereas responses of individual species may be more specific. Examples of applicable fish guilds are those based on feeding and trophic relations, reproduction, or habitat affinity (Simon 1999b). Furthermore, higher levels of fish taxa (e.g., families or genera, such as Salmonidae or darters including genera *Ammocrypta*, *Etheostoma*, and *Percina*) may be considered indicator taxa.

Disadvantages of the indicator fishes approach lie primarily in its subjectivity and ecological basis. Although several lists partitioning fish guilds exist (e.g., Karr et al. 1986; Halliwell et al. 1999; Simon 1999c), standard criteria for guild delineation are lacking. Another problem is that mechanisms unrelated to ecological integrity may influence occurrence or abundance of fishes, including zoogeography, biotic interactions, or harvest (Fausch et al. 1990). Further complicating the use of fishes as ecological indicators is that response to environmental conditions in fishes can vary with space, time, life history stage, and type or degree of environmental stress, which could confound comparisons among sites or ecosystems.

While indicator taxa, such as common carp or salmonids, have been recognized in fisheries science for decades, the first widespread application of a formal biotic index based on fishes was the index of biotic integrity (IBI: Karr et al. 1986). Since its development for wadeable, warmwater streams in the midwestern USA, the IBI has been modified for coldwater streams, large rivers, lakes, wetlands, estuaries, and highly modified habitats, such as reservoirs and tailraces, in various regions of the USA and other countries (Simon 1999b). Today, IBIs are widely applied and provide a conceptual and procedural framework for assessing biological integrity based on fish assemblages.

The IBI was designed as an index that integrates attributes of the assemblage, population, and individual based on relative abundance of fish species and condition of individuals in a representative sample of the assemblage. The primary advantage of the IBI over community structural indices is that it was developed based on ecological relationships of fishes to assess ecological integrity and anthropogenic disturbance. The original IBI framework included 12 metrics that described various aspects of fish species composition, trophic composition, abundance, and condition (Table 12.2). Although metrics have been omitted, augmented, or modified in applying the IBI to other regions and habitats, the original framework is typically maintained (Miller et al. 1988). Increasingly, metrics are developed systematically for a specific region (Hughes et al. 1998; Angermeier et al. 2000). A number rating, or score (5, 3, or 1), based on ecological expectations (i.e., based on reference or least-disturbed conditions) is assigned to each metric, and metric scores are summed to yield a composite index score. The IBI scores may be compared, or scores may be categorized from "very poor" to "excellent."

Metrics developed for any fish IBI should follow a common ecological rationale and represent empirical relations with anthropogenic impact for the entire assemblage. Metrics describing species composition are intended to characterize biological integrity through measures of fish species richness and occurrence of relatively tolerant and intolerant species. Trophic composition metrics are based on the premise that alterations in food resources and productivity, influenced by water and habitat quality, are reflected in the trophic structure of the fish assemblage. As habitat degrades, food resources fluctuate more and omnivores may replace more specialized feeders. The presence of piscivores, or other top carnivores, indi-

Table 12.2. Generalized fish assemblage metrics and scoring criteria for an index of biotic integrity (IBI) applied to streams (modified from Karr et al. 1986). Scores are assigned to each metric based on the sample deviation from that expected from a least-disturbed reference system, with 5 representing the highest integrity and 1 the lowest.

Attribute category and metric	Scoring criteria		
	5	3	1
Species richness and composition			
Total number of fish species			
Number and identity of darter species	Expectations vary with stream size, region, and basin		
Number and identity of sunfish species			
Number and identity of sucker species			
Number and identity of intolerant species			
Percent individuals as tolerant	<5%	5–20%	>20%
Trophic composition			
Percent individuals as omnivores	<20%	20–45%	>45%
Percent individuals as insectivorous cyprinids	>45%	20–45%	<20%
Percent individuals as piscivores	>5%	1–5%	<1%
Fish abundance and condition			
Number of individuals sampled	Expectations vary with stream size, region, and basin		
Percent individuals as hybrids	0	>0–1%	>1%
Percent individuals diseased or with anomalies	0–2%	>2–5%	>5%
Total IBI score (sum of 12 metrics)	60 ←——————→ 12		
Integrity class	Excellent – Good – Fair – Poor – Very Poor		

cates a more complex food web. Metrics describing fish abundance and condition were designed to incorporate population- and individual-level effects of environmental degradation. Metrics for hybrid individuals and for disease and anomalies are among those most difficult to apply and are frequently omitted or replaced (Miller et al. 1988). Metrics related to introduced species abundance or reproductive guilds have been substituted for the hybrid metric to represent a similar ecological rationale.

The IBI was an important advance in biotic index development that has evolved to a useful assessment framework for local and regional assessments. Fisheries and aquatic scientists apply their own experience, perspective, and creativity into assessing ecological integrity based on fish and invertebrate assemblages, as well as physical attributes of aquatic habitats, riparian zones, and landscapes. Development of an IBI is an intensive endeavor that requires sampling a substantial number of sites and conducting careful deliberation and analyses regarding metric and criteria development and refinement. However, investigators undertaking more limited assessments may utilize individual IBI metrics or related indices singly or in aggregate as quantitative characteristics for comparison and assessment. For example, IBI development would not be practical to compare a small number of sites or to monitor trends over time at a single site. In such situations, managers may apply an existing IBI, select a subset of the most relevant IBI metrics for the region, or choose several community indices (Box 12.2) and compare them among sites or over time.

12.2.4 Quantitative Comparisons

A variety of approaches are available for describing, comparing, and seeking associations in ecological integrity assessment. Many approaches for data presentation and analysis designed for community-level investigations are applicable to ecological assessment (see Krebs 1998; Kwak and Peterson 2007). Most results related to assessment are relative values that are meaningful in a comparative context rather than in an absolute sense. Thus, absolute probabilities (P-values) and statistical significance (α-levels), to which many fisheries scientists are accustomed, are less applicable, and reliance upon them may confuse interpretation. This is not to say that subjective or qualitative approaches are optimal; on the contrary, intensive data description, exploration, and quantitative comparison are the most objective, defensible, and scientifically-relevant means to describe and understand ecological integrity.

Simple descriptive statistics and graphical displays of data can be effective for expressing results of biotic assessment data. Simple biplots of sample means or medians with a measure of dispersion (e.g., SD, SE, and 95% confidence intervals), especially in a series or among sites, can convey the magnitude and direction of ecologically relevant findings. More complex data may require multivariate statistical analyses (i.e., simultaneous observation and analysis of more than one statistical variable) and are often graphically displayed to detect trends that may not be apparent by significance testing (Johnson 1998). Such graphical techniques are helpful in examining broad trends among samples, detecting relationships among assemblages, and identifying outliers. Many univariate statistical procedures have underlying assumptions about data distributions (i.e., parametric tests), whereas others are available without distribution assumptions (i.e., nonparametric tests). Both types of univariate methods are useful for comparing single variables among sites or over time. However, the unusual statistical properties of many habitat or biotic indices do not conform to those required for use of traditional statistical procedures and statistical treatment of index values should be interpreted with caution (e.g., Jackson and Somers 1991).

Ecological field assessment and monitoring results may be amenable to analysis in an experimental design to detect ecological impacts. While rarely planned or designed as manipulative field experiments, ecological monitoring data can be analyzed as a time series or in a before-after-control-impact (BACI) design framework (Underwood 1994). In a BACI design, reference sites or least-disturbed conditions serve as an experimental control to detect relative changes in response variables over time as an interaction of time and site effects, and results may be more relevant with multiple controls or when analyzed in a hierarchical framework.

If a multimetric index approach is adopted, there are several important steps to follow before application. Metrics likely need to be calibrated to account for influences of geology, physiography, or other factors unrelated to integrity and to define expectations in the fish assemblage that may vary with stream size, basin, or region (Karr et al. 1986; Barbour et al. 1999). Fausch et al. (1984) presented a graphical calibration technique to demonstrate the effect of stream longitudinal position on fish species richness and to approximate expected criteria values for IBI applications. When species richness is plotted against stream order or watershed area, the distribution of sites forms a positive-sloped line of maximum species richness for a river system or region, with lower richness values scattered below the maximum line. The line of maximum species richness is used in IBI practice to define "excellent" species richness (metric score = 5) that varies with stream size; similarly, sites with richness falling below the maximum expected may be rated depending on the degree of deviation below the line for a given stream order or watershed area. The line of maximum expected values can also be quantitatively derived by calculating the 95th-percentile regression (Blackburn et al. 1992) rather than by visually fitting a line to perceived maximum values. This technique can be applied to other metrics describing species composition (Table 12.2) and may be less relevant to metrics that vary less due to stream size (Karr et al. 1986). Furthermore, such spatial and size effects apply to limnetic zones of lakes and reservoirs and may be considered in analogous lentic comparisons among sites.

Multivariate statistical procedures offer advantages over univariate approaches for pattern analysis associated with ecological integrity assessments. Much of the relevant empirical information may be in the form of community level data or multicollinear variables (i.e., correlated independent variables), which may not be easily interpreted using univariate techniques. Furthermore, multivariate approaches may detect subtle trends or effects that are not apparent when examining individual parameters. One shortcoming of many multivariate approaches, however, is that the ecological interpretation of results is more abstract and difficult (e.g., what does a principal component score mean in terms of habitat suitability or community structure?), but inclusion of reference conditions in the analysis can form a basis for comparison among conditions. Some applicable multivariate procedures include various forms of clustering, ordination techniques (e.g., principal component analysis and multidimensional scaling), discriminant analysis, and time series analysis (Norris and Georges 1993; Kwak and Peterson 2007).

12.3 MANAGEMENT FOR ECOLOGICAL INTEGRITY

12.3.1 Problem Identification

Two types of management issues are commonly confronted by natural resource managers. In the case of the first type, one or more water bodies in a manager's jurisdiction fail to meet designated goals for ecological integrity. These water bodies may not meet water quality criteria for designated uses or may score low on standardized habitat assessments or fish IBIs. The manager is then challenged with determining likely causes of the apparent ecological impairment and identifying the most appropriate remedies.

The second type of management issue involves avoiding the loss of integrity in aquatic systems that are valued for their natural and biological resources. The manager must then determine how to sustain ecological integrity in streams and lakes that are considered high quality (e.g., designated outstanding resource waters as defined by the antidegradation provision of the CWA), or that support extraordinary biodiversity or species at risk of imperilment (Master et al. 1998; Moyle and Randall 1998) given foreseeable changes in regional land use, human population growth, and human demands on water resources.

In either case, before developing a management plan, the manager must have a clear idea of either what caused the loss of integrity or what is likely to cause a future loss of integrity. The difficulty, of course, is that seldom does loss of ecological integrity result from a single, easily-identified cause. To complicate matters, land use or management practices that occurred in the past also may limit biological condition in a stream or river. These persistent legacy effects of land use practices have been termed the "ghost of land use past" (Harding et al. 1998), and common examples include the following.

- Agricultural legacies—historically intense agriculture in some regions left incised channels and large sediment deposits in stream valleys (Jackson et al. 2005) with the result that currently reforested watersheds may continue to exhibit depressed ecological integrity (Harding et al. 1998; Wenger et al. 2008a).
- Logging legacies—the historical practice of transporting timber in floods of logs scoured rivers with lingering effects on present-day habitat (Wohl and Merritts 2007).
- Mining and industrial discharge legacies—toxic metals from mine tailings are associated with species losses in streams even where watersheds are nearly entirely forested (Maret and MacCoy 2002); elevated mercury levels in sediments are a legacy of 19th-century gold mining (Leigh 1997) and may be responsible for loss of multiple mussel species (Burkhead et al. 1997).
- Dam legacies—loss of diadromous and migratory species in streams and lakes upstream of dams that impede fish passage alters present-day faunal and ecosystem condition (Freeman et al. 2003); perhaps less obvious are effects of thousands of smaller dams built in the 17th through early 20th centuries on the geomorphic structure of streams today (Walter and Merritts 2008).
- Stream channelization and snag removal legacies—efforts to facilitate navigation and reduce flooding have involved extensive channel straightening and removal of wood (snags) in major rivers across North America (e.g., Wohl and Merritts 2007). Smaller streams were also extensively channelized during the past century, typically to improve drainage of agricultural lands, with multiple effects on present-day habitat and ecosystem functions (NRC 1992).

It is important to recognize potential legacy effects of historical land and water uses because these legacies may also limit what can be obtained as a management goal under current conditions.

Given the multiple potential causes, both contemporary and historical, of species losses and depressed ecological integrity, developing management approaches to protect or restore integrity poses a substantial challenge. Progressing with the most efficient use of resources and best possibility of success requires explicit identification of management objectives and the most influential factors impairing or threatening ecological integrity. Two basic components of effective management are (1) a statement of objectives and (2) a statement of how the system is thought to function. Objectives are tied to values and state how a value is defined in a particular context (Clemen 1996). Objectives may be separated into two types: fundamental and means. Fundamental objectives are important components of ecological integrity, such as maintaining native species richness, water quality, and natural habitat dynamics; means objectives help achieve those desired, fundamental outcomes (Peterson and Evans 2003). Without clearly-stated objectives, it is impossible to measure success. By separately-identifying means objectives (e.g., limiting excessive water diversions, watershed erosion, or sediment inputs; restoring connectivity between river and floodplain habitats), the manager can articulate hypotheses about how management actions may achieve desired outcomes. Identifying a variety of means objectives in relation to fundamental objectives can also help to reveal and develop management options (Clemen 1996).

A statement of beliefs regarding how environmental drivers (e.g., streamflow alteration, sediment loads from historic land uses, or barriers to dispersal) relate to ecological integrity (defined by fundamental and means objectives) provides a basis for evaluating the probable success of particular management actions. Managers commonly use conceptual models to formulate ideas of how stressors are related to ecological integrity and what actions could effectively protect or restore integrity. A hypothesis about how drivers relate to objectives can be used to build a parameterized decision model (Reiman et al. 2001; Peterson and Evans 2003) in which process models (e.g., population dynamics in relation to streamflow patterns) or expert judgments are used to derive quantitative predictions of outcomes for alternative decisions. A sensitivity analysis of a decision model, by which inputs are varied to determine their relative effect on output, elucidates what factors (including sources of uncertainty, such as future climate or land use) most strongly influence outcomes of alternative decisions.

The process of specifying objectives for a particular management context, and then constructing and analyzing sensitivity of a decision model based on hypothesized causal relations between factors and outcomes, is the basis of decision analysis (Peterson and Evans 2003). Decision analysis is discussed to emphasize that success in managing for ecological integrity will always require, at a minimum, clearly-stated objectives and explicit hypotheses of how different ecological factors relate to attainment of those objectives. The manager will return to these statements of what management is trying to achieve and why management thinks a particular action will be effective, to develop measures of success and to evaluate why a particular objective was or was not met.

12.3.2 Management Options

Innovative approaches and partnerships are keys to successful management for ecological integrity in aquatic systems, in both short and long timeframes. Short-term actions address proximal causes of species declines and often are taken at local scales. Examples include habitat restoration or species propagation. Long-term actions address ecological processes, with a goal of restoring or sustaining integrity at larger scales, such as watersheds, stream networks, large rivers, or lakes.

12.3.2.1 Near-Term, Local-Scale Actions

Habitat Restoration. Habitat restoration aims to correct or replace habitat losses with the goal of improving suitability for desired aquatic communities. Many actions of the past have been taken under the general notion of "improvement": for example, stream poisoning to remove "rough fish" prior to stocking introduced sport fishes. However, in managing for ecological integrity, the intent generally is to restore ecological processes (e.g., fish movement onto floodplains) and habitat features hypothesized to promote viability of native species.

Stream restoration, in particular, is a major activity in the USA (Bernhardt et al. 2005; Kondolf 2006; Chapter 10) and is often undertaken as mitigation for other human-caused habitat loss (Sudduth et al. 2007). Stream restoration encompasses a variety of activities (Table 12.3). These restoration activities generally target stream segments, sometimes long segments, as in the case of a river reach downstream from a hydropower dam. Habitat restoration in lakes pres-

Table 12.3. Common stream and river restoration activities with examples of management goals and applications. Potential drawbacks (in italics) are noted for selected activities and highlight ecological complexity associated with restoration.

Restoration activities	Goals and applications
Riparian plantings and protection	Goals—improve water quality by reducing nonpoint source pollutant input; maintain naturally cool or cold water regimes (via shading); maintain wood and detritus inputs to streams • Maintenance or restoration of riparian vegetation • Fencing to exclude livestock
Streambank stabilization	Goal—reduce bank erosion and associated sediment inputs to streams • Bank reshaping and strengthening (e.g., with rock, logs, and live vegetation) • *Bank hardening enhances flood conveyance and downstream erosion and impedes natural channel dynamics* (Florsheim et al. 2008)
Channel redesign	Goal—restore channel to natural geometry and improve instream habitat • Channel reshaping; bank stabilization; meander creation; placement of habitat features (e.g., boulders and rootwads) and channel control (e.g., rock weirs) • *Project failures are common; typical applications attempt to create single-thread, symmetrically-meandering channels, even where channels are naturally braided or irregularly meandering* (Kondolf 2006)
Fish passage provision	Goal—restore biological integrity by removing impediments to faunal migrations • Culvert replacement with bridges or other road crossings that do not impede movements by aquatic organisms • Installation of fish ladders or fishways at dams • Operation of navigation locks to facilitate fish movements past dams • *Fish devices at dams fail to pass fish of all species and sizes; impoundments remain barriers to fluvial fishes and impede or prevent reproduction*
Dam removal	Goal—restore free-flowing river habitat and upstream–downstream species dispersal and migrations • Removal of structures that are no longer safe, functional, or economical as designed, including many small dams built in the last 100 years • *Mobilization of sediments stored behind dams can result in mortality and exposure to contaminants in downstream aquatic communities; dam removals may facilitate spread of exotic species* (Hart et al. 2002; Stanley and Doyle 2003)

Table 12.3. Continued.

Restoration activities	Goals and applications
Instream flow provisions	Goals—protect aquatic communities from deleterious low flows downstream from dams and diversions; improve instream habitat conditions and ecological functions downstream from dams • Flow requirements above a prescribed minimum flow at dams and water diversions to protect downstream aquatic communities during low-flow periods • Water releases at dams regulated so as to ameliorate effects on downstream aquatic communities; may be accompanied by dam retrofits to raise temperature and dissolved oxygen in water released from reservoir hypolimnia • *Minimum flow provisions alone cannot protect the range of flow-dependent ecological functions necessary for ecological integrity* (Poff et al. 1997)

ents a different challenge. Lakes tend to retain what we put into them, and contaminant remediation is thus a major concern, as is controlling invasive species. Most extensively, however, lake restoration has addressed excess nutrient loading, particularly by diverting or treating waste inputs and by modifying watershed land uses (NRC 1992).

Restoration activities may achieve specific management goals by alleviating a particular cause of poor water quality, habitat degradation, or low survival, reproduction, or colonization by native species. It is important to state from the outset the specific management goals and how those targets are believed to depend on the processes or habitat features that are being restored. For example, if providing higher flows below a dam or reconfiguring a channel to a more natural shape fails to result in intended improvement in ecological integrity, then from a management perspective, we have just tested a specific hypothesis about what limits integrity in that situation. It should be equally clear that implementing habitat restoration without monitoring is, at best, incomplete management and, at worst, a waste of resources. Monitoring is necessary to ascertain how well the restoration is implemented (Bernhardt et al. 2007). Ecological outcomes might be most efficiently monitored at a subset of similar projects in a region (Bernhardt et al. 2007) to provide the feedback needed to know if what was intended by a given restoration activity is achieved (Palmer et al. 2005).

Species propagation and reintroduction, or translocation of wild-caught individuals, to formerly-occupied habitats offers another tool for maintaining natural biological diversity. Despite a long history in North America of aquaculture to support fisheries based on introduced species or to replace decimated fisheries (e.g., Pacific salmon), propagation to restore declining nongame aquatic species has been relatively more restricted (Chapter 9). Examples vary from captive propagation and reintroduction of small-bodied, imperiled stream fishes in the southeastern USA (Shute et al. 2005) and southwestern USA and northern Mexico (Johnson and Jensen 1991) to captive breeding to supplement existing populations of riverine fishes such as the robust redhorse (Fiumera et al. 2004) and pallid sturgeon (Webb et al. 2005). For the razorback sucker, capturing and transferring wild-spawned larvae to isolated backwaters that are free of nonnative predators appears essential for avoiding near-term extinction of this critically endangered fish native to the Colorado River (Marsh et al. 2005). For other species that are critically-endangered by habitat loss, holding and rearing "ark" populations in captivity may represent the best hedge against extinction, provided there is a chance that environmental conditions can eventually improve enough to allow species reintroduction to native habitats.

Translocations involve moving wild-caught individuals into habitats within the species' historic distribution (Minckley 1995; Harig and Fausch 2002), in some cases following improvement in environmental conditions. For example, mollusks and most fishes were eliminated from 100 km of the Pigeon River in North Carolina and Tennessee by toxic chemicals discharged by a paper mill for most of the 20th century. Following modernization of the mill and drastic pollution reduction in the late 1990s, some fishes began recolonizing the river, but many other species lacked connected sources or sufficient mobility to re-establish on their own. To further ecological recovery, biologists have been reintroducing various snails and fishes from nearby streams in the same drainage (Coombs et al. 2004).

Conserving native species by holding and propagating individuals in artificial systems is an expensive approach that is most appropriate for species that have declined to levels of near extinction (Helfman 2007). Propagation and reintroduction have in some cases (such as the smoky madtom in the southern Appalachian Mountains) restored species to parts of their native distribution where natural recolonization or recovery was unlikely (Shute et al. 2005). On the other hand, relatively large expenditures on other species, such as razorback sucker in the lower Colorado River (Minckley et al. 2003) and pallid sturgeon in the upper Missouri River (Webb et al. 2005), have resulted in only limited contributions to species recovery in the wild. Additionally, maintaining genetic diversity will always be a major concern with captive propagation programs. Ultimately, whether the goals are conservation of imperiled species in particular or management for ecological integrity more generally, success will require attention to large-scale and long-term impediments to natural ecological processes.

12.3.2.2 Long-Term, Large-Scale Actions

Short-term actions to maintain or restore ecological integrity will be most successful when they are part of long-term strategies. Opportunities for long-term management arise, for example, in dam relicensing, habitat conservation planning, and local land use and watershed planning. These activities frequently target time frames of two or more decades—short in ecological terms but long compared with typical management time scales. However, what also sets long-term, larger-scale actions apart from those more tactical is the consideration of multiple stressors and their interactions in framing management alternatives and the integration of human uses and values into management.

Managing for ecological integrity over the long term entails working at large spatial scales, such as watersheds or river basins. Large spatial scales become linked to long-term management because of the interconnection of freshwater systems with their surrounding catchments. Over years and decades, changes in land uses and associated inputs to streams and lakes in thermal or hydrologic regimes or in connectivity to upstream and downstream systems can overwhelm local-scale efforts to maintain ecological integrity. Inland waters are viewed as components of landscapes that are embedded in climatic regimes and spatially integrated by transport of water, sediments, nutrients and energy between terrestrial and aquatic systems, headwaters and downstream habitats, surface waters and aquifers, as well as to and from floodplains (Ward 1989). Connectivity in aquatic landscapes is accomplished by water flow and material transport but also by fishes and other migratory fauna and by organisms (plants and animals) for which dispersal is critical for periodic population recolonization (e.g., following droughts or floods). Thus, the occurrence of introduced species is an impact that must be managed at the landscape level.

Long-term management for ecological integrity focuses on ways to avoid, ameliorate, or mitigate large-scale changes in natural patterns of discharge, nutrient and sediment inputs, thermal and chemical regimes, and organism movements. Importantly, long-term management seldom means "stabilizing" an aquatic system in an optimum state but rather protecting natural levels of variability. For example, geomorphologists emphasize dynamism as a critical component of physical integrity of streams and rivers (Graf 2001), meaning, in this case, that channels and riparian areas must be allowed to undergo natural cycles of erosion and deposition, storage and transport. Stream ecologists similarly recognize the critical importance of natural flow variability to ecological integrity (Poff et al. 1997) and that ecological communities are themselves naturally dynamic (Palmer et al. 2005).

Managing for improved ecological integrity addresses processes at watershed and landscape scales, such as those below.

- Hydrologic factors—minimizing stormwater runoff in urban and suburban areas, most effectively by managing near the source and using approaches (such as infiltration) that reduce pollutant loads. Innovative instream-flow management includes policies that minimize alteration of a suite of ecologically important flow variables and that incorporate water conservation into human population growth and landscape planning.
- Nutrient and sediment export—implementing best management practices in agriculture, urban development, and other land-disturbing activities to minimize nutrient and sediment mobilization.
- Riparian management—protecting (e.g., through local ordinances, exotic species control, and natural flow restoration) naturally-vegetated buffers adjacent to streams and lakes, including natural riparian processes such as water filtration, flooding, channel migration, and canopy dynamics.
- Connectivity—using landscape level planning to ensure that new reservoirs are placed to minimize effects on connectivity among and within high-quality streams; constructing "passage-friendly" road crossings at streams, such as bridges or bottomless or embedded culverts; and constructing effective fish passes at dams.

Ideally, management options are considered in the context of landscape level projections (e.g., where growth is likely to occur), options for water supply and wastewater management, and predicted effects of alternative scenarios on integrity of aquatic systems. As with short-term actions, integrity must be defined by specific quantifiable objectives to assess if management efforts are achieving desired outcomes.

A sobering fact is that the global climate is changing (Palmer et al. 2008). We are no longer in a position where managing to minimize direct human impacts on natural systems will result in long-term maintenance of a system. Although major uncertainties remain about precise rates of change and some of the mechanisms that may be involved, there is certainty that the global temperature is rising, glaciers and polar ice are melting, sea levels are rising, and precipitation patterns are changing. At a minimum, changes in local climate anywhere on the globe will overwhelm efforts to manage for ecological integrity if specific actions are not taken to halt human emissions of greenhouse gases. Even with our best efforts to slow it, we know that some degree of climate change is certain and will affect management options (Palmer et al. 2008).

An implication of our changing world is that environmental management will increasingly require close integration between measures of ecological integrity and benefits of ecologically healthy systems to human communities (Rapport et al. 1998). It will also be critical to maintain flexibility, so that management can be adjusted in response to changing environmental conditions and as management outcomes become better understood. These are characteristics of adaptive management (Williams et al. 2007), which is appropriate to situations in which management decisions are made repeatedly and can strongly affect outcomes but there also is high uncertainty regarding future conditions and system dynamics. Adaptive management requires monitoring to track outcomes relative to management objectives (e.g., biological monitoring to assess changes in ecological integrity) and a framework for using information gained from monitoring to improve understanding of the system (i.e., explicit objectives and hypotheses) as a basis for subsequent management decisions. Adaptive management also entails engagement of the multiple stakeholders affected by management decisions to identify values and objectives and differing beliefs about how the system works. Broad engagement of stakeholders will generally be a key element for effective large-scale ecological management.

12.4 POLITICS AND ENVIRONMENTAL LEGISLATION

12.4.1 Importance of Legislation

Legislation provides agencies the authority to manage for ecological integrity. Of course, legislation generally reflects societal values and needs of the time, and diverse and changing values can result in conflicting objectives and mandates. For example, sport fish restoration mandates often result in stocking of nonnative sport fishes that are detrimental to ecological integrity; in turn, pollution control mandates for reducing nutrient discharges to surface waters can lower productivity of managed fisheries (see section 12.1.2). Adding to the complexity of conflicting goals and authorities among government programs is the sheer number of agencies involved in managing water resources (e.g., approximately 20 federal agencies alone in the USA; Graf 2001).

Reviewing the full body of legislation relevant to ecological status of freshwater systems is beyond the scope of this chapter (see Chapter 4). However, understanding key environmental legislation can be useful in creating innovative management options. To this end, the following section provides an overview of two legislative mandates (the CWA and the Endangered Species Act) that open opportunities for ecological integrity management in inland waters of the USA. States also play important roles in implementing these mandates, and many have analogous state level legislation. Similarly, laws and policies aimed at protecting water resources and imperiled species may also provide management opportunities in Canada (e.g., Federal Water Policy and Species at Risk Act) and Mexico (e.g., National Water Law).

12.4.2 Federal and State Protective Programs and Legislation

12.4.2.1 The Clean Water Act

The CWA is the primary U.S. federal mandate for managing for ecological integrity of aquatic systems. As previously discussed (section 12.1), Section 101 of the CWA establishes as a national priority the restoration and maintenance of the "physical, chemical, and biological integrity of the Nation's waters." Application of the CWA has evolved from an emphasis on regulating point discharges of pollutants to including a focus on addressing nonpoint or diffuse sources of pollutants. Over the same time frame, regard for the physical and biological components of integrity has replaced a narrower focus on water quality.

The CWA is administered by the EPA, which delegates certain authorities to states and tribes (e.g., developing water quality standards and permitting discharges). Along with the EPA, the U.S. Army Corps of Engineers (ACE) administers Section 404 of the CWA, which requires a permit be obtained before discharging dredged or fill material (e.g., dams and fill for road crossings) into wetlands and other "jurisdictional" waters. The jurisdictional scope of federal authority to regulate water quality has been an ongoing source of controversy and legal challenges (NRC 1992; Downing et al. 2003). Generally, provisions of the CWA extend to all interstate waters and to streams, rivers, and lakes that do or could contribute to interstate commerce (e.g., be used by interstate travelers for recreation), as well as to tributaries and adjacent wetlands to these waters. Court challenges have addressed the applicability of CWA authority to isolated intrastate wetlands and to headwater, intermittent, and ephemeral streams, despite multiple ecological connections between headwaters and wetlands and the integrity of downstream waters (Nadeau and Rains 2007). Controversy will probably continue, but the CWA remains the major legislative basis for managing freshwaters for ecological integrity.

The CWA establishes responsibility for establishing water quality standards and regulations, which are largely implemented by states. Under the CWA, states have the responsibility for designating "use classifications" (e.g., drinking water, recreation, fishing or aquatic habitat, or industrial) for individual water bodies and water quality standards for each use. Three sections of the CWA are particularly relevant to management of ecological integrity.

Section 305(b) requires that each state prepare a report every 2 years describing the status of streams, lakes, and reservoirs with respect to the standards established for their designated uses. To prepare the 305(b) report, states must implement monitoring and assessment programs. Although assessments have traditionally focused on water quality, many states are incorporating biological assessments and criteria, particularly using macroinvertebrates as indicators of water quality. Increasingly, states are also using fish assemblage data, typically IBI-type measurements, to assess biological status of aquatic systems (see Box 12.1).

Section 303(d) requires that states prepare a list of impaired waters—those on the 305(b) list that fail to meet the standards for their designated uses. Once a water body is 303(d) listed, it is the state's responsibility to develop a total maximum daily load (TMDL) for the pollutant(s) that are responsible for impairment. This requires identifying the sources of impairment (could be historic or even natural, as well as point and nonpoint sources) and strategies for eliminating or managing these sources to meet the TMDL.

Section 319 establishes a program for addressing nonpoint source pollution, requiring states to examine carefully their major nonpoint sources and to develop a plan to reduce pollution inputs. The federal government provides funding that states match to support plan implementation; a portion of these funds are designated for watershed-based programs to restore waters on the state's 303(d) list.

Admittedly, state assessment programs are often resource limited and the TMDL focus on pollutants makes it hard to address the ecological reality of multiple interacting stressors (e.g., pollutants, hydrologic alteration, and habitat disturbances; Karr and Yoder 2004). However, the CWA mandates provide a basis for improving state programs by increasing the emphasis on biological monitoring and assessment and by fostering more in-depth and holistic considerations of ecological linkages and interactions in the analysis of sources of impairment for 303(d) listed waters (Karr and Yoder 2004). Section 319 programs provide an avenue for innovative collaborations with diverse stakeholders to implement management strategies, often at watershed scales (Hardy and Koontz 2008).

Mexico's National Water Law addresses use of surface water and groundwater for purposes including ecosystem protection. Individual states also issue water supply and environmental laws. In Canada, the Federal Water Policy provides a framework for supporting research, development of water quality guidelines, and public outreach related to water management issues; however, provincial governments have authority to legislate water use and pollution control. Provincial legislation, such as Ontario's Clean Water Act passed in 2006 and targeting protection of sources of drinking water, may provide avenues for managing watersheds in ways that generally benefit ecological integrity of streams and lakes (Davies and Mazumder 2003).

12.4.2.2 The Endangered Species Act

The Endangered Species Act (ESA) was passed by Congress in 1973. The ESA aims to "provide a means whereby the ecosystems on which endangered species and threatened species depend may be conserved" as well as the more familiar goal of providing a program for conserving threatened and endangered species. Mexico also has laws protecting endangered taxa and maintains an official list of species at risk (Norma Oficial Mexicana NOM-059-ECOL-2001). Canada's Species at Risk Act, passed in 2003, provides habitat protection and recovery planning in cooperation with provincial and territorial ministers for listed species on federal lands. In addition, provincial legislation, such as Ontario's Endangered Species Act (2007) provides opportunities for protection, recovery, and stewardship of freshwater habitats that are not on federal lands and that harbor listed species.

In the USA, at least 275 freshwater species, primarily fishes and mussels, are presently listed as threatened or endangered under the ESA, which is administered by the U.S. Fish and Wildlife Service (FWS; for terrestrial and freshwater species) and the National Marine Fisheries Service (NMFS; for marine and anadromous species). The ESA (Section 9 and regulations for implementing the ESA) protects listed species through a prohibition on "take" (defined as killing, harming, harassing, and so on, including habitat degradation that impairs species behavior or reproduction). Additionally, Section 7 of the ESA requires insurance that actions authorized, carried out, or funded by a U.S. government agency are not likely to jeopardize the existence of a listed species or to modify critical habitat for a listed species adversely, if critical habitat has been designated. Federal agencies comply with Section 7 by consulting with FWS or NMFS whenever they fund or authorize an activity that could affect a listed species. If consultation FWS or NMFS determines that the activity is likely to jeopardize the species or adversely affect critical habitat, then FWS or NMFS specifies "reasonable and prudent alternatives" for the proposed action. Even without finding a likelihood of jeopardy to the species, FWS or NMFS can recommend reasonable or prudent measures to reduce impact to the species and its habitat. Thus Section 9 prevents harm to listed individuals or their habitats, and Section 7 provides a mechanism for avoiding extinction of listed species as a result of federal actions.

The reasonable and prudent alternatives or measures specified for federal actions as a result of Section 7 consultation represent opportunities to modify activities in ways that lessen effects on ecological integrity of aquatic systems that harbor ESA-listed species. For example, through consultation with ACE regarding a CWA Section 404 permit to install a culvert at a new road crossing on a stream with a listed species, FWS may require that the project be modified (e.g., by using an appropriately-sized culvert embedded into the streambed). The result could minimize effects of the new culvert on habitat and passage of aquatic species generally.

A third section of the ESA, Section 10, may offer particular opportunities for innovative ecological management. Section 10 applies when an activity by an individual is lawful and does not involve a federal action (e.g., no federal permit is needed), but is likely to result in take of a listed species. In this case, Section 10 provides for issuance of an "incidental take permit" to the individual to allow the anticipated take, provided that the individual prepares and FWS approves a plan for minimizing and mitigating the anticipated take. The plan is termed a "habitat conservation plan" (HCP). Many HCPs have been prepared over the last 20 years, mostly for small-scale activities that degrade habitat locally for a listed species. However, applying habitat conservation planning at watershed scales offers the opportunity to engage in long-range, large-scale management to protect ecological integrity. For example, an HCP prepared to address anticipated take of three ESA-protected stream fishes endemic to the upper Coosa River basin, Georgia, has included requirements intended to minimize hydrologic change, erosion and stream sedimentation, riparian disturbance, and stream system fragmentation from new development (Wenger et al. 2008a, 2008b). Collaboration among county and municipal governments and planning departments, academic and agency biologists, and local environmental groups is key in developing a workable, watershed scale HCP.

Section 6 of the ESA provides for cooperation between federal and state governments to promote protection of threatened and endangered species. Many states have legislation establishing processes for listing imperiled species and programs for research and recovery activities. State wildlife and wildflower protection acts may complement ESA regulations, for example, by preventing actions on state lands that would harm state-listed species. State environmental and fish and wildlife agencies are also important partners in habitat conservation planning and receive funding under Section 6 to support research and other activities targeting recovery of ESA-listed species.

12.4.2.3 Other Legislation

There are additional opportunities to manage for ecological integrity arising from federal and state mandates other than the CWA and ESA. Prominent examples include the following.

- Hydropower licensing and relicensing of nonfederal projects by the Federal Energy Regulatory Commission and operation updates for federally-owned dams (e.g., updates of water control plans for dams owned and operated by ACE). Both situations provide opportunities for environmental stakeholders, including state and federal agencies, to advise on how dams might be operated to reduce ecological impacts, particularly of downstream river systems.

- Annual U.S. Farm Bills, which set provisions for agriculture policies including multiple programs that provide economic incentives for erosion and nutrient control, wetland conservation, water use efficiency, organic farming, wildlife habitat protection, and other conservation practices.
- State and provincial water resource development and protection programs, including policies addressing water conservation and drought management, reservoir development, and instream flow requirements.

Clearly, the ensemble of federal, state, and provincial activities creates a large and complex framework in which an innovative manager may find multiple approaches to improve and protect ecological systems. An important take-home message is that collaboration with experts in law and policy may be as important as any partnership.

12.4.3 Respective Roles of Agencies, Nongovernmental Organizations, and Grass Roots

Natural resource agencies are primary institutions for protecting the ecological integrity of freshwater systems as these agencies are charged with implementing and enforcing legislative requirements for permitting, monitoring, reporting on, and, in many respects, managing publicly held natural resources. However, resource agencies differ from one another in their core missions (e.g., programs that promote small impoundment construction that may be detrimental to stream habitat for imperiled or special-concern species); are individually charged with sometimes conflicting mandates (e.g., developing productive reservoir sport fisheries while reducing nutrient loads to surface waters); and typically have smaller budgetary resources than are needed to meet all management objectives fully. Thus, government agencies can do only so much, leaving room for others to contribute substantively to management for ecological integrity.

Nongovernmental organizations (NGOs) play a large role in creating opportunities for public involvement in resource management. A number of national and international NGOs are involved in protecting ecological integrity of freshwaters in North America (Table 12.4). The list in Table 12.4 is by no means exhaustive but illustrates the variety of NGO activities related to environmental and habitat management. These organizations engage in various activities such as

- public outreach and education (e.g., concerning water conservation and low-impact development);
- advocacy for environmentally-relevant legislation (e.g., CWA renewal);
- litigation concerning policy implementation (e.g., law suits challenging EPA administration of the CWA);
- representation of environmental interests in controversial projects (e.g., dam removal campaigns);
- synthesis and dissemination of information relevant to ecological integrity (e.g., conservation status of freshwater systems); and
- development of partnerships to promote innovative, ecologically-based management of freshwaters (e.g., The Nature Conservancy's Sustainable Rivers Program; Richter et al. 2006).

Table 12.4. Examples of nongovernmental organizations with large influences on managing freshwater systems for ecological integrity. This list is not exhaustive but illustrates a variety of common activities. Abbreviations are EPA (U.S. Environmental Protection Agency), CWA (U.S. Clean Water Act), TMDL (total maximum daily load), and ACE (U.S. Army Corps of Engineers).

Organization	Scope and activities	Example relevant to freshwater ecological integrity
American Rivers	National organization with offices across the USA; advocates for legislation and policies to restore and protect free-flowing rivers, including improving and sustaining water quality, encouraging water conservation, and restoring natural flow regimes	"America's Most Endangered Rivers," an annual report, highlights rivers facing near-term policy and management decisions that could affect future ecological integrity
The Sierra Club	National organization with chapters across the USA; advocates and litigates for environmental protection	Sierra Club chapters and partners have filed lawsuits in multiple states challenging the EPA's administration of the CWA and state agencies' failures to establish TMDLs for rivers that do not meet water quality standards
The Nature Conservancy	International organization with offices across the USA and in over 30 other countries; broadly involved in nature conservation through land acquisition and management, scientific research, outreach and education, policy development, and collaborations with governmental agencies and others	The "Sustainable Rivers Project" is a partnership with ACE to explore innovative management options that would restore habitat and ecological functions downstream from ACE dams while also meeting other societal benefits from the dams
Waterkeeper Alliance	International organization; supports local Waterkeeper programs that advocate and build community support for watershed and water quality protection in specific aquatic systems	"Riverkeeper" organizations across the USA, Canada, and other countries engage in legal action, community outreach and education, monitoring, and political advocacy to promote policies that protect water quality, instream flows, and habitat in specific river systems
World Wildlife Fund	International organization engaged in wildlife conservation in 100 countries through research, natural area protection, and promotion of sustainability through community initiatives and corporate partnerships	The "Southeast Rivers and Streams" program has identified priority watersheds for protecting biodiversity in three major southeastern U.S. river basins, advocates for environmental protection, and collaborates with corporations and other foundations to fund local groups working to conserve aquatic species and habitats

Grass-root environmental organizations include local watershed coalitions, land trusts, and similar community-based groups organized around shared goals of protecting or restoring elements of nature that people value. There are thousands of such groups across North America, with members participating in activities ranging from removing trash from streams to adopt-a-stream-type monitoring programs to advocating for local ordinances and state legislation protective of streams and water quality. Angler organizations (e.g., fishing clubs) can also be important allies in advocating for water quality and habitat protection programs locally. The scope and reach of these locally-organized citizens groups can be vast, as when organizations coordinate their efforts to lobby for statewide legislation, but also in effecting change (e.g., in building codes to promote low-impact development) at the most local levels of government.

Professional scientific organizations may also include in their mission efforts to comment on, or provide resolutions and policy statements germane to protecting ecological integrity or improving ecological condition of surface waters. Such organizations include the American Fisheries Society (see Table 12.1), North American Benthological Society, American Society of Limnology and Oceanography, North American Lake Management Society, Society of Wetland Scientists, and Ecological Society of America. The Web sites of these organizations provide examples of past and current actions.

Meyer (1997) stressed the necessity of studying stream ecology in the context of the human attitudes and institutions that form a stream's "societal watershed," especially if the intent of the research is to contribute to improving management and ecological condition. Politics, legislation, and individuals from regulators and biologists to local landowners and business owners are part of an aquatic system's societal watershed, and managing for ecological integrity clearly requires skillful interaction with these societal drivers. As Meyer concluded, "A pristine stream in a politically unstable setting or with no supporters is not a healthy stream because it is not sustainable."

12.5 CONCLUSIONS

Ecological integrity has been a mainstream component of environmental science and policy since the 1970s, and it is an important aspect of modern fisheries management. It may be beyond the scope of routine fisheries management to conduct thorough ecological integrity assessments, but the management actions of fisheries biologists can result in profound effects on the integrity of aquatic ecosystems. The involvement of fisheries biologists in planning for and evaluating ecological integrity may range from examination of a subset of habitat, community, or integrity indices or metrics to regional ecological integrity assessments. Most importantly, integrity effects should be included in the management planning process.

The effects of proposed management actions on ecological integrity are equally valid considerations as are those outcomes related to sport fish populations or constituent desires. Incorporating ecological integrity concerns into the management process compels a broad, integrated, and holistic view. Integrating management approaches, such as adaptive management and structured decision making, is leading to improvement in the management process, results, and public support. The role of fisheries managers in educating constituents and the public on sustainable resource use and ecological integrity goals is paramount to effecting long-term change. The inclusion of a chapter on ecological integrity in a fisheries management textbook is a clear indication that the topic is relevant to the manager, resources, and society, and we are confident that the importance of this topic will continue to grow as managers seek to include all physical, biotic, and human components of fisheries into their goals.

12.6 REFERENCES

Allan, J. D. 2004. Landscapes and riverscapes: the influence of land use on stream ecosystems. Annual Reviews of Ecology, Evolution and Systematics 35:257–284.

Allan, J. D., and M. M. Castillo. 2007. Stream ecology: structure and function of running waters, second edition. Springer, Dordrecht, Netherlands.

Angermeier, P. L., R. A. Smogor, and J. R. Stauffer. 2000. Regional frameworks and candidate metrics for assessing biotic integrity in mid-Atlantic highland streams. Transactions of the American Fisheries Society 129:962–981.

Bain, M. B., and N. J. Stevenson, editors. 1999. Aquatic habitat assessment: common methods. American Fisheries Society, Bethesda, Maryland.

Barbour, M. T., J. Gerritsen, B. D. Snyder, and J. B. Stribling. 1999. Rapid bioassessment protocols for use in streams and wadeable rivers: periphyton, benthic macroinvertebrates, and fish, second edition. U.S. Environmental Protection Agency, Office of Water, EPA 841-B-99–002, Washington, D.C.

Benke, A. C. 1990. A perspective on America's vanishing streams. Journal of the North American Benthological Society 9:77–88.

Bernhardt, E. S., M. A. Palmer, J. D. Allan, G. Alexander, K. Barnas, S. Brooks, J. Carr, S. Clayton, C. Dahm, J. Follstad-Shah, D. Galat, S. Gloss, P. Goodwin, D. Hart, B. Hassett, R. Jenkinson, S. Katz, G. M. Kondolf, P. S. Lake, R. Lave, J. L. Meyer, T. K. O'Donnell, L. Pagano, B. Powell, and E. Sudduth. 2005. Synthesizing U.S. river restoration efforts. Science 308:636–637.

Bernhardt, E. S., E. B. Sudduth, M. A. Palmer, J. D. Allan, J. L. Meyer, G. Alexander, J. Follstad-Shah, B. Hassett, R. Jenkinson, R. Lave, J. Rumps, and L. Pagano. 2007. Restoring rivers one reach at a time: results from a survey of U.S. river restoration practitioners. Restoration Ecology 15:482–493.

Binns, N. A., and F. M. Eiserman. 1979. Quantification of fluvial trout habitat in Wyoming. Transactions of the American Fisheries Society 108:215–228.

Blackburn, T. M., J. H. Lawton, and J. N. Perry. 1992. A method for estimating the slope of upper bounds of plots of body size and abundance in natural animal assemblages. Oikos 65:107–112.

Burkhead, N. M., S. J. Walsh, B. J. Freeman, and J. D. Williams. 1997. Status and restoration of the Etowah River, an imperiled southern Appalachian ecosystem. Pages 375–441 in G. W. Benz and D. E. Collins, editors. Aquatic fauna in peril: the southeastern perspective. Lenz Design and Communications, Decatur, Georgia.

Carpenter, S. R., N. F. Caraco, D. L. Correll, R. W. Howarth, A. N. Sharpley, and V. H. Smith. 1998. Nonpoint pollution of surface waters with phosphorus and nitrogen. Ecological Applications 8:559–568.

CEQ (Council on Environmental Quality). 2005. Regulations for implementing the procedural provisions of the National Environmental Policy Act. Council on Environmental Quality, Executive Office of the President, Washington, D.C.

Chandler, J. R. 1970. A biological approach to water quality management. Water Pollution Control 69:415–421.

Chessman, B. C. 2006. Prediction of riverine fish assemblages through the concept of environmental filters. Marine and Freshwater Research 57:601–609.

Chessman, B. C., and M. J. Royal. 2004. Bioassessment without reference sites: use of environmental filters to predict natural assemblages of river macroinvertebrates. Journal of the North American Benthological Society 23:599–615.

Clemen, R. T. 1996. Making hard decisions, second edition. Duxbury Press, Pacific Grove, California.

Clements, F. E. 1916. Plant succession. Carnegie Institution of Washington Publication 242, Washington, D.C.

Collier, M., R. H. Webb, and J. C. Schmidt. 1996. Dams and rivers: a primer on the downstream effects of dams. U.S. Geological Survey Circular 1126, Denver, Colorado.

Contreras-B., S., and M. A. Escalante-C. 1984. Distribution and known impacts of exotic fishes in Mexico. Pages 102–130 in W. R. Courtenay Jr. and J. R. Stauffer Jr., editors. Distribution, biology, and management of exotic fishes. Johns Hopkins University Press, Baltimore, Maryland.

Coombs, J. A., L. Wilson, B. Tracy, and V. Harrison. 2004. Pigeon River revival. Wildlife in North Carolina 68(12):26–29.

Crossman, E. J. 1984. Introduction of exotic fishes into Canada. Pages 78–101 in W. R. Courtenay Jr. and J. R. Stauffer Jr., editors. Distribution, biology, and management of exotic fishes. Johns Hopkins University Press, Baltimore, Maryland.

Davies, J. M., and A. Mazumder. 2003. Health and environmental policy issues in Canada: the role of watershed management in sustaining clean drinking water quality at surface sources. Journal of Environmental Management 68:273–286.

Davis, W. S. 1995. Biological assessment and criteria: building on the past. Pages 15–29 in W. S. Davis and T. P. Simon, editors. Biological assessment and criteria: tools for water resource planning and decision making. Lewis Publishers, Boca Raton, Florida.

Downing, D. M., C. Winer, and L. D. Wood. 2003. Navigating through Clean Water Act jurisdiction: a legal review. Wetlands 23:475–493.

Fausch, K. D., J. R. Karr, and P. R. Yant. 1984. Regional application of an index of biotic integrity based on stream fish communities. Transactions of the American Fisheries Society 113:39–55.

Fausch, K. D., J. Lyons, J. R. Karr, and P. L. Angermeier. 1990. Fish communities as indicators of environmental degradation. Pages 123–144 in S. M. Adams, editor. Biological indicators of stress in fish. American Fisheries Society, Symposium 8, Bethesda, Maryland.

Fausch, K. D., C. E. Torgersen, C. V. Baxter, and H. W. Li. 2002. Landscapes to riverscapes: bridging the gap between research and conservation of stream fishes. BioScience 52:483–498.

Fisher, W. L., and F. J. Rahel, editors. 2004. Geographic information systems in fisheries. American Fisheries Society, Bethesda, Maryland.

Fiumera, A. C., B. A. Porter, G. Looney, M. A. Asmussen, and J. C. Avise. 2004. Maximizing offspring production while maintaining genetic diversity in supplemental breeding programs of highly fecund managed species. Conservation Biology 18:94–101.

Florsheim, J. L., J. F. Mount, and A. Chin. 2008. Bank erosion as a desirable attribute of rivers. Bioscience 58:519–529.

Freeman, M. C., C. M. Pringle, E. A. Greathouse, and B. J. Freeman. 2003. Ecosystem-level consequences of migratory faunal depletion caused by dams. Pages 255–266 in K. E. Limburg and J. R. Waldman, editors. Biodiversity, status and conservation of the world's shads. American Fisheries Society, Symposium 35, Bethesda, Maryland.

Fuller, P. L., L. G. Nico, and J. D. Williams. 1999. Nonindigenous fishes introduced into inland waters of the United States. American Fisheries Society, Special Publication 27, Bethesda, Maryland.

Graf, W. L. 2001. Damage control: restoring the physical integrity of America's rivers. Annals of the Association of American Geographers 91:1–27.

Grossman, G. D., R. E. Ratajczak, M. Crawford, and M. C. Freeman. 1998. Assemblage organization in stream fishes: effects of environmental variation and interspecific interactions. Ecological Monographs 68:395–420.

Halliwell, D. B., R. W. Langdon, R. A. Daniels, J. P. Kurtenbach, and R. A. Jacobson. 1999. Classification of freshwater fish species of the northeastern United States for use in the development of indices of biological integrity, with regional applications. Pages 301–337 in T. P. Simon, editor. Assessing the sustainability and biological integrity of water resources using fish communities. CRC Press, Boca Raton, Florida.

Harding, J. S., E. F. Benfield, P. V. Bolstad, G. S. Helfman, and E. B. D. Jones III. 1998. Stream biodiversity: the ghost of land use past. Proceedings of the National Academy of Sciences 95:14843–14847.

Hardy, S. D., and T. M. Koontz. 2008. Reducing nonpoint source pollution through collaboration: policies and programs across the U.S. states. Environmental Management 41:301–310.

Harig, A. L., and K. D. Fausch. 2002. Minimum habitat requirements for establishing translocated cutthroat trout populations. Ecological Applications 12:535–555.

Hart, D. D., T. E. Johnson, K. L. Bushaw-Newton, R. J. Horwitz, A. T. Bednarek, D. F. Charles, D. A. Kreeger, and D. J. Velinsky. 2002. Dam removal: challenges and opportunities for ecological research and river restoration. Bioscience 52:669–681.

Hawkins, C. P. 2006. Quantifying biological integrity by taxonomic completeness: its utility in regional and global assessments. Ecological Applications 16:1277–1294.

Helfman, G. S. 2007. Fish conservation: a guide to understanding and restoring global aquatic biodiversity and fishery resources. Island Press, Washington, D.C.

Heltsley, R. M., W. G. Cope, D. Shea, R. B. Bringolf, T. J. Kwak, and E. G. Malindzak. 2005. Assessing organic contaminants in fish: comparison of a nonlethal tissue sampling technique to mobile and stationary passive sampling devices. Environmental Science and Technology 39:7601–7608.

Hilsenhoff, W. L. 1987. An improved index of organic stream pollution. The Great Lakes Entomologist 20:31–39.

Horwitz, R. J. 1978. Temporal variability patterns and the distributional patterns of stream fishes. Ecological Monographs 48:307–321.

Hughes, R. M. 1995. Defining acceptable biological status by comparing with reference conditions. Pages 31–47 *in* W. S. Davis and T. P. Simon, editors. Biological assessment and criteria: tools for water resource planning and decision making. Lewis Publishers, Boca Raton, Florida.

Hughes, R. M., P. R. Kaufmann, A. T. Herlihy, T. M. Kincaid, L. Reynolds, and D. P. Larsen. 1998. A process for developing and evaluating indices of fish assemblage integrity. Canadian Journal of Fisheries and Aquatic Sciences 55:1618–1631.

Hynes, H. B. N. 1970. The ecology of running waters. Liverpool University Press, Liverpool, UK.

Hynes, H. B. N. 1975. The stream and its valley. Verhandlungen der Internationalen Vereinigung fur Theoretische und Angewandte Limnologie 19:1–15.

Jackson, C. R., J. K. Martin, D. S. Leigh, and L. T. West. 2005. A southeastern piedmont watershed sediment budget: evidence for a multi-millennial agricultural legacy. Journal of Soil and Water Conservation 60:298–310.

Jackson, D. A., and K. M. Somers. 1991. The spectre of 'spurious' correlations. Oecologia 86:147–151.

Jelks, H. L., S. J. Walsh, N. M. Burkhead, S. Contreras-Balderas, E. Díaz-Pardo, D. A. Hendrickson, J. Lyons, N. E. Mandrak, F. McCormick, J. S. Nelson, S. P. Platania, B. A. Porter, C. B. Renaud, J. Jacobo Schmitter-Soto, E. B. Taylor, and M. L. Warren Jr. 2008. Conservation status of imperiled North American freshwater and diadromous fishes. Fisheries 33(8):372–407.

Jeschke, J. M., and D. L. Strayer. 2005. Invasion success of vertebrates in Europe and North America. Proceedings of the National Academy of Sciences 102:7198–7202.

Johnson, D. E. 1998. Applied multivariate methods for data analysts. Duxbury Press, Pacific Grove, California.

Johnson, J. E., and B. L. Jensen. 1991. Hatcheries for endangered freshwater fishes. Pages 199–217 *in* W. L. Minckley and J. E. Deacon, editors. Battle against extinction. University of Arizona Press, Tucson.

Johnson, P. T. J., J. D. Olden, and M. J. Vander Zanden. 2008. Dam invaders: impoundments facilitate biological invasions into freshwaters. Frontiers in Ecology and the Environment 6:357–363.

Jorgensen, S. E., F.-L. Xu, and J. C. Marques. 2005. Application of indicators for the assessment of ecosystem health. Pages 5–66 *in* S. E. Jorgensen, R. Costanza, and F.-L. Xu, editors. Handbook of ecological indicators for assessment of ecosystem health. CRC Press, Boca Raton, Florida.

Junk, W. L., P. B. Bayley, and R. E. Sparks. 1989. The flood pulse concept in river–floodplain systems. Pages 110–127 *in* D. P. Dodge, editor. Proceedings of the international large river symposium. Canadian Special Publication of Fisheries and Aquatic Sciences 106, Department of Fisheries and Oceans, Ottawa.

Karr, J. R., J. D. Allan, and A. C. Benke. 2000. River conservation in the United States and Canada. Pages 3–39 *in* P. J. Boon, B. R. Davies, and G. E. Petts, editors. Global perspectives on river conservation: science, policy, and practice. John Wiley and Sons, Chichester, UK.

Karr, J. R., and E. W. Chu. 1999. Restoring life in running waters. Island Press, Washington, D.C.

Karr, J. R., and D. R. Dudley. 1981. Ecological perspective on water quality goals. Environmental Management 5:55–68.

Karr, J. R., K. D. Fausch, P. L. Angermeier, P. R. Yant, and I. J. Schlosser. 1986. Assessing biological integrity in running waters: a method and its rationale. Illinois Natural History Survey Special Publication 5, Champaign.

Karr, J. R., and C. O. Yoder. 2004. Biological assessment and criteria improve total maximum daily load decision making. Journal of Environmental Engineering 130:594–604.

Kolkwitz, R., and M. Marsson. 1908. Ökologie der pflanzlichen Saprobien. [Ecology of plant saprobia.] Berichte der Deutschen Botanischen Gesellsschaft 26a:505–519. (In German.) English translation 1967. Pages 47–52 *in* L. E. Keup, W. M. Ingram, and K. M. Mackenthum, editors. Biology of water pollution. U.S. Department of Interior, Federal Water Pollution Control Administration, Washington, D.C.

Kondolf, G. M. 2000. Assessing salmonid spawning gravel quality. Transactions of the American Fisheries Society 129:262–281.

Kondolf, G. M. 2006. River restoration and meanders. Ecology and Society 11:42. Available: www.ecologyandsociety.org/vol11/iss2/art42/. (December 2008).

Krebs, C. J. 1998. Ecological methodology, second edition. Benjamin/Cummings, Menlo Park, California.

Kwak, T. J., and J. T. Peterson. 2007. Community indices, parameters, and comparisons. Pages 677–763 *in* C. S. Guy and M. L. Brown, editors. Analysis and interpretation of freshwater fisheries data. American Fisheries Society, Bethesda, Maryland.

Leigh, D. S. 1997. Mercury-tainted overbank sediment from past gold mining in north Georgia, USA. Environmental Geology 30:244–251.

Lodge, D. M., C. A. Taylor, D. M. Holdich, and J. Skurdal. 2000. Nonindigenous crayfishes threaten North American freshwater biodiversity: lessons from Europe. Fisheries 25(8):7–20.

Lyons, J. 1992. Using the index of biotic integrity (IBI) to measure environmental quality in warmwater streams of Wisconsin. U.S. Forest Service, North Central Forest Experiment Station, General Technical Report No. 149, St. Paul, Minnesota.

Lyons, J. 2006. A fish-based index of biotic integrity to assess intermittent headwater streams in Wisconsin, USA. Environmental Monitoring and Assessment 122:239–258.

Lyons, J., R. R. Piette, and K. W. Niermeyer. 2001. Development, validation, and application of a fish-based index of biotic integrity for Wisconsin's large warmwater rivers. Transactions of the American Fisheries Society 130:1077–1094.

Lyons, J., L. Wang, and T. D. Simonson. 1996. Development and validation of an index of biotic integrity for coldwater streams in Wisconsin. North American Journal of Fisheries Management 16:241–256.

Maret, T. R., and D. E. MacCoy. 2002. Fish assemblages and environmental variables associated with hard-rock mining in the Coeur d'Alene River basin, Idaho. Transactions of the American Fisheries Society 131:865–884.

Marsh, G. P. 1867. Man and nature; or, physical geography as modified by human action. Charles Scribner, New York.

Marsh, P. C., B. R. Kesner, and C. A. Pacey. 2005. Repatriation as a management strategy to conserve a critically imperiled fish species. North American Journal of Fisheries Management 25:547–556.

Master, L. L., S. R. Flack, and B. A. Stein. 1998. Rivers of life: critical watersheds for protecting freshwater biodiversity. The Nature Conservancy, Arlington, Virginia.

Matthews, W. J. 1998. Patterns in freshwater fish ecology. Chapman and Hall, New York.

Mattson, K. M., and P. L. Angermeier. 2007. Integrating human impacts and ecological integrity into a risk-based protocol for conservation planning. Environmental Management 39:125–138.

McMahon, T. E., A. V. Zale, and D. J. Orth. 1996. Aquatic habitat measurements. Pages 83–120 in B. R. Murphy and D. W. Willis, editors. Fisheries techniques, second edition. American Fisheries Society, Bethesda, Maryland.

Meyer, J. L. 1997. Stream health: incorporating the human dimension to advance stream ecology. Journal of the North American Benthological Society 16:439–447.

Miller, D. L., R. M. Hughes, J. R. Karr, P. M. Leonard, P. B. Moyle, L. H. Schrader, B. A. Thompson, R. A. Daniels, K. D. Fausch, G. A. Fitzhugh, J. R. Gammon, D. B. Halliwell, P. L. Angermeier, and D. J. Orth. 1988. Regional applications of an index of biotic integrity for use in water resource management. Fisheries 13(5):12–20.

Minckley, W. L. 1995. Translocation as a tool for conserving imperiled fishes: experience in western United States. Biological Conservation 72:297–309.

Minckley, W. L., P. C. Marsh, J. E. Deacon, T. E. Dowling, P. W. Hedrick, W. J. Matthews, and G. Mueller. 2003. A conservation plan for native fishes of the lower Colorado River. Bioscience 53:219–234.

Moyle, P. B., and M. P. Marchetti. 2006. Predicting invasion success: freshwater fishes in California as a model. BioScience 56:515–524.

Moyle, P. B., and T. Light. 1996. Biological invasions of freshwater: empirical rules and assembly theory. Biological Conservation 78:149–161.

Moyle, P. B., and P. J. Randall. 1998. Evaluating the biotic integrity of watersheds in the Sierra Nevada, California. Conservation Biology 12:1318–1326.

Nadeau, T.-L., and M. C. Rains. 2007. Hydrological connectivity of headwaters to downstream waters: introduction to the featured collection. Journal of the American Water Resources Association 43:1–4.

Norris, R. H., and A. Georges. 1993. Analysis and interpretation of benthic macroinvertebrate surveys. Pages 234–286 in D. M. Rosenberg and V. H. Resh, editors. Freshwater biomonitoring and benthic macroinvertebrates. Chapman and Hall, New York.

NRC (National Research Council). 1992. Restoration of aquatic ecosystems: science, technology, and public policy. National Academy Press, Washington, D.C.

O'Connor, J. S., and R. T. Dewling. 1986. Indices of marine degradation: their utility. Environmental Management 10:335–343.

Omernik, J. M. 1995. Ecoregions: a spatial framework for environmental management. Pages 49–62 in W. S. Davis and T. P. Simon, editors. Biological assessment and criteria: tools for water resource planning and decision making. Lewis Publishers, Boca Raton, Florida.

Omernik, J. M., and R. G. Bailey. 1997. Distinguishing between watersheds and ecoregions. Journal of the American Water Resources Association 33:935–949.

Palmer, M. A., E. S. Bernhardt, J. D. Allan, P. S. Lake, G. Alexander, S. Brooks, J. Carr, S. Clayton, C. N. Dahm, J. Follstad-Shah, D. L. Galat, S. G. Loss, P. Goodwin, D. D. Hart, B. Hassett, R. Jenkinson, G. M. Kondolf, R. Lave, J. L. Meyer, T. K. O'Donnell, L. Pagano, and E. Sudduth. 2005. Standards for ecologically successful river restoration. Journal of Applied Ecology 42:208–217.

Palmer, M. A., C. A. R. Liermann, C. Nilsson, M. Florke, J. Alcamo, P. S. Lake, and N. Bond. 2008. Climate change and the world's river basins: anticipating management options. Frontiers in Ecology and the Environment 6:81–89.

Paulsen, S. G., A. Mayio, D. V. Peck, J. L. Stoddard, E. Tarquinio, S. M. Holdsworth, J. Van Sickle, L. L. Yuan, C. P. Hawkins, A. T. Herlihy, P. R. Kaufmann, M. T. Barbour, D. P. Larsen, and A. R. Olsen. 2008. Condition of stream ecosystems in the U.S.: an overview of the first national assessment. Journal of the North American Benthological Society 27:812–821.

Peterson, J. T., and J. W. Evans. 2003. Quantitative decision analysis for sport fisheries management. Fisheries 28(1):10–21.

Peterson, J. T., and C. F. Rabeni. 1996. Natural thermal refugia for temperate warmwater stream fishes. North American Journal of Fisheries Management 16:738–746.

Pimentel, D., R. Zuniga, and D. Morrison. 2005. Update on the environmental and economic costs associated with alien-invasive species in the United States. Ecological Economics 52:273–288.

Poff, N. L., J. D. Allan, M. B. Bain, J. R. Karr, K. L. Prestegaard, B. D. Richter, R. E. Sparks, and J. C. Stromberg. 1997. The natural flow regime: a paradigm for river conservation and restoration. BioScience 47:769–784.

Pringle, C. M., M. C. Freeman, and B. J. Freeman. 2000. Regional effects of hydrologic alterations on riverine macrobiota in the New World: tropical–temperate comparisons. BioScience 50:807–823.

Radwell A. J., and T. J. Kwak. 2005. Assessing ecological integrity of Ozark rivers to determine suitability for protective status. Environmental Management 35:799–810.

Rapport, D. J., R. Costanza, and A. J. McMichael. 1998. Assessing ecosystem health. Trends in Ecology and Evolution 13:397–402.

Reiman, B., J. T. Peterson, J. Clayton, P. Howell, R. Thurow, W. Thompson, and D. Lee. 2001. Evaluation of potential effects of federal land management alternatives on trends of salmonids and their habitats in the interior Columbia River basin. Forest Ecology and Management 153:43–62.

Resh, V. H., and J. K. Jackson. 1993. Rapid assessment approaches to biomonitoring using benthic macroinvertebrates. Pages 195–233 *in* D. M. Rosenberg and V. H. Resh, editors. Freshwater biomonitoring and benthic macroinvertebrates. Chapman and Hall, New York.

Richter, B. D., A. T. Warner, J. L. Meyer, and K. Lutz. 2006. A collaborative and adaptive process for developing environmental flow recommendations. River Research and Applications 22:297–318.

Rosenberg, D. M., V. H. Resh, and R. S. King. 2008. Use of aquatic insects in biomonitoring. Pages 123–137 *in* R. W. Merritt, K. W. Cummins, and M. B. Berg, editors. An introduction to the aquatic insects of North America, fourth edition. Kendall/Hunt, Dubuque, Iowa.

Ruesink, J. L. 2005. Global analysis of factors affecting the outcome of freshwater fish introductions. Conservation Biology 19:1883–1893.

Scheffer, M., S. Carpenter, J. A. Foley, C. Folke, and B. Walker. 2001. Catastrophic shifts in ecosystems. Nature 413:591–596.

Shannon, C. E., and W. Weaver. 1949. The mathematical theory of communication. University of Illinois Press, Urbana.

Shea, D. 2004. Transport and fate of toxicants in the environment. Pages 479–499 *in* E. Hodgson, editor. A textbook of modern toxicology, third edition. John Wiley and Sons, Hoboken, New Jersey.

Shute, J. R., P. L. Rakes, and P. W. Shute. 2005. Reintroduction of four imperiled fishes in Abrams Creek, Tennessee. Southeastern Naturalist 4:93–110.

Simon, T. P. 1999a. Introduction: biological integrity and use of ecological health concepts for application to water resource characterization. Pages 3–16 *in* T. P. Simon, editor. Assessing the sustainability and biological integrity of water resources using fish communities. CRC Press, Boca Raton, Florida.

Simon, T. P., editor. 1999b. Assessing the sustainability and biological integrity of water resources using fish communities. CRC Press, Boca Raton, Florida.

Simon, T. P. 1999c. Assessment of Balon's reproductive guilds with application to midwestern North American freshwater fishes. Pages 97–121 *in* T. P. Simon, editor. Assessing the sustainability and biological integrity of water resources using fish communities. CRC Press, Boca Raton, Florida.

Smock, L. A. 1996. Macroinvertebrate movements: drift, colonization, and emergence. Pages 371–390 *in* F. R. Hauer and G. A. Lamberti, editors. Methods in stream ecology. Academic Press, San Diego, California.

Stanley, E. H., and M. W. Doyle. 2003. Trading off: the ecological effects of dam removal. Frontiers in Ecology and the Environment 1:15–30.

Sudduth, E. B., J. L. Meyer, and E. S. Bernhardt. 2007. Stream restoration practices in the southeastern United States. Restoration Ecology 15:573–583.

Suter, G. W., II. 1993. A critique of ecosystem health concepts and indexes. Environmental Toxicology and Chemistry 12:1533–1539.

Taylor, C. A., G. A. Schuster, J. E. Cooper, R. J. DiStefano, A. G. Eversole, P. Hamr, H. H. Hobbs III, H. W. Robison, C. E. Skelton, and R. F. Thoma. 2007. A reassessment of the conservation status of crayfishes of the United States and Canada after 10+ years of increased awareness. Fisheries 32(8):372–389.

Underwood, A. J. 1994. On beyond BACI: sampling designs that might reliably detect environmental disturbances. Ecological Applications 4:3–15.

USACE (U.S. Army Corps of Engineers). 2005. National inventory of dams methodology: state and federal manual, version 3.0. U.S. Army Corps of Engineers, Civil Works Engineering Division, Washington, D.C.; Association of State Dam Safety Officials, Lexington, Kentucky; and U.S. Army Topographic Engineering Center, Alexandria, Virginia.

USEPA (U.S. Environmental Protection Agency). 1990. Biological criteria: national program guidance for surface waters. U.S. Environmental Protection Agency, Office of Water Regulations and Standards, EPA 440–5-90–04, Washington, D.C.

USEPA (U.S. Environmental Protection Agency). 1994. Water quality standards handbook, second edition. U.S. Environmental Protection Agency, Water Quality Standards Branch, Office of Science and Technology, EPA 823-B-94–005, Washington, D.C.

Vannote, R. L., G. W. Minshall, K. W. Cummins, J. R. Sedell, and C. E. Cushing. 1980. The river continuum concept. Canadian Journal of Fisheries and Aquatic Sciences 37:130–137.

Vitousek, P. M., H. A. Mooney, J. Lubchenco, and J. M. Melillo. 1997. Human domination of earth's ecosystems. Science 277:494–499.

Walter, R. C., and D. J. Merritts. 2008. Natural streams and the legacy of water-powered mills. Science 319:299–304.

Wang, L., T. Brenden, P. Seelbach, A. Cooper, D. Allan, R. Clark Jr., and M. Wiley. 2008. Landscape based identification of human disturbance gradients and reference conditions for Michigan streams. Environmental Monitoring and Assessment 141:1–17.

Ward, J. V. 1989. The four-dimensional nature of lotic ecosystems. Journal of the North American Benthological Society 8:2–8.

Washington, H. G. 1984. Diversity, biotic and similarity indices: a review with special relevance to aquatic ecosystems. Water Research (Great Britain) 18:653–694.

Waters, T. F. 1995. Sediment in streams: sources, biological effects, and control. American Fisheries Society, Monograph 7, Bethesda, Maryland.

Webb, M. A. H., J. E. Williams, and L. R. Hildebrand. 2005. Recovery program review for endangered pallid sturgeon in the upper Missouri River basin. Reviews in Fisheries Science 13:165–176.

Weigel, B. M. 2003. Development of stream macroinvertebrate models that predict watershed and local stressors in Wisconsin. Journal of the North American Benthological Society 22:123–142.

Wenger, S. J., T. L. Carter, L. A. Fowler, and R. A. Vick. 2008b. Runoff limits: an ecologically-based stormwater management program. Stormwater 9:45–58.

Wenger, S. J., J. T. Peterson, M. C. Freeman, B. J. Freeman, and D. D. Homans. 2008a. Stream fish occurrence in response to impervious cover, historic land use, and hydrogeomorphic factors. Canadian Journal of Fisheries and Aquatic Sciences 65:1250–1264.

Wheeler, A. P., P. L. Angermeier, and A. E. Rosenberger. 2005. Impacts of new highways and subsequent landscape urbanization on stream habitat and biota. Reviews in Fisheries Science 13:141–164.

Wilcove, D. S., D. Rothstein, J. Dubow, A. Phillips, and E. Losos. 1998. Quantifying threats to imperiled species in the United States: assessing the relative importance of habitat destruction, alien species, pollution, over exploitation, and disease. BioScience 48:607–615.

Wiley, M. J., L. L. Osborne, and R. W. Larimore. 1990. Longitudinal structure of an agricultural prairie river system and its relationship to current stream ecosystem theory. Canadian Journal of Fisheries and Aquatic Sciences 47:373–384.

Williams, B. K., R. C. Szaro, and C. D. Shapiro. 2007. Adaptive management: the U.S. Department of the Interior technical guide. U.S. Department of the Interior, Adaptive Management Working Group, Washington, D.C.

Williams J. D., M. L. Warren Jr., K. S. Cummings, J. L. Harris, and R. J. Neves, 1993. Conservation status of freshwater mussels of the United States and Canada. Fisheries 18(9):6–22.

Williamson, M. 1996. Biological invasions. Chapman and Hall, New York.

Wohl, E., and D. J. Merritts. 2007. What is a natural river? Geography Compass 1(4):871–900.

Yuan, L. L. 2006. Theoretical predictions of observed to expected ratios in RIVPACS-type predictive model assessments of stream biological condition. Journal of the North American Benthological Society 25:841–850.

Chapter 13

Ecology and Management of Lake Food Webs

STEVEN R. CHIPPS AND BRIAN D. S. GRAEB

13.1 INTRODUCTION

Knowledge of factors that affect lake productivity, species interactions, and energy flow in aquatic food webs has increased appreciably in the last half century. Contributions from the fields of fisheries science and limnology, along with technical and analytical advancements, have provided fisheries managers with new insights and tools to evaluate fish populations. Resource managers now recognize that stocking or removing certain fishes (i.e., biomanipulation) can have an important influence on food web structure and resulting water quality. Similarly, energy-based analytical techniques, such as stable isotope analysis, are increasingly used to supplement traditional diet studies.

In many ways, movement toward food web-based research and management strategies represents a shift in the traditional paradigms of fisheries biologists and limnologists. Not long ago, fisheries management and limnology were guided by different paradigms (Rigler 1982). Fisheries managers, because of their emphasis on stock–recruitment relationships, adult survival, and fishing pressure, often ignored factors such as lake productivity or species interactions. In many cases, fishing mortality was the driving variable of their predictive models. Limnologists, on the other hand, believed that the appropriate scale to study was the whole lake or drainage basin. In their view, useful theories for predicting fish production must contain physical, chemical, and biological components of the system. Unlike the quantitative models used by fisheries managers, the limnological view was more conceptual because quantifying food web interactions required intensive, whole-lake studies (Rigler 1982). Similar studies were generally lacking in fisheries science.

Fortunately, we have come a long way in the last 50 years. Limnologists and fisheries biologists recognize that they share a common goal to understand the structure and function of aquatic systems. Indeed, fisheries management challenges are often linked to water quality issues and (or) poor growth and survival of fishes. Thus, understanding food web interactions can help direct management decisions about water quality improvement, fish stocking, macrophyte management, species introductions, or other food web components. In this chapter, factors that influence the structure and function of aquatic food webs are discussed and the implications of food web interactions for fisheries management are explored.

13.2 NUTRIENTS AND PRODUCTIVITY

Nutrient availability in lakes has long been an important issue facing fisheries managers. It is estimated that environmental problems associated with nutrient enrichment (eutrophication) cost US$2.2 billion per year in the USA due to reduced property values, diminished recreation, water treatment expenses, and lost biodiversity (Dodds et al. 2009). Similarly, low nutrient availability, or oligotrophication, limits fish production and can require expensive management actions such as whole-lake fertilization.

In the early 1970s, there was much debate about which nutrients "limited" productivity in aquatic environments. Although much of the early research focused on carbon (C) as a limiting nutrient, biologists now recognize that phosphorus (P) and nitrogen (N) are the primary nutrients that limit productivity and contribute to eutrophication of lakes and streams (Schindler 1978). Since the later 1970s, much effort has gone into studying the sources and dynamics of these nutrients in aquatic environments.

13.2.1 Phosphorus

Phosphorus has received considerable attention as a limiting nutrient in aquatic ecosystems. Because P occurs in a variety of forms, it is important to recognize what constitutes the available P "pool" and how different forms relate to primary productivity. Soluble reactive P (SRP) is measured from filtered water samples and is considered to be the most biologically-available form of P. However, accurate measures of SRP can be difficult to obtain because of low concentrations and variability across space and time. Total P (TP) is measured from unfiltered water samples and includes SRP plus particulate P (PP), which is bound in plant or animal tissue. Because a major part of TP can be contained in algae (e.g., PP), it is not surprising that algal biomass (i.e., measured as chlorophyll a) is often positively correlated with TP concentration (Scheffer 1998). Nonetheless, TP has been correlated to lake productivity and is the most frequently reported value of P availability.

The availability of P is driven primarily by three sources: external inputs, internal loading, and biological activity and cycling. External loading refers to P inputs from inflowing water, runoff, and weathering of rocks and soils. Internal loading refers to P derived from lake sediments, primarily through geochemical processes. Invertebrates and fishes can also influence P availability via behavior (e.g., disturbance of sediments) or metabolic processes (e.g., excretion). All of these sources can influence P availability and resulting productivity of lakes and streams.

Efforts to reduce eutrophication of freshwaters often focus on reducing external inputs to inflowing waters. By altering land use practices in watersheds, managers have been successful in reducing nutrient inputs and reversing the effects of eutrophication in lakes. Examples of these practices include establishing buffer zones along stream riparian areas, limiting cattle grazing or feedlot operations, reducing fertilizer application on lawns or crops, constructing storm water retention ponds, or a combination of these approaches (Roni 2005). Although reduction of external P inputs can lead to improvements in water quality, improvements may take many years to occur. The reason centers on the importance of sediment as a P buffer and the mechanisms that govern sediment P release. Internal P loading can be affected by several factors, including dissolved oxygen (DO) concentration, iron (Fe) availability, water temperature, and pH. Dissolved oxygen concentration is the most important mechanism regulating P release from sediments, and Fe is the major agent responsible for binding SRP under oxygenated (aerobic) conditions. Because Fe can bind about 10% of its own weight in P, Fe availability can influence internal nutrient loading (Jensen et al. 1992). Under aerobic conditions, sediment acts as a buffer to P release. However, under anaerobic conditions, the capacity of Fe to bind SRP is reduced, and P is released to the water. Lakes that are anoxic near the water–sediment interface are prone to internal P loading and, as a result, may take many years to respond to lowered external P inputs (Figure 13.1). In Lake Søbygaard, Denmark, P concentration remained elevated after 8 years of reduced external P inputs. The large P pool that accumulated in the sediments from years of external P inputs continued to influence P availability and lake productivity despite 80–90% reductions in external loading (Søndergaard et al. 1993).

Phosphorus concentration in water can also be influenced by the composition and abundance of aquatic organisms. Feeding and burrowing activities of benthic organisms can release bound P from sediment to overlying waters, a process referred to as translocation. In general, the rate of P release increases with the density of benthic organisms (Wisniewski and Planter 1985). Similarly, excretion of soluble nutrients by fishes and invertebrates can contribute to nutrient availability in lake water.

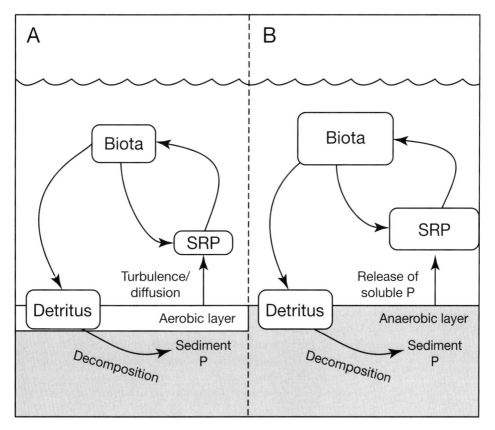

Figure 13.1. Internal phosphorus (P) loading in lakes: (**A**) under aerobic conditions, iron (Fe) binds with sediment P in the aerobic layer, preventing its release to overlying water. Some sediment P (soluble reactive P; SRP) can be released by turbulence and (or) diffusion; (**B**) under anaerobic conditions, Fe loses its capacity to bind P and SRP is released to the overlying water, thereby contributing to increased nutrient availability (adapted from Scheffer 1998).

13.2.2 Nitrogen

Processes that govern N cycling are different from those that regulate P cycling. Unlike P, N does not accumulate appreciably in lake sediments, can be released as N_2 into the atmosphere under reducing (i.e., anaerobic) conditions, and can be used in gaseous form (as N_2) as a nutrient by cyanobacteria (Scheffer 1998; Figure 13.2). Decomposition of organic matter occurs at the sediment–water interface resulting in production of ammonium (NH_4^+), which diffuses into the water and becomes readily available to algae. In aerobic sediments, ammonium can be converted to nitrate (NO_3^-) via microbial activity—a process called nitrification. Under anaerobic conditions, nitrate is converted to nitrite (NO_2^-; denitrification) and ultimately to N_2, which is not readily available to green algae. Because denitrification occurs under anaerobic conditions, the process is common where conditions alternate between aerobic and anaerobic environments (e.g., sediments). Hence, DO plays a critical role in denitrification, much like its role in P availability.

Nitrogen can be a limiting nutrient to primary production in all types of lake environments, from shallow wetlands to large, oligotrophic lakes. Because phytoplankton cells contain about 10 times more N than P, the N:P ratio in water can be used as an index of potential nutrient limitation (Scheffer 1998). Generally speaking, N is more likely to be a limiting nutrient when the N:P ratio is less than 10 (Smith 1982). Conversely, waters with N:P ratios greater than 20 are generally considered to be P limited.

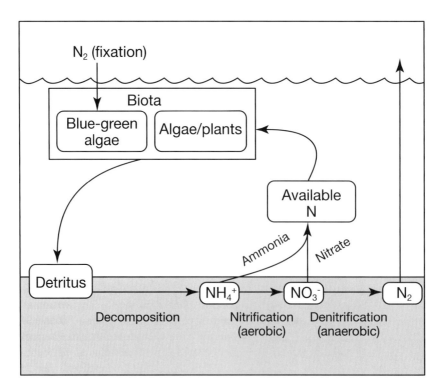

Figure 13.2. Generalized nitrogen (N) cycle in lakes. Cyanobacteria (blue-green algae) have the ability to fix atmospheric N by means of specialized cells called heterocysts (adapted from Scheffer 1998).

13.2.3 Productivity Index

Agencies charged with managing aquatic resources routinely monitor nutrient concentrations along with other water quality attributes such as water clarity (i.e., Secchi disk depth or total suspended solids), algal biomass, and DO concentration. Because nutrient concentrations can affect water quality, information should be easily summarized and communicated to the public. Many laypeople, for example, may not recognize that a TP concentration of 125 µg/L would be indicative of eutrophic conditions. To remedy this, Carlson (1977) developed the trophic state index (TSI) based on a numerical scale varying from 0 to 100, for which each 10 units represents a doubling in algal biomass. The index is calculated from TP concentration (TSI_{TP}), algal biomass (as chlorophyll a, $TSI_{Chl\,a}$), or Secchi depth (TSI_{Secchi}):

$$TSI_{TP} = 10\left|6 - \frac{\log_e \frac{48}{TP}}{\log_e 2}\right|; \qquad (13.1)$$

$$TSI_{Chl\,a} = 10\left|6 - \frac{2.04 - 0.68 \log_e Chl\,a}{\log_e 2}\right|; \text{ and} \qquad (13.2)$$

$$TSI_{Secchi} = 10\left|6 - \frac{\log_e Secchi}{\log_e 2}\right|; \qquad (13.3)$$

where TP is measured in the surface waters as µg/L, chlorophyll a is measured as µg/L), and Secchi depth is measure in meters (Carlson 1977). Carlson (1977) recommended the use of $TSI_{Chl\,a}$ for samples taken in summer and TSI_{TP} if sampling is conducted in spring or winter when algae abundance may be limited by factors other than nutrient availability. Values of TSI less than 30 are characteristic of oligotrophic conditions, values of 50–70 indicate eutrophic conditions, and values greater than 70 indicate hypereutrophic conditions. When data on Secchi depth, algal biomass, and TP concentration are available, deviations between TSI values can be examined graphically for factors affecting water clarity (Figure 13.3).

13.3 LAKE FOOD WEBS

13.3.1 Fish Production

In most aquatic food webs, the energy that supports higher trophic levels is derived primarily from the energy fixed by primary producers. As a result, fisheries managers are often interested in predicting fish production (or biomass) as a function of lake productivity (Figure 13.4). The morphoedaphic index (MEI) is one such model that predicts fisheries yield based on lake morphometry (mean water depth, z) and productivity (indexed by total dissolved solids, TDS):

$$MEI = TDS/z, \qquad (13.4)$$

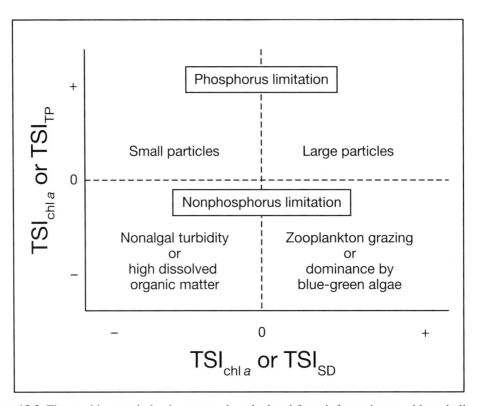

Figure 13.3. The trophic state index is commonly calculated from information on chlorophyll-*a* concentration ($TSI_{chl\,a}$), total P (TSI_{TP}), or Secchi depth (TSI_{SD}). In some lakes, TSI values change seasonally owing to grazing pressure by herbivorous zooplankton, nutrient availability, or nonalgal sources of turbidity (Wetzel 2001). By graphically examining deviations among TSI values, the causes of nutrient-limited and nonnutrient-limited deviations can be explored. Values above the *x*-axis (positioned at zero) indicate P limitation, whereas values below the *x*-axis indicate that trophic status is regulated by something other than P availability. Points in the lower left quadrant imply that water transparency is influenced more by nonalgal sources of turbidity, such as high concentrations of dissolved organic matter or small, nonalgal particles (e.g., clay). Points in the lower right quadrant imply that water transparency is greater than predicted by chlorophyll-*a* concentrations, owing to large cyanobacteria or zooplankton grazing and the relative reduction of smaller particles (adapted from Carlson 1992).

where fish yield (kg/ha/y) = $2\sqrt{MEI}$ (Ryder 1982). The MEI has been successfully used to relate reservoir productivity to fish harvest and standing crop (Oglesby and Jenkins 1982). Based on data from 290 U.S. reservoirs, maximum standing crop of fishes would be expected at MEI values between 50 and 200 (Oglesby and Jenkins 1982). Similarly, growth and body condition of reservoir sport fishes have been linked to MEI values for Alabama and Texas reservoirs (DiCenzo et al. 1995; Wilde and Muoneke 2001). In some studies, MEI values have been shown to be unrelated or only weakly correlated to fishery attributes, prompting researchers to refine the index or develop models using other parameters. Youngs and Heimbuch (1982), for example, suggest that lake surface area is a better predictor of fish yield than is mean depth. Similarly, models based on phytoplankton production or biomass (e.g., chlorophyll *a*) appear to be more accurate for predicting fish yield than those based on morphoedaphic factors (Oglesby 1977) and should be considered when these data are available.

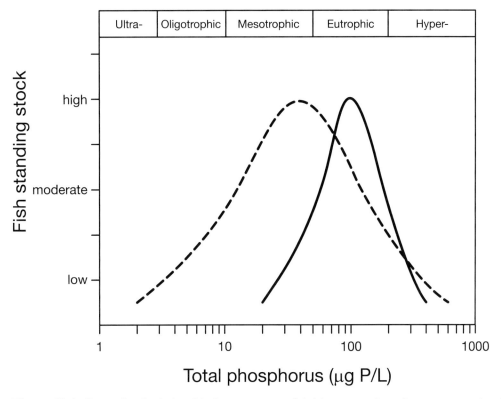

Figure 13.4. Generalized relationship between sport fish biomass and nutrient concentration in lakes (dotted line) and reservoirs (solid line) (adapted from Ney [1996] and Stockner et al. [2000]).

13.3.2. Food Web Conceptualizations

Describing and quantifying food web interactions provide important information to fisheries managers. Conceptually, food webs can be characterized three ways (Box 13.1). A *descriptive food web* is the simplest approach and is based on observations of "who eats what." Here, basic diet information (e.g., presence–absence) can be used to construct a general food web by linking predators and their prey. A descriptive food web, while useful for identifying food web linkages, relays little information about relative energy flow or interaction strengths among food web components. *Energy flow webs* attempt to characterize the relative energy flow from producer to consumer. To evaluate patterns in energy flow, quantitative approaches are needed that include detailed diet information, bioenergetics modeling, stable isotope analysis, and (or) experimental data. By quantifying the relative energetic contribution of different prey to the diet of a consumer, we gain an understanding of important energy pathways (Box 13.2). Finally, *interaction webs* draw upon knowledge of species interactions and emphasize connections that have a large influence on the dynamics of food web structure and function (Paine 1980). Here, concerns are centered on understanding the influence of removing (or adding) an individual species on the rest of the food web. This approach formed the foundation of the keystone species concept (Paine 1980) and later served as a cornerstone of the trophic cascade hypothesis (Carpenter and Kitchell 1993).

Box 13.1. Aquatic Food Webs

Representative food web configurations for a large oligotrophic lake with introduced opossum shrimp. Panel **A** depicts a *descriptive food web*, showing general feeding relationships. Panel **B** depicts an *energy flow diagram*, showing relative energetic contribution of different prey to consumer's diets; width of arrows is proportional to energy transfer. Panel **C** depicts relative *interaction strengths* among food web components. In this example, the establishment of opossum shrimp had negative impacts on cladoceran zooplankton, but positive impacts on lake trout, which ultimately reduced the abundance of kokanee, which, in turn, affected fish and avian predators (Spencer et al. 1991; Martinez et al. 2009).

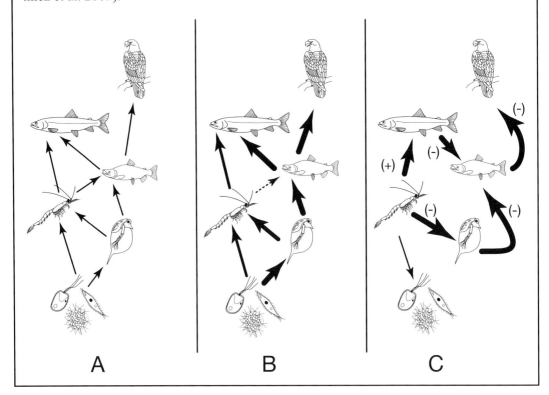

13.3.3 Food Web Components

13.3.3.1 Plants

Phytoplankton. Phytoplankton represent the primary source of production in pelagic food webs. By converting sunlight and nutrients into usable (i.e., edible) forms of energy, phytoplankton provide the food base for most aquatic food webs. As a group, phytoplankton consist of a diverse array of taxa that are tolerant to a wide range of environmental conditions. Although extreme variation exists in phytoplankton composition among lakes, some generalities have emerged with regard to phytoplankton associations in lakes with differing nutrient

Box 13.2. Bioenergetics Modeling

Bioenergetics models are commonly used to estimate food consumption by fishes. A generalized bioenergetics model can be expressed as $C = M + A + SDA + F + U + G$, where food consumption (C) is balanced by standard (M) and active (A) metabolism, specific dynamic action (SDA), egestion (F), excretion (U), and somatic plus gonadal growth (G). Fish Bioenergetics 3.0 is a popular software package that can be used to model food consumption by fishes and can be obtained from the University of Wisconsin Sea Grant College Program (see Hanson et al. 1997).

Estimating prey contribution

When food consumption is expressed in energy units such as joules or calories, the relative energetic contribution of different prey can be evaluated. The table below shows prey taxa found in the diets of age-0 and age-1 western mosquitofish (adapted from Chipps and Wahl 2004). Food consumption was estimated from observed growth rates by means of Fish Bioenergetics 3.0 (Hanson et al. 1997). Prey-specific food consumption (kJ) was obtained as the product of prey energy density (kJ/g dry weight [wt]) and consumption (g dry wt). The energetic contribution of individual prey (P_i) can then be calculated as (prey$_i$ consumption [kJ]/total consumption [kJ]) × 100 to estimate the relative contribution of different prey to the growth of fish.

(Box continues)

availability. In temperate, oligotrophic lakes, chrysophycean algae (golden-brown algae), diatoms (especially *Cyclotella* and *Tabellaria*), and to a lesser extent Chlorophyta (green algae) represent dominant taxa in phytoplankton communities. As nutrient availability increases (mesotrophic), dinoflagellates become a dominant taxon in the phytoplankton communities. In eutrophic lakes, diatoms (especially *Asterionella* and *Fragilaria*) are present much of the year—along with many other algae. During warmer months, green algae and cyanobacteria (formerly blue-green algae) can become dominant taxa in the phytoplankton communities of eutrophic lakes (Wetzel 2001). Cyanobacteria, in particular, can present water quality problems by reducing light penetration and affecting the potability of water.

Seasonal succession in phytoplankton composition, though variable from lake to lake, can change dramatically. Small flagellates (e.g., cytomonads and dinoflagellates) that are adapted to low light and water temperature generally dominate phytoplankton biomass during winter in temperate lakes. In spring, the biomass of diatoms increases rapidly followed by development of green algae in early summer. Depending on lake trophic status, a second development of diatoms may occur in late summer or early autumn (less productive lakes) or, in the case of more productive lakes, cyanobacteria can develop in summer and dominate phytoplankton biomass (Wetzel 2001). Seasonal changes in phytoplankton composition and abundance, from spring to summer, have been linked to reduced nutrient availability, changes in light availability, and grazing pressure by zooplankton.

Box 13.2. Continued.

Table. Prey-specific consumption estimates for age-0 and age-1 western mosquitofish from a southern Illinois wetland. Bioenergetics modeling was used to estimate consumption for age-0 (August–October) or age-1 (March–May) fish. The relative energetic contribution of individual prey (P_i, %) was then calculated as prey-specific consumption (kJ) divided by total energy consumed (kJ).

Prey taxon	Energy density (kJ/g dry wt)	Age-0 Consumption (g dry wt)	Age-0 (kJ)[a]	Age-0 Energetic contribution (P_i, %)	Age-1 Consumption (g dry wt)	Age-1 (kJ)[a]	Age-1 Energetic contribution (P_i, %)
Amphipoda	18.5	0.002	0.04	1	0.01	0.18	3
Chironomidae	22.8	0.130	2.96	61	0	0	0
Cladocera	25.1	0.016	0.40	8	0.09	2.26	33
Copepoda	23.2	0.010	0.23	5	0.16	3.71	53
Culicidae	20.4	0.047	0.96	20	0	0	0
Ephemeroptera	16.8	0	0	0	0.046	0.77	10
Gastropoda	4.7	0.021	0.1	2	0.003	0.01	1
Notonectidae	17.9	0.007	0.12	3	0	0	0
Total		0.233	4.81	100	0.309	6.93	100

[a] In all versions of Fish Bioenergetics software, the calculation of prey consumption in either joules or calories is incorrect. To obtain correct values, either multiply prey consumption (in g) by prey energy density (J/g wet wt) or multiply model output (joules or calories consumed) by the square of prey energy density.

Macrophytes. Macrophytes and associated epiphytic (attached) algae represent an important energy source in littoral zones. Like phytoplankton, macrophytes convert solar energy and nutrients into primary production. However, unlike phytoplankton, macrophytes are large, sessile plants that provide an important structural component to aquatic environments. Macrophytes provide food, shelter, and spawning habitat for many fishes and help to increase water clarity by reducing erosion.

Macrophytes also have an important influence on environmental conditions in lakes and rivers (Carpenter and Lodge 1986). Macrophyte stands can influence local conditions such as water temperature and DO concentrations. In dense macrophyte beds, vertical gradients in water temperature can vary up to 10°C/m. Similarly, diel changes in DO can be as large as 8 mg/L in dense macrophyte stands, with submersed plants being more effective at generating oxygen than are floating plants. Macrophytes also have important effects on nutrient fluxes. Actively-growing plants remove inorganic C from the surrounding water and can either assimilate it into organic matter (biomass) or precipitate it as carbonate salts on the surface of leaves. Plants release photosynthetically-fixed C as dissolved organic C to the water, where most is then used by the epiphytic community (e.g., microbes and algae). Abundant macrophytes also influence sedimentation processes by enhancing deposition of sediments that would otherwise be eroded. This, in turn, can have a strong influence on water clarity and resultant light penetration that favors macrophyte growth (see alternative stable state, section 13.4.3). Conversely, a decline in phytoplankton abundance can lead to increased water transparency (light penetration) and contribute to improved macrophyte growth and coverage.

Macrophytes can represent either a source or sink for dissolved nutrients, such as P and organic C, depending on whether they are actively growing or senescing. During the growing season, macrophytes and associated epiphytic complexes generally represent a sink for dissolved P. When macrophytes senesce and begin to decompose they represent a net source of dissolved P to the sediments and overlying water. Decomposition of organic matter (i.e., plants) also results in the release of N, usually as ammonium, that can be quickly used by algae and plants. However, unlike P, which tends to accumulate in lake sediments, significant amounts of N can be lost in shallow, macrophyte-dominated lakes via denitrification. In shallow Danish lakes, over 75% of N loss occurs through denitrification (Jensen et al. 1991; Scheffer 1998).

Because algae and macrophytes often compete for the same nutrients (e.g., P and N), it is not surprising that the abundance of one can suppress that of the other. It is suspected that macrophytes inhibit the growth of green algae and cyanobacteria through excretion of chemical substances—a process called allelopathy (Scheffer 1998). Although the chemical substances associated with allelopathy are poorly known, empirical studies show that dense macrophyte stands can suppress phytoplankton abundance, especially cyanobacteria. Studies of *Ceratophyllum* and *Myriophyllum* showed that cyanobacteria density declined nearly 90% in the presence of these plants (Kogan and Chinnova 1972).

Macrophytes provide important habitat for a variety of consumers, including microbes (e.g., bacteria), epiphytic algae, invertebrate consumers (e.g., grazers), and vertebrate predators (e.g., fishes and waterfowl). Epiphytic organisms rely on macrophytes for substrate, nutrients, refugia, breeding sites, and other factors that enhance their production. In turn, many macrophytes have evolved to take advantage of epiphytic complexes to enhance their own production. Epiphytes are thought to benefit macrophytes by distracting grazers and, as a result, macrophytes have evolved to benefit epiphytes by providing substrate and nutrients

(Hutchinson 1975). However, dense epiphytic growth can inhibit macrophyte photosynthesis (Carpenter and Lodge 1986). Coontail, for example, has been shown to grow better in the presence of grazing snails than in their absence (Brönmark 1985). While the epiphytic complex represents an important energy source for most grazing organisms, some animals directly consume macrophytes. Fishes such as rudd and grass carp are known to consume macrophytes and can have an important influence on macrophyte abundance in lakes (Carpenter and Lodge 1986). Similarly, some species of waterfowl and crayfishes are dependent on macrophytes as a food source.

13.3.3.2 Macroinvertebrates

Benthic macroinvertebrates perform a variety of important functions that influence energy flow and food web structure in aquatic environments (Palmer et al. 1997; Covich et al. 1999). Burrowing and feeding activities of benthic invertebrates release bound nutrients from sediment into overlying water, often in the form of SRP that is quickly taken up by algae (Gallepp 1979). Benthic invertebrates also contribute to nutrient cycling via excretion of soluble nutrients. Moreover, by converting dead organic material into secondary production, benthic invertebrates accelerate detrital decomposition. Lastly, because many benthic invertebrates are predators, they influence food web structure by regulating the abundance, distribution, and size of their prey (Figure 13.5).

The recent spread of a number of exotic and nonnative benthic invertebrates has increased our awareness of how invertebrates influence trophic interactions. For example, expansion of the rusty crayfish into northern Wisconsin demonstrates how grazing by crayfish can influence fish populations. The establishment of rusty crayfish in shallow lakes resulted in declines of the native northern crayfish. Rusty crayfish are able to displace northern crayfish by excluding them from shelters and outcompeting them for limited food resources (Garvey et al. 1994; Hill and Lodge 1994). They can also increase predation mortality for northern crayfish by displacing them from refuge areas, making them more vulnerable to fish predators (Garvey et al. 1994). Because rusty crayfish feed on detritus and macrophytes, once established, they can significantly reduce submerged macrophytes, which ultimately leads to reduced abundance of gastropods and fishes (Lodge et al. 1994; Covich et al. 1999).

Intentional introductions have taught us how invertebrate behaviors and life history attributes can influence food web dynamics. Large-scale introductions of the opossum shrimp in western North American lakes were intended to provide additional forage for salmonids (Nesler and Bergersen 1991). Once populations of opossum shrimp became established, fisheries biologists recognized changes in the food web, most notably reduced abundance of cladoceran zooplankton and declines in the abundance and growth of salmonids. Because opossum shrimp undergo extensive diel vertical migrations, spending much of the day on the lake bottom, they were unavailable to visually-feeding pelagic fishes. This, combined with their omnivorous feeding habits, provided them a unique niche and allowed their populations to flourish in lakes where they were introduced (Chipps and Bennett 2000). In Flathead Lake, Montana, opossum shrimp introductions resulted in dramatic declines of kokanee and native fishes, whereas nonnative lake trout abundance increased. These changes "cascaded" up the food web by severing the link with the terrestrial–riparian food web, resulting in fewer bald eagles and bears in the watershed (Spencer et al. 1991; Box 13.1).

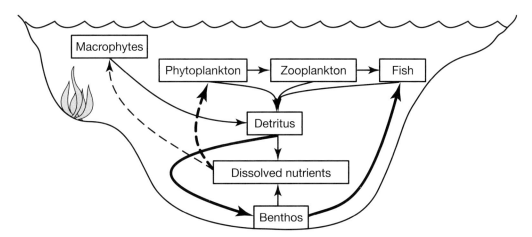

Figure 13.5. Influence of benthic invertebrates on detrital processing, nutrient recycling, and food availability (fishes) in lake food webs (adapted from Covich et al. 1999).

13.3.3.3 Pelagic Zooplankton

Herbivorous zooplankton (e.g., *Daphnia*) are often considered the nexus of pelagic food webs because they serve a critical link between primary production and higher trophic levels. They can also play an important role in regulating algal biomass, particularly the more edible forms of green algae. The clear-water phase that occurs in many temperate lakes is a good example of how zooplankton can limit algal densities. During late spring, grazing pressure by dense *Daphnia* populations can reduce algal biomass and noticeably increase water clarity. Seasonal succession in phytoplankton composition (i.e., food quality), combined with increased predation pressure from juvenile fishes, are often related to reduced *Daphnia* abundance by midsummer.

The composition, morphology, and behavior of pelagic zooplankton can provide important information about fish assemblages in lakes and reservoirs (Mills and Schiavone 1982). In lakes dominated by planktivorous fishes (i.e., plankton-feeding fishes), smaller cladocerans usually predominate because of size-selective predation by fishes (Brooks and Dodson 1965). Larger zooplankton, such as *Daphnia*, are generally preferred over smaller taxa, such as *Bosmina*. Smaller zooplankton also tend to be less efficient filter feeders than are larger zooplankton and as a result are less efficient at regulating algal biomass.

Morphological features of cladoceran zooplankton can vary among populations as a result of fish predation. In populations exposed to relatively high predation, long tail spines and enlarged "helmets" become more conspicuous and are believed to be an adaptation for reducing predation. Experimental studies have shown that long tail spines and enlarged helmets increase the handling time and reduce capture success by small predatory fishes and invertebrates (Kolar and Wahl 1998). An interesting example of this can be seen among glacial lakes in the northern Great Plains that experience winter fish kills. A comparison of *Daphnia pulex* populations in these lakes showed that spine lengths were shorter in lakes that experience frequent winterkills, presumably due to reduced predation pressure from fishes (Isermann et al. 2004).

In many populations, zooplankton undergo diel vertical migrations, ascending to surface waters at night to feed and then descending to deeper depths during the day. Patterns in diel vertical migration have been linked to relative predation pressure by fishes (Gliwicz 1986). In lakes and reservoirs containing planktivorous fishes, the magnitude of diel vertical migrations is more pronounced than it is in lakes with little to no predation pressure. This behavior is believed to be an adaptation to predation pressure: by migrating to deeper depths during the day, zooplankton are less vulnerable to visually feeding fishes. Although not well documented, this fish-induced behavior may have indirect effects on phytoplankton abundance by reducing the grazing pressure by herbivorous zooplankton.

13.3.3.4 Fishes

Fishes represent an important component of aquatic food webs. As a group, fishes span a broad range of trophic levels from herbivores to apex predators. However, most fishes undergo diet shifts as they grow by switching prey types. Because many fish hatch at small sizes, the first food item is often zooplankton, which are easier to capture and fit within the gape width of larval fishes. The availability of zooplankton during the larval and juvenile stage of many fishes is related to recruitment success. Laboratory and field studies have shown that larval yellow perch (<12 mm) grow faster and have higher survival when small copepods are abundant (Graeb et al. 2004). Furthermore, juvenile yellow perch have greater growth, survival, and recruitment when larger cladocerans (e.g., *Daphnia*) are abundant.

Although recruitment in fishes is influenced by many factors, food availability is a critical component that can affect survival of young fishes. If larval fish emerge when their preferred zooplankton prey are abundant, their survival is generally higher than if they emerge before or after zooplankton blooms. Because of the importance of zooplankton to growth and survival of larval fish, many fisheries biologists monitor zooplankton abundance and species composition to assess potential fish growth and recruitment.

As fish grow, they develop larger body sizes, wider gape widths, and increased swimming ability that allows them to forage on a broader range of prey types. For example, largemouth bass, yellow perch, and walleye often switch as small juveniles from a zooplankton-dominated diet to a macroinvertebrate diet. Benthic macroinvertebrates are an important energy source to fishes because these organisms convert organic matter (e.g., plants and detritus) into usable energy. Common macroinvertebrates consumed by fishes include aquatic insects (midges, mayflies, caddisflies, and dragonflies), mollusks (mussels and snails), and crustaceans (amphipods and crayfishes). For piscivorous fishes, the shift from invertebrates to fish prey is an important process because it often confers faster growth and may increase survival (Ludsin and DeVries 1997). Some generalist fishes (e.g., yellow perch) may maintain an invertebrate diet through most of their life as long as invertebrate prey remain available, but will readily shift to piscivory when fish prey are abundant.

Zooplanktivory. Some fishes feed almost exclusively on zooplankton throughout their life cycle. Kokanee, rainbow smelt, and alewife are examples of sport and prey fishes that are planktivorous as both juveniles and adults. For these species, monitoring zooplankton communities is critical because changes in zooplankton species composition and (or) size structure can indicate changes in food web structure (e.g., abundance of planktivores). For example, Johnson and Goettl (1999) showed that the abundance of rainbow smelt could be tracked in Horsetooth Reservoir, Colorado, by monitoring the abundance and species composition of zooplankton. After rainbow smelt became established in Horsetooth Reservoir, the abundance of zooplankton decreased, particularly larger *Daphnia*, and remained low for several years. Subsequent zooplankton monitoring showed another shift in the zooplankton community in which the abundance of large-bodied zooplankton began increasing. Indeed, managers later learned that rainbow smelt abundance was decreasing during this time period, which reduced predation pressure on zooplankton and allowed the zooplankton community to recover. Routine monitoring of zooplankton allowed managers to detect a decline in the abundance of rainbow smelt.

Omnivory. Omnivorous fishes derive their energy by consuming a variety of plant and animal material. Common carp is a well-known omnivore that has molar-like surfaces on its pharyngeal teeth that allow it to grind plant material prior to ingestion. The benthic feeding behavior of common carp is characterized by sucking up lake sediment, expelling it into the water, and then filtering food items (Scott and Crossman 1973). A variety of aquatic animals are reported from the diets of common carp and include aquatic insects, mollusks, annelids, and crustaceans. Sediment resuspension, nutrient cycling, and omnivorous feeding by common carp can have profound effects on lake water quality. High common carp densities have been linked to reduced water clarity and submerged aquatic vegetation and represent a significant challenge to fisheries managers (Weber and Brown 2009).

Channel catfish, another omnivorous fish, is also known to consume fruits and berries of certain plants (i.e., frugivory). Studies in the Mississippi River have shown that channel catfish consume red mulberry and eastern swamp privet fruit (Chick et al. 2003). Fruit consumption by channel catfish improved the germination success of seeds, implying that seed dispersal by channel catfish could have a positive influence on recruitment of riparian plants.

Piscivory. Fishes that obtain most of their energy by consuming other fishes are piscivores. Piscivores often are apex predators in aquatic systems and can affect the entire food web. For example, an abundant predator population can reduce the abundance of fishes at lower trophic levels (planktivores or macroinvertebrate predators). Many of the most popular sport fish in North America are piscivores (e.g., largemouth bass, Chinook salmon, muskellunge, walleye, striped bass, and flathead catfish). Management efforts for most of these predators involve maintaining balance between predator demands (abundance of piscivores) and prey supply (abundance of planktivores and macroinvertebrate predators).

Newly-hatched larval fish are highly vulnerable to many organisms because they hatch at fairly small sizes and typically cannot swim well enough to avoid most predators. So even top-level piscivores are prey items to other fishes during larval and juvenile stages. One result of this relationship is that potential competitors can also prey on each other. The relationship in which potential competitors can act as predators and prey is called intraguild predation (Gotelli 2001). Northern pike and muskellunge are top-level piscivores common in the northern USA and Canada and provide an excellent example of intraguild predation. Northern pike spawn and hatch earlier in the season than do muskellunge. As such, young northern pike that emerge prior to muskellunge can readily prey on them and may suppress recruitment of muskellunge (Inskip 1986). Thus, we should be cognitive of the relative hatch timing of potential competitors and how this may influence predation by one species on another.

The abundance of one species may also influence relative predation on another species.

In an earlier example (section 13.3.3.2), the introduction of opossum shrimp and the subsequent decline of kokanee were initially thought to result from competition for zooplankton. However, opossum shrimp are important prey for juvenile lake trout and in lakes where they were introduced, they contributed to increased growth and recruitment of this top piscivore. In turn, lake trout predation on kokanee contributed significantly to declines in kokanee populations. Such examples are termed "apparent competition" (Holt 1977) because the mechanism influencing kokanee abundance appeared to be competition with opossum shrimp, when in fact predation by lake trout was important.

13.4 FOOD WEB THEORY

13.4.1 Trophic Cascade Hypothesis

The direct effects of predation are fairly predictable—as predation increases, prey density decreases. Often these effects indirectly influence other trophic levels. Hrbacek et al.'s (1961) study was one of the first to document the indirect effects of fish predation on zooplankton and algae. They showed that when fish density was low, large-bodied zooplankton (e.g., *Daphnia* spp.) dominated the plankton community, resulting in less algae in fish ponds. Later studies quantified these interactions, and the term "trophic cascade" was coined (Paine 1980) to characterize how predation by higher trophic levels cascades to lower trophic levels (Brooks and Dodson 1965; Carpenter et al. 1985). A simple model of the trophic cascade hypothesis predicts that increasing piscivore abundance reduces abundance of planktivores by direct predation. The predatory effect cascades to lower trophic levels because reduced abundance of planktivores decreases predation pressure on zooplankton, allowing zooplankton densities to increase or taxa composition to shift toward larger, more efficient grazers. Increased grazing pressure from zooplankton reduces phytoplankton density, ultimately reducing algal biomass and increasing water clarity.

Trophic cascades can influence the composition and size structure of fish and zooplankton populations that, in turn, influence algal biomass. Because mass-specific metabolism decreases with body size, nutrient recycling (i.e., excretion) can be strongly influenced by the size composition of aquatic organisms (Figure 13.6). As a result, dense populations of smaller fishes (i.e., planktivores) and zooplankton can enhance nutrient availability via excretion of soluble nutrients. In lakes dominated by planktivores and smaller zooplankton, algal abundance is enhanced by both reduced grazing pressure and increased nutrient subsidies (Vanni and Layne 1997).

Empirical evidence and whole lake manipulations have shown that while food web effects are real, they are not always as straightforward as theory predicts owing to complex ecosystem responses (Carpenter and Kitchell 1992; DeMelo et al. 1992). Many changes can occur that decrease the effectiveness of trophic cascades, particularly at lower trophic levels. One example of how trophic cascades break down involves the role of gizzard shad in Midwestern reservoirs. Gizzard shad can attain very high densities in eutrophic systems, and much research and management efforts have been devoted to reducing gizzard shad density through trophic interactions (Stein et al. 1995). For example, largemouth bass and hybrid striped bass were stocked to increase predation on gizzard shad, but researchers quickly learned that gizzard shad, and not piscivores, drove these food webs. Gizzard shad had very fast growth rates

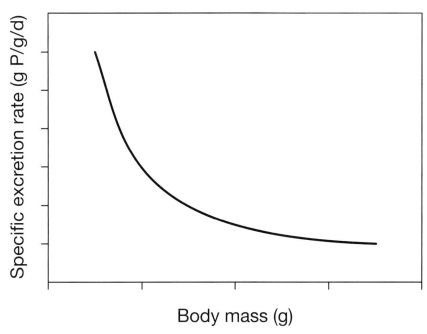

Figure 13.6. General relationship between specific excretion rate and body mass for aquatic organisms. At the same biomass, a population of smaller-bodied fish or invertebrates excrete more soluble nutrients than do larger-bodied populations.

and quickly outgrew the gape of many piscivores. Thus, predation pressure was typically over a short window during the first couple of months after gizzard shad hatch. Gizzard shad also competed directly with largemouth bass when both fishes were reliant on zooplankton prey (larval and early juvenile), but gizzard shad have the ability to switch to detritus after zooplankton densities decline, giving them a competitive advantage over their predators. Thus, in an effort to control gizzard shad abundance through trophic cascades, we learned that gizzard shad actually structure food webs from the "middle." Managers need to predict the direct and indirect effect of management actions across multiple trophic levels.

13.4.2 Benthic–Pelagic Links

The influence of benthic productivity is often overlooked in studies of lake food webs, even though benthic organisms are important to the diets of many fishes. For piscivorous fishes such as walleye, northern pike, and largemouth bass, the contribution of benthic prey to the diet can be substantial (>50%; Vander Zanden et al. 1997; Vadeboncoeur et al. 2002). The use of benthic prey by fishes can vary among populations and is often influenced by lake size. The production and contribution of benthic organisms to fish diets is generally more pronounced in small, shallow lakes and decreases as mean depth and lake area increase (Lindegaard 1994). Thus, interactions between pelagic and benthic communities are affected by lake size.

The transfer and direction of energy flow between pelagic and benthic environments is strongly influenced by nutrient concentration in the water column (Vadeboncoeur et al. 2002). In oligotrophic lakes, whole-lake primary productivity can be dominated by the production of benthic plants (e.g., periphyton and submerged vegetation). As nutrient concentration increases, a transition from benthic to pelagic productivity occurs because high phytoplankton biomass reduces light availability to benthic algae and submerged vegetation. In shallow lakes, the relative influence of benthic or pelagic primary productivity can alternate at intermediate nutrient levels, resulting in dramatic changes in water quality and energy flow. Such nonlinear changes in ecosystem states are referred to as alternative stable states and have important implications for habitat management and fish production.

13.4.3 Alternative Stable State Hypothesis

Many ecosystems can alternate between alternative stable states. The most common example of this in aquatic systems is the clear-water and turbid-water states found in many shallow lakes (Scheffer et al. 1993; Scheffer 1998). The clear-water state is characterized by high water transparency and primary production dominated by submerged aquatic vegetation. This state tends to be stable as long as nutrient- or fish-induced turbidity remains below a certain threshold. Conversely, lakes in the turbid-water state have low water transparency and primary production dominated by pelagic phytoplankton such as green algae or cyanobacteria. Although nutrient concentration has an important effect on water clarity, it is often difficult to change turbid lakes into clear-water systems by reducing nutrient concentration alone. The reasons for this center on trophic interactions and sediment resuspension that influence the light climate needed by submerged vegetation. By manipulating fish abundance, resource managers have been successful in "switching" lakes from a turbid state to a more desirable, clear-water state. Reductions in fish biomass can influence water turbidity in two important ways. First, as discussed earlier, a reduction of planktivorous fish abundance can cascade down the food web and subsequently lead to increased water clarity because of reduced predation on herbivorous zooplankton. Second, in lakes dominated by benthic feeding fishes (e.g., common carp), a reduction in fish biomass often results in improved water clarity because the feeding activities of these species resuspend bottom sediments and contribute to increased turbidity (Scheffer 1998).

13.5 FOOD WEB MANAGEMENT

13.5.1 Nutrient Management

Excessive algal blooms are common in many lakes and reservoirs during summer months. Dense blooms of algae can degrade water quality, making water unsuitable for drinking and impacting recreational opportunities (Welch and Cooke 1999). These problems are often linked to high nutrient concentrations and present a pragmatic challenge to resource managers. Reductions in external nutrient loading are a primary goal in all lake restoration efforts; however, not all lakes respond similarly to reduced nutrient inputs. In some lakes, significant reductions in external P loading can have little influence on short-term improvements to water quality owing to the importance of the sediment P pool that has accumulated over time. Here, internal nutrient loading from the sediments can contribute to high P concentrations and algal blooms.

Application of aluminum salts, such as aluminum sulfate ($Al_2[SO_4]_3$), is one method used to reduce internal nutrient loading. Aluminum sulfate, or alum for short, is usually applied in dissolved form and the application dose is dependent on a number of factors including P release rate and alkalinity of the water (Figure 13.7). By binding with inorganic sediment P, alum can be effective in reducing nutrient concentrations and improving water quality. Chlorophyll *a* has been reduced by about 40% after 5–18 years following treatments (Welch and Cooke 1999). Moreover, the abundance of cyanobacteria has declined significantly after alum treatments, whereas macrophyte coverage has increased in many lakes due to enhanced water clarity. The use of alum to switch turbid lakes to clear, macrophyte-dominant lakes may be as effective as the removal of undesirable fishes (e.g., common carp); however, case studies of alum-treated lakes suggests that removal of fish may be necessary to achieve adequate control of internal loading when fish densities are high (Welch and Cooke 1999).

The longevity and effectiveness of alum treatments can vary among lake types. Because shallow lakes are often well mixed (i.e., polymictic), they do not exhibit large vertical gradients in water temperature or oxygen concentration. In contrast, deeper lakes that thermally stratify in summer months (e.g., dimictic) often experience anoxia in hypolimnetic waters, which can contribute to sediment P release. Comparative studies have shown that polymictic lakes generally exhibit less of a response to alum application than do dimictic lakes. Moreover, dense stands of macrophytes or wind-driven sediment resuspension typical of many shallow lakes can reduce the effectiveness of alum treatments. When successful, follow-up studies have shown that longevity of alum treatments was about 10 years in polymictic lakes compared with 15 years in most dimictic lakes (Welch and Cooke 1999).

Figure 13.7 Application of aluminum sulfate (commonly referred to as alum) to reduce internal P loading in Lake Mitchell, South Dakota (photo courtesy of T. St. Sauver, South Dakota Department of Game, Fish, and Parks).

Unlike eutrophic lakes, oligotrophic lakes are characterized by low nutrient concentration. Although desirable from a water quality perspective, oligotrophic lakes, by definition, are low-productivity systems. Low primary productivity results in slow growth of fishes, which can present challenges for fisheries managers. In such cases, it is sometimes desirable to increase nutrient availability as a way to increase fish growth. The addition of inorganic fertilizer, such as ammonium polyphosphate and urea ammonium, can increase growth rate of sport fishes in oligotrophic lakes and streams (Stockner and MacIsaac 1996; Bradford et al. 2000) and provide an important management tool to fisheries managers. Given widespread concern regarding eutrophication, intentional fertilization of lakes should be approached cautiously on a lake-by-lake basis. Detailed knowledge of limnological conditions, particularly nutrient concentration (N and P) and N:P ratio, are required prior to developing a fertilization treatment (Stockner et al. 2000). Maintaining a high N:P ratio (>20) is important because as P increases, the N:P ratio declines and phytoplankton production becomes N-limited, potentially leading to cyanobacteria blooms. Thus, a well-balanced N:P supply is important in the development and application of inorganic fertilizers (Stockner et al. 2000).

13.5.2 Macrophyte Management

Submerged aquatic vegetation provides important rearing and refuge areas for fishes and is often associated with the clear-water state typical of many shallow lakes. While the clear-water state is often a desirable condition, high densities of submerged aquatic vegetation can pose problems. The spread of nonnative plants such as Eurasian watermilfoil provides a good example of the problems often encountered with dense vegetation (Olson et al. 1998). High densities of submerged vegetation can impair recreational opportunities and negatively affect fisheries.

Growth of fishes can be influenced by dense vegetation in several ways. First, by providing refuge for small fish, dense vegetation impedes foraging efficiency by piscivores, leading to reduced growth of piscivores (Savino and Stein 1982; Savino et al. 1992). Second, dense vegetation enhances survival of small fishes, leading to overcrowding and reduced growth, particularly among panfishes. Finally, dense macrophytes can lead to water quality degradation, most notably diel fluctuations in DO concentration (Frodge et al. 1990). A variety of approaches, from herbicide application to mechanical harvest of macrophytes, have been used to reduce macrophyte abundance in shallow lakes. Multi-lake studies with largemouth bass and bluegill populations have shown that growth rates of some age classes increased following a 20% removal of littoral vegetation (Olson et al. 1998). Similarly, a modeling study of vegetation removal patterns suggested that bluegill growth would increase following a 20–40% removal of littoral vegetation (Trebitz and Nibbelink 1996). In the latter study, the authors recommended that edge habitat be maximized by clearing narrow channels through vegetation rather than clear-cutting large areas.

13.5.3 Biomanipulation

Today, fisheries managers recognize that stocking or removing certain fishes (i.e., biomanipulation) can have an important influence on food web structure and resulting water quality. Biomanipulation is increasingly used as a lake restoration technique in large part because the ability to enhance water quality and sport fish populations is attractive to a variety of lake users (Mehner et al. 2002). Although biomanipulation research has contributed substantially to our understanding of lake food webs, successful application of these techniques is not a "one-size fits all" approach. In a review of biomanipulation research, Mehner et al. (2002) discussed factors affecting food web complexity and success of biomanipulation efforts. They made the following observations: (1) nutrient recycling by aquatic organisms contributes to bottom-up effects on lake productivity, but the magnitude of this effect varies greatly among lakes; (2) the complexity of food web interactions is enhanced by size-dependent interactions of fishes and can limit our ability to predict the outcome of a biomanipulation event successfully; (3) it is important to consider the temporal and spatial scales in biomanipulation research—repeated interventions may be necessary to maintain the desired outcome in lakes; and (4) the appropriate balance among piscivorous, planktivorous, and benthivorous fishes is needed to achieve the desired outcome of a biomanipulation, but can present a challenge due to a general lack of quantitative assessment.

When practical, the removal of undesirable fishes seems to have a larger impact on water quality than does stocking piscivorous fishes. A synthesis of biomanipulation studies involving 39 lakes, varying in size from 0.18 to 2,650 ha (mean depth 1–23 m), showed that changes in phytoplankton biomass and water transparency were most successful in small, shallow lakes (<25 ha, mean depth < 3 m; Drenner and Hambright (1999). Changes in water quality were also influenced by the type of biomanipulation implemented. Ninety percent of studies using partial fish removal were successful in improving water quality, followed by (1) elimination and restocking of fish (67%), (2) partial fish removal plus piscivore stocking (60%), (3) elimination of fish (40%), and (4) piscivore stocking (27%). Moreover, only about 15% of the studies that used biomanipulation techniques were unsuccessful in enhancing water quality for at least 1 year (Drenner and Hambright 1999). Although the mechanisms responsible for these changes can be varied, evidence suggests that in lakes dominated by benthivorous fishes (e.g., common carp), over 50% of the turbidity can be caused by sediment resuspension (Meijer et al. 1994). Reduction of benthivorous fishes can also indirectly reduce algal biomass by triggering a shift to the clear, macrophyte-dominated state characteristic of many shallow lakes.

13.6 CASE HISTORIES: FOOD WEB DYNAMICS

13.6.1 Predator–Prey Balance

Balancing the consumptive demands of predators with the dynamics of prey populations is critical to managing sport fisheries (Swingle 1950; Ney 1990). A variety of approaches can be used to assess balance between predator and prey populations, including traditional indices such as proportional size distribution or relative weight. Because "balance" implies an equilibrium condition that rarely occurs in predator–prey populations, Ney (1990) coined a more appropriate phrase, "trophic economics," to reflect the dynamics between predators and prey. Understanding food web linkages that help explain fisheries resources (e.g., growth, body condition, and size structure) is an important responsibility of biologists.

One example that highlights the importance of trophic economics occurred in Colorado, where managers sought to create trophy lake trout fisheries in several high-elevation reservoirs (Johnson and Martinez 2000; Martinez et al. 2009). Because lake trout are long-lived species that typically maintain low densities, managers sought to increase length structure through restrictive harvest regulations (e.g., protected slot length limits and low bag limits) to reduce exploitation and allow lake trout to grow to quality length. Harvest regulations succeeded in allowing lake trout populations to increase; however, the cost of this expansion soon became apparent. Lake trout are apex piscivores that naturally reproduce in these reservoirs whereas their primary prey in Colorado reservoirs are kokanee and rainbow trout, neither of which naturally recruit in the reservoirs and are maintained by stocking. As the restrictive regulations were enacted, managers started examining food web dynamics by means of a variety of tools. Hydroacoustic surveys were used to determine the standing stock of lake trout, kokanee, and rainbow trout. Managers also determined growth rate and size structure through standardized gill netting and age estimation. Finally, diets and stable isotope signatures were used to determine diet composition of lake trout. These data were incorporated into a bioenergetics model to estimate total consumption demand by lake trout. In some of the reservoirs, predator consumption exceeded the standing stock of prey during some months. Lake trout were overshooting their prey supply!

The regulations designed to improve lake trout length structure resulted in a large imbalance in the food web. This was particularly troublesome because lake trout were competing with anglers for kokanee and rainbow trout. Moreover, it cost money to produce fish in a hatchery, and analysis of predator demands suggested that it could cost more than US$1,000 to produce one trophy lake trout. Finally, as lake trout populations expanded and prey became limited, lake trout growth slowed dramatically and negated the effectiveness of harvest regulations.

As the food web imbalance became apparent, mangers shifted focus to reducing lake trout populations through liberalized regulations (e.g., four or eight fish daily harvest limit and no length limit). Not only did this approach reduce lake trout predation pressure on rainbow trout and kokanee, but lake trout growth in some systems increased substantially. In fact, lake trout in Blue Mesa Reservoir, which had the most liberalized regulations at eight fish per day and no length restrictions, grew to 760 mm in 10 years. Thus, by understanding food web dynamics, managers were able to determine that the best way to grow trophy lake trout in these systems was to maintain low-density populations through liberalized regulations.

13.6.2 Nutrient Cycling in Salmonid Ecosystems

Stable isotope analysis is being increasingly used to track energy flow and identify trophic relationships in aquatic and terrestrial environments (Kling et al. 1992; Cabana and Rasmussen 1994; Fry 2006). This approach has been used to address a variety of topics that include quantifying diet shifts from pelagic to benthic prey (Vander Zanden et al. 1998), modeling contaminant uptake in piscivorous fishes (Kidd et al. 1996), and monitoring the effects of eutrophication on aquatic food webs (Cabana and Rasmussen 1996). Unlike gut content analysis that provides a snapshot of recent feeding events, stable isotope analysis provides time-integrated information about feeding history. The stable isotope approach has several advantages as a method for quantifying energy flow: (1) it is economical, often requiring less time and fewer fish (<10 per species or life stage) than do traditional diet studies; (2) it reflects materials actually assimilated by fish; (3) it allows more efficient use of sampled fish because there is no loss of information when stomachs are empty; and (4) it can be used to evaluate variation in intra-population feeding patterns.

Carbon (δ^{13}C) and N (δ^{15}N) are the most commonly used isotopes in aquatic food web studies. In general, C signatures of consumers are similar to those of their prey and can be used to identify C sources at the base of the food chain. Conversely, the stable isotope of N exhibits a stepwise increase from prey to predator. A 3.4‰ enrichment of the heavy N isotope (δ^{15}N) is often used to represent a trophic-level increment of one (e.g., from zooplankton to fish; Post 2002).

The use of stable isotope analysis has been particularly helpful in tracking energy flow in streams that support anadromous salmon. Anadromous salmon spend most of their lives in marine environments before returning to freshwater rivers where they spawn. After spawning, salmon die and their carcasses provide an important nutrient subsidy to freshwater environments. Because the isotopic signature of marine-derived N (MDN) is generally much different than that found in freshwater systems, stable isotopes can be used to track the contribution of MDN to aquatic and terrestrial organisms.

A comparison of stream reaches with and without spawning salmon in the Indian and Kadashan rivers, Alaska, showed that nutrients from salmon carcasses significantly influence riparian vegetation (Helfield and Naiman 2001). At spawning sites, the percentage of MDN in riparian foliage varied from 22% in ferns (i.e., *Dryopteris* and *Athyrium*) to 24% in Sitka spruce (Box 13.3). Moreover, basal area growth of Sitka spruce was significantly higher near spawning sites (~2,250 mm^2/y) than in nonspawning areas (~660 mm^2/y; Helfield and Naiman 2001). By using stable isotope analysis, researchers have shown that nutrient subsidies contributed by salmon carcasses enhance riparian production, thus contributing to increased shading, sediment filtration, and production of large woody debris. Enhancement of riparian vegetation provides a positive feedback mechanism that improves spawning and rearing habitat for young salmon (Helfield and Naiman 2001).

13.6.3 Fish Removal

Lake Christina is a large, shallow lake in central Minnesota that has a long history of waterfowl hunting. Prolonged high water and an abundance of planktivorous fishes caused turbid conditions in Lake Christina, with low densities of aquatic plants. The turbid water and lack of aquatic vegetation made Lake Christina unappealing to nesting and migrating waterfowl. To bring back waterfowl, managers decided to shift Lake Christina back to its former clear-water state by initiating a shift in the food web. They attempted to remove all fish (primarily common carp and black bullhead) from the lake by applying rotenone three different times. Reclamation was successful after the two earlier rotenone treatments. Aquatic vegetation density increased after each treatment and waterfowl usage increased (Hanson and Butler 1994; Hansel-Welch et al. 2003). However, Lake Christina gradually shifted back to a turbid-water state after each treatment. The most recent reclamation has also shown positive results as aquatic vegetation and waterfowl use have increased dramatically in the years following reclamation. Time will tell if Lake Christina is able to maintain a clear-water state.

13.6.4 Lake Fertilization

In contrast to eutrophication, oligotrophication is a process that occurs when lakes and reservoirs become less productive through time. A variety of factors can contribute to the cultural oligotrophication of freshwaters, including nutrient abatement, impoundment (e.g., res-

> **Box 13.3. Stable Isotope Analysis**
>
> In a detailed study of forested riparian areas, Helfield and Naiman (2001) measured stable nitrogen isotope values ($\delta^{15}N$) of Pacific salmon carcasses and riparian vegetation (Sitka spruce) at sites with and without spawning salmon.
>
> **Table.** Mean nitrogen isotope values ($\delta^{15}N$) of Pacific salmon and Sitka spruce.
>
> | Taxon | $\delta^{15}N$ | |
> |---|---|---|
> | | Reference sites (no salmon) | Salmon spawning sites |
> | Sitka spruce | −3.3 | 0.63 |
> | Pacific salmon | — | 13.39 |
>
> Using a two-source mixing model, Helfield and Naiman (2001) calculated the percent contribution of marine-derived nitrogen (MDN) to riparian trees as
>
> $$\% \text{ MDN} = [(\text{SAMPLE} - \text{REF})/(\text{FISH} - \text{REF})] \times 100,$$
>
> where SAMPLE is the mean $\delta^{15}N$ value of Sitka spruce from spawning areas, REF is the terrestrial end member (i.e., 0% MDN) measured as the mean $\delta^{15}N$ value of Sitka spruce from reference sites, and FISH is the marine end member (i.e., 100% MDN) measured as the mean $\delta^{15}N$ value from salmon carcass tissue. Using this equation, we obtain
>
> $$\% \text{ MDN} = [(0.63 + 3.3)/(13.39 + 3.3)] \times 100$$
>
> $$= (3.93/16.69) \times 100$$
>
> $$= 0.235 \times 100 = 24\%.$$
>
> Thus, Sitka spruce trees near spawning sites derive approximately 24% of their foliar nitrogen from Pacific salmon carcasses.

ervoir aging), wetland drainage, fish reduction, acidification and liming, logging, and climate change (Ney 1996; Stockner et al. 2000). Reservoirs, in particular, can act as nutrient sinks, and as they age become progressively less productive (see Chapter 17). In the southeastern USA, oligotrophication of reservoirs has been linked to low planktivorous fish biomass and reduced growth and harvest of sport fishes (Ney 1996).

The fertilization of nutrient-limited systems is an effective tool for increasing fish production. Chilko Lake in eastern British Columbia is a large, oligotrophic lake that represents an important rearing environment for sockeye salmon. Because of low P concentration (<3 µg/L), experimental fertilization was initiated in the late 1980s and early 1990s to increase sockeye salmon production (Bradford et al. 2000). Aerial application of inorganic fertilizer

(N:P = 25) was applied at 4 mg P/m^2 at weekly intervals from June to September for 5 years. Using 39 years of pretreatment data plus data from other local populations, Bradford et al. (2000) found that mean productivity (recruits per spawner) increased 73% following fertilization. Similarly, primary productivity and zooplankton biomass were higher following fertilization and likely contributed to increased growth of age-1 (+34%) and age-2 (+58%) sockeye salmon smolts. Fertilization of Chilko Lake appeared to increase adult production by improving growth and survival of sockeye salmon smolts (Bradford et al. 2000).

13.7 CONCLUSIONS

More than a century ago, Forbes (1887) highlighted the importance of food webs in his classic work *The Lake as a Microcosm*. Since then, evidence from whole-lake studies, combined with theoretical and analytical advances, have shown that trophic interactions have an important influence on energy flow and resultant fish production in freshwater systems. Increasingly, fisheries managers recognize how knowledge derived from studies of trophic cascades, benthic–pelagic coupling, and alternative stable states can be used to manage aquatic ecosystems. The challenge is not so much in applying this knowledge, but in deciding when and where to use it given limited time and resources. Removing fish, reducing nutrient inputs, or fertilizing lake basins require more time and funding than do traditional management activities (e.g., fish stocking, evaluating regulations, and monitoring population trends). Nonetheless, by manipulating entire lakes we gain insight into how aquatic systems work. Because most of us learn by doing, these techniques lend themselves to the adaptive management approach; only by applying these principles, monitoring their outcomes, and reporting results will we begin to appreciate fully their usefulness in contemporary management applications.

13.8 REFERENCES

Bradford, M. J., B. J. Pyper, and K. S. Shortreed. 2000. Biological responses of sockeye salmon to the fertilization of Chilko Lake, a large lake in the interior of British Columbia. North American Journal of Fisheries Management 20:661–671.
Brönmark, C. 1985. Interactions between macrophytes, epiphytes and herbivores: an experimental approach. Oikos 45:26–30.
Brooks, J. L., and S. I. Dodson. 1965. Predation, body size, and composition of plankton. Science 150:28–35.
Cabana, G., and J. B. Rasmussen. 1994. Modeling food chain structure and contaminant bioaccumulation using stable nitrogen isotopes. Nature 372:255–257.
Cabana, G., and J. B. Rasmussen. 1996. Comparison of aquatic food chains using nitrogen isotopes. Proceedings of the National Academy of Sciences 93:10844–10847.
Carlson, R. E. 1977. A trophic state index for lakes. Limnology and Oceanography 22:361–369.
Carlson, R. E. 1992. Expanding the trophic state concept to identify non-nutrient limited lakes and reservoirs. Pages 59–71 *in* Proceedings of a National Conference on enhancing the states' lake management programs, monitoring and lake impact assessment. Chicago.
Carpenter, S. R., and J. F. Kitchell. 1992. Trophic cascade and biomanipulation: interface of research and management—a reply to the comment by DeMelo et al. Limnology and Oceanograhpy 37:208–213.
Carpenter, S. R., and J. F. Kitchell. 1993. The trophic cascade in lakes. Cambridge University Press, New York.

Carpenter, S. R., J. F. Kitchell, and J. R. Hodgson. 1985. Cascading trophic interactions and lake productivity. BioScience 35:634–639.

Carpenter, S. R., and D. M. Lodge. 1986. Effects of submersed macrophytes on ecosystem processes. Aquatic Botany 26:341–370.

Chick, J. H., R. J. Cosgriff, and L. S. Gittinger. 2003. Fish as potential dispersal agents for floodplain plants: first evidence in North America. Canadian Journal of Fisheries and Aquatic Sciences 60:1437–1439.

Chipps S. R., and D. H. Bennett. 2000. Zooplanktivory and nutrient regeneration by invertebrate and vertebrate planktivores: implications for trophic interactions in oligotrophic lakes. Transactions of the American Fisheries Society 129:569–583.

Chipps, S. R. and D. H. Wahl. 2004. Development and evaluation of a western mosquitofish bioenergetics model. Transactions of the American Fisheries Society 133:1150–1162.

Covich, A. P., M. A. Palmer, and T. A. Crowl. 1999. The role of benthic invertebrate species in freshwater ecosystems. BioScience 49:119–127.

DeMelo, R., R. France, and D. J. McQueen. 1992. Biomanipulation: hit or myth? Limnology and Oceanography 37:192–207.

DiCenzo, V. J., M. J. Maceina, and W. C. Reeves. 1995. Factors related to growth and condition of the Alabama subspecies of spotted bass in reservoirs. North American Journal of Fisheries Management 15:794–798.

Dodds, W. K., W. W. Bouska, J. L. Eitzmann, T. J. Pilger, K. L. Pitts, A. J. Riley, J. T. Schloesser, and D. J. Thornbrugh. 2009. Eutrophication of U.S. freshwaters: analysis of potential economic damages. Environmental Science and Technology 43:12–19.

Drenner, R. W., and K. D. Hambright. 1999. Biomanipulation of fish assemblages as a lake restoration technique. Archiv für Hydrobiologie 146:129–165.

Forbes, S. A. 1887. The lake as a microcosm. Bulletin of the Scientific Association (Peoria, IL) 1887:77–87. Reprinted 1925, Bulletin of the Illinois Natural History Survey 15:537–550.

Frodge, J., G. L. Thomas, and G. Pauley. 1990. Effects of canopy formation by floating and submergent aquatic macrophytes on the water quality of two shallow Pacific Northwest lakes. Aquatic Botany 38:231–248.

Fry, B. 2006. Stable isotope ecology. Springer, New York.

Gallepp, G. W. 1979. Chironomid influence on phosphorus release in sediment–water microcosms. Ecology 60:547–556.

Garvey, J. E., R. A. Stein, and H. M. Thomas. 1994. Assessing how fish predation and interspecific prey competition influence a crayfish assemblage. Ecology 75:532–547.

Gliwicz, M. Z. 1986. Predation and the evolution of vertical migration in zooplankton. Nature 320:746–748.

Gotelli, N. J. 2001. A primer of ecology. Sinauer Associates, Sunderland, Massachusetts.

Graeb, B. D. S., J. M. Dettmers, D. H. Wahl, and C. E. Cáceres. 2004. Fish size and prey availability affect growth, survival, and foraging behavior of larval yellow perch. Transactions of the American Fisheries Society 113:504–514.

Hansel-Welch, N., M. G. Butler, T. J. Carlson, and M. A. Hanson. 2003. Changes in macrophyte community structure in Lake Christina (Minnesota), a large shallow lake, following biomanipulation. Aquatic Botany 75:323–337.

Hanson, M. A., and M. G. Butler. 1994. Responses of plankton, turbidity, and macrophytes to biomanipulation in a shallow prairie lake. Canadian Journal of Fisheries and Aquatic Sciences 51:1180–1188.

Hanson, P.C., T. B. Johnson, D. E. Schindler, and J. F. Kitchell.1997. Fish bioenergetics 3.0. University of Wisconsin System, Sea Grant Institute, WISCU-T-97-001, Madison. Available: http://limnology.wisc.edu/research/bioenergetics/bioenergetics.html (June 2010).

Helfield, J. M., and R. J. Naiman. 2001. Effects of salmon-derived nitrogen on riparian forest growth and implications for stream productivity. Ecology 82:2403–2409.

Hill, A. M. and D. M. Lodge. 1994. Diel changes in resource demand: competition and predation in species replacement among crayfishes. Ecology 75:2118–2126.

Holt, R. D. 1977. Predation, apparent competition, and the structure of prey communities. Theoretical Population Biology 12:197–229.

Hrbacek, J., M. Dvorakova, V. Korinek, and L. Prochazkova. 1961. Demonstration of the effect of the fish stock on the species composition of zooplankton and the intensity of metabolism of the whole lake plankton assemblage. Mitteilungen Internationale Vereinigung für theoretische und angewandte Limnologie 14:192–195.

Hutchinson, G. E. 1975. A treatise on limnology. Volume III: limnological botany. Wiley, New York.

Inskip, P. D. 1986. Negative associations between abundances of muskellunge and northern pike: evidence and possible explanations. Pages 135–150 in G. E. Hall, editor. Managing muskies: a treatise on the biology and propagation of muskellunge in North America. American Fisheries Society, Special Publication 15, Bethesda, Maryland.

Isermann, D. A., S. R. Chipps, and M. L. Brown. 2004. Seasonal *Daphnia* biomass in winterkill and nonwinterkill glacial lakes of South Dakota. North American Journal of Fisheries Management 24:287–292.

Jensen, J. P., P. Kristensen, and E. Jeppesen. 1991. Relationships between N loading and in-lake N concentrations in shallow Danish lakes. Verhandlungen Internationale Vereinigung für theoretisch und angewandte Limnologie 24:201–204.

Jensen, J. P., P. Kristensen, E. Jeppesen, and A. Skytthe. 1992. Iron : phosphorus ratio in surface sediment as an indicator of phosphate release from aerobic sediments in shallow lakes. Hydrobiologia 235–236:731–743.

Johnson, B. M., and J. P. Goettl Jr. 1999. Food web changes over fourteen years following introduction of rainbow smelt into a Colorado reservoir. North American Journal of Fisheries Management 19:629–642.

Johnson, B. M., and P. J. Martinez. 2000. Trophic economics of lake trout management in reservoirs of differing productivity. North American Journal of Fisheries Management 20:127–143.

Kidd, K. A., R. H. Hesslein, B. J. Ross, K. Koczanski, G. R. Stephens, and D. C. G. Muir 1996. Bioaccumulation of organochlorines through a remote freshwater food web in the Canadian Arctic. Environmental Pollution 102:91–103.

Kling, G. W., B. Fry, and W. J. O'Brien. 1992. Stable isotopes and planktonic trophic structure in arctic lakes. Ecology 73:561–566.

Kogan, S. I., and G. A. Chinnova. 1972. Relations between *Ceratophyllum demersum* and some blue-green algae. Hydrobiological Journal 5:14–19. Translation of Gidrobiol. Zh. 8:21–27.

Kolar, C. S., and D. H. Wahl. 1998. Daphnid morphology deters fish predators. Oecologia 116:556–564.

Lindegaard, C. 1994. The role of zoobenthos in energy flow in two shallow lakes. Hydrobiologia 275/276:313–322.

Lodge, D. M., M. W. Kershner, J. E. Aloi, and P. A. Covich. 1994. Effects of an omnivorous crayfish (*Orconectes rusticus*) on a freshwater littoral food web. Ecology 75:1265–1281.

Ludsin, S. A., and D. R. DeVries. 1997. First-year recruitment of largemouth bass: the interdependency of early life stages. Ecological Applications 7:1024–1038.

Martinez, P. J., P. E. Bigelow, M. A. Deleray, W. A. Fredenberg, B. S. Hansen, N. J. Horner, S. K. Lehr, R. W. Schneidervin, S. A. Tolentino, and A. E. Viola. 2009. Western lake trout woes. Fisheries 34:424–442.

Mehner, T., J. Benndorf, P. Kasprzak, and R. Koschel. 2002. Biomanipulation of lake ecosystems: successful applications and expanding complexity in the underlying science. Freshwater Biology 47:2453–2465.

Meijer, M. L., E. Jeppesen, E. Van Donk, and B. Moss. 1994. Long term responses to fish-stock reduction in small shallow lakes—interpretation of five year results of four biomanipulation cases in the Netherlands and Denmark. Hydrobiologia 276:457–466.

Mills, E. L., and A. Schiavone Jr. 1982. Evaluation of fish communities through assessment of zooplankton populations and measures of lake productivity. North American Journal of Fisheries Management 2:14–27.

Nesler, T. P., and E. P. Bergersen. 1991. Mysids in fisheries: hard lessons in headlong introductions. American Fisheries Society, Symposium 9, Bethesda, Maryland.

Ney, J. J. 1990. Trophic economics in fisheries: assessment of demand–supply relationships between predators and prey. Reviews in Aquatic Sciences 2:55–81.

Ney, J. J. 1996. Oligotrophication and its discontents: effects of reduced nutrient loading on reservoir fisheries. Pages 285–295 in L.E. Miranda and D.R. DeVries, editors. Multidimensional approaches to reservoir fisheries management. American Fisheries Society, Symposium 16, Bethesda, Maryland.

Oglesby, R. T. 1977. Relationships of fish yield to lake phytoplankton standing crop, production and morphoedaphic factors. Journal of the Fisheries Research Board of Canada 34:2271–2279.

Oglesby, R. T., and R. M. Jenkins. 1982. The morphoedaphic index and reservoir fish production. Transactions of the American Fisheries Society 111:133–140.

Olson, M. H., S. R. Carpenter, P. Cunningham, S. Gafny, B. R. Herwig, N. P. Nibbelink, T. Pellett, C. Storlie, A. S. Trebitz, and K. A. Wilson. 1998. Managing macrophytes to improve fish growth: a multi-lake experiment. Fisheries 23(2):6–11.

Paine, R. T. 1980. Food webs: linkage, interaction strength, and community infrastructure. Journal of Animal Ecology 49:667–685.

Palmer, M., A. P. Covich , B. J. Finlay, J. Gibert, K. D. Hyde, R. K. Johnson, T. Kairesalo, P. S. Lake, C. R. Lovell, R. J. Naiman, C. Ricci, F. Sabater, and D. Strayer. 1997. Biodiversity and ecosystem processes in freshwater sediments. Ambio 26:571–577.

Post, D. M. 2002. Using stable isotopes to estimate trophic position: models, methods, and assumptions. Ecology 83:703–718.

Rigler, F. H. 1982. The relation between fisheries management and limnology. Transactions of the American Fisheries Society 111:121–132.

Roni, P. 2005. Monitoring stream and watershed restoration. American Fisheries Society, Bethesda, Maryland.

Ryder, R. A. 1982. The morphoredaphic index—use, abuse, and fundamental concepts. Transactions of the American Fisheries Society 111:154–164.

Savino, J. F., E. A. Marschall, and R. A. Stein. 1992. Bluegill growth as modified by plant density: an exploration of underlying mechanisms. Oecologia 89:153–160.

Savino, J. F., and R. A. Stein. 1982. Predator–prey interactions between largemouth bass and bluegills as influenced by simulated, submersed, vegetation. Transactions of the American Fisheries Society 111: 255–266.

Scheffer, M. H. 1998. Ecology of shallow lakes. Chapman and Hall, London, UK.

Scheffer, M. H., S. H. Hosper, M.-L. Meijer, B. Moss, and E. Jeppeson. 1993. Alternative equilibria in shallow lakes. Trends in Ecology and Evolution 8:275–279.

Schindler, D. W. 1978. Factors regulating phytoplankton production and standing crop in the world's freshwaters. Limnology and Oceanography 23:478–486.

Scott, W. B., and E. J. Crossman. 1973. Freshwater fishes of Canada. Fisheries Research Board of Canada Bulletin 183.

Smith, V. H. 1982. The nitrogen and phosphorus dependence of algal biomass in lakes: an empirical and theoretical analysis. Limnology and Oceanography 27:1101–1112.

Søndergaard, M., P. Kristensen, and E. Jeppesen. 1993. Eight years of internal phosphorus loading and changes in the sediment phosphorus profile in Lake Søbygaard Denmark. Hydrobiologia 253:345–356.

Spencer, C. N., B. R. McClelland, and J. A. Stanford. 1991. Shrimp stocking, salmon collapse, and eagle displacement. Bioscience 41:14–21.

Stein, R. A., D. R. DeVries, and J. M. Dettmers. 1995. Food-web regulation by a planktivore: exploring the generality of the trophic cascade hypothesis. Canadian Journal of Fisheries and Aquatic Sciences 52:2518–2526.

Stockner, J. G., and E. A. MacIsaac. 1996. British Columbia lake enrichment programme: two decades of habitat enhancement for sockeye salmon. Regulated Rivers: Research and Management 12:547–561.

Stockner, J. G., E. Rydin, and P. Hyenstrand. 2000. Cultural oligotrophication: causes and consequences for fisheries resources. Fisheries 25(5):7–14.

Swingle, H. S. 1950. Relationships and dynamics of balanced and unbalanced fish populations. Alabama Agricultural Experiment Station, Auburn University, Bulletin 274.

Trebitz, A. S., and N. Nibbelink. 1996. Effect of pattern of vegetation removal on growth of bluegill: a simple model. Canadian Journal of Fisheries and Aquatic Sciences 53:1844–1851.

Vadeboncoeur, Y., M. J. Vander Zanden, and D. M. Lodge. 2002. Putting the lake back together: reintegrating benthic pathways into lake food web models. Bioscience 52:44–54.

Vander Zanden, M. J., G. Cabana, and J. B. Rasmussen. 1997. Comparing trophic position of freshwater fish calculated using stable nitrogen isotope ratios and literature dietary data. Canadian Journal of Fisheries and Aquatic Sciences 54:1142–1158.

Vander Zanden, M. J., M. Hulshof, M. S. Ridgway, and J. B. Rasmussen. 1998. Application of stable isotope techniques to trophic studies of age-0 smallmouth bass. Transactions of the American Fisheries Society 127:729–739.

Vanni, M. J., and C. D. Layne. 1997. Nutrient recycling and herbivory as mechanisms in the "top-down" effect of fish on algae in lakes. Ecology 78:21–40.

Weber, M. J., and M. L. Brown. 2009. Effects of common carp on aquatic ecosystems 80 years after "Carp as a Dominant": ecological insights for fisheries management. Reviews in Fisheries Science 17:524–537.

Welch, E. B., and D. C. Cooke. 1999. Effectiveness and longevity of phosphorus inactivation with alum. Journal of Lake and Reservoir Management 15:5–27.

Wetzel, R. G. 2001. Limnology, 3rd edition. Academic Press, New York.

Wilde, G. R., and M. I. Muoneke. 2001. Climate-related and morphoedaphic correlates of growth in white bass. Journal of Fish Biology 58:453–461.

Wisniewski, R., and M. Planter. 1985. Exchange of phosphorus across the sediment–water interface in several lakes of different trophic status. Verhandlungen Internationale Vereinigung für theoretisch und angewandte Limnologie 22:3345–3349.

Youngs, W. D., and D. G. Heimbuch. 1982. Another consideration of the morphoedaphic index. Transactions of the American Fisheries Society 111:151–153.

Chapter 14

Use of Social and Economic Information in Fisheries Assessments

KEVIN M. HUNT AND STEPHEN C. GRADO

14.1 INTRODUCTION

Fisheries management can be defined as the process of working with a given aquatic habitat and community of organisms for the benefit of people in a recreational or commercial setting (Weithman 1999). This process depends on numerous inputs to decision-making with strong emphasis on scientifically-based information (Decker et al. 2001). For many years, fisheries managers collected biological and ecological information to support management decisions, but information from resource users and other constituents was typically collected in an informal manner or through public hearings and responses to public notices. Nevertheless, it has often been said that any policy or regulation, no matter how scientifically sound, will fail if it is not in accord with the fundamental views of the public. Many seasoned fisheries managers will admit, often reluctantly, that fisheries management is as much or more about people management as it is about the fish and to be effective they must have information about those with an interest in the fate of aquatic resources. These individuals and groups are often referred to as stakeholders (see Chapter 5). Therefore, it is important that fisheries managers collect scientifically-based information from their stakeholders as well as the fishes and their habitats.

The information needed from stakeholders is diverse and involves numerous fields related to the study of humans, such as psychology, sociology, demography, anthropology, public administration and policy, geography, and economics. Each disciplinary perspective considers a different dimension (or different perspective on the same dimension) of the complex of social phenomena that is fisheries management. Collectively, these disciplines are more commonly referred to as the "human dimensions" of fisheries management and their consideration is essential for fisheries administrators and managers to make more informed fisheries management decisions. Compared with biological and ecological studies of lakes, reservoirs, rivers, and streams, human dimensions studies are relatively recent arrivals to the inland fisheries management process, primarily taking root in the past 40 years. However, understanding human dimensions has become important because the angling (and nonangling) public is increasingly demanding that the fisheries management process be open and transparent and that decisions be based on a fair process that considers the best available scientific information.

The intent of this chapter is to introduce the stakeholders involved in inland fisheries management and the general uses of social and economic information in the management process. The types and uses of social and economic information that is needed from recreational anglers in fisheries assessments are discussed and the various levels of human dimensions research that provide this information are described. The numerous types of social and economic information needed in fisheries assessments are usually collected with one of the various survey methodologies available to researchers: personal interviews, telephone interviews, self-administered mail questionnaires, Internet-based surveys, or secondary analysis of existing data. Specific rationales and techniques for survey research methods are beyond the scope of this chapter and readers are encouraged to see Knuth et al. (in press) for an in-depth treatment of the use of these methodologies in fisheries management.

14.2 SOCIAL AND ECONOMIC RESEARCH

During the course of a fisheries manager's day, important questions may arise about stakeholders and the infrastructure surrounding inland recreational fisheries. Unless management wants to rely on anecdotal information, it is necessary to have valid and reliable data. The information needed from stakeholders can be "broad" for long-range planning purposes; "comprehensive" for short-range planning decisions, commitment of resources, or establishment of goals and objectives at the program level; or "focused" for more immediate action decisions and implementation of activities (Brown 1987). Social research involves identifying stakeholders, their trip origins, species and experience preferences, and their attitudes toward fisheries resources and management (Decker et al. 2001). Social research can provide information to guide long-range planning decisions and measure what constitutes satisfying fishing experiences so that managers can develop the best possible products, services, marketing strategies, and educational programs. Economic research can assess expenditures made by stakeholders in pursuit of activities associated with fisheries resources, resultant economic impacts to local and regional economies, and what aquatic resources and fishing is "worth" to individuals and to society. Social and economic information is used to provide justification and guidance for policies and programs at various levels of the fisheries management process, and it gives agencies multidisciplinary problem-solving capabilities. Both social and economic information is critically needed but is often lacking in fisheries assessments.

14.2.1 Stakeholders in Fisheries Management

A fishery can be defined as a social system that includes fish, harvesters, and the entire support industry whose long-term success rests with sustainable fishery resources (Ditton 1997). Thus, one of the first steps in human dimensions research is identifying stakeholders who will be affected by possible changes to management practices designed to alter a fishery. Stakeholders are individuals or groups who may be affected by, or may influence, fisheries management decisions and actions—they are interested parties who have a stake in the decision. Likely stakeholders in inland fisheries management decisions include licensed recreational anglers, unlicensed recreational anglers (e.g., those exempt from license requirements like youth and the elderly), recreational fishing guides, charter and headboat operators, commercial fishers, private landowners, fishing tackle producers, local businesses that cater to anglers, Native Americans who often have fishing rights protected by certain treaties, nongovernmental organizations, other state and federal regulatory agencies, and the general public.

Although much of the general public does not currently use many fisheries resources, they may still be considered stakeholders as they can value fisheries resources and opportunities intrinsically or for their use potential. This can be considered "option value," meaning that individuals like to know they have the opportunity to use aquatic resources if they so desire. In addition to recreational opportunities, some stakeholders may value educational and research uses (e.g., scientific value). "Existence value" describes when people value fisheries despite the fact that they do not currently use aquatic resources or ever plan to in the future (Weithman 1999). Knowing that resources are out there providing important ecological functions (i.e., ecological value), valuing resources for the benefit of future generations (i.e., bequest value), and valuing the survival of species (i.e., altruistic value) are examples of existence values (Loomis and White 1996).

In addition to identifying stakeholders, fisheries managers must be prepared to weigh various viewpoints in their decision-making process because the diversity and number of stakeholders may be quite large (Krueger and Decker 1999). Often, those with a vested social or economic interest in fisheries resources receive the largest consideration in the decision-making process. Nevertheless, fisheries managers can benefit from collecting information from a variety of stakeholders before decisions are made. If managers consistently gather information from only one stakeholder group, their views of how people are affected by management decisions will be limited and biased. Often, debates about potential fisheries management goals, funding mechanisms, effects on people, and effects on other resources may be germane to more than just anglers and commercial fishers. The failure to account for all stakeholder groups on a particular decision could result in underrepresented or dissatisfied stakeholder groups challenging the legitimacy of the final decision (e.g., initiating litigation), which may delay proposed changes to management regulations or practices and sometimes can derail them altogether.

An important stakeholder group in inland fisheries management is licensed recreational anglers. Recreational fishing is important to national economies. In the USA, anglers spent nearly US$45.3 billion in 2006 on recreational freshwater fishing (Southwick Associates 2007). Canadians spent nearly CAN$2.5 billion in 2005 on fishing-related expenditures (Fisheries and Oceans Canada 2007). These expenditures have a significant effect on national economies. For example, in the USA in 2006, the US$45.3 billion spent on recreational fishing was associated with a total economic output of nearly US$125 billion and supported over 1 million jobs (Southwick Associates 2007). Sizeable portions of local and regional economies, aquatic resource conservation programs, numerous nongovernmental organizations, and fishing tackle and boating manufacturers rely on recreational fishing remaining a viable outdoor recreational activity. Additionally, many of the U.S. aquatic resource conservation efforts are funded by angler expenditures by means of license fees and excise taxes paid on fishing equipment and motorboat fuel collected through the Federal Aid in Sport Fish Restoration Act. From 1950 to 2000 in the USA, anglers contributed over US$12.0 billion for state-level fisheries conservation as a result of this act (Bohnsack and Sousa 2000). On average, 83% of funding for state fish and wildlife agencies' aquatic resource management budgets is supported by sportsmen and sportswomen (Southwick Associates 2002). Therefore, considerable social and economic research conducted for inland fisheries assessments has focused on this stakeholder group and the remainder of this chapter uses angler research as the central example to explore the application of social and economic research in inland fisheries management.

Although the chapter focuses largely on recreational anglers, that should not diminish the need to consider other stakeholder groups in decision-making processes. Fisheries stakeholder groups are diverse and may include fish watchers (e.g., salmon migration viewing), underwater photographers, scientists, members of animal protection groups, water resource dependent industries, and private property rights organizations. Varying opinions on the appropriate uses of fisheries resources are also pervading the fisheries profession (Muth et al. 1998). As a general rule, the more the inland resource is used for multiple purposes the more likely managers will encounter diverse groups with differing opinions on what constitutes appropriate use of resources. Resultantly, managers will need to seek feedback from those other than recreational anglers.

14.2.2 General Uses of Social and Economic Information

To determine how human dimensions research is used by U.S. fisheries management agencies, Simoes (2009) conducted telephone interviews with agency contact persons for human dimensions from each state and the District of Columbia. When respondents were asked to report on the ways in which human dimensions data were used by their agency from a list of five items, the majority of respondents reported that human dimensions data were used in the design of fisheries regulations (89%); local resource management plans (84%); statewide resource management plans (82%); angler educational and outreach programs and materials (69%); and other uses (38%; see Box 14.1). Most of the comments offered by respondents in the "other" category could be grouped into one of two broad categories: fiscal justification or outreach (e.g., information for legislature, public relations, economic impacts, or other fiscal justification) or, to a lesser extent, recruitment and retention of anglers (e.g., angler marketing or angler motivations). These results indicated that angler human dimensions data are being used to communicate the mission of fisheries management agencies and the economic and other societal benefits of angling activities.

Economic information is needed in fisheries assessments for several reasons. First, angler expenditures provide important revenue and employment for local communities, states, provinces, and nations. Second, many communities and their businesses, especially those in rural areas, are dependent on users of local resources for tax generation and retail sales revenues. Third, because of these benefits, there likely will be economic consequences to fisheries legislation and management decisions. Fourth, economic dependency can help to justify the need for protection or conservation of fisheries resources. Fifth, economic information can show the value of resources over time, which can reflect the changing quality of the fisheries resource and (or) fishing experiences. Sixth, economic information can aid in determining compensation in the event of environmental damage to fisheries resources through negligent land use practices or blatant criminal activity (e.g., dumping). Finally, economic information is useful in setting license and permit fee structures.

14.2.3 Specific Types and Uses of Social Information

The information needed for various uses of social information falls into six general categories: (1) angler characteristics; (2) participation patterns; (3) opinions and preferences; (4)

Box 14.1. Collection, Use, and Importance of Angler Human Dimensions Data: a Survey of U.S. Fisheries Management Agencies

Table A. Frequency and percent of U.S. fisheries management agencies ($N = 55$) that indicated they used human dimensions information for the listed reasons.

Human dimensions use	N	Percent
Design of fishery regulations	49	89
Local resource management plans	46	84
Statewide strategic resource management plans	44	81
Development of angler educational and outreach programs and materials	38	69
Other	19	35

Those who responded to the "other" category were asked to indicate how human dimensions data were used by their agency. Respondents reiterated each of the four original response categories, with several other dominant themes also emerging. Most of the 19 respondents added that angler human dimensions data were used in developing resource management plans; less than half indicated that data were used in conducting public relations and outreach, informing legislature or validating programs, and developing regulations.

Table B. Frequency of open-ended responses by U.S. fisheries management agencies that indicated "other" uses of human dimensions information in their agency ($N = 19$).

Human dimensions use	Frequency
Developing resource management plans	14
Conducting public relations and outreach	9
Informing legislature or validating program	8
Developing regulations	7
Researching angler motivations or behavior or profiling	7
Obtaining economic information or assessing impacts and valuation	6
Marketing to anglers and increasing recruitment retention	5
Evaluating programs and services	3
Evaluating fiscal justification (state and federal funding)	3

perceptions, beliefs, and attitudes; (5) motivations, expectations, and satisfaction; and (6) culture and value orientations. As natural resource agencies continue to conduct human dimensions research, they can develop a perspective on trends in each of these areas. Together with biological, economic, and policy information, agency personnel are better able to develop an integrated management perspective that allows them to make more justifiable recommendations to fisheries decision-makers (Brown 1987). For example, collecting social information may be paired with the collection of biological data to understand the interdependence of human well-being and ecosystem health and services, and to inform fisheries managers. Further, such comprehensive studies help inform decisions about species management and regulation alternatives. For example, social research addressing whether the angling public is satisfied with current fishing opportunities, fish populations, composition of catches, or fishing regulations can be compared with fish stock assessments to determine whether different regulations may be possible or whether manipulations of the organisms (e.g., stocking) or habitat (e.g., fish-attracting structures) should be considered as possible management alternatives. If management goals may be achieved through different regulations or organism or habitat manipulations, angler surveys can help inform managers about the course of action likely to be desired most by anglers.

14.2.3.1 Angler characteristics

Collecting information to describe an angler population is similar to collecting water temperature, dissolved oxygen, and pH data each time a biologist goes to the lake. Information about angler populations provides managers with basic information needed to put other findings in context. By understanding the relationships of human descriptors with other social and economic information, managers can design management programs that are responsive to the needs and abilities of a variety of angler groups or stakeholders. Information that characterizes participants includes demographic and social information.

Demographic information characterizes who anglers are and their trip origins, and includes data such as age, income level, education level, race, ethnicity, and gender. Anticipated changes in the U.S. general population, particularly an aging population, increased immigration, and an increased percentage of minority populations, will likely have future effects on fisheries management (Murdock et al. 1996). For example, some states in the USA, such as California, Florida, Arizona, New Mexico, and Texas, are currently witnessing drastic increases in their Hispanic-Latino populations. Tracking changes in the demographic composition of anglers is important in determining which strategies to use to ensure recreational fisheries management will continue to be relevant to all of society. Additionally, a more diverse population will require more diverse amenities at fisheries resources to meet their needs and expectations.

Coupled with demographic information and anglers' trip origins, social characteristics of anglers can further inform the manager about this stakeholder group. For example, the proportions of anglers who belong to fishing clubs or organizations (e.g., Trout Unlimited or Bass Anglers Sportsman Society), participate in fishing tournaments, or subscribe to fishing magazines, and with which organizations, tournaments, or magazines they are affiliated, can assist fisheries managers in determining the types of clientele visiting a particular resource. This can assist managers in determining where to relay fishing and marketing information about agency programs and services. The incorporation of geographic information system tools into social research also enables managers to visualize better where angler trips originate, provided zip code, city, or county of home residence information is collected.

14.2.3.2 Participation patterns

Many fisheries management objectives are targeted toward achieving a particular level of angler-days and increasing overall participation in recreational fishing at particular resources or statewide or province-wide. Therefore, it is important to gather information about fishing frequency and participation rates in recreational fishing. For instance, fisheries management agencies track angling participation rates (the proportion of the public that participates in fishing) for determining short- and long-term demands that will be placed on fisheries resources. Agencies and researchers have also recognized that angler populations are not composed of the same people year after year and therefore investigate the rate at which anglers enter or drop out of the customer base in terms of purchasing or not purchasing fishing licenses; this is termed the angler "churn rate" (Strouse 1999). Understanding participation is critical to the financial side of fisheries management. The number of fishing licenses sold is part of the equation for determining each state's allotment of funds from the Federal Aid in Sport Fish Restoration Program (see Box 4.3) and provides much of the matching funds necessary to receive that allotment. Such data enable fisheries agencies to document the values of fisheries management to policy makers and the general public.

Whereas the number of anglers in the USA appears to be relatively stable, participation rates have decreased nationwide in the past 25 years (USDI 2007). Similarly, resident angler participation rates in Canada have shown a downward trend in most provinces and territories since 1995 (Fisheries and Oceans Canada 2007). It should be noted that a rise in the number of anglers does not necessarily result in a rise in the participation rate. In fact, the participation rate can still decline if the number of new recruits to fishing does not increase at the same rate as the general population increases. Whereas a large number of anglers can show legislators that fishing is important to many people, agencies may be more effective if they can show that the percentage of the population participating in fishing is increasing as well.

Recognizing declining participation rates, many fisheries management agencies are trying to improve angler retention by producing more satisfying recreational experiences and are instituting recruitment efforts designed primarily to attract youth. However, most of the increases in the U.S. population, and most of the future recruits to recreational fishing, are projected to come from nontraditional groups (e.g., those other than Caucasian males) (Murdock et al. 1996). It is important to know whether nontraditional groups differ in their resource use and socialization patterns so agencies can develop recruitment efforts to attract them for further financial and political support. For example, Hunt and Ditton (2002) found that the average African-American male angler in Texas did not start fishing until his teens and the average Hispanic male angler did not start until his early twenties. Current agency efforts aimed at reaching out to youth alone may not be as effective at attracting new participants from an increasingly diverse cultural population. Fisheries managers need to be able to track and recognize participation trends by various segments of a population so they can be proactive in developing strategies to address these new circumstances. In the future, it will be difficult for agencies to maintain their support from increasingly diverse state legislatures if they cannot provide equitable benefits and services to a diverse citizenry.

14.2.3.3 Opinions and preferences

Social research helps managers assess the likely effects of management decisions on people. Understanding opinions and preferences about alternative management approaches enables managers to judge the probable political and social acceptability of various sets of actions. Managers are able to select management actions that have a high degree of probable acceptance, effectiveness, and desirable human outcomes in an affected community, which in turn enhance compliance. Those who interact with fisheries resources often have novel ideas that may assist managers in providing quality recreational experiences (Ditton 2004).

Three common approaches for obtaining preference information from anglers include: (1) the traditional single-item question approach, (2) revealed-preference models, and (3) stated-preference choice models (Louviere and Timmermans 1990). The traditional single-item question approach involves asking anglers to indicate whether they support or oppose each of several management options, usually as stand-alone items (e.g., a proposed 305-mm minimum length limit or a five-fish-daily bag limit on rainbow trout). Although this has been the traditional approach used in fisheries assessments, it does not convey the relative importance of each of the options to anglers and the tradeoffs they are willing to make when considering restrictions jointly (Ditton 2004). The revealed-preference approach looks at actual behavior to determine angler preferences for regulations. It is assumed that anglers will choose fishing locations with the regulations that they prefer. To determine preference, anglers are surveyed as to where and how often they fish at various locations and what the regulations (and other attributes) were at those locations. The stated-preference choice approach makes use of hypothetical scenarios to derive individuals' preferences by measuring their choice of preferred scenarios. This approach assumes that complex decisions are not based on one factor, but on several considered jointly. Results allow managers to understand how anglers combine their preferences for various management measures under consideration and the relative influence of each management measure (e.g., a combination of 305-mm minimum length on rainbow trout and a five-fish-daily bag with the bag limit contributing most to the angler's support rating). Because of its ability to use hypothetical scenarios, the stated-preference choice approach has been used increasingly in angler surveys as more researchers and agencies discover its benefits (Aas et al. 2000; Gillis and Ditton 2002; Oh et al. 2006).

14.2.3.4 Perceptions, beliefs, and attitudes

A complete understanding of participation patterns and preferences often requires understanding angler perceptions, beliefs, and attitudes. Information about people's perceptions, beliefs, and attitudes helps managers understand what people think about a fishery resource, its importance, and how stakeholders would like to interact with the resource. Individuals' perceptions about what is real (whether it is or not) often influence their beliefs (e.g., whether something is good or bad) and resultant attitudes (e.g., positive or negative evaluation) toward particular behaviors and management actions. For example, where most individuals are no longer relying on fish and wildlife resources for subsistence purposes (i.e., living directly off of the land), many individuals use fisheries resources for sustenance (i.e., supplementing their diet). Often, those who rely more on fisheries resources for sustenance include lower-income anglers, and some ethnic groups have been found to have attitudes more consistent with higher consumption (Burger 2000; Hunt et al. 2007). We would expect these groups to respond unfavorably to reductions in bag limits, even if reductions may protect anglers from high levels of toxic substances increasingly being found in inland waters. We would expect this response because their beliefs are consistent with their perception that fish are a healthy and inexpensive source of protein. As a result, they have developed positive attitudes toward consuming fish, especially if they have not witnessed any negative health effects. Where contamination risks and these populations coincide, a stronger propensity to consume fish puts them at greater health risk than others (Burger 2000). In this situation, knowing the perceptions, beliefs, and attitudes held by anglers about consuming fish beforehand can assist agencies in developing information and education programs designed to produce voluntary changes in behavior that may limit their risks and (or) create more compliance with necessary regulations.

14.2.3.5 Motivations, expectations, and satisfactions

The reasons why people participate in recreational fishing have been studied extensively by human dimensions researchers. Research in this area began when Bultena and Taves (1961) observed that anglers returning from fishing trips in the Quetico Superior area of Minnesota were not dissatisfied with their visit to the area despite not having any fish in their creels. Bultena and Taves hypothesized that there must be multiple motivations for fishing, and researchers since have sought to investigate reasons for fishing aside from the catch. Most of the non-catch-related motivations for fishing have been found to be to relax, to experience natural settings, to explore or achieve, to escape temporarily from the regular routine, to be with family and friends, or to get away from family and friends (Fedler and Ditton 1994). Although many believe that fulfillment of these motivations are out the fisheries manager's control, it has helped to educate fisheries managers that the fishing experience is more than catching fish and to lead managers to look for better ways to improve the esthetic and social settings surrounding fisheries resources.

In social science research, satisfaction is defined as the fulfillment of expectations (or motivations) and is ideally measured by subtracting a posttrip or postseason measure of performance from a pretrip or preseason measure of expectation (Brunke and Hunt 2007). Nevertheless, only posttrip or postseason satisfaction ratings have been found to provide managers with useful information about fishing trips (Arlinghaus 2006). Measuring satisfaction allows fisheries managers to determine to what extent the needs and desires of people are met through a fishery resource and a fishing experience. Fisheries management actions may be developed to increase satisfaction, either by manipulating the biota (e.g., stock different species), altering the physical environment (e.g., provide more observation or access points), or informing or governing people (e.g., setting realistic expectations by providing factual information or shifting fishing pressure by regulating angling behavior). Fly-fishing-only, catch-and-release-only, limited-use, or family-oriented areas that offer various amenities are just a few examples of managing resources to attract similarly motivated anglers. Thus, multiple uses can be managed in a manner that maximizes overall satisfaction.

The definition of what constitutes a satisfying fishing experience is highly subjective. Each angler derives somewhat different benefits from a fishery resource and has a different set of preferences and opinions of how a fishery should be managed. Because preferences vary widely among individual anglers, the capability to manage resources based on specific angler desires is in its infancy, and some may even argue that it is impractical. Therefore, most agencies focus on providing a diversity of recreational fishing opportunities with the hope that each angler will seek out the experience that meets his or her particular expectations. According to Weithman (1999), fisheries managers will be successful in meeting expectations if they: (1) carefully consider the entire array of benefits that fisheries resources offer, (2) make the effort to determine what anglers want, and (3) develop a process to explain to anglers the differences between their expectations and a manager's ability to affect change at a particular resource. A fourth item should be added to Weithman's list: communicating the types of available resources so the angler can make a more informed decision in choosing a fishing locale from the diversity of opportunities.

14.2.3.6 Culture and value orientations

Culture embodies a system of shared beliefs, values, customs, symbols, and behaviors that members of society use to cope with their world and with one another (Bates and Fratkin 2002). Cultural patterns or value orientations contribute to the way people think about the world and the manner in which they behave. Four cultural patterns have been identified as key descriptors of differences in leisure and environmental orientation: (1) humankind-nature orientation (utilitarian, harmonic, or fatalistic), (2) time orientation (past, present, or future), (3) activity orientation (doing or being), and (4) relational orientation (individualistic or collectivistic; Simcox 1993). U.S. society has traditionally been dominated by the Anglo or European culture, which is seen as utilitarian and individualistic and in which people are future oriented and goal driven to achieve desired end states. Other cultures around the world and some subcultures within the USA have been found to have different value orientations (Bates and Fratkin 2002). This is important because value orientations combine to represent a collective feeling toward fish and wildlife as well as recreational opportunities (Weithman 1999).

Recent research indicates that wildlife and fisheries value orientations are changing as the USA becomes a more diverse society (Teel et al. 2007), and some U.S. subcultures have been found to be different with respect to fishing behavior, motivations, and attitudes (Toth and Brown 1997; Hunt and Ditton 2001, 2002; Hunt et al. 2007). Additionally, recent declines in fishing participation, increased attention to the actions of animal rights organizations, and increasing multiuse conflicts over wildlife- and fisheries-related issues are all signs of changing value orientations. For example, catch-and-release fishing is becoming increasingly popular among the traditional clientele. However, some cultures view fish predominately as a food source and argue that this practice is akin to playing with fish and results in waste. Studies of culture and values can help managers understand why people use fisheries in certain ways.

Information about the social world or culture surrounding the human element of a fishery also provides a context for the manager to understand the sources of human beliefs about a fishery and the importance of a fishery to local or regional communities. For example, in many Hispanic-Latino communities in the USA outdoor recreation participation occurs in large groups consisting of family and extended family (Hunt and Ditton 2002). Thus, their selection of resources will most likely consist of areas that can support larger groups and may be inconsistent with some current management practices that are designed to minimize participation to help anglers "get away from it all." This latter philosophy is based primarily on the needs and desires of the traditional angler clientele, not necessarily those that managers will be increasingly encountering in the future. Thus, understanding differences in cultural value orientations and how they relate to uses of fisheries resources will be an increasingly important component to fisheries management.

14.2.4 Specific Types and Uses of Economic Information

To gather economic information related to fisheries assessments, economic research focuses on two primary areas: (1) economic impacts, often referred to as input-output analysis, and (2) economic valuation.

14.2.4.1 Input-output analysis

Economic impact analysis often focuses on fishing activities and economics associated with an economy of interest (typically a county or combination of counties, parishes, a state, states, or provinces) and provides measures of fishing activities' contribution to these economies. Economic impacts derived from each separate or collective set of counties, states, or provinces are increasingly modeled using impact analysis for planning (IMPLAN) software originally developed by the U.S. Forest Service to assess its forest management plans on a local economy. This software uses economic data from an area of interest (e.g., individual counties or a specific state or province) to construct a model of the economy. Expenditure profiles of fishing trip participants, coupled with respective angler use in terms of activity days, are needed to perform an economic impact analysis. Economic impacts of expenditures from angling activities (e.g., sporting gear) and associated trip activities (e.g., fuel and food) can be generated from county or statewide models derived from IMPLAN software.

Expenditures can be identified during a survey process (e.g., mail or on-site) by type of purchase made in a specific location on behalf of fishing activity and the expenses associated with all fishing trip activities. These expenditures can then be organized into final demands on county, state, or provincial industries and businesses. An activity day is the presence of one person for a portion of a day at a resource where the activity is taking place. As a result an itemized participant expenditure profile (U.S. dollar/participant/activity day) is often used as an input in the IMPLAN model, in which each item is entered separately and aligned with its appropriate economic sector. Once all expenses are entered, they are matched with activity days for a site or activity.

Input-output models for each county, county combination, or state (or province) economy can be built to generate direct and secondary economic impacts resulting from in-economy expenditures and coinciding activity days. Table 14.1 is a typical economic impact table that presents results of a local input-output analysis which fisheries managers will likely come across during their professional career. Direct impacts include sales, salaries, wages, and jobs created by the initial purchases of recreational anglers and represent that portion of these expenditures retained by local businesses in their business operations. Secondary impacts are composed of indirect and induced impacts. Indirect economic impacts occur when industries or businesses the local economy sell their products to those making direct sales. Indirect impacts are created through purchases made by directly engaged businesses or individuals that have supporting businesses in the local economy. Indirect impacts will include the same categories as direct impacts since industries or businesses then purchase additional inputs such as materials and labor from other economy sectors (Grado et al. 2001). Induced impacts occur

Table 14.1. Economic impacts of fishing trips associated with the 2007 recreational boat fishery at Sardis Lake, Mississippi, to the three-county region surrounding the reservoir. All dollar amounts are in U.S. Dollars.

Industry	Direct impacts ($)	Secondary impacts ($)	Total economic impacts ($)	Value added ($)	Employment number
Agriculture, forestry, and fisheries	1,198	110,763	111,961	49,785	1
Mining	0	499,059	499,059	307,711	3
Construction	0	703	703	92	0
Manufacturing	9,333,182	4,397,459	13,730,641	6,245,044	163
Transportation, communication, and utilities	0	434,824	434,824	216,956	4
Trade	0	382,833	382,833	197,448	5
Finance, insurance, and real estate	0	129,123	129,123	63,286	3
Services	6,052,332	2,027,145	8,079,477	4,242,441	105
Total	15,386,712	7,981,909	23,368,621	11,322,763	283

from household consumption generated by employment tied to direct and indirect economic impacts that generate sales, salaries, wages, and jobs. An example is contributions to the local economy from wages spent by hotel and lodging employees catering to anglers visiting the resource. The sum of direct and secondary impacts is the total economic impact to the economy of interest as a result of angler expenditures.

In addition to total economic impacts from employment and income, value added and total full- or part-time jobs are often presented in economic impact tables. An important component of value added is the employee income garnered by the local labor force. This benefit is measured by the number of jobs supported annually on a full- and part-time basis. Additionally, economic multipliers are derived and used to evaluate incremental contributions to each economy from changes in final demand for commodities associated with fishing and fishing trip-related activities. Results enable researchers to determine the extent to which other industries of a local economy (e.g., manufacturing, government, and services) benefit from resource-related activities and (or) may be underserving current clientele through their scarcity or absence in a local economy. Leakages (i.e., expenditures leaving the local economy due to its lack of capacity to supply the good or service) from both direct and indirect purchases do occur and are taken into account in the IMPLAN model.

Economic multipliers derived from input-output analysis are used to explain a respective economy's (e.g., local, region, state, or province) ability to absorb and use in-region fishing-related expenditures. Several key ratios or multipliers are developed from IMPLAN outputs for each resource-based activity. Social accounting matrix (SAM) multipliers are used to evaluate incremental contributions to an economy from per unit changes in activity-based expenditure levels. A SAM multiplier (often referred to as TYPE SAM) is computed by dividing total economic impacts by direct economic impacts (Olson and Lindall 2000). Figure 14.1 shows the relationship between direct economic and total economic impacts. As can be seen from the figure, two reservoirs that have the same amount of direct impacts from angler expenditures can have different total economic impacts. Dividing total economic impacts by direct economic impacts shows that the SAM multiplier for reservoir 2 is 1.6 whereas that of reservoir 1 is 1.4. Therefore, money

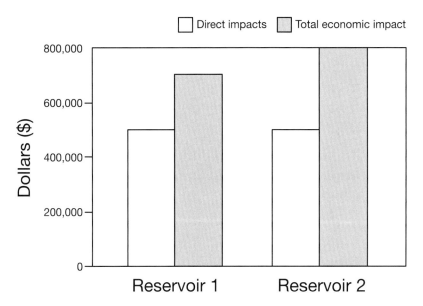

Figure 14.1. Economic impact of two reservoirs illustrating the relationship of direct impacts (equal in this case) to total economic impact. Money spent at reservoir 1 "leaks" out of the local economy at a faster rate than it does at reservoir 2.

spent at reservoir 1 "leaks" out of the local economy at a faster rate than it does at reservoir 2. Multiplier size may be related to the areal size of an economy because value added within an area has the potential to increase as its geographic area increases: more than likely a smaller proportion of expenditures are purchased outside the region (Loomis and Walsh 1997). Also, the extent of development within an economy is a factor in multiplier size. Typical recreational expenditure multipliers vary from 1.5 to 2.7 in the USA (Loomis and Walsh 1997).

The above discussion of economic impact analysis made no distinction between where the angler expenditures used in the input-output models originated. For nonresidents who do not live in the impact area surrounding the resources in question, their expenditures represent an influx of new money to the county, state, or provincial industrial and commercial bases and are always used in total in economic impact analyses. A debate among economists surrounds how to treat resident expenditures. Some researchers have discounted using resident expenditures to derive in-economy impacts because they are viewed as a redistribution of money within a respective economy. Specifically, it is argued that if residents did not go fishing at a particular resource in a county, state, or province, they would likely spend their money on something else in that same area. However, it is also likely that some resident anglers would fish elsewhere (e.g., another county, state, or province) and some of their expenditures would leave the economy of interest. It is up to the researcher to decide how to parse out this breakdown and determine the appropriate use of resident expenditures. This determination begins by asking appropriate questions on this issue in a survey process.

As understood from the preceding discussion on economic impacts, recreational fishing expenditures can have a pronounced effect on a particular economy from a solely monetary perspective (Johnson and Moore 1993). Regional expenditures can generate millions of dollars in sales and taxes and can be related to the number of jobs supported in the public sector and private industry (Burger et al. 1999; Steinback 1999). Recreational expenditures can have a positive effect on the natural setting in which they take place since the availability of fisheries, fisheries habitat, and off-site accommodations (e.g., lodging and food) have been known to affect user interest and participation in fishing and could limit revenues to an affected economy unless there is some level of stewardship. Whereas information generated from economic impact analysis can provide fisheries administrators with a great deal of power to garner support from politicians and the business community, it must be emphasized that this method of tallying expenditures results in a monetary value to a local or regional economy and not to the recreation participant or society. Nor does it relate to the on-site value of the activity. Expenditure data are frequently misused by laypersons to represent the value of fishing trips to anglers (Pollock et al. 1994). The economic value of fishing trips to the individual angler and to society is presented in the next section.

14.2.4.2 Economic valuation research

Since the 1970s, economists have become increasingly interested in placing monetary values on goods and services not exchanged in the marketplace. These values are called "nonmarket values" and include recreational opportunities such as fishing. The impetus to develop appropriate methods of valuation arose from the need for benefit-cost assessments of public goods (Swanson and Loomis 1996; Davis et al. 2001). An economically efficient mix of market and nonmarket goods can be determined if public good values (e.g., recreational fishing opportunities) can be estimated in a manner that makes them directly comparable with market prices.

Whereas expenditures and resultant economic impacts discussed in the previous section are useful indicators of the importance of recreational fishing to local, state, provincial, and national economies, they do not measure the economic benefit to either the individual participant or society (beyond the impacts on the economy; Boyle et al. 1998; Aiken and Pullis La Rouche 2003). Expenditures and net economic values are two widely-used but distinctly-different measures of the economic value of recreational fishing. Net willingness to pay (WTP) is usually referred to as "consumer surplus" and represents an individual's WTP for fishing over and above what they actually spend to participate. The summation of all recreational angler consumer surplus, for example, represents the benefit of recreational fishing to society. Figure 14.2 is adapted from an economic valuation addendum to the *1996 National Survey of Fishing, Hunting, and Wildlife-Associated Recreation* (Boyle et al. 1998) and simplifies the concept of economic valuation and its relationship to expenditures and economic impacts.

As per Figure 14.2, the previous section on economic impact is concerned with only expenditures (rectangle a-b-d-e). Net economic value measures participants' WTP for recreational fishing over and above what they actually spend to participate (triangle b-c-d). The benefit to society is the summation of WTP across all individuals. However, there is a direct relationship between expenditures and net economic value. A demand curve for a representative angler is shown in the figure. An individual angler's demand curve gives the number of trips the angler would take per year for each different cost per trip. The downward sloping demand curve represents marginal or additional WTP per trip and indicates that each additional trip is valued less by the angler than is the preceding trip. All other factors being equal, the lower the cost per trip (vertical axis) the more trips the angler will take (horizontal axis). The cost of a fishing trip serves as an implicit price for fishing since a market price generally does not exist for this activity. At $60 per trip, the angler would choose not to fish, but if fishing trips were free, the angler would take 20 fishing trips. At a cost per trip of $25 the angler will

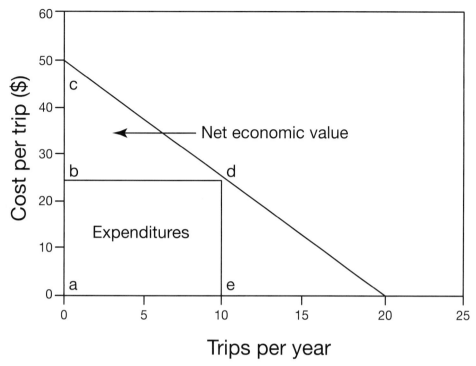

Figure 14.2. Individual angler's demand curve for fishing trips (adapted from Boyle et al. [1998]). The curve represents marginal or additional willingness to pay per trip and indicates that each additional trip is valued less by the angler than the preceding trip. The difference between what the angler is willing to pay and what is actually paid is net economic value.

take 10 trips, with a total WTP of $375 (area a-c-d-e). Total WTP is the total value the angler places on participation. The angler will not take more than 10 trips because the cost per trip ($25) exceeds what they would pay for an additional trip. For each trip between 0 and 10, however, the angler would actually have been willing to pay more than $25 (i.e., the demand curve, showing marginal WTP, lies above $25; Boyle et al. 1998).

The difference between what the angler is willing to pay and what is actually paid is net economic value. In this simple example, net economic value is $125 ([$50 − $25] × 10 ÷ 2; i.e., area of triangle b-c-d in Figure 14.2) and angler expenditures are $250 ($25 × 10; area of rectangle a-b-d-e in Figure 14.2). Thus, the angler's total WTP is composed of net economic value and total expenditures. Net economic value is simply total WTP minus expenditures.

The relationship between net economic value and expenditures is the basis for asserting that net economic value is an appropriate measure of the benefit an individual derives from participation in an activity and that expenditures are not the appropriate benefit measure. Expenditures are out-of-pocket expenses on items an angler purchases in order to fish. The remaining value, net WTP (net economic value), is the economic measure of an individual's satisfaction after all costs of participation have been paid. Summing the net economic values of all individuals who participate in recreational fishing derives the value to society. For our example, let us assume that there are 100 anglers who fish at a particular reservoir and all have demand curves identical to that of our typical angler presented in Figure 14.2. The total value per year of this reservoir to society is $12,500 ($125 × 100; Boyle et al. 1998).

Data for estimating net economic values are expensive to collect, challenging to analyze, and difficult to put into useful form for management decisions; they also present difficulties when trying to make convincing arguments as to their validity. Numerous methodologies (i.e., travel cost method, hedonic pricing method, and contingent valuation method [CVM]) can be used to estimate net economic values to achieve a common ground for comparison with market values, but it is not our intent to discuss these methodologies in-depth. A brief description of the CVM follows as this is the methodology that is used by the U.S. Fish and Wildlife Service (FWS) and is increasingly being used by fisheries economists to determine angler WTP. For a thorough discussion of economic valuation techniques, the reader is referred to Weithman (1999) and Haab and McConnell (2002).

The CVM has become a commonly-used tool for valuing natural resources (e.g., aquatic habitats and air and water quality) and other public goods not traded in the marketplace (e.g., recreation activities; Loomis and Walsh 1997; Groothuis 2005; Oh 2005). The CVM has primarily been directed toward the benefit–cost analyses of public goods such as recreation and has been defined as any approach to valuation that relies upon individual responses to contingent circumstances in an artificially-structured market. It typically uses a bidding approach (Swanson and Loomis 1996). The CVM uses simulated markets in an interview process to estimate WTP. The resource to be valued is often described orally with possible on-site inspections or it is presented through the use of self-administered mail surveys, with descriptions, photographs, or drawings to describe changes in characteristics like water, fishing, or habitat quality (Klemperer 1996). In CVM, a sample of the affected population is asked about their WTP contingent on hypothetical changes in existing environmental qualities, settings, or recreational opportunities. Alternatively, as per the previous discussion, CVM questions can be directed to the maximum amount a respondent is willing to pay to experience a resource or recreational fishing opportunity. While these formats are tailored to open-ended questions, closed-ended questions can also be formulated in this process. Regardless, a key assumption in the CVM is that consumers are able and willing to answer such questions truthfully and they have the knowledge to do so accurately. Studies done to test the validity of CVM research found this method to be a reasonable approach to valuation, particularly in the area of natural resources (Loomis and Walsh 1997).

In addition to WTP for recreational fishing opportunities, fisheries management agencies have also used CVM to make management assessments on license fees (Enck et al. 2000). Generally, license fees have been kept low and have not increased at the same rate as the cost of living (Sutton et al. 2001). With the current erosion of fishing and hunting participation across the USA (Mehmood et al. 2003; Miller and Vaske 2003), agencies have been forced to increase license fees or introduce new fees to maintain the current levels of management, programs, and funding. However, increasing license fees or introducing new fees can have negative effects on fish and wildlife agencies in the long run by decreasing participant satisfaction or causing them to drop the activity (Sutton et al. 2001). Sutton et al. (2001) used the CVM to determine angler WTP for license fees at a fishery near Fort Hood, Texas, and to assess how prices could be set to "reduce access," "maximize profits," or "maximize access." The study should be reviewed for a more thorough description. However, Box 14.2 shows a hypothetical example of angler WTP for increases to a reservoir fishing permit. With participation rates decreasing, the box illustrates how results of a CVM question can be used to keep loss of anglers to a minimum while achieving funding goals rather than to maximize revenues.

Box 14.2. Economic Analysis to Set License Fee Increases to Meet Increased Management Costs and Maximize Angler Retention

Imagine you work at a lake where about 26,500 anglers purchase an annual fishing permit for $10.00. This permit covers your salary, a technician's salary, and management costs. For a couple of years, everything is going well and anglers are happy with their fishing and the facilities. After a few years, however, you have witnessed the influx of exotic vegetation and your facilities and boat ramps are in dire need of repair. You need more money for the removal of the vegetation and building materials for facility repair. Your only way to increase revenue is to increase the cost of the fishing permit, but at the same time you do not want to lose a lot of anglers. To accomplish this, you need to determine what price increase will satisfy your need for revenue while minimizing attrition. From a contingent valuation method question you asked in a survey of a sample of anglers using the lake, the logistic regression analysis determined that the median willingness to pay for a permit increase was around $20. Increasing the permit fee to $30 would result in over $250,000 in added revenue, more than enough for your management needs, but at this cost, you would lose 13,250 (i.e., half) the anglers and probably would not win you much favor in the local community. Because you really need only $150,000 for your activities, the prudent thing to do is to determine where the total revenue increase needed (i.e., management costs) intersects the total revenue curve (illustrated by the circle in the figure below) and drop down to the demand curve. Drawing a straight line to the permit cost increase indicates a permit price increase of roughly $8.00 meets your revenue needs and minimizes attrition by only a few thousand anglers.

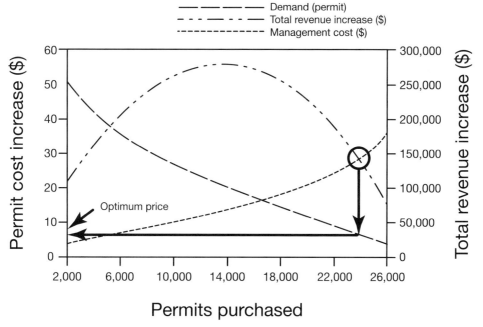

Figure. A graphical demonstration of how to use willingness to pay for increases to a $10 permit to set permit fees at an optimum price. An optimum price maximizes participation while producing sufficient revenues to meet management costs.

14.2.5 Levels of Analysis in Human Dimensions Studies

Evaluation is a key component of the fisheries management process (Krueger and Decker 1999). Most social and economic research starts in an effort to inform, evaluate, and provide a sound scientific basis for broad policies and goals for fisheries management. Many fisheries management objectives focus on human-related outcomes (e.g., number of angler-hours, measures of angler satisfaction, angler approval, or generation of economic impacts), requiring social and economic information to determine if the objective has been met. If objectives for managing a fishery have not been met, human dimensions information may help a manager understand why the objectives were not met. In developing and evaluating management plans, human dimensions information can help inform whether the objectives from the current management cycle are appropriate to continue in the next management cycle or whether modifications or further discussion about the reasons for specific objectives may be necessary.

Social and economic evaluations are often progressive endeavors with many fisheries management agencies. Regardless of the methodology chosen to collect social and economic information, formalized social and economic studies of anglers usually begin at the national level and then evolve to state- or province-wide, species, and resource level studies.

14.2.5.1 National level

The national survey of hunting, fishing, and wildlife-associated recreation conducted every 5 years since 1955 by the FWS (USDI 2007) often is the basis of U.S. management agencies' understanding of their angler populations. Fisheries and Oceans Canada (2007) has conducted a survey of recreational fishing in Canada since 1995. The FWS study is funded by Federal Aid in Sport Fish Restoration Program monies that are removed prior to distribution to the states because all states benefit from the information. The national survey focuses on all participants in fishing and not just licensed anglers. This study gives a broad view of each state's angler population (in-state and out-of-state), their fishing participation patterns, species preferences, and expenditure levels. Similar province-wide information is generated from the Canadian study. Some states and provinces rely solely on the information available in these reports and public hearings for their social and economic information needs. However, many states and provinces use this information as a starting point and embark on more in-depth state or provincial level studies for more precise estimates about their licensed anglers and other human dimensions information not collected in national studies (Wilde et al. 1996).

14.2.5.2 State- or province-wide level

State- or province-wide licensed angler studies are designed to give a broad view of the respective jurisdiction's licensed fishing population. These surveys are conducted annually by some states or provinces or in 3- or 5-year intervals (Wilde et al. 1996). Statewide surveys gather information in a variety of areas: demographics, use patterns, preferred species, participation in clubs and tournaments, reasons for fishing, and attitudes about fish, fisheries resources, and general management tools (e.g., support for the idea of stocking and bag limits). This information provides a rationale for current fisheries policy and management goals and guides their future modification. For example, statewide angler surveys reveal species preferences and species sought by licensed freshwater anglers in a particular state. Identification of preferred species identifies to agencies which species their management efforts should target. For example, if survey results show that over 60% of all angler-days are targeted at largemouth bass in reservoirs, these results reaffirm to decision-makers that freshwater recreational fisheries management should primarily focus on largemouth bass and that one of the primary management goals should be to provide satisfying largemouth bass fishing experiences in reservoirs.

14.2.5.3 Species level

State- or province-wide surveys are designed to examine fishing activity in general, thus they offer little insight into the fishing behavior of anglers seeking a particular species. Because some agencies are interested in learning more about fishing in the state or province for a particular species (e.g., smallmouth bass), angler responses to species preference questions create a list of species-specific anglers with whom to follow up about their fishing behavior. Since the anglers in state- or province-wide angler surveys are randomly selected from license files, this group can adequately represent smallmouth bass anglers in those areas as well. Currently, in the absence of specific tags or licenses for a specific fish, this is often the most cost-effective way to identify anglers fishing for a particular species for survey research purposes. However, many U.S. states require specific species licenses (e.g., trout stamps, hand-fishing permits, or paddlefish snagging permits), so random sampling of license holders from the databases associated with these licenses is possible.

While the information from state- or province-wide surveys helps justify and guide policies and goals, species level studies shape management plans or can provide economic impact and valuation for a particular species. A smallmouth bass angler study, for example, allows decision makers to determine the popularity of customized smallmouth bass fishing regulations versus uniform state- or province-wide regulations, which of the various management alternatives are most palatable to anglers in particular areas of those boundaries, and how expenditure levels may vary under customized regulations. This type of information is needed to guide a management approach based on providing diverse opportunities not only to anglers and local businesses but also to law enforcement administrators. Law enforcement becomes a more difficult task under a system with customized regulations for each reservoir or stretch of river. Amid this concern, feedback from species level surveys gives decision makers a better idea of the right mix of regulations needed on water bodies throughout the state or province and where they should be implemented to match angler distribution and preference.

14.2.5.4 Resource level

Resource level studies are directed at those anglers who use a particular resource (e.g., lake, reservoir, stream, or stretch of river). They obtain information about fishing success at that resource, attitudes toward its management and issues of local concern, expenditure levels, and degree of support for possible management alternatives. Resource level angler studies usually are conducted as follow-ups of anglers intercepted during creel surveys designed to estimate fish catch and fishing effort. These are called "add-on" surveys in human dimensions research (Pollock et al. 1994). Resource level studies provide a multitude of benefits to fisheries administrators and managers. They evaluate management goals and objectives, equip local managers with the information they need to become more effective, and provide feedback to improve fishing experiences.

Resource level studies allow administrators to determine whether diverse products are indeed attracting the expected clientele and expected economic consequences. They also give local fisheries managers information they need to be more effective. First, they quantify some of the anecdotal information managers receive through observation and daily interactions, making managers more informed when dealing with the public. Second, information from volunteered comments calls attention to potential problems and is useful because managers can relate to issues and locations mentioned and can investigate. Making changes based on feedback from anglers is extremely important for the agency and local managers because it demonstrates that the agency is attempting to be responsive. This enhances credibility of agencies, increases the legitimacy of management decisions and actions, builds public trust with the agency and managers, and translates into more satisfying fishing experiences.

14.3 CONCLUSIONS

Regardless of the level at which social and economic assessments are conducted in fisheries management, this chapter has hopefully enlightened the reader in their uses and importance in inland fisheries assessments. Nevertheless, the use of social and economic information is still in its infancy. The fisheries profession has long relied on traditional father-son socialization processes to create new customers, and communication by word of mouth and the fishing tackle and outdoor media industry to guide anglers to desired resources. Until recently, there has not been much collaboration between marketing and fisheries departments within natural resource agencies. However, amid high turnover in the fishing public and stagnant or declining rates of participation throughout the USA and Canada, fisheries and marketing departments are realizing they must team up so they can compete with other recreational providers for the public's leisure time. With social and economic data becoming more prevalent, valuable information is available for marketing strategies to guide anglers to desired resources and (or) to market products to targeted clientele. Guiding and attracting anglers to desired resources will create more satisfying fishing trips. In turn, this will aid in retaining anglers in the activity. Although it will take some time for social and economic information to be fully incorporated into fisheries management programs, efforts are underway to integrate social and economic information better into the decision-making process of fisheries agencies.

14.4 REFERENCES

Aas, O., W. Haider, and L. Hunt. 2000. Angler responses to harvest regulations in Engerdal, Norway: a conjoint-based choice modeling approach. North American Journal of Fisheries Management 20:940–950.

Aiken, R., and G. Pullis La Rouche. 2003. Net economic values for wildlife-related recreation in 2001: addendum to the 2001 national survey of fishing, hunting, and wildlife-associated recreation. U.S. Fish and Wildlife Service, Report 2001–3, Washington, D.C.

Arlinghaus, R. 2006. On the apparently striking disconnect between motivation and satisfaction in recreational fishing: the case of catch orientation of German anglers. North American Journal of Fisheries Management 26:592–605.

Bates, D. G., and E. M. Fratkin. 2002. Cultural anthropology, 3rd edition. Allyn and Bacon, New York.

Bohnsack, B. L., and R. J. Sousa. 2000. Sport fish restoration: a conservation funding success story. Fisheries 25(7):54–56.

Boyle, K. J., B. Roach, and D. J. Waddington. 1998. 1996 net economic values for bass, trout, and walleye fishing, deer, elk and moose hunting, and wildlife watching. Addendum to the 1996 national survey of fishing, hunting, and wildlife-associated recreation. U.S. Fish and Wildlife Service, Report 96-2, Washington, D.C.

Brown, T. L. 1987. Typology of human dimensions information needed for Great Lakes sport-fisheries management. Transactions of the American Fisheries Society 116:320–324.

Brunke, K. D., and K. M. Hunt. 2007. Comparison of two approaches for the measurement of waterfowl hunter satisfaction. Human Dimensions of Wildlife 12:1–15.

Bultena, G., and M. J. Taves. 1961. Changing wilderness images and forest policy. Journal of Forestry 59:167–171.

Burger, J. 2000. Consumption advisories and compliance: the fishing public and the deamplification of risk. Journal of Environmental Planning and Management 43:471–488.

Burger, L. W., D. A. Miller, and R. L. Southwick. 1999. Economic impact of northern bobwhite hunting in the southeastern United States. Wildlife Society Bulletin 27:1010–1018.

Davis L. S., K. N. Johnson, P. Bettinger, and T .E. Howard. 2001. Forest management, 4th edition. McGraw-Hill, New York.

Decker, D. J., T. L. Brown, and W. F. Siemer. 2001. Human dimensions of wildlife management in North America. The Wildlife Society, Bethesda, Maryland.

Ditton, R. B. 1997. Choosing our words more carefully. Fisheries 4(10):4.

Ditton, R. B. 2004. Human dimensions of fisheries. Pages 199–208 *in* M. J. Manfredo, J. J. Vaske, B. L. Bruyere, D. R. Field, and P. J. Brown, editors. Society and natural resources: a summary of knowledge prepared for the 10th international symposium on society and resource management. Modern Litho, Jefferson City, Missouri.

Enck, J. W., D. J. Decker, and T. L. Brown. 2000. Status of hunter recruitment and retention in the United States. Wildlife Society Bulletin 28:817–824.

Fedler, A. J., and R. D. Ditton. 1994. Understanding angler motivations in fisheries management. Fisheries 19(4):6–13.

Fisheries and Oceans Canada. 2007. Survey of recreational fishing in Canada 2005. Fisheries and Oceans Canada, Economic Analysis and Statistics Policy Sector, Ottawa.

Gillis, K. S., and R. B. Ditton. 2002. A conjoint analysis of the U.S. Atlantic billfish fishery management alternatives. North American Journal of Fisheries Management 22:1218–1228.

Grado, S. C., R. M. Kaminski, I. A. Munn, and T. A. Tullos. 2001. Economic impacts of waterfowl hunting on public lands and at private lodges in the Mississippi Delta. Wildlife Society Bulletin 29:846–855.

Groothuis, P. A. 2005. Benefit transfer: a comparison of approaches. Growth and Change 36:551–564.

Haab, T., and K. McConnell. 2002. Valuing environmental and natural resources: the econometrics of nonmarket valuation. Edward Elgar, North Hampton, Massachusetts.

Hunt, K. M., and R. B. Ditton. 2001. Perceived benefits of recreational fishing to Hispanic-American and Anglo anglers. Human Dimensions of Wildlife 6:153–172.

Hunt, K. M., and R. B. Ditton. 2002. Freshwater fishing participation patterns of racial and ethnic groups in Texas. North American Journal of Fisheries Management 22:52–65.

Hunt, K. M., M. F. Floyd, and R. B. Ditton. 2007. African-American and Anglo anglers' attitudes toward the catch-related aspects of fishing. Human Dimensions of Wildlife 12:227–239.

Johnson, R. L., and E. Moore. 1993. Tourism impact estimation. Annals of Tourism Research 20:279–288.

Klemperer, W. D. 1996. Forest resource economics and finance. McGraw-Hill, New York.

Knuth, B. A., T. L. Brown, and K. M. Hunt. In press. Measuring the human dimensions of recreational fisheries. *In* A. V. Zale, D. L. Parrish, and T. M. Sutton, editors. Fisheries techniques, 3rd edition. American Fisheries Society, Bethesda, Maryland.

Krueger, C. C., and D. J. Decker. 1999. The process of fisheries management. Pages 31–59 *in* C. C. Kohler and W. A. Hubert, editors. Inland fisheries management in North America, 2nd edition. American Fisheries Society, Bethesda, Maryland.

Loomis J. B., and R. G. Walsh. 1997. Recreation economic decisions: comparing benefits and costs, 2nd edition. Venture Publishing, State College, Pennsylvania.

Loomis, J. B., and D. S. White. 1996. Economic values of increasingly rare and endangered fish. Fisheries 21(11):6–11.

Louviere, J. J., and H. Timmermans. 1990. Stated preference and choice models applied to recreation research: a review. Leisure Sciences 12:9–12.

Mehmood, S., D. Zhang, and J. Armstrong. 2003. Factors associated with declining hunting license sales in Alabama. Human Dimensions of Wildlife 8:243–262.

Miller, C. A., and J. J. Vaske. 2003. Individual and situational influences on declining hunter effort in Illinois. Human Dimensions of Wildlife 8:263–276.

Murdock, S. H., D. K. Loomis, R. B. Ditton, and M. N. Hoque. 1996. The implications of demographic change for recreational fisheries management in the United States. Human Dimensions of Wildlife 1:14–37.

Muth, R. M., D. A. Hamilton, J. F. Ogden, D. J. Witter, M. E. Mather, and J. J. Daigle. 1998. The future of wildlife and fisheries policy and management: assessing the attitudes and values of wildlife and fisheries professionals. Transactions of the North American Wildlife and Natural Resources Conference 63:604–627.

Oh, C. O. 2005. Understanding differences in nonmarket valuation by angler specialization level. Leisure Sciences 27:263–277.

Oh, C. O., R. B. Ditton, , B. Genter, and R. Riechers. 2006. A stated discrete choice approach to understanding angler preferences for management options. Human Dimensions of Wildlife 10:173–186.

Olson, D., and S. Lindall. 2000. IMPLAN professional, 2nd edition. Minnesota IMPLAN Group, Stillwater.

Pollock, K. H., C. M. Jones, and T. L. Brown. 1994. Angler survey methods and their applications in fisheries management. American Fisheries Society, Special Publication 25, Bethesda, Maryland.

Simcox, D. E. 1993. Cultural foundations for leisure preferences, behavior, and environmental orientation. Pages 267–280 *in* A.W. Ewert, D. Chavez, and A. W. Magill, editors. Culture, conflict, and communication in the wildland–urban interface. Westview Press, Boulder, Colorado.

Simoes, J. C. 2009. Recreational angler surveys: their role and importance national and the 2008 Michigan angler survey. Master's thesis, Michigan State University, East Lansing.

Steinback, S. R. 1999. Regional economic impact assessments of recreational fisheries: an application of the IMPLAN modeling system to marine party and charter boat fishing in Maine. North American Journal of Fisheries Management 19:724–736.

Strouse, K.G. 1999. Marketing telecommunications services: new approaches for a changing environment. Artech House, Norwood, Massachusetts.

Southwick Associates. 2002. Sportfishing in America: values of our traditional pastime. Produced for the American Sportfishing Association with funding from the Multistate Conservation Grant Program, Alexandria, Virginia.

Southwick Associates. 2007. Sportfishing in America: an economic engine and conservation powerhouse. Produced for the American Sportfishing Association with funding from the Multistate Conservation Grant Program, Alexandria, Virginia.

Sutton, S. G., J. R. Stoll, and R. B. Ditton. 2001. Understanding anglers' willingness to pay increased fishing license fees. Human Dimensions of Wildlife 6:115–130.

Swanson, C. S., and J. B. Loomis. 1996. Role of nonmarket economic values in benefit–cost analysis of public forest management options. U.S. Department of Agriculture, Forest Service,

Pacific Northwest Research Station, General Technical Report PNW-GTR- 361, Portland, Oregon.

Teel T. L, M. J. Manfredo, and H. M. Stinchfield. 2007. The need and theoretical basis for exploring wildlife value orientations cross-culturally. Human Dimensions of Wildlife 12:297–305.

Toth, J. F., and R. B. Brown. 1997. Racial and gender meanings of why people participate in recreational fishing. Leisure Sciences 19:129–146.

USDI (U.S. Department of the Interior). 2007. 2006 national survey of fishing, hunting, and wildlife-associated recreation. U.S. Department of the Interior, Fish and Wildlife Service, and U.S. Department of Commerce, Census Bureau, Washington, D.C.

Weithman, S. A. 1999. Socioeconomic benefits of fisheries. Pages 193–213 *in* C. C. Kohler and W. A. Hubert, editors. Inland fisheries management in North America, 2nd edition. American Fisheries Society, Bethesda, Maryland.

Wilde, G. R., R. B. Ditton, S. R. Grimes, and R. K Riechers. 1996. Status of human dimensions surveys sponsored by state and provincial fisheries management agencies in North America. Fisheries 21(11):12–23.

Chapter 15

Natural Lakes

MICHAEL J. HANSEN, NIGEL P. LESTER, AND CHARLES C. KRUEGER

15.1 INTRODUCTION

Natural lakes are important resources throughout North America and contribute substantial economic benefits to people through fishing, boating, swimming, and other recreational uses. Recreational, commercial, and subsistence fisheries are among the most important, though rarely exclusive, social and economic values of natural lakes. Therefore, fisheries managers of natural lakes must balance demands for fishing with demands for boating, swimming, or other uses, which requires managers to be skilled in fishery science, resource economics, and public policy.

This chapter focuses on the unique attributes of natural lakes that challenge managers within the larger environment of fisheries management. The chapter begins with an overview of natural lakes in North America. Next, types of fisheries found in natural lakes are described (section 15.2), which serves as a background for methods of predicting fishery potential in natural lakes (section 15.3) and common management goals for different types of fisheries in natural lakes (section 15.4). Next, methods for evaluating fishery potential in natural lakes (section 15.5) and strategies for managing fisheries in natural lakes are presented (section 15.6).

15.1.1 Natural Lakes in North America

Natural lakes are formed by the interplay between climate and geologic forces (Wetzel 2001). First, geologic forces must form a basin in which water can be held, and second, precipitation must be adequate to supply enough water to offset evaporative, seepage, and outlet losses, thereby sustaining water in the lake basin throughout the year. As a consequence of geology and the interplay between availability of water and lake basins, most natural lakes are in the northern and eastern parts of North America, where annual precipitation is adequate and glaciers left numerous basins for storage of surface waters. In contrast, lakes are less common in western and southern North America, either because few lake basins are present or precipitation is insufficient to overcome evaporation or seepage losses.

Regardless of the forces that form natural lakes, they can be classified based on the presence or absence of inlets or outlets, which correlates to landscape position (high to low elevation) and connectivity with other lakes or rivers (low to high connectivity). Seepage lakes have neither inlets nor outlets, are often positioned highest in landscapes, and have small watersheds. Drained lakes have one or more outlets, are often positioned at intermediate altitudes in landscapes, and have intermediate-size watersheds. Drainage lakes have at least one inlet and one outlet, are positioned lowest in landscapes, and have large watersheds. Such lake categories follow gradients of landscape position and watershed size that directly influence the magnitude of watershed influence on lake productivity.

15.1.2 Distribution of Natural Lakes in North America

The last continental glacier left North America with a large number of freshwater natural lakes that extended over the northern half of the continent mostly north of 40° north latitude (Cole 1994). The number of lakes in the glaciated region of North America is uncertain because numbers of small lakes are difficult to estimate (Downing et al. 2006). For example, Meybeck (1995) estimated about 49,000 lakes (725,000 km^2) in Canada and 4,000 lakes (106,000 km^2) in the USA that were larger than 1 km^2 in surface area, whereas Lehner and Döll (2004) estimated about 90,000 lakes (868,000 km^2) in Canada and 13,000 lakes (160,000 km^2) in the USA. Regardless of estimation method, natural lakes are more densely concentrated in Canada than they are in the USA because more of Canada was glaciated (Meybeck 1995). Consequently, lakes in Canada are 12.5-fold more numerous, 7.3-fold greater in total surface area, and 11.7-fold greater in density than they are in the USA (Table 15.1). Globally, Canada (8%, Meybeck 1995; 9%, Lehner and Döll 2004) is rivaled only by Scandinavia (9%, Meybeck 1995; 6%, Lehner and Döll 2004) in the percentage of total land mass covered by lakes, followed by the USA (1%, Meybeck 1995; 2%, Lehner and Döll 2004), and the former Union of Soviet Socialist Republics (2%, Meybeck; 2%, Lehner and Döll 2004).

North America contains 8 of the 20 largest lakes of the world, defined as having more than 10,000-km^2 surface area, all of which were formed by glaciers (Kalff 2002). Five of these "Great Lakes" are in the Laurentian system, that is, the St. Lawrence River watershed (Superior, Huron, Michigan, Erie, and Ontario), and three others lie entirely in Canada (Great Bear, Great Slave, and Winnipeg; Table 15.2). Based on volume, the eight largest lakes in North America change their order of size because of large differences in depth (Table 15.2).

Table 15.1. Number, total surface area, and density of natural lakes in Canada and the USA, North America (adapted from Meybeck 1995).

Lake area (km^2)	Number		Total surface area (km^2)		Density (number/Mkm^2)	
	Canada	USA	Canada	USA	Canada	USA
0.1	440,000	35,000	114,000	9,100	44,000	3,750
1	44,000	3,500	114,000	9,100	4,400	375
10	4,500	450	117,000	11,700	450	48
100	523	55	136,000	14,219	52	5.9
1,000	31	7	88,000	13,064	3.1	0.75
10,000[a]	7	1	270,000	57,750	0.7	0.11
Sum	489,061	39,013	839,000	114,933	48,906	4,180

[a] Lakes Superior, Huron, Erie and Ontario are tallied in Canada.

Table 15.2. Surface area, volume, elevation, average depth, maximum depth, shoreline length, and drainage area of eight North American Great Lakes. Rank represents position among the 20 Great Lakes of the world, including the Caspian Sea (from Kalff 2002).

Metric (units)	Lake							
	Superior	Huron	Michigan	Great Bear	Great Slave	Erie	Winnipeg	Ontario
Surface area (km^2)	82,400	59,596	58,000	31,153	27,200	25,744	24,514	19,500
Rank	2	5	6	9	11	13	14	17
Water volume (km^3)	12,000	3,540	4,900	2,236	1,580	480	294	1,640
Rank	4	7	6	9	10	18	20	12
Elevation (m)	186	176	176	186	156	174	217	75
Average depth (m)	147	59	85	72	41	19	12	86
Maximum depth (m)	406	229	281	446	614	64	36	246
Shore length (m)	2,725	6,157	2,636	2,719	3,057	1,370	1,858	1,146
Drainage area (km^2)	127,700	118,100	133,900	114,717	971,000	58,800	984,200	70,000

The Laurentian system contains more surface freshwater than anywhere else in the world, and Lake Superior contains more than half of the total volume in the Laurentian system. Lake Superior is second largest in surface area to only the Caspian Sea and is fourth largest in total volume (first is Caspian Sea, second is Lake Baikal, and third is Lake Tanganyika; Kalff 2002). Other notable glacial lakes include Reindeer Lake, Saskatchewan–Manitoba, which has the 17th largest volume in the world (585 km^3) and Quesnel Lake, British Columbia, the 13th deepest lake in the world (475 m; Kalff 2002).

Elsewhere in North America, natural lakes were formed by geologic forces other than glaciers. For example, a notable lake formed by volcanic activity is Crater Lake, Oregon, the eighth deepest lake in the world (589 m; Kalff 2002). Notable lakes formed by tectonic activity include Lake Tahoe, California–Nevada, 11th deepest in the world (501 m); Lake Chelan, Washington, 13th deepest lake in the world (489 m); and Adams Lake, British Columbia, 16th deepest lake in the world (457 m; Kalff 2002). Solution (karst) lakes are formed in regions of highly soluble rock, mostly limestone, and are relatively common in Kentucky, Indiana, Tennessee, and Florida (Kalff 2002).

15.1.3 Lake Types and Habitat Zones

Fisheries managers must acknowledge the limitations of lake type when making informed decisions because of strong relationships between lake type and nutrient concentration, primary production, secondary production, and aquatic community structure. Naumann (1919, cited in Kalff 2002) developed a classification of lakes based on relative supply of nutrients (phosphorus, nitrogen, and calcium) and resulting phytoplankton. Oligotrophic lakes are unproductive (poorly nourished), eutrophic lakes are productive (well nourished), and mesotrophic lakes are intermediate in productivity (medium nourished; Naumann 1919, cited in Kalff 2002; Table 15.3). Numerous attributes of lakes correlate to the trophic-state continuum (oligotrophic → eutrophic), including depth (deep → shallow), bank slope (steep → gradual), epilimnetic : hypolimnetic ratio (small → large), transparency (deep → shallow), organic sediments (low → high), hypolimnetic oxygen (high → low), rooted plants (sparse → abun-

Table 15.3. Chemical and biological characteristics of oligotrophic, mesotrophic, and eutrophic lakes based on a classification system by Busch and Sly (1992).

Parameter (unit)	Lake classification		
	Oligotrophic	Mesotrophic	Eutrophic
Dissolved oxygen deficit (mg O$_2$/cm^2/month)	<0.75	0.75–1.65	>1.65
Primary production (g C/m^2)	<25	25–75	>75
Total phosphorus (mg P/m^3)	<8	8–23	>23
Total nitrogen (mg N/m3)	<300	300–650	>650
Chlorophyll a (mg chlorophyll *a*/m^3)	<2.9	2.9–5.6	>5.6
Secchi disc transparency (m)	>4.0	2.5–4.0	<2.5
Morphoedaphic index (total dissolved solids/mean depth)	<6.0	6.0–7.0	>7.0

dant), phytoplankton abundance (low → high), benthic invertebrate abundance (low → high), and fish production (low → high; compiled from Cole 1994; Wetzel 2001). Assemblages of organisms also vary across the trophic-state continuum, with oligotrophic lakes typified by diatom algae, *Tanytarsus*-type midges, and coregonid and salmonid fishes; mesotrophic lakes characterized by green algae, mixed midge assemblages, and percid fishes; and eutrophic lakes typified by blue-green algae, *Chironomus* midges, and centrarchid fishes (compiled from Cole 1994; Wetzel 2001).

Fisheries managers must also recognize the mix of habitats present in individual lakes, regardless of trophic state, because the relative amount of inshore versus offshore area and volume of water within the photic zone largely determine management prescriptions. Habitat zones based on temperature stratification during summer include the epilimnion, where warm surface waters are well mixed by winds; metalimnion (thermocline), where temperature transitions from warm to cold; and hypolimnion, where cold deep waters are of relatively uniform temperature (Kalff 2002). The depth of light penetration defines habitat zones based on primary production that overlap with temperature strata to cause oxygen stratification (Cole 1994). The photic zone often coincides with the epilimnion, includes all water shallower than the depth of maximum light penetration, and is divided between the inshore littoral zone, which includes all areas where light reaches the bottom of the lake, and the offshore limnetic zone, which includes all areas where light does not reach bottom of the lake. Rooted aquatic plants are important sources of primary production in the littoral zone, whereas phytoplankton provide all primary production in the limnetic zone. Lakes with large fractions of their surface area in the littoral zone and large fractions of their volume in the photic zone are typically eutrophic, whereas lakes with small littoral and photic zones are typically oligotrophic. The profundal (aphotic) zone often coincides with the hypolimnion and includes all waters deeper than the maximum depth of light penetration where primary production is absent and only respiration and decomposition occurs. Lakes with low hypolimnetic volume are typically eutrophic, whereas lakes with high hypolimnetic volume are typically oligotrophic. The benthic zone includes all water that touches the lake bottom, including littoral and profundal zones.

15.1.4 Watershed Influences

The relative influence of nutrient inputs from atmospheric versus watershed sources ultimately limits fish production and depends on the ratio of watershed area to lake surface area (drainage ratio), geology, and land use (Kalff 2002). Lakes with small drainage ratios, typically located at higher elevation in the landscape, are usually affected more by atmospheric inputs than by watershed inputs. In contrast, lakes with large drainage ratios, typically located at lower elevation in the landscape, are usually affected more by watershed inputs than by atmospheric inputs. The combined influence of watershed size and land use on nutrient loading, primary production, and secondary production in natural lakes is predictable: (1) terrestrial nutrient loading increases with drainage size; (2) nutrient loading is much higher in agricultural drainages than in natural drainages; (3) nutrient loading from atmospheric sources is substantial only in lakes with small natural drainages; and (4) algal production and fish yield increase with drainage size and are both higher in agricultural than in natural watersheds (Kalff 2002).

Landscape position and connectivity affect many attributes of lakes, including fish assemblage structure. The position of a lake in the landscape constrains its physical, chemical, and biological attributes (Webster et al. 1996; Kratz et al. 1997; Soranno et al. 1999; Riera et al. 2000 Webster et al. 2000). For example, surface area, ion concentration, and fish species richness vary predictably with landscape position in northern Wisconsin lakes: lakes higher in the drainage network typically are smaller in surface area, lower in ion concentration, and lower in fish species richness than are lakes lower in the drainage network (Webster et al. 1996; Kratz et al. 1997; Riera et al. 2000; Webster et al. 2000). Connectivity also explains variation in fish assemblage structure among lakes. For instance, in 18 northern Wisconsin lakes, two fish assemblage types were differentiated primarily based on connectedness and minimum oxygen conditions (Tonn and Magnuson 1982). Further, an inlet or outlet provides a migratory corridor for species to gain access from other systems and may provide winter refuge from low oxygen concentrations or predators, thereby supporting greater species diversity.

15.1.5 Fish Assemblage Structure

The number of fish species present in natural lakes (species richness) increases with lake size, presumably because lake size is correlated to habitat diversity (Matuszek and Beggs 1988; Kalff 2002). Species richness in natural lakes is also affected by physical (latitude and altitude) and chemical (e.g., pH) attributes of individual lakes (Matuszek and Beggs 1988). Native species occupying any lake were derived by a process of "hierarchical filters" operating at sequentially lower spatial scales, including continental, regional, lake type, and local scales (Tonn 1990). Consequently, the particular assemblage of fishes occurring naturally in any lake is a consequence of the fish fauna with access to the lake, either during lake formation or through connectedness to other rivers and lakes, that can live in the habitat present in the lake (e.g., tolerance to ambient temperature; Tonn 1990). Through this spatial and temporal sequence of filters, fish assemblages in natural lakes of North America have developed into characteristic assemblages for cold water (such as cisco and lake trout), cool water (yellow perch and walleye), or warm water (bluegill and largemouth bass). Large lakes may be complex enough to harbor multiple assemblages, such as when cisco and lake trout occupy deep hypolimnetic cold basins, yellow perch and walleye occupy the shallower, cooler metalimnetic fringes of deep basins, and bluegill and largemouth bass occupy the shallow epilimnetic peripheral bays of the same lake.

Fish assemblage structure varies with latitude and altitude across natural lakes in North America largely because of thermal tolerances of fish species. Along a south to north latitudinal gradient, fish assemblage structure trends from a predominance of warmwater species (black basses, centrarchid panfishes, and catfishes) in southern states to a predominance of coolwater species (walleye, yellow perch, smallmouth bass, and northern pike) in north-central states and to coldwater species (Arctic char, Arctic grayling, and whitefishes) in northern Canada. In western North America, where natural lakes are few in number and mostly found at high elevations, fish assemblages are dominated by coldwater species (Arctic char, salmon, and sculpins), though warmwater and coolwater species from eastern North America have been widely introduced. Coldwater fish assemblages in lakes of eastern and western North America differ in species within the same taxonomic groups, rather than between taxonomic groups (e.g., bull trout in western North America versus lake trout in eastern North America), because of differences in geologic history and long-term separation of fish assemblages by mountain ranges.

15.2. TYPES OF FISHERIES

Recreational fisheries predominate in natural lakes of North America, but commercial and subsistence fisheries also occur in some systems. Recreational fisheries typically rely on angling as the capture method, whereas commercial and subsistence fisheries rely on a greater variety of capture methods, including gill netting, trap-netting, angling, long-lining, and spearing. The extent of each type of fishery in natural lakes of North America is of interest to fisheries managers. For recreational fishing, national surveys in Canada during 2005 (FOC 2007) and in the USA during 2006 (USDI 2006) provide information. Because of differences in survey and questionnaire design, recreational fishing participation and effort in Canada could not be assigned to natural lakes. Nonetheless, the number and density of natural lakes in Canada indicated that most participation and effort in freshwater was aimed at natural lakes.

15.2.1 Recreational Fisheries

In Canada, 39.8 million angler-days were spent on freshwater (FOC 2007; NRC 2008). The top five taxonomic groups accounting for 84% of the total harvest were walleye (24%), trout (20%), yellow perch (16%), black basses (13%), and northern pike (11%), all of which are species that are common in natural lakes.

In the USA, 83% of all freshwater anglers (25.4 million freshwater anglers) fished in lakes and reservoirs in 2006 (USDI 2006). Angling effort in lakes and reservoirs was 70% of all freshwater angling effort (in total, 433.3 million days). In the Great Lakes, 1.4 million anglers spent 18 million days fishing. On the Great Lakes, walleye and sauger were the most popular fishes, followed by yellow perch, salmon, lake trout, black basses, and steelhead. On freshwaters not including the Great Lakes, black basses were the most popular fishes sought, followed by panfishes, catfishes and bullheads, trout, and crappies.

In Mexico, recreational angling in natural lakes is popular, especially for largemouth bass in the northern part of the country. Notable natural lakes for angling fisheries include Lake Guerrero, Lake Huites, Lake Materos, Lake El Salto, and Lake Baccarac.

15.2.2 Commercial Fisheries

Commercial fisheries exist in many natural lakes of North America and sometimes represent important local industries. The most notable commercial fisheries are in the Laurentian Great Lakes (Brown et al. 1999) and many lakes of northwestern Canada. In all five Great Lakes, commercial fisheries exist for coregonids (whitefishes and ciscoes). Other commercially-important species include lake trout in Lake Superior, yellow perch in Lake Michigan, and yellow perch and walleye in Lake Erie (Kinnunen 2003). The most valued species based on total commercial landings from the Great Lakes in the USA and Canada are lake whitefish, yellow perch, and walleye (Table 15.4).

Lake Winnipeg, Lake Manitoba, and Great Slave Lake are three of the largest lakes supporting commercial fisheries among natural lakes of Canada. Fish from these lakes are bought, processed, and marketed through the Freshwater Fish Marketing Corporation (FFMC) in Winnipeg, Manitoba. The corporation is a self-sustaining federal Crown corporation created in 1969 to handle freshwater fishes caught from waters of Manitoba, Saskatchewan, Alberta, Northwest Territories, and part of northwestern Ontario. Profits, in the form of final payments,

Table 15.4. Total landings (kg) and value (US$) of commercially caught fish from Canadian and U.S. waters of the Great Lakes in 2000 (from Kinnunen 2003).

Species	USA Landings	USA Value	Canada Landings	Canada Value	Total Landings	Total Value
Lake whitefish	4,484,355	10,256,122	5,065,266	8,379,717	9,549,621	18,635,839
Yellow perch	530,441	3,034,896	1,816,184	7,887,079	2,346,625	10,921,975
Walleye	10,383	38,851	3,297,163	10,081,376	3,307,546	10,120,227
Chubs	737,366	1,588,906	136,078	301,500	873,444	1,890,406
Rainbow smelt	209,034	751,793	3,261,329	1,107,979	3,470,364	1,859,772
Lake trout	450,910	531,462	255,373	230,098	706,283	761,560
Channel catfish	230,105	299,270	14,061	9,762	244,166	309,032
Carps	591,506	140,837	89,358	19,799	680,864	160,636

are distributed annually to participating fishers. The FFMC processes and markets most of western Canada's wild-caught inland commercial fishes and distributes these fishes to Canada, the USA, and 13 other countries. Marketed species include walleye, sauger, lake whitefish, northern pike, lake trout, common carp, yellow perch, inconnu (sheefish), white bass, and cisco. Product forms include fresh and frozen whole fish, fillets, portions, and caviar, with total sales exceeding Can$50 million (FFMC 2008).

Smaller commercial fisheries, often associated with aboriginal treaty rights or defined by statute to be directed at nongame fishes (e.g., suckers and common carp in Minnesota [Minnesota Statute 97C.827] and Kansas [Kansas Commercial Harvest Regulation 115–17–12]), also exist in some lakes in the USA. For example, whitefishes, ciscoes, suckers, and bullheads are commercially harvested in Leech Lake, Minnesota (Leech Lake Band of Ojibwe 2008), and commercial fishing for walleye reopened in 2007 in Red Lake, Minnesota (Melmer 2007). In Flathead Lake, Montana, lake whitefish are commercially harvested by angling (Flathead Lakers 2001).

Small-scale commercial fisheries occur in Mexican lakes with sales mostly to local markets of live, eviscerated, filleted, or dried fish. For example, Lake Chapala in west-central Mexico, the country's largest natural lake (1,100 km^2), supports an important commercial fishery (Macias and Lind 1990). The fishery involves approximately 2,500 fishers, two-thirds of whom are organized into 59 unions and eight cooperatives, while the remainder comprises unregistered pescadores libres, or "free fishers" (Pomeroy 1994). The lake supports commercial fisheries ranging 300–8,900 metric tons for native charal (five similar small-bodied [<125 mm] species of silversides), exotic common carp, and exotic tilapia (Lyons et al. 1998).

15.2.3 Subsistence Fisheries

The term "subsistence fisheries" refers to local, noncommercial, nonrecreational fisheries, oriented toward procurement of fish for consumption by fishers, their families, and their community (Berkes 1988). In Alaska, subsistence is defined by statute as follows:

Subsistence uses of wild resources are defined as "noncommercial, customary and traditional uses" for a variety of purposes. These include: direct personal or family consumption as food, shelter, fuel, clothing, tools, or transportation, for the making and selling of handicraft articles out of nonedible by-products of fish and wildlife resources taken for personal or family consumption, and for the customary trade, barter, or sharing for personal or family consumption. [Alaska Statute 16.05.940(32)]

Subsistence fisheries primarily occur in northern North America (Canada and Alaska) and to a lesser extent in the Great Lakes region. Depending on locality, subsistence fisheries are conducted by rural residents or are the exclusive opportunity of aboriginal peoples (First Nations). Especially among aboriginal peoples, subsistence fishing often has a multigenerational tradition, woven into the fabric of culture and guiding seasonal activities of local communities (Burch 1998; Helm 2000; Cleland 2001).

Management of subsistence fisheries can be complicated because of complex jurisdictional arrangements, treaties, and federal legislation. For example, subsistence fisheries in Alaska involve dual management between the state of Alaska and the U.S. federal government (Buklis 2002). Subsistence fishing often occurs in remote localities such as in Great Slave Lake and Great Bear Lake.

15.3 FISHERY POTENTIAL OF NATURAL LAKES

The importance of understanding fish production in natural lakes is obvious. Fisheries management deals with "two really different issues: what nature can produce, and what we can do to manage the activities of those who would capture that production. Presumably we can do a better job with the management issue if we understand the production issue" (Walters and Martell 2004). In this section, determinants of fish productivity and potential yield are summarized and then related to management of natural lakes. We discuss the theory of maximum sustained yield (MSY) as a reference point for planning a fishery, not as a goal for managing a fishery, because knowledge of MSY is critical for addressing goals related to fishing quality. For example, if MSY is low, a lake can support quality fishing for only a few fishers. The link between MSY, fishing effort, and fishing quality seems obvious and is presented as a background for understanding how to develop goals for fisheries.

15.3.1. Lake Productivity and Potential Fish Yield

The fishery potential of a lake depends on its productivity, that is, the ability of any trophic level to produce biomass. Because lower trophic levels are the food base of higher trophic levels, productivity at different levels tends to be positively correlated. Therefore, productivity of a fish assemblage is constrained by factors that limit primary production in any ecosystem, including natural lakes.

Rates of annual phytoplankton production in lakes range over three orders of magnitude (Brylinsky 1980; Kalff 2002). Primary production generally declines with increasing latitude and altitude but varies greatly among lakes at any latitude or altitude (Figure 15.1). Production is lower at high latitudes and altitudes than at low latitudes and altitudes because the growing season is shorter and water temperature is lower. Variation in production at any latitude or altitude is caused mainly by differences in availability of limiting nutrients (Kalff 2002).

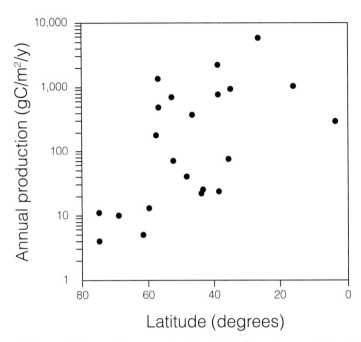

Figure 15.1. Annual phytoplankton primary production (g C/m²/year) versus latitude for selected lakes of the world. The figure is drawn from data in Table 21-4 of Kalff (2002). The latitude scale is reversed to depict a range from cold (left) to warm (right).

The limiting nutrient in lakes is usually phosphorus (Dillon et al. 2004). Concern about eutrophication of lakes led to development of many empirical models to predict phytoplankton biomass from nutrient concentrations (e.g., Schindler 1978; Kalff 2002). Chlorophyll-a concentration is usually measured as a surrogate for biomass and increases roughly in proportion to total phosphorus (TP) concentration. For example, one of the earliest models showed that yearly average chlorophyll-a concentration in the euphotic zone (μg/L) was strongly correlated to yearly average TP concentration (μg/L) in 77 temperate lakes ($r^2 = 0.77$; Vollenweider and Kerekes 1982).

Secondary production refers to production of consumers, the various trophic levels (including fish) that use food based on primary production (e.g., zooplankton, planktivores, and piscivores). If ecological efficiency, the ratio of predator production to prey production, is constant at each step of the food chain, then each level of secondary production will be proportional to primary production. Ecological efficiency is often assumed to be about 10% (range = 5–20%), and empirical models reveal a correlation between primary and secondary production because variation in primary production is much larger than is variation in ecological efficiency (Lampert and Sommer 1997). For example, Downing et al. (1990) showed that annual fish production in lakes ranged over three orders of magnitude (1.2–400 kg/ha/year), the same range as primary production. Fish production was strongly correlated with TP concentration in 14 lakes ($r^2 = 0.67$) and even more strongly correlated to primary production in 19 lakes ($r^2 = 0.79$).

An alternative means of studying fish production has been the use of fishery yield as an index of production (kg/ha/year). Fishery yield is defined as the consumption by the next highest trophic level, in this case, fish harvest. Long-term average yield from moderately to intensively exploited fisheries provides an approximate estimate of maximum sustainable yield (Ryder et al. 1974; Leach et al. 1987). Fishery yield underestimates total fish production of a lake because it excludes production of (1) fish that are smaller than harvested size, (2) fish that die of natural causes, and (3) other tissue losses, such as gametes through spawning. Therefore, potential yield of a fishery may be as low as 10% of total production (Morgan 1980; Downing et al. 1990).

Most fish production models have used sustained yield to examine how physical and chemical attributes of lakes influence fish production. Researchers have identified correlations with many limnological variables, including mean depth, total dissolved solids, TP, algal biomass, macrobenthos density, and temperature (Leach et al. 1987; Downing et al. 1990; Kalff 2002). This large number of variables is not surprising because many of these variables co-vary in lakes (Duarte and Kalff 1989).

The most well-known yield model is the morphoedaphic index (MEI; Ryder 1965), calculated as the concentration of total dissolved solids (TDS) divided by mean depth of the lake. This model indicates that potential fishery yield is proportional to the square root of MEI. The simplicity of the MEI likely contributed to its popularity, but it has also been criticized (Downing et al. 1990; Jackson et al. 1990; Rempel and Colby 1991). Although the MEI was derived empirically (Ryder 1965), the conceptual framework and assumptions underlying the model were later described (Ryder et al. 1974; Ryder 1982). The assumptions are (1) bedrock geology largely determines the concentration of TDS entering lakes; (2) TDS is a surrogate for essential nutrients, such as phosphorus, that control lake productivity; and (3) mean depth is a surrogate for hydrological characteristics such as thermal stratification, nutrient circulation, and dilution, all of which affect how energy is processed within the water column. Empirical support of these assumptions was provided by Chow-Fraser (1991), who showed that MEI accounted for 83% of the variation in TP concentration among 73 Canadian lakes.

An important variable not explicit in the original MEI is climate because the model was developed from a set of lakes largely in one climatic zone of North America (Ryder 1965), and one condition for its application was that lakes in the data set must be subject to homogeneous climatic conditions (Ryder et al. 1974; Ryder 1982). The MEI was subsequently expanded to incorporate temperature effects (Schlesinger and Regier 1982). Using data that spanned diverse climates (e.g., mean air temperatures of −4.4°C to 25.6°C), Schlesinger and Regier (1982) showed that, on a global scale, temperature was more important than MEI in predicting sustained fisheries yield.

The success of MEI probably depends on its ability to predict TP. The MEI was developed before measurement of TP was a standard procedure in lake assessment. The MEI is a surrogate for TP, based on bedrock geology, but may not account for anthropogenic phosphorus inputs. Therefore, phosphorus is likely a better predictor of fish yield. If phosphorus levels are outside the range expected from MEI (Chow-Fraser 1991), then MEI is probably not appropriate.

The climatic-MEI model (Schlesinger and Regier 1982) is a first approximation of how climate and nutrient factors interact to determine potential annual fishery yield and shows that climate plays a major role in fish production across a broad latitudinal range (Figure 15.2). Given the diversity of climate in North America, potential yield in natural lakes should vary by several orders of magnitude. When nutrient level is high (MEI = 40), this model predicts that potential annual fishery yield will range from 1 kg/ha/year in the north to 100 kg/ha/year in the south. Within climatic zones, observed variation in MEI predicts about a threefold

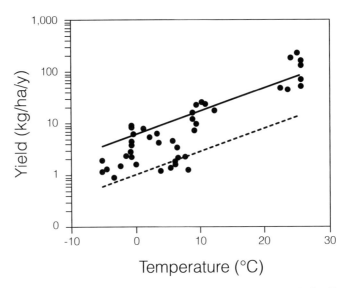

Figure 15.2. Fish yield (kg/ha/year) versus annual mean air temperature (°C) in 43 natural lakes of the world (circles) for two levels of the morphoedaphic index (MEI = 40, dashed upper line; MEI = 1, dotted lower line). The figure is drawn from data shown in Table 1 and Table 3 (equation 3) of Schlesinger and Regier (1982).

range in potential annual fish yield. Therefore, potential annual fishery yield should vary from 0.33–1 kg/ha/year in the north to 33–100 kg/ha/year in the south.

15.3.2. Production and Potential Yield of Fish Populations

Fisheries management is usually concerned with the potential yield of a few selected species that are targeted by fisheries. Recreational fisheries in North America typically target large fishes of the fish assemblage, such as lake trout, walleye, black bass, and northern pike, although smaller fishes that attain edible size, such as yellow perch and sunfishes, also attract angler interest. Commercial fisheries, which are less frequent on North American lakes, are less specialized and focus on edible-size fish of any species for which a market exists or can be developed.

Empirical studies have contributed to our understanding of production and potential yield of targeted species. Long-term median yields from moderately to intensively exploited fisheries has been used to estimate MSY and indicate that MSY varies widely among lakes for several species, such as walleye, lake trout, lake whitefish, and northern pike (Figure 15.3). Empirical models have shown that habitat differences account for much of this variation (e.g., Christie and Regier 1988; Marshall 1996; Shuter et al. 1998; Lester et al. 2004a). Temperature, oxygen, and water clarity are important habitat variables affecting species abundance and potential yield. Production studies indicate that life history traits are also important determinants of potential yield and have shown that the ratio of production to biomass (P/B) is higher for smaller species (e.g., Downing and Plante 1993; Randall et al. 1995). Downing and Plante (1993) also found that P/B increased with mean annual air temperature. Natural mortality decreases with body size and increases with water temperature (Pauly 1980), indicating that P/B (and hence MSY) increases with natural mortality.

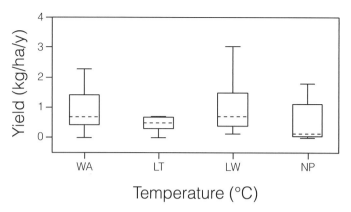

Figure 15.3. Variation in sustained yield of walleye (WA), lake trout (LT), lake whitefish (LW), and northern pike (NP). Dashed line depicts the median, large box depicts the 25th and 75th percentiles, and error bars depict the nonoutlier range. Data are from Christie and Regier (1988), Shuter et al. (1998), and Lester et al. (2004a).

A useful rule of thumb for understanding fishery potential is

$$\text{MSY} = p \times M \times B_{max} \qquad (15.1)$$

where p is an empirical factor, M is the instantaneous rate of natural mortality, and B_{max} is the biomass that would exist if the stock were not fished (Walters and Martell 2004). In other words, MSY is proportional to production of the unexploited stock, which is measured as $M \times B_{max}$ (Dickie et al. 1987; Mertz and Myers 1998). Early models suggested $p \approx 0.5$, but recent experiences suggest this value may be an upper bound (see Die and Caddy 1997; Quinn and Deriso 1999; Walters and Martell 2002). The appropriate value of p depends on the size at which fish become vulnerable to fishing in relation to the size at maturity. Low values of p apply when stocks are harvested well before maturity, and higher values (e.g., 0.5) apply when stocks are harvested at maturity. Because exploitation of a stock generally begins when fish reach a size similar to the size at maturity (Pauly 1984; Leach et al. 1987; Shuter et al. 1998), we suggest that $0.5 \times M \times B_{max}$ is a good first approximation of MSY.

If a fishery exists, parameters of this production model can be estimated from surveys that estimate biomass and mortality rate of a targeted fish population. Parameters of unexploited populations can still be estimated on remote lakes, and should be undertaken, especially in areas where access to lakes is likely to change because of road development (Hunt and Lester 2009). A historic unexploited benchmark may be valuable for planning and evaluating the impact of fishery development (e.g., Miller 1999). In many parts of North America such lakes do not exist, so estimates of surplus production parameters must be determined long after a fishery has developed.

Understanding mortality rates of species in their natural environments has been a central focus of ecological and evolutionary theory and the topic of many empirical studies. Across a broad range of animals, the natural mortality rate changes predictably with body size and temperature. For example, Pauly (1980) showed that the natural mortality rate of fish populations decreased with maximum body length and increased with mean annual temperature of the waters inhabited by the population (Figure 15.4). Other models indicate that natural mor-

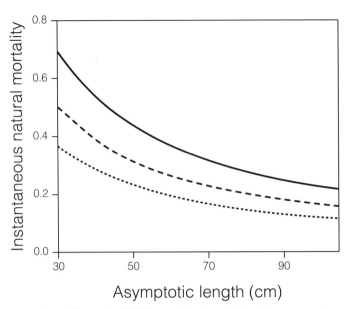

Figure 15.4. Demonstration of Pauly's (1980) empirical model relating natural mortality rate (M) to asymptotic total length (L_∞) of a fish population and mean annual water temperature in which it lives. Three temperature levels are shown (24, 12, and 6°C from top to bottom respectively). Curves were generated using an alternative version of Pauly's model, as in Table 2 of Shuter et al. (1998): $\log_e(M) = -0.0238 - 0.9326[\log_e(L_\infty)] + 0.6551[\log_e(\omega)] + 0.4646[\log_e(T)]$, where ω is the product of the von Bertalanffy growth parameters (i.e., $\omega = k \times L_\infty$). This example assumes $\omega = 8$ cm/year.

tality is related to life history features of fishes (Hoenig 1983; Peterson and Wroblewski 1984; Chen and Watanabae 1989; Jensen 1996; Quinn and Deriso 1999; Lester et al. 2004b; Shuter et al. 2005; Miranda and Bettoli 2007). Further, natural mortality of fishes is consistent with a general model of natural mortality across a broad range of animals and plants (Gillooly et al. 2001; McCoy and Gillooly 2008).

Given that MSY increases in proportion to instantaneous natural mortality (equation 15.1; Walters and Martell 2004), Pauly's (1980) mortality model indicates two important generalizations about fish species in natural lakes: (1) populations of large-bodied species will support lower potential yield than will populations of small-bodied species; and (2) coldwater species will support lower potential yield than will warmwater species. Application of Pauly's (1980) mortality model to natural lakes may explain some variation in potential yield. First, for a given species, L_∞ (theoretical maximum length; Walters and Martell 2004) may vary widely among lakes, so some variation in yield among lakes is likely due to differences in life history. For lake trout, as an example, a threefold variation in L_∞ implies at least a twofold variation in M and thus MSY. Because L_∞ increases with lake area, small lakes should support higher potential yield (kg/ha) than large lakes. This effect has been overlooked as an explanation for why yield is higher in small lakes (Marshall 1996; Shuter et al. 1998). Second, for a species that inhabits the littoral zone and spans a broad latitudinal distribution, the water temperature in which fish live varies widely among lakes. For example, southern populations of walleyes tend to have higher M and therefore should support higher potential yield than do northern populations.

The other population parameter that influences MSY is B_{max}, the carrying capacity of a fish stock in its natural unexploited condition. The relevant measure of B_{max} is the biomass of the segment of the fish population that is vulnerable to exploitation. Because exploitation directed on a stock generally begins when fish reach a size similar to the size at sexual maturity (Pauly 1984; Leach et al. 1987; Shuter et al. 1998), B_{max} is roughly equal to adult biomass. Therefore, attributes of lakes that affect adult biomass of unexploited populations (adult carrying capacity) are important to understand. In particular, adult unexploited biomass depends on (1) the availability of suitable habitat to support growth and reproduction of adult fish and (2) habitat and community factors that affect survival and growth of progeny from the egg to adult stage (i.e., recruitment).

A life cycle approach for understanding adult carrying capacity is important because bottlenecks at earlier life stages can control adult abundance (Shuter 1990) and render analysis of adult habitat irrelevant. Alternatively, effort to increase carrying capacity by creating new spawning habitat will be fruitless when habitat for other life stages is limiting (Minns et al. 1996). This comprehensive approach, which calls for an assessment of how habitat and community factors affect growth and survival at each life stage, is described by Hayes et al. (2009), though examples of its application are few (Minns et al. 1996; Chu et al. 2006). Most models of carrying capacity previously focused on habitat requirements of adult fish and assumed that adult habitat was the bottleneck constraining abundance.

Ryder and Kerr (1989) proposed a hierarchical framework for evaluating habitat needs of fish species. First, key survival determinants, such as oxygen, temperature, light, and nutrients, must be addressed. Then, limiting effects of structural habitat should be considered. Most models developed for predicting potential yield of populations have followed this direction by focusing on one or more key survival determinants. Marshall (1996) described how this approach can be used to assess potential yield of lake trout in natural lakes. For lake trout, optimal habitat is defined by water temperature (<10°C) and dissolved oxygen concentration (>6 mg O_2/L; Evans et al. 1991). These criteria imply that many North American lakes are not suitable for lake trout because summer water temperatures are higher than 10°C. Lake trout thrive in only deep lakes where thermal stratification during summer provides a coldwater refuge and small lakes in the Arctic where air temperature is never warm (Martin and Olver 1980).

Lake depth is an important environmental variable because it governs the thermal regime of a lake and the amount of oxygen stored in the hypolimnion during stratification (Cornett and Rigler 1979; Walker 1979). Because hypolimnetic oxygen concentration declines during summer, the initial supply of oxygen affects whether hypolimnetic oxygen levels later in summer are sufficient to support a species such as lake trout (Ryan and Marshall 1994; Clark et al. 2004; Dillon et al. 2004; Evans 2007). Because water temperature and dissolved oxygen define the living space for lake trout, abundance and potential yield should be closely linked to measures of this space. For example, Christie and Regier (1988) showed that lake trout harvest (kg) was related to thermal habitat volume (or area), defined as the volume (or area) of water bounded by 8°C and 12°C temperature isotherms, averaged over summer. Harvest of lake whitefish, walleye, and northern pike was also related to measures of thermal habitat (Christie and Regier 1988).

This "living space" approach acknowledges that surface area of a lake is not a precise measure of how much habitat exists to support production of a species. Suitable area is often correlated with lake area, but lakes with the same surface area can differ widely in the amount of habitat suited to different species (Figure 15.5). Consequently, the density (biomass/surface area) and potential yield (harvest/surface area/year) of a species can vary greatly among lakes.

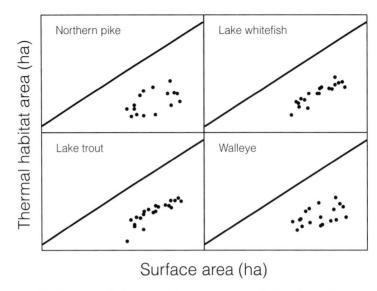

Figure 15.5. Thermal habitat area (ha) versus lake surface area (ha) estimated for northern pike, lake whitefish, lake trout, and walleye in 21 lakes (Christie and Regier 1988). Axes are both logarithmic scales; the solid line depicts the line of equality to show that lake surface area is usually larger than the area suited for any individual species.

When lake size is measured in terms of suitable habitat, measures of density (biomass/suitable area) and potential yield (harvest/suitable area/year) are less variable.

Community attributes may also account for variation in density and yield of fish species in natural lakes, so habitat alone may not be sufficient to account for variation in abundance. For example, changes in the fish assemblage may affect survival of young fish and reduce adult abundance to less than would be expected from measures of habitat suitability. Such effects argue for using a full life cycle assessment when attempting to predict or describe fish density or yield.

15.4 MANAGEMENT GOALS AND OBJECTIVES

The difference between a goal and an objective is not always obvious because both words seem to describe the same concept. Goals are general statements about what managers desire to achieve in the long term, and they explain the purpose of management. To set goals, the basic questions to answer are "where are we?" and "where do we want to go?" On the other hand, objectives specify measurable, expected outcomes that indicate achievement or progress toward goals and when those outcomes are to be achieved.

For example, a recreational fishery goal for a warmwater lake in Illinois might be "to provide a largemouth bass fishery with the opportunity to catch a trophy largemouth bass and a sustained harvest fishery for bluegills." Three objectives could be used to define this goal: by the year 2020, (1) increase the catch rate of largemouth bass from 1 bass per 4 h of angling to 3 bass per hour, (2) develop a fishery where at least one 61-cm largemouth bass is caught for every 100 bass angled, and (3) maintain a catch rate for bluegills of 4 fish per hour with an average size of 15 cm. This example illustrates that goals are broad statements of intent about the purpose of management whereas objectives define exactly what management intends to accomplish and when. Approaches that can be used to define goals and objectives for recreational, commercial, and subsistence fisheries in natural lakes are described below.

15.4.1 Recreational Fisheries

The first step in setting goals and objectives for recreational fisheries is to define the fishery! Goals and objectives for recreational fisheries depend on an exact definition of the fishery. Recreational fisheries in lakes can be defined in four ways: by spatial, temporal, taxonomic, and special definitions. Spatial definitions of fisheries may be at a state or provincial level (e.g., Florida or Ontario), at a regional level (e.g., Catskill or Uinta mountains), by water type (e.g., lakes, farm ponds, or Great Lakes), or for a specific water (e.g., Cayuga Lake or Lake Simcoe). Water-specific designation is typically used for only lakes of special significance. Temporal definitions of fisheries are most often used to define periods of special opportunities (e.g., special winter trout fisheries). Taxonomic definitions are often used to specify fisheries for single species (e.g., walleye) or groups of species (e.g., salmonids or panfishes). Special definitions of fisheries are sometimes used to define size of fish (e.g., trophy fishery), method of fishing (e.g., fly fishing), or a unique feature of a fishery (e.g., urban, wilderness, handicapped, or children-only fishing).

Clearly-defined fisheries enable development of meaningful and understandable goals and objectives for managers and stakeholders. Statewide or province-wide plans with stated goals and objectives can be crucial for coordinating management of several different fisheries or across jurisdictional boundaries. Nested under these broad geographic plans are management plans with goals and objectives for fisheries that are more discretely defined by more than one of the categories described above (e.g., brook trout lakes of Ontario, walleye lakes of north-central Minnesota, or salmonids in Lake Ontario). Most of the ensuing discussion focuses on discretely-defined goals and objectives for specific fisheries.

Goals for recreational fisheries in natural lakes often focus on attributes of angler opportunity or satisfaction, such as consistent year-round fishing opportunities, a high level of user satisfaction, needs of special users, or simply the opportunity to fish. Each of these purposes can be shaped into a goal statement. Choosing the right goal for management of a lake fishery is critically important because the goal for a fishery sets the context under which to choose actions (e.g., regulations) to achieve the goal. Fisheries managers should enlist the help of stakeholders in defining and choosing goals. Goals should reflect the values and interests of society, while not foreclosing opportunities for future generations of anglers.

Objectives for recreational fisheries in natural lakes can focus on the actual parameters that measure angler opportunity or satisfaction or provide a measure of a "good" fishery. For example, harvest and catch relate to angler satisfaction because most anglers prefer to catch more fish than fewer fish. Total annual harvest or catch is typically specified in numbers, rather than weight of fish, and can be set to be "ecologically realistic" in relation to system productivity. However, harvest and catch are usually insufficient as single measures of fishing success because stunted populations (i.e., small individual size of fish) could provide high numbers of fish caught or harvested, but these fish would be unattractive to most anglers. Therefore, harvest and catch are usually specified as parts of more complex objectives that also include harvest or catch rates and average size of fish caught.

Catch per unit effort (C/f), often defined as the average number of fish caught per hour of fishing, is commonly used in objectives to specify success rate of recreational fishing. Management plans often use harvest in combination with average *C*/*f*, which is an element of a successful fishery but also varies among anglers. For example, anglers who fish only a few times each year would likely rank a high *C*/*f* as crucial to their perception of success or satisfaction. Alternatively, anglers who fish many times each year may view *C*/*f* as less important than the species and (or) size of fish caught. Therefore, *C*/*f* by itself is not usually adequate for describing a successful recreational fishery, as illustrated by a stunted panfish population that supports a high *C*/*f* but likely induces low angler satisfaction. Another problem with relying on *C*/*f* as the sole parameter in a fishery objective is that harvest and catch can only attain a maximum level that is set by the biological productivity of a system. For example, if a lake can produce an annual harvest of only 10,000 fish to a fishery and the management objective specifies 0.5 fish per angler-hour (2 angler-hours/fish), then effort cannot exceed a total of 20,000 angler-hours per year. Managers may be successful in managing the lake to produce 10,000 fish consistently, but fisheries managers in North America rarely attempt to control angler effort in recreational fisheries. Therefore, *C*/*f* objectives assume that effort will not exceed a maximum level.

Average length of fish harvested is often used as a parameter in recreational fishery objectives for lakes in combination with harvest and *C*/*f*. Average length helps to avoid the problem illustrated by the stunted panfish example described above. If an objective specifies that average size of bluegill caught should be 20 cm, most people would agree that this average length would provide a satisfactory fishery. Average-length parameters stated in objectives should be set based on the size variability observed locally and should be intended to meet local average length (or higher) of fish caught because angler expectations and values reflect their past fishing experiences.

Combining all three metrics (harvest or catch, success rate, and average length) is better for defining recreational fishery management objectives than is using a single metric. For example, an objective could be stated as over the next 5 years the total harvest of northern pike will be 10,000 fish annually at an average length of 64 cm and caught at an average rate of 0.5 fish per angler-hour. This objective captures several different elements of a fishing experience. However, many other variables contribute to successful fishing trips, such as good companions or spectacular sunsets, which are difficult to measure.

The limitation of using average length or *C*/*f* is that all attributes of recreational fishing vary through time and among lakes. Most fisheries are complex, with a variety of species, sizes, and catch rates, all of which appeal to a variety of anglers. If an angler catches the same number of 64-cm northern pike at the same rate of 1 fish per angler-hour every time out fishing, this predictability could make the fishery uninteresting (Borgeson 1978). However, if once each season the same angler experiences a fishing rate of 5 fish per hour or catches one 114-cm fish, then such exceptional days of fishing may increase overall satisfaction. The simplest way to capture inherent variation in recreational fisheries is to specify management objectives as ranges around average values (e.g., 38–127 cm) rather than single target values (e.g., 51 cm).

A more sophisticated approach used by some agencies to capture variation within management objectives is called proportional size distribution (PSD, formerly called proportional stock density, and RSD, formerly relative stock density; see Guy et al. 2006, 2007 for discussion on the change in terminology; see Gabelhouse 1984, Gabelhouse et al. 1992, and Anderson and Neumann 1996 for a description of the development and use of the index). By means of this approach, fish populations in lakes (not angler catch) are evaluated by estimating the proportion of individuals in a population above a certain percentage of the length of the all-tackle angler-caught world record fish. The assumption here is that by managing for length variation in the population, the fishery catch will reflect that length variation and generate fisheries acceptable to the public. Based on a survey conducted during 1985–1986, the province of Quebec and 33 U.S. states were using PSD to evaluate fish populations (Gabelhouse et al. 1992). Management agencies continue to use PSD concepts to set management objectives and evaluate populations (Illinois Department of Natural Resources 2007; Marteney et al. 2007; Stewart et al. 2008). A potential disadvantage of PSD is that indexing a population length frequency against the angling world record length may be inappropriate wherever expectations are better measured against another length. In such fisheries, the population length frequency can be indexed against a more useful length rather than published lengths that are indexed against world record length. Using PSD as a parameter specified in a management objective infers a commitment to assess populations regularly to estimate their length structure. Fishery-dependent metrics, such as catch, harvest, C/f, and angling effort, depend on creel surveys that are more costly than are fishery-independent metrics, such as PSD.

Names used to describe recreational fisheries often infer goals, such as trophy fisheries, catch-and-release fisheries, urban fisheries, special-user-group fisheries, or wilderness fisheries. Trophy management objectives could rely on PSD or choose a length that is considered a trophy in the local area. Length limits could then be used as an action to control lengths of fish harvested (see Chapter 7). For example, muskellunge fisheries are often managed as trophy fisheries, and some popular trout fisheries are managed as catch-and-release fisheries. In these cases, management objectives will likely be stated as higher C/f and larger average lengths than exist in other local waters. Trophy and catch-and-release fisheries avoid the problem earlier mentioned for C/f objectives (i.e., that effort may need to be restricted) because catch is not limited by the production of a population because harvest is negligible (assuming survival after release). This type of management, recycling fish within the population, is one means to accommodate rising angler effort. Objectives for urban fisheries and special-user groups (e.g., children, elderly, or disabled) may focus on acquisition and maintenance of accessible places to fish combined with C/f. Objectives for wilderness fisheries could be to minimize the number of interactions among individuals (Kennedy and Brown 1976). Actions to achieve wilderness fishery objectives could be not to stock fish, advertise, or undertake any management action that would increase angler participation.

15.4.2 Commercial Fisheries

Commercial fisheries goals and objectives in natural lakes have a history of theoretical development derived through the management of marine fisheries. A critical theoretical development occurred when Baranov (1918) made the case that sometime in the life of a year-class (cohort) of fish, a peak in total biomass should occur as a function of growth and natural mortality rate, so, for a cohort, optimal periods should exist when the greatest harvest could be taken. Over the decades that followed, various approaches for managing commercial fisheries were developed with goals including sustainable yield (Beverton and Holt 1957; Ricker 1975; Quinn and Deriso 1999; Walters and Martell 2004), economic yield (Christy and Scott 1965; Clark 1985: Grafton et al. 2006), and (or) the precautionary principle (Garcia 1994; O'Riordan and Cameron 1994; Restrepo et al. 1999; Fisher et al. 2006).

Goal statements for commercial fisheries typically include the idea that the fishery is to be managed to maximize weight or economic benefit of the annual commercial harvest in perpetuity. Therefore, commercial fisheries objectives are often articulated as specific maximum harvest or economic levels believed to be sustainable. Sustainable yield theory applies to "maximum weight," economic yield theory applies to "maximum economic benefit," and "in perpetuity" or "sustainable" is addressed in the concept of the precautionary principle.

Commercial fisheries objectives for natural lakes can be specified as the fish biomass that can be continually harvested from a population without diminishing its viability. If a population is stable (i.e., births replace deaths), then harvest from the population should not cause a decline in abundance. Fishery harvest can be maintained partly because populations compensate for harvest through density-dependent increases in growth, reproduction, and survival. A population that is fished responds by increasing its productivity through increased growth, which then increases reproductive capacity (births or recruitment) and survival (Tyler and Gallucci 1980). Any theoretical equilibrium yield can serve as an objective for a commercial fishery in a lake, so equilibrium yield is more conservative or precautionary than is MSY at population levels lower than those that would support MSY. However, MSY has often been used as an objective for commercial fisheries.

Management of fisheries for MSY is problematic for many reasons (Larkin 1977). One problem with MSY is that survival varies over time because of environmental variation. Environments of natural lakes are prone to changes in physical (e.g., climate change) and biotic (e.g., nonnative species invasions) forces, which can alter predator–prey relationships, competition, mortality, and recruitment or reproduction. Therefore, MSY must be re-estimated whenever the ecosystem changes. In the Great Lakes, for example, new species have invaded so frequently that these ecosystems have not stabilized at equilibrium conditions long enough to enable reliable estimations of MSY. In addition, estimation of MSY from data for a number of populations causes MSY to reflect the equilibrium yield for an average population. Therefore, less productive populations may be overfished while more productive populations may be underfished (foregoing economic benefit). A more conservative approach would be to choose an equilibrium yield for a commercial fishery objective that is less than MSY and at higher population abundance than the level needed to support MSY.

Maximum economic yield (MEY) for a commercial fishery in a lake may not coincide with maximum sustainable yield (Christy and Scott 1965; Grafton et al. 2006). Managing for economic yield instead of biological yield is sensible because commercial fisheries are aimed at selling fish for profit (Grahm 1943). In a hypothetical example for lake whitefish in Lake Huron (Figure 15.6), MSY results in the maximum total (gross) revenue from the sale of fish, if the price for fish is constant and demand–supply effects do not occur. Costs increase as the number of fishers increase, who individually and collectively expend more effort to catch more lake whitefish. At some point, total cost intersects total revenue at the breakeven point, beyond which the commercial fishery loses money. To the left of this breakeven point, positive economic yield occurs because the total costs of fishing are lower than are total revenues from sale of lake whitefish. In this example, the maximum difference between the two curves, CD (= MEY), is less than MSY.

Economic forces of fisheries and fish markets are much more complicated than illustrated in Figure 15.6 because cost curves are much more complex and revenues are a function of the interaction between catch and price and, hence, are affected by the forces of supply and demand. Consequently, problems with management for MEY are numerous. In our lake white-

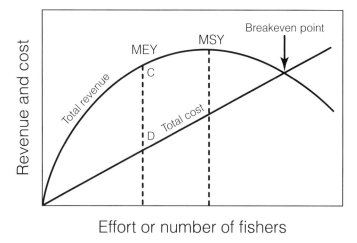

Figure 15.6. Relationship between number of fishers and total costs of fishing and total revenue from the sale of fish (adapted from Christy and Scott 1965). The difference between total revenue and total costs define a set of economic yields from the fishery. The line CD represents the largest difference between total revenue and total cost and defines maximum economic yield (MEY), which occurs at a lower level of effort than does maximum sustainable yield (MSY).

fish example, if management is successful and MEY is achieved as an objective then new fishers and more boats and gill nets are attracted to the profits, which reduces MEY over time. These forces would tend to drive the lake whitefish fishery to the break-even point. To solve this problem and to manage for economic yield, fishing effort must be controlled by limiting participation in the fishery (e.g., limited entry regulations).

The precautionary principle states that when fishing may cause irreversible harm to the long-term well-being of the resource, managers should establish and implement objectives to minimize risk in the face of scientific uncertainty (Garcia 1994). The precautionary principle states that fishery management should "do no harm" and it is "better to be safe than sorry." Proponents of more aggressive fishing or high harvest levels are required to prove that risk is minimal to long-term sustainability of fishery resources. Therefore, if fishers or managers desire to raise a quota limit for lake whitefish on Lake Huron, they must prove that increased harvest and mortality will not harm the population. This approach may seem like common sense, but the reality of fisheries management in the past has been the reverse. Precautionary management objectives were not adopted unless proof could be supplied that higher levels of fishing posed a risk or threat to populations. The precautionary principle shifts the burden of proof from those advocating precautionary harvest levels to those desiring higher harvest and fishing levels. The precautionary principle infers that the primary goal is sustainability, so yield or harvest must be specified at levels that do not pose risk to sustainability.

The precautionary principle in fisheries management was stimulated by recognizing that political and biological forces surrounding commercial fisheries management often led to depletion, collapse, or extinction of fish populations (Ludwig et al. 1993). Though long-term sustainability as a concept is typically part of commercial fisheries goals and objectives in lakes, management has often failed to achieve long-term sustainability of fish populations and associated fisheries. First, safe-yield limits are often estimated with great uncertainty, and second, fisheries are difficult to regulate within safe-yield limits, even if safe yield is known with great certainty. Overfishing can cause the collapse of populations, so fishing must be managed to prevent abundance from falling to potentially irreversible low levels. Managers must recognize uncertainty in their knowledge of fish populations and fisheries and take action in the face of this risk.

Precautionary management usually relies on a combination of stock size (B) and fishing mortality rate (F) to define overfishing and then to establish specific reference points (e.g., limit or targets) at which management actions are triggered. For example, B_{lim} is the spawning biomass or population size beyond which reproduction is unable to replace fish caught, F_{lim} is the fishing mortality rate beyond which removals from a population are higher than the rate of reproduction can compensate for losses (recruitment overfishing), and B_{lim} and F_{lim} define boundaries at which a population is in danger of collapse (Figure 15.7). Precautionary levels for B and F are also defined to provide boundaries within which to fish a population safely without undue risk to its long-term sustainability. The precautionary biomass (abundance) level, B_{pa}, is determined with reasonable certainty to sustain the population above B_{lim} in the face of interannual fluctuations, and the precautionary fishing mortality rate, F_{pa}, is set sufficiently below F_{lim} to have a low probability of stock collapse. These four variables can be used to define three zones of fishing (Figure 15.7): danger of population collapse, overfish-

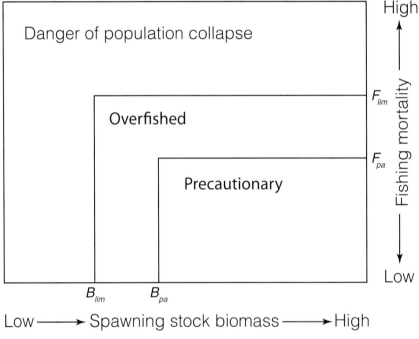

Figure 15.7. Graphical representation of precautionary levels of population abundance and fishing mortality rate to avoid overfishing and population collapse (adapted from Restrepo et al. 1999). The spawning biomass or population size beyond which reproduction is unable to replace those fish caught is designated as B_{lim}. The fishing mortality rate beyond which removals from a population are so high the rate of reproduction is unable to compensate for the losses (recruitment overfishing) is F_{lim}. The precautionary biomass (abundance) level (B_{pa}) is determined with reasonable certainty that even with year-to-year fluctuations, the population will stay above B_{lim}. The precautionary fishing mortality rate (F_{pa}) is sufficiently below F_{lim} so that a low probability of stock collapse exists.

ing, and precautionary. Precautionary management objectives for commercial fishing would be defined at levels below B_{pa} and F_{pa}. Herein lies a major difficulty, because no objective method has been developed to decide how large a difference should exist between F_{lim} and F_{pa}. Implementing precautionary objectives for commercial fisheries management is also difficult because of the need to estimate population size (B) and fishing mortality rates (F), both of which are often estimated with great uncertainty but both of which are also needed to develop precautionary reference points and to evaluate success (Essington 2001).

15.4.3 Subsistence Fisheries

Goals for subsistence fisheries management in natural lakes usually focus on ensuring the opportunity for subsistence fishing. In Alaska, one stated goal for subsistence management is as follows:

> [I]f the harvestable portion of the stock or population is sufficient to provide for subsistence uses and some, but not all, other consumptive uses, the appropriate board shall adopt regulations that provide a reasonable opportunity for subsistence uses of those stocks or populations …. [Alaska Statute 16.05.258]

In this example, the primary purpose or goal of subsistence fisheries management is to provide a "reasonable opportunity" for subsistence fisheries. Therefore, the role of subsistence objectives would be to provide a specific definition of a reasonable opportunity. A possible objective might be that fish population abundance would be managed at levels to permit fishing a minimum of 6 d/week with all types of gear. Choosing specific metrics for subsistence fisheries objectives to measure progress toward goals should be done by consulting with subsistence users.

Goals and objectives for subsistence fisheries should not be chosen in isolation from recreational and commercial fisheries or other subsistence activities. Subsistence use of fish may have a legal priority over recreational and commercial uses, so goals and objectives for all fisheries may be intertwined. In Alaska, if populations of fish can support a limited harvest, then subsistence fisheries must be allowed to provide a reasonable opportunity to harvest quantities of fish deemed necessary for subsistence purposes (Alaska Statute 16.05.258, Subsistence Use and Allocation of Fish and Game). If potential yield exceeds the amount deemed necessary for subsistence, then other fisheries may be opened. Fish provide one source of food for rural communities and must be considered in conjunction with other subsistence food gathering of wildlife and plants. Subsistence fisheries often have a special place in a calendar of subsistence activities (e.g., linked to weather appropriate for preservation drying) and a seasonality in diets of residents. Goals or objectives must be carefully chosen so that management actions consider cultural effects of fisheries management.

15.4.4 Species Restoration

Fisheries management is sometimes focused on goals not associated with fisheries but rather on re-establishment of fish populations or species within their native distributions. For example, goals were established to restore self-sustaining populations of lake trout into each of the five Laurentian Great Lakes (Hansen 1999; Krueger and Ebener 2004). In the USA, the

Endangered Species Act can require that plans be developed and implemented to prevent extinction of populations or species of fish (e.g., sockeye salmon in Redfish Lake, Idaho; Flagg et al. 1992; Selbie et al. 2007). Goals describe the scope and scale expected for successful species restoration. For Lake Michigan, as an example, Bronte et al. (2008) recommended a goal for lake trout management as follows: "in targeted rehabilitation areas, reestablish genetically diverse populations of lake trout composed predominately of wild fish able to sustain fisheries." This goal specifies that rehabilitation of the species should be focused in certain areas of the lake, that some level of genetic diversity should be a part of management (Reisenbichler et al. 2003), and that the populations should be able to support fisheries.

Objectives defining measurable progress toward the goal should be established based on attributes of populations similar to the target population. Metrics for use in management objectives should be those that define self-sustaining wild populations. For Lake Michigan, objectives for lake trout restoration were based on (1) total annual mortality rate (<50%); (2) numbers of mature fish in spawning areas (C/f in fall assessments should consistently exceed 50 lake trout/1,000 ft [305 m] of graded-mesh gill net); (3) numbers of eggs deposited in spawning areas (500 viable eggs/m^2); (4) numbers of spawning age-classes (>5 year-classes); (5) density of naturally reproduced year-classes (100,000 naturally produced yearlings); and (6) establishment of more than one morphotype. Morphotype refers to different forms of lake trout that exist as semireproductively isolated populations occupying unique niches in lakes (Krueger and Ihssen 1995; Zimmerman et al. 2009).

Objectives for restoration of species should also have specific deadlines for accomplishment. Dates specified for measurement of each metric must reflect the life history of the species, including age at maturity, number of spawning age-classes required, and maximum life span. For long-lived species such as lake trout or lake sturgeon, dates specified in objectives tend to be many years into the future (10–50 years) to reflect the long life span of these species. An example objective recommended by Bronte et al. (2008) for lake trout in Lake Michigan is as follows: by 2024, spawning populations in targeted rehabilitation areas stocked prior to 2008 should be at least 25% female and contain 10 or more age-groups older than age 7.

15.5 STOCK ASSESSMENT DESIGNS

Stock assessment is an important activity of fisheries management agencies to gauge progress toward achieving goals and objectives for fisheries. Stock assessment is powerful when both fishery-dependent surveys and fishery-independent surveys are conducted. Fishery-dependent surveys rely on people participating in the fishery to report catch and effort, whereas fishery-independent surveys are conducted by management agencies to investigate characteristics of fished populations. Fishery-dependent surveys provide subjective, self-reported indices of fishery attributes, so effort, catch, and locations fished are often not verifiable. The accuracy of reporting in fishery-dependent surveys can be muddled by several external factors (e.g., income tax obligation, mistrust of management agency, among other factors). For example, in Lake Superior, unreported changes in gear and fishing locations masked underlying changes in lake trout stock density during a period of stock collapse (Wilberg et al. 2004). To reduce such bias, agencies must incorporate on-board monitoring (often termed observers) to provide more accurate reporting.

In contrast to fishery-dependent surveys, fishery-independent surveys provide objective indices of stock status. Fishery-independent surveys are typically conducted by a management agency with its own vessels and crews. Fishery-independent surveys use fixed or random sampling stations and do not target areas of high catch or effort. If fishery-independent surveys are undertaken objectively, then C/f will reflect underlying trends in stock density and abundance. For example, fishery-independent surveys were corrected for changes in fishery attributes and then blended with fishery-dependent surveys to reconstruct the history of lake trout stock collapse and recovery in Lake Superior from 1929 to 1999 (Wilberg et al. 2003).

15.5.1 Indicators and Biological Reference Points

An indicator is a type of metric that can be used as an objective, whereas a reference point is a specific "benchmark" value of an indicator that triggers or guides management decisions to achieve an objective. Perhaps the most famous reference point in fisheries management is F_{MSY}, the fishing mortality rate that results in maximum sustainable yield of a stock. Biomass (B) and yield change as F increases from 0 to $F_{Extinction}$, the point at which B is 0 (Figure 15.8). Therefore, if assessment indicates that F is less than F_{MSY}, fishery expansion is justified. Alternatively, if F is greater than F_{MSY}, then F should be reduced. An alternative application of this model could be based on stock biomass. Because equilibrium biomass decreases with F relative to F_{MSY} (Figure 15.8), biomass at MSY also provides a reference point (B_{MSY}). Therefore, comparison of B to B_{MSY} could trigger the same management decision. The precautionary approach discussed above (Figure 15.7) was developed from these concepts.

Although MSY is no longer widely viewed as a primary objective in fisheries management, this example demonstrates several features of a reference point framework. First, in-

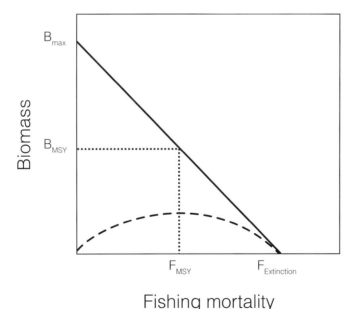

Figure 15.8. Illustration of MSY and associated reference points for a hypothetical fish stock. The *x*-axis is fishing mortality rate. The relationship between fishing mortality and equilibrium biomass is depicted by the solid line; the relationship between fishing mortality and equilibrium yield is depicted by the dashed line. The MSY reference levels are depicted by dotted lines.

dicators and reference points are closely linked to an objective, so the value of an indicator in relation to a reference point dictates the management direction that will help to meet an objective. Second, establishment of reference points requires a model of system performance. In this example, the model describes how fishing affects stock biomass. Third, several indicators and their associated reference points can be used to guide management decisions aimed at the same objective. In this example, the same decision results when either F or B is used. In practice, measurement of indicators and their associated model-based estimates of reference points are subject to error, so indicators may not agree in directing management action. For this reason, multiple indicators are preferred, and choice of action is based on the weight of evidence. Lastly, the framework is transparent because decision criteria are identified before collection of data and can be communicated to resource users.

Fisheries management entered a new realm of complexity with the demise of MSY (Larkin 1977). Optimum benefits from fisheries do not result from attempting to maximize yield, so ecologically sustainable fisheries must consider impacts of fishing on the aquatic ecosystems that support harvested stocks. More conservative yield objectives for fisheries have resulted not only because they are expected to optimize short-term benefits, but also because they are more likely sustainable. This shift of emphasis toward ecologically sustainable fisheries has resulted in a new role for MSY and an increased number of indicators that are needed to manage a fishery.

Maximum sustainable yield is now viewed as a limit reference point that specifies a limit to exploitation to safeguard long-term productivity of a fish stock (e.g., NOAA 1998; Mace 2001). Therefore, biomass and fishing mortality reference points are still used but now bound an acceptable zone of operation within which targets must be set. Straying outside of the zone should initiate prompt, strong actions to reduce fishing mortality and restore fish biomass. This response is the essence of precautionary management.

In addition to indicators related to harvested stocks, fisheries management that includes ecosystem-related objectives requires indicators or parameters that measure impacts on other parts of the ecosystem. Depletion of stocks by fishing affects the structure and dynamics of aquatic ecosystems, so changes in fish assemblage structure and long-term productivity of harvested stocks must be assessed. In recent years, many variables have been proposed as candidates for assessing ecosystem effects of marine fishing (e.g., Sainsbury and Sumaila 2001; Rochet and Trenkel 2003; Methratta and Link 2006). The list includes the same population measures that apply in single-species management (e.g., abundance, exploitation rate, and mean fish size), as well as a long list of assemblage-related variables (Tables 15.5, 15.6). Many of these indicators may be redundant (Methratta and Link 2006), so the final list may shrink. For population indicators, a strong theoretical basis supports setting reference points, but this is not the case for assemblage indicators. Translating these community indicators into decision criteria with reference points and associated control rules remains a challenge (Link 2005).

Development of a reference point framework for natural lake fisheries is less advanced than is the evolving framework for marine fisheries. The perceived need for a rigorous framework has been less, perhaps because recreational, not commercial, fishing is the predominant form of exploitation on most natural lake waters. Recreational fisheries are often viewed as different from commercial fisheries in that they are assumed to be self-sustaining and not controlled by the same social and economic forces that have driven many fish populations and associated commercial fisheries to collapse (Post et al. 2002). Consequently, many recreational fisheries in North American lakes are managed passively, without specific management plans,

Table 15.5. Population indicators and reference points for assessing the effect of fishing (after Rochet and Trenkel 2003). Indicators can apply to all species but "effect of fishing" applies to only species directly impacted by fishing (i.e., harvest and bycatch). The following abbreviations are used: maximum sustained yield (MSY); catchability coefficient (q); parameters of the von Bertalanffy growth model (K and L_∞); length at first maturity (L_m); and length at first capture (L_c).

Indicator	Effect of fishing	Reference point
Biomass or abundance (B)	–	B_{MSY}
Catch per unit effort (C/f)	–	$q \times B_{MSY}$
Total mortality rate (Z)	+	$K(L_\infty - L_m) / (L_m - L_c)$
Fishing mortality rate (F)	+	F_{MSY}
F/Z	+	0.5
Mean length of population	–	
Median length of catch	–	Median length at maturity
Proportional size distribution (PSD)	–	
Growth	+	

Table 15.6. Indicators related to the fish assemblage that might be affected by fishing (after Rochet and Trenkel 2003).

Indicator	Effect of fishing
Average growth rate	+
Average maximum length	–
Average age at maturity	–
Average size at maturity	–
Mean length distribution	Shift left
Size–abundance relationship	Slope decrease
Ordination of species traits	Smaller and faster traits
Biomass variability	+
Proportion of piscivorous fish	–
Pelagic to demersal ratio	+
Mean trophic level	–
Fishing in balance	–
Proportion of nonharvested species in community	+
Average weight in the community	–
Size spectrum	Fewer large fish Less total biomass

goals, objectives, or methods to control exploitation within prescribed limits. Further, if recreational fisheries have management plans, the plans are often vague or generically applied to many water bodies (Pereira and Hansen 2003). Active management of fisheries implies that a complete management procedure is in place, with clear goals and objectives, management strategies and actions aimed at meeting the objectives, and methods for assessing and evaluating whether the objectives have been met. Active management is practiced on large North American lakes that support commercial fisheries and to restore native fishes but has been applied less frequently to recreational fisheries.

15.5.2 Case Studies of Active Management

In the three sections that follow, examples of active management applied to North American lakes are provided. In each case, the key ingredients of an active management system are described.

15.5.2.1 Lake Trout in the Great Lakes

An example of intensive sampling in a large lake is the monitoring of lake trout stocks in Lake Superior (Hansen 1996). The goal of lake trout restoration in Lake Superior is to restore self-sustaining stocks that can provide an annual catch of 2 million kilograms, which was the average annual yield during 1929–1943 when yield was thought to be sustainable (Busiahn 1990). Since lake trout stocks collapsed in the 1950s, fishery management has sought to restore stocks by restricting harvest, controlling sea lampreys, and stocking hatchery-reared lake trout (Hansen 1996). Effectiveness of management actions is monitored through fishery-dependent and fishery-independent annual surveys, from which indices of lake trout harvest, abundance, recruitment, mortality, and sea lamprey wounding rates are developed for comparison to measurable objectives.

Lake trout abundance, mortality, wounding, recruitment, and harvest are monitored annually to track progress toward restoration objectives, develop management strategies, and implement management actions. Collection, analysis, and reporting of data for each metric are standardized and reported for individual lake trout management areas. Abundance of stocked (fin-clipped) and wild (not fin-clipped) adult lake trout is indexed from C/f in standardized gill nets fished for one night (or corrected for soak time; Hansen et al. 1998) in late April through May. Total annual mortality (A) of stocked and wild lake trout is estimated from catch curves based on data for fish sampled during the spring gill-net surveys. The mortality rate estimated from a single gill-net mesh size is recognized to be an index of the true mortality rate because of strong size selectivity (Hansen et al. 1997). Sea lamprey wounding rate is indexed as the mean number of fresh wounds per 100 lake trout caught in spring gill nets. Recruitment of wild lake trout is indexed as the C/f in standardized gill nets fished for one night from late July through August. Catches of lake trout in recreational, commercial, and subsistence fisheries are estimated for comparison with allowable-harvest limits. Recreational fishing effort and harvest are monitored through creel surveys of recreational fishing effort (counts of anglers and boat trailers) and catch rates (angler interviews), commercial fishing effort and harvest are monitored through mandatory reporting and on-board monitoring, and subsistence fishing effort and harvest are reported by users of this fishery.

Statistical catch-at-age models are used to integrate survey data and to estimate adult abundance, recruitment, and mortality from harvest at age in commercial and recreational fisheries and C/f at age in large-mesh and small-mesh gill-net assessment fisheries (e.g., Linton et al. 2007). A projection model is then used to estimate future harvest quotas from trends in recruitment and abundance at age with fixed natural mortality (including sea lamprey mortality). The rate of fishing mortality (F) is set at a level that does not induce a decline in abundance (N), and harvest quotas (C) are then estimated from abundance and total mortality (Z = total instantaneous mortality; A = total annual mortality) by use of Baranov's equation, $C = NFA/Z$ (Ricker 1975). The biological reference point specified as a measurable objective for managing fishery harvest was taken from a meta-analysis of lake trout in North America that showed lake trout subjected to total annual mortality (A) exceeding 50% tended to decline (Healey 1978). Therefore, the annual fishing mortality rate must be set at a level so that when added to the annual natural and sea lamprey mortality rates, the total annual mortality rate does not exceed 50%.

15.5.2.2 Walleye in Wisconsin

Monitoring of walleye populations and associated fisheries in the northern third of Wisconsin is carried out under the aegis of federal court decisions that affirmed tribal rights to hunt, fish, and gather in lands ceded in treaties of 1836 and 1842 (Hansen et al. 1991; Beard et al. 1997). Each year from among lakes containing walleye, lakes that were subjected to state-licensed angling and tribal-licensed spearing are selected at random (Hansen et al. 1991). Tribal spearing occurs during a few weeks in spring, while sexually mature walleyes are attempting to spawn, whereas angling occurs from the first Saturday in May through the first of March in the next year. Tribal spearing targets spawning walleyes, so numbers of sexually mature walleyes are estimated during the spawning period. Angling occurs while mature and immature walleyes are intermingled, so total numbers of walleyes are estimated 2–3 weeks after spawning when immature and mature fish are mixed in the total population.

Abundances of adult walleyes and all walleyes (immature and mature) are estimated during spring of each year by use of Chapman's modification of the Petersen estimator (Ricker 1975). Fyke nets are set shortly after ice-out, when mature walleyes congregate nearshore for spawning, to capture walleyes for marking (Hansen et al. 1991; Beard et al. 1997). Ten percent of mature walleyes in each lake are targeted for marking. Mature walleyes are defined as all fish for which the sex can be determined or fish longer than 38 cm, the length at which both males and females are likely to be sexually mature. All fish are marked by partial removal of a fin. To estimate adult walleye abundance, fish are recaptured 1–2 d after netting while adults are still congregated near shore. Because the interval between marking and recapturing is short, the entire shoreline is electrofished to ensure marked and unmarked fish are equally vulnerable to capture. In order to estimate the total number of walleyes in the population, immature and mature fish that are present during the first electrofishing recapture run, but which have not been previously marked, are marked by removal of one or more fins. To estimate total walleye abundance, fish are captured about 2–3 weeks after the first recapture sample. Again, the entire shoreline of each lake is electrofished to ensure marked and unmarked fish are equally vulnerable to capture. Marked fish from the fyke netting and first electrofishing runs are combined as the marked sample, and the ratio of marked fish in the population is estimated from the second electrofishing run.

Abundance and variance are estimated for four length classes (≤30 cm, 30–38 cm, 30–51 cm, and ≥51 cm) and then summed to estimate adult and total abundance and variance for each lake. Prior to estimating abundance, walleyes speared after the start of the marking period are subtracted from the number of marked fish at large during the recapture period. These fish are then added to the number of fish estimated to have been present at the time of marking for the populations of interest.

Based on an estimated nightly quota for each lake and the number of interested individuals, spearing is permitted for only a limited number of tribal members on each lake each night (USDI 1991). Each spearer is allowed a quota for the evening (commonly, 25 fish/spearer), which must be landed at a specified location on each lake. Spearing is pursued to take the allotted quota in as little time as possible. Clerks record the starting and ending times of each spearing trip, count all fish speared, and record the length of each fish. Maximum length of walleyes speared is limited by allowing only one fish to be between 51 and 61 cm, one additional fish to be of any length, and the balance to be of lengths shorter than 51 cm.

Angling harvest is estimated from creel surveys conducted during the entire angling season for walleye (Beard et al. 1997). Creel surveys follow a stratified-random roving-access design (Pollock et al. 1994; Rasmussen et al. 1998). Angling catch rates are estimated from catch and effort data collected during interviews of anglers as they complete fishing each day. Data are stratified by month and day type (weekdays versus weekends). The mean annual walleye catch rate for each lake in a year is calculated as the weighted mean of the monthly estimates.

The biological reference point for limiting spearing and angling exploitation was derived from long-term observation of the walleye population in Escanaba Lake, which was sustainable at an annual exploitation rate not exceeding $u = 0.35$. By federal court order, the sum of spearing and angling exploitation cannot exceed $u = 0.35$ more than 1 in 40 times (Beard et al. 2003). Spearing harvest is limited by lake-specific safe-harvest levels that account for uncertainty in the method by which the total allowable catch was estimated (Hansen et al. 1991). Angling bag limits are altered on individual lakes each year in response to spearing harvest to ensure that combined spearing and angling exploitation does not exceed the biological reference point. Spearing exploitation is estimated as the ratio of spearing harvest to adult walleye abundance, whereas angling exploitation is estimated as the ratio of angling harvest to total walleye abundance. During 1980–1998, the combined spearing and angling exploitation rate exceeded $u = 0.35$ only slightly more than 1 in 40 times, with total exploitation averaging 11.8%, angling exploitation averaging 8.4%, and spearing exploitation averaging 3.5% (Beard et al. 2003).

15.5.2.3 Inland Fisheries in Ontario

Traditional methods of managing fisheries call for annual or frequent monitoring of a fishery so that regulations can be changed if exploitation exceeds a desired level. This approach is not feasible for managing inland lake fisheries in Ontario because lakes are too numerous. Lakes in Ontario offer diverse angling opportunities for many species, including walleye, lake trout, brook trout, northern pike, and smallmouth bass. Overall, the size and economic importance of this resource are large but are divided into so many small isolated populations that detailed stock assessment on each lake is not feasible. Because management on a lake-by-lake basis is not feasible, a passive management approach was applied. Except for a few large lakes, management was conducted without a detailed plan that included assessment and criteria for directing management action. A common regulatory scheme (i.e., seasons, bag

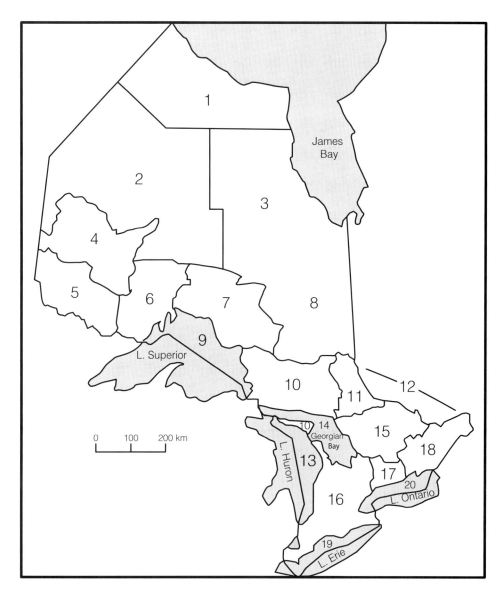

Figure 15.9. Fishery management zones in Ontario, Canada. Within each zone, a stratified random sample of lakes is selected to monitor fisheries.

limits, and length limits) was applied to all lakes in a geographic region (Figure 15.9), with exceptions to regional regulations for individual lakes when overexploitation was detected, often when anglers complained about deteriorating fishing quality.

Evaluation of this management approach identified several flaws that led to development of a more active form of management (Lester et al. 2003). The most important flaw related to the scale of management, by which management decisions were predominated by an individual-lake approach. This approach is not tenable when the resource is widely dispersed among so many lakes and the angling population is free to move among lakes. Because regulations applied to individual lakes may influence the distribution of angling effort, a broader spatial scale of management is needed and should incorporate four elements: (1) consensus on ecologically achievable objectives; (2) minimally biased assessment of the state of the resource; (3) periodic evaluation of whether current management practices are meeting objectives; and (4) an adaptive management approach in choosing among alternative management actions. In 2008, the Ontario Ministry of Natural Resources initiated a new approach for managing inland lake fisheries (Lester et al. 2003; Lester and Dunlop 2004). Components of this approach are summarized to illustrate how a rigorous management framework can be applied to a landscape of lakes.

Provincial fisheries management objectives were developed within the context of high-level goals to address a broad range of themes, including sustainable use, healthy aquatic ecosystems supporting biodiversity, optimization of social and economic benefits, and improvement of partnerships. The goals are important because they communicate to the public what fisheries managers are striving to achieve in Ontario. The goals are also important because they provide a linkage between high-level strategic directions and indicators specified in objectives. The hierarchy that links goals to indicators in objectives identifies a set of variables that must be measured on lakes (Figure 15.10). Some variables are indicators for evaluating resource status (e.g., fish abundance, mortality rate, and contaminant levels) and some are needed to calculate indicators of reference points (e.g., lake morphometry).

Sampling methods used to measure variables include large-fish netting, small-fish netting, zooplankton sampling during summer, water sampling in the spring, and a seasonal aerial survey of fishing effort (Figure 15.10; Sandstrom et al. 2008). Large-fish netting surveys use a standard gill netting protocol proposed for North American lakes (Bonar et al.2009). Because a North American standard for sampling small fish does not yet exist, small-fish netting surveys use a gill-netting method adapted from a European standard (Appelberg 2000).

Fish stocks	Habitat	Community	Exploitation
Parameters			
• Abundance • Size structure • Age structure • Mortality rate • Maturation • Disease	• Bathymetry • Temperature • Oxygen • Water clarity • Water chemistry	• Fish species • Zooplankton	• Angling effort
Sampling methods			
• Large fish netting	• Spring water chemistry survey • Other data collected during fish surveys	• Small fish netting • Plankton hauls	• Aerial effort survey

Figure 15.10. Variables measured on each lake sampled for fishery attributes in Ontario, Canada. For each category of variables, the sampling method is indicated in the bottom half of the figure.

These standard netting methods are used to obtain *C/f* indices of fish biomass, which can be converted to population biomass if catchability of the gear is known. However, research is needed to measure catchability of these gears so that biomass estimates of harvested species can be compared with biomass reference points. Research is also needed to develop biomass and mortality reference points for harvested species. These reference points vary depending on lake characteristics that affect carrying capacity and life history traits (Shuter et al. 1998; Lester et al. 2004a).

Because sampling cannot be conducted on all lakes each year, implementation of the management assessment scheme involves decisions about where and when to sample. The temporal scale is a 5-year cycle, and sampling intensity is 5% of all lakes larger than 20 ha. This design implies that 1,660 lakes will be sampled in each 5-year cycle (332 lakes/year), half of which will be monitored as fixed sites once in each 5-year cycle and half of which will be re-selected as variable sites within each 5-year cycle. This mixed-site design is a compromise between trend detection, favored by fixed sites, and status reporting, favored by variable sites (Urquhart et al. 1998). Lakes are selected for sampling based on stratified random-sampling principles. Strata include geographical zones and lake size. Although lakes of all sizes (20–250,000 ha) are included, a higher proportion of large lakes are sampled because large lakes are deemed more valuable than are small lakes for supporting regional fisheries.

Evaluation of current management occurs at the end of each 5-year cycle and includes reporting of indicator values (objectives) and performance assessment based on reference points. For example, if the ratio of B/B_{MSY} exceeds 1 for all lake trout populations, the current management system is meeting the objective of sustaining lake trout populations at desired levels of abundance. The geometric mean of these ratios offers a suitable statistic for scoring the performance of a set of lakes. The same scoring method applies to each indicator, such as biomass, mortality rate, and fishing effort, and scores are produced for each harvested species in each management zone. Results obtained at the end of each 5-year cycle can be compared with results from previous assessments to monitor trends and to assess exploitation in different parts of Ontario.

A graphical example of this evaluation process is based on data from southern Ontario during the 1980s (Figure 15.11). Angling *C/f*, an index of lake trout abundance, effort, an index of fishing mortality, and MSY reference points were developed with the model described by Shuter et al. (1998). Panels A and B show results for individual lakes plotted against lake area. The line on each graph specifies MSY reference points to indicate that reference points for lake trout depend on lake size. In small lakes, lake trout are usually small but density can be high, whereas in large lakes, lake trout grow larger but density is lower. The data indicate that lake trout are overexploited because effort is above and abundance is below MSY levels in most lakes. Managers of Ontario have four options for reducing overexploitation: (1) reduce effort by changing fishing seasons or restricting access; (2) reduce angling efficiency by imposing gear restrictions; (3) limit harvest by implementing annual quotas or by changing creel limits and length limits; or (4) stock fish.

15.6 MANAGEMENT STRATEGIES

Based on the fisheries management paradigm, fisheries management strategies aimed at protecting, rehabilitating, or enhancing fisheries in natural lakes can be categorized into habitat, population, or people management (see Chapter 5) to organize the issues or problems that

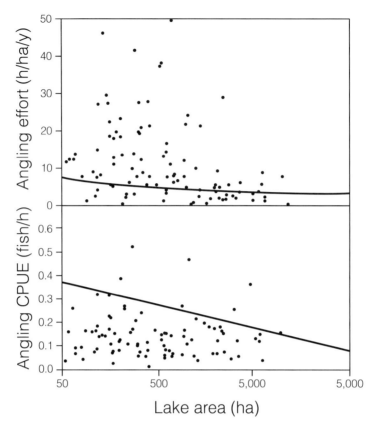

Figure 15.11. Illustration of how the state of the lake trout resource can be evaluated using angling effort (h/ha/year) and catch rate (fish/h) (redrawn from Lester and Dunlop 2004). The solid line in each graph is the MSY reference level. Lakes that fall above the line in the upper panel (angling effort versus lake size) or below the line in the lower panel (angling catch rate versus lake size) are considered to be overfished.

need to be addressed to achieve goals and objectives. Habitat issues in natural lakes are usually related to water quality or alteration of physical habitat. Fish population issues in natural lakes are usually related to low abundance of a target species, negative effects of exotic species, or imbalance of predator–prey relationships. People issues in natural lakes are usually related to overharvest or perceptions of quality.

The interconnectedness of a fishery often means that a single symptom, such as reduced size structure, may be caused by multiple forces, such as high recruitment that leads to high population density, high fishing mortality that removes large individuals from the population, and (or) slow growth that is caused by poor habitat quality. Diagnosis of causal forces for problem symptoms is difficult, but necessary, for prescribing management actions to remedy the problem.

15.6.1 Habitat Management

Problems with habitat in natural lakes are numerous but are often related to nutrient enrichment (cultural eutrophication), infestations of aquatic plants (often exotic species), shoreline alteration, acidification, or contaminants.

15.6.1.1 Cultural Eutrophication

Cultural eutrophication is likely the most prevalent habitat problem in natural lakes (NAS 1969) and occurs when nutrients, primarily phosphorus and nitrogen, enter lakes from runoff or streams draining feedlots, sewage outfalls, urbanized landscapes, or fertilized fields at rates that exceed the decomposition rate in the lake (Kalff 2002). Phosphorus was long thought to limit primary production in most lakes, but the N/P ratio is tightly coupled in most lakes, so both nutrients must be added to stimulate increased algal production (Elser et al. 1990). Algal production is strongly correlated to nutrient concentration (Sakamoto 1966, referenced in Kalff 2002), and increased algal production leads to increased enrichment of the hypolimnion where aerobic decomposition strips away dissolved oxygen to produce an oxygen deficit (Kalff 2002). Lake size and form influence transport processes such as sedimentation, re-suspension, and mixing, which, in turn, influence chemical concentrations, algal production, and water clarity (Håkanson 2005). Oxygen deficit leads to fish kills during summer when stratification separates the hypolimnion from oxygen sources (photosynthesis and atmospheric oxygen) and in winter when ice and snow prevent photosynthesis and oxygen replenishment from the atmosphere. During periods of summer stratification, fish that depend on cold water will either die or be forced into warm surface waters that are less suitable for growth and survival. Shallow, nutrient-rich lakes are most susceptible to summer and winter fish kills. Eutrophication has been associated with oxygen supersaturation and depletion, toxic algae, and high pH, each of which can impact fishes at various life stages (Müller and Stadelmann 2004).

Methods for reducing nutrient levels can be categorized into methods that reduce (1) nutrient inputs (sewage treatment, wastewater diversion, wetland protection or construction, and buffer strips), (2) in-lake nutrient loads (inactivating phosphorus by adding aluminum sulfite, increasing phosphorus oxidation by adding lime, dredging nutrient-rich sediments, and harvesting aquatic macrophytes), or (3) oxygen-deficient water (pumping or aerating hypolimnetic water; Kalff 2002). Effects of nutrient abatement programs are often difficult to quantify, likely because nutrient loads must be reduced substantially while in-lake nutrients are flushed. Nutrients are therefore more difficult to reduce in lakes with long retention times, even if inputs are abated. For example in Lake Washington wastewater diversion dramatically reduced phosphorus levels because Lake Washington has a short retention time for flushing of in-lake nutrients and most nutrient loading was from point sources. Effects of in-lake nutrient reduction treatments are either not yet known (adding aluminum sulfate or lime) or too small in relation to total loads to produce positive effects (e.g., dredging sediments and harvesting aquatic plants). Aeration of the water column by pumping air or water into the hypolimnion replenishes oxygen by circulating water during summer and by maintaining open water during winter so that oxygen is replenished by contact with the atmosphere. However, aeration is only a symptomatic treatment and does not solve underlying problems. Aeration is costly on a large scale but more feasible for small lakes (see Chapter 16).

15.6.1.2 Aquatic Plant Infestation

Dense aquatic vegetation is a nuisance in many natural lakes of North America, especially when exotic plant species such as Eurasian watermilfoil, hydrilla, or curly-leaf pondweed invade and proliferate. Submersed aquatic plants influence both physical and chemical environments for fishes living in the littoral zone (Crowder and Cooper 1982; Wiley et al. 1984;

Carpenter and Lodge 1986) and thereby influence fish abundance (Miranda and Pugh 1997; Trebitz et al. 1997), survival (Miranda and Pugh 1997), and growth (Crowder and Cooper 1979; Olson et al. 1998). Aquatic vegetation affects fish growth rates by altering behavior and distribution (Savino et al. 1992) because predators are less able to capture prey at high vegetation density (Savino and Stein 1982), so foraging efficiency is low when plant stem density is high (Crowder and Cooper 1982). Conversely, when plant stem density is sparse, predators may overexploit their prey (Heck and Crowder 1991).

Stands of aquatic plants in natural lakes are often manipulated by mechanical harvesting, application of herbicides, and stocking of fish or invertebrates (Kalff 2002). Use of mechanical harvesters or herbicides permits plants to be removed from all or selected areas of lakes, often depending on stakeholder needs unrelated to fisheries, such as beach esthetics and swimming. Mechanical harvesting has the advantage of removing plant biomass from the lake, whereas herbicides leave plant biomass in the lake for recycling of nutrients. However, this apparent advantage may be trivial if the amount of nutrients removed through plant harvesting is small in relation to the total amount of nutrients available in the lake. Use of biological agents requires stocking of plant-eating fish, such as grass carp (illegal in many states) that feed on all plants, or plant-eating insects, such as beetles that feed on single or a few target species of plants. However, stocking fishes and insects as biological agents for plant control is risky because lakes are never closed, so stocked fish and insects, most of which are nonnative species in North America, are likely to escape to other lakes and rivers.

If aquatic plant management is focused on fishery benefits, rather than other uses, reducing rather than eradicating plants should be the objective of management actions regardless of the control tactic used. Dense aquatic vegetation reduces angling effectiveness for larger largemouth bass (≥ 2.27 kg; Maceina and Reeves 1996), so anglers may fish less (Colle et al. 1987). Therefore, reducing plant density may lead to increased angling effort (Mitzner 1978). Overall catch rates may be positively correlated with plant density cover because average fish size is often lower when plant density is high (Maceina and Reeves 1996; Wrenn et al. 1996). The amount of plant bed edge seems particularly important for increasing angling catch rates (Smith 1993), so cutting channels through dense plant beds, rather than removing all plants, may improve fish growth and survival and also improve angling (Unmuth et al. 1999, 2001). Maintaining intermediate plant density, rather than high or low plant density, is important for maximizing foraging efficiency and thereby increasing size structure (Crowder and Cooper 1979), which may increase production of harvestable-size fish (Moxely and Langford 1985).

15.6.1.3 Acidification

Acid rain or deposition is caused when burning of coal produces constituents of sulfuric acid and nitric acid that are transported by the atmosphere to a lake basin (Haines 1981; Clair and Hindar 2005). The acid neutralizing capacity (ANC) of a lake, usually indexed by alkalinity (mg HCO_3^-/L), determines vulnerability to acid rain, the effect of which is highest from Tennessee and Kentucky northeast to New York and also the Precambrian Shield lakes of Canada. Lakes in this region have low ANC and high inputs of acid rain. In contrast, areas underlain by limestone are less vulnerable to acidification even in the presence of acid rain. If acid inputs saturate the ANC of the water, lakes can become too acidic for fish to survive, particularly eggs and larvae (Hulsman et al. 1983; McCormick and Leino 1999). Species richness is also positively correlated to pH, which suggests that acidification may negatively affect the number of species occupying lakes (Rago and Wiener 1986; Minns 1989; Graham 1993).

Short-term remediation of acidification is possible through addition of lime ($CaCO_3$) directly to the lake or watershed, whereas long-term remediation is only possible by reductions in emissions. Liming quickly increases ANC and thereby neutralizes free hydrogen ions, though the effects are short-term and costs are high (Kalff 2002; Clair and Hindar 2005). Lime can be added to the entire lake surface, and application of lime to specific areas may adequately increase pH enough to increase survival of eggs and young of fishes (e.g., Gunn and Keller 1984; Hasselrot and Hultberg 1984; Booth et al. 1993). The only long-term solution to acidification is to reduce use of fossil fuels that produce acid constituents during combustion or to make combustion cleaner (Haines 1981; Kalff 2002). The effect of reduced emissions is illustrated by lakes near Sudbury, Ontario, where an 80% reduction of SO_2 emissions from copper smelting led to increased pH and subsequent reproduction by lake trout. In North America, air quality standards implemented by federal governments may not be sufficient to increase pH of surface freshwaters above the threshold (pH = 5.5–6.0) at which fish species diversity can be restored to original levels (Doka et al. 2003; Grupert 2003).

15.6.1.4 Contaminants

Contaminants of greatest concern in natural lakes, such as organichlorines, heavy metals, and mercury, are problematic because they are toxic, persistent, and often lipophilic, so they biomagnify and bioaccumulate in the biota (Kalff 2002). Long-lived fatty fishes that are high in food chains, such as lake trout, are likely to carry higher concentrations of contaminants than are other biota in affected aquatic communities. Contaminants have been linked to increased incidence of tumors in fish and possibly reproductive inhibition, but the most serious effects are felt in fish-eating birds and humans. Therefore, birds and humans that eat contaminated fish are more likely to suffer chronic effects related to toxicity of such substances, which are usually related to reproduction. Because eating contaminated fish poses health hazards to humans, fisheries managers of natural lakes most often face problems associated with human health advisories from eating contaminated fish rather than direct effects of contaminants on the health of fish. For example, fish consumption advisories apply to 30% of Wisconsin lakes, 90% of northeastern Minnesota lakes, 90% of Ontario lakes, and more than 1,000 Michigan lakes. The only remedy is to eliminate sources, although aerial transport makes elimination extremely difficult.

15.6.1.5 Shoreline Alteration

Alterations to the shorelines of natural lakes through natural or anthropogenic forces may affect fish assemblages in a variety of ways. Shorelines erode naturally through wind and wave action. Aquatic and riparian vegetation along shorelines stabilizes soil, buffers the impacts of waves, provides structure for fishes, and acts as a sink for incoming nutrients (Wetzel 2001). Vegetated or gravel shorelines may be important spawning, rearing, or refuge areas for many species. Human alterations to the riparian and littoral zones of lakes include aquatic and terrestrial vegetation removal, construction, dredging or filling on or near the shoreline, sand deposition, submerged woody debris removal, or seawall construction. Alterations such as these lead to accelerated rates of erosion and sedimentation, increased eutrophication, loss of aquatic plants and plant diversity, invasion of nonnative plant species, and exploitation of sport fishes. As fish habitats are altered by humans, fish assemblages and fish populations may be positively or negatively affected depending on the perturbation and species. In general, along a gradient of increasing shoreline development and thus human impacts, the spatial aggregation of fishes decreased (Scheuerell and Schindler 2004), fish assemblage composition changed (Jeppesen et al. 2000), trophic interactions were altered (Jeppesen et al. 1997), and fish growth rates declined (Schindler et al. 2000).

Shorelines in natural lakes should be protected from human impacts through landowner education, permitting, and zoning, each of which is a viable management action. Maintaining lower boating speed laws near shorelines (i.e., to reduce the effects of wakes), maintaining wide buffer strips and installing erosion control fences for construction projects, discouraging the removal of vegetation, inspecting septic systems and limiting effluent discharge, and limiting the number of piers are among the measures taken voluntarily or by enforcement to protect natural lake shorelines. Once altered, shorelines may be restored using a variety of methods. Shoreline stabilization most often involves planting the riparian zone, riprapping eroded shorelines, or constructing seawalls. Planting the riparian area with natural vegetation will help to stabilize banks and buffer against wind and waves. Windbreaks may reduce erosion of gravel crucial for walleye spawning. Riprapping with large rock will help to stabilize the wave zone.

15.6.1.6 Manipulations of Physical Habitat

Techniques to manipulate physical habitat for fishes in natural lakes are less fully developed than they are in other aquatic ecosystems such as trout streams, though models for relating fish habitat to fish production are emerging (Hayes et al. 1996; Minns et al. 1996a). Such models should enable development of habitat management techniques for natural lakes, which historically emerged mostly by trial and error. For example, littoral zone habitat structure in natural lakes was greatly altered by humans through removal of large woody debris, which eliminated critical inshore habitat and thereby reduced recruitment and growth of some fish species (Sass et al. 2006). To overcome widespread loss of inshore habitat, the addition of brush piles, log cribs, stumps, and whole trees have been used to restore woody debris in natural lakes, with an ultimate objective to attract fish for anglers or provide spawning habitat (Bassett 1994). However, effects of such habitat manipulations have only rarely been quantified, perhaps because power to detect effects of habitat manipulations is low (Lester et al. 1996; Minns et al. 1996b).

Some physical habitat manipulations, such as logs or spawning reefs, may serve as spawning habitat and increase fish production, whereas other manipulations, such as log cribs, may only aggregate fish rather than increase fish production. Both of these habitat manipulations may serve to achieve a fishery management objective, one to increase fish production and the other to increase fishery harvest. Choice of an appropriate habitat manipulation requires knowledge of the fishery objective (the easy part of the problem) and diagnosis of the habitat feature or problem that limits achievement of the fishery objective (the hard part of the problem). For example, raising lake levels in early spring will allow northern pike to reach spawning marshes, which would lead to increased population density if access to spawning habitat was limiting. However, if spawning habitat was not limiting northern pike production, then raising lake levels could lead to increased population densities and ultimately to stunted populations. Obviously, manipulating lake levels is only possible in natural drainage lakes where a dam or other structure has been installed on the lake outlet.

15.6.2 Population and Assemblage Management

Problems with fish populations and assemblages in natural lakes are often related to low abundance of a target species, negative effects of exotic species, or imbalance of predator–prey relationships. Low abundance of target species, often related to low recruitment or overfishing, can be partly addressed through stocking but usually requires management of habitat and people. Negative effects of exotic species are usually addressed through removal of such species. Imbalance of predator–prey relationships is complicated but is often addressed by stocking of new predator or prey species, habitat management, or people management. As with habitat problems, fish species and assemblage problems are not unique to natural lakes, so discussion is focused on ways in which these problems affect natural lakes in particular and how resource managers have attempted to mitigate effects in natural lakes.

15.6.2.1 Stocking

Fish stocking in natural lakes is often used to restore, maintain, or enhance fishery benefits (see Chapter 9). Restoration stocking can accelerate species restoration when natural recolonization is unlikely to occur. Stocking to maintain or enhance fisheries requires the understanding that stocking must occur indefinitely and will not solve problems associated with low recruitment. Lake trout restoration in the Great Lakes is an example of large-scale stocking for the purpose of restoring a key species to ecosystems where overfishing, predation by exotic species, and habitat degradation led to near extirpation of the species (Hansen 1999; Krueger and Ebener 2004). Early attempts to stock lake trout were poorly planned but led to widespread reproduction in eastern Lake Superior, where yearlings from adult fish captured near Marquette, Michigan, were stocked at high density near shorelines. To date, lake trout have been restored to self-sustaining status only in Lake Superior and remote areas of Lake Huron, whereas large-scale stocking now sustains populations elsewhere in the basin. Elsewhere in the basin, lake trout failed to reproduce at sustainable levels when stocked likely because of differences in available spawning habitat, prevalence of alewife that prey upon newly hatched lake trout, and mortality from fishing and sea lamprey predation. The disparity in effectiveness of stocking among the Great Lakes illustrates how stocking produces differing results in ecosystems because of differences in limiting factors.

Stocking to maintain or enhance fisheries is widespread in natural lakes, though stocking programs vary widely across North America. Fisheries supported by such stocking are not naturally self-sustaining but rather depend on continued stocking. In the USA, rainbow trout and kokanee are stocked predominantly in the West, where natural lakes are few. Walleye are stocked predominantly in the midwestern and northeastern USA, where natural lakes are numerous, and striped bass, hybrid striped bass, largemouth bass, and bluegill are stocked predominantly in the south, where natural lakes are few and stocking is focused on farm ponds and reservoirs (Halverson 2008). In natural lakes, stocking for enhancement of self-sustaining populations is often much less successful than is stocking for maintenance of non-self-sustaining populations. An example of successful maintenance stocking is the Great Lakes, where stocking of coldwater species of salmonids produced world-class fisheries (Hansen and Holey 2002). In Lake Michigan, the average rate of return to creel of stocked Chinook salmon (12–13%) was much higher than it was in western North America within their native distribution (usually ≤1%). The walleye serves as an example of a species for which survival is difficult to predict in natural lakes, especially when trying to supplement walleye populations that are partly sustained by natural reproduction (e.g., Li et al. 1996a, 1996b; Jennings et al. 2005). The largemouth bass serves as an example of a species for which stocking can disrupt fitness of native populations by outbreeding depression (Phillip et al. 2002). The genetic consequences of stocking fish should always be carefully considered when contemplating whether to implement this management action (e.g., Reisenbichler et al. 2003).

15.6.2.2 Community Manipulations

Manipulation of fish assemblages usually focuses on removal of undesirable species or addition of new species. Removal of undesirable species has typically been aimed at nonnative species, such as common carp in warmwater lakes and sea lamprey in the Great Lakes. Common carp have been the subject of chemical and physical control programs for decades. Of 250 fish control projects reviewed by Meronek et al. (1996), the most successful projects targeted rough fish (usually common carp). Chemical treatments were equally successful for rotenone and antimycin; physical removal with nets, traps, seines, electrofishing, and drawdowns was successful for 33–57% of all projects; and combined chemical and physical methods were successful in 4 of 6 projects (Meronek et al. 1996). Similarly, the sea lamprey has been the subject of an intensive control program in the Laurentian Great Lakes since the mid-1950s. Sea lamprey colonized the Great Lakes in a well-documented invasion that led to large-scale changes of native fish assemblages (Hansen 1999; Krueger and Ebener 2004). Efforts to restore or rehabilitate native fish assemblages and associated fisheries therefore depended on successful control of sea lamprey populations. The sea lamprey control program began with the discovery of a relatively specific chemical toxicant that targeted sea lampreys but evolved through time to include physical controls (barriers and traps) and biological controls (releases of sterile-male sea lampreys). At present, the control program suppresses sea lamprey populations to levels that are about 10–15% of precontrol levels but seeks to reduce populations even further by use of pheromones to increase trapping efficiency.

Manipulation of fish assemblages by addition of new species to natural lakes was widespread in the past when species were distributed widely across North America but has slowed greatly in modern times. Prey fish, primarily gizzard shad and threadfin shad, were widely introduced into reservoirs to provide prey for highly valued predators, but this practice has not been as widespread in natural lakes to manipulate predator–prey balance. More often, population density of predators has been manipulated by stocking or harvest regulations to manipulate predator–prey balance. In some natural western lakes, kokanee were introduced to support native predators such as rainbow trout or bull trout and nonnative predators such as lake trout. However, nonnative lake trout seem to destabilize food webs in some western lakes, where they prey upon or compete with native predators (e.g., Bowles et al. 1991; Ruzycki et al. 2001; Stafford et al. 2002; Ruzycki et al. 2003; Vander Zanden et al. 2003; Hansen et al. 2008). Lake trout are presently the subject of control programs in Yellowstone Lake and Lake Pend Oreille, Idaho.

15.6.3 People Management

Problems with people in the management of natural lakes are often related to overharvest that either erodes quality (human perception of the fishing experience—quality overfishing), yield (growth overfishing), or sustainability (recruitment overfishing). Quality overfishing is based in human perceptions so must be addressed through regulations that are aimed at providing better perceived fishery benefits for stakeholders. Growth overfishing fails to maximize yield, so must be addressed by changing physical attributes of capture gears to harvest fish at a size that maximizes growth potential in the face of natural mortality. Recruitment overfishing leads to stock collapse and must be addressed by controlling harvest. As with habitat and fish problems, human problems are not unique to natural lakes, so we focus our discussion on ways in which these problems affect natural lakes in particular and how resource managers have attempted to mitigate effects in natural lakes.

15.6.3.1 Input Controls

Input controls on harvest is a term used to describe the control of the number of access points to a fishery or control of the number of participants who can fish (limited entry). Input controls are not common for regulating human use in natural lakes, though the number of access points in remote areas may serve the purpose of limiting effort and harvest. For example, in many remote areas of Canada access to natural lakes is limited (not by management) to primitive roads or airplanes, which inadvertently protects such lakes from excessive fishing effort (Hunt and Lester 2009). In such lakes, angling participation is limited by the number of individuals who are willing or can afford to access the fishery or by fishing resorts that have limited numbers of openings for clients. Management goals and objectives for such fisheries are appropriately aimed at providing opportunities to harvest trophy fish by avoiding quality overfishing. Few areas of the USA are so limited in access, so examples of input controls on fishing are limited to caps on the number of licenses issued to commercial fisheries (e.g., limited entry in the Great Lakes). Management goals and objectives for such fisheries are often aimed at allocating fishing opportunities between recreational and commercial users while avoiding overfishing by both constituencies (quality, growth, and recruitment overfishing). However, angling fisheries are becoming increasingly efficient in North America, so active controls on access or fishing effort may be needed in the future (Pereira and Hansen 2003).

15.6.3.2 Output Controls

Output controls refers to restrictions on harvest commonly applied to fisheries in natural lakes, such as the use of numerical limits or quotas, seasons, or size restrictions (often, all three types of limitations are applied to the same fishery). When applied to commercial fisheries, as in the Great Lakes, output controls are designed to limit numerically the number and sizes of fish harvested so constitute active controls on harvest. When applied to recreational fisheries, as in most natural lakes, output controls are usually used to allocate harvest among users or regulate size of fish harvested while only passively controlling total harvest. As with input controls, we suspect that active output controls on fish harvest will become more common for regulating recreational fisheries of North America in the future, as was suggested by Pereira and Hansen (2003).

15.7 CONCLUSIONS

Throughout North America natural lakes are important public resources that support socially and economically valuable recreational, commercial, and subsistence fisheries. The fishery potential of natural lakes depends on productivity, which limits the potential yield of species targeted by fisheries. Goals and objectives for fisheries in natural lakes often focus on angler opportunity or satisfaction for recreational fisheries, maximum sustainable yield or maximum economic yield for commercial fisheries, and ensuring opportunities for subsistence fisheries. Fishery-dependent and fishery-independent surveys provide managers of natural lake fisheries with information for setting management goals and objectives, assessing the status of fish populations, and evaluating achievement of goals and objectives. Fishery management strategies for natural lakes rely on management strategies to overcome problems with lake habitat (water quality or alteration of physical habitat), fish populations or fish assemblages (low abundance of a target species, negative effects of exotic species, or imbalance of predator–prey relationships), or people (overharvest or perceptions of quality).

15.8 REFERENCES

Anderson, R. O., and R. M. Neumann. 1996. Length, weight, and associated structural indices. Pages 447–482 *in* B. R. Murphy and D. W. Willis, editors. Fisheries techniques, 2nd edition. American Fisheries Society, Bethesda, Maryland.

Appelberg, M. 2000. Swedish standard methods for sampling freshwater fish with multi-mesh gillnets. Fiskeriverket Information 2000:1. Fiskeriverket, Göteborg, Sweden.

Baranov, T. I. 1918. On the question of the biological basis of fisheries. Nauchnyi Issledovatelskii Ikhtiologicheskii Institut Isvestia 1(1):81–128. [Reports from the Division of Fish Management and Scientific Study of the Fishing Industry.] (English translation by W. E. Ricker, 1945. Mimeographed.)

Bassett, C. E. 1994. Use and evaluation of fish habitat structures in lakes of the eastern United States by the USDA Forest Service. Bulletin of Marine Science 55:2–3.

Beard, T. D., Jr., S. P. Cox, and S. R. Carpenter. 2003. Impacts of daily bag limit reductions on angler effort in Wisconsin walleyes lakes. North American Journal of Fisheries Management 23: 1283–1293.

Beard, T. D., Jr., S. W. Hewett, Q. Yang, R. M. King, and S. J. Gilbert. 1997. Prediction of angler catch rates based on walleye population density. North American Journal of Fisheries Management 17:621–627.

Beard, T. D., Jr., P. W. Rasmussen, S. Cox, and S. R. Carpenter. 2003. Evaluation of a management system for a mixed walleye spearing and angling fishery in Northern Wisconsin. North American Journal of Fisheries Management 23:481–491.

Berkes, F. 1988. Subsistence fishing in Canada: a note on terminology. Arctic 41:319–320.

Beverton, R. J. H., and S. J. Holt. 1957. On the dynamics of exploited fish populations. United Kingdom Ministry of Agriculture and Fisheries, Fisheries Investigations (Series 2) 19. Reprint 1993, Chapman and Hall, London.

Bonar, S. A., W. A. Hubert, and D. W. Willis, editors. 2009. Standard methods for sampling North American freshwater fishes. American Fisheries Society, Bethesda, Maryland.

Booth, G. M., C. D. Wren, and J. M. Gunn. 1993. Efficacy of shoal liming for rehabilitation of lake trout populations in acid-stressed lakes. North American Journal of Fisheries Management 13:766–774.

Borgeson, D. P. 1978. The anatomy of wild trout fishing. Pages 61–66 *in* K. Hashagen, editor. A national symposium on wild trout management. California Trout, San Francisco.

Bowles, E. C., B. E. Rieman, G. R. Mauser, and D. H. Bennett. 1991. Effects of introductions of *Mysis relicta* on fisheries in northern Idaho. Pages 65–74 *in* T. P. Nesler and E. P. Bergersen, editors. Mysids in fisheries: hard lessons from headlong introductions. American Fisheries Society Symposium 9, Bethesda, Maryland.

Bronte, C. R., C. C. Krueger, M. E. Holey, M. L. Toneys, R. L. Eshenroder, and J. L. Jonas. 2008. A guide for the rehabilitation of lake trout in Lake Michigan. Great Lakes Fishery Commission Miscellaneous Publication 2008–01. Great Lakes Fishery Commission, Ann Arbor, Michigan.

Brown, R. W., M. Ebener, and T. Gorenflo. 1999. Great Lakes commercial fisheries: historical overview and prognosis for the future. Pages 307–354 *in* W. W. Taylor and C. P. Ferreri, editors. Great Lakes fishery policy and management. Michigan State University Press, East Lansing.

Brylinsky, M. 1980. Estimating the productivity of lakes and reservoirs. Pages 411–454 *in* E. D. Le Cren and R. H. Lowe-McConnell, editors. The functioning of freshwater ecosystems. Cambridge University Press, New York.

Buklis, L. S. 2002. Subsistence fisheries management on federal public lands in Alaska. Fisheries 27(7):10–18.

Burch, E. S., Jr. 1998. The Inupiaq Eskimo nations of northwest Alaska. University of Alaska Press, Fairbanks.

Busch, W. D. N., and P. G. Sly. 1992. The development of an aquatic habitat classification system for lakes. CRC Press, Boca Raton, Florida.

Busiahn, T. R. 1990. Fish community objectives for Lake Superior. Great Lakes Fishery Commission Special Publication 90–1. Great Lakes Fishery Commission, Ann Arbor, Michigan.

Carpenter, S. R., and D. M. Lodge. 1986. Effects of submersed macrophytes on ecosystem processes. Aquatic Botany 26:341–370.

Chen, S., and S. Watanabae. 1989. Age dependence and natural mortality coefficient in fish population dynamics. Nippon Suisan Gakkaishi 55:205–208.

Christie, G. C., and H. A. Regier. 1988. Measures of optimal thermal habitat and their relationship to yields for four commercial fish species. Canadian Journal of Fisheries and Aquatic Sciences 45:301–314.

Christy, F. T., Jr., and A. Scott. 1965. The common wealth in ocean fisheries. John Hopkins Press, Baltimore, Maryland.

Chow-Fraser, P. 1991. Use of the morphoedaphic index to predict nutrient status and algal biomass in some Canadian lakes. Canadian Journal of Fisheries and Aquatic Sciences 48:1909–1918.

Chu, C., N. C. Collins, N. P. Lester, and B. J. Shuter. 2006. Population dynamics of smallmouth bass in response to habitat supply. Ecological Modeling 195:349–362.

Clair, T. A., and A. Hindar. 2005. Liming for the mitigation of acid rain effects in freshwaters: a review of recent results. Environmental Reviews 13:91–128.

Clark, B. J, P. J. Dillon, and L. A. Molot. 2004. Lake trout (*Salvelinus namaycush*) habitat volumes and boundaries in Canadian Shield lakes. Pages 111–117 *in* J. Gunn, R. Steedman and R. Ryder, editors. Boreal shield watersheds: lake trout ecosystems in a changing environment. CRC Press, Boca Raton, Florida.

Clark, C. W. 1985. Bioeconomic modeling and fishery management. John Wiley & Sons, New York.

Cleland, C. E. 2001. The place of the pike (Gnoozhekaaning). University of Michigan Press, Ann Arbor.

Cole, G. A. 1994. Textbook of limnology, 4th edition. Waveland Press, Prospect Heights, Illinois.

Colle, D. E., J. V. Shireman, W. T. Haller, J. C. Joyce, and D. E. Canfield Jr. 1987. Influence of hydrilla on harvestable sport fish populations, angler use, and angler expenditures at Orange Lake, Florida. North American Journal of Fish Management 7:410–417.

Cornett, R. J., and F. H. Rigler. 1979. Hypolimnetic oxygen deficits—their prediction and interpretation. Science 205:580–581.

Crowder, L. B., and W. E. Cooper. 1979. Structural complexity and fish–prey interactions in ponds:

a point of view. Pages 2–10 *in* D. L. Johnson and R. A. Stein, editors. Response of fish to habitat structure in standing water. American Fisheries Society, North Central Division, Special Publication 6, Bethesda, Maryland.

Crowder, L. B., and W. E. Cooper. 1982. Habitat structural complexity and the interaction between bluegills and their prey. Ecology 63:1802–1813.

Dickie, L. M., S. R. Kerr, and P. Schwinghamer. 1987. An ecological approach to fisheries assessment. Canadian Journal of Fisheries and Aquatic Sciences 44:68–74.

Die, D. J., and J. F. Caddy. 1997. Sustainable yield indicators from biomass: are there appropriate reference points for use in tropical fisheries? Fisheries Research 32:69–79.

Dillon, P. J., B. J. Clark, and H. E. Evans. 2004. The effects of phosphorus and nitrogen on lake trout (*Salvelinus namaycush*) production and habitat. Pages 119–131 *in* J. Gunn, R. Steedman and R. Ryder, editors. Boreal shield watersheds: lake trout ecosystems in a changing environment. CRC Press, Boca Raton, Florida.

Doka, S. E., D. K. McNicol, M. L. Mallory, I. Wong, C. K. Minns, and N. D. Yan. 2003. Assessing potential for recovery of biotic richness and indicator species due to changes in acid deposition and lake pH in five areas of southeastern Canada. Environmental Monitoring and Assessment 88:53–101.

Downing, J. A., and C. Plante. 1993. Production of fish populations in lakes. Canadian Journal of Fisheries and Aquatic Sciences 50:110–120.

Downing, J. A., C. Plante, and S. Lalonde. 1990. Fish production correlated with primary productivity, not the morphoedaphic index. Canadian Journal of Fisheries and Aquatic Sciences 47:1929–1936.

Downing, J. A., Y. T. Prairie, J. J. Cole, C. M. Duarte, L. J. Tranvik, R. G. Striegl, W. H. McDowell, P. Kortelainen, N. F. Caraco, J. M. Melack, and J. J. Middelburg. 2006. The global abundance and size distribution of lakes, ponds, and impoundments. Limnology and Oceanography 51:2388–2397.

Duarte, C. M., and J. Kalff. 1989. The influence of catchment geology and lake depth on phytoplankton biomass. Archives of Hydrobiology 115:27–40.

Elser, J. J., E. R. Marzolf, and C. R. Goldman. 1990. Phosphorus and nitrogen limitation of phytoplankton growth in the freshwaters of North America: a review and critique of experimental enrichments. Canadian Journal of Fisheries and Aquatic Sciences 47:1468–1477.

Essington, T. E. 2001. The precautionary approach in fisheries management: the devil is in the details. Trends in Ecology & Evolution 16:121–122.

Evans, D. O. 2007. Effects of hypoxia on scope-for-activity and power capacity of lake trout (*Salvelinus namaycush*). Canadian Journal of Fisheries and Aquatic Sciences 64:345–361.

Evans, D. O., J. Brisbane, J. M. Casselman, K. E. Coleman, C. A. Lewis, P. G. Sly, D. L. Wales, and C. C. Willox. 1991. Anthropogenic stressors and diagnosis of their effects on lake trout populations in Ontario lakes. Ontario Ministry of Natural Resources, Lake Trout Synthesis, Toronto.

FFMC (Freshwater Fish Marketing Corporation). 2008. About the freshwater fish marketing corporation. Available: www.freshwaterfish.com/english.htm. (July 2008).

Fisher, E., E. C. Fisher, J. S. Jones, and R. von Schomber. 2006. Implementing the precautionary principle: perspectives and prospects. Edward Elgar Publishing, Northampton, Massachusetts.

Flagg, T., W. McAuley, D. Frost, M. Wastel, W. Fairgrieve, and C. Mahnken. 1992. Redfish Lake sockeye salmon captive broodstock rearing and research. BPA (Bonneville Power Association) Report DOE (Department of Energy)/BP-41841-4, Portland, Oregon.

Flathead Lakers. 2001. 25 September draft of the Flathead Lake and River fisheries co-management plan released. Available: www.flatheadlakers.org/HOTISSUE/f_process.htm. (July 2008).

FOC (Fisheries and Oceans Canada). 2007. Survey of recreational fishing in Canada 2005. Fisheries and Occans Canada, Economic Analysis and Statistics Policy Sector, Ottawa.

Gabelhouse, D. W., Jr. 1984. A length-categorization system to assess fish stocks. North American Journal of Fisheries Management 4:273–285.

Gabelhouse, D. W., Jr., R. Anderson, L. Aggus, D. Austen, R. Bruch, J. Dean, F. Doherty, D. Dunning, D. Green, M. Hoeft, B. Hollender, K. Kurzawski, A. LaRoche III, G. Matlock, P. McKeown, B. Schonhoff, D. Stang, G. Tichacek, D. Willis, C. Wooley, and D. Workman. 1992. Fish sampling and data analysis techniques used by conservation agencies in the U. S. and Canada. American Fisheries Society, Fisheries Techniques Standardization Committee, Fisheries Management Section, Bethesda Maryland.

Garcia, S. M. 1994. The precautionary approach to fisheries with reference to straddling fish stocks and highly migratory fish stocks. FAO (Food and Agriculture Organization of the United Nations) Fisheries Circular 871, Rome.

Gillooly, J. F., J. H. Brown, G. B. West, V. M. Savage, and E. L. Charnov. 2001. Effects of size and temperature on metabolic rate. Science 293:2248–2251.

Grafton, R. Q., J. Kirkley, T. Kompas, and D. Squires. 2006. Economics for fishery management. Ashgate Publishing, Burlington, Vermont.

Graham, J. H. 1993. Species diversity of fishes in naturally acidic lakes in New Jersey. Transactions of the American Fisheries Society 122:1043–1057.

Grahm, M. 1943. The fish gate. Faber and Faber, London.

Grupert, J. P. 2003. Acid deposition in the eastern United States and neural network predictions in the future. Journal of Environmental Engineering and Science 2:99–109.

Gulland, J. A. 1971. Science and fishery management. Journal du Conseil International pour l'Exploration de la Mer 33:471–477.

Gunn, J. M., and W. Keller. 1984. In situ manipulation of water chemistry using crushed limestone and observed effects on fish. Fisheries 9(1):19–24.

Guy, C. S., R. M. Neumann, and D. W. Willis. 2006. New terminology for proportional stock density (PSD) and relative stock density RSD: proportional size structure (PSS). Fisheries 31(2):86–87.

Guy, C. S., R. M. Neumann, D. W. Willis, and R. O. Anderson. 2007. Proportional size distribution (PSD): a further refinement of population size structure index terminology. Fisheries 32(7):348.

Haines, T. A. 1981. Acid precipitation and its consequences for aquatic ecosystems: a review. Transactions of the American Fisheries Society 110:669–707.

Håkanson, L. 2005. The importance of lake morphometry for the structure and function of lakes. International Review of Hydrobiology 90(4):433–461.

Halverson, M. A. 2008. Stocking trends: a quantitative review of governmental fish stocking in the United States, 1931 to 2004. Fisheries 33(2):69–75.

Hansen, M. J., editor. 1996. A lake trout restoration plan for Lake Superior. Great Lakes Fishery Commission, Ann Arbor, Michigan.

Hansen, M. J. 1999. Lake trout in the Great Lakes: basin-wide stock collapse and binational restoration. Pages 417–453 *in* W. W. Taylor and C. P. Ferreri, editors. Great Lakes fishery policy and management: a binational perspective. Michigan State University Press, East Lansing.

Hansen, M. J., and M. E. Holey. 2002. Ecological factors affecting the sustainability of Chinook and coho salmon populations in the Great Lakes, especially Lake Michigan. Pages 155–179 *in* K. D. Lynch, M. L. Jones, and W. W. Taylor, editors. Sustaining North American salmon: perspectives across regions and disciplines. American Fisheries Society, Bethesda, Maryland.

Hansen, M. J., N. J. Horner, M. Liter, M. P. Peterson, and M. A. Maiolie. 2008. Dynamics of an increasing lake trout population in Lake Pend Oreille, Idaho, USA. North American Journal of Fisheries Management 28:1160–1171.

Hansen, M. J., C. P. Madenjian, T. E. Helser, and J. H. Selgeby. 1997. Gillnet selectivity of lake trout (*Salvelinus namaycush*) in Lake Superior. Canadian Journal of Fisheries and Aquatic Sciences 54:2483–2490.

Hansen, M. J., R. G. Schorfhaar, and J. H. Selgeby. 1998. Gill-net saturation by lake trout in Michigan waters of Lake Superior. North American Journal of Fisheries Management 18:847–853.

Hansen, M. J., M. D. Staggs, and M. H. Hoff. 1991. Derivation of safety factors for setting harvest quotas on adult walleye from past estimates of abundance. Transactions of the American Fisheries Society 120:620–628.

Hasselrot, B., and H Hultberg. 1984. Liming of acidified Swedish lakes and streams and its consequences for aquatic ecosystems. Fisheries 9(1):4–9.

Hayes, D. B., C. P. Ferreri, and W. W. Taylor. 1996. Linking fish habitat to their population dynamics. Canadian Journal of Fisheries and Aquatic Sciences 53 (Supplement 1):383–390.

Hayes, D., M. Jones, N. Lester, C. Chu, S. Doka, J. Netto, J. Stockwell, B. Thompson, C. K. Minns, B. Shuter, and N. Collins. 2009. Linking fish population dynamics to habitat conditions: insights from the application of a process-oriented approach to several Great Lakes species. Reviews in Fish Biology and Fisheries 19:295–312.

Healey, M. C. 1978. The dynamics of exploited lake trout populations and implications for management. Journal of Wildlife Management 42:307–328.

Heck, K. L., Jr., and L. B. Crowder. 1991. Habitat structure and predator–prey interactions in vegetated aquatic systems. Pages 281–299 *in* S. S. Bell, E. D. McCoy, and H. R. Mushinsky, editors. Habitat structure and the physical arrangement of objects in space. Chapman and Hall, New York.

Helm, J. 2000. The people of the Denendeh. Ethnohistory of the Indians of Canada's Northwest Territories. McGill-Queen's University Press, Montreal.

Hoenig, J. M. 1983. Empirical use of longevity data to estimate mortality rates. Fishery Bulletin 82:898–903.

Hulsman, P. F., P. M. Powles, and J. M. Gunn. 1983. Mortality of walleye eggs and rainbow trout yolk sac larvae in low-pH waters of the LaCloche Mountain area, Ontario. Transactions of the American Fisheries Society 112:680–688.

Hunt, L. M., and N. P. Lester. 2009. The effect of forestry roads on access to remote fishing lakes in northern Ontario, Canada. North American Journal of Fisheries Management 29:586–597.

Illinois Department of Natural Resources. 2007. Status of the walleye and sauger fishery. Illinois Department of Natural Resources, Springfield.

Jackson, D. A., H. H. Harvey, and K. M. Somers. 1990. Ratios in aquatic sciences—statistical shortcomings with mean depth and the morphoedaphic index. Canadian Journal of Fisheries and Aquatic Sciences 47:1788–1795.

Jennings, M. J., J. M. Kampa, G. R. Hatzenbeler, and E. E. Emmons. 2005. Evaluation of supplemental walleye stocking in northern Wisconsin lakes. North American Journal of Fisheries Management 25:1171–1178.

Jensen, A. L. 1996. Beverton and Holt life history invariants result in optimal trade-off of reproduction and survival. Canadian Journal of Fisheries and Aquatic Sciences 53:820–822.

Jeppesen, E., J. P. Jensen, M. Søndergaard, T. Lauridsen, and F. Landkildehus. 2000. Trophic structure, species richness, and biodiversity in Danish lakes: changes along a phosphorus gradient. Freshwater Biology 45:201–218.

Jeppesen, E., J. P. Jensen, M. Søndergaard, T. Lauridsen, L. J. Pedersen, and L. Jensen. 1997. Top-down control of freshwater lakes: the role of nutrient state, submerged macrophytes and water depth. Hydrobiologia 342-343:151–164.

Kalff, J. 2002. Limnology: inland water ecosystems. Prentice Hall, Upper Saddle River, New Jersey.

Kennedy, J. J., and P. J. Brown. 1976. Attitudes and behavior of fishermen in Utah's Uinta Primitive Area. Fisheries 1(6):15–17, 30–31.

Kinnunen, R. E. 2003. Great Lakes commercial fisheries. Michigan Sea Grant Extension, Marquette. Available: www.miseagrant.umich.edu/downloads/fisheries/GLCommercialFinal.pdf. (December 2009).

Kratz, T. K., K. E. Webster, C. J. Bowser, J. J. Magnuson, and B. J. Benson. 1997. The influence of landscape position on lakes in northern Wisconsin. Freshwater Biology 37:209–217.

Krueger, C. C., and M. Ebener. 2004. Rehabilitation of lake trout in the Great Lakes: past lessons and future challenges. Pages 37–56 *in* J. M. Gunn, R. J. Stedman, and R. A. Ryder, editors. Boreal shield watersheds: lake trout ecosystems in a changing environment. CRC Press, Boca Raton, Florida.

Krueger, C. C., and P. E. Ihssen. 1995. Review of genetics of lake trout in the Great Lakes: history, molecular genetics, physiology, strain comparisons, and restoration management. Journal of Great Lakes Research 21 (Supplement 1):348–363.

Lampert, W., and U. Sommer. 1997. Limnoecology: the ecology of lakes and streams. Oxford University Press, New York.

Larkin, P. A. 1977. An epitaph for the concept of maximum sustained yield. Transactions of the American Fisheries Society 106:1–11.

Leach, J. H., L. M. Dickie, B. J. Shuter, U. Borgmann, J. Hyman, and W. Lysack. 1987. A review of methods for prediction of potential fish production with application to the Great Lakes and Lake Winnipeg. Canadian Journal of Fisheries and Aquatic Sciences 44:471–485.

Leech Lake Band of Ojibwe. 2008. Fisheries program. Leech Lake Division of Resource Management, Fish, Wildlife, and Plant Resources Department, Minnesota. Available: www.lldrm.org/fish.html. (July 2008).

Lehner, B., and P. Döll. 2004. Development and validation of a global database of lakes, reservoirs and wetlands. Journal of Hydrology 296:1–22.

Lester, N. P., A. J. Dextrase, R. S. Kushneriuk, M. R. Rawson, and P. A. Ryan. 2004a. Light and temperature: key factors affecting walleye abundance and production. Transactions of the American Fisheries Society 133:588–605.

Lester, N. P., and W. I. Dunlop. 2004. Monitoring the state of the lake trout resource: a landscape approach. Pages 293–321 in J. M. Gunn, R. J. Stedman, and R. A. Ryder, editors. Boreal shield watersheds: lake trout ecosystems in a changing environment. CRC Press, Boca Raton, Florida.

Lester, N. P., W. I. Dunlop, and C. C. Willox. 1996. Detecting changes in the nearshore fish community. Canadian Journal of Fisheries and Aquatic Sciences 53 (Supplement 1):391–402.

Lester, N. P., T. R. Marshall, K. Armstrong, W. I. Dunlop, and B. Ritchie. 2003. A broad-scale approach to management of Ontario's recreational fisheries. North American Journal of Fisheries Management 23:1312–1328.

Lester, N. P., B. J. Shuter, and P. A. Abrams. 2004b. Interpreting the von Bertalanffy model of somatic growth in fish: the cost of reproduction. Proceedings of the Royal Society B 271:1625–1631.

Li, J., Y. Cohen, D. H. Schupp, and I. R. Adelman. 1996a. Effects of walleye stocking on population abundance and fish size. North American Journal of Fisheries Management 16:830–839.

Li, J., Y. Cohen, D. H. Schupp, and I. R. Adelman. 1996b. Effects of walleye stocking on year-class strength. North American Journal of Fisheries Management 16:840–850.

Link, J. S. 2005. Translating ecosystem indicators into decision criteria. ICES Journal of Marine Science 62:569–576.

Linton, B. C., M. J. Hansen, S. T. Schram, and S. P. Sitar. 2007. Dynamics of a recovering lake trout population in eastern Wisconsin waters of Lake Superior, 1980–2001. North American Journal of Fisheries Management 27:940–954.

Ludwig, D., R. Hilborn, and C. Walters. 1993. Uncertainty, resource exploitation, and conservation: lessons from history. Science 260:17–36.

Lyons, J., G. Gonzalez-Hernandez, E. Soto-Galera, and M. Guzman-Arroyo. 1998. Decline of freshwater fishes and fisheries in selected drainages of west-central Mexico. Fisheries 23(4):10–18.

Mace, P. M. 2001. A new role for MSY in single-species and ecosystem approaches to fisheries stock assessment and management. Fish and Fisheries 2:2–32.

Maceina, M. J., and W. C. Reeves. 1996. Relations between submersed macrophyte abundance and largemouth bass tournament success on two Tennessee River impoundments. Journal of Aquatic Plant Management 34:33–38.

Macias, J. G. L., and O. T. Lind. 1990. The management of Lake Chapala (Mexico): considerations after significant changes in the water regime. Lake and Reservoir Management 6:61–70.

Marshall, T. R. 1996. A hierarchical approach to assessing habitat suitability and yield potential of lake trout. Canadian Journal of Fisheries and Aquatic Sciences 53:332–341.

Marteney, R., L. Aberson, C. Johnson, S. Lynott, R. Sanders, J. Stephen, J. Vajnar, and S. Waters. 2007. Largemouth bass management plan. Kansas Department of Wildlife and Parks, Pratt.

Martin, N. V., and C. H. Olver. 1980. The lake char, *Salvelinus namycush*. Pages 209–277 *in* E. K. Balon, editor. Charrs: salmonid fishes of the genus Salvelinus. Perspectives in vertebrate science, volume 1. Dr. W. Junk Publishers, The Hague, Neatherlands.

Matuszek, J. E., and G. L. Beggs. 1988. Fish species richness in relation to lake area, pH, and other abiotic factors in Ontario lakes. Canadian Journal of Fisheries and Aquatic Sciences 45:1931–1941.

McCormick, J. H., and R. L. Leino. 1999. Factors contributing to first-year recruitment failure of fishes in acidified waters with some implications for environmental research. Transactions of the American Fisheries Society 128:265–277.

McCoy, M. W., and J. F. Gillooly. 2008. Predicting natural mortality rates of plants and animals. Ecology Letters 11:710–716.

Melmer, D. 2007. Red Lake walleye make comeback to retail market. Indian Country Today. Available: www.indiancountry.com/content.cfm?id = 1096415026. (July 2008).

Meronek, T. G., P. M. Bouchard, E. R. Buckner, T. M. Burri, K. K. Demmerly, D. C. Hatlelli, R. A. Klumb, S. H. Schmidt, and D. W. Coble. 1996. A review of fish control projects. North American Journal of Fisheries Management 16:63–74.

Mertz, G., and R. A. Myers. 1998. A simplified formulation for fish production. Canadian Journal of Fisheries and Aquatic Sciences 55:478–484.

Methratta, E. T., and J. S. Link. 2006. Evaluation of quantitative indicators for marine fish communities. Ecological Indicators 6:575–588.

Meybeck, M. 1995. Global distribution of lakes. Pages 1–35 *in* A. Lerman, D. M. Imboden, and J. R. Gat, editors. Physics and chemistry of lakes. Springer-Verlag, Berlin.

Miller, R. J. 1999. Courage and the management of developing fisheries. Canadian Journal of Fisheries and Aquatic Sciences 56:897–905.

Minns, C. K. 1989. Factors affecting fish species richness in Ontario Lakes. Transactions of the American Fisheries Society 118:533–545.

Minns, C. K., J. R. M. Kelso, and R. G. Randall. 1996b. Detecting the response of fish to habitat alterations in freshwater ecosystems. Canadian Journal of Fisheries and Aquatic Sciences 53 (Supplement 1):403–414.

Minns, C. K., R. G. Randall, J. E. Moore, and V. W. Cairns. 1996a. A model simulating the impact of habitat supply limits on northern pike, *Esox lucius*, in Hamilton Harbour, Lake Ontario. Canadian Journal of Fisheries and Aquatic Sciences 53 (Supplement 1):20–34.

Miranda, L. E., and P. W. Bettoli. 2007. Mortality. Pages 229–277 *in* C. S. Guy and M. L. Brown, editors. Analysis and interpretation of freshwater fisheries data. American Fisheries Society, Bethesda, Maryland.

Miranda, L. E., and L. L. Pugh. 1997. Relationship between vegetation coverage and abundance, size, and diet of juvenile largemouth bass during winter. North American Journal of Fisheries Management 17:601–610.

Mitzner, L. 1978. Evaluation of biological control of nuisance aquatic vegetation by grass carp. Transactions of the American Fisheries Society 107:135–145.

Morgan, N. C. 1980. Secondary production. Pages 247–340 *in* E. D. Le Cren and R. H. Lowe-McConnell, editors. The functioning of freshwater ecosystems. International Biological Programme Synthesis Series 22. Cambridge University Press, New York.

Moxely, D. J., and F. H. Langford. 1985. Beneficial effects of hydrilla on two eutrophic lakes in Central Florida. Proceedings of the Annual Conference Southeastern Association of Fish and Wildlife Agencies 36(1982):280–286.

Müller, R., and P. Stadelmann. 2004. Fish habitat requirements as the basis for rehabilitation of eutrophic lakes by oxygenation. Fisheries Management and Ecology 11:251–260.

NAS (National Academy of Sciences). 1969. Eutrophication: causes, consequences, correctives. National Academy of Sciences, Washington, D.C.

NOAA (National Oceanic and Atmospheric Administration). 1998. Technical guidance on the use of precautionary approaches to implementing national standard 1 of the Magnuson–Stevens Fishery Conservation and Management Act. U.S. Department of Commerce, NOAA Technical Memorandum NMFS (National Marine Fisheries Service)-F/SPO.

NRC (Natural Resources Canada). 2008. The atlas of Canada: recreational fishing. Natural Resources Canada. Available: http://atlas.nrcan.gc.ca/sites/english/maps/freshwater/recreational/fishing. (June 2008).

Olson, M. H., S. R. Carpenter, P. Cunningham, S. Gafny, B. R. Herwig, N. P. Nibbelink, T. Pellett, C. Storlie, A. S. Trebitz, and K. A. Wilson. 1998. Managing macrophytes to improve fish growth: a multi-lake experiment. Fisheries 23(2):6–12.

O'Riordan, T., and J. Cameron, editors. 1994. Interpreting the precautionary principle. Earthscan Publications Limited, London.

Pauly, D. 1980. On the interrelationships between natural mortality, growth parameters, and mean environmental temperature in 175 fish stocks. Journal du Conseil International pour l'Exploration de la Mer 39:175–192.

Pauly, D. 1984. Fish population dynamics in tropical waters: a manual for use with programmable calculators. International Center for Living Aquatic Resources Management, Manila.

Pereira, D. L., and M. J. Hansen. 2003. A perspective on challenges to recreational fisheries management: summary of the symposium on active management of recreational fisheries. North American Journal of Fisheries Management 23:1276–1282.

Peterson, I., and J. S. Wroblewski. 1984. Mortality rate of fishes in the pelagic ecosystem. Canadian Journal of Fisheries and Aquatic Sciences 41:1117–1120.

Phillip, D. P., J. E. Claussen, T. W. Kassler, and J. M. Epifanio. 2002. Mixing stocks of largemouth bass reduces fitness through outbreeding depression. Pages 349–363 *in* D. P. Phillip and M. S. Ridgway, editors. Black bass: ecology, conservation, and management. American Fisheries Society, Symposium 31, Bethesda, Maryland.

Pollock, K. H., C. M. Jones, and T. L. Brown. 1994. Angler survey methods and their application in fisheries management. American Fisheries Society, Special Publication 25, Bethesda, Maryland.

Pomeroy, C. 1994. Obstacles to institutional development in the fishery of Lake Chapala, Mexico. Pages 17–39 *in* C. L. Dyer and J. R. McGoodwin, editors. Folk management in the world's fisheries. University Press of Colorado, Niwot.

Post J. R., M. Sullivan, S. Cox, N. P. Lester, C. J. Walters, E. A. Parkinson, A. J. Paul, L. Jackson, and B. J. Shuter. 2002. Canada's recreational fisheries: the invisible collapse? Fisheries 27(1):6–17.

Quinn, T. J., and R. B. Deriso. 1999. Quantitative fish dynamics. Oxford University Press, New York.

Rago, P. J., and J. G. Wiener. 1986. Does pH affect fish species richness when lake area is considered? Transactions of the American Fisheries Society 115:438–447.

Randall, R. G., J. R. M. Kelso, and C. K. Minns. 1995. Fish production in fresh waters—are rivers more productive than lakes? Canadian Journal of Fisheries and Aquatic Sciences 52:631–643.

Rasmussen, P. W., M. D. Staggs, T. D. Beard Jr., and S. P. Newman. 1998. Bias and confidence interval coverage of creel survey estimators evaluated by simulation. Transactions of the American Fisheries Society 127:469–480.

Reisenbichler, R. R., F. M. Utter, and C. C. Krueger. 2003. Genetic concepts and uncertainties in restoring fish populations and species. Pages 149–183 *in* R. C. Wissmar and P. A. Bisson, editors. Strategies for restoring river ecosystems: sources of variability and uncertainty in natural and managed systems. America Fisheries Society, Bethesda, Maryland.

Rempel, R. S., and P. J. Colby. 1991. A statistically valid model of the morphoedaphic index. Canadian Journal of Fisheries and Aquatic Sciences 48:1937–1943.

Restrepo, V. R., P. M. Mace, and F. M. Serchuk. 1999. The precautionary approach: a new paradigm or business as usual? Pages 61–70 *in* Our living oceans. Report on the status of U.S. living marine resources, 1999. U.S. Department of Commerce, NOAA (National Oceanic and Atmospheric Administration)Technical Memorandum NMFS (National Marine Fisheries Service)-F/SPO-41, Washington, D.C.

Ricker, W. E. 1975. Computation and interpretation of biological statistics of fish populations. Fishery Research Board of Canada Bulletin 191.

Riera, J. L., J. J. Magnuson, T. K. Kratz, and K. E. Webster. 2000. A geomorphic template for the analysis of lake districts applied to the Northern Highland Lake District, Wisconsin, U.S.A. Freshwater Biology 43:301–318.

Rochet, M. J., and V. M. Trenkel. 2003. Which community indicators can measure the impact of fishing? A review and proposals. Canadian Journal of Fisheries and Aquatic Sciences 60:86–99.

Ruzycki, J. R., D. A. Beauchamp, and D. L. Yule. 2003. Effects of introduced lake trout on native cutthroat trout in Yellowstone Lake. Ecological Applications 13:23–37.

Ruzycki, J. R., W. A. Wurtsbaugh, and C. Luecke. 2001. Salmonine consumption and competition for endemic prey fishes in Bear Lake, Utah–Idaho. Transactions of the American Fisheries Society 130:1175–1189.

Ryan, P. A., and T. R. Marshall. 1994. A niche definition for lake trout (*Savelinus namaycush*) and its use to identify populations at risk. Canadian Journal of Fisheries and Aquatic Sciences 51:2513–2519.

Ryder, R. A. 1965. A method for estimating potential fish production of north-temperate lakes. Transactions of the American Fisheries Society 94:214–218.

Ryder, R. A. 1982. The morphoedaphic index—use, abuse, and fundamental concepts. Transactions of the American Fisheries Society 111:154–164.

Ryder, R. A., and S. R. Kerr. 1989. Environmental priorities: placing habitat in hierarchic perspective. Pages 2–12 *in* C. D. Levings, L. B. Holtby, and M. A. Henderson, editors. Proceedings of the national workshop of effects of habitat alteration on salmonid stocks. Canadian Special Publication in Fisheries and Aquatic Sciences 105.

Ryder, R. A., S. R. Kerr, K. H. Loftus, and H. A. Regier. 1974. Morphoedaphic index, a fish yield estimator—review and evaluation. Journal of the Fisheries Research Board of Canada 31:663–688.

Sainsbury, K., and U. R. Sumaila. 2001. Incorporating ecosystem objectives into management of sustainable marine fisheries including "best practice" reference points and use of marine protected areas. In Summary of the Reykjavik Conference on Responsible Fisheries in the Marine Ecosystem. Available: www.iisd.ca/sd/sdice/sdvol61num1.html. (January 2010).

Sakamoto, M. 1966. Primary production by phytoplankton community in some Japanese lakes and its dependence on lake depth. Archives of Hydrobiology 62:1–28.

Sandstrom, S., M. Rawson, and N. Lester. 2008. Manual for broad-scale fish community monitoring using large mesh gillnets and small mesh gillnets. Ontario Ministry of Natural Resources, Peterborough, Ontario.

Sass, G. G., J. F. Kitchell, S. R. Carpenter, T. R. Hrabik, A. E. Marburg, and M. G. Turner. 2006. Fish community and food web responses to a whole-lake removal of coarse woody habitat. Fisheries 31(7):321–330.

Savino, J. F., E. A. Marschall, and R. A. Stein. 1992. Bluegill growth as modified by plant density: an exploration of underlying mechanisms. Oecologia 89:153–160.

Savino, J. F., and R. A. Stein. 1982. Predator–prey interaction between largemouth bass and bluegills as influenced by simulated, submersed vegetation. Transactions of the American Fisheries Society 111:255–265.

Scheuerell, M. D., and D. E. Schindler. 2004. Changes in the spatial distribution of fishes in lakes along a residential development gradient. Ecosystems 7:98–106.

Schindler, D. W. 1978. Factors regulating phytoplankton production and standing crop in the world's freshwaters. Limnology and Oceanography 23:478–486.

Schindler, D. E., S. I. Geib, and M. R. Williams. 2000. Patterns of fish growth along a residential development gradient in north temperate lakes. Ecosystems 3:229–237.

Schlesinger, D. A., and H. A. Regier. 1982. Climatic and morphoedaphic indexes of fish yield from natural lakes. Transactions of the American Fisheries Society 111:141–150.

Selbie, D. T., B. A. Lewis, J. P. Smol, and B. P. Finney. 2007. Long-term population dynamics of the endangered Snake River sockeye salmon: evidence of past influences on stock decline and impediments to recovery. Transactions of the American Fisheries Society 136:800–821.

Shuter, B. J. 1990. Population-level indicators of stress. Pages 145–166 in S. M. Adams, editor. Biological indicators of stress in fish. American Fisheries Society, Symposium 8, Bethesda, Maryland.

Shuter, B. J., M. L. Jones, R. M. Korver, and N. P. Lester. 1998. A general, life history based model for regional management of fish stocks: the inland lake trout (*Salvelinus namaycush*) fisheries of Ontario. Canadian Journal of Fisheries and Aquatic Sciences 55:2161–2177.

Shuter, B. J., N. P. Lester, J. La Rose, C. Purchase, K. Vascotto, G. Morgan, N. Collins, and P. Abrams. 2005. Optimal life histories and food web position: linkages between somatic growth, reproductive investment and mortality. Canadian Journal of Fisheries and Aquatic Sciences 62:725–729.

Smith, K. D. 1993. Vegetation–open water interface and the predator–prey interaction between largemouth bass and bluegills. Doctoral dissertation, University of Michigan, Ann Arbor.

Soranno, P. A., K. E. Webster, J. L. Riera, T. R. Kratz, J. S. Baron, P. A. Bukaveckas, G. W. Kling, D. S. White, N. Caine, R. C. Lathrop, and P. R. Leavitt. 1999. Spatial variation among lakes within landscapes: ecological organization along lake chains. Ecosytems 2:395–410.

Stafford, C. P., J. A. Stanford, F. R. Hauer, and E. B. Brothers. 2002. Changes in lake trout growth associated with *Mysis relicta* establishment: a retrospective analysis using otoliths. Transactions of the American Fisheries Society 131:994–1003.

Stewart, B., M. Meding, and D. Rogers. 2008. Lake Pleasant striped bass. Arizona Game and Fish Department Technical Guidance Bulletin 11, Phoenix.

Tonn, W. M. 1990. Climate change and fish communities: a conceptual framework. Transactions of the American Fisheries Society 119:337–352

Tonn, W. M., and J. J. Magnuson. 1982. Patterns in the species composition and richness of fish assemblages in northern Wisconsin lakes. Ecology 63:1149–1166.

Trebitz, A., S. Carpenter, P. Cunningham, B. Johnson, R. Lillie, D. Marshall, T. Martin, R. Narf, T. Pellett, S. Stewart, C. Storlie, and J. Unmuth. 1997. A model of bluegill–largemouth bass interactions in relation to aquatic vegetation and its management. Ecological Modeling 94:139–156.

Tyler, A.V., and V. F. Gallucci. 1980. Dynamics of fished stocks. Pages 111–147 in R. T. Lackey and L. A. Nielsen, editors. Fisheries management. Blackwell Scientific Publications, Boston.

Unmuth, J. M. L., M. J. Hansen, and T. D. Pellett. 1999. Effects of mechanical harvesting of Eurasian watermilfoil on largemouth bass and bluegill populations in Fish Lake, Wisconsin. North American Journal of Fisheries Management 19:1089–1098.

Unmuth, J. M. L., M. J. Hansen, P. W. Rasmussen, and T. D. Pellett. 2001. Effects of mechanical harvesting of Eurasian watermilfoil on angling in Fish Lake, Wisconsin. North American Journal of Fisheries Management 21:448–454.

Urquhart, N. S., S. G. Paulsen, and D. P. Larsen. 1998. Monitoring for policy-relevant regional trends over time. Ecological Applications 8:246–257.

USDI (U.S. Department of the Interior). 1991. Casting light upon the waters: a joint fishery assessment of the Wisconsin ceded territory. U.S. Department of the Interior, Bureau of Indian Affairs, Minneapolis, Minnesota.

USDI (U.S. Department of the Interior), Fish and Wildlife Service, and U.S. Department of Commerce,

Census Bureau. 2006. National survey of fishing, hunting, and wildlife-associated recreation. U.S. Department of the Interior, Fish and Wildlife Service, and U.S. Department of Commerce, Census Bureau, Washington, D.C.

Vander Zanden, M. J., S. Chandra, B. C. Allen, J. E. Reuter, and C. R. Goldman. 2003. Historical food web structure and restoration of native aquatic communities in the Lake Tahoe basin. Ecosystems 6:274–288.

Vollenweider, R., and J. Kerekes. 1982. Eutrophication of waters: monitoring, assessment and control. Organisation for Economic Co-operation and Development, Paris.

Walker, W. W. 1979. Use of hypolimnetic oxygen depletion rate as a trophic state index for lakes. Water Resources Research 15:1463–1470.

Walters, C. J., and S. J. D. Martell. 2002. Stock assessment needs for sustainable fisheries management. Bulletin of Marine Science 70:629–638.

Walters, C. J., and S. J. D. Martell. 2004. Fisheries ecology and management. Princeton University Press, Princeton, New Jersey.

Webster, K. E., T. K. Kratz, C. J. Browser, J. J. Magnuson, and W. J. Rose. 1996. The influence of landscape position on lake chemical responses to drought in Northern Wisconsin. Limnology and Oceanography 41:977–984.

Webster, K. E., P. A. Soranno, S. B. Baines, T. K. Kratz, C. J. Bowser, P. J. Dillon, P. Campbell, E. J. Fee, and R. E. Hecky. 2000. Structuring features of lake districts: landscape controls on lake chemical response to drought. Freshwater Biology 43:499–515.

Wetzel, R. G. 2001. Limnology: lake and river ecosystems, 3rd edition. Academic Press, San Diego, California.

Wilberg, M. J., C. R. Bronte, and M. J. Hansen. 2004. Fleet dynamics of the commercial lake trout fishery in Michigan waters of Lake Superior during 1929–1961. Journal of Great Lakes Research 30:252–266.

Wilberg, M. J., M. J. Hansen, and C. R. Bronte. 2003. Historic and modern density of wild lean lake trout in Michigan waters of Lake Superior. North American Journal of Fisheries Management 23:100–108.

Wiley, M. J., R. W. Gorden, S. W. Waite, and T. Powless. 1984. The relationship between aquatic macrophytes and sport fish production in Illinois ponds: a simple model. North American Journal of Fisheries Management 4:111–119.

Wrenn, W. B., D. R. Lowery, M. J. Maceina, and W. C. Reeves. 1996. Relationships between largemouth bass and aquatic plants in Guntersville Reservoir, Alabama. Pages 382–393 *in* L. E. Miranda and D. R. DeVries, editors. Multidimensional approaches to reservoir fisheries management. American Fisheries Society, Symposium 16, Bethesda, Maryland.

Zimmerman, M. S, S. N. Schmidt, C. C. Krueger, M. J. Vander Zanden, and R. L. Eshenroder. 2009. Ontogenetic niche shifts and resource partitioning of lake trout morphotypes. Canadian Journal of Fisheries and Aquatic Science 66:1007–1018.

Chapter 16

Farm Ponds and Small Impoundments

DAVID W. WILLIS, ROBERT D. LUSK, AND JEFFREY W. SLIPKE

16.1 INTRODUCTION

A wealth of recreational fishing opportunities exists in farm ponds and small impoundments, and many anglers have had their first angling experience on such small waters. Recent estimates indicate that there are at least 2.6 million small, constructed water bodies in the USA (Smith et al. 2002). In 2006, 25.4 million anglers fished in freshwater, and 84% of the overall angling effort occurred on lakes, reservoirs, and ponds (USDI 2007). The last year that the National Survey of Fishing, Hunting, and Wildlife-Associated Recreation further categorized angling by water type was in 1991 (USDI 1993), when 35% of 30.1 million anglers fished in ponds smaller than 4.2 ha.

Ponds and small impoundments are defined as waters less than 40 ha in surface area. However, such a size definition certainly is arbitrary, and other authors have used various size definitions. Predator–prey interactions tend to be similar in small ponds of 1 ha and small impoundments up to approximately 40 ha. Larger waters tend to have more diversity in both habitat and predator–prey relations.

Ponds can easily and consistently be manipulated by a fishery manager to produce desired results, especially in comparison with other habitat types. In larger reservoirs and natural lakes, environmental conditions can override fishery management attempts. In large rivers, which are open rather than closed systems, results from management efforts can be difficult to discern. In contrast, various management strategies can be attempted in pond management with relatively high confidence for success. Many urban fishing opportunities occur in ponds and small impoundments (Box 16.1).

Despite the small size of ponds, pond management still involves the same three primary management components as all fisheries: habitat, biota, and humans (Nielsen 1999; Willis et al. 2009). Thus, these three management components are emphasized throughout this chapter. In addition, whether developing a pond for personal use, helping a private landowner, or managing a public water body, clear management objectives must precede pond construction, stocking, and habitat management. This chapter provides the management options; managers or pond owners need to decide which are feasible and, of those, which are desired.

A paradigm shift has occurred in pond management, and the shift also applies to management of public small impoundments. First, views on fisheries management in ponds are changing from the older, more simple and inexpensive model of fingerling stockings with largemouth bass, bluegill, and channel catfish (Dauwalter and Jackson 2005) to a model of

Box 16.1. Managing Fisheries in the Urban Environment

Richard T. Eades[1] and Thomas J. Lang[2]

Many state agencies have specific programs focusing on management of urban fisheries (Hunt et al. 2008). Urban fisheries management is rooted in traditional fisheries management, and many of the same techniques (e.g., fish stocking, creel limits, and habitat improvement) and strategies used on other waters are applicable in urban areas. However, urban fisheries require more intensive management to maintain successful fisheries. From a physical standpoint, a 2-ha pond in a rural area on private property or in a state park may be very similar to one in a public park in the middle of Chicago or St. Louis, but the pond in the city will likely require much more intensive management to establish and maintain a similar fish population.

Water quality is a primary concern in the urban setting. Urban runoff can carry a variety of pollutants that can be detrimental to both fish and humans. After years of contaminant, sediment, and organic matter accumulation, urban impoundments may be too degraded to support desirable fish assemblages. Restoring these waters to productive fisheries is possible but costly.

Angler use can be substantial in urban areas. Urban ponds may experience angling effort as high as 30,000 h/ha/year (Alcorn 1981). While this high rate is uncommon, angling effort on urban waters is often considerably higher than that at other lakes (Lang et al. 2008). Even if anglers keep only a portion of their catch, a fish assemblage may quickly become out of balance or dominated by less desirable species under these conditions. Therefore, many states stock their urban lakes on a regular basis with catchable-length channel catfish or rainbow trout to keep catch rates high. Providing channel catfish and rainbow trout for anglers to harvest also can be a method of reducing pressure on species such as largemouth bass and bluegill, which are often managed as self-sustaining populations with catch-and-release or restricted creel limits. Unfortunately, some states are finding angler compliance with regulations in urban areas to be a problem, and even a combination of repeated stockings and catch-and-release regulations does not always result in the creation of a desirable fishery (Lang 2007; Eades et al. 2008).

Most urban lakes are owned by local municipalities and controlled by their parks and recreation departments or similar city departments. City park managers are typically eager to work with fisheries agencies to improve fishing opportunities, but they also have their own set of priorities and responsibilities that do not always dovetail with fisheries goals. Many park managers are more concerned with park aesthetics than with fish productivity. For instance, they may prefer fountains in their lakes rather than more efficient and desirable aeration systems. Many park managers prefer their parks to have as much mowed grass as possible, including any and all vegetation growing along the lake shoreline. Algal growth is an eyesore for urban residents, and many park managers will chemically treat lakes with an unsightly green tint to restore them to the clarity of a neighborhood swimming pool. All this leaves the fisheries manager having to deal with

[1] Nebraska Game and Parks Commission, Lincoln.
[2] Kansas Department of Wildlife and Parks, Pratt.

(Box continues)

Box 16.1. Continued.

extra constraints not faced in more rural areas or on lakes under state ownership. Fortunately, a growing number of communities are recognizing the value of local fishing opportunities, and cooperation and management is improving in many locations.

Access can be another constraint for urban fisheries managers. Most small urban impoundments do not allow boating, for safety reasons, and thus boat ramps are usually not present. Shorelines may be covered with riprap or constructed of seawall, which inhibit boat access. This may make it impossible for a manager to launch an electrofishing boat to conduct a population assessment. Likewise, setting nets may be impossible without boat access. In these high-use settings, managers may be reluctant to set nets in a lake and leave them for a sufficient time period to obtain a sample. In these situations, managers must rely on creel surveys as their primary indicator of fish population structure. Poor vehicle access (e.g., landscaping, fences, and high curbs) can hamper fish-stocking efforts if large fish-hauling trailers loaded with channel catfish or rainbow trout cannot get to the water's edge. Poor access may also prevent good fish stocking practices, such as acclimating fish to the pond.

More so than other biologists, urban fisheries managers should embrace the role of agency representative. They may be the only employee of their agency that a member of the urban public has ever encountered. They need to be willing to take a break from sampling or stocking fish and explain to onlookers what it is they are doing and why it is important. Urban fisheries managers can take advantage of these opportunities to develop strong relationships with their urban clienteles while developing their fisheries.

Urban fisheries managers faced with the task of trying to maintain a good fishery under these circumstances have completed numerous research studies in attempts to find the most cost-effective stocking methods for urban waters, looking at sizes at stocking and stocking densities and frequencies. Other researchers have considered "new" species for urban waters that may be more cost-effective to stock and manage while also offering diversity in angling opportunities. Recently, hybrid striped bass have been stocked in urban lakes in several states and have shown promise in providing an exciting sport fish for urban anglers, as well as improving panfish size structure through predation.

The primary goals of urban fisheries programs are typically to increase fishing opportunities, recruit and retain anglers, and educate the public about aquatic resources. Fishing participation rates in the USA have been declining over the past 25 years, and many states have implemented urban fisheries programs in an attempt to reverse this trend (Pajak 1994; Aiken 1999; Hunt et al. 2008). Urban fisheries managers are increasingly turning to education, marketing, and other public outreach methods to boost fishing participation. Establishing and maintaining good, fishable lakes in urban environments is a real challenge but only half the battle. The second, and often more difficult, part is making the public aware of enhanced fishing opportunities and providing a safe, comfortable environment for them to enjoy. Urban ponds must be convenient and safe if they are to attract and recruit new anglers. Finally, it is imperative that fisheries agencies invest time and resources in urban fisheries and increase and diversify their clientele if these fisheries are to remain socially relevant in the future.

innovation, species and size diversity of fishes, and a sense of investment. Overharvest of fishes, especially predators, was a primary concern for both private and public waters in past years, but underharvest or lack of harvest can cause as many or more problems. Traditional pond management strategies are biologically sound but can take more time to accomplish than some private pond owners are willing to accept. Thus innovative techniques are being used to increase species diversity in ponds, improve the length structure or abundance of target fishes, or speed the process for developing desirable fish assemblages. Second, views on water management are shifting and many of the modifications in management philosophy involve a change of focus from fish production to water management. Landowners are learning that water is borrowed and they should take better care of it while they use it.

This chapter begins with pond habitat development and management because available habitat determines which management strategies are viable. Then, pond management strategies for managing the biota (i.e., fish species options) and human users (i.e., harvest regulation) are discussed. Many of the examples and information focus on warmwater systems because much less research has been conducted in coldwater habitats.

16.2 HABITAT DEVELOPMENT

The quality of pond habitat is the deciding factor as to which management alternatives can be used; without clean water and quality habitat, management options are limited. Coldwater fishes cannot be stocked into warmwater habitats. Sight-feeding predators will not be successful in ponds with high colloidal clay turbidity. Only after creating new pond habitat through construction, or by assessing current habitat for an existing pond, can management strategies such as species combinations and harvest regulations be developed.

16.2.1 Building a Pond

Most ponds are embankment type (Figure 16.1), also known as hill ponds, meaning a dam is constructed to impound water from the watershed. However, some ponds are excavated, which allows them to be constructed in nearly any location and filled via a source such as pumping. Excavated ponds typically have a bottom seal of some type (e.g., liner or clay) to maintain water levels. However, some excavated ponds are deep enough to tap into the water table. Common examples include sand and gravel pits (e.g., borrow pits along interstate highways) or rock quarries.

Building a pond is an orderly, step-by-step process, and several considerations are involved in planning and execution: permitting; site selection, including an evaluation of watershed size; and evaluation of soil types and quantities. These are tasks often best suited for experienced engineers and contractors. The process and important considerations are presented here, with greater detail available in Deal et al. (1997).

A key consideration in pond construction is the esthetics of site selection. Esthetics can be particularly important for decisions made by private pond owners because most ponds have recreational value. However, similar considerations obviously apply to ponds built and managed by public funds. Advice for land shopping by potential pond owners was provided by Lusk et al. (2007).

Whether a pond will be privately developed or constructed and managed with public funds, the initial step in site selection is to learn the regulations that apply to pond construction in that geographic location. What are the laws involving water rights and impounding water

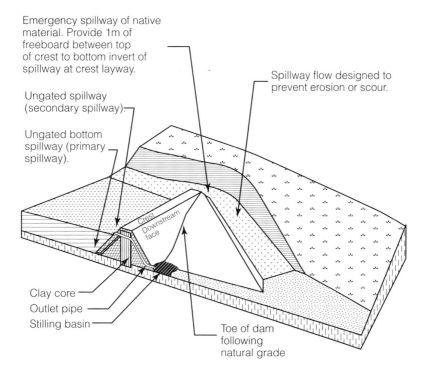

Figure 16.1. Cross-sectional view of the dam for an embankment pond and an emergency spillway design commonly used in pond construction (from Blaser and Eades 2006; courtesy of Chris Lemke).

that may affect another person farther down in the watershed? What permitting is required for that state or province? What county zoning laws might be involved? Will an environmental impact statement be required? It is important to learn the answers to all such questions before initiating pond construction.

Knowing the drainage area and the average annual precipitation and runoff expected at a geographic location given the soil types help determine the area and depth of a prospective pond. For example, consider a pond with a surface area of 1 ha, a maximum depth of 4 m, and a mean depth of 2 m. Such a pond would have a volume of 2 ha-m. In southern Mississippi where rainfall is plentiful, approximately 13.4 ha of watershed would be required to maintain the pond water level (Deal et al. 1997). In eastern Wyoming, the same size pond would require a watershed of approximately 234 ha due to lower average annual precipitation and runoff.

Ponds should be constructed at locations that allow sufficient water depth. Two primary considerations affect planned water depth. First, the pond should be deep enough to account for seepage (influenced by soil type) and evaporation (influenced by local climate) and still maintain sufficient depth for fish production or other landowner goals. Thus, ponds in more arid locations typically are planned for greater depth than those in locations where average annual precipitation is higher. Second, those at northern latitudes typically are planned for greater depths because more water volume is needed to prevent winterkill during periods of ice cover (see section 16.3.2). Therefore, ponds in northern or arid western portions of North America should be built to depths of 3.7–6.1 m. However, in the southern portions of North America, depths of no more than 1.8–2.4 m are sufficient because most biological productivity occurs within the top 2 m of the water column.

Prior to the start of pond construction, a site survey will determine pond characteristics and the expected final elevation of the pond water. Such surveys allow planning and design of the dam and spillways and determination of water levels and projected shoreline. Site surveys require the use of surveying equipment and should be conducted by experienced professionals. Site surveys must also include a thorough assessment of soils. Contractors typically dig a series of test holes around the proposed pond location. Highly porous soils may indicate that the site is a poor location or some type of seal is needed to prevent excessive seepage of water. Similarly, building a pond over a location with a subsurface seam of fractured rock may lead to leaks. Locations with subsurface limestone formations are notorious for pond leakage.

To maintain pond depth over time, inflowing water should not carry excessive sediment. Thus, protection of the watershed is important in pond construction. Runoff from grasslands is low in sediment, whereas runoff from agricultural row crops carries substantial sediment. If a watershed cannot be protected with appropriate vegetation plantings, then a small check dam can be constructed just upstream of the pond to allow sediment to settle, decreasing the volume of sediment that reaches the pond. Sediment can be removed from check dams at intervals to maintain their capacities. Blaser and Eades (2006) described an alternative method in which the outlet structure for a pond is a bottom withdrawal spillway, which discharges water from the bottom of the pond. This structure is designed to carry much of the incoming muddy water through the pond and dam, releasing it downstream.

Construction of dams to impound water is both a science and an art. Experienced contractors know the best fill material that can be used to build the dam. Borrow material taken from the pond itself is typically the least expensive option. However, if the soil type in the pond is inadequate for proper dam construction, materials may need to be transported from another location. The best soils for earthen dams contain particles ranging from small gravel to fine sand and clay (Deal et al. 1997). The soil should contain a minimum 20% by weight of clay particles. If ideal soils are not available, then a clay core (Figures 16.1, 16.2) is often constructed within the dam. The clay core acts as an underground barrier to water movement. Earthen dams commonly have constructed slopes of 3:1 on the upstream side and 2:1 on the downstream side (Deal et al. 1997). The width for the top of a dam increases with dam height. For example, a 3.1-m-high dam should have a top width of 1.8 m, whereas a 6.1- to 7.3-m-high dam should have a 3.7-m top width.

When choosing a pond location, there should be sufficient space for an emergency spillway. An emergency spillway is designed to allow surface runoff to overflow safely from a pond and prevent water from breaching the dam or eroding the dam (Figure 16.1). The volume of water coming from a watershed will vary with geographic location (i.e., peak rainfall events) and hydrologic characteristics of a watershed. Some soil types (e.g., sand or gravel) have a high infiltration rate, whereas others (e.g., clay soils or impervious materials) have a low infiltration rate. Thus, consultation with local U.S. Natural Resource Conservation Service personnel or experienced pond construction engineers is desirable.

The pond outlet, also called a primary spillway, is typically constructed to allow water from normal runoff to pass through the dam. Drop-type outlet structures (Figure 16.3) are commonly used to maintain water levels in ponds. Wire baskets (Figure 16.4), called trash racks, are often installed to prevent objects from clogging the outlet pipe. Another type of outlet structure is the trickle tube (Figure 16.4). In addition, siphon systems or bottom outlets can be used. Regardless of the type of outlet, anti-seep collars should be added to prevent leakage (Figures 16.4, 16.5).

Figure 16.2. This core trench was excavated in sandy soils in western Tennessee. After the trench is filled with clay, it will prevent leakage through the pond dam (image courtesy of Robert Lusk).

Figure 16.3. A drop-inlet structure maintains the water level in this Texas pond (image courtesy of Mike Otto).

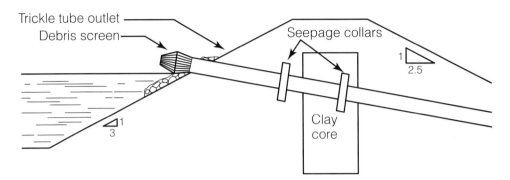

Figure 16.4. Trickle-tube outlet system constructed through the dam during construction. Note the anti-seep collars that are added to prevent water movement and subsequent erosion around the outlet pipe. Also note the debris screen, or trash rack, used to prevent tree branches or other large debris from clogging the outlet pipe (from Blaser and Eades 2006; courtesy of Chris Lemke).

Figure 16.5. Installation of a pipe with anti-seep collars designed to prevent future leakage in this pond (image courtesy of Mike Otto).

All trees and brush should be removed from the dam site and emergency spillway prior to construction and prevented from growing thereafter. If any trees are pushed over, the roots must not penetrate to a permeable soil or rock layer that will create a leak.

Excavated ponds are built with knowledge of the water table depth at that location. One option is to excavate a pond on flat ground and then use either natural clay soils or another type of sealant such as an impermeable liner to prevent leakage. Water for such ponds typically is supplied via pumping or a similar alternative. If a pond is excavated at a location with a watershed above it, an emergency spillway system should be constructed. Some ponds are excavated into the groundwater table, such as sand and gravel pits that are dug using draglines. The water level in these ponds will fluctuate with the water table. Thus, what appears to be a viable habitat for fish during wet years may not be suitable during a drought.

The effort that goes into sealing a pond depends primarily on soil types. Most pond construction sites will have a variety of soils. Clay soils are best for building a dam and blanketing a pond bottom. Porous soils such as sand and rock can be used on the backside slope of a dam during construction but are avoided in a pond basin where clay soils are necessary to seal a pond. The most common causes of pond leaks are improper site selection, poor construction methods, permeable soils such as sand and gravel, fissures in bedrock, and compromises in the integrity of the structure, such as tree roots, improperly installed outlets, and damage from muskrat or nutria burrows.

Compaction of soils during pond construction helps prevent or minimize leaks, especially if the soils contain sufficient clay content. If not, clay can be hauled from nearby sources and applied as a blanket. If content of the imported blanket is at least 20% clay, then approximately 30 cm of clay blanket should be added for ponds that are 3.1 m deep or shallower. Deeper ponds require a thicker clay blanket. Bentonite is a colloidal clay that swells upon contact with water and can also be used to minimize pond leakage. Bentonite needs to be thoroughly mixed with existing soils to a depth of at least 30 cm. A sheepsfoot roller can help accomplish such a task (Figure 16.6). Alternatively, waterproof linings can be used to seal a pond bottom. Such liners are made of polyethylene, polyvinyl chloride, or rubber. Liners are expensive and require installation by experienced personnel to avoid cuts or tears.

16.2.2 Fish Habitat Development

The ideal time to plan fish habitat is prior to pond construction because habitat choices vary depending on the species of fish to be stocked into a pond. Some fishes prefer rock habitat, others are better suited to large woody habitat, and others prefer submergent or emergent aquatic macrophytes. Habitat selection thus depends on the set of goals developed for that fish assemblage.

During pond construction, irregularities in pond perimeter and bottom contours should be created. Bays can be dug and islands (Figure 16.7) or peninsulas can be constructed during earth moving. Most fishes prefer drop-offs, humps, or bumps rather than a flat, smooth bottom. If rocks are encountered during construction, they can be set aside and later placed to enhance fish habitat. For example, a line of rocks can run from the shore to deep water on one or more of the points in a pond. Also, rocks can be used to create such points. During construction of larger ponds, the dam face is often rip-rapped to prevent erosion, and these rocks also provide fish habitat.

Figure 16.6. Sheepsfoot roller being used to apply a bentonite seal to a pond bottom at a site with porous soils (image courtesy of Robert Lusk).

Figure 16.7. Fish habitat should be considered during the pond construction phase. Here, an earthen hump or island was created with earthmoving equipment, and a wood structure was constructed on top of the hump (image courtesy of Robert Lusk).

Although trees should be removed from a dam site and spillway, a small number of trees can be left standing in the upper end of a pond to provide fish habitat. If trees are cut near a dam, they can be used to create sunken brush piles elsewhere in a pond. Small trees can be weighted with concrete blocks or held down with steel cable prior to a pond being filled. Once trees are waterlogged, they no longer float and remain in place. Trees and brush piles last much longer if they are continually submerged rather than if they are alternately exposed and submerged by varying water levels (Figure 16.8). Care must be taken not to breach the lining of a pond when installing habitat. Ninety percent of the fish will live in 10–15% of the pond, so strategically planned habitat provides structure and cover not only for fish but also for anglers to direct fishing efforts.

Aquatic vegetation should be considered when a pond is constructed. Shallow areas (1–2 m) of most ponds will eventually develop aquatic vegetation. While some vegetation is desirable for many fish assemblages (e.g., Wiley et al. 1984; Engel 1988; Bettoli et al. 1993), overly abundant vegetation can interfere with both recreational uses (e.g., angling and swimming) and predator–prey interactions. Thus, creating proper basin slopes for a pond during earthmoving can reduce later abundance of rooted aquatic vegetation.

16.2.3 Watershed Influences

The concept of pond ecology is considerably broader than simply the fish. Whatever is in a watershed above a pond will eventually find its way to a pond. Thus, stewardship of the entire watershed should be a management goal (Schramm and Hubert 1999; Iowa Department of Natural Resources 2007). When someone wants to build a pond, the mission is usually sharply focused. But, as they consider their plan, most people do not think about the impact of a pond on upstream or downstream ecosystems. Where a pond exists, it influences its surroundings by its fish assemblage, level of eutrophication, and depth. Not only does it become an ecological system, it also influences surrounding systems.

Decisions made during construction affect the pond environment for a minimum of four future generations of humans. Thus, wise planning in conjunction with eventual pond management objectives need to occur before heavy equipment arrives.

16.3 COMMON HABITAT MANAGEMENT TECHNIQUES

In established ponds, habitat determines which fish management strategies are feasible. As such, this section considers commonly-used habitat management techniques.

16.3.1 Fertilization and Liming

In many locations where soil fertility is low, fertilizers are added to ponds to increase fish production by stimulating phytoplankton growth. Ponds in fertile watersheds, well-managed pastures with high densities of cattle, northern latitudes where winterkill is a problem, or areas where clay turbidity is a persistent problem, as well as ponds that receive very little fishing effort, generally should not be fertilized. Fertilization is typically used in Alabama, Arkansas, some parts of northern Florida, Georgia, Kentucky, Mississippi, North Carolina, South Carolina, Tennessee, Texas, and Virginia. With a few exceptions, fertilization is generally not recommended elsewhere.

Figure 16.8. Christmas trees placed into concrete bases can provide fish habitat in a pond. Longevity of the structures is greatest when they are continuously submerged. The tree on the left was submerged for 3 years, the center one for 5 years, and those on the right for 10 years (images courtesy of Todd St. Sauver, South Dakota Department of Game, Fish and Parks).

Swingle and Smith (1938) found that fertilized ponds in Alabama supported four to five times the weight of largemouth bass and bluegill as did unfertilized ones. The shading effect of a phytoplankton bloom also aids in control of submergent aquatic macrophytes, and this beneficial turbidity increases fishing success. However, most fertilizers are expensive, and the pond owner should determine whether the increased production justifies the expense. Bennett (1971) cautioned about the dangers of summerkill at any latitude and winterkill at northern latitudes when abundant nutrients result in excessive plant growth.

Inorganic fertilizers are most often used as pond fertilizers in the USA. A common fertilizer is granular 20–20–5: 20% nitrogen (as N), 20% phosphorus (as P_2O_5), and 5% potassium (as K2O; also called potash). Secondary nutrients in fertilizers include calcium, magnesium, and sulfur. Minor or trace nutrients such as copper and zinc may also be present. In recent years, powdered or crystalline fertilizers and liquid formulations have become popular with pond owners because they are easy to apply. A standard fertilizer application procedure for the southeastern USA can be found in Boyd (1979).

Swingle et al. (1965) found that old ponds which had been fertilized for many years had adequate supplies of nitrogen and potassium through nitrogen fixation by bacteria or algae and through decomposition of bottom organic materials. For such ponds, they recommended a standard fertilization rate with phosphorus-only fertilizer: 45 kg/ha of superphosphate (16–20%) or 20 kg/ha of triple (or treble) superphosphate (44–54%). If a satisfactory phytoplankton bloom cannot be maintained, use of a complete fertilizer can be resumed.

Granular fertilizers applied to waters over 1 m deep may sink into bottom sediment, be locked into bottom water and sediment, and be unavailable to phytoplankton. Therefore, granular fertilizers should be broadcast over shallow water or be placed on underwater platforms in the euphotic zone (zone with sufficient light penetration to support plant growth). A platform at a depth of about 0.3 m and with an area of 4 m^2 is adequate for 2–4 ha of pond area. The mixing of liquid fertilizers (i.e., after first diluting) in the prop wash of a boat motor is another method of distributing fertilizer. The advent of powdered and crystalline formulations has nearly eliminated the need for platforms or mixing of liquid fertilizers in most cases.

In certain geographic locations, pond waters have low alkalinity and may be acidic. These waters may not respond to fertilization unless alkalinity is increased by adding lime to the water. Lime has been routinely used to increase fish production in ponds where naturally low levels of calcium result in low pH and low buffering capacity. Liming has several effects on water quality and productivity, including (1) increased pH of bottom sediment and thereby increased availability of phosphorus; (2) increased benthic invertebrate production via increased nutrient availability; (3) increased alkalinity and thereby availability of carbon dioxide for photosynthesis; (4) increased microbial activity by means of increased pH; (5) improved buffering against pH shifts common in soft-water ponds; and (6) reduced humic stains of vegetative origin and clearer water that allows for increased light penetration. The net effect is an increase in phytoplankton productivity thereby increasing food production for fishes.

Liming needs are determined by total alkalinity measurements of water. Boyd (1979) described methods used in analysis of agricultural soils and pond sediments to determine the lime requirement. Agricultural limestones, $CaCO_3$, or $CaMg(CO_3)_2$, are the materials most commonly used in ponds. New ponds can be limed by spreading limestone over the pond bottom and disking it with farm equipment. Established ponds can be limed by spreading limestone over the entire pond surface (Figure 16.9). It is practically impossible to overap-

Figure 16.9. Agricultural limestone being applied to a pond. Lime increases production in impoundments with acidic bottom sediments and soft water (less than 20 mg $CaCO_3$/L [20 ppm] total alkalinity). Lime is especially important in making more phosphorus available for phytoplankton production. (Image courtesy of Jeffrey Slipke).

ply lime to ponds with naturally low alkalinity soils. For such ponds, application rates of lime of 11.2–22.4 metric tons per hectare, or even higher, rarely increase total alkalinity to levels observed in naturally high alkalinity ponds (>50 mg/L $CaCO_3$). Liming at a high rate reduces the frequency at which lime needs to be re-applied. Addition of limestone or other liming materials is best accomplished during winter before the fertilization season. The length of time a treatment is effective will depend on the rate of water loss through overflow and seepage.

16.3.2 Water Quality Maintenance

Summer aeration can be used to improve water quality in ponds, especially those that thermally stratify. A stratified pond that is fertile will accumulate decaying organic matter in the hypolimnion (the lowermost water layer in a stratified water body), and bacterial respiration typically results in a loss of dissolved oxygen (DO), in an anoxic benthic zone. This bottom layer then becomes a nutrient sink, as nitrogen and phosphorous from organic decomposition are trapped in the bottom water layer and are not available for use by algae and higher vascular plants. Aeration can circulate the entire pond and keep nutrients available for use in the aquatic food web. Circulation systems should be initiated before or soon after summer thermal stratification occurs. If circulation begins after thermal stratification, mixing of anoxic waters can result in a sufficiently reduced DO level to kill fish.

Aeration can also help decrease the likelihood of winterkill (Box 16.2) in northern ponds (Figure 16.10). Traditionally, the air stone or diffuser is placed on the pond bottom in the deepest water. The air bubbles lift 4°C bottom water to the surface, melting a hole in the ice. Sunlight penetration allows phytoplankton and plants to maintain photosynthesis and increase DO levels. Recent evidence indicates that diffusers should be placed in shallow waters because circulating the entire pond from bottom (4°C) to top (0°C) at sufficient volume may result in cooling of the entire pond, resulting in a lack of 4°C refuge for fishes. While the process is not entirely understood, the lack of the 4°C water for an extended period may be stressful on fishes at a time of year when additional stress is not desirable. In addition, fish may not feed at such low temperatures, inhibiting winter angling opportunities. Winter aeration should be initiated soon after initial ice formation. If aeration begins after oxygen depletion has occurred, which happens from the bottom up in ponds during winter, then the circulation system start-up can mix the entire water column, resulting in a decline in overall DO that may cause a fish kill.

A variety of aerators are commercially available, with the most popular type being a shore-mounted electrical compressor that pumps air to diffusers in a pond. When electricity is not available, windmills can be used to create an air pumping system; however, they obviously work only when the wind is blowing. Compressed air cylinders, another aeration option, can be weighted and sunk, with the valve slightly open to create a bubble stream to lift water and open a hole in the ice above; cylinder size depends on length of ice cover at a given location. Finally, some circulation units are mechanical in nature, rather than air compressors. For example, one method to keep an open area in pond ice during winter is through horizontal water movement near the surface, often with a propeller system that mechanically moves waters. Such a system would be desirable when cooling of bottom waters is a concern.

Box 16.2. Winterkill

Winterkill is a phenomenon that occurs in lakes during winter with loss of dissolved oxygen (DO) under ice. If oxygen loss is severe, then a fish kill will result. Other forms of aquatic life, such as zooplankton, are also susceptible to winterkill.

Shallow, eutrophic lakes at northern latitudes are most susceptible to winterkill. The longer the duration of ice and snow cover, the more likely a winterkill. The more productive a water body, the greater the decomposition of organic material such as dying algae and other plants. The higher the organic load in a water body, the more DO will be respired by the bacteria that decompose the organic material. Dissolved oxygen is replenished in lakes by photosynthesis by algae and other aquatic plants. The rate of photosynthesis is least when snow prevents light penetration through ice. When the amount of oxygen used during decomposition is greater than the amount of oxygen replenished during photosynthesis, then DO levels in a water body decline. If oxygen levels reach sufficiently low levels, then fish begin to die. Finally, gases such as hydrogen sulfide and methane may accumulate with decomposition and can contribute to winterkill because they are toxic to fishes at high concentrations.

Different fishes have varying levels of susceptibility to low levels of DO in water. Some fishes, such as salmonids, are highly sensitive to low DO concentrations and will die before DO levels reach 2 mg/L. Other fishes, such as largemouth bass and walleye, have an intermediate tolerance to low DO levels and typically will die when DO levels are 1–2 mg/L. Tolerant fishes, such as northern pike, yellow perch, and fathead minnow, will not begin to die until DO levels decline well below 1 mg/L.

Figure 16.10. Winter aeration can decrease the likelihood of winterkill under the ice in northern ponds (photograph courtesy of Ted Lea, ForeverGreen, Inc.).

16.3.3 Supplemental Feeding

Supplemental feeding can be useful when harvest is high or where large fish are desired. Channel catfish and rainbow trout have been the species most commonly fed formulated feeds, and their nutritional requirements are well documented (National Research Council 1993). In addition, interest in feeding bluegill, hybrid sunfishes, hybrid striped bass, and largemouth bass has increased over time (Lewis and Heidinger 1971; Nail and Powell 1975; Brown et al. 1993; Kerby et al. 2002; Lane and Morris 2002; Porath and Hurley 2005). Use of floating feed is advisable because fish can be observed feeding and feeding rates can be appropriately adjusted. Because nutritional requirements vary by species, use an appropriate diet for selected target fishes.

Seasonal timing, quantity, and frequency of feeding vary with fish species, fish sizes, water temperature, water quality, weather conditions, and quality of feed. For example, seasonal feeding of channel catfish typically begins when water temperatures are over 15°C. They may be fed at 3% of their body weight per day. The addition of large amounts of feed increases the danger of oxygen depletion, and feeding should be discontinued when surface DO levels drop below 5.0 mg/L. See Barrows and Hardy (2001) for more extensive information on feeding rates and frequencies for various fishes.

16.3.4 Turbidity

Pond water needs to be reasonably clear for production of sight-feeding fishes. In addition to limiting food production, turbid water can reduce the success of fish feeding and reproduction. Many ponds become turbid following high rainfall, but silt typically settles within a week. To cure a poor water clarity problem, the source needs to be identified. A simple approach is to collect a jar of water from the pond and put it on a shelf. If the suspended silt settles within a week and the water becomes relatively clear, the turbidity is probably due to wind action or the activities of livestock, fishes such as common carp, or crayfish. If the water in the jar remains turbid after a week, the problem is due to the chemistry of the soil suspended in the water.

Turbidity caused by soil type is the most difficult water clarity problem to manage. The turbidity is caused by suspension of negatively charged colloidal clay particles that electrically repel each other and will not clump together to form a particle large enough to settle. This problem can be treated by adding material that causes these particles to clump together and settle. Agricultural grade gypsum (hydrated calcium sulfate), available from most fertilizer dealers, can clear colloidal clay problems temporarily. Powdered gypsum is scattered evenly over the surface of a pond at 1,930 kg/ha-m of water volume. If a pond does not clear within 4 weeks and there is no other source of turbidity, one-quarter the original amount of gypsum can then be added.

Another material that can be used to clear clay turbidity is aluminum sulfate (alum); this material causes clay particles to flocculate and settle. An application of about 185 kg/ha-m of water will clear most ponds within a week. Alum is dissolved in water and quickly sprayed over the entire pond surface. This should be done on a calm day because wave action will break up the floc and it will not settle. Alum has an acid reaction with water. If a pond is acidic or has very soft water, about 73 kg of hydrated lime (calcium hydroxide) per hectare-meter of water should first be added. Sometimes the liming itself will cause the to settle.

Organic matter can also be added to water to settle clay particles. Organic matter provides food for desirable bacteria. As bacteria break down organic matter, byproducts allow clay particles to clump together and settle. Manure, hay, and cottonseed meal all work. However, too much organic matter can cause oxygen deficiency in a pond. If organic matter is added, it is best to use something that decomposes rather slowly (e.g., dry hay).

These treatments are temporary and likely will have to be repeated each year (usually at lower application levels) and after periods of substantial water inflow. Ponds with chronic clay turbidity may be best managed for channel catfish (see section 16.4.1.5), in which case treatment of the turbidity problem is unnecessary.

High winds can cause shoreline erosion and wave action that keep soil particles in suspension. The effect of wind can be minimized by use of windbreaks and shoreline protection. A windbreak can be planted on the upwind side of a pond to dissipate prevailing summer winds. If a dam is eroding badly, it should be protected by rock rip-rap. Erosion of the rest of the shoreline can be lessened by deepening the shoreline during construction (or by draining and basin re-sculpturing), thus eliminating mud flats.

Livestock with access to a pond can trample shoreline vegetation and wade in the water, especially during summer. These activities add feces that negatively affect pond water quality and stir sediment that can then be carried over an entire pond by wind and wave action. Livestock should be fenced out of a pond if fish production is important. If livestock water is needed, a pipe through the dam to a tank below a dam can be used. If this is not possible, all but a small corner of the pond can be fenced. Although limited livestock access may cause some reduction in water clarity, the effect is less than if livestock had access to an entire pond.

High abundance of some fishes (e.g., common carp) can cause water to be turbid because of their feeding behaviors. Renovation, a management strategy of planned partial or complete removal of fishes from a pond, is discussed in section 16.3.6. A dense crayfish population can also cause pond water to be turbid due to their burrowing and bottom-feeding activities that stir up sediment. The introduction of predatory fish such as largemouth bass or channel catfish usually solves this problem. In addition, crayfish can readily be trapped and used as a food source for humans.

16.3.5 Aquatic Vegetation

Primary production by aquatic plants, especially phytoplankton, is the means by which a portion of the sun's energy is fixed into forms available to other aquatic organisms. A variety of invertebrates and a few vertebrates feed on living and dead aquatic plants. Many aquatic organisms use plants for shelter and attachment. Isolated plant beds attract fish. However, overly abundant vegetation can create unpleasant conditions for anglers, swimmers, and boaters. Aquatic plant respiration and decomposition deplete oxygen and can result in a fish kill.

Vegetation may need to be reduced but not eliminated. Durocher et al. (1984) found a positive relationship between submerged vegetation (up to 20% of total lake coverage) in Texas reservoirs and both standing stock of largemouth bass and numbers being recruited to harvestable size. Reduction in submerged vegetation below 20% coverage resulted in decreases in both recruitment and standing stock of largemouth bass. The optimal vegetation coverage for largemouth bass production in Illinois ponds was approximately 36% (Wiley et al. 1984).

To control aquatic plants, it is essential to know what type is causing problems. Aquatic plants can be grouped into four general categories: algae, floating plants, submergent plants, or emergent plants.

Phytoplankton is microscopic, single-celled, free-floating algae. Filamentous algae are often called "moss" by pond owners and consist of masses of long, stringy strands that float on top of or just under the water surface. The two common genera for filamentous algae are *Cladophora* and *Spirogyra*. Some types of algae, such as *Chara* spp., or muskgrass, and *Nitella* spp., or stonewort, look very much like vascular, submergent plants, explaining why correct identification is essential. A macrophyte may be an emergent, submergent or floating type of aquatic plant. Floating plants have leaves that float on the water and have roots that hang down without being connected to the bottom. A common example is duckweed. Submergent plants grow under the water, are rooted to the bottom, have stems and leaves, and produce seeds. Some species have leaves that reach to the surface and are a different shape than those below the surface. Common examples of submergent plants are the various pondweeds and milfoils. Emergent plants generally grow around the margin of a pond, are rooted to the bottom, and have parts extending above the water surface. Common examples of such plants are the cattails and bulrushes.

Control methods for aquatic vegetation are generally categorized as mechanical, chemical, and biological. Mechanical methods such as raking and cutting are labor intensive and produce short-term results. Properly timed drawdowns can temporarily reduce aquatic macrophytes, but there is also a risk of creating conditions that allow macrophytes to colonize areas that previously were too deep for establishment.

The use of herbicides requires that appropriate chemicals be selected for the targeted vegetation (Masser et al. 2006). Check the label on the herbicide container to be certain that the chemical has been cleared for the intended use (Avery 2003); both the U.S. Environmental Protection Agency (EPA) and the U.S. Food and Drug Administration (FDA) control the registration of applied chemicals. In addition, most states now require that chemicals be applied by a certified applicator. Herbicides can vary seasonally in their effectiveness, many chemicals are expensive, and decomposition of dead plants depletes DO. Thus, expertise is required to undertake any type of chemical control program.

Biological control of aquatic vegetation is receiving considerable attention largely because of environmental concerns over chemicals. However, biological control, specifically the use of grass carp, has generated other environmental concerns. Some of these concerns include nontarget effects, downstream migration of grass carp, and change from a macrophyte-dominated plant community to a phytoplankton community that is often dominated by blue-green algae. Use of grass carp is illegal in some states, and other states require the use of triploid (sterile) grass carp or have other restrictions. Grass carp are long lived (Hill 1986), which should be considered prior to any introduction. Grass carp most often are too effective and completely eliminate vegetation; however, Blackwell and Murphy (1996) reported success in partial vegetation control in Texas.

Prevention is always the best control method. Plants are common in ponds that have clear water, high fertility, and extensive shallow areas. Plant problems can be minimized through proper pond construction. Any shallow mud flats should be eliminated by digging the shore area to at least 1 m deep with a 3:1 slope. Existing ponds with extensive shallow areas can be dug deeper during periods of low water.

16.3.6 Renovation

At times, an undesirable fish assemblage may develop in a pond. For example, in the absence of predators such as largemouth bass, bullheads and common carp sometimes develop such dense populations that their feeding activities create substantial physical turbidity. Their feeding actions may also resuspend nutrients that cause algal blooms and thereby further reduce water clarity. Even if largemouth bass were stocked in such a pond, they could not see to feed, and their impact on bullheads and common carp may be negligible. Draining a pond is the most economical alternative for removing unwanted fish. Before a pond is drained, consider that private and public waters downstream could be damaged by fishes released from a pond. In many states, it is illegal to stock public or private waters without permission.

If a pond cannot be drained, fishes can be chemically removed. Liquid rotenone, historically available in a 5% liquid formulation or a 2.5% synergized formulation, is the chemical commonly used. In the near future, relabeling by the EPA is likely to render the 2.5% synergized formula unavailable. A 5% rotenone formulation, CFT Legumine, contains an emulsifier and solvent package that reduces the presence of petroleum solvents, making it virtually odor free yet retaining its efficacy. Rotenone kills only animals with gills and is not harmful to any warm-blooded animals except pigs. The 5% rotenone formulation typically is applied at a rate of 30.7 L/ha-m of water to obtain a concentration of approximately 3 mg/L. Freshwater springs or abundant aquatic vegetation may create a need for a higher treatment rate. The total amount of rotenone required may be reduced if the water volume of a pond can be lowered through siphoning or pumping, which is desirable because the chemical is expensive.

Treatment typically occurs when water temperature is 16°C or above. In ponds smaller than 1 ha, the chemical can be mixed into the water using the propeller wash of a stationary outboard motor. The front end of a small boat should be fixed firmly to the pond bank for safety, typically securing the bow to t-bar fence posts pounded into the shoreline, and the motor should be run in forward gear. The rotenone is then poured slowly into the prop wash. It is best to dilute rotenone with water before it is poured into a pond so that the treatment is done gradually. The propeller wash circulates the chemical to all depths of the pond. Change the boat location several times to ensure that the mixing action reaches all areas of the pond. Shallow areas should be treated with a hand sprayer or by applying the chemical by bucket.

Rotenone may not reach all areas of large ponds or ponds deeper than 3 m when the chemical is mixed with an outboard motor, especially during warm weather. In such cases, it should be pumped into areas not reached by motor mixing and into the deepest water. In ponds that are thermally stratified, rotenone should be pumped into the cold bottom water if the pond is treated in the summer. Fish can be stocked into a pond within 2–3 weeks after rotenone has been applied in the summer. It is best to wait until the following spring to stock ponds treated in the fall. Toxicity of a pond can be tested by putting a few fish in a cage.

Fishes can also be eliminated using rotenone application under ice. A rotenone concentration of 0.5 mg/L can be used because the chemical remains toxic as long as ice cover remains because rotenone decomposes slowly at cold temperatures. This technique should be considered experimental as few evaluations have been completed.

16.4 TRADITIONAL MANAGEMENT STRATEGIES

Historically, one of the most common pond management problems was overharvest of largemouth bass (Funk 1974). Up to 70% of the adult largemouth bass can be caught and harvested within a few days of angling in a new pond (Redmond 1974). If that many largemouth bass are harvested, the predator–prey balance will be impaired. Once largemouth bass are overharvested, panfish species are likely to overpopulate with an associated decline in population quality (i.e., a decline in panfish length structure). Thus, traditional pond management strategies sought to protect largemouth bass. Minimum length limits were used and required that all largemouth bass less than a specified length (e.g., 38 cm) be immediately released. If obeyed, this regulation protects largemouth bass and other fish species from overharvest. However, if habitat is appropriate for high rates of largemouth bass recruitment and high resulting population abundance, a minimum length limit may result in overpopulation of largemouth bass and reduced growth and length structure. This condition is desirable in some pond management options but not in others. A protected slot length limit is a better regulation choice when larger largemouth bass are desired (Novinger 1990). However, anglers must harvest smaller largemouth bass below the slot; otherwise the regulation will function like a minimum length limit (see Chapter 7).

Today, the catch-and-release ethic is so ingrained in anglers that overpopulation of largemouth bass in ponds is far more common than overpopulation of panfishes. Selective harvest of largemouth bass thus is often necessary for pond management. Selective harvest refers to the appropriate harvest of an abundant or targeted segment of a population. In some cases, overly abundant small fishes might be harvested. In others, low recruitment rates call for protection of small and mid-size fish, with only large fish being harvested. Combined with selective harvest, traditional pond management strategies remain viable today.

16.4.1 Traditional Stocking and Management Strategies

Several species combinations and management strategies have been recommended along with stocking strategies. Warmwater fishes include largemouth bass, bluegill, and channel catfish, and coolwater fishes include walleye, yellow perch, smallmouth bass, and northern pike. Coldwater fishes are primarily salmonids of differing species.

Fish species combinations vary according to management objectives and geographic location. The traditional warmwater largemouth bass—bluegill stocking combination (Dillard and Novinger 1975; Modde 1980) is widely used in the southern USA and to some extent in more northerly locations. Rainbow trout are most commonly used in coldwater ponds, although other salmonid species may be used in some locations. Native species are increasingly used for coldwater pond management today. Examples include brook trout in the eastern USA and Canada and native subspecies of cutthroat trout in western North America.

Gabelhouse et al. (1982) presented five options for stocking and management strategies to meet specific objectives in warmwater ponds (all-purpose option, harvest quota option, panfish option, big bass option, and catfish-only option). We will begin with those five options and then discuss two additional options, including coldwater pond options.

16.4.1.1 All-purpose option

This option allows the harvest of largemouth bass, bluegill, and channel catfish of a variety of sizes. It uses a protected slot limit of 30–38 cm for largemouth bass once the fish assemblage has had time to become established (perhaps 3–4 years after stocking, with shorter times in the south and longer times in the north). To reduce competition and allow largemouth bass consistently to grow over 38 cm, about 75 largemouth bass per hectare (length range 20–30 cm) should be harvested annually after the fourth year following largemouth bass fingerling stocking (Gabelhouse et al. 1982) in ponds of average fertility. Initiation of harvest can be sooner or later depending on growth rates at any particular geographic location. Release of all 30–38 cm largemouth bass ensures that some of the catchable-size largemouth bass survive to lengths of 38 cm and longer. Maintaining moderate largemouth bass abundance ensures predation on small bluegills and allows some bluegills to exceed 20 cm.

Bluegills and channel catfish can be harvested as desired, but subsequently stocked channel catfish must be 20 cm or longer to reduce predation by largemouth bass (Storck and Newman 1988; Shaner et al. 1996). With the all-purpose option, if anglers do not adhere to the slot limit, overharvest of largemouth bass can result in overpopulation of bluegills. On the other hand, if anglers release largemouth bass in the slot limit but do not harvest 20–30 cm largemouth bass under the slot, overpopulation and stunting of largemouth bass can occur. Selective harvest of the right lengths and numbers of fish is essential for proper pond management, and determining harvest rates can be challenging. Ponds managed under the all-purpose option might also be considered as a "balance-seeking" strategy (see section 16.6).

16.4.1.2 Harvest quota option

Quotas have been proposed as an alternative to length limits for regulating harvest of largemouth bass. This option involves the harvest of a given number or weight of largemouth bass annually, regardless of size. Catch-and-release fishing is possible after the quota has been reached. Few or no largemouth bass should be harvested the first few years after stocking, followed by an annual harvest of about 125 individuals or 22 kg of largemouth bass per hectare without regard for length. This is a general guideline that should be adjusted for the fertility of a particular pond. For example, ponds in the southeastern USA tend to be infertile due to highly leached soils. Standing stocks of largemouth bass may be only 20 kg/ha in such waters without fertilization programs (see section 16.3.1). In contrast, largemouth bass standing stocks in ponds built on fertile Midwestern soils can be 70–100 kg/ha (Hackney 1978; Hill and Willis 1993).

With this option, there is a tendency to overharvest large largemouth bass and underharvest small ones. Channel catfish harvest is unrestricted, but those harvested should be replaced with 20 cm or larger individuals. Very accurate harvest records will be needed for successful use of the quota management option.

16.4.1.3 Panfish option

This option emphasizes production of large panfishes instead of largemouth bass by imposing a 38-cm minimum length limit on largemouth bass. Few largemouth bass will grow beyond 38 cm because of their expected overpopulation. The high density of 20–38 cm largemouth bass will reduce bluegill density and allow faster growth of bluegills to 20 cm and beyond. Other panfish species used under this strategy include black crappie, white crappie, black bullhead, and yellow perch (Gabelhouse 1984a; Boxrucker 1987; Saffel et al. 1990; Guy and Willis 1991). Because crappies, black bullhead, and yellow perch tend to overpopulate, addition of these species requires release of nearly all largemouth bass and maintenance of water clarity to depths greater than 46 cm to enhance predation by sight-feeding largemouth bass. Consensus on the use of crappies in pond management has not been reached within the profession, and some managers will not include them as pond stocking options because of their propensity to overpopulate.

With adherence to the length restriction on harvest of largemouth bass, the panfish option is unlikely to fail and actually provides a relatively safe management strategy. However, pond owners may be disappointed in the small size of largemouth bass caught. It is also possible to have too many small largemouth bass, which can then compete with bluegills (Gabelhouse 1987). In such a case it would be desirable to remove some small largemouth bass periodically or temporarily impose a slot limit of 30–38 cm (Gabelhouse 1987; Novinger 1990; Neumann et al. 1994). Finding an appropriate balance between harvest and sufficient numbers remaining for effective predation can be a challenge because of differences in productivity and standing stock among ponds. Most private pond managers monitor changes in fish body condition (e.g., relative weight; see section 16.6.3) to help determine when sufficient harvest has occurred.

An alternative approach for panfish fishing is to stock hybrid sunfish and largemouth bass. Several sunfish crosses (bluegill × green sunfish and redear sunfish × green sunfish) produce offspring that are primarily male (Childers 1967; Lewis and Heidinger 1978). With limited reproductive potential, hybrid sunfishes are less likely to overpopulate, and thereby achieve large sizes. The F1 generation of hybrid sunfishes exhibits hybrid vigor and higher vulnerability to angling. Subsequent generations of hybrid sunfish tend to exhibit a range of characteristics between the two parental species and often do not reach the larger sizes exhibited by the F1 generation, and hybrids will backcross with parental species. Because of reduced reproduction, hybrids do not provide as much prey for predators as do the parental species; thus, many pond managers prefer not to use hybrids.

Smallmouth bass can be used in ponds (e.g., Bennett and Childers 1972). Traditional advice has been not to rely on smallmouth bass as a primary predator in pond fisheries. For example, the smallmouth bass–bluegill combination typically results in overpopulated bluegills because of insufficient predation by the smallmouth bass in ponds containing submergent aquatic vegetation. However, the smallmouth bass–redear sunfish combination has been recommended for pond management (Gabelhouse 1978) because of the limited reproductive potential of redear sunfish compared with bluegill.

16.4.1.4 Big bass option

The objective of this option is consistent production of trophy largemouth bass without regard to bluegill size. Catch rate for largemouth bass will be low with this option, but the largemouth bass caught should be large. After 4 years, densities of 20–38-cm largemouth bass should be greatly reduced to allow for rapid growth of those remaining. In a pond of average fertility, 75 largemouth bass 20–30 cm long and about 12 largemouth bass 30–38 cm long should be harvested per hectare each year. All largemouth bass over 38 cm should continue to be released, other than an occasional trophy largemouth bass. Expect the abundance of panfishes such as bluegills to be high under this option, with concomitant declines in panfish growth, length structure, and maximum length.

The key to development of the big bass option is availability of a diverse and abundant prey base. Large prey items such as gizzard shad are essential to produce trophy largemouth bass consistently. However, because gizzard shad can grow quickly to a size too large for all but the largest largemouth bass in a pond to consume, they should not be introduced into a pond until the largemouth bass population is dominated by fish longer than 41 cm. Growth of gizzard shad is often inversely related to density; therefore a fertile environment that lends itself to consistent production of age-0 gizzard shad will help reduce their growth and allow them to remain vulnerable for a longer time (DiCenzo et al. 1996). If large adult gizzard shad become overly abundant in a pond, reproduction by gizzard shad will be reduced because of density-dependent mechanisms (Kim and DeVries 2000). Partial rotenone treatment to reduce gizzard shad abundance selectively every few years can maintain an actively reproducing gizzard shad population (Irwin et al. 2003). At southern latitudes, threadfin shad can be used as a prey species for trophy largemouth bass. Because of their smaller maximum size, they remain vulnerable to predation by adult largemouth bass.

Rainbow trout can be stocked as an alternative to gizzard shad for providing trophy largemouth bass with a large prey item. In warmer climates, rainbow trout only persist in ponds during that portion of the year when water temperatures are less than approximately 22°C. The advantage of rainbow trout is that their abundance can be controlled because they will not reproduce in a pond, and all biomass can be directed toward trophy largemouth bass production. In addition, rainbow trout also provide an alternative angling opportunity.

16.4.1.5 Catfish-only option

The catfish-only option is especially desirable for turbid ponds where sight-feeding largemouth bass and bluegill would not do well. The pond should be free of any structures, such as hollow logs and rock ledges, which would provide seclusion and cavities required for channel catfish spawning. Natural recruitment by channel catfish in small ponds without predators leads to overpopulation and decreased growth rates and length structure. Fathead minnows can be stocked as prey, and harvest of channel catfish is generally unrestricted. Replacement stocking of channel catfish over 20 cm should take place during cool weather in spring or fall; the number harvested plus an additional 10% to account for natural mortality should be stocked. A density of 250–500 channel catfish per hectare should be maintained in ponds receiving no supplemental feeding. When supplemental feeding is planned, stock 500–1,500 per hectare but be prepared to harvest channel catfish as they reach 45 cm. Densities in excess of 2,540 channel catfish per hectare can be sustained with supplemental feeding, as long as water quality is maintained with sufficient aeration.

16.4.1.6 Black bass only option

In ponds where owners are not interested in panfishes or in very shallow ponds where excessive aquatic vegetation provides too much cover for bluegill, black bass alone is a viable option. Bennett (1952) found that spotted bass, smallmouth bass, or largemouth bass, when stocked alone in Illinois ponds, were able to persist by feeding on crayfish, large aquatic insects, and their own young. Swingle (1952) experimented with stocking largemouth bass alone but considered this to be an inefficient use of available food resources of a pond. A largemouth bass only pond is often is considered a "safe" management option as no panfish species are present to overpopulate if largemouth bass are overharvested. Thus, some pond managers use this strategy when a private pond owner is disinterested and unlikely to spend time managing a pond.

Overpopulation is quite common with the black bass only option, but this can be corrected with harvest under a protected slot length limit (Gabelhouse 1987; Neumann et al.

1994). While some biologists believe that larger largemouth bass control recruitment of small largemouth bass, experience indicates that overpopulation should be expected in ponds with appropriate habitat for largemouth bass reproduction and recruitment.

The largemouth bass–golden shiner combination is a slight variation to the black bass only option that can be used in ponds that contain sufficient macrophytes to protect some adult golden shiners. Regier (1963) suggested this combination for the pond owner in northern climates who is not interested in bluegill and whose pond is small and shallow and has surface water temperatures that exceed 22°C in summer. Golden shiners can also be used as a prey source for smallmouth bass in ponds.

16.4.1.7 Trout options

Historically, most coldwater pond management has involved rainbow trout. Because rainbow trout will not spawn in standing water, growth rates and ultimate lengths are easily controlled by stocking rates. In addition, rainbow trout readily accept formulated feeds if supplemental feeding is desired. However, many strains of hatchery-reared rainbow trout have become so domesticated that stocked trout do not live much beyond 2–3 years, but they grow very quickly in a pond environment.

Small ponds at high elevations or at high latitudes typically are used for trout management. Generally speaking, water temperatures should not exceed 22°C for extended periods if that pond is to support rainbow trout. Brown trout can withstand slightly warmer water, and brook trout typically require cooler water. In northern portions of the USA and Canada, some ponds that thermally stratify during summer can support trout. Although the surface waters may exceed 22°C, the trout apparently retreat to the thermocline to escape the warm temperatures but come to the surface to feed on insects at sunrise and sunset.

For some anglers, knowing which species and the available size of trout that will be caught reduces the quality of the angling experience. The experience can be enhanced by adding a few large fish or by stocking more than one species or a color variation. Hatchery broodstock from private dealers become available for stocking from time to time. Several state and private trout hatcheries have golden-colored strains of rainbow trout that can provide an unusual catch when stocked in low numbers. Brown trout generally live longer and grow larger than do rainbow trout, but they can be cannibalistic and harder to catch. Interest in cutthroat trout in western states and brook trout in eastern states is increasing as knowledge of native trout distributions becomes more commonplace.

16.4.2 Stocking Rates

Initial stocking strategies should be designed to provide a desirable pond fishery in the shortest time possible within physical and fiscal limits (Dillard and Novinger 1975). This desire dictates a diversity of stocking strategies rather than one standard strategy. Fisheries managers sometimes become overly concerned with adjustments to stocking rates to achieve desired results. In reality, harvest has more to do with the long-term development of a pond fishery than with stocking rates.

Ponds should not be stocked with fish caught elsewhere by anglers. In some states, it is illegal for anglers to transport live fishes. In addition, stocking a few individuals of sometimes erroneously identified species is not likely to produce a desirable fish assemblage, and diseases and parasites might also be inadvertently introduced. Similarly, shipments of fish should be checked for unwanted species before stocking. Most warmwater fish-rearing facilities raise many species, and it is not uncommon for a few stray fish to find their way into a shipment of another species. Sometimes a shipment contains desirable species like fathead minnows mixed with largemouth bass, but other times the mixture may contain undesirable species such as goldfish with channel catfish. In addition, most states now require health certificates for interstate shipment of pond fishes for stocking because of concerns over diseases such as viral hemorrhagic septicemia (Elsayed et al. 2006; Jones and Dettmers 2007).

16.4.2.1 Warmwater ponds

Introductory stocking rates vary geographically, but 125–250 largemouth bass fingerlings per hectare and 1,250–2,500 sunfish fingerlings (either all bluegill or a combination of bluegill and redear sunfish at southern latitudes) per hectare remain the most common initial stocking rates. Lower initial stocking rates for largemouth bass can increase their growth rates. Stocking of fingerlings into a pond that already contains fishes is not effective because predation by fish present in the pond greatly reduces fingerling survival.

Successive, or split, stocking involves stocking fingerling sunfishes in the fall and fingerling largemouth bass the following spring. By stocking bluegills before largemouth bass, bluegills have an opportunity to reach sexual maturity by the time largemouth bass are stocked. Bluegill spawning then produces appropriate-size prey for stocked fingerling largemouth bass. In northern regions, this sequence is reversed or both are stocked at the same time. Due to the slow growth of largemouth bass at northern latitudes, largemouth bass may not reproduce until age 2, whereas bluegills typically reproduce at age 1. Thus, some biologists add a second stocking of fingerling largemouth bass the year after largemouth bass and bluegills originally were stocked.

When channel catfish are desired, they should be stocked initially at a rate of 250 fish per hectare when stocked with other species and up to 500 fish per hectare when stocked alone and when no supplemental feeding is provided. Periodic restocking is necessary for perpetuation of this species in most ponds. Alternate-year stocking ensures a steady supply of channel catfish for harvest. If adult largemouth bass are present, fingerling channel catfish longer than 20 cm should be stocked to minimize predation.

Over the past decade, private pond managers, particularly in the southern USA, have begun experimenting with innovative stocking rates for new ponds. This typically has involved stocking a higher rate of sunfish fingerlings (e.g., 5,000 per hectare) and a lower rate of largemouth bass fingerlings (e.g., 125 per hectare). In addition, the introduction of alternative prey species such as threadfin shad and golden shiner concurrent with the traditional mixture of sunfish has become more common. The objective of this innovative strategy is to create a prey-rich pond environment in terms of both density and species diversity before introducing largemouth bass fingerlings. The ultimate goal is to maximize early growth and minimize subsequent recruitment of largemouth bass to prevent or at least delay the common problem of largemouth bass overpopulation. Southern ponds stocked with this strategy have commonly produced largemouth bass in excess of 0.6, 1.3, 2.7, 3.6 and 4.5 kg in their first through fifth years of life, respectively, after initially being stocked as fingerlings. Reduced recruitment of largemouth bass through the first 5 years of many new ponds has been observed as well, thereby reducing the level of largemouth bass harvest required to maintain balance in these ponds.

16.4.2.2 Coldwater ponds

Most available information on coldwater fish stocking in ponds involves rainbow trout. Recommended stocking rates with 5–10 cm rainbow trout are 600 fish per hectare in western states and 1,500 fish per hectare in eastern states (Marriage et al. 1971). If larger rainbow trout are used, fewer should be stocked. Fingerling rainbow trout less than 10 cm are adequate for initial stocking, but subsequent stockings should be done with fish greater than 10 cm. Recommendations for initial stocking of rainbow trout in northern plains states and provinces range from 375 to 600 fish per hectare (Willis et al. 1990; Blaser and Eades 2006), with re-stockings scheduled for every second or third year. Record keeping aids in determining the need for more or fewer rainbow trout in subsequent stockings. These rainbow trout stocking rates should be used for brook trout and cutthroat trout, until such time as species-specific evaluations are completed. Cutthroat trout fingerlings are not yet highly available for pond stocking programs, but that may change as interest in native species conservation continues to increase. Because of their piscivorous nature, brown trout have not been extensively used in pond management.

16.5 INNOVATIVE MANAGEMENT STRATEGIES

Expectations of many pond owners have shifted over the past decade or two, with concurrent changes in pond management strategies. High-quality fisheries may now be the expected norm rather than the exception. Many pond owners are now willing to invest substantially in their ponds, and they expect a high return on their investment in the form of high catch rates, above-average fish lengths, and impressive fishing experiences.

This section covers some innovative approaches to pond management to achieve high-quality fisheries. However, much of the information supporting the approaches in this section is anecdotal because relatively little research has been conducted on pond management methods since the 1970s.

16.5.1 Coppernose Bluegill

Perhaps the largest change in pond management, at least in the southeastern USA, is the widespread use of coppernose bluegill. The coppernose bluegill, a subspecies of bluegill native to central and southern Florida (Hubbs and Allen 1943), is considered by many fisheries managers and pond enthusiasts to be superior to native bluegill due to its rapid growth and aggressiveness. However, little research has been conducted to support these claims. One notable exception was Prentice and Schlechte (2000), who found that coppernose bluegills grew faster than native bluegill strains in Texas ponds through their second year of life. Coppernose bluegills are the preferred subspecies for use in ponds in the southeastern USA. Coppernose bluegill culture began in Arkansas in the late 1980s and early 1990s. Since then, their popularity has grown and they have been widely stocked in ponds throughout the southeastern USA and as far north as Missouri, Kentucky, West Virginia, and Ohio.

Supplemental feeding has become a common practice in the southeastern states and the observation that coppernose bluegill more readily consume artificial feeds compared with native bluegill has contributed to its popularity. Coppernose bluegill growth can be increased substantially when its diet is supplemented with an artificial feed. For example, fish stocked at

5 g can grow to 225 g in 1 year when provided a supplemental diet. This rapid growth enables ponds to support harvestable populations of coppernose bluegill within 1 year of initial stocking. Additionally, rapid growth enables this bluegill subspecies to begin spawning earlier, thereby enhancing the prey base for largemouth bass.

Regardless of the subspecies of bluegill inhabiting a pond, bluegill harvest strategies can be adjusted to maximize length structure. For example, harvest of only female bluegills allows large parental males (Drake et al. 1997; Jennings et al. 1997; Jacobson 2005) to remain in a pond. Due to behavioral plasticity, smaller, younger males will then continue to grow rather than maturing and may attain another several centimeters of growth prior to maturation and the associated slowing of growth that occurs when energy is diverted to gonads rather than somatic growth. A strategy of female-only harvest has the exciting potential to produce a fish assemblage with the largest possible male bluegills for that pond.

16.5.2 Feed-Trained Largemouth Bass

Pond owners are increasingly concerned with the genetic composition of the largemouth bass they stock. Widely publicized accounts of huge largemouth bass produced in southern California and Texas reservoirs have created a demand for largemouth bass with a similar genetic composition. This often leads to pond owners stocking Florida largemouth bass in ponds far removed from their native range. From the 1970s to 1990s, Florida largemouth bass were stocked into southern and southwestern ponds in the USA to maximize the production of trophy-size fish. However, as these populations began to age, many pond owners became disgruntled by the low angling vulnerability expressed by Florida-strain largemouth bass. Anglers began questioning the practicality of growing trophy largemouth bass that were difficult to catch.

In addition, many pond owners observed a decline in largemouth bass catch rates as their ponds aged, even in cases where Florida largemouth bass were not stocked. Presumably, the reliance on angling as the sole means of attaining an annual harvest quota (see section 16.4) leads to the selective removal of the most aggressive fish from the population. Because angling vulnerability has been shown to be a heritable trait (Garrett 2002), such a practice is one plausible explanation for a reduction in angling vulnerability over time in a small impoundment, as would be learned behavior from catch-and-release angling.

To meet the demand for increased angling vulnerability, the private fish culture industry has developed strains of largemouth bass that are trained to feed on artificial diets. These feed-trained largemouth bass are produced in a hatchery by exposing young largemouth bass to prepared diets soon after they hatch. Not all largemouth bass accept a prepared diet, but those that do are separated and reared exclusively on a high-protein feed to adulthood, after which time they are stocked into ponds. The result is an aggressive fish that is vulnerable to angling. Many pond owners claim that stocking these fish in their pond has revolutionized their angling experience because a day of fishing turned into a day of catching. Feed-trained largemouth bass have become so popular that the fish culture industry has had difficulty keeping up with demand.

There are drawbacks to stocking feed-trained largemouth bass. Fish tend to prefer the prepared diet, even in the presence of abundant natural prey. Although feed-trained largemouth bass generally maintain relative weight (see section 16.6.3) values in excess of 100, anecdotal reports have shown that they rarely reach trophy length in ponds. Whether this is a result of them being reared to adulthood on a strictly artificial diet or their reliance on a prepared diet after being stocked into a pond is unknown. Another drawback of stocking feed-trained largemouth bass is that they can be too aggressive. Some pond owners have reported mixed feelings after stocking them; initial excitement at the increased angler catch rates is followed by disappointment at the ease of catching the feed-trained largemouth bass.

16.5.3 Female-Only Largemouth Bass

In ponds where the management objective is to grow largemouth bass as large as possible, even at the expense of sacrificing angler catch rates, female-only largemouth bass stocking is an option. Females grow faster and larger than do males and when they are stocked in the absence of males to preclude reproduction, energy otherwise lost to the production of gonads and spawning is directed into somatic growth. In addition, the common problem of largemouth bass overpopulation is eliminated by stocking one gender of fish because managers can precisely control abundance.

This stocking option has the potential to revolutionize pond management, but its use is currently limited by the supply of female-only largemouth bass for stocking. The use of methyltestosterone to mass produce monosex fish (Garrett 1989) is prohibited by the FDA, and visual determination of gender is currently the only commercially viable method of obtaining female-only largemouth bass. Visual determination of gender is time and labor intensive and generally cost prohibitive for stocking even moderate-size ponds.

16.5.4 Hybrid Striped Bass

Hybrid striped bass, or wipers as they are sometimes known, are a cross between a striped bass and a white bass. The American Fisheries Society (Nelson et al. 2004) recognizes the cross between male white bass and female striped bass as the palmetto bass, while the cross between female white bass and male striped bass is termed the sunshine bass. Both crosses are considered wipers or hybrid striped bass as used in this chapter.

Hybrid striped bass are popular as an alternative predator for diversifying pond fisheries in many geographic locations. Attributes that make hybrid striped bass popular with pond owners include their aggressiveness and willingness to strike artificial lures, fast growth, ability to train to consume an artificial diet, and lack of reproduction in ponds. While hybrid striped bass will not produce an F2 generation in ponds, they can backcross with either parental species in some situations, which is a potential problem where they escape from a pond and move to an environment where the parental species occur. Consequently, they should not be stocked into ponds with high water exchange rates because they apparently are attracted to current and readily escape through emergency spillways and other types of outlets.

Hybrid striped bass can be an effective predator for controlling crappie recruitment in small ponds. Although crappies are extremely popular with pond owners, problems caused by excessive recruitment in some locales have prompted many managers to discourage their use. However, the limnetic nature of larval and juvenile crappies makes them vulnerable to predation by hybrid striped bass, a limnetic predator (Neal et al. 1999). For example, several Alabama ponds stocked with hybrid striped bass, black crappie, threadfin shad, and golden shiners successfully produced trophy fisheries for both hybrid striped bass and crappie along with the suppression of crappie recruitment in each of the 4 years following initial stocking in an ongoing study conducted by one of the authors.

16.5.5 Northern Pike

Northern pike typically are not recommended in pond management because they can have substantial effects on fish assemblages through predation on relatively large prey. In contrast to largemouth bass control of bluegill recruitment through cropping of small bluegills, northern pike tend to feed on larger prey items and may even compete directly with anglers for harvestable-size panfishes (e.g., Johnson 1966; Anderson and Schupp 1986; DeBates et al. 2003; Jolley et al. 2008). Thus, a high biomass of northern pike certainly is not conducive to management of a pond fishery. However, both Gurtin et al. (1996) and DeBates et al. (2003) reported decreased abundance and increased length structure of largemouth bass due to northern pike predation on small largemouth bass in small (4–25 ha) water bodies. Thus, an innovative strategy of stocking low numbers of female-only northern pike in ponds to avoid reproduction and subsequent overabundance could result in improved size structure of high-density largemouth bass populations and likely without loss of quality (i.e., length structure) for bluegills (Gurtin et al. 1996). Such a strategy is considered experimental at this time.

16.5.6 Smallmouth Bass

Smallmouth bass have been used in pond management for decades (e.g., Bennett and Childers 1972). Traditional advice has been not to rely on smallmouth bass as a primary predator in pond fish assemblages. For example, the smallmouth bass–bluegill combination typically results in overpopulated bluegills because of insufficient predation by smallmouth bass. However, in ponds with limited coverage of submergent aquatic vegetation, smallmouth bass may be an effective predator, although further research is needed to confirm this.

In the southern USA, smallmouth bass typically exhibit erratic recruitment. Thus maintenance of populations through natural reproduction can be problematic. However, erratic recruitment can also be a positive attribute by preventing the development of high-density, slow-growing populations. In northern states, smallmouth bass commonly overpopulate when used alone in ponds and require substantial selective harvest of smaller individuals to maintain high-quality populations (see section 16.4.1.6).

16.5.6 Walleye

The walleye is a highly popular sport fish across the northern USA and Canada, and many pond owners thus want to include it as part of a fish assemblage. Traditional advice has been not to use walleyes in small ponds because they typically will not reproduce in this habitat. However, reproduction can occur in some instances, especially in ponds that exceed 6 ha or in sand and gravel pits. The reproduction may not be sufficient to support a harvestable population but does maintain species presence.

The best use of walleyes in ponds may be as a "bonus" fish, such as in a northern pond being managed with a largemouth bass–bluegill combination, although maintenance stockings will likely be required for the walleyes. Larger sizes of walleye fingerlings will also be necessary for stocking in ponds with established fish communities. Many suppliers in northern states can provide 15–20-cm walleyes by their first fall of life. An initial stocking rate might be 25 fish per hectare each year, with adjustments depending on survival and the amount of harvest.

16.6 POPULATION AND ASSEMBLAGE ANALYSIS

The concept of balanced fish populations was proposed by Swingle (1950) and further developed by Anderson (1973). Historically, ponds were usually stocked with species combinations that were likely to attain balance. A balanced system is a dynamic one characterized by continual reproduction of predator and prey species, diverse length composition of prey species so that food is available for all lengths of predators, high growth rates of predators and prey, and an annual yield of harvestable-size fish in proportion to basic fertility.

Bennett (1971) refuted the concept of balance largely on the basis that ponds represent artificial ecosystems that cannot be expected to show stability. Gabelhouse et al. (1982) and Gabelhouse (1984b) then proposed panfish (section 16.4.1.3) and big bass (section 16.4.1.4) management options that would be considered out of balance by some biologists. The all-purpose option (section 16.4.1.1) does fit the concept of balance and is the management strategy that both Swingle and Anderson originally had in mind. Many of the innovative pond management strategies being applied today seek to provide more and bigger fishes and similarly push the bounds of what traditionally was considered balance (see section 16.4).

No one has applied the concept of balance to coldwater ponds. To be sure, lack of reproduction and no prey fish are major deviations from Swingle's (1950) definition of balance. However, the objective of sustained crops of harvestable-size fish in proportion to pond fertility is equally valid for coldwater systems.

16.6.1 Biomass Indices

In an attempt to describe balanced and unbalanced populations, Swingle (1950) analyzed biomass data from 55 balanced and 34 unbalanced ponds and computed certain biomass ratios or indices. His most commonly used biomass indices are the F:C ratio, the Y:C ratio, and the A_T value. The F:C ratio is the total weight of all forage fishes (F) divided by the total weight of all carnivorous fishes (C). Ratios from 3.0 to 6.0 are considered the most desirable in the balanced range (Figure 16.11). The Y:C ratio is the total weight of all forage fishes that are small enough to be eaten by the average-size individual in the C group divided by the C value. The most desirable range for Y:C ratios in balanced populations is between 1.0 and 3.0. The A_T value (total availability value) is the percentage of the total weight of a fish population composed of fishes of harvestable size. Before an A_T value could be calculated, minimum weights suitable for harvest had to be defined, and Swingle (1950) did this for common pond species (e.g., 45 g for sunfishes, 180 g for largemouth bass, and 230 g for channel catfish). Swingle suggested an A_T range of 60–85% as most desirable for balanced populations.

Swingle (1956) also presented a method of analysis that was based on use of a minnow seine to sample for evidence of reproduction and a larger seine to sample intermediate and harvestable-size fishes. His description of possible catches and interpretations that may be drawn are summarized in Box 16.3. Because so much of the analysis is based on presence or absence of young fishes, sampling should be done after spawning has occurred. However, Swingle (1956) reported that his method was valid from June through October in Alabama ponds.

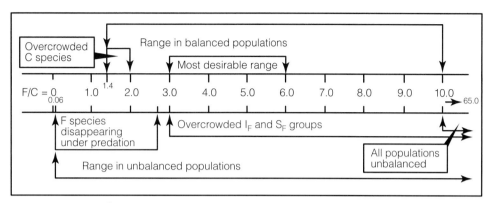

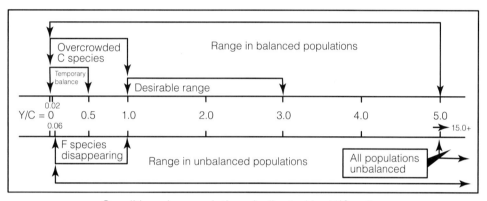

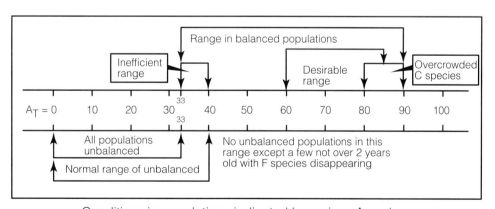

Figure 16.11. Status of fish populations as indicated by $F:C$ ratios, $Y:C$ ratios, and A_T values (see section 16.6.1 for definitions). Additional abbreviations are I_F (intermediate forage) and S_F (small forage) groups. (Figure taken from Swingle 1950).

Box 16.3. Swingle's (1956) Method of Pond Analysis Based on Seining

1. No young largemouth bass present.

    A. Many recently hatched bluegills; no or very few intermediate bluegills
       (Temporary balance with bass overcrowded)

    B. No recent hatch of bluegills; many intermediate bluegills
       (Unbalanced population with overcrowded bluegills and insufficient bass)

    C. No recent hatch of bluegills; many intermediate bluegills; many tadpoles, minnows, or crayfish
       (Unbalanced population with overcrowded bluegills and very few bass)

    D. No recent hatch of bluegills; few intermediate bluegills
       (Unbalanced population, crowding due to species competitive with bluegills)

    E. No recent hatch of bluegills; few intermediate bluegills; many intermediate fish of a species competitive with bluegills
       (Unbalanced population due to crowding by competitive species)

    F. No recent hatch of bluegills; no intermediate bluegills
       (Unbalanced population; possibly no fish present or water unsuitable for bass–bluegill reproduction)

2. Young largemouth bass present:

    A. Many recently hatched bluegills; few intermediate bluegills
       (Balanced population)

    B. Many recently hatched bluegills; very few or no intermediate bluegills
       (Balanced population with slightly crowded bass)

    C. No recent hatch of bluegills; no intermediate bluegills
       (Unbalanced population; bluegills prevented from spawning by some extrinsic factor such as temperature or water salinity)

    D. No recent hatch of bluegills; few intermediate bluegills
       (Temporary balance with possibility of imbalance developing due to a reduction of the food available to bluegill or overcrowding by a species growing to a competitive size)

    E. No recent hatch of bluegill; many intermediate bluegills
       (Unbalanced population similar to [1][B] but less severely overcrowded)

16.6.2 Length Structure Indices

It is difficult to sample with equal effort the different species and lengths of fishes needed to obtain biomass data required for Swingle's (1950) pond analysis system. Consequently, Anderson (1976) recognized that length-frequency samples were more easily obtained than was biomass information. He introduced an analysis method based on fish population length structure that was initially termed proportional stock density (PSD) and now is known as proportional size distribution (still PSD; Guy et al. 2007). The primary advantage of PSD is that it can be calculated from samples of pond fish populations (Anderson and Neumann 1996) rather than requiring that an entire pond be drained or treated with rotenone, as with the Swingle indices. Length structure indices reflect an interaction of the rates of recruitment, growth, and mortality of the age-groups present in a population (Anderson and Weithman 1978; Gabelhouse 1984a).

The PSD index is calculated by dividing the number of fish of a given species greater than or equal to quality length by the number that are greater than or equal to stock length and then multiplying by 100 (Neumann et al., in press). Stock- and quality-length categories are based on a percentage of world record length for various fishes (Gabelhouse 1984b), and the lengths are standard for anyone using them (Table 16.1). In contrast, Swingle's (1950) notations of small (S), intermediate (I), and large (A) had enough latitude that different biologists could put different lengths of a given species into the categories. Such a practice without standards can hinder comparisons and communication among biologists (Bonar et al. 2009).

Proportional size distribution is similar to A_T (Swingle 1950) in that both indices represent a percentage of fish that reach a size that interests anglers. However, PSD is based on length frequency instead of total weight. In addition, fish below stock length are not included in PSD; therefore, no special effort is needed to sample small fish. For balanced ponds, largemouth bass PSD should be between 40 and 70 (Gabelhouse 1984b) and bluegill PSD should be between 20 and 60 (Anderson 1985). Sequential sampling for PSD may shorten the amount of time spent sampling (Weithman et al. 1980). Methods to calculate confidence limits for PSD estimates can be found in Gustafson (1988).

Proportional size distribution is a relatively coarse index of the length frequency for a fish population sample. Thus, PSD can also be further subdivided into additional length-groups (Table 16.1) such as PSD-P, the percentage of preferred-length (P) fish that also exceed stock length, and PSD-M, the percentage of memorable-length (M) fish that also exceed stock length (Guy et al. 2007). See Neumann et al. (in press) for further information on calculation of various PSD indices and Gabelhouse (1984b) for the basis of this five-cell analysis for fish lengths. The PSD indices do allow a biologist or pond owner to set measurable objectives for different management strategies (Table 16.2).

16.6.3 Abundance and Weight Indices

The indices presented so far do not provide an indication of fish abundance. Desirable ratios might exist with very low abundance, giving the impression that good fishing is available when catch rates probably would be low. A measure of fish abundance that is sometimes used is the number of stock-length and longer fish captured per hour of electrofishing. For example, a very general consensus of most fisheries biologists conducting elec-

Table 16.1. Proposed maximum total length in English (E; in) and metric (M; cm) units for minimum stock, quality, preferred, memorable, and trophy lengths for selected fish species based on percentages of world record lengths (Gabelhouse 1984b; Neumann et al. in press).

Species	Length									
	Stock		Quality		Preferred		Memorable		Trophy	
	E	M	E	M	E	M	E	M	E	M
Black bullhead	6	15	9	23	12	30	15	39	18	46
Black crappie	5	13	8	20	10	25	12	30	15	38
Bluegill	3	8	6	15	8	20	10	25	12	30
Brown trout, lentic	8	20	12	30	16	40	20	50	24	60
Channel catfish	11	28	16	41	24	61	28	71	36	91
Flathead catfish	14	35	20	51	28	71	34	86	40	102
Green sunfish	3	8	6	15	8	20	10	25	12	30
Largemouth bass	8	20	12	30	15	38	20	51	25	63
Pumpkinseed	3	8	6	15	8	20	10	25	12	30
Rainbow trout	10	25	16	40	20	50	26	65	31	80
Redear sunfish	4	10	7	18	9	23	11	28	13	33
Rock bass	4	10	7	18	9	23	11	28	13	33
Smallmouth bass	7	18	11	28	14	35	17	43	20	51
Spotted bass	7	18	11	28	14	35	17	43	20	51
Walleye	10	25	15	38	20	51	25	63	30	76
White bass x striped bass	8	20	12	30	15	38	20	51	25	63
White crappie	5	13	8	20	10	25	12	30	15	38
Yellow perch	5	13	8	20	10	25	12	30	15	38

Table 16.2. Proportional size distribution (PSD; Q = quality length, P = preferred length, M = memorable length) values for largemouth bass and bluegill under three different management strategies.

Management strategy	Largemouth bass			Bluegill	
	PSD	PSD-P	PSD-M	PSD	PSD-P
Panfish	20–40	0–10		50–80	10–30
Balance	40–70	10–40	0–10	20–60	5–20
Big bass	50–80	30–60	10–25	10–50	0–10

trofishing under reasonable sampling conditions would be that 100 stock-length (20 cm) and longer largemouth bass captured per hour of electrofishing reflects a dense largemouth bass population. Other catch-per-unit-effort methods, such as number of target species per angler-hour, can be used as indices of population density given that gear, methods, and sampling design are standardized (Hubert 1996; Malvestuto 1996; Bonar et al. 2009).

Relative weight (W_r) is the actual weight of a fish divided by a standard weight (W_s) for the same length for that species multiplied by 100 (Wege and Anderson 1978; Anderson and Neumann 1996). Equations for calculating W_s have been developed for selected species (Table 16.3). The key to appropriate use of W_r as an assessment tool is the availability of appropriate standards. Recent research (Gerow et al. 2004, 2005; Rennie and Verdon 2008) indicated valid concerns over the use of currently accepted standards; however, we recommend the continued use of the standards in Table 16.3 until such time as improved standards are developed (see Neumann et al. [in press] for further information).

A Wr of 100 is not average, but for most W_s equations it represents the 75th percentile of weights attained by that species across its range. Because pond fish with a W_r of 100 are at the 75th percentile, a typical conclusion is that food or access to food is not limiting the population. In contrast, fish with values below 85 are probably near the 25th percentile and may be too abundant for their food supply. Low W_r values cannot solely be attributed to food supply but can also be the result of factors such as recent spawning, environmental problems causing stress, parasites or disease, and excessive turbidity.

Mean W_r should not be calculated for an entire population because such averaging may obscure important patterns in condition across fish lengths. Individuals of the same species but of different lengths often have different food habits, which can affect consumption and prey availability; consequently, fish of different lengths could have considerably different W_r values. A plot of W_r as a function of length may indicate when food is abundant or limiting. Another useful technique is to calculate mean W_r values for fish grouped by the five-cell length categories proposed by Gabelhouse (1984b).

The progression of material presented here may lead readers to the conclusion that Swingle's concepts of balance are outdated. This is not at all true, and there is no better discussion of the structure and dynamics of warmwater pond fish populations than Swingle (1950). Furthermore, Swingle provided an understanding of pond analysis from a biomass perspective rather than one of relative numbers, providing an ecological perspective on pond fish interactions.

Table 16.3. Recommended equations for calculating standard weight (W_s) for a given length of a given fish species. Standard weights are then used in relative weight (W_r) computations. Modified from Blackwell et al. (2000) and Neumann et al. (in press). Intercepts are given in metric (M; mm, g) and English (E; in, lb) measurements; all lengths are maximum total length.

Species	Intercept (a) M	Intercept (a) E	Slope (b)	Minimum applicable length (mm)
Black bullhead	−4.974	−3.297	3.085	130
Black crappie	−5.618	−3.576	3.345	100
Bluegill	−5.374	−3.371	3.316	80
Brown trout, lentic	−5.422	−3.592	3.194	140
Channel catfish	−5.800	−3.829	3.294	70
Flathead catfish	−5.542	−3.661	3.230	130
Gizzard shad	−5.376	−3.580	3.170	180
Golden shiner	−5.593	−3.611	3.302	50
Green sunfish	−4.915	−3.216	3.101	60
Largemouth bass	−5.528	−3.587	3.273	150
Pumpkinseed	−5.179	−3.289	3.237	50
Rainbow trout, lentic	−4.898	−3.354	2.990	120
Redear sunfish	−4.968	−3.263	3.119	70
Rock bass	−4.827	−3.166	3.074	80
Smallmouth bass	−5.329	−3.491	3.200	150
Spotted bass	−5.392	−3.533	3.215	100
Walleye	−5.453	−3.642	3.180	150
White bass x striped bass	−5.201	−3.448	3.139	115
White crappie	−5.642	−3.618	3.332	100
Yellow perch	−5.386	−3.506	3.230	100

16.6.4 Angler-Collected Data

Swingle (1956) also developed a method of distinguishing balanced and unbalanced ponds from angler catches. Assuming that anglers have been fishing for both largemouth bass and bluegill, the following interpretations can be made: (1) in balanced ponds most bluegills caught will be larger than 15 cm, and the average largemouth bass will be 500–1,000 g, although smaller and larger ones also will be caught; (2) in unbalanced ponds the catch will be principally bluegills varying from 7.5 to 12.5 cm, and the few largemouth bass caught will be larger than 1,000 g (note the similarity to the big bass option, section 16.4.1.4); and (3) in ponds crowded with largemouth bass (unbalanced condition), bluegills will average more than 150 g and largemouth bass will be less than 500 g and in poor condition (note similarity to the panfish option, section 16.4.1.3). New York biologists used angler-collected data to provide recommendations to private pond owners (Green et al. 1993). In fact, a pond management bulletin for New York (Eipper et al. 1997) recommended that pond owners use angling catch rates and size structure of fishes that are caught to assess their pond fishery.

Managers should consider size selectivity of the angler catch when angler-collected data are used to assess populations. The type of lure or bait used (e.g., Payer et al. 1989) and the size of lures used (e.g., Gabelhouse and Willis 1986) can influence the resulting length structure of samples and could lead to improper recommendations. For example, the use of large lures to capture largemouth bass could result in a sample dominated by large (e.g., >30 cm) individuals, even if a population was dominated by smaller individuals. Managers can alternatively use angling gear themselves to obtain useful samples from ponds. Isaak et al. (1992) purposely varied the lure sizes used to collect largemouth bass in small impoundments and found that catch per unit effort and size structure were highly correlated between angling and electrofishing samples.

Private pond owners in many geographic locations use W_r to assess their pond fish populations. For example, overly abundant largemouth bass populations can be recognized by high angling catch rates, small length structure of fish, and low W_r values. Pond owners can purchase scales to weigh fish, and W_s equations are commonly shared among pond owners, private sector biologists, and many Web sites. Given that fish condition is likely to reflect population dynamics (i.e., recruitment, growth, and mortality) in small waters and simple fish assemblages (Blackwell et al. 2000), such analyses are appropriate.

16.7 CONCLUSIONS

There has been a proliferation of information available on pond fish management and aquaculture on the internet. Many of the topics in this chapter are covered in greater depth in other sources. In addition, regional differences in pond management strategies typically are valid, and thus using local information sources can be quite important. Because of the ever-changing nature of internet addresses, lists of such sites are not included here. However, internet searches involving the five Regional Aquaculture Centers (northeastern, north-central, southern, tropical–subtropical, and western) provide good starting locations. These centers were established by the U.S. Congress and are administered by the U.S. Department of Agriculture's National Institute of Food and Agriculture. Most state conservation agencies have pond management information on their websites, as do private pond management organizations and many universities.

16.8 REFERENCES

Aiken, R. 1999. 1980–1995 participation in fishing, hunting, and wildlife watching: national and regional demographic trends. U.S. Fish and Wildlife Service, Division of Federal Aid, Washington, D.C.

Alcorn, S. R. 1981. Fishing quality in two urban fishing lakes, St. Louis, Missouri. North American Journal of Fisheries Management 1:80–84.

Anderson, D. W., and D. H. Schupp. 1986. Fish community responses to northern pike stocking in Horseshoe Lake, Minnesota. Minnesota Department of Natural Resources, Division of Fish and Wildlife, Investigational Report 387, St. Paul.

Anderson, R. O. 1973. Application of theory and research to management of warmwater fish populations. Transactions of the American Fisheries Society 102:164–171.

Anderson, R. O. 1976. Management of small warmwater impoundments. Fisheries 1(6):5–7, 26–28.

Anderson, R. O. 1985. Managing ponds for good fishing. University of Missouri Extension Division, Agricultural Guide 9410, Columbia.

Anderson, R. O., and R. M. Neumann. 1996. Length, weight, and associated structural indices. Pages 447–482 *in* B. R. Murphy and D. W. Willis, editors. Fisheries techniques, 2nd edition. American Fisheries Society, Bethesda, Maryland.

Anderson, R. O., and A. S. Weithman. 1978. The concept of balance for coolwater fish populations. Pages 371–381 *in* R. L. Kendall, editor. Selected coolwater fishes of North America. American Fisheries Society, Special Publication 11, Bethesda, Maryland.

Avery, J. L. 2003. Aquatic vegetation management: herbicide safety, technology, and application techniques. Texas A&M University, Southern Regional Aquaculture Center Publication 3601, College Station.

Barrows, F. T., and R. W. Hardy. 2001. Nutrition and feeding. Pages 483–558 *in* G. A. Wedemeyer, editor. Fish hatchery management, 2nd edition. American Fisheries Society, Bethesda, Maryland.

Bennett, G. W. 1952. Pond management in Illinois. Journal of Wildlife Management 16:249–253.

Bennett, G. W. 1971. Management of lakes and ponds. Van Nostrand Reinhold, New York.

Bennett, G. W., and W. F. Childers. 1972. Thirteen-year yield of smallmouth bass from a gravel pit pond. Journal of Wildlife Management 36:1249–1253.

Bettoli, P. W., M. J. Maceina, R. L. Noble, and R. K. Betsill. 1993. Response of a reservoir fish community to aquatic vegetation control. North American Journal of Fisheries Management 13:110–124.

Blackwell, B. G., M. L. Brown, and D. W. Willis. 2000. Relative weight (W_r) status and current use in fisheries assessment and management. Reviews in Fisheries Science 8:1–44.

Blackwell, B. G., and B. R. Murphy. 1996. Low-density triploid grass carp stockings for submersed vegetation control in small impoundments. Journal of Freshwater Ecology 11:475–484.

Blaser, J., and R. Eades. 2006. Nebraska pond management. Nebraska Game and Parks Commission, Lincoln.

Bonar, S. A., W. A. Hubert, and D. W. Willis, editors. 2009. Standard methods for sampling North American freshwater fishes. American Fisheries Society, Bethesda, Maryland.

Boxrucker, J. 1987. Largemouth bass influence on size structure of crappie populations in small Oklahoma impoundments. North American Journal of Fisheries Management 7:273–278.

Boyd, C. E. 1979. Water quality in warmwater fish ponds. Alabama Agricultural Experiment Station, Auburn University, Auburn.

Brown, P. B., M. E. Griffin, and M. R. White. 1993. Experimental and practical diet evaluations with juvenile hybrid striped bass. Journal of the World Aquaculture Society 24:80–89.

Childers, W. F. 1967. Hybridization of four species of sunfishes (Centrarchidae). Illinois Natural History Survey Bulletin 29:159–214.

Dauwalter, D. C., and J. R. Jackson. 2005. A re-evaluation of U.S. state fish-stocking recommendations for small, private warmwater impoundments. Fisheries 30(8):18–27.

Deal, C., J. Edwards, N. Pellmann, R. W. Tuttle, and D. Woodward. 1997. Ponds—planning, design, construction. U.S. Department of Agriculture, Natural Resources Conservation Service, Agriculture Handbook 590, Washington, D.C.

DeBates, T. J., C. P. Paukert, and D. W. Willis. 2003. Fish community responses to the establishment of a piscivore, northern pike (*Esox lucius*) in a Nebraska Sandhill lake. Journal of Freshwater Ecology 18:353–359.

DiCenzo, V. J., M. J. Maceina, and M. R. Stimpert. 1996. Relations between reservoir trophic state and gizzard shad population characteristics in Alabama reservoirs. North American Journal of Fisheries Management 16:888–895.

Dillard, J. G., and G. D. Novinger. 1975. Stocking largemouth bass in small impoundments. Pages 459–474 *in* H. Clepper, editor. Black bass biology and management. Sport Fishing Institute, Washington, D.C.

Drake, M. T., J. E. Claussen, D. P. Philipp, and D. L. Pereira. 1997. A comparison of bluegill reproductive strategies and growth among lakes with different fishing intensities. North American Journal of Fisheries Management 17:496–507.

Durocher, P. P., W. C. Provine, and J. E. Kraai. 1984. Relationship between abundance of largemouth bass and submerged vegetation in Texas reservoirs. North American Journal of Fisheries Management 4:84–88.

Eades, R. T., L. D. Pape, and K. M. Hunt. 2008. The role of law enforcement in urban fisheries. Pages 41–51 in R. T. Eades, J. W. Neal, T. J. Lang, K. M. Hunt, and P. Pajak, editors. Urban and community fisheries programs: development, management, and evaluation. American Fisheries Society, Symposium 67, Bethesda, Maryland.

Eipper, A. W., H. A. Reiger, and D. M. Green. 1997. Fish management in New York ponds. Cornell Cooperative Extension Information Bulletin 116, Ithaca, New York.

Elsayed, E., M. Faisal, M. Thomas, G. Whelan, W. Batts, and J. Winton. 2006. Isolation of viral hemorrhagic septicaemia virus from muskellunge, *Esox masquinongy* (Mitchill), in Lake St. Clair, Michigan, USA reveals a new sublineage of the North American genotype. Journal of Fish Diseases 29:611–619.

Engel, S. 1988. The role and interactions of submersed macrophytes in a shallow Wisconsin lake. Journal of Freshwater Ecology 4:329–341.

Flickinger, S. A., and F. J. Bulow. 1993. Small impoundments. Pages 469–492, 587 in C. C. Kohler and W. A. Hubert, editors. Inland fisheries management in North America. American Fisheries Society, Bethesda, Maryland.

Flickinger, S. A., F. J. Bulow, and D. W. Willis. 1999. Small impoundments. Pages 561–587 in C. C. Kohler and W. A. Hubert, editors. Inland fisheries management in North America, 2nd edition. American Fisheries Society, Bethesda, Maryland.

Funk, J. L. 1974. Symposium on overharvest and management of largemouth bass in small impoundments. American Fisheries Society, North Central Division, Special Publication 3, Bethesda, Maryland.

Gabelhouse, D. W., Jr. 1978. Redear sunfish for small impoundments? Pages 109–123 in G. D. Novinger and J. D. Dillard, editors. New approaches to the management of small impoundments. American Fisheries Society, North Central Division, Special Publication 5, Bethesda, Maryland.

Gabelhouse, D. W., Jr. 1984a. An assessment of crappie stocks in small Midwestern private impoundments. North American Journal of Fisheries Management 4:371–384.

Gabelhouse, D. W., Jr. 1984b. A length-categorization system to assess fish stocks. North American Journal of Fisheries Management 4:273–285.

Gabelhouse, D. W., Jr. 1987. Responses of largemouth bass and bluegills to removal of surplus largemouth bass from a Kansas pond. North American Journal of Fisheries Management 7:81–90.

Gabelhouse, D. W., Jr., R. . Hager, and H. E. Klaassen. 1982. Producing fish and wildlife from Kansas ponds. Kansas Fish and Game Commission, Pratt.

Gabelhouse, D. W., Jr., and D. W. Willis. 1986. Biases and utility of angler catch data for assessing size structure and density of largemouth bass. North American Journal of Fisheries Management 6:481–489.

Garrett, G. P. 1989. Hormonal sex control of largemouth bass. The Progressive Fish-Culturist 51:146–148.

Garrett, G. P. 2002. Behavioral modification of angling vulnerability in largemouth bass through selective breeding. Pages 387–392 in D. P. Philipp and M. S. Ridgway, editors. Black bass: ecology, conservation, and management. American Fisheries Society, Symposium 31, Bethesda, Maryland.

Gerow, K. G., R. C. Anderson-Sprecher, and W. A. Hubert. 2005. A new method to compute standard-weight equations that reduces length-related bias. North American Journal of Fisheries Management 25:1288–1300.

Gerow, K. G., W. A. Hubert, and R. C. Anderson-Sprecher. 2004. An alternative approach to detection of length-related biases in standard weight equations. North American Journal of Fisheries Management 24:903–910.

Green, D. M., E. L. Mills, and D. J. Decker. 1993. Participatory learning in natural resource education: a pilot study in private fishery management. Journal of Extension 31:13–15.

Gurtin, S. D., M. L. Brown, and C. G. Scalet. 1996. Dynamics of sympatric northern pike and largemouth bass populations in small prairie impoundments. Pages 64–72 in R. Soderberg, editor. Warmwater workshop proceedings: esocid management and culture. American Fisheries Society, Northeast Division, Bethesda, Maryland.

Gustafson, K. A. 1988. Approximating confidence intervals for indices of fish population size structure. North American Journal of Fisheries Management 8:139–141.

Guy, C. S., R. M. Neumann, D. W. Willis, and R. O. Anderson. 2007. Proportional size distribution (PSD): a further refinement of population size structure index terminology. Fisheries 32(7):348.

Guy, C. S., and D. W. Willis. 1991. Evaluation of largemouth bass–yellow perch communities in small South Dakota impoundments. North American Journal of Fisheries Management 11:43–49.

Hackney, P. A. 1978. Fish community biomass relationships. Pages 25–36 in G. D. Novinger and J. D. Dillard, editors. New approaches to the management of small impoundments. American Fisheries Society, North Central Division, Special Publication 5, Bethesda, Maryland.

Hill, K. R. 1986. Mortality and standing stocks of grass carp planted in two Iowa lakes. North American Journal of Fisheries Management 6:449–451.

Hill, T. D., and D. W. Willis. 1993. Largemouth bass biomass, density, and size structure in small South Dakota impoundments. Proceedings of the South Dakota Academy of Science 72:31–39.

Hubbs, C. L., and E. R. Allen. 1943. Fishes of Silver Springs, Florida. Proceedings of the Florida Academy of Science 6:110–130.

Hubert, W. A. 1996. Passive capture techniques. Pages 157–181 in B. R. Murphy and D. W. Willis, editors. Fisheries techniques, 2nd edition. American Fisheries Society, Bethesda, Maryland.

Hunt, K. M., H. L. Schramm Jr., T. J. Lang, J. W. Neal, and C. P. Hutt. 2008. Status of urban and community fishing programs nationwide. Pages 177–202 in R. T. Eades, J. W. Neal, T. J. Lang, K. M. Hunt, and P. Pajak, editors. Urban and community fisheries programs: development, management, and evaluation. American Fisheries Society, Symposium 67, Bethesda, Maryland.

Iowa Department of Natural Resources. 2007. Working for clean water: 2007 watershed improvement successes in Iowa. Iowa Department of Natural Resources, Des Moines.

Irwin, B. J., D. R. DeVries, and R. A. Wright. 2003. Evaluating the potential for predatory control of gizzard shad by largemouth bass in small impoundments: a bioenergetics approach. Transactions of the American Fisheries Society 132:913–924.

Isaak, D. J., T. D. Hill, and D. W. Willis. 1992. Comparison of size structure and catch rate for largemouth bass samples collected by electrofishing and angling. Prairie Naturalist 24:89–96.

Jacobson, P. C. 2005. Experimental analysis of a reduced daily bluegill limit in Minnesota. North American Journal of Fisheries Management 25:203–210.

Jennings, M. J., J. E. Claussen, and D. P. Philipp. 1997. Effect of population size structure on reproductive investment of male bluegill. North American Journal of Fisheries Management 17:516–525.

Johnson, L. 1966. Consumption of food by the resident population of pike *Esox lucius* in Lake Windermere. Journal of the Fisheries Research Board of Canada 23:1523–1535.

Jolley, J. C., D. W. Willis, T. J. DeBates, and D. D. Graham. 2008. The effects of mechanically reducing northern pike density on the sport fish community of West Long Lake, Nebraska, USA. Fisheries Management and Ecology 15:251–258.

Jones, M. L., and J. M. Dettmers. 2007. Making wise decisions about transferring fish among lakes within the Great Lakes basin. Journal of Great Lakes Research 33:930–934.

Kerby, J. H., J. M. Everson, R. M. Harrell, J. G. Geiger, C. C. Starling, and H. Revels. 2002. Perfor-

mance comparisons between diploid and triploid sunshine bass in freshwater ponds. Aquaculture 211:91–108.
Kim, G. W., and D. R. DeVries. 2000. Effects of a gizzard shad reduction on trophic interactions and age-0 fishes in Walker County Lake, Alabama. North American Journal of Fisheries Management 20:860–872.
Lane, R. L., and J. E. Morris. 2002. Comparison of prepared feed versus natural food ingestion between pond-cultured bluegill and hybrid sunfish. Journal of the World Aquaculture Society 33:517–519.
Lang, T. J. 2007. Impacts of stocking frequency on fishing quality in the Arkansas urban fishing program and evaluation of Arkansas fishing derby events. Master's thesis, University of Arkansas–Pine Bluff.
Lang, T. J., J. W. Neal, and C. P. Hutt. 2008. Stocking frequency and fishing quality in an urban fishing program in Arkansas. Pages 379–389 *in* R. Eades, W. Neal, T. Lang, K. Hunt, and P. Pajak, editors. Urban and community fisheries programs: development, management, and evaluation. American Fisheries Society, Symposium 67, Bethesda, Maryland.
Lewis, W. M., and R. C. Heidinger. 1971. Supplemental feeding of hybrid sunfish populations. Transactions of the American Fisheries Society 100:619–623.
Lewis, W. M., and R. C. Heidinger. 1978. Use of hybrid sunfishes in the management of small impoundments. Pages 104–108 *in* G. D. Novinger and J. G. Dillard, editors. New approaches to the management of small impoundments. American Fisheries Society, North Central Division, Special Publication 5, Bethesda, Maryland.
Lusk, B., M. Otto, and M. McDonald. 2007. Perfect pond ... want one? Pond Boss, Sadler, Texas.
Malvestuto, S. P. 1996. Sampling the recreational creel. Pages 591–623 *in* B. R. Murphy and D. W. Willis, editors. Fisheries techniques, 2nd edition. American Fisheries Society, Bethesda, Maryland.
Marriage, L. D., A. E. Borell, and P. M. Scheffer. 1971. Trout ponds for recreation. U.S. Department of Agriculture, Farmers' Bulletin 2249, Washington, D.C.
Masser, M. P., T. R. Murphy, and J. L. Shelton. 2006. Aquatic weed management: herbicides. Texas A&M University, Southern Regional Aquaculture Center Publication 361, College Station.
Modde, T. 1980. State stocking policies for small warmwater impoundments. Fisheries 5(5):13–17.
Nail, M. L., and D. H. Powell. 1975. Observations on supplemental feeding of a 75-acre lake stocked with largemouth bass, bluegill, redear, and channel catfish. Proceedings of the Annual Conference Southeastern Association of Game and Fish Commissioners 28(1974):378–384.
National Research Council. 1993. Nutrient requirements of fish. National Academy Press, Washington, D.C.
Neal, J. W., R. L. Noble, and J. A. Rice. 1999. Fish community response to hybrid striped bass introduction in small warmwater impoundments. North American Journal of Fisheries Management 19:1044–1053.
Nelson, J. S., E. J. Crossman, H. Espinosa-Perez, L. T. Findley, C. R. Gilbert, R. N. Lea, and J. D. Williams. 2004. Common and scientific names of fishes from the United States, Canada, and Mexico, 6th edition. American Fisheries Society, Special Publication 29, Bethesda, Maryland.
Neumann, R. M., C. S. Guy, and D. W. Willis. In press. Length, weight, and associated structural indices. Pages 000–000 *in* A. V. Zale, D. L. Parrish, and T. M. Sutton, editors. Fisheries techniques, 3rd edition. American Fisheries Society, Bethesda, Maryland.
Neumann, R. M., D. W. Willis, and D. D. Mann. 1994. Evaluation of largemouth bass slot length limits in two small South Dakota impoundments. Prairie Naturalist 26:15–32.
Nielsen, L.A. 1999. History of inland fisheries management in North America. Pages 3–30 *in* C. C. Kohler and W. A. Hubert, editors. Inland fisheries management in North America, 2nd edition. American Fisheries Society, Bethesda, Maryland.
Novinger, G. D. 1990. Slot length limits for largemouth bass in small private impoundments. North American Journal of Fisheries Management 10:330–337.

Pajak, P. 1994. Urban outreach: fishery management's next frontier? Fisheries 19(10):6–7.

Payer, R. D., R. B. Pierce, and D. L. Pereira. 1989. Hooking mortality of walleyes caught on live and artificial baits. North American Journal of Fisheries Management 9:188–192.

Porath, M. T., and K. L. Hurley. 2005. Effects of waterbody type and management actions on bluegill growth rates. North American Journal of Fisheries Management 25:1041–1050.

Prentice, J. A., and J. W. Schlechte. 2000. Performance comparisons between coppernose and native Texas bluegill populations. Proceedings of the Annual Conference Southeastern Association of Fish and Wildlife Agencies 54:196–206.

Redmond, L. C. 1974. Prevention of overharvest of largemouth bass in Missouri impoundments. Pages 54–68 in J. L. Funk, editor. Symposium on overharvest and management of largemouth bass in small impoundments. American Fisheries Society, North Central Division, Special Publication 3, Bethesda, Maryland.

Regier, H. A. 1963. Ecology and management of largemouth bass and golden shiners in farm ponds in New York. New York Fish and Game Journal 10:139–169.

Rennie, M. D., and R. Verdon. 2008. Development and evaluation of condition indices for the lake whitefish. North American Journal of Fisheries Management 28:1270–1293.

Saffel, P. D., C. S. Guy, and D. W. Willis. 1990. Population structure of largemouth bass and black bullheads in South Dakota ponds. Prairie Naturalist 22:113–118.

Schramm, H. L., Jr., and W. A. Hubert. 1999. Ecosystem management. Pages 111–123 in C. C. Kohler and W. A. Hubert, editors. Inland fisheries management in North America, 2nd edition. American Fisheries Society, Bethesda, Maryland.

Shaner, B. L., M. J. Maceina, J J. McHugh, and S. E. Cook. 1996. Assessment of catfish stocking in public fishing lakes in Alabama. North American Journal of Fisheries Management 16:880–887.

Smith, S. V., W. H. Renwick, J. D. Bartley, and R. W. Buddemeier. 2002. Distribution and significance of small, artificial water bodies across the United States landscape. The Science of the Total Environment 299(1–3):21–36.

Storck, T., and D. Newman. 1988. Effects of size at stocking on survival and harvest of channel catfish. North American Journal of Fisheries Management 8:98–101.

Swingle, H. S. 1950. Relationships and dynamics of balanced and unbalanced fish populations. Alabama Polytechnic Institute, Agricultural Experiment Station Bulletin 274, Auburn.

Swingle, H. S. 1952. Farm pond investigations in Alabama. Journal of Wildlife Management 16:243–249.

Swingle, H. S. 1956. Appraisal of methods of fish population study, part 4: determination of balance in farm fish ponds. Transactions of the North American Wildlife Conference 21:298–322.

Swingle, H. S., B. C. Gooch, and H. R. Rabanal. 1965. Phosphate fertilization of ponds. Proceedings of the Annual Conference Southeastern Association of Game and Fish Commissioners 17(1963):213–218.

Swingle, H. S., and E. V. Smith. 1938. Fertilizers for increasing the natural food for fish in ponds. Transactions of the American Fisheries Society 68:126–135.

USDI (U.S. Department of the Interior). 1993. 1991 national survey of fishing, hunting, and wildlife-associated recreation.. U.S. Department of the Interior, Fish and Wildlife Service, and U.S. Department of Commerce, Census Bureau, Washington, D.C.

USDI (U.S. Department of the Interior. 2007. 2006 national survey of fishing, hunting, and wildlife-associated recreation. U.S. Department of the Interior, Fish and Wildlife Service, and U.S. Department of Commerce, Census Bureau, Washington, D.C.

Wege, G. J., and R. O. Anderson. 1978. Relative weight (W_r): a new index of condition for largemouth bass. Pages 79–91 in G. D. Novinger and J. G. Dillard, editors. New approaches to the management of small impoundments. American Fisheries Society, North Central Division. Special Publication 5, Bethesda, Maryland.

Weithman, S. A., J. B. Reynolds, and D. E. Simpson. 1980. Assessment of structure of largemouth bass stocks by sequential sampling. Proceedings of the Annual Conference Southeastern Association of Fish and Wildlife Agencies 33(1979):415–424.

Wiley, M. J., R. W. Gordon, S. W. Waite, and T. Powless. 1984. The relationship between aquatic macrophytes and sport fish production in Illinois ponds: a simple model. North American Journal of Fisheries Management 4:111–119.

Willis, D. W., M. D. Beem, and R. L. Hanten. 1990. Managing South Dakota ponds for fish and wildlife. South Dakota Department of Game, Fish and Parks, Pierre.

Willis, D. W., C. G. Scalet, and L. D. Flake. 2009. Introduction to wildlife and fisheries: an integrated approach, 2nd edition. W. H. Freeman and Company, New York.

Chapter 17

Large Reservoirs

LEANDRO E. MIRANDA AND PHILLIP W. BETTOLI

17.1 INTRODUCTION

Large impoundments, defined as those with surface area of 200 ha or greater, are relatively new aquatic ecosystems in the global landscape. They represent important economic and environmental resources that provide benefits such as flood control, hydropower generation, navigation, water supply, commercial and recreational fisheries, and various other recreational and esthetic values. Construction of large impoundments was initially driven by economic needs, and ecological consequences received little consideration. However, in recent decades environmental issues have come to the forefront. In the closing decades of the 20th century societal values began to shift, especially in the developed world. Society is no longer willing to accept environmental damage as an inevitable consequence of human development, and it is now recognized that continued environmental degradation is unsustainable. Consequently, construction of large reservoirs has virtually stopped in North America. Nevertheless, in other parts of the world construction of large reservoirs continues.

The emergence of systematic reservoir management in the early 20th century was guided by concepts developed for natural lakes (Miranda 1996). However, we now recognize that reservoirs are different and that reservoirs are not independent aquatic systems inasmuch as they are connected to upstream rivers and streams, the downstream river, other reservoirs in the basin, and the watershed. Reservoir systems exhibit longitudinal patterns both within and among reservoirs. Reservoirs are typically arranged sequentially as elements of an interacting network, filter water collected throughout their watersheds, and form a mosaic of predictable patterns.

Traditional approaches to fisheries management such as stocking, regulating harvest, and in-lake habitat management do not always produce desired effects in reservoirs. As a result, managers may expend resources with little benefit to either fish or fishing. Some locally expressed effects, such as turbidity and water quality, zooplankton density and size composition, or fish growth rates and assemblage composition, are the upshot of large-scale factors operating outside reservoirs and not under the direct control of reservoir managers. Realistically, abiotic and biotic conditions in reservoirs are shaped by factors working inside and outside reservoirs, with the relative importance of external factors differing among reservoirs.

With this perspective, large reservoirs are viewed from a habitat standpoint within the framework of a conceptual model in which individual reservoir characteristics are influenced by both local- and landscape-scale factors (Figure 17.1). In the sections that follow, how

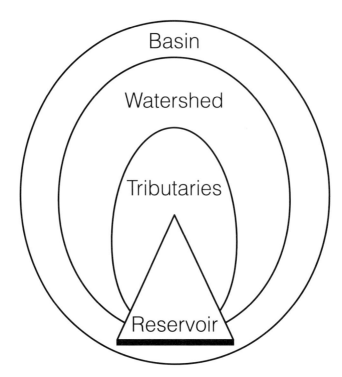

Figure 17.1. Abiotic and biotic conditions in reservoirs are shaped by factors working inside (section 17.3) and outside the reservoir (tributaries, section 17.4; watershed, section 17.5; and basin, section 17.6), with their relative importance differing among reservoirs.

each element of this hierarchical model influences habitat and fish assemblages in reservoirs is considered. Important in-reservoir habitat issues and reservoirs as part of larger systems, where reservoir management requires looking for real solutions outside individual reservoirs are described

17.2 THE DIVERSITY OF RESERVOIRS

The geographic distribution of reservoirs reflects a complex interaction among topography, climate, and economic needs to control water passage through river basins. Some large river systems have been transformed into chains of reservoirs by stacking reservoirs in sequence. Altogether, there are over 2,000 large reservoirs in the USA, and most are described as multipurpose (NID 2008). The uses of these reservoirs include hydropower (26.9%), irrigation (16.4%), water supply (16.3%), recreation (15.7%), flood control (13.0%), navigation (6.3%), or other uses (5.4%). Canada boasts several of the world's largest reservoirs, particularly in the eastern provinces of Quebec (e.g., Manicouagan Reservoir; 195,000 ha), Newfoundland, and Labrador (e.g., Smallwood Reservoir; 652,700 ha). Most (78%) of the large dams built in Canada have been constructed for hydropower (Prowse et al. 2004); dams built to aid navigation or control flooding are far less common in Canada than they are in the USA. The James Bay Project in northwestern Quebec is one of the largest hydroelectric projects in the world. In fact, hydroelectric dams meet nearly all of Quebec's electrical needs, and they often produce electricity to export to the USA. In the United Mexican States (Mexico), from the arid north to the subtropical south, the Mexican government has built a patchwork of dams, primarily for irrigation and power generation, that supply nearly 30% of the energy flowing on the nation's electric power grid (Robinson 2000).

Marked patterns in the geographic distribution of large reservoirs in North America are evident. In general, hydropower reservoirs are regionalized in mountainous areas (e.g., the western slopes of the Rocky Mountains, Cascade Mountains, and Sierra Nevada) and along the edge of the Laurentian Plateau (i.e., the Canadian Shield). Precipitation and topography in these areas provide ideal inflows and required head conditions to generate hydropower. Irrigation reservoirs in the USA are common in drier areas of the central plains and the western coastal valleys. Irrigation reservoirs in the central plains have low-relief topography and are often shallow and turbid owing to the effects of wind action and inputs from highly-erodible soils. Demands for drinking and industrial water in densely populated areas have led to the development of water supply reservoirs throughout the arid central and populated eastern USA and near urban centers in western coastal regions. Flood control reservoirs have been constructed in areas where strong, short-duration rainfall events and allied runoff are common, especially in the central USA. Maintaining navigable channels for transportation has encouraged the impoundment of low-gradient rivers to retain passage throughout most of the year in the western (e.g., Columbia and Snake rivers) and central USA (e.g., Mississippi, Missouri, and Tennessee rivers). Normally, reservoirs in the USA constructed for navigation and hydropower tend to be very large due to their location on large rivers. Some large hydroelectric reservoirs in Canada have been created not by a single dam on a large river, but by damming the outflow of natural lakes and impounding other lakes in the watershed (e.g., Caniapiscau Reservoir in Quebec). Other large Canadian hydroelectric reservoirs have been created by building dikes to flood numerous lakes in a watershed to create a single, large reservoir (e.g., 88 dikes spanning 64 km created Smallwood Reservoir in Labrador).

A reservoir's purpose has much to do with how its water regime is operated (Kennedy 1999). Hydropower and navigation uses require that reservoirs be maintained near 100% of the storage capacity and no less than about 70%. Hydropower is most efficient at maximum hydraulic head, whereas navigation, which often occurs in shallow channels, requires stable water levels. Water level fluctuations are often minimal and may exhibit diel patterns. Although there is great variability, reservoirs with their main purpose being water supply for agriculture are generally maintained between 40% and 80% of their maximum capacity. Water level often varies on a scale of years, as the reservoir may be used to store water during wet years to compensate for dry years. Conversely, flood control reservoirs experience seasonal drawdowns to 10–30% of capacity due to their requirement to store excessive runoff. Water levels in flood control reservoirs generally follow sharp annual patterns.

Large impoundments may also be classified as tributary storage, main-stem storage, or run-of-river reservoirs (Table 17.1). Tributary storage reservoirs are generally located on low-order, high-gradient streams and are often small to moderate in surface area because they have relatively small watersheds. Main-stem storage reservoirs provide moderate to large storage capacity and are impounded on mid-order rivers. Because they inundate more of the floodplain than do tributary storage reservoirs, main-stem storage reservoirs tend to have more shallow littoral areas. Run-of-river reservoirs often occur in series over hundreds of kilometers and are most common on large rivers; therefore, run-of-river reservoirs are typically shallow with low retention time. These reservoirs tend to be long and narrow, with minimal lateral expansion. Tributary storage reservoirs are generally deep relative to their area, presenting vertical abi-

Table 17.1. Physical and operational characteristics of tributary storage, main-stem storage, and run-of-river reservoirs. Peaking hydropower differs from baseline hydropower in that water is released (and electricity is generated) for short periods of time when demand on the regional power grid peaks, typically in the morning and early evening. Peaking hydropower is also available to supply electricity quickly to a power grid in the event of a problem or emergency shutdown at a baseline power production facility (e.g., a coal-fired plant or nuclear facility).

	Reservoir type		
Attribute	Tributary storage	Main-stem storage	Run-of-river
Primary uses	Peaking hydropower flood control	Navigation, hydropower flood control	Navigation, hydropower flood control
Basin morphology	Steep banks; deep basin; small littoral zone	Expansive overbank (littoral areas); relatively shallow	Long and narrow
Typical water retention time	Months or years	Weeks	Days
Water level fluctuations	Large	Modest	Slight

otic and biotic diversity. In contrast, run-of-river reservoirs can be much larger in surface area (although perhaps not volume), tend to be shallow relative to their area, and show longitudinal diversity. Main-stem storage reservoirs are intermediate in diversity relative to vertical and longitudinal gradients. It is important to know why a particular reservoir was built and how it is operated because fish habitat and primary production are profoundly influenced by reservoir operations. Also, it is important to understand that promoting recreational or commercial fisheries is usually not the reason why large reservoir projects were proposed, funded, and built. This fact implies that reservoir fisheries management usually occurs within constraints imposed by uses that take priority over fisheries.

As much as possible, considering the vagaries of precipitation, reservoir operators follow pre-established guide curves, which dictate what the water level should be on each day of the year (Figure 17.2). Guide curves are formal (often legal) descriptions of where the water level in a given reservoir should be on any given day of the year to meet the reservoir's primary purposes (e.g., flood control, power generation, or navigation) within the constraints posed by droughts, floods, and other unforeseen circumstances. Regional authorities or regulatory bodies such as the U.S. Army Corps of Engineers, U.S. Bureau of Reclamation, and Tennessee Valley Authority (TVA) develop guide curves to optimize water storage benefits among reservoirs in a river basin. For instance, the TVA manages 41 reservoirs on the main-stem Tennessee River and its numerous tributaries, and the operations of each dam affect the operations of other dams in that system.

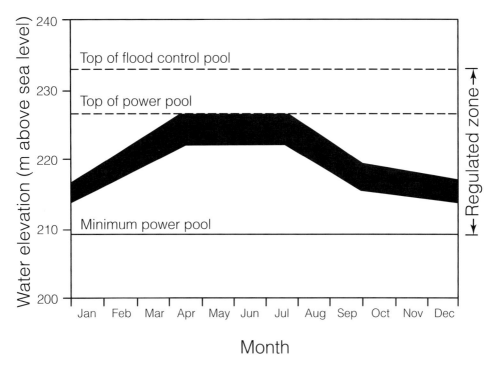

Figure 17.2. Guide curve for a hypothetical tributary storage impoundment. The shaded polygon is the guide curve, and it represents the range of water elevations that are targeted on each day of the year and is established to meet local and regional demands for hydropower, flood control, navigation, and recreation. Extreme floods or droughts can move actual reservoir elevations outside the guide curve.

The geographical, hydrological, and morphological diversity of reservoirs has created a diversity of fish assemblages and fisheries. Most fishes that flourish in reservoirs are generalists and have wide native distributions. Salmon, important to western reservoirs (mainly coho salmon, sockeye salmon, and Chinook salmon), are native to the western states. Trout are also important to western reservoirs and have been introduced into selected deep eastern U.S. reservoirs. Native trout in western reservoirs include mainly cutthroat trout and rainbow trout; introduced trout include brook trout (native to the central and eastern USA), lake trout (native to northern North America), and brown trout (native to Europe). Other major fishes in reservoirs are native to the central and eastern parts of North America but have been introduced into many western reservoirs. In fact, reservoir fisheries in the southwest are maintained mainly or entirely by species introduced from east of the Rocky Mountains. Two species, northern pike and muskellunge, and their hybrid (tiger muskellunge) support most of the esocid fisheries in reservoirs, but these species are mostly absent from southern U.S. reservoirs. Percids, which include walleye and sauger, that are native to the central states west of the Appalachian Mountains and their hybrid, the saugeye, have been introduced into many Midwest reservoirs. Another percid is the yellow perch, which is native to the northern portions of North America, although native populations can also be found east of the Appalachian Mountains and in some southern reservoirs. Catfishes include the *Ameiurus* species that are native east of the Rocky Mountains, and the *Ictalurus* and *Pylodictis* catfishes that are native to central North America west of the Appalachian Mountains and east of the Rocky

Mountains, although they have been introduced elsewhere. Of four *Morone* species, in general two are limited to central drainages (white bass and yellow bass), one to Atlantic drainages (white perch), and one to Atlantic drainages and Gulf of Mexico drainages east of the Mississippi River (striped bass), although striped bass and white bass have been introduced in the west coast. Both hybrid crosses between the striped bass and white bass (palmetto bass and sunshine bass) are commonly introduced into southern reservoirs. Of the centrarchids, several species of sunfishes, three species of black basses, and both species of crappies contribute to reservoir fisheries in North America. Distributions of centrarchid species appear to be influenced largely by latitude, the Rocky Mountains, and the Appalachian Mountains. Representation of these major groups of fishes in reservoir fisheries vary geographically as well as within geographical regions due to peculiarities associated with the different types of reservoirs (Miranda 1999).

Considering reservoir fish assemblages from a trophic perspective, fishes in North American reservoirs are often organized into herbivore, detritivore, planktivore, invertivore, and carnivore guilds. Many species operate in multiple trophic guilds as their niche shifts over life history stages or their diets shift due to changes in prey availability. Representation by these groups varies greatly among reservoirs depending on abiotic characters of reservoirs (Miranda et al. 2008). In warmwater reservoirs the most abundant species in term of biomass are usually filter-feeding herbivores–detritivores (e.g., gizzard shad) and planktivores (e.g., threadfin shad). Other guilds common to warmwater reservoirs include invertivores–carnivores (e.g., black basses, crappies, and catfishes), invertivores (e.g., sunfishes and minnows), invertivores–detritivores (e.g., common carp), invertivores–herbivores (e.g., buffaloes), planktivores, carnivores (e.g., gars), and detritivores–planktivores (e.g., carpsuckers).

17.3 THE RESERVOIR

In-lake characteristics of reservoirs exert major influences on fish assemblages and fisheries. Many chemical, physical, and biological factors characterize reservoirs, and their relative importance varies geographically. Some of the most universal influences include suspended sediments and sedimentation, nutrients and water quality, water retention and fluctuation, and submerged structure and vegetation. Moreover, many of these factors show marked longitudinal variation along the length of a reservoir.

17.3.1 Suspended Sediments and Sedimentation

Suspended sediments enter reservoirs principally through tributaries but also through overland runoff from their watersheds. Sedimentation typically occurs in natural lakes on a geologic time scale, but it occurs much faster (i.e., over several decades) in reservoirs. This effect is due to differences in incoming volumes of water caused by differences in watershed sizes—given a lake and a reservoir of similar size, the watershed for the reservoir will almost always be larger. That is, the ratio of watershed area to water surface area will usually be higher (sometimes several times higher) for reservoirs because reservoirs are designed to capture as much water as possible by means of the smallest possible dams.

Reservoirs are efficient sediment traps. As sediments settle, they alter the surface of coarse rock substrates into a homogenous surface of fine silt and clay particles. In fact, depending on the geology of the watershed and basin morphology, the useful lifespan of tributary storage reservoirs (where retention times are high) may be measured in decades or a few centuries because of sediment accumulation. Extreme examples of sedimentation problems in reservoirs can be found in Africa, Australia, and Puerto Rico, where humid tropical environments and highly erodible soils promote high sedimentation rates and storage capacity losses exceeding 1% per year. Sedimentation is a major concern of reservoir fisheries managers because lithophilic fishes (i.e., those that prefer rock substrates for spawning or feeding) will eventually decline in abundance and be replaced by species with broader habitat requirements. The "carpeting" of reservoir substrates by layers of fine sediment eliminates substrates for invertebrates and periphyton, both of which are important biotic components of reservoir food webs. Of all the processes regulating reservoir aging, sedimentation is perhaps the most dominant.

As reservoirs age, sedimentation and shoreline erosion produce a loss of littoral habitat. Within a few years after impoundment, aging is apparent in the littoral zones (Agostinho et al. 1999). Reservoir aging received scant attention from fisheries managers before the 1970s, perhaps because the construction of new reservoirs proceeded at a rapid pace between the 1920s and 1960s, and new reservoirs, along with "boom" fisheries, were constantly coming on line. The 1950s and 1960s were considered the "golden age of dam building" in the USA (Doyle et al. 2003). Robert Jenkins, an early pioneer of reservoir ecology, was one of the first to note that sport fish harvest was inversely related to reservoir age (Jenkins 1967). In many parts of North America it was easy for anglers dissatisfied with the fishing in an "old" reservoir experiencing trophic depression to travel to a new nearby reservoir to exploit fisheries in the boom phase associated with trophic upsurge (see section 17.3.2). However, construction of large reservoirs in the USA declined precipitously in the 1970s because dams had been constructed at most prime reservoir construction sites. Although new, large reservoir projects in the USA are now relatively rare, construction of large reservoirs in other parts of the world is proceeding at a rapid pace, particularly in developing countries such as China and Turkey (World Wildlife Fund 2004), as well as in remote regions of Canada (e.g., northern Quebec, Manitoba, and Northwest Territories), which may be experiencing a golden age of large reservoir projects (Prowse et al. 2004). Mexico began building dams in the late 1930s and continued at a fast pace until the 1980s, but large dams are still being constructed in that country. Given the current prevalence of aging reservoirs in the USA (and in the future in Canada and Mexico), fisheries managers can no longer rely on new reservoirs to satisfy the demand for quality fishing. Instead, managers will have to rely on innovative ways to manage the habitat, the fish assemblage, and the fisheries in aging reservoirs.

Sediments and suspended fine materials can have a major influence on fish assemblages that develop in reservoirs or individual embayments of reservoirs. Suspended fine materials can limit light penetration and photosynthesis, diminish plant biomass, alter zooplankton assemblages, reduce visibility, reduce fish growth, decrease fish size at sexual maturity, limit maximum fish size, and produce a shift in habitat use by fishes (Bruton 1985). The influence of suspended fine materials entering reservoirs is exacerbated by loss of depth from sedimentation. Loss of depth encourages resuspension of sediments through wave action (Hamilton and Lewis 1990) as well as through stirring action by benthivorous fishes searching for food (Scheffer 2001). Decreases in water clarity driven by suspended fine materials interfere with the feeding of large zooplankton but not of smaller zooplankton such as rotifers (Kirk and Gilbert 1990), favoring dominance by small zooplankton and fish that feed on small zooplankton. Excessive sedimentation and suspended fine materials also reduce benthic production, which is reflected by diminished representation of fishes that depend on benthic plants or invertebrates.

Foraging by visual piscivores is limited in turbid reservoirs, such that their representation in fish assemblages declines. However, tactile, nonvisual species that forage by ingesting sediment often thrive, including common carp, adult gizzard shad, some catfishes, and buffaloes. Turbid environments may also allow prey fishes to expand because turbidity decreases vulnerability to predation. In advanced stages of sedimentation, fish assemblages in reservoirs generally include few predators and many species that thrive in turbid, shallow systems.

17.3.2 Nutrients and Water Quality

Another major factor influencing reservoir function is nutrient loading (i.e., the amount of nutrients entering the water body). Kimmel and Groeger (1986) were among the first to describe in limnological terms what fisheries managers had long known, namely, that the quality of sport fisheries and overall fish production are initially high after dams are closed but decline after a decade or so. Part of the decline can be attributed to loss of high-quality habitats due to sedimentation, but much of the decline can be attributed to changes in internal nutrient loading. Internal nutrient loading refers to organic detritus and inorganic nutrients liberated following inundation of soils and terrestrial vegetation. Conversely, external nutrient loading refers to input of nutrients into reservoirs that originated from outside reservoirs; that is, externally loaded nutrients are carried downstream into reservoirs by tributary streams or overland from surrounding watersheds. Both forms of nutrient loading are important, but it is the internal nutrient loading flux that drives the characteristic boom–bust fisheries among new reservoirs. As depicted in Figure 17.3, rates of external nutrient loading remain essentially unchanged in the absence of human-induced changes in watersheds. Changes in the rates of internal nutrient loading assume overarching significance to fisheries managers because of the tight linkage between nutrient concentrations (principally nitrogen and phosphorous), fish biomass, and production.

The limnology of reservoirs is greatly affected by trophic state, which ultimately refers to the amount of primary production in a reservoir during a particular time, usually summer. Primary production can be represented by rooted or floating macrophytes or, more commonly, microscopic algae in the water column (i.e., phytoplankton). Reservoirs fall along a trophic state continuum ranging from unproductive (oligotrophic) to moderately productive (mesotrophic) to extremely productive (eutrophic or hypereutrophic). Trophic state refers to primary production but also provides insight into how productive a particular water body will be in terms of fish. Classifications of reservoir trophic state diverge from lake-based classification schemes that rely on the presence or absence of oxygen in the hypolimnion. For instance, it is not uncommon for oligotrophic reservoirs to have depleted concentrations of dissolved oxygen in the hypolimnion (which indicates eutrophy in natural lakes) because they experience high rates of nutrient and sediment loading.

Several indices of trophic state based on readily obtainable water quality data are used to describe the trophic state of reservoirs. Two of the most widely used are those by Carlson (1977) and Forsberg and Ryding (1980). Both indices assign trophic states according to the algal biomass present during summer (indexed by chlorophyll-a biomass), the concentrations of key nutrients (phosphorous and nitrogen), and water transparency as measured with a Secchi disk (Table 17.2). Chlorophyll a is a photosynthetic pigment common to all photosynthetic organisms in freshwater (e.g., green algae, diatoms, and cyanobacteria); thus, high levels of chlorophyll a (the most common form of chlorophyll) in a filtered water sample generally equates to high standing crops of algae.

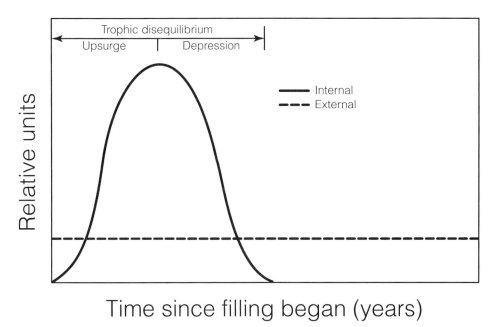

Figure 17.3. The "boom-and-bust" aspect of most reservoir fisheries is driven in large measure by fluxes in nutrient loading; specifically, internal nutrient loading. Rates of primary production and fish production will lag slightly behind the pulse in internal nutrient loading. External nutrient loading rates will remain unchanged in the absence of any cultural eutrophication activities (Adapted from Kimmel and Groeger 1986).

Table 17.2. Trophic state classification scheme proposed by Forsberg and Ryding (1980). Nitrogen (N), phosphorous (P), and chlorophyll-a concentrations are expressed in mg/m^3; water transparency is expressed as Secchi disk depth in m. To account for chlorophyll and nutrients sequestered in aquatic macrophytes, see Canfield et al. (1983).

Trophic state	Total N	Total P	Chlorophyll a	Transparency
Oligotrophic	<400	<15	<3	> 4.0
Mesotrophic	400–600	15–25	3–7	2.5–4.0
Eutrophic	600–1,500	25–100	7–40	1.0–2.5

Hypolimnetic oxygen concentrations are not used to assign reservoirs to a particular trophic state; nevertheless, the depiction of dissolved oxygen (DO) and water temperature in a vertical profile tells a great deal about a reservoir's trophic state. In particular, DO–temperature profiles show what habitat (especially offshore) is available to fishes with differing DO and thermal preferences and tolerances. Clinograde DO profiles (Figure 17.4, panel A) are common in eutrophic reservoirs and result from high DO consumption in the hypolimnion due to excessive biological oxygen demand (e.g., algal respiration and decomposition of organic matter), chemical oxygen demand (e.g., anoxic groundwater containing reduced chemicals such as Fe^{+2} will lower DO concentrations), or both processes. Heterograde curves can be either positive or negative depending on whether or not the metalimnion, with its higher phytoplankton concentration, is in or out of the photic zone. The metalimnion (also known as the thermocline) serves as a barrier to mixing between the warm epilimnion and colder hypolimnion; differences in density of water at different temperatures slow the settling of phytoplankton and leads to their accumulation in the metalimnion. The photic zone refers to the layer of water where photosynthesis can occur; it extends from the surface down to a depth where light intensity is approximately 1% of that at the water's surface. In Dale Hollow Reservoir (Figure 17.4, panel B) water clarity is sufficient to allow photosynthesis as deep as 10 m (and oxygen is generated by photosynthetic activity in the metalimnion). However, in Center Hill Reservoir (Figure 17.4, panel C) phytoplankton accumulated in the metalimnion cannot photosynthesize because too little light reaches that depth; thus, oxygen is consumed via bacterial decomposition and algal respiration. It is common in mesotrophic reservoirs for zones of hypoxia to extend up from the bottom and merge with hypoxic metalimnetic waters in late summer and early fall (before fall overturn), resulting in clinograde DO profiles. In riverine reservoirs with short retention times turbulent flows prevent strong stratification. The water column remains mixed throughout the year, so water temperatures are nearly isothermal and DO concentrations usually vary little from top to bottom.

The amount of DO available in the water column does not fully describe how much (or where) habitat is available to reservoir fishes. Being poikilothermic, all fishes exhibit preferences for a particular range of water temperatures that is mediated by numerous factors such as fish size, genetics, and acclimation state. The thermal acclimation state of a fish refers to the physiological status it assumes in response to the thermal environment it inhabits. The environment (e.g., presence of structure) also mediates temperature selection (Bevelhimer 1996). Eurythermic species such as centrarchids are capable of occupying waters with broad ranges of temperatures and DO concentrations relative to stenothermic species such as salmonids. Nevertheless, all fish species will perform best bioenergetically at particular temperatures and with some minimum DO. When habitats with optimal combinations of temperatures and DO are not present or availability is limited, fish will experience a "temperature–DO squeeze," a long-recognized concept in reservoir fisheries management (Coutant 1985). To visualize how much habitat is available to coolwater fishes such as striped bass, a profile of temperatures in summer (when habitat is most limited) can be overlain onto a DO profile (Figure 17.5). For example, Zale et al. (1990) noted that striped bass in an Oklahoma reservoir selected the coolest water where DO concentrations were at least 2 mg/L; if the temperature in that layer was 27–28°C for more than 1 month, fish would die (at warmer temperatures they would die sooner).

The temperature–DO squeeze phenomenon is not limited to coldwater or coolwater species. Hale (1999) observed the occurrence of late-summer growth depression of crappies that

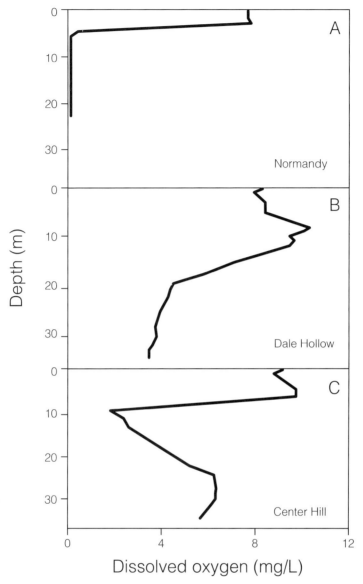

Figure 17.4. Midsummer dissolved oxygen profiles for three middle Tennessee reservoirs exhibiting (**A**) a clinograde curve, (**B**) a positive heterograde curve, and (**C**) a negative heterograde curve. Note that panels (B) and (C) also reveal the presence of oxygen-consuming processes in the hypolimnion.

were forced by low DO concentrations to inhabit water too warm for growth in a Kentucky reservoir. The problems experienced by crappies and other species forced to inhabit waters that are either too warm or too cold can be explained with bioenergetics principles (e.g., Hayward and Arnold 1996). The thermal preferences and tolerances of most freshwater fish species that occur in reservoirs have been described. Conversely, the specific DO requirements of fishes are not understood nearly as well as their temperature requirements; however, U.S. Environmental Protection Agency and state or provincial water quality standards for DO are typically around 5 or 6 mg/L. All fish species can tolerate low DO concentrations for short

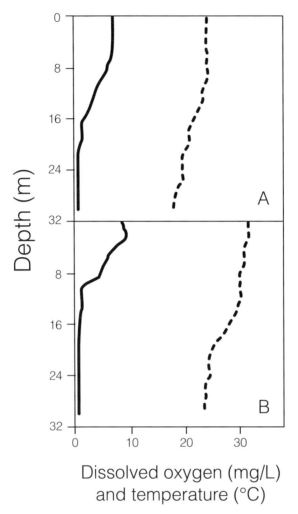

Figure 17.5. Dissolved oxygen (solid line) and temperature (dashed line) profiles in the forebay of Lake Texoma, Oklahoma–Texas, 2003, a eutrophic reservoir with a naturally-reproducing population of striped bass. Good habitat for large (>5 kg) striped bass (defined as water cooler than 25°C with at least 2 mg DO/L) is restricted to the upper 15 m of the water column in June **(A)**. By August **(B)**, the coolest water with at least 2 mg DO/L is 27.8°C; thus, there is no good habitat for large striped bass in the reservoir forebay by midsummer. Reservoirs with similar water quality can support exceptional fisheries for small striped bass because they tolerate warmer temperatures, but the lack of cool, well-oxygenated water will limit the abundance of large adults. (Data courtesy of Matt Mauck, Oklahoma Department of Wildlife Conservation).

periods of time (coldwater species less so than warmwater species), but the physiology and health of a fish will be compromised if it is forced to inhabit waters with less DO than it requires to meet minimum metabolic needs, which will vary among species.

If well-oxygenated, cold (<20°C) water is present in a reservoir, the ability to develop and maintain a "two-story" fishery is available to fisheries managers, especially in deep storage impoundments in the lower latitudes of North America. A classic two-story reservoir fishery consists of warmwater species (e.g., centrarchids and ictalurids) inhabiting the warm epilimnion and coldwater species (salmonids) inhabiting the cold hypolimnion. Spawning habitat for salmonids is often lacking in reservoir environments and many salmonid fisheries are supported wholly by stocking programs. Black bass, crappie, and sunfish populations are usually self-sustaining in two-story fisheries. In practice, two-story fisheries are often managed as three-story fisheries because the thermal and oxygen requirements of coolwater species such as walleyes, muskellunge, and striped bass are usually present in two-story reservoirs. As with salmonids, such coolwater species may not be able to reproduce and may have to be maintained through stocking (see Chapter 9). The advent and popularity of two-story fisheries beginning in the 1950s and 1960s served to justify the indiscriminate stocking of coldwater prey fishes to satisfy new predator demands, often with unexpected consequences. For instance, the introduction of alewives into many eastern U.S. water bodies (as prey for percids and salmonids) invariably led to the collapse of some native fisheries via mechanisms that are still being debated, such as competition with, or predation on, early life history stages of sport fishes (Brooking et al. 1998) and dietary deficiencies when alewives are the primary forage (Honeyfield et al. 2005). Introductions of rainbow smelt to serve as a coldwater prey species have caused similar problems; for instance, their introduction into a Colorado reservoir to boost growth rates of walleyes and smallmouth bass ultimately led to recruitment failure by walleyes (Johnson and Goettl 1999). Fisheries managers at the present time are unlikely to stock coldwater prey species to improve growth rates by sport fishes when the specter of such negative consequences exists; unfortunately, many prey species that *could* be stocked *were* stocked through the 1990s, and managers now have little control over the species assemblage present in a particular reservoir.

The discussion of trophic states in reservoirs is important to fisheries managers because the amount of fish biomass a system can support is linked to its primary productivity (i.e., trophic state). Researchers have investigated the relationship between surrogate measures of trophic state (e.g., phosphorous or chlorophyll-*a* concentrations) and fish biomass in and harvest from reservoirs. Some of the earliest work focused on the relationship between Canadian fisheries and the morphoedaphic index (MEI), which incorporated a measure of a system's nutrient availability and its ability to process those nutrients (Ryder 1965). Specifically, the MEI was the ratio of total dissolved solids (mg/L) over mean depth (m). Despite many field investigations and published papers, the MEI eventually fell out of favor because of statistical concerns over the spurious correlations between ratios and their denominators (Jackson et al. 1990). Nevertheless, early work on MEI spurred research into the statistical relationship between indices of primary production and fish biomass and harvest. For instance, Jones and Hoyer (1982) explained more than 80% of the variability in biomass of sport fish harvest in midwestern U.S. lakes and reservoirs as a simple linear function of chlorophyll-*a* concentrations. Yurk and Ney (1989) explained 75% of the variation in fish community standing crops in southeastern reservoirs as a function of the concentration of phosphorous. The 1980s was a period when the paradigm of fisheries management (especially in reservoirs) shifted (Rigler 1982), and fish populations were no longer viewed or managed as separate entities in aquatic ecosystems but as components of complex systems; such research is now commonplace around the world (e.g., Gomes et al. 2002). Reservoir fisheries biologists need to be aware of the relationships between limnology, trophic state, and fisheries because it is likely that reservoirs they manage will undergo shifts in trophic state through time on a human scale. A solid working knowledge of these relationships allows biologists to communicate with environmental engineers and others tasked with maintaining quantity and quality of water to meet societal needs.

The fear that all freshwater systems were in danger of undergoing cultural eutrophication (i.e., the addition of nutrients into a water body due to direct human action) was one of many catalysts to the U.S. Clean Water Act of 1972. Although early stages of eutrophication may enhance fish growth and biomass and seem to be desirable from a fisheries perspective (i.e., more nutrients = more fish), water quality changes associated with higher trophic states (e.g., hypoxia, denser algal blooms, reduced water clarity, and altered fish fauna) usually argue against promoting higher trophic states because of changes in fish food habits, spatial distribution, and community composition. In fact, extreme cases of hypereutrophication promote dense, noxious algal blooms that can cause fish kills. Moreover, phytoplankton communities in eutrophic reservoirs can shift from domination by green algae to potentially noxious cyanobacteria (i.e., blue–green algae, which are actually bacteria that photosynthesize). While this dominance may shift seasonally in many reservoirs, cyanobacteria tend to dominate for an increasingly longer segment of the year in eutrophic and hypereutrophic reservoirs (Smith 1998) and are considered "sentinels" of eutrophication (Stockner et al. 2000). In turn, zooplankton composition is affected by phytoplankton availability because macrofiltrators (usually large-bodied zooplankton) that are more abundant in oligotrophic reservoirs give way to low-efficiency, small-bodied, algal and bacterial feeders as nutrients increase (Taylor and Carter 1998). In hypereutrophic reservoirs, the food supply for zooplankton may actually decrease because of the dominance by cyanobacteria. Eutrophication is particularly relevant in reservoirs because of their large watersheds relative to their surface areas and corresponding high sediment and nutrient inputs. As a result of the U.S. Clean Water Act, sweeping changes in infrastructure have been undertaken to reduce nutrient inputs and prevent eutrophication in aquatic systems. In 1970, Canada passed the Canada Water Act. One of its key provisions was to regulate nutrients in cleansing products (to combat cultural eutrophication). Although still on the books, the Canada Water Act has been supplanted by provincial laws aimed at protecting water quality in rivers, lakes, and reservoirs (e.g., 2005 Water Protection Act of Manitoba and 2006 Clean Water Act of Ontario). A similar program in Mexico, the *Programa Agua Limpia* instituted by the *Comisión Nacional del Agua* in 1991, aims to assure that water resources are of adequate quality for a variety of uses by society.

Concerted efforts by government agencies and private citizens to reverse cultural eutrophication (e.g., promoting or mandating the use of phosphorous-free laundry detergents and building new wastewater treatment plants) led to unexpected consequences. Specifically, nutrient loading rates into reservoirs were reduced at a time when many reservoirs were experiencing (or were about to experience) reduced internal nutrient loading rates and trophic depression. Rates of nutrient loading and trophic states were changing so abruptly in some systems that a new word entered the lexicon of reservoir and lake managers: *oligotrophication*. Moving from eutrophy to mesotrophy or from mesotrophy to oligotrophy usually resulted in clearer water because of reduced phytoplankton biomass, which most citizens equated to "cleaner" water. However, the costs to fisheries were largely ignored for decades until the trade-offs between "clean water" and productive fisheries began to be discussed by fisheries biologists (e.g., Ney 1996; Stockner et al. 2000), and those discussions have continued to the present (Anders and Ashley 2007). The tight linkage between algal standing crops or phosphorous concentrations and fish biomass and sport fish harvest meant that fisheries could suffer in reservoirs that shifted to a lower trophic state. Such shifts in trophic state are particularly important in southern and western U.S. reservoirs. Oligotrophication also receives considerable attention in Canada, where reservoir trophic states often are low (i.e., oligotrophic), even during the period of trophic upsurge and can drop even lower into ultra-oligotrophy as reservoirs age (Stockner et al. 2000). The trade-offs between cleaner water and popular sport fisheries were subsequently investigated; for instance, Maceina et al. (1996) showed that modest shifts in trophic state could achieve cleaner (i.e., clearer) water while still maintaining good black bass fisheries.

In the mid-1990s, the TVA sought to improve releases from its 41 reservoirs. Discharging water with low DO was a chronic, widespread problem, and various approaches were investigated including aerating the forebay (i.e., immediately upstream of the dam) by pumping liquid oxygen through submersed, porous, diffuser lines. In Cherokee Lake, a 11,000-ha reservoir in east Tennessee, nearly 15 km of porous line were suspended just off the bottom in the reservoir's forebay in 1995, and as predicted, DO in discharges improved. In some systems it makes more sense from an engineering perspective to pump compressed air into the hoses. The oxygen (or air) bubbles aerated the cool, anoxic waters in the turbine withdrawal layer above the diffuser lines, but they did not destratify the forebay. This approach to aerating Cherokee Dam's discharge provided summer refuges for fishes such as striped bass, a species that historically suffered from a temperature–DO squeeze every summer. The forebay refuge was used so extensively by striped bass (and anglers that pursued them) that emergency legislation was enacted to prohibit fishing in the forebay during summer to prevent overfishing and excessive catch-and-release mortality (Bettoli and Osborne 1998). The success of this approach to aerating downstream discharges led to the installation of similar systems in eight other TVA reservoirs with an unintended benefit, pelagic reservoir habitat was created in those reservoirs.

17.3.3 Water Retention and Water Level Fluctuations

Hydraulic retention time influences the trophic state of reservoirs. Phytoplankton communities reach their full production potential in reservoirs with high retention times that behave more like lakes and less like rivers. In a study of lakes and reservoirs across the USA, Soballe and Kimmel (1987) noted that algal communities needed about 60–100 d of retention time to realize their full potential at any given level of nutrients. In Alabama reservoirs, the relationship between retention time and algal production (and trophic state) was confirmed, but the threshold retention time was thought to be closer to 35 d (Maceina and Bayne 2003). The retention time in any reservoir is driven by the amount of rain falling in the watershed, and the amount of rainfall in region varies seasonally and annually. Thus, a reservoir's trophic state could vary within and between years.

Although the amount of precipitation falling into a reservoir's watershed cannot be controlled, hydraulic retention times and water levels are under the direct control of reservoir operators, and both of these aspects of reservoir hydrology are known to influence reservoir fisheries. Whereas relationships between retention times and fisheries can be subtle or complex, the impact of water levels on fish populations can be direct and dramatic. For instance, rapid changes in water levels can disrupt or eliminate spawning by some species (e.g., littoral nesting species), and statistical relations are regularly observed between water levels and recruitment by many species (e.g., gizzard shad, Michaletz 1997; largemouth bass, Sammons et al. 1999; white bass, DiCenzo and Duval 2002; crappies, Maceina 2003). The ecological mechanisms are not well understood, but responses of many fish species to hydrologic factors are well established. Unfortunately, it is difficult to alter the manner in which reservoirs are operated (i.e., change the guide curve), even when clear relationships between fish production and hydrology are identified.

Water level fluctuations within the regulated zone (i.e., between the minimum pool and top of the flood control pool) may partially or fully encompass the original floodplain near inlets or may occur entirely within bands occupied by upland vegetation in regions toward the dam. Substitution of a naturally variable flooding pattern with a standardized one, and loss of wet–dry cycles, has lasting ecological effects. Biotic communities in littoral areas of reservoirs become dominated by species adapted to lakes and standing water and may be less diverse or abundant than those in floodplain backwaters. Floodplain vegetation often dies with prolonged flooding, removing a major component of floodplain ecosystems. Flooding upland terrestrial vegetation in the regulated zone also creates habitat diversity, shelter, and a favorable environment for littoral fishes. Nevertheless, uniform fluctuations over time controlled by an engineered hydrograph (i.e., guide curve) will limit vegetation growth and produce barren slopes and mudflats, except perhaps near inlets where the regulated zone straddles the original floodplain. Guide curves are often justified and maintained by the misperception that a fixed guide curve will benefit all sport fish species (Miranda and Lowery 2007).

Managed floods within the regulated zone are often used to simulate a flood pulse and maintain ecological processes associated with random floods. During high water in free-flowing rivers, nutrients are exchanged between rivers and their floodplain, generally resulting in a net nutrient gain in floodplains. In reservoirs, nutrients normally dissolved in the incoming water may have already settled in upstream reservoirs, or may settle soon after entering reservoirs, resulting in reduced particulate organic matter and nutrient deposition in regulated zones. In fact, a net reduction in nutrients may occur, as these are extracted from regulated zones by nutrient-impoverished water (Thomaz et al. 2004).

Enhancement of regulated zones to recreate floodplains is central to reservoir habitat management. Unvarying, standardized guide curves produce regulated zones devoid of vegetation and of limited ecological value to floodplain species and do not provide a diversity of floods needed to maintain a diverse fish assemblage (Miranda and Lowery 2007). Some year-to-year variation in the flood pattern is necessary to maintain a full complement of species, periodically enhance nearly all species, and produce extreme events that have a rejuvenating function by connecting the basin to distant backwaters. Operational flexibility can often be found to produce floods artificially of suitable extent and duration, but institutionalizing a realistic level of flood-regime randomness into highly engineered systems is challenging. In developing water management plans, regulatory agencies could consider incorporating managed randomness into guide curves.

Many hydropower reservoirs in the USA operate under licenses issued by the Federal Energy Regulatory Commission, and those licenses periodically come up for renewal (see Box 4.5). Dams in the USA can be relicensed for up to 50 years; therefore, agencies tasked with managing reservoir fisheries devote substantial resources to providing input into the relicensing process when those opportunities arise to ensure that fishery resources are given consideration along with hydropower, flood control, and navigation needs. The relicensing process is complex and time consuming but provides a rare opportunity to influence reservoir operations for decades because a consideration in relicensing is mitigation of environmental damage. Under the auspices of the U.S. Environmental Protection Act, an environmental impact statement is required for any major federal action that may have a significant impact on the environment. Such mitigation can take many forms, including altering guide curves to promote spawning in a reservoir, improving the quality and (or) quantity of water discharged through a dam, and developing management plans to protect littoral habitats and riparian zones. The Canadian Environmental Assessment Agency, which administers the Canadian Environmental Assessment Act of 1995, does not permit dam operations per se. However, upgrades to hydroelectric facilities in Canada or changes in how reservoirs are operated will trigger mandated federal assessment of environmental impacts, and fisheries managers have ample opportunity to participate in that process.

17.3.4 Submerged Structure and Vegetation

The lack of submerged woody or vegetative structure in reservoirs, particularly old reservoirs and those that experience substantial winter drawdowns, has prompted collaboration among many agencies and citizens groups to add structure. According to a recent survey (Tugend et al. 2002), most fish and game agencies in the USA devote money and personnel to adding structure and evaluating its impact in otherwise barren littoral zones. The response of fish assemblages to artificial structures placed in littoral zones is also being investigated in neotropical reservoirs such as Lajes Reservoir, Brazil, an oligotrophic reservoir built for hydropower production (Santos et al. 2008). Enhancement structures are typically made of either natural materials (e.g., brush or discarded trees) or manufactured products such as automobile tires (Figure 17.6). Structures made of artificial materials are not as effective as those constructed of logs or brush in providing refuge areas or serving as fish attractors (Roni et al. 2005). Stake beds (i.e., lengths of slender lumber driven into soft substrates to create a matrix of vertical structure) are common in reservoirs supporting crappie fisheries. Such management efforts fall under the category of habitat enhancement, and the primary objective is usually to attract fish to improve angler catch rates, not to boost population density. In some locales, the placement of half-log structures (Hoff 1991) to attract spawning black bass species, especially smallmouth bass, is commonplace (Figure 17.6). In most instances, only demersal or structure-oriented species (e.g., centrarchids and cichlids) will use structures placed in barren littoral zones; there should be no expectations that open-water, pelagic species will benefit (Santos et al. 2008).

Brown (1986) stated that few evaluations of the effects of fish attractors or other habitat enhancements on fisheries production had been published, and that is still true over two decades later (Roni et al. 2005). Similarly, few publications documenting the positive influences of such habitat enhancements on angler catch rates are available, despite the ubiquity of this management activity. Wills et al. (2004) provided a review of the rationale behind reservoir habitat enhancement techniques and noted the pervasive role that habitat variables (e.g., substrate) play in determining whether fish use artificial structures. For instance, the proximity of complex physical structures to habitat enhancement structures affects the use of enhancement structures by spawning largemouth bass (Hunt et al. 2002).

Although the efficacy of habitat enhancement techniques is weakly established and additional studies are needed to determine population level responses to habitat enhancements (Roni et al. 2005), there is little doubt that reservoir managers will promote such activities because they are embraced by the angling public, regardless of the public's species preferences. For example, in Norris Lake, Tennessee, collaborating anglers and biologists have placed more than 21,000 structures along the shoreline of the reservoir since 1992 to enhance fish habitat. It took decades for anglers to accept that the hatchery truck could not solve all fishing woes and that the key, invariably, to healthy fish populations and good fishing was

Figure 17.6. Barren littoral zones often prompt agency biologists and anglers to add natural and artificial structures to increase habitat complexity. Christmas trees and wooden pallets (upper photo) attract various species. Spawning benches (also known as half-logs, lower photo) placed along an exposed shoreline provide spawning habitat for black bass, particularly smallmouth bass. Stake beds (particularly attractive to crappies) and downed trees are visible in the background of the lower photo.

good habitat (e.g., Quinn 1992). As with planting vegetation or seeding shorelines, habitat enhancement projects provide opportunity to forge relationships and establish lines of communication among stakeholders and biologists, which promote mutual trust and help resolve conflicts (Box 17.1; see also Box 5.2). If the ecological impact of collaborative, in-reservoir habitat enhancement techniques is subsequently shown to be modest (or absent), the credibility of biologists is not lost if they acknowledged early in the process that the science is not well established and that such enhancements may not result in improved fisheries.

If soil substrate exists and water level fluctuations are not too severe, aquatic plants will inevitably colonize the littoral zone of reservoirs. Aquatic vegetation management has been debated for decades, and the responses of reservoir ecosystems to vegetation colonization and vegetation control are broadly understood (e.g., Bettoli et al. 1993). Most biologists would agree that some aquatic vegetation is desirable because studies have demonstrated that sport fish production is maximized at intermediate levels of vegetation (e.g., Wiley et al. 1984; Miranda and Pugh 1997). Nevertheless, controversies over vegetation management routinely besiege fisheries professionals because different stakeholders have different opinions regarding what constitutes desirable levels of vegetation (Wilde et al. 1992; Henderson 1996). When reservoir shorelines are urbanized and shoreline homeowners and developers enter the arena, the potential for conflict escalates. The likelihood for conflict rises even further when reservoirs are colonized by exotic species such as hydrilla or water hyacinth. In the absence of natural pathogens or grazers, these and other exotic species are much more likely to reach nuisance levels. In our experience, if vegetation coverage reaches 40–50% of the surface area of a reservoir, the ability to manage the vegetation is compromised and vegetation eradication may become the only option, even in large reservoirs. Ideally, the decision to manage vegetation will be made early when more options are available.

Four approaches are often used to manage nuisance aquatic vegetation: chemical, biological, mechanical, or water level manipulation. Reservoir managers often find themselves weighing the pros and cons of chemical and biological control. With few exceptions, biological control of aquatic vegetation is synonymous with stocking grass carp (either diploid or sterile triploid fish). The use of grass carp to control aquatic vegetation has been well studied for more than 30 years. Grass carp are long-lived, obligate herbivores capable of consuming all submersed and floating vegetation if stocked at a large-enough size (>300 mm total length) and at high-enough rates (20–70 fish/ha of vegetation; Martyn et al. 1986; Bonar et al. 2002). The ability of grass carp to control a wide array of plant species is not debated, nor is their inability to reproduce in reservoirs and small impoundments questioned; rather, the challenge is to devise strategies to control, rather than eliminate, vegetation with grass carp. It was once thought that incremental grass carp stockings combined with close monitoring of vegetation might achieve desirable, intermediate plant densities (e.g., Bain 1993); however, the responses of individual ecosystems to grass carp stocking vary widely. A consensus has been reached that grass carp should be stocked only in water bodies where most stakeholders can tolerate the complete elimination of submersed plants (Bonar et al. 2002).

Whereas the amount of plant biomass consumed by grass carp cannot be controlled once the fish are stocked, herbicides have long been used to control vegetation in reservoir ecosystems. A voluminous literature exists on chemical control of nuisance aquatic vegetation, and information on herbicide use is readily available on the Internet. In most locales, herbicides and pesticides can be dispensed only by certified applicators. Although chemical treatment of aquatic vegetation can be much more expensive than introducing grass carp, the ability of

Box 17.1. The Costs of Ignoring Human Dimensions in Reservoir Fisheries Management: The Norris Lake Story[1]

The abundance of nutrients released from the flooded reservoir basin (i.e., trophic upsurge), combined with the creation of expansive and unoccupied lacustrine habitat allowed sauger, walleye, smallmouth bass, and other native species to flourish in Norris Lake, Tennessee, in the two decades after its impoundment in 1936. In fact, at one time reservoir biologists advocated letting private citizens use gill nets to catch fish that would otherwise go to waste! The boom fishery did not last, as the trophic upsurge phase was followed by trophic depression (i.e., a decline in the rate of internal nutrient loading) and by loss of high-quality habitats. In the 1960s, biologists introduced striped bass and a popular fishery quickly developed—but not among those anglers who grew up learning to fish for species native to the river system. Those "native-species" anglers soon voiced their concern that striped bass were being promoted and managed to the exclusion of native species (walleye, black basses, and crappies), most of which were less abundant in the late 1960s than during the two decades following impoundment. The response of the state fisheries management agency to such angler concerns through the 1970s was to continue and eventually to expand the striped bass stocking program. The agency also conducted field studies to "prove" that striped bass did not compete with, or prey upon, native sport fishes to any appreciable degree. The response of biologists to growing complaints was to do what they had been trained to do—establish new fisheries to take advantage of pelagic habitats and abundant prey resources in reservoirs and collect biological data to defend their management activities.

A state-agency-sponsored task force was created in 1992 to address the concerns of organized, increasingly vocal opposition to striped bass management in Norris Lake. Unfortunately, no amount of persuasion or data could convince opponents to abide by the findings and recommendations of the task force; instead, they sought redress through legislative mandates. Five bills were introduced by state legislators on behalf of these constituents, but none passed and became law. Anglers on each side of the striped bass issue assumed hardened positions that made resolution of the problem difficult.

Thousands of dollars were spent on research (e.g., Raborn et al. 2002) and an advisory committee was created in 1998 to move opposing factions toward common ground regarding management of the reservoir's diverse fisheries. The balancing role of a diverse group of stakeholders on the Norris Lake Advisory Committee played an important part in achieving compromises and resolving most of the major issues regarding stocking of various species, including native species such as crappies and walleye. Ten years later in 2008, anglers from diverse backgrounds and with different attitudes toward fishing on Norris Lake regularly interact with state biologists and collaborate on projects beneficial to all parties (Figure 17.6).

Failure to consider the importance of human dimensions, a much more common component of fisheries education in the 21st century, led to the creation and prolonging of the Norris Lake controversy, and at one time or another threatened the entire structure and existence of the Tennessee Wildlife Resources Agency. As populations grow and public pressure for water and quality fishing increase in this century, reservoir biologists will find more and more opportunities and needs to consider human dimensions in their regular management activities.

[1] Adapted from Churchill et al. 2002.

herbicides to kill plants only where and when desired justifies their use in many situations. Advances in herbicide formulations and delivery systems have made large-scale plant control using chemicals more cost-effective in recent years. For example, a drip-delivery system of Sonar™ (a fluridone herbicide that inhibits carotenoid photosynthesis) in the headwaters of an embayment in Lake Seminole, Georgia, eliminated 1,200 ha of hydrilla (Sammons et al. 2003). Fluridone was considered an excellent herbicide to treat hydrilla for nearly two decades because the dose that killed hydrilla had minimal impacts on native emergent and submersed plant species such as *Vallisneria* spp., *Potamogeton* spp., and *Scirpus* spp. (Hoyer et al. 2005). However, fluridone-resistant hydrilla was discovered in Florida in 2000, and the higher doses and longer exposure times necessary to control resistant hydrilla negatively affect native plant species. Efforts are underway to develop and license new herbicides to control hydrilla (and other noxious plants) economically and safely and to determine whether hydrilla biotypes maintain resistance to fluridone in the absence of fluridone-selective pressure (Puri et al. 2007). Reservoir managers concerned with maintaining ecological function and habitat diversity face significant challenges if and when fluridone-resistant hydrilla spreads into new reservoirs.

The topic of native vegetation establishment has received far less attention than has vegetation control, but interest has increased in recent years. Field studies have usually entailed planting native vegetation (e.g., American pondweed and wild celery) in exclosures to serve as founder colonies (Smart et al. 1996). Exclosures are critical because small patches of transplanted plants or propagules growing along a barren shoreline will quickly be grazed by terrestrial, aquatic, amphibian, and avian herbivores (Smart et al. 1998). Although the prospect of establishing native plants is appealing to stakeholders, the efficacy of such programs has not been well demonstrated. Exclosures are prone to failure when they are forcibly entered by turtles and other grazers (Bettoli and Gordon 1990), and plants that expand outside the exclosures are often cropped by herbivores. Nevertheless, successful establishment of aquatic macrophytes in selected reservoirs may be possible (Smart et al. 1996).

There are many obstacles to establishing aquatic plants (with or without exclosures) in reservoirs that lack suitable substrate or experience large water level fluctuations. This fact prompted biologists to investigate the efficacy of seeding exposed shorelines with annual terrestrial grasses (e.g., winter wheat, millet, and ryegrass) at reservoirs that experience winter drawdowns. After numerous investigations, the general consensus is that any benefits to fisheries resulting from seeding shorelines are modest and transitory (Strange et al. 1982). Annual grasses will grow readily on exposed shorelines with little or no preparation of the soil, and lush stands of vegetation can grow in lower latitudes through fall and winter. However, the vegetation often does not persist once inundated in spring; thus, no vegetation is available to serve as nursery habitat for juvenile fishes later in spring and early summer. A benefit of shoreline seeding projects is that lush grass growing along dewatered shorelines will prevent erosion when the soil is exposed to rainfall.

American water willow, an emergent plant species, has been extensively planted along reservoir shorelines throughout North America for decades, with mixed success. This species has a semirigid, fibrous stem, and it is capable of rapidly colonizing new habitats via rhizomatous growth. Although American water willow is not readily grazed by herbivores (e.g., Dick et al. 2004), transplanted shoots and established plants will not survive if they experience extensive periods of inundation or desiccation, although they tolerate the latter more than the former (Strakosh et al. 2005). Ongoing research seeks to define reservoirs, habitats, and water-level-fluctuation regimes for which American water willow and other transplanted species might have a good chance of colonizing reservoir shorelines.

The importance of woody vegetation to reservoir ecosystems has prompted many investigations into the feasibility of establishing pioneer riparian tree species such as willows and cottonwoods, and water-tolerant trees such as baldcypress, in the drawdown zone of reservoirs. The same environmental hurdles limiting herbaceous vegetation re-establishment (e.g., herbivores and periods of desiccation and freezing followed by periods of inundation) can be present when saplings or seedlings are planted along reservoir shorelines. A general consensus is being reached that woody species (as well as herbaceous species) are more likely to become established if the pattern of water level fluctuations is altered (i.e., change the guide curves) to promote plant survival in the drawdown zone. However, altering guide curves of hydropower reservoirs to favor plants along shorelines will likely incur steep costs in lost power generation (BC Hydro 2007).

Caveats not withstanding, shoreline seeding and efforts to establish aquatic vegetation or riparian trees invariably garner instant and enthusiastic public support. Such projects serve to inform stakeholders that their fisheries are regulated in large part by the quality of the habitat. The following excerpt from an article that appeared in *BASS Times* (December 2004), a popular sportfishing magazine, illustrates the point that planting vegetation is enthusiastically received by the fishing public—even when it is not successful.

> "The grass plantings also served to strengthen the relationship between the West Virginia BASS Federation and the DNR [West Virginia Department of Natural Resources], in addition to helping to generate interest in this ongoing project. While setbacks are inevitable, the establishment of a successful aquatic vegetation program in West Virginia will take both time and effort. But the DNR has made a solid commitment to its anglers that quality fish habitat remains a priority in the Mountain State."

Although the ecological merits, cost-effectiveness, and procedures for shoreline seeding and native aquatic plant establishment projects are still being debated, the public relations impact of such programs is substantial and cannot be dismissed. Thus, managers should view favorably any activity that does no harm to the resource, fosters interactions among fishery biologists and stakeholder groups, and may, through future breakthroughs, potentially reap important benefits. As mentioned earlier, biologists should clearly establish realistic expectations for any such habitat projects in order to maintain their subsequent credibility, in the event that enhancement activities provide no tangible results.

17.3.5 Longitudinal Patterns along the Reservoir

Spatial patterns occur in reservoirs due to longitudinal changes in reservoir morphology, flow velocity, suspended solids, light penetration, and nutrient dynamics (Kimmel et al. 1990). The upper sections of a reservoir and major bays are often characterized by a lotic-like environment that is generally shallower and narrower than are downstream sections and is influenced by the original river's geomorphology and basin contour (Figure 17.7). This lotic region exhibits higher flows, shorter water retention times, higher nutrient levels, lower light penetration, and greater sedimentation relative to downstream regions. Like the upstream river, the water is well mixed and oxygenated and is often turbid. Primary production is lim-

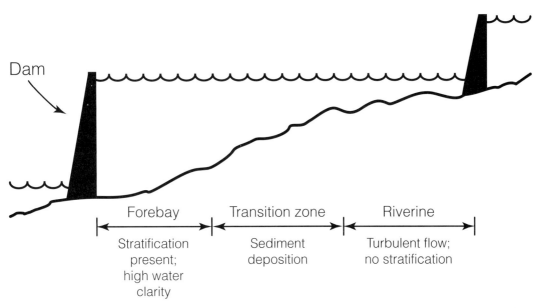

Figure 17.7. Reservoirs of all types and sizes typically display longitudinal gradients in terms of the physical habitat and water clarity. The relative lengths of each of the three zones distinguish one reservoir from another. For instance, the transition zone can be quite short in large tributary storage impoundments; whereas, the forebay can be short (or nearly nonexistent) in main-stem, run-of-river reservoirs. If there is a reservoir upstream, the water clarity in the riverine reach can be high; otherwise, the riverine reach is usually the most turbid zone of the reservoir. Sediment deposition is highest at the upper end of the transition zone, where water velocity drops and suspended particles settle to the bottom.

ited despite high levels of nutrients due to low transparency and low water residence time (Bernot et al. 2004). Further downstream, the reservoir becomes wider and deeper, showing signs of reduced flow, suspended fine materials, and turbidity but augmented light availability. In this transitional zone nutrient levels are lower than those upstream, but because more light is available due to settling of fine sediments, algal photosynthesis increases and might be the highest in the reservoir. In the transitional zone, more of a reservoir's production originates from autochthonous inputs rather than allochthonous inputs from the river. Near the dam, a reservoir is broadest and deepest, flow is generally unnoticeable unless large volumes of waters are being discharged, and water is the clearest. Nutrient concentrations in this lacustrine-like section are low relative to upstream, reducing volumetric primary production but perhaps not total production given a deeper photic zone. It is not uncommon for these three sections to be classified at different trophic states due to differences in nutrient concentrations, sediment loads, and depth. For instance, Center Hill Lake, a 7,400-ha tributary storage impoundment in Tennessee, is classified as eutrophic in the transition zone and embayments but mesotrophic in the lacustrine zone near the dam. Nevertheless, the intensity of the longitudinal gradient observed in reservoirs depends on factors such as size and layout of tributary streams, reservoir basin morphology, and position of the reservoir along the river basin.

Fish assemblages and fisheries may also change in a longitudinal manner. For example, in Itaipu Reservoir, Brazil–Paraguay, fish species richness decreased toward the dam (De Oliveira et al. 2005). Abundances of shads and other major species common to reservoirs in the southeastern USA tend to be lowest near the dam and increase upstream in more riverine environments (Michaletz and Gale 1999). In John Day Reservoir, Oregon, introduced walleyes were largely restricted to the upper third of the reservoir, whereas the number of introduced smallmouth bass increased progressively down lake (Beamesderfer and Rieman 1991). In Lake Texoma, Texas–Oklahoma, Gido et al. (2002) found the littoral fish assemblage to be highly predictable along the length of the reservoir and along the length of tributary arms. The basis for this predictability was shaped largely by differential responses of individual species to physical and chemical gradients in the reservoir. In general, many species associated with the upper reservoir require flowing water during at least a portion of their life history, are species that orient toward the banks or bottom, or require sand or gravel substrates. These are frequently migratory (i.e., potamodromous) species whose spawning habitat lies in tributaries or upstream floodplains. In contrast, near the dam pelagic species adapted to lake environments are most common, although periodically the deeper layers may not be occupied by fishes because of thermal and chemical stratification. Lentic and lotic species may coexist in transitional segments between the upper and lower reservoir, and thus species richness and diversity may be highest in these segments (Agostinho et al. 1999).

Strong longitudinal patterns observed in many reservoirs can influence management within a single reservoir. For instance, the concept of a cline provided the foundation required to ask what factors affect longitudinal microbial community composition, production, metabolism, and biomass accumulation, facts needed to gain insights into eutrophication processes (Lind 2002). The longitudinal patterns framework has also been used to guide sampling programs to evaluate water quality (Davis and Reeder 2001), sedimentation (Pagioro and Thomaz 2002), total maximum daily loads, and fish populations. For example, longitudinal patterns along Itaipu Reservoir showed dominances by different fish species and disparate fishery yields requiring recognition in fishery management plans (Okada et al. 2005). In Cave Run Lake (a Kentucky reservoir), the intra-lake distributional gradients of three black bass species linked to clinal changes in habitat and nutrients indicated the need to stratify sampling and assess population characteristics separately according to zones (Buynak et al. 1989). In long reservoirs with strong longitudinal patterns, fishery management may be divided according to zones, with different habitat enhancement, harvest regulations, and infrastructure development strategies in each zone. Management alternatives that are appropriate for a particular zone of a reservoir may not be the best choice in other zones with different environmental conditions.

17.4 TRIBUTARIES

Tributary streams are elements of all reservoirs and include the main-stem river impounded by the reservoir and inlets associated with reservoir coves and arms. The discharge, width, and length of tributaries vary greatly, ranging from small creeks to major rivers. Thus, the influence of tributaries over reservoir fish assemblages varies from a minimal effect on reservoirs with small watersheds to a sizeable effect on reservoirs with large tributaries. This range of tributary conditions results in fish assemblages varying between those that are dominated by pool and backwater fish species to those dominated by riverine species.

17.4.1 Tributary Size and the Reservoir Fish Community

Reservoir fishes are mostly of riverine origin (Fernando and Holčík 1982) and many are obligatory tributary spawners. However, many species are generalists able to spawn in tributaries as well as in a reservoir, although their abundance in a reservoir is often enhanced by access to tributaries. Warmwater species such as longnose gar, paddlefish, white bass, and various catostomids migrate from reservoirs to spawn in tributary rivers and creeks (Colvin 1993; Johnson and Noltie 1996; Hoxmeier and DeVries 1997). Moreover, most salmonids use tributaries for reproduction due to the availability of suitable substrates, current velocities, and temperatures (Parsons and Hubert 1988; Crisp et al. 1990; Stables et al. 1990). In most cases, preferred spawning sites are large gravel bars, but pre-spawn staging occurs in adjacent pools. Gravel bars are used by multiple species, concurrently and segregated spatially and temporally. After hatching, juveniles may migrate immediately back into a reservoir (e.g., kokanee) or rear for months to years in the streams (e.g., trout) and backwaters (e.g., gars, paddlefish, and catostomids) before returning to a reservoir.

In reservoirs with long riverine stretches upstream, large lateral tributaries, or extensive floodplains upriver, a large fraction of the fish assemblages will be composed of potamodromous species. For example, upriver from Itaipu Reservoir, a long stretch of the free-flowing Paraná River connects to an extensive floodplain providing spawning and rearing habitats for many species in ephemeral floodplain lagoons (Agostinho et al. 2001). In fact, 6 out of the 10 species that sustain Itaipu's subsistence and commercial fisheries develop in the floodplains upstream (Okada et al. 2005). At the urging of conservationists and fishers, this section of the river above the reservoir was set aside as a national park by the Brazilian government to prevent impoundment and degradation and maintain fisheries in the reservoir downstream. Among several flood control reservoirs in Mississippi that provide quality crappie fisheries, juvenile crappie densities were one to two orders of magnitude higher in sloughs, backwaters, and oxbow lakes immediately upstream of the reservoirs than in the reservoirs (Meals and Miranda 1991). Those backwaters flooded annually or semiannually, simulating river floodplains, and crappie populations that developed there probably supplemented or even sustained populations in the reservoirs. Backwaters serve as nursery areas for juveniles of many species that spawn in rivers and have larval development in lentic lateral pools. In reservoirs with riverine floodplain habitats upstream, restoration and conservation of such habitats should be a reservoir management priority.

Whereas many short-distance migrators persist in impounded rivers by using the reservoir and associated tributaries, long-distance migrators may be gravely affected by the barrier presented by the dam and the reservoir (Larinier 2001). Elimination or reduction of spawning grounds or delayed access to spawning areas have been significant effects of reservoirs. Dams and reservoirs can affect fish by blocking upstream and downstream passage. These movements are most important to anadromous and catadromous fishes such as salmon and eels, which spend part of their life cycles in rivers and part in oceans or other large water bodies, but also to potamodromous species that, during a certain phase of their life cycle, depend on longitudinal movements in river systems (e.g., paddlefish and sauger). For adult fish trying to move upstream, a dam and associated reservoir can pose an impassable barrier unless passage is provided, and juvenile and adult fish moving downstream are at high risk of being entrained in a turbine or preyed upon in the relatively still waters of a reservoir (Lucas and Baras 2001). The deleterious impact of dams and reservoirs on migrating fishes is clearly illustrated by the plight of many imperiled stocks of Pacific salmon in the Columbia and Snake river watersheds, where 18 large main-stem dams (and many smaller tributary dams) impede upstream migrations of adults and the downstream out-migrations of juveniles. Fish passage facilities have been provided in many impounded rivers (particularly in the Pacific Northwest) where migratory species are major components of fish assemblages; fish passage facilities are lacking at many other dams, most notably in eastern North America.

17.4.2 Management of Tributaries to Benefit Reservoir Fish

The need to manage tributaries to enhance reservoir species depends on a reservoir's fish assemblage but, in general, includes protecting gravel bars, maintaining bank stability, manipulating access to adjoining sloughs and oxbow lakes, developing artificial backwaters, or providing proper flows and water levels during key periods. A first step in the management process is to inventory tributary habitats and rate their quality relative to reservoir species that use them. There is a large body of literature relevant to stream habitat maintenance and restoration (e.g., FISRWG 1998; also see Chapter 10). This literature has concentrated on river restoration to benefit riverine species rather than to benefit reservoir species that use rivers during part of their life cycle, although these two aims overlap. Depending on a reservoir's position along a river basin (section 17.6), management of fish assemblages requires attention to tributary protection and restoration.

17.5 THE WATERSHED

We define a watershed as an area that drains into a reservoir and its tributaries. Land cover in a watershed is a major determinant of water quality and consequent fish assemblage composition. As described earlier, watersheds contribute nutrients that influence primary production in reservoirs. Nutrients flow to reservoirs from their watersheds by way of streams, groundwater, and runoff, often embedded in organic and inorganic particles. Watersheds typically experience various levels of deforestation, agricultural development, industrial growth, urban expansion, surface and subsurface mining activities, water diversion, and road construction. These changes destabilize runoff, change annual amplitudes and spatial distributions of flow, and enhance downstream movements of nutrients, sediments, and detritus that are ultimately trapped by reservoirs and regulate primary productivity, species composition, and food web interactions.

Sediments are a major watershed export into reservoirs and produce suspended fine material and sedimentation discussed earlier. Mean total suspended solids varying from 1.2 to 47 mg/L in 135 Missouri reservoirs were directly related to the proportion of cropland and inversely related to the proportion of forest cover in the watershed (Jones and Knowlton 2005). Sedimentation rates in reservoirs are higher in agricultural watersheds and are affected by agricultural land management. Sedimentation not only affects backwaters of reservoirs, but as backwaters fill, sedimentation extends upwards beyond reservoir into tributaries.

Watershed practices are directly linked to eutrophication (Carpenter et al. 1998). Row-crop agriculture with frequent tillage and fertilizer application is a major disturbance in watersheds (Novotny 2003). Nutrient exports from croplands are several times higher than from grasslands and forestlands (Beaulac and Reckhow 1982). In Missouri reservoirs, phosphorus and nitrogen levels were high in reservoirs surrounded by croplands and lower in reservoirs surrounded by forests, resulting in a sevenfold minimum difference in nutrients between reservoirs with watersheds dominated by forests and ones with watersheds dominated by croplands (Jones et al. 2004). Similar relations have been reported for lakes and reservoirs in Connecticut (Field et al. 1996), Iowa (Arbuckle and Downing 2001), and Ohio (Knoll et al. 2003). Nutrient input per unit of land area from urban watersheds often equals or exceeds that from agriculture as impervious surfaces enhance runoff (Beaulac and Reckhow 1982).

Critical portions of watersheds are the strips of land immediately adjacent to reservoirs, termed riparian zones, that begin at the shorelines and move inland a loosely defined distance. In tributaries, riparian zones have been defined as encompassing the terrestrial landscape from the high-water marks upland to where vegetation may be influenced by elevated water tables and flooding (Naiman and Decamps 1997). In reservoirs, riparian zones resemble those of tributaries only near the entrance of tributaries. Reservoirs lack true riparian zones in their lower reaches (toward the dam) because original river channels have been submerged and the shoreline contours consist of upland vegetation that provides buffer zones, although not true riparian zones.

Riparian zones are key ecotones for regulating aquatic–terrestrial interactions (Correll 1997). In tributaries, major roles of riparian zones include thermal buffering, shading, contribution of woody debris, bank stability, and sediment and nutrient interception (Pusey and Arthington 2003). These roles remain relevant in reservoirs, but protection from strong winds also becomes an important feature. Furthermore, riparian buffers present esthetic visual barriers that help maintain quality of recreational fishing experiences.

17.5.1 Links between Watersheds and Reservoir Fisheries

The effects on reservoirs of sediment and nutrients imported from watersheds have been described, but in addition to nitrogen and phosphorus, watersheds can contribute large quantities of particulate organic matter to reservoirs. The fish assemblages of many reservoirs in North America are dominated by gizzard shad, a clupeid that depends on small zooplankton at larval stages (Miranda and Gu 1998) but is capable of consuming large amounts of organic detritus during postlarval stages (Mundahl and Wissing 1987). According to Vanni et al. (2005), gizzard shad represent a key link between reservoir fish assemblages and watersheds. Agricultural watersheds tend to export greater quantities of particulate organic matter than do forested watersheds and reservoirs in agricultural watersheds support higher abundances of gizzard shad, probably through mechanisms operating on larval and adult stages. Thus, reliance on watershed exports gives species such as gizzard shad and buffaloes a large advantage over other reservoir fishes because they can utilize this exogenous food resource.

Riparian zones have multiple effects on fish. Without suitable riparian buffers, fine sediments are transferred from watersheds to shallow reservoir environments where they can affect littoral fish species. Increased turbidity and sedimentation alter food availability (e.g., benthic invertebrates and algae; Berkman and Rabeni 1987), affect fish foraging behavior and efficiency (Bruton 1985), and alter intraspecific interactions. Other effects include reductions in habitat suitability for spawning (Walser and Bart 1999) and increased egg mortality, as well as reductions in rates of larval development and survival (Jeric et al. 1995). As banks and associated littoral habitats degrade, densities of fish that rely on littoral zones during all or part of their life history are likely to decrease. Fish assemblages may shift toward dominance by species that depend less on substrate-based resources and can exploit pelagic resources.

With few exceptions, research on the influence of riparian zones to lake and reservoir ecosystems has focused on filtration and its positive effects on water quality; direct influences on fish assemblages are rarely inferred. In reservoirs of the southern USA, species richness and centrarchid abundance are generally higher in coarse woody habitat (Barwick 2004) provided by forests surrounding reservoirs. In a lake in Wisconsin, experimental removal of coarse woody habitat originating from riparian zones resulted in largemouth bass consuming less fish and more terrestrial prey and growing more slowly (Sass et al. 2006). Moreover, yellow perch declined to extremely low densities as a consequence of predation with little or no recruitment.

17.5.2 Watershed Management

The goal of watershed management is to facilitate self-sustaining natural processes and linkages among terrestrial, riparian, and reservoir environments. Watershed management involves controlling the quantity, makeup, and timing of runoff flowing into reservoirs or tributaries from the surrounding terrain (Box 17.2). The first and most critical step must be halting or eliminating anthropogenic practices causing degradation of reservoirs. Such approaches can involve a wide range of adjustments to human activities. For example, they may involve increasing widths of buffer strips around fields, altering livestock grazing strategies to minimize impacts, moving tillage operations in fields farther away from riparian systems and water, changing tillage methods and timing, or stopping the release of industrial wastes that cause water pollution. To this end, various protocols dubbed best management practices have been developed to target and minimize impacts from nonpoint sources in watersheds. Best management practices are usually applied as systems of practices because one practice rarely solves all problems and one practice will not work everywhere. Reservoir managers should be acquainted with the large body of literature on watershed management; however, watershed management is generally not the direct responsibility of fisheries managers (section 17.7).

Buffer strips with multiple vegetation types may protect water bodies against the negative effects of agriculture. This concept uses three interactive zones that are in a consecutive upslope order from shore: a strip of permanent forest, a strip of shrubs and trees, and a strip of herbaceous vegetation (Schultz et al. 1995). Width and composition of the strips are adapted to the geographical variability of terrestrial plant communities and riparian conditions. The first strip influences the aquatic environment directly (e.g., temperature, shading, bank stability, wind break, and source of coarse woody habitat). The second strip controls pollutants in subsurface flow and surface runoff and is where biological and chemical transformations, storage in woody vegetation, infiltration, and sediment deposits are maximized. The first two strips contribute to nitrogen, phosphorus, and sediment removal. Grasses in the third strip spread the overland flow, thus facilitating deposition of coarse sediments. Grassy riparian areas trapped more than 50% of sediments from uplands when overland water flows were less than 5 cm deep (Magette et al. 1989). In North Carolina, riparian areas removed 80–90% of the sediments leaving agricultural fields (Daniels and Gilliam 1997). Riparian buffer zones accumulate nutrients and absorb them into plant biomass, serving as nutrient filters. In Vermont, reductions of approximately 20% in mean total phosphorus concentrations and 20–50% in mean total phosphorus loads were observed (Meals and Hopkins 2002). In Lake Rotorua, New Zealand, riparian management reduced fine sediment loads by 85%, particulate phosphorus and soluble phosphorus by about 25%, particulate nitrogen by 40%, and soluble nitrogen by 26% (Williamson et al. 1996). These reductions reduced the chlorophyll-*a* concentrations in the lake and helped shift the lake's trophic state from eutrophic to mesotrophic.

Box 17.2. Iowa's Comprehensive Lake and Watershed Management Program[1]

Iowa leads the nation in the proportion of its land area converted to cropland: 72%. An additional 10% of its land is pastureland and 5% is urbanized. Consequently, 87% of Iowa's land area is directly disturbed by humans. As a result, many natural and constructed lakes in Iowa are impaired with poor water quality, fisheries, and recreational value. Over the years, fish assemblages in many lakes were renovated, some multiple times, resulting in improved fisheries that were eventually degraded because the underlying problems of sedimentation, excessive nutrients, and poor water quality were not addressed.

A lake classification system was developed by the Iowa Department of Natural Resources (IDNR) based on systematic assessment of both lake water quality and watersheds. This classification, combined with socioeconomic factors, resulted in a priority ranking of lakes and watersheds for restoration. Once local commitments are demonstrated and feasibility verified, comprehensive restoration is initiated to address both watershed and in-lake issues. Watershed models are used to simulate hydrologic processes and pinpoint the major sources of sediment and nutrient loading. These loads are reduced to acceptable levels through land-use changes and application of best management practices (BMPs).

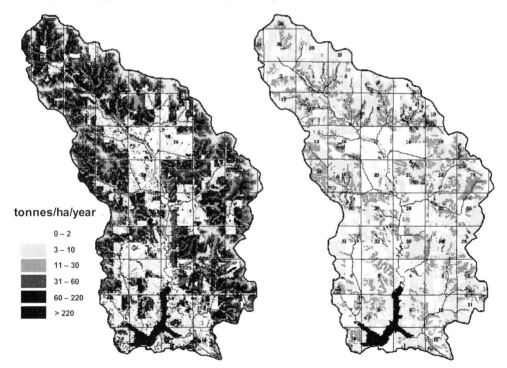

Figure. Geographical information system representation of the Rock Creek Lake, a reservoir in central Iowa. The left plate shows soil erosion estimated by the revised universal soil loss equation assuming no conservation efforts (32.3 metric tons/ha/year) and the right plate shows soil erosion assuming various BMPs (6.5 metric tons/ha/year). Shades of gray identify an array of erosion rates (as per the accompanying scale). (Plates courtesy of IDNR)

[1] Most of the information used in developing this box was contributed by Don Bonneau, Iowa Department of Natural Resources.

(Box continues)

> **Box 17.2. Continued.**
>
> Fisheries managers work in partnerships comprising government agencies, landowners, and nongovernment organizations and invest 25–30% of their time on watershed work associated with lake and stream projects. Fisheries managers work in various capacities within partnerships, often as leaders in providing technical details for specific projects. Although the approach may be intimidating to managers at first, it works by producing success stories that get the public support necessary to obtain funding needed to work at a watershed scale. Restorations are expensive and require years to complete but are also an investment in the local economy, fishing quality, and natural resources as a whole. Consequently, the Iowa legislature approved a state lakes program and allocated US$8.5 million in 2007 and similar amounts from 2008 to 2010. This lake protection and improvement program is administered by the IDNR with local and other matching funds expanding the program by over $5.5 million each year.

First and foremost, vegetation in riparian zones surrounding reservoirs serves to stabilize shorelines and reduce erosion and the flow of sediments into reservoirs. As sedimentation is considered the most important factor contributing to reservoir aging and habitat degradation, anything that can be done to protect riparian zones is desirable. Some reservoir management agencies, such as the TVA, have been active in establishing programs to protect and enhance riparian zones on private lands bordering their reservoirs, especially on main-stem reservoirs where water level fluctuations are less dramatic compared with tributary impoundments. It is generally accepted that retaining walls (i.e., bulkheads) are the worst form of shoreline stabilization because they result in impoverished littoral habitat and fish assemblages (Trial et al. 2001). From a fisheries perspective, the placement of riprap (with native plant establishment above the riprap zone) is a better way to stabilize shorelines subject to severe erosion. Regardless of what form bank stabilization takes, it will reduce the positive effects of the riparian forest (if present) on the distribution and abundance of large woody debris known to benefit freshwater fisheries (Angradi et al. 2004).

17.6 THE RIVER BASIN

A river basin is the portion of land drained by a river and its tributaries and includes multiple watersheds both upstream and downstream from a reservoir. Broad patterns of reservoir characteristics are evident at the river basin scale. In large river basins, variability in climate and physical characteristics among geographical sections of the basin influence diversity of hydrology. Patterns are also evident within river basins in relation to longitudinal gradients along chains of reservoirs. Basin scale variables are rarely controllable, but they constrain the expression of processes at smaller scales. Thus, an appreciation of basin patterns helps set limits for smaller-scale determinants and thereby helps managers understand the potentials and limits of reservoir management.

17.6.1 Longitudinal Gradients among Reservoirs in River Basins

The river continuum concept (RCC; Vannote et al. 1980) proposed a clinal view of rivers. According to the RCC, the physical character of a river has a gradient of conditions from headwaters to mouth, with upstream processes affecting downstream processes (see Chapters 18–21). The RCC does not apply directly to a reservoir chain in a basin. However, the notion of clinal change along a basin does apply to a reservoir chain. Clinal trends in reservoir attributes are basin specific yet exhibit broad common patterns. In general, the upper reaches of most basins tend to be forested, whereas the lower reaches tend to have higher levels of modifications due to agriculture. Characteristics such as mean depth, relative size of the limnetic zone, water retention time, oxygen and thermal stratification, substrate size, and water level fluctuations tend to increase in upstream reservoirs. Conversely, reservoir area, extent of the riverine and littoral zones, access to floodplains and associated wetlands, habitat diversity, and nutrient and sediment inputs tend to increase in downstream reservoirs. Many of these patterns are dictated by landscape characteristics and are also evident in chains of natural lakes (Martin and Soranno 2006), but exceptions are common given the diversity of landscapes.

Nutrient trapping by reservoirs along a basin reduces productivity down a series of reservoirs, although in reservoirs with large tributaries, nutrients and productivity may actually increase downstream. Lake Mead experienced a drastic drop in productivity after the impoundment of Lake Powell upstream on the Colorado River (Vaux et al. 1995). Similarly, in the Tietê River, Brazil, the uppermost reservoir in a chain of nine impoundments captured most of the nutrients released from São Paulo, the largest city in South America (Barbosa et al. 1999). In reservoirs on large rivers with low retention and (or) multiple influential tributaries, the effects of upstream reservoirs may not be as pronounced as in the above examples (Bruns et al. 1984; Agostinho et al. 2004). In the Tennessee River, upstream reservoirs retain a greater portion of inflowing nutrients owing to greater water retention, although their net loads are lower owing to smaller watersheds with different geomorphology and land cover (Voigtlander and Poppe 1989). Thus, nutrients and associated primary productivity, and likely many water quality variables, show spatial gradients among reservoirs within basins so that conditions in a given reservoir are predictable based on its position in the basin.

The RCC postulates that fish assemblages change along lotic systems in response to physical and nutrient gradients. Analogously, in impounded basins, reservoirs higher in a chain of reservoirs tend to have largely lacustrine, generalist fishes characteristic of sluggish upper reaches of basins (McDonough and Barr 1979). The reduction of riverine species is particularly evident for large migratory fishes stopped by dams that lack passage or interrupted by multiple dams with passages (Agostinho et al. 1999). Depending on latitude, upstream reservoirs in high elevations may include coolwater and coldwater species assemblages, and reservoirs lower in the series may transition into warmwater species assemblages. Riverine species become more common in downstream reservoirs, an effect that is especially evident in reservoirs below long, unimpounded stretches, with unimpounded tributaries or with extensive upstream floodplains.

In reservoirs of the Tennessee River, fish species richness, composition, and biomass changed longitudinally along the basin. Number of species increased from a low of less than 20 in high-elevation impoundments to nearly 70 in the lowermost reservoir (Miranda et al. 2008). Similarly, fish abundance increased in reservoirs further downstream. Additionally, species composition showed strong organization relative to position in the chain. Reservoirs high in the basin were characterized by a greater composition of bluegill, smallmouth bass, walleye, largemouth bass, river redhorse, and white bass. On the lower end of the basin, reservoir fish assemblages included greater representation by shads, blue catfish, buffaloes, gars, yellow bass, and redear sunfish. A relatively linear cline existed in between the two extremes. Trophic guild composition also tends to change along reservoir chains, with percentage composition by number of detritivores and planktivores increasing down basins and that of invertivores, invertivores–carnivores, and invertivores–detritivores increasing up basins (Miranda et al. 2008).

17.6.2 The Basin Perspective

Considering impoundments at a basin scale and viewing them as reaches in a river or links in a chain may generate management insight not available when considering them as isolated entities. An obvious feature of reservoir chains is a predictable spectrum of fish assemblages that can provide a diversity of recreational and commercial fisheries. Traditional management approaches may be organized relative to features of the reservoir series. For example, the effectiveness with which typical management efforts influence fish assemblages is likely to decrease downstream because reservoir size, species richness, and fish assemblage stability increase. Correspondingly, stocking, harvest regulations, and habitat manipulation programs are likely to be increasingly more effective in upstream reservoirs. Efforts to foster diverse commercial, subsistence, or recreational fisheries, and to provide multispecies fish-passage facilities to increase connectivity among impounded reaches separated by dams, are likely to be more effective in downstream reservoirs because those reservoirs tend to have more diversity of habitats and water regimes. These principles apply whether a basin has one or many reservoirs. Thus, a basin perspective professed by the RCC can serve as a template for considering reservoirs because they generally show longitudinal gradients at the scale of a single reservoir as well as at the scale of a chain of reservoirs constructed along a river.

17.7 CONCLUSIONS

Management of large reservoirs has emphasized solving reservoir-level problems. Expanding this view of reservoirs to include tributaries, watersheds, and river basin enhances a manager's abilities to influence reservoir fish populations and fish assemblages and can increase the effectiveness of in-reservoir management measures such as stocking (Box 17.3). Given a potentially overwhelming expansion in management problems, there is a need for reservoir biologists to expand the level of human resources involved in management through partnering among state, provincial, and federal agencies, local governments, universities, nongovernment organizations, corporations, and the public. Within this environment, the traditional control exerted by fisheries managers over a resource is diminished, but the potential to bring big, long-lasting changes to reservoir environments and biota is increased.

The importance of looking for solutions to reservoir fisheries problems beyond a reservoir itself is likely to increase with the level of human disturbance in a watershed. Reservoirs in relatively undisturbed watersheds with high-quality tributaries and riparian zones are likely to require mainly watershed protection and traditional in-reservoir management approaches. In contrast, reservoirs in highly-disturbed watersheds with highly engineered tributaries may require considerable out-of-reservoir attention before in-reservoir efforts become effective. In this latter class of reservoirs, a focus on traditional management activities such as regulations, stocking, and littoral zone improvement may be shortsighted and represent only short-term fixes to complicated landscape issues that are the underlying problems to maladies afflicting

Box 17.3. Introductions to Create New Fisheries and Stocking to Augment Native Fisheries

There is no such thing as a "reservoir species," and biologists have leeway when manipulating reservoir fish assemblages. The need to introduce species can arise because of the lack of pre-adapted species to colonize habitats such as the pelagic zone. Stocking existing species is often necessary because aging reservoirs lose their ability to sustain robust, exploitable fish populations. For all of these reasons, reservoir biologists routinely stock fish. A thorough discussion of when these activities are called for and how to determine which species (and how many) to use is presented in Chapter 9, but some comments specific to reservoir programs are appropriate.

Reservoir construction in the 20th century created millions of hectares of lacustrine habitat where aquatic resources were scarce. With the creation of vast expanses of offshore, deepwater habitats, new species were introduced to occupy those habitats. Nonnative lake trout were stocked into many western U.S. reservoirs to create trophy fisheries where none would otherwise have existed. Similarly, nonnative smallmouth bass and walleyes were stocked in impoundments in the Columbia River watershed to provide sport fisheries in new lacustrine habitats that were created when that major river system was regulated by more than two dozen dams (and the wisdom of stocking potential predators of juvenile salmon, some of which are endangered species, has been roundly questioned). One of the most successful reservoir stocking programs in the 20th century involved striped bass, an anadromous species native to the Atlantic and Gulf coasts of North America. Striped bass are stocked widely in the USA and are prized because of the large sizes they can achieve (20+ kg) and their pelagic habits. Their large size and preference for clupeid prey such as gizzard shad that often dominate the fish biomass in reservoirs render them particularly suitable in large reservoirs. Other species that are widely introduced and stocked in reservoirs to create offshore fisheries include hybrid striped bass (the usual cross is the palmetto bass), blue catfish, and several species of trout and salmon.

Whereas introductions of striped bass and other species have created sport fisheries where none previously existed, other reservoir stocking programs seek to replace declining recruitment by species native to impounded watersheds that have declined as reservoirs aged. Some examples include stocking crappies and walleyes in Tennessee tributary storage reservoirs (Isermann et al. 2002; Vandergoot and Bettoli 2003), walleyes in large main-stem Missouri River impoundments (Fielder 1992), and muskellunge in Ohio reservoirs (Bevelhimer et al. 1985). Conservation biologists stock various subspecies of native cutthroat trout (e.g., Bonneville and Rio Grande) into western U.S. reservoirs to preserve those native fishes and their unique genomes. Stocking the nonnative Florida subspecies of largemouth bass into reservoirs containing only native northern subspecies to promote introgression is still common, particularly in Texas (Buckmeier et al. 2003).

The introduction of new prey species into reservoirs was once widespread but has slowed because most candidate species have already been introduced in most reservoirs. The realization that the ecological costs of stocking nonnative prey species sometimes outweigh possible benefits also halted many planned introductions. For instance, nonnative alewives are readily preyed upon by pelagic predators, but alewives have caused the collapse of resident

(Box continues)

> **Box 17.3. Continued.**
>
> fish populations through various mechanisms (e.g., preying on larval sport fish and causing a vitamin deficiency in sport fish). One of the most widely stocked prey species in southern and western U.S. reservoirs is the threadfin shad because it assumes a pelagic existence in reservoirs (unlike many native forage fish species) and has many attributes of the ideal prey species (e.g., small maximum size). Another widely-introduced prey species is the emerald shiner, which has been stocked in reservoirs in Canada and the western and midwestern USA to provide forage for pelagic predators.

reservoir fish assemblages. The science of reservoir management is relatively new, and it is unclear how watershed improvements might improve fisheries (or reverse declines) in old reservoirs that have sustained decades of sedimentation, loss of woody debris, and general habitat degradation. Nevertheless, protecting watersheds is important even for old reservoirs to prevent further habitat degradation and unwanted shifts in fish assemblage structure.

Reservoir managers wishing to engage in out-of-reservoir activities to protect and enhance water quality, physical habitat, and fish assemblages may lack jurisdiction or expertise to operate beyond reservoir shores. Thus, landscape level partnerships must be forged. Partnerships provide the organization needed to plan, fund, and complete restoration work and give reservoir managers the political clout they may not have outside the reservoir basin. As partnership members, managers must be prepared to show linkages between a reservoir and a watershed and to be advocates for change that benefits fish in a reservoir. Managers should be equipped to contribute information suitable for developing restoration and protection plans, particularly relevant to how specific actions may affect reservoir water quality and biotic communities. To this end, river basin and watershed inventories documenting features important to reservoir condition are essential, focusing on critical areas representing major sources of problems likely to have large effects on the reservoir. Such problems might include large stretches of channelized tributaries that discharge excessive sediment loads and lack adequate habitat for reservoir species that spawn in tributaries, mistimed discharges from upstream impoundments, major tracts of wetlands disconnected from adjacent tributaries or the reservoir, agricultural ventures stretching down to the banks, and forest clear-cutting operations.

Considering that reservoir managers have traditionally focused on in-lake processes, links between reservoir fish assemblages and watersheds have not received sufficient attention and are likely to require research emphasis to build the capacity of managers to participate in landscape level partnerships. Fisheries researchers have established links between eutrophication and fish assemblage composition, concluded that oligotrophication can reduce fishery yields, and developed target ranges for optimum nutrient levels in some watersheds. However, the associations between watershed imports and reservoir fish assemblages are tenuous at best and are only beginning to be ascertained. The importance of riparian and buffer zones as filters has been studied in streams, but their contribution to littoral habitats in reservoirs has largely been ignored. Although reservoir managers know that some reservoir fishes use tributaries, the relationships between tributaries, their backwaters, river discharges, and fish assemblages that develop in reservoirs have received little or no attention. Furthermore, the natural gradient in abiotic and biotic features of reservoirs along a river basin deserves greater consideration when developing local or large-scale reservoir management plans.

17.8 REFERENCES

Agostinho, A. A., L. C. Gomes, S. Verissimo, and E. K. Okada. 2004. Flood regime, dam regulation and fish in the upper Paraná River: effects on assemblage attributes, reproduction, and recruitment. Reviews in Fish Biology and Fisheries 14:11–19.

Agostinho, A. A., L. C. Gomes, and M. Zalewski. 2001. The importance of floodplains for the dynamics of fish communities of the upper River Parana. International Journal of Ecohydrology & Hydrobiology 1:209–217.

Agostinho, A. A., L. E. Miranda, L. M. Bini, L. C. Gomes, S. M. Thomaz, and H. I. Susuki. 1999. Patterns of colonization in neotropical reservoirs, and prognoses on aging. Pages 227–265 in J. G. Tundisi and M. Straškraba, editors. Theoretical reservoir ecology and its applications. Backhuys Publishers, Leiden, The Netherlands.

Anders, P. J., and K. I. Ashley. 2007. The clear-water paradox of aquatic ecosystem restoration. Fisheries 32(3):125–128.

Angradi, T. R., E. W. Schweiger, D. W. Bolgrien, P. Imert, and T. Selle. 2004. Bank stabilization, riparian land use and the distribution of large woody debris in a regulated reach of the upper Missouri River, North Dakota, USA. River Research and Applications 20:829–846.

Arbuckle, K. E., and J. A. Downing. 2001. The influence of watershed land use on lake N:P in a predominantly agricultural landscape. Limnology and Oceanography 46:970–975.

Bain, M. B. 1993. Assessing impacts of introduced aquatic species: grass carp in large systems. Environmental Management 17:211–224.

Barbosa, F. A. R., J. Padisak, E. L. G. Espindola, G. Borics, and O. Rocha. 1999. The cascading reservoir continuum concept (CRCC) and its application to the River Tietê, São Paulo State, Brazil. Pages 425–437 in J. G. Tundisi and M. Straškraba, editors. Theoretical reservoir ecology and its applications. Backhuys Publishers, Leiden, The Netherlands.

Barwick, D. H. 2004. Species richness and centrarchid abundance in littoral habitats of three southern U.S. reservoirs. North American Journal of Fisheries Management 24:76–81.

BC (British Columbia) Hydro. 2007. Columbia River Project water use plan, Kinbasket and Arrow Lakes reservoirs revegetation management plan. Unpublished internal report. BC Hydro, Vancouver, British Columbia.

Beamesderfer, R. C., and B. E. Rieman. 1991. Abundance and distribution of northern squawfish, walleyes, and smallmouth bass in John Day Reservoir, Columbia River. Transactions of the American Fisheries Society 120:439–447.

Beaulac, M. N., and R. H. Reckhow. 1982. An examination of nutrient export relationships. Water Research Bulletin 18:1013–1024.

Berkman, H. E., and C. F. Rabeni. 1987. Effect of siltation on stream fish communities. Environmental Biology of Fishes 18:285–294.

Bernot, R. J., W. K. Dodds, M. C. Quist, and C. S. Guy. 2004. Spatial and temporal variability of zooplankton in a Great Plains reservoir. Hydrobiologia 525:101–112.

Bettoli, P. W., and J. A. Gordon. 1990. Aquatic macrophyte studies on Woods Reservoir, Tennessee. Journal of the Tennessee Academy of Science 65:4–8.

Bettoli, P. W., M. J. Maceina, R. L. Noble, and R. K. Betsill. 1993. Response of a reservoir fish community to large-scale aquatic vegetation removal. North American Journal of Fisheries Management 13:110–124.

Bettoli, P. W., and R. S. Osborne. 1998. Hooking mortality and behavior of striped bass following catch and release angling. North American Journal of Fisheries Management 18:609–615.

Bevelhimer, M. S. 1996. Relative importance of temperature, food, and physical structure to habitat choice by smallmouth bass in laboratory experiments. Transactions of the American Fisheries Society 125:274–283.

Bevelhimer, M. S., R. A. Stein, and R. F. Carline. 1985. Assessing significance of physiological differences among three esocids with a bioenergetics model. Canadian Journal of Fisheries and Aquatic Sciences 42:57–69.

Bonar, S. A., B. Bolding, and M. Divens. 2002. Effects of triploid grass carp on aquatic plants, water quality, and public satisfaction in Washington State. North American Journal of Fisheries Management 22:96–105.

Brooking, T. E., L. G. Rudstam, M. H. Olson, and A. J. VanDeValk. 1998. Size-dependent alewife predation on larval walleyes in laboratory experiments. North American Journal of Fisheries Management 18:960–965.

Brown, A. M. 1986. Modifying reservoir fish habitat with artificial structures. Pages 98–102 in G. E. Hall and M. J. Van Den Avyle, editors. Reservoir fisheries management: strategies for the 80's. American Fisheries Society, Southern Division, Reservoir Committee, Bethesda, Maryland.

Bruns, D. A., G. W. Minshall, C. E. Cushing, K. W. Cummins, J. T. Brock, and R. L. Vannote. 1984. Tributaries as modifiers of the river-continuum concept: analysis by polar ordination and regression models. Archiv für Hydrobiologie 99:208–220.

Bruton, M. N. 1985. Effects of suspensoids on fish. Hydrobiologia 125:221–241.

Buckmeier, D. L., J. W. Schlechte, and R. K. Betsill. 2003. Stocking fingerling largemouth bass to alter genetic composition: efficacy and efficiency of three stocking rates. North American Journal of Fisheries Management 23:523–529.

Buynak, G. L., L. E. Kornman, A. Surmont, and B. Mitchell. 1989. Longitudinal differences in electrofishing catch rates and angler catches of black bass in Cave Run Lake, Kentucky. North American Journal of Fisheries Management 9:226–230.

Canfield, D. E., Jr., K. A. Langeland, M. Maceina, W. T. Haller, and J. V. Shireman. 1983. Trophic state classification of lakes with aquatic macrophytes. Canadian Journal of Fisheries and Aquatic Sciences 40:1713–1718.

Carlson, R. E. 1977. A trophic state index for lakes. Limnology and Oceanography 22:361–369.

Carpenter, S. R., N. F. Caraco, D. L. Correll, R. W. Howarth, A. N. Sharpley, and V. H. Smith. 1998. Nonpoint pollution of surface waters with phosphorous and nitrogen. Ecological Applications 8:559–568.

Churchill, T. N., P. W. Bettoli, D. C. Peterson, W. C. Reeves, and B. Hodges. 2002. Angler conflicts in fisheries management: a case study of the striped bass controversy at Norris Reservoir, Tennessee. Fisheries 27(2):10–19.

Colvin, M. A. 1993. Ecology and management of white bass: a literature review. Missouri Department of Conservation, Federal Aid in Sport Fish Restoration, Project F-1-R-42, Study I-31, Final Report, Jefferson City.

Correll, D. L. 1997. Buffer zones and water quality protection: general principles. Pages 7–20 in N. E. Haycock, T. P. Burt, K. W. T. Goulding, and G. Pinay, editors. Buffer zones: their processes and potential in water protection. Quest Environmental, Harpendon, UK.

Coutant, C. C. 1985. Striped bass, temperature, and dissolved oxygen: a speculative hypothesis for environmental risk. Transactions of the American Fisheries Society 114:31–61.

Crisp, D. T., R. H. K. Mann, P. R. Cubby, and S. Robson. 1990. Effects of impoundment upon trout (*Salmo trutta*) in the basin of Cow Green Reservoir. Journal of Applied Ecology 27:1020–1041.

Daniels, R. B., and J. W. Gilliam. 1997. Sediment and chemical load reduction by grass and riparian filters. Soil Sciences Society of America Journal 60:246–251.

Davis, S. E., III, and B. C. Reeder. 2001. Spatial characterization of water quality in seven eastern Kentucky reservoirs using multivariate analyses. Aquatic Ecosystem Health & Management 4:463–477.

De Oliveira, E. F., C. V. Minte-Vera, and E. Goulart. 2005. Structure of fish assemblages along spatial gradients in a deep subtropical reservoir (Itaipu Reservoir, Brazil–Paraguay border). Environmental Biology of Fishes 72:283–304.

DiCenzo, V. J., and M. C. Duval. 2002. Importance of reservoir inflow in determining white bass year-class strength in three Virginia reservoirs. North American Journal of Fisheries Management 22:620–626.

Dick, G. O., R. M. Smart, and J. R. Snow. 2004. Aquatic vegetation restoration in Drakes Creek, Tennessee. U.S. Army Corps of Engineers, Engineer Research and Development Center, Waterways Experiment Station, Aquatic Plant Control Research Program Bulletin A-04-1, Vicksburg, Mississippi.

Doyle, M. W., E. H. Stanley, J. M. Harbor, and G. S. Gordon. 2003. Dam removal in the United States: emerging needs for science and policy. Eos 84:29, 32–33.

Fernando, C. H., and J. Holčík. 1982. The nature of fish community: a factor influencing the fishery potential and yields of tropical lakes and reservoir. Hydrobiologia 97:127–140.

Field, C. K., P. A. Siver, and A. M. Lott. 1996. Estimating the effects of changing land use patterns on Connecticut lakes. Journal of Environmental Quality 25:325–333.

Fielder, D. G. 1992. Evaluation of stocking walleye fry and fingerlings and factors affecting their success in lower Lake Oahe, South Dakota. North American Journal of Fisheries Management 12:336–345.

FISRWG (Federal Interagency Stream Restoration Working Group). 1998. Stream corridor restoration: principles, processes, and practices. Federal Interagency Stream Restoration Working Group, Government Printing Office Item 0120-A, Washington, D.C.

Forsberg, C., and S. O. Ryding. 1980. Eutrophication parameters and trophic state indices in 30 Swedish waste-receiving lakes. Archiv für Hydrobiologie 89:189–207.

Gido, K. B., C. W. Hargrave, W. J. Matthews, G. D. Schnell, D. W. Pogue, and G. W. Sewell. 2002. Structure of littoral-zone fish communities in relation to habitat, physical, and chemical gradients in a southern reservoir. Environmental Biology of Fishes 63:253–263.

Gomes, L. C., L. E. Miranda, and A. A. Agostinho. 2002. Fishery yield in relation to phytoplankton biomass in reservoirs of the upper Paraná River, Brazil. Fisheries Research 55:335–340.

Hale, R. S. 1999. Growth of white crappies in response to temperature and dissolved oxygen conditions in a Kentucky Reservoir. North American Journal of Fisheries Management 19:591–598.

Hamilton, S. K., and W. M. Lewis, Jr. 1990. Basin morphology in relation to chemical and ecological characteristics of lakes on the Orinoco River floodplain, Venezuela. Archiv für Hydrobiologie 119:393–425.

Hayward, R. S., and E. Arnold. 1996. Temperature dependence of maximum daily consumption in white crappie: implications for fisheries management. Transactions of the American Fisheries Society 125:132–138.

Henderson, J. E. 1996. Management of nonnative aquatic vegetation in large impoundments: balancing preferences and economic values of angling and nonangling groups. Pages 373–381 in L. E. Miranda and D. R. DeVries, editors. Multidimensional approaches to reservoir fisheries management. American Fisheries Society, Special Publication 16, Bethesda, Maryland.

Hoff, M. H. 1991. Effects of increased nesting cover on nesting and reproduction of smallmouth bass in northern Wisconsin lakes. Pages 39–43 in D. C. Jackson, editor. Proceedings of the first international smallmouth bass symposium. Mississippi State University, Starkville.

Honeyfield, D. C., J. P. Hinterkopf, J. D. Fitzsimmons, D. E. Tillit, J. L. Zajicek, and S. B. Brown. 2005. Development of thiamine deficiencies and early mortality syndrome in lake trout by feeding experimental and feral fish diets containing thiaminase. Journal of Aquatic Animal Health 17:4–12.

Hoxmeier, R. J. H., and D. R. DeVries. 1997. Habitat use, diet, and population size of adult and juvenile paddlefish in the lower Alabama River. Transactions of the American Fisheries Society 126:288–301.

Hoyer, M. V., M. D. Netherland, M. S. Allen, and D. E. Canfield Jr. 2005. Hydrilla management in

Florida: a summary and discussion of issues identified by professionals with future management recommendations. University of Florida, Institute of Food and Agricultural Sciences, Department of Fisheries and Aquatic Sciences, Florida LAKEWATCH, Gainesville.

Hunt, J., N. Bacheler, D. Wilson, E. Videan, and C. A. Annett. 2002. Enhancing largemouth bass spawning: behavioral and habitat considerations. Pages 277–290 in D. P. Philipp and M. S. Ridgway, editors. Black bass: ecology, conservation, and management. American Fisheries Society, Symposium 31, Bethesda, Maryland.

Isermann, D. A., P. W. Bettoli, S. M. Sammons, and T. N. Churchill. 2002. Initial poststocking mortality, oxytetracycline marking, and year-class contribution of black-nosed crappies stocked into Tennessee reservoirs. North American Journal of Fisheries Management 22:1399–1408.

Jackson, D. A., H. H. Harvey, and K. M. Somers. 1990. Ratios in aquatic sciences: statistical shortcomings with mean depth and the morphoedaphic index. Canadian Journal of Fisheries and Aquatic Sciences 47:1788–1795.

Jenkins, R. M. 1967. The influence of some environmental factors on standing crop and harvest of fishes in U.S. reservoirs. Pages 298–321 in Reservoir Committee. Reservoir fishery resources symposium. American Fisheries Society, Southern Division, Reservoir Committee, Bethesda, Maryland.

Jeric, R. J., T. Modde, and J. M. Godfrey. 1995. Evaluation of a method for measuring intragravel dissolved oxygen concentrations and survival to emergence in shore-spawned salmonids. North American Journal of Fisheries Management 15:185–192.

Johnson, B. L., and D. B. Noltie. 1996. Migratory dynamics of stream-spawning longnose gar (*Lepisosteus osseus*). Ecology of Freshwater Fishes 5:97–107.

Johnson, B. M., and J. P. Goettl Jr. 1999. Food web changes over fourteen years following introduction of rainbow smelt in a Colorado reservoir. North American Journal of Fisheries Management 19:629–642.

Jones, J. J., and M. V. Hoyer. 1982. Sportfish harvest predicted by summer chlorophyll-*a* concentrations in midwestern lakes and reservoirs. Transactions of the American Fisheries Society 111:176–179.

Jones, J. R., and M. F. Knowlton. 2005. Suspended solids in Missouri reservoirs in relation to catchment features and internal processes. Water Research 39:3629–3635.

Jones, J. R., M. F. Knowlton, D. V. Obrecht, and E. A. Cook. 2004. Importance of landscape variables and morphology on nutrients in Missouri reservoirs. Canadian Journal of Fisheries and Aquatic Sciences 61:1503–1512.

Kennedy, R. H. 1999. Reservoir design and operation: limnological implications and management opportunities. Pages 1–28 in J. G. Tundisi and M. Straškraba, editors. Theoretical reservoir ecology and its applications. Backhuys Publishers, Leiden, The Netherlands.

Kimmel, B. L. and A. W. Groeger. 1986. Limnological and ecological changes associated with reservoir aging. Pages 103–109 in G. E. Hall and M. J. Van Den Avyle, editors. Reservoir fisheries management: strategies for the 80's. American Fisheries Society, Southern Division, Reservoir Committee, Bethesda, Maryland.

Kimmel, B. L., O. T. Lind, and L. J. Paulson. 1990. Reservoir primary production. Pages 133–193 in K. W. Thornton, B. L. Kimmel, and E. E. Payne, editors. Reservoir limnology: ecological perspectives. Wiley, New York.

Kirk, K. L., and J. J. Gilbert. 1990. Suspended clay and the population dynamics of planktonic rotifers and cladocerans. Ecology 71:1741–1755.

Knoll, L. B., M. J. Vanni, and W. H. Renwick. 2003. Phytoplankton primary production and photosynthetic parameters in reservoirs along a gradient of watershed land use. Limnology and Oceanography 48:608–617.

Larinier, M. 2001. Environmental issues, dams and fish migration. Pages 45–90 in G. Marmulla, editor. Dams, fish and fisheries: opportunities, challenges and conflict resolution. FAO (Food and Agriculture Organization of the United Nations) Fisheries Technical Paper 419, Rome.

Lind, O. T. 2002. Reservoir zones: microbial production and trophic state. Lake and Reservoir Management 18:129–137.

Lucas, M. C., and E. Baras, editors. 2001. Migration of freshwater fishes. Blackwell Science, Oxford, UK.

Maceina, M. J. 2003. Verification of the influence of hydrologic factors on crappie recruitment in Alabama reservoirs. North American Journal of Fisheries Management 23:470–480.

Maceina, M. J., and D. R. Bayne. 2003. The potential impact of water reallocation on retention and chlorophyll a in Weiss Lake, Alabama. Lake and Reservoir Management 19:200–207.

Maceina, M. J., D. R. Bayne, A. S. Hendricks, W. C. Reeves, W. P. Black, and V. J. DiCenzo. 1996. Compatibility between water clarity and quality black bass and crappie fisheries in Alabama. Pages 296–305 in L. E. Miranda and D. R. DeVries, editors. Multidimensional approaches to reservoir fisheries management. American Fisheries Society, Symposium 16, Bethesda, Maryland.

Magette, W. L., R. B. Brinsfield, R. E. Palmer, and J. D. Wood. 1989. Nutrient and sediment removal by vegetated filter strips. Transactions of the American Society of Agricultural Engineering 32:663–667.

Martin, S., and P. A. Soranno. 2006. Lake landscape position: relationships to hydrologic connectivity and landscape features. Limnology and Oceanography 51:801–814.

Martyn, R. D., R. L. Noble, P. W. Bettoli, and R. C. Maggio. 1986. Mapping aquatic weeds and evaluating their control by grass carp with aerial infrared photography. Journal of Aquatic Plant Management 24:45–56.

McDonough, T. A., and W. C. Barr. 1979. An analysis of fish associations in Tennessee and Cumberland drainage impoundments. Proceedings of the Annual Conference of the Southeastern Association of Fish and Wildlife Agencies 31(1977):555–563.

Meals, D. W., and R. B. Hopkins. 2002. Phosphorus reductions following riparian restoration in two agricultural watersheds in Vermont, USA. Water Science and Technology 45(9):51–60.

Meals, K. O., and L. E. Miranda. 1991. Abundance of age-0 centrarchids in littoral habitats of flood control reservoirs in Mississippi. North American Journal of Fisheries Management 11:298–304.

Michaletz, P. H. 1997. Factors affecting abundance, growth, and survival of age-0 gizzard shad. Transactions of the American Fisheries Society 126:84–100.

Michaletz, P. H., and C. M. Gale. 1999. Longitudinal gradients in age-0 gizzard shad density in large Missouri reservoirs. North American Journal of Fisheries Management 19:765–773.

Miranda, L. E. 1996. Development of reservoir fisheries management paradigms in the 20th century. Pages 3–11 in L. E. Miranda and D. R. DeVries, editors. Multidimensional approaches to reservoir fisheries management. American Fisheries Society, Symposium 16, Bethesda, Maryland.

Miranda, L. E. 1999. A typology of fisheries in large reservoirs of the United States. North American Journal of Fisheries Management 19:536–550.

Miranda, L. E., and H. Gu. 1998. Dietary shifts of a dominant reservoir planktivore during early life stages. Hydrobiologia 377:73–83.

Miranda, L. E., M. D. Habrat, and S. Miyazono. 2008. Longitudinal gradients along a reservoir cascade. Transactions of the American Fisheries Society 137:1851–1865.

Miranda, L. E., and D. R. Lowery. 2007. Juvenile densities relative to water regime in mainstem reservoirs of the Tennessee River. Lakes & Reservoirs: Research and Management 12:89–98.

Miranda, L. E., and L. L. Pugh. 1997. Relationship between vegetation coverage and abundance, size, and diet of juvenile largemouth bass during winter. North American Journal of Fisheries Management 17:601–610.

Mundahl, N. D., and T. E. Wissing. 1987. Nutritional importance of detritivory in the growth and condition of gizzard shad in an Ohio reservoir. Environmental Biology of Fishes 20:129–142.

Naiman R. J., and H. Decamps. 1997. The ecology of interfaces: riparian zones. Annual Review of Ecology and Systematics 28:621–658.

Ney, J. J. 1996. Oligotrophication and its discontents: effects of reduced nutrient loading on reservoir fisheries. Pages 285–295 *in* L. E. Miranda and D. R. Devries, editors. Multidimensional approaches to reservoir fisheries management. American Fisheries Society, Special Publication 16, Bethesda, Maryland.

NID (National Inventory on Dams). 2008. National inventory on dams. U.S. Army Corps of Engineers, Alexandria, Virginia. Available: http://crunch.tec.army.mil/nidpublic/webpages/nid.cfm (January 2010).

Novotny V. 2003. Water quality: diffuse pollution and watershed management. Wiley, Hoboken, New Jersey.

Okada, E. K., A. A. Agostinho, and L. C. Gomes. 2005. Spatial and temporal gradients in artisanal fisheries of a large Neotropical reservoir, the Itaipu Reservoir, Brazil. Canadian Journal of Fisheries and Aquatic Sciences 62:714–724.

Pagioro, T. A., and S. M. Thomaz. 2002. Longitudinal patterns of sedimentation in a deep, monomictic subtropical reservoir (Itaipu, Brazil–Paraguay). Archiv für Hydrobiologie 154:515–528.

Parsons, B. G. M., and W. A. Hubert. 1988. Influence of habitat availability on spawning site selection by kokanees in streams. North American Journal of Fisheries Management 8:426–431.

Prowse, T. D., F. J. Wrona, and G. Power. 2004. Dams, reservoirs and flow regulation. Pages 9–19 *in* Environment Canada. Threats to water availability in Canada. NWRI (National Water Research Institute) Scientific Assessment Report Series 3 and ACSD (Atmospheric and Climate Science Directorate) Science Assessment Series 1, Burlington, Ontario.

Puri, A., G. E. MacDonald, and W. T. Haller. 2007. Stability of fluridone-resistant hydrilla (*Hydrilla verticallata*) biotypes over time. Weed Science 55:12–15.

Pusey, B., and A. H. Arthington. 2003. Importance of the riparian zone to the conservation and management of freshwater fish: a review. Marine and Freshwater Research 54:1–16.

Quinn, S. P. 1992. Angler perspectives on walleye management. North American Journal of Fisheries Management 12:367–378.

Raborn, S. W., L. E. Miranda, and T. M. Driscoll. 2002. Effects of simulated removal of striped bass from a southeastern reservoir. North American Journal of Fisheries Management 22:406–417.

Rigler, F. H. 1982. The relation between fisheries management and limnology. Transactions of the American Fisheries Society 111:121–132.

Robinson, S. 2000. The experience with dams and resettlement in Mexico. Contributing paper to Displacement, resettlement, rehabilitation, reparation and development. World Commission on Dams Thematic Review Social Issues 1.3, Cape Town, South Africa.

Roni, P., K. Hanson, T. Beechie, G. Pess, M. Pollock, and D. M. Bartley. 2005. Habitat rehabilitation for inland fisheries: global review of effectiveness and guidance for rehabilitation of freshwater ecosystems. FAO (Food and Agriculture Organization of the United Nations) Fisheries Technical Paper 484, Rome.

Ryder, R. A. 1965. A method for estimating the potential fish production of north temperate lakes. Transactions of the American Fisheries Society 94:214–218.

Sammons, S. M., L. G. Dorsey, P. W. Bettoli, and F. C. Fiss. 1999. Effects of reservoir hydrology on reproduction by largemouth bass and spotted bass in Normandy Reservoir, Tennessee. North American Journal of Fisheries Management 19:78–88.

Sammons, S. M., M. J. Maceina, and D. G. Partridge. 2003. Changes in behavior, movement, and home ranges of largemouth bass following large-scale hydrilla removal in Lake Seminole, Georgia. Journal of Aquatic Plant Management 41:31–38.

Santos, L. N., F. G. Araújo, and D. S. Brotto. 2008. Artificial structures as tools for fish habitat rehabilitation in a neotropical reservoir. Aquatic conservation: marine and freshwater ecosystems 18(6):896–908. Available: www.interscience.wiley.com. (February 2008).

Sass, G. G., J. F. Kitchell, S. R. Carpenter, T. R. Hrabik, A. E. Marburg, and M. G. Turner. 2006. Fish community and food web responses to a whole-lake removal of coarse woody habitat. Fisheries 31(7):321–330.

Scheffer, M. 2001. Ecology of shallow lakes. Kluwer Academic Publishers, London.

Schultz, R. C., J. P. Colletti, T. M. Isenhart, W. W. Simpkins, C. W. Mizc, and M. L. Thompson. 1995. Design and placement of a multispecies riparian buffer strip system. Agroforestry Systems 29:201–226.

Smart, R. M., G. O. Dick, and R. D. Doyle. 1998. Techniques for establishing native aquatic plants. Journal of Aquatic Plant Management 36:44–49.

Smart, R. M., R. D. Doyle, J. D. Madsen, and G. O. Dick. 1996. Establishing native submersed aquatic plant communities for fish habitat. Pages 347–356 in L. E. Miranda and D. R. DeVries, editors. Multidimensional approaches to reservoir fisheries management. American Fisheries Society, Symposium 16, Bethesda, Maryland.

Smith, V. H. 1998. Cultural eutrophication of inland, estuarine and coastal waters. Pages 7–49 in M. L. Pace and P. M. Groffman, editors. Successes, limitations and frontiers in ecosystem science. Springer, New York.

Soballe, D. M., and B. L. Kimmel. 1987. A large-scale comparison of factors influencing phytoplankton abundance in rivers, lakes, and impoundments. Ecology 68:1943–1954.

Stables, T. B., G. L. Thomas, S.L.Thiesfeld, G. B. Pauley, and M. A. Wert. 1990. Effects of reservoir enlargement and other factors on the yield of wild rainbow and cutthroat trout in Spada Lake, Washington. North American Journal of Fisheries Management 10:305–314.

Strakosh, T. R., J. L. Eitzmann, K. B. Gido, and C. S. Guy. 2005. The response of water willow *Justicia americana* to different water inundation and desiccation regimes. North American Journal of Fisheries Management 25:1476–1485.

Strange, R. J., W. B. Kittrell, and T. D. Broadbent. 1982. Effects of seeding reservoir fluctuation zones on young-of-the-year black bass and associated species. North American Journal of Fisheries Management 2:307–315.

Stockner, J. G., E. Rydin, and P. Hyenstrand. 2000. Cultural oligotrophication: causes and consequences for fisheries resources. Fisheries 25(5):7–114.

Taylor, W. D., and J. C. H. Carter. 1998. Zooplankton size and its relationship to trophic status in deep Ontario lakes. Canadian Journal of Fisheries and Aquatic Sciences 54:2691–2699.

Thomaz, S. M., T. A. Pagioro, L. M. Bini, M. D. Roberto, and R. R. de Araújo Rocha. 2004. Limnological characterization of aquatic environments and the influence of hydrometric levels. Pages 75–102 in S. M. Thomaz, A. A. Agostinho, and N. S. Hahn, editors. The upper Paraná River and its floodplain: physical aspects, ecology and conservation. Backhuys Publishers, Leiden, The Netherlands.

Trial, P. F., F. P. Gelwick, and M. A. Webb. 2001. Effects of shoreline urbanization on littoral fish assemblages. Lake and Reservoir Management 17:127–138.

Tugend, K. I., M. S. Allen, and M. A. Webb. 2002. Use of artificial habitat structures in U. S. lakes and reservoirs: a survey from the Southern Division AFS Reservoir Committee. Fisheries 27(5):22–26.

Vandergoot, C. S., and P. W. Bettoli. 2003. Relative contribution of stocked walleyes in Tennessee reservoirs. North American Journal of Fisheries Management 23:1036–1041.

Vanni, M. J., K. Arend, M. T Bremigan, D. B. Bunnell, J. E. Garvey, M. J. González, W. H. Renwick, P. A. Soranno, and R. A. Stein. 2005. Linking landscapes and food webs: effects of omnivorous fish and watersheds on reservoir ecosystems. BioScience 55:155–167.

Vannote, R. L., J. V. Minshall, K. W. Cummins, J. R. Seddell, and C. E. Cushing. 1980. The river continuum concept. Canadian Journal of Fisheries and Aquatic Sciences 37:130–137.

Vaux, P., L. Paulsen, R. Axler, and S. Leavitt. 1995. Water quality implications of artificially fertilizing a large desert reservoir for fisheries enhancement. Water Environment Research 67:189–200.

Voigtlander, C.W., and W. L. Poppe. 1989. The Tennessee River. Pages 372–384 *in* C. P. Dodge, editor. Proceedings of the international large river symposium. Canadian Special Publications in Fisheries and Aquatic Sciences 106.

Walser C. A., and H. L. Bart Jr. 1999. Influence of agriculture on in-stream habitat and fish community structure in Piedmont watersheds of the Chattahoochee River system. Ecology of Freshwater Fish 8:237–246.

Wilde, G. R., R. K. Reichers, and J. Johnson. 1992. Angler attitudes towards control of freshwater vegetation. Journal of Aquatic Plant Management 30:77–79.

Wiley, M. J., R. W. Gorden, S. W. Waite, and T. Powless. 1984. The relationship between aquatic macrophytes and sport fish production in Illinois ponds: a simple model. North American Journal of Fisheries Management 4:111–119.

Williamson, R. B., C. M. Smith, and A. B. Cooper. 1996. Watershed riparian management and its benefits to a eutrophic lake. Journal of Water Resources Planning and Management 122:24–32.

Wills, T. C., M. T. Bremigan, and D. B. Hayes. 2004. Variable effect of habitat enhancement structures across species and habitats in Michigan reservoirs. Transactions of the American Fisheries Society 133:399–411.

World Wildlife Fund. 2004. Rivers at risk: dams and the future of freshwater ecosystems. World Wildlife Fund, Surrey, England.

Yurk, J. J., and J. J. Ney. 1989. Phosphorous-fish community biomass relationships in southern Appalachian reservoirs: can lakes be too clean for fish? Lake and Reservoir Management 5:83–90.

Zale, A. V., J. D. Wiechman, R. L. Lochmiller, and J. Burroughs. 1990. Limnological conditions associated with summer mortality of striped bass in Keystone Reservoir, Oklahoma. Transactions of the American Fisheries Society 119:72–76.

Chapter 18

Coldwater Streams

ROBERT E. GRESSWELL AND BRUCE VONDRACEK

18.1 INTRODUCTION

Coldwater streams are typically found in headwater areas across North America. These systems tend to have channel slopes of greater than 2%, pool–riffle sequences that promote aeration, and riparian canopies that moderate temperatures. Environmental gradients and processes often produce continuous and predictable changes in habitat from headwaters downstream, and species assemblages (e.g., macroinvertebrates, amphibians, and fish) generally reflect the gradients.

Maximum daily mean water temperature is usually less than 22°C in coldwater streams. Water temperature is maintained by groundwater inputs and (or) weather conditions in high-elevation and temperate areas. Most coldwater streams occur in snowmelt-dominated drainages, but in regions that are more temperate, coldwater streams can occur in rain-dominated systems where groundwater inputs are common.

Productivity and faunal diversity in coldwater streams are low (especially in western North America) compared with warmwater streams. In Yellowstone National Park, for example, there were only 13 native fishes in almost 4,300 km of coldwater streams. At the same time, the proportion of coldwater streams occupied by fishes was great. Only high-elevation coldwater streams isolated above barriers were historically devoid of fish, apparently because they were not invaded following late-Pleistocene glaciation (Smith et al. 2002). Since the latter part of the 19th century, however, salmonids have been introduced into most all of these formerly fishless streams in North America.

Salmonids, cottids, and cyprinids are the dominant fish taxa in coldwater streams, and salmonids support highly-valued recreational fisheries. In fact, coldwater streams in North America attract anglers from around the world who seek opportunity to catch native and nonnative salmonids. In this chapter, abiotic and biotic characteristics of coldwater streams with emphasis on factors that influence fisheries management are discussed. Although historical and current approaches are noted, an emphasis is maintained on emerging management trends, concepts, and approaches. The reader is encouraged to seek detailed information concerning specific topics from preceding chapters in this book and cited literature. We have limited the discussion to potamodromous (migrating only in freshwater) and nonmigratory fishes; Chapter 19 provides information on anadromous (feeding and growing in the ocean or an estuary, but reproducing in freshwater) fishes and tailwater habitats.

18.2 CHARACTERISTICS OF COLDWATER STREAMS

In general, the river continuum concept (Vannote et al. 1980) provides insight into the general organization of coldwater stream systems from headwater tributaries to large main-stem rivers. Streams are viewed as systems in which the physical characteristics and co-occurring biotic communities change along a gradient from source to mouth. This gradient is generally reflected by increasing size and complexity moving downstream. Production in upstream portions of coldwater streams is generally from allochthonous (coming from outside the system) sources, and streams are tightly linked to the bordering riparian and terrestrial systems. Fish in these areas often spend their entire lives in a limited portion of stream, so they are susceptible to changes in terrestrial habitats (Allan 1995). As the channel widens and discharge increases, the riparian canopy has less direct affect, and increased light and water temperature lead to greater autochthonous (instream) production. Diversity of both macroinvertebrates and fishes generally increases in a downstream direction.

The river continuum concept has provided insight into coldwater stream systems, but it has not been useful for finer-scale questions concerning the distribution and abundance of specific stream biota. There has been growing awareness about the effects of spatial and temporal landscape dynamics on habitat complexity (Frissell et al. 1986; Pickett and Cadenasso 1995) and habitat–fish relationships (Fausch et al. 2002; Gresswell et al. 2006). Consequently, a hierarchical view of stream systems in the context of the watershed has been promoted (Frissell et al. 1986; Figure 18.1). This integrated multiscale approach incorporates spatial variation from microhabitats to watersheds that may persist from minutes to millennia. Linkage between spatial extent and temporal persistence is especially valuable for managing salmonid fisheries and stream habitats that support them.

The structure and composition of fish assemblages in coldwater streams are influenced by a complex set of interacting factors. In the broadest sense, these factors can be divided into two categories: (1) abiotic factors, including physical and chemical attributes that affect biological activity, and (2) biotic factors, such as competition and predation. Abiotic factors generally control the distribution and abundance of species at broad spatial and temporal scales (e.g., Rieman and McIntyre 1995), and biotic factors generally influence fishes at finer scales (Quist and Hubert 2005). Although the following sections address individual factors, it is virtually impossible to separate discrete factors in the natural world (Warren and Liss 1980). Abiotic and biotic processes are difficult to discuss individually because some of the best examples are a result of their interactions. Biotic interactions are strongest where fish abundance is high, and this situation is most often related to a relatively benign and predicable abiotic environment (Allan 1995). Because of these structuring constraints, abiotic factors provide a good starting point for discussing factors that influence coldwater fish assemblages.

18.2.1 The Physicochemical Template

Physical processes shape fish assemblages in coldwater streams through formation of suitable habitat space. Climate and geology are two of the primary physical factors influencing habitat space and constraining fish assemblages (Montgomery and Buffington 1998). Major climate-related factors are temperature and precipitation. At the landscape scale, water temperature can be used to predict the presence of thermally-sensitive fish species that com-

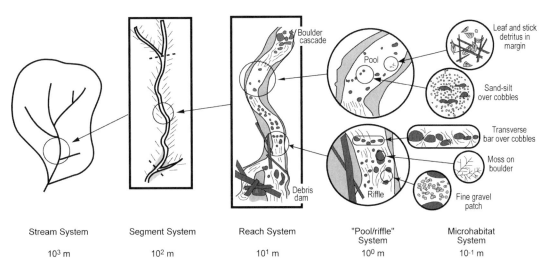

Figure 18.1. Conceptualization of the hierarchical organization of a stream system and associated habitat subsystems (from Frissell et al. 1986).

monly occur in coldwater streams (Rieman and McIntyre 1995). For example, Dunham et al. (2003a) found that when water temperature was combined in analyses with other environmental variables (instream cover, channel form, substrate, and the abundance of native and nonnative salmonid fishes), water temperature was the only parameter that was strongly associated with the distribution of bull trout across the landscape. Within individual streams, water temperature can influence the dynamics between native and nonnative fishes (Peterson et al. 2004; Coleman and Fausch 2007).

Precipitation is ultimately responsible for supply of water to a stream system, either directly as runoff or indirectly through groundwater. The resulting hydrological regime, specifically peak discharges (Montgomery and Buffington 1998), interacts with bedrock geology through the transport of sediment and with large woody debris to shape and maintain habitat for aquatic and riparian species (Frissell et al. 1986; Reeves et al. 1995; Poff et al. 1997). At the watershed, stream segment, and reach scales, geologic processes form a physical habitat template (sensu Southwood 1977; Poff and Ward 1990) that interacts with climatic factors to influence habitat space and constrain fish assemblages through variations in stream depth, width, gradient, sediment type and availability, and microhabitat. Finer-scale features at the habitat unit (e.g., individual pool, run, or riffle) and microhabitat scale are also influenced by geology.

The interaction of geology and hydrology results in the formation and distribution of riffles and pools in coldwater streams, which affect water depth, water velocity, and substrate type. These variables contribute to a diversity of potential habitat spaces, ultimately contributing to variation in salmonid assemblages and distributions (Hicks and Hall 2003; Ganio et al. 2005). Increasing habitat diversity and complexity generally lead to increased assemblage diversity (Flebbe and Dolloff 1995).

Typically, coldwater fishes require streams ample rocky substrate and an adequate proportion of pool habitat. Most coldwater fishes use gravels as substrate for spawning, and the size of the gravels used is largely a function of the species and size of sexually-mature fish (Bateman and Li 2001; Crisp 2000; Mundahl and Sagan 2005). Additionally, the presence of downwelling and upwelling currents through gravels can be important for spawning and survival of embryos and fry (Kondolf 2000). Optimal water velocities where fishes forage most efficiently vary by species, age, and life stage (Crisp 1993). Because habitat is influenced by discharge regimes, human manipulation of stream discharge can have substantial effects on fishes.

Pool habitat is important for salmonids throughout the year and movements among pools are common (Young 1996; Gresswell and Hendricks 2007). The linear range of movements varies by species, life stage, and habitat. For example, adult brown trout often display high site fidelity for a single pool or pool–riffle combination (Northcote 1997; Burrell et al. 2000). In rain-dominated systems, habitat may be limited in late summer and fall during periods of low discharge when connectivity among pools is reduced. However, brown trout in the southeastern USA may be more active in fall and winter than in spring and summer (Burrell et al. 2000). During winter in more northern latitudes, salmonid behavior often changes from feeding and defending territories to hiding and schooling (Northcote 1978), and large-scale movements decrease (Hilderbrand and Kershner 2000; Gresswell and Hendricks 2007). Large, deep pools immediately adjacent to the channel and connected to groundwater can be important overwinter habitat (Harper and Farag 2004). These examples underscore the importance of recognizing the variability of habitat through time and space and the importance of incorporating this recognition into management strategies.

18.2.2 Biotic Factors

18.2.2.1 Food and feeding

Abundance and production of fish are directly related to growth, mortality, and reproduction, processes that are influenced by both abiotic and biotic factors and their interactions. The biotic community is dependent on energy inputs from autotrophic periphyton (primarily diatoms), coarse particulate organic matter (e.g., leaves, wood, and grass) that is decomposed by microbes and some macroinvertebrates (detritivores), and dissolved and fine particulate organic matter that originates in adjacent riparian areas or from upstream (Allan 1995). Herbivores and larger detritivores utilize these energy sources, and, in turn, they are consumed by predators, both invertebrates and vertebrates. Linkages among trophic levels in food webs of coldwater streams are complex and vary through time in relation to changes in the physical environment (e.g., water quality and discharge).

In small coldwater streams, fish generally feed on drift (terrestrial or aquatic invertebrates in the water column), benthic invertebrates, and (or) fish. There is a strong terrestrial influence on drift in headwater streams because of the linkage to adjacent riparian areas (Allan 1995; Romero et al. 2005). Some fishes are dependent on smaller fish as prey, and in larger coldwater streams, piscivory is common.

Foraging is often site specific with substantial fidelity. For example, as young fish become free swimming, they move to refuge and feeding areas. In some species, salmonid fry establish and defend a territory to maximize potential energy intake and increase their growth and survival (Grant and Kramer 1990). Most young salmonids feed on drift and those with the best locations, usually the upper portion of a riffle, encounter food first. A dominance hierarchy directly related to fish size can occur (Elliott 1994). Greater habitat diversity tends to increase the availability of prey species at feeding locations.

Adaptation to variable food availability is common among fishes and periods of starvation may be followed by periods of high food abundance. Many fishes appear to be opportunistic feeders, and food items are often consumed according to availability (Romero et al. 2005). Individuals of piscivorous species initially feed on invertebrates, but growth usually results in a shift from insectivory to piscivory. As fish age, food selectivity may increase (Grant 1990).

18.2.2.2 Mortality

Mortality rates generally differ among life stages and are highest during the egg stage. During early life stages, mortality rates are often density dependent and vary with the carrying capacity of sites. Although mortality is high immediately after emergence, mortality rates tend to decrease as fish grow. Mortality of juvenile salmonids during winter may approach 50% of the fall population. Because fry often seek shelter from strong currents at stream margins and in or near the substrate, mortality at this life stage may be related to mechanical injury due to bed movement or impingement on stream substrate, or may be due to stranding in impounded areas related to ice accumulation (Griffith 1993). In rain-dominated systems, mortality may be greatest in autumn during periods of low discharge (Berger and Gresswell 2009).

Disease may cause mortality in salmonid populations and negatively affect fisheries. For example, *Flavobacterium psychrophilum* causes "coldwater disease" in populations of wild salmonids and can cause up to 50% mortality in juveniles (Bratovich et al. 2004). Similarly, whirling disease, caused by the exotic parasite *Myxobolus cerebralis*, has resulted in declines of rainbow trout recruitment in rivers of the western USA (Vincent 1996). The effects of whirling disease vary among species and with the size of fish. Brown trout are somewhat resistant, whereas brook trout and cutthroat trout are susceptible (Thompson et al. 1999). Young salmonids are more vulnerable than adults (Vincent 1996).

Predation can be a significant factor in mortality of fishes in coldwater streams (Quist and Hubert 2005). In many cases, life history organization and habitat use are directly affected by predation (Gilliam and Fraser 2001). For example, Bardonnet and Heland (1994) observed that in the absence of predator fishes (age-1 and older trout and sculpins), emerging brown trout were common at water depths of 20–30 cm with a pebble substrate, but when predators were present, most remained hidden in water less than 10 cm deep. Predation in coldwater streams is often temporally variable, and in some cases there are substantial seasonal differences in predation related to the developmental stage of fish (of both predator and prey) or habitat availability (Berger and Gresswell 2009).

Angling mortality, primarily associated with harvest, can cause declines in sport fish populations (Gresswell 1988; Post et al. 2002). Substantial mortality can occur even with low harvest rates and modest levels of angling effort (Gresswell 1995), so most state fisheries management agencies attempt to control angling mortality through fishing regulations (see Chapter 7). Regulations designed to maintain or rebuild naturally-reproducing fish populations (i.e., special regulations) include creel limits, size limits, terminal gear specifications, and season-length restrictions (Gresswell and Harding 1997). Used either singly or in combination, special regulations have been effective for protecting and rebuilding fisheries in many regions of North America; however, they are not without limitations. For instance, hooking mortality must be low. If angler harvest does not represent a major portion of total mortality, or if natural mortality is compensatory, regulations aimed at reducing angler harvest will be ineffective (Shetter and Alexander 1967).

Where environmental conditions limit fish growth, modifications to population size structure may not occur even in the absence of angling (Clark and Alexander 1985). Therefore, fish size targets may not be attainable even when a fish population is protected from excessive angler mortality. Furthermore, some fish species (or even segments of a population) may not be vulnerable to angling, and therefore, angling quality may be low even when population density is high. Unequal availability of fish to anglers can influence the response to regulations, especially in mixed-species fisheries (Gresswell and Harding 1997).

18.2.2.3 Life history characteristics

Life history characteristics of fishes include numerous physiological and behavioral qualities associated with maturation and reproduction, such as age and size at maturity, fecundity, migration characteristics, reproductive life span, and parental care. Life history variations exist among and within fish species. In fact, life history variation can occur at several levels, including species, subspecies, metapopulations (a group of populations linked by episodic movements of individuals among populations), populations, or individuals (Gresswell et al. 1994). Individual life history characteristics, however, can occur at more than one level of organization (i.e., groups of local populations and metapopulations may share similar life histories). For example, anadromy and potamodromy can occur in the same species, and in some species (e.g., brown trout), the same individual may exhibit both life history strategies during its life (Elliott 1994).

Although specifics of spawning vary, a general characteristic of salmonid spawning migrations is natal homing (return of adult spawners to the area of their birth). Migratory behavior of potamodromous salmonids and the environmental factors that influence them are frequently the subjects of research related to movements of salmonid fishes (e.g., Northcote 1978; Gresswell et al. 1997). Definitions related to spawning migrations (Varley and Gresswell 1988) have been used to identify broad life history categories (Northcote 1997). Four migratory spawning patterns have been observed for Yellowstone cutthroat trout: (1) fluvial (stream residents dispersing locally within the home range), (2) fluvial–adfluvial (fluvial residents moving into tributaries to spawn), (3) lacustrine–adfluvial (lake residents moving into lake tributaries to spawn), and (4) allacustrine (lake residents moving into the outlet stream; Varley and Gresswell 1988). Northcote (1997) reported fluvial and fluvial–adfluvial migrations were the most common forms of potamodromy for salmonids in general, but lacustrine–adfluvial migrations were also common. Of the four patterns, allacustrine migrations were the least common (Northcote 1997). Fluvial life history types can include those fishes in headwater streams where true migrations do not occur and reproductive movements are to local areas where suitable spawning substrates are available (Gresswell and Hendricks 2007).

Although the specifics of spawning vary, completion of the life cycle involves many complex behaviors. During spawning, eggs of most Salmoninae (trout, salmon, and *Salvelinus* spp.) are buried in redds (nests), and the creation, choice, and guarding of redds by the female is common. However, Coregoninae (whitefishes and ciscoes) and Arctic grayling are broadcast spawners, and males defend territories visited by females. Alevins (yolk sac fry) hatch after an incubation period that varies from weeks to months. Alevins of species that spawn in the fall (e.g., brook trout and bull trout) emerge in spring. While in the gravel, larvae receive nourishment from the yolk sac. Following emergence they obtain food from the gravel surface, then disperse and establish feeding territories upon absorption of the yolk sac.

Unlike salmonids, male slimy sculpin guard a nest and provide parental care for offspring, often from more than one female. Females produce about 100 eggs (Keeler and Cunjak 2007). During the reproductive period, nests built by males using cobble substrate in shallow water can be sensitive to changes in discharge. Adult mottled sculpin have restricted home ranges and are territorial, exhibiting little overlap with neighboring sculpins (Petty and Grossman 2007). In contrast, juvenile mottled sculpin are not territorial and occupy overlapping home ranges along stream margins. Reticulate sculpin in the Coast Range of Oregon exhibit positive selection for moderately-embedded cobble substrate even when the availability of such habitat varies among streams (Bateman and Li 2001). Apparently, cobble provides cover for males guarding nests.

Life history characteristics of coldwater fishes are linked to both abiotic and biotic components of the environment (Northcote 1978; Gresswell et al. 1997); therefore, changes in the environment resulting from human activities can have negative consequences for coldwater fishes. Barriers to movements resulting from dams or water diversion structures can block migrations or alter discharge patterns that act as cues for migrating spawners. Moreover, migratory life history types are suppressed by habitat fragmentation (Rieman and McIntyre 1995). Additionally, reduced sediment inputs and increased embeddedness can limit spawning and rearing habitats below dams (Van Kirk and Benjamin 2001). In some cases, physical characteristics that promote reproductive isolation related to the timing of reproduction are altered and the probability of hybridization with nonnative fishes can increase (Henderson et al. 2000). Habitat fragmentation negatively affects persistence by reducing total available habitat, inhibiting dispersal behaviors, simplifying habitat structure, and limiting resilience to stochastic disturbance. Increased water temperature related to riparian zone management and altered discharge patterns can negatively affect native species, and in some cases such changes can favor expansion of nonnative fishes with "generalist" habitat requirements (e.g., brown trout) and reductions in species with narrow habitat requirements (Dunham et al. 2003b).

18.2.2.4 Biotic interactions

Perhaps the clearest and most common example of biotic interactions in coldwater streams is predation. When predator abundance is high, prey species composition and abundance may be regulated, regardless of abiotic conditions (Quist and Hubert 2005). Predation is often implicated when native salmonids are replaced by invasive species (Kruse et al. 2000). Less obvious indirect effects of predation may also occur. For example, to reduce predation risk, prey species may seek poorer quality habitat that limits abundance (Gilliam and Fraser 2001) and growth rates of larval fish may be impeded (Bardonnet and Heland 1994)

Competitive interactions affect fish assemblage structure in some coldwater streams, and some level of both interspecific and intraspecific competition is likely where habitat is suitable for multiple species. Interspecific competition occurs between individuals of different species and habitat partitioning, rather than competitive exclusion, appears to be the most frequent outcome (Freeman and Grossman 1992; Jackson et al. 2001). Interspecific competition can involve competition among native species or competition between native and nonnative species, but in many cases, niche separation may reduce interspecific competition among co-evolved native species. For example, native brook trout and slimy sculpin can coexist without evidence of competition (Zimmerman and Vondracek 2006).

Species often inhabit areas that are associated with species-specific habitat requirements or are simply free of competition. In one study, riffle sculpin and speckled dace were observed in similar habitats at opposite ends of a 12.5-km reach of stream (Baltz et al. 1982). Riffle sculpin did not occur at sites in downstream portions of the section that were warm, whereas speckled dace were not found in upper, colder parts of the reach. Among three species that exhibit longitudinal replacement in streams of the Rocky Mountains (brook trout at high elevations, introduced brown trout at mid-elevations, and creek chub at lower elevations), Taniguchi et al. (1998) reported that competitive capacity varied along a temperature gradient of 3–26°C. In these examples, it appears that interactions among the thermal optima of individual coldwater fish species and spatial and temporal variations in water temperature result in habitat segregation.

Interspecific competition often appears to be greatest between native and nonnative fishes. For example, Zimmerman and Vondracek (2006) found no evidence of competition between native brook trout and slimy sculpin, but it appeared that competition between introduced brown trout and slimy sculpin did occur. Indeed, interactions may be most intense among species with similar habitat requirements and life history characteristics. In a controlled experiment, interspecific competition resulted in slower growth of brook trout in the presence of brown trout (Dewald and Wilzbach 1992). In the wild, it appears that interspecific competition has resulted in proliferation of introduced brown trout and exclusion of native cutthroat trout in many streams (McHugh and Budy 2005).

Intraspecific competition between adults and juveniles of the same species may be decreased through size-structured habitat use or ontogenetic niche separation (Heggenes et al. 1999). Although competition may have negative consequences for some individuals, especially for small fish (Jenkins et al. 1999), populations ultimately benefit because abundance is maintained within the capacity of the habitat. In fact, intraspecific competition for limited space and food resources can affect the number of individual fish supported in a given environment (i.e., carrying capacity; Grant and Kramer 1990). With an increase in density or relative body size, "self-regulation" of a population is likely (Keeley 2003) and emigration from an area is common.

18.3 MANAGEMENT OF COLDWATER STREAMS IN THE 21ST CENTURY

Since the early 1970s, fisheries management in coldwater streams has shifted from stocking and providing for human consumption to a greater focus on native fish conservation and habitat restoration. This shift reflects a change of values associated with angling and preservation of native fish assemblages (Gresswell and Liss 1995). Furthermore, there has been a growing recognition that streams cannot be managed as isolated entities independent of their watershed (Williams et al. 1997). In many cases, threats to the persistence of native coldwater fishes are the result of past management activities that included widespread introductions of nonnative species, especially salmonids such as brook trout, brown trout, and rainbow trout (Thurow et al. 1997).

Management of coldwater fisheries has traditionally been separated among groups (state versus federal, but also nongovernmental land management entities in some cases) that have primary jurisdictions for habitat or fish populations. This dichotomy is rooted in legislative and administrative processes associated with the creation of these entities; however, it has also led to disjunct and unfocused management of coldwater fisheries and stream habitat. In recent decades, however, there has been increased cooperation and coordination among entities, especially in the management of native fishes. Furthermore, it is increasingly apparent that collaborations are critical for the persistence of fisheries in coldwater streams.

18.3.1 Threats to the Persistence of Coldwater Fishes

18.3.1.1 Disturbance

Disturbances, either anthropogenic or natural, have been described as pulse, press, or ramp disturbances (Lake 2000). A pulse disturbance is an abrupt change that progressively dissipates, whereas a press disturbance begins quickly and reaches a level that is maintained for an extended period. Ramp disturbance, a less-commonly-discussed type of disturbance, increases through time and space. Press and ramp disturbances may provide the greatest challenges to management of coldwater streams.

Examples of pulse disturbances are natural events (e.g., fire, floods, and windstorms), but some anthropogenic activities (e.g., a chemical spill) can be categorized as pulse disturbances. Although individual fish may be killed or dislocated by a pulse disturbance, population scale effects are generally brief. Reoccupation and (or) recolonization commence soon after habitat becomes available again.

Anthropogenic activities, such as grazing, row-crop agriculture, road construction, and mining, are commonly categorized as press disturbances. Roads have immediate and long-lasting effects on erosion patterns, watershed fragmentation, and water quality (Trombulak and Frissell 2000). Responses to press disturbances by fishes are linked to intensity, extent, and duration of the disturbances, and although fish may endure in altered environment, demographics of populations may remain depressed for extended periods. Such depressed populations are vulnerable to replacement by fishes that may be better adapted to altered environments (Grossman et al. 1998; Dunham et al. 2003b).

Long-term droughts (extended temporal periods of declines in rainfall) are examples of ramp disturbances. This term is also applicable to anthropogenic activities, such as suburban development, and it may be especially appropriate for describing effects of climate change. Furthermore, continued degradation of stream habitat leads to corresponding declines in fish populations and increases the probability of shifts in fish assemblages (Grossman et al. 1998). For instance, the cumulative effects of increasing water temperatures, changing hydrological patterns, more frequent and widespread wildfires, and human development may interact to increase negative consequences of habitat degradation and introduced species. Although these changes may not be linear, they can be expected to change steadily for decades, and management plans and long-term strategies should anticipate these changes.

In reality, disturbance events often exhibit characteristics of more than one disturbance category. For example, both pulse and press disturbances are often attributed to wildfires and the magnitude of effects is generally related to the temporal and spatial scale of disturbance events (Gresswell 1999). A rapid increase in water temperature associated with burning vegetation near a stream is an example of a pulse disturbance that will not be noticeable after a few hours, but longer-term press disturbance of increased summer water temperatures may accompany the removal of riparian vegetation and resulting lack of shade. Anthropogenic activities (e.g., timber harvesting) may also exhibit characteristics of more than one category of disturbance. At the local scale, effects of clear-cutting and an extreme fire event may be similar when most of the biomass is removed, but most fires are small. At the landscape scale, the affected area (i.e., portion converted to an earlier successional state) has historically been greater for timber harvesting. Clear-cutting occurred on about 20% of 4.6 million hectares of land in western Oregon between 1972 and 1995 (Cohen et al. 2002). The proportion of the landscape affected by large fires was much less; however, evidence suggests the effects of large fires may increase during the coming decades (Westerling et al. 2006). Management strategies that focus on protecting robust coldwater stream communities and restoring habitat structure and life history complexity may provide the most effective means to protect the capacity of coldwater streams from disturbances (Ebersole et al. 1997).

18.3.1.2 Introduced and invasive nonnative species

Nonnative species can be divided into those that have been deliberately introduced by management agencies, such as a rainbow trout, and invasive nonnative species, such as sea lamprey, that have gained access unintentionally or illegally. Introduced species may replace native fishes, but they are often important to recreational anglers (Quist and Hubert 2004). In contrast, invasive species are valued negatively by humans. Regardless of the introduction mechanism, nonnative species generally have negative (often unanticipated) consequences for native communities and ecosystems.

The primary threat to native coldwater fishes resulting from nonnative species is related to introduced fishes (Behnke 1992), both exotic (naturally occurring outside North America) and those founded by interbasin transfers of fishes native to North America. Major continental scale introductions of nonnative fishes have been common (frequently associated with government programs) since the latter part of the 19th century (Rahel 1997). The perceived scarcity of native fishes suitable for food or fishing was a frequent justification for early introductions. In general, the pattern of nonnative salmonid introductions first occurred in eastern North America and then proceeded westward, but rainbow trout, initially found in western coastal states, were introduced across the continent (Nico and Fuller 1999).

Predicting the outcome of nonnative salmonid introductions and invasions is not easy because few studies have been conducted at the population scale (Peterson and Fausch 2003). Hybridization is likely when invasive species interbreed with native fish, such as introductions of rainbow trout into streams supporting native cutthroat trout (Gresswell 1988). Competition and predation have been documented for individual species, but outcomes are frequently altered by abiotic factors that influence adaptation to new environments by the invader (Dunson and Travis 1991).

Nonnative fishes often expand from areas where they were first introduced. Barriers to movements may restrict access by invasive species, but the probability of interbasin transfers by humans exists (Rahel and Olden 2008). Major continental scale introductions of nonnative fishes have occurred in conjunction with official government programs (Behnke 1992; Rahel 1997), but unofficial introductions commonly occur. In Montana alone, 375 unauthorized introductions of fishes were documented through the mid-1990s, and 45 different species were illegally introduced into 224 different waters (Vashro 1995). Furthermore, anthropogenic activities, such as eutrophication and removal of apex predators, can increase the probability of successful establishment of nonnative species (Byers 2002). Habitat that has been degraded may be more vulnerable to invasion and establishment of nonnative species because abiotic and biotic conditions of the system may have become more favorable for introduced rather than native fishes (Thurow et al. 1997; Dunham et al. 2003b; Rahel and Olden 2008).

18.3.1.3 Habitat degradation

Habitat degradation associated with surface water diversions, dam construction, grazing, mineral extraction, timber harvest, or road construction is common among coldwater streams, and these activities frequently have negative consequences for the distribution and abundance of coldwater fishes. Barriers to migration, reduced flows, fine sediment deposition, streambank instability, erosion, increased water temperatures, and pollution are all associated with human activities (McIntosh et al. 1994). Impoundments have altered fish migration patterns, and reductions of peak flows, rapid fluctuations in discharges related to hydropower generation, and sediment loss immediately downstream from dams have changed habitat downstream from dams. Reduced coarse sediment inputs and increased embeddedness limit spawning and rearing habitats downstream from dams, and these problems are exacerbated by changes in the timing and magnitude of discharge (Van Kirk and Benjamin 2001).

Water diversions related to hydroelectric power, industry, and irrigated agriculture affect coldwater streams and have been significant factors in the decline of many native trout populations (McIntosh et al. 1994). Degraded water quality and unscreened irrigation ditches contribute to problems associated with water diversions. Thousands of salmonids, as well as large numbers of nongame fishes, can be entrained in poorly-designed diversions or fish screening facilities (Post et al. 2006). Water diversions may also provide new routes for species invasions when ditches or tunnels traverse watershed boundaries.

Habitat fragmentation can negatively affect the persistence of coldwater fish populations by reducing available habitats, inhibiting dispersal behaviors, simplifying habitat structure, and limiting resilience to stochastic disturbances. Road culverts often form barriers to fish movements and play a role in habitat fragmentation (Belford and Gould 1989). Wofford et al. (2005) found genetic diversity and allelic richness of coastal cutthroat trout were lowest in small tributaries where immigration had been blocked by culverts. Similar genetic effects have been reported for bull trout in larger systems where dams have fragmented stream networks (Neraas and Spruell 2001). Fragmentation can also reduce dispersal pathways among fish populations, inhibiting repopulation following local extirpations (Guy et al. 2008). The message for managers is that genetic variability is linked to the number of successful spawners and, regardless of potential genetic effects on persistence, the probability of extirpation increases if population abundance is reduced (Hilderbrand and Kershner 2000; Kruse et al. 2001).

Although the effects of excessive livestock grazing on riparian habitats (e.g., streambank sloughing, channel instability, erosion, and siltation) are widely documented (Platts 1991), consequences to distributions and abundances of fishes in coldwater streams can vary. Bank erosion and fine sediment in the streambed can be reduced by altering grazing management along streams (Platts 1991; Lyons et al. 2000), and in many cases, livestock grazing may be less of a threat to native salmonids than are hybridization, competition with introduced fishes, or dewatering (Varley and Gresswell 1988).

Mineral extraction does not appear to have altered the distribution of native coldwater fishes substantially, but, local extirpations of native cutthroat trout and bull trout associated with toxic heavy metals have occurred in numerous coldwater streams (Woodward et al. 1997; Farag et al. 2003). Furthermore, deposition of waste materials from dredging and hydraulic mining can alter sediment dynamics of streams (Nelson et al. 1991). In many cases, mine tailings continue to act as point sources for acid mine drainage and associated heavy metal pollution, factors that inhibit local populations of coldwater fishes (Woodward et al. 1989). Dredging can cause direct mortality to fish eggs and fry (Griffith and Andrews 1981), and these activities continue for extraction of both precious metals and gravel (Brown et al. 1998; Harvey and Lisle 1998).

18.3.1.4 Climate change

Climate change may be the greatest threat to persistence of fishes in coldwater streams because of synergistic relationships among climate, invasive aquatic species, and habitat degradation. During the past 100 years, mean global air temperature has increased about 0.6°C and it is expected to increase from 1.4 to 5.8°C during this century (IPCC 2007). Water temperatures increase 0.6–0.8°C for each degree rise in air temperature, so a 3–5°C increase in air temperature equates to a 2–3°C increase in water temperature (Morrill et al. 2005).

As water temperatures increase, the current ranges of fishes in coldwater streams are anticipated to shift up in elevation and northward in latitude. Based on an upper temperature threshold of 22°C for brook trout, cutthroat trout, and brown trout, Keleher and Rahel (1996) predicted that an increase in water temperature from 1°C to 5°C would produce a 7.5–43.3% decrease in the length of streams occupied by coldwater fishes in Wyoming. Even considering increases in suitable habitats at higher elevation as water temperature rises, the overall distribution of salmonids will probably diminish (Keleher and Rahel 1996; see Box 18.1). Rieman et al. (2007) argued that model results predicting 18–92% declines of thermally-suitable natal habitat for bull trout and 27–99% declines of large habitat patches (>10,000 ha) indicate that population effects of climate warming over the range of anticipated changes may be disproportionate to the simple loss of habitat area. Downstream from dams the effects of climate change on stream temperatures may be reduced if water is released from the hypolimnion (deep water), but if releases are taken from the epilimnion (reservoir surface), model estimates indicate that the negative consequences of climate change (complete loss of coldwater habitat) are unaffected (Sinokrot et al. 1995). It appears, however, that coldwater streams with high groundwater discharge are less sensitive to climate change than are streams with low groundwater discharge (Chu et al. 2008).

Water temperatures can influence fish directly though alteration of metabolism, feeding, and growth rates and indirectly by altering prey availability and mediating competitive interactions (Wehrly et al. 2007). Furthermore, water temperatures can influence species interactions (Rahel and Olden 2008). For instance, when growth rates of bull trout and brook trout were compared in the laboratory at a range of water temperatures from 8°C to 20°C, brook trout grew faster than bull trout at higher temperatures, but there was no growth advantage for bull trout at cooler temperatures (McMahon et al. 2007).

The effect of climate change on precipitation patterns is more complex, but changes in precipitation patterns will subsequently affect discharge regimes. Global trends cannot be

Box 18.1. Climate Change Impacts on Stream Fishes: Exploring Management Implications

Jack E. Williams[1] and Amy L. Haak[2]

Warmer waters, reduced snowpacks, earlier peak runoffs, lower summer flows, and increased frequency and intensity of disturbances are some of the factors associated with climate change that are likely to impact native salmonid populations in western North America (Poff et al. 2002). Currently, many inland salmonid species and subspecies occupy only 10–30% of their historic distributions because of habitat degradation and introduced species (Young 1995). Physical instream barriers (e.g., culverts and dams), invasions of nonnative salmonids, habitat degradations, and management strategies isolating native populations in headwater reaches above artificial barriers all have contributed to highly-fragmented landscapes with many small, isolated populations of native cutthroat trout. Larger, interconnected populations that were important to persistence are now quite rare (Colyer et al. 2005). Increased stress from climate change is likely to compound these existing problems.

Isolated cutthroat trout populations are increasingly at risk of extinction from two primary causes. First, their small stream habitats are vulnerable to disturbances such as wildfire, flood, or prolonged drought. Second, small isolated populations are at increased risk of extinction because of demographic and genetic factors associated with reduced population sizes and loss of interpopulation connectivity (Neville et al. 2006a; Guy et al. 2008).

How might a more detailed understanding of climate change impacts alter management strategies? We examined this question by modeling three factors associated with climate change (i.e., increased summer temperatures, uncharacteristic winter flooding, and increased wildfires) on Bonneville cutthroat trout populations. Determining the risk from climate change may alter management priorities by demonstrating the value of larger metapopulations that are more resilient to disturbance (Dunham et al. 2003b), at least in areas where problems associated with nonnative species can be addressed (Fausch et al. 2006).

Effects of a rapidly-changing climate are apparent in streams and watersheds of western North America. In Colorado, earlier emergence of the mayfly *Baetis bicaudatus* has been observed since 2001 because of earlier peak stream runoff associated with warmer stream temperatures during dryer years (Harper and Peckarsky 2006). Since the mid-1980s, there has been a 60% increase in the frequency of large wildfires in the northern Rocky Mountains associated with warmer spring and summer temperatures and earlier spring snowmelt (Westerling et al. 2006).

What are the implications of such changes to stream-dwelling salmonid populations? As a prerequisite to answering that question, we need to understand existing management and population status. State and federal management agencies divide the distribution of Bonneville cutthroat trout into four discrete management areas: the Bear River drainage, Northern Bonneville, Southern Bonneville, and West Desert. The amount of habitat currently occupied in these four management areas varies widely from only 94 km of stream

---

[1] Trout Unlimited, Medford, Oregon.
[2] Trout Unlimited and Conservation Geography, Boise, Idaho.

(Box continues)

Box 18.1. Continued

stream habitat in the West Desert to 1,752 km in the Bear River drainage. Larger, interconnected populations are restricted to the Bear River drainage and Northern Bonneville, whereas populations in the Southern Bonneville and West Desert are best characterized as small and isolated.

Our models indicate that Bonneville cutthroat trout populations are at a relatively high risk from climate change despite the fact that the Bear River drainage and Northern Bonneville management areas include several large, interconnected populations that are inherently resilient to disturbance. A small portion of increased risk is from higher summer temperatures, which may disproportionally affect populations in the West Desert and Southern Bonneville areas. Most of the increased risk is associated with greater likelihood of winter flooding. As measured by subwatersheds in the historic range, watersheds in nearly 50% of current and historic range face high risks of winter flooding. Increased wildfire risk affects fewer subwatersheds than does flood risk, but wildfire risk is greatest in the Bear River and Northern Bonneville areas. When areas subject to increased summer temperatures, winter flooding, and wildfire are combined, 73% of current habitat ranks at high risk (see Figure).

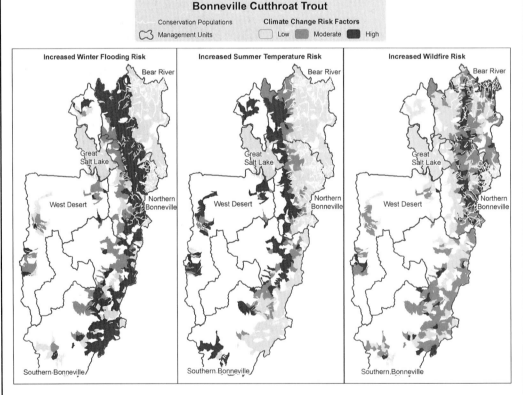

Figure. Predicted risk, associated with increased winter flooding, increased summer temperatures, and increased wildfire, to stream-dwelling populations of Bonneville cutthroat trout. Analysis unit is the subwatershed (4th level hydrologic unit code [indicates size of drainage area]; see Williams et al. 2007 for method details). *(Box continues)*

> **Box 18.1. Continued**
>
> Our results indicate that remaining populations in the West Desert and Southern Bonneville areas are more susceptible to near-term loss. This threat should not lead to despair but rather to action. Many proactive measures can be taken to improve resistance and resiliency of these populations to climate change and help ensure their future persistence (Williams et al. 2007). For instance, initial efforts should focus on expanding small isolated populations by increasing available downstream habitat and improving existing habitat quality. Salmonids will have a much better chance of persisting in the face of increasing environmental threats if they have access to heterogeneous habitat and refugia, both seasonally and during disturbance. Second, ecological and life history diversity should be restored by providing instream flows and reconnecting fragmented stream systems by removing instream barriers.
>
> What about the larger, more interconnected populations in the Bear River and Northern Bonneville areas? Protection of existing high-quality habitats, restoration of valley bottoms, and monitoring to detect changes offer the best prescription. Although the impacts of climate change appear dire, stream populations will have the best chance to survive rapid environmental changes if we act sooner rather than later to remove external stressors and maintain remaining genetic and ecological diversity.

accurately predicted because precipitation is so variable in time and space and because there are few reliable long-term records. In general, the effects of climate change are expected to differ regionally due to variation in intensity, frequency, duration, and magnitude of precipitation (Trenberth et al. 2003). Furthermore, rising air temperatures can alter stream discharge regime by diminishing snowpack and increasing evaporation (Field et al. 2007). Ultimately, changes in magnitude, frequency, duration, timing, and rate of change of discharge patterns are likely to reduce spatial distributions and sizes of coldwater fish populations (Jager et al. 1999).

Climate change will also affect persistence of coldwater fishes through complex behavioral responses to shifts in water temperatures and precipitation. Where species that can hybridize are sympatric, the probability of introgression may increase if migration cues are altered by changing hydrological patterns (Henderson et al. 2000). Other interspecific interactions, such as competition and predation, may be modified as a result of changing physical conditions. Understanding effects of climate change on interactions among co-occurring fishes, or those residing in close proximity, is especially important for determining future management options (Rahel and Olden 2008).

18.3.2. Current and Emerging Management Trends

18.3.2.1 Angler harvest

Providing quality angling experiences is still a major component of most coldwater-stream management programs. Angling is a social–psychological activity and the quality of recreational fishing depends on individual motivations and preferences for recreational experiences (Schroeder et al. 2006; Anderson et al. 2007). Satisfactory angling experiences generally include both social (e.g., spending time with family or away from crowds) and catch-related (e.g., harvesting fish to eat or catching a certain number of fish per trip; Kyle et al. 2007) aspects. The opportunity to test angling skills can also contribute to angler satisfaction. Agency programs that vary angling regulations among waters are acknowledgments of the variety of motivations among the angling public. In recognition of the cultural and social values applied to coldwater species, many states have designated heritage species or state fish to elevate public awareness and create cultural values for designated species (Epifanio 2000).

Although ability to harvest fish is important to some anglers, it is broadly recognized that overharvest can cause substantial declines in fish populations where vulnerability to angling is high. In fact, the number of anglers has reached a point in some areas where even limited levels of harvest can be detrimental. Anglers are often attracted to a fishery by high catchability, but substantial declines in abundance can occur when harvest is not restricted (Gresswell and Liss 1995). Because nonnative fishes are often less vulnerable to angling than are native fishes, unequal mortality can result in the decline of native species (Moyle and Vondracek 1985).

In some areas, angler harvests in coldwater streams are sustained or supplemented by hatchery-raised fish; however, this management strategy has become less common in recent decades (see Chapter 9, this volume). Where habitat can sustain naturally-reproducing populations, the repeated stocking of cultured fishes has few positive effects on fish population abundance or angling quality (Benson et al. 1959; Vincent 1987). Furthermore, it is broadly recognized that nonnative fish introductions have resulted in detrimental consequences for native coldwater fishes. Hatchery stocking is still used to support put-and-take fishing in streams, but this practice is more commonly associated with reservoir and lake fisheries.

18.3.2.2 Species and habitat assessment

Historically, fish management focused on individual fisheries that comprised a single body of water (e.g., lake or reservoir) or section of stream, often delineated geographically by species composition or access. Since the early 1970s, there has been a shift to a more broadly-based approach. These changes are related to continued declines in native species and habitat quality and national legislation such as the U.S. Endangered Species Act, Clean Water Act, and National Forest Management Act. In response to petitions for listing a particular species under the Endangered Species Act, individual agencies or groups of agencies have conducted status reviews across the historic distribution of numerous species of interest. These efforts often include status determinations, viability analyses, and risk assessments.

In other cases, status assessments that inform management are related to social or ecological concerns about declining populations and efforts to reverse declines or restore populations to viable levels. There is growing awareness of the recreational and economic importance of both native and nonnative fisheries in coldwater streams and the value of collaborations among various stakeholder groups, managers, and agencies (Gresswell and Liss 1995; Granek et al. 2008). Assessments for management activities are generally focused on assembling data on populations, habitat condition, and threats to populations and habitats.

Standardized protocols have been developed to summarize information concerning abundance and distribution of native salmonids (e.g., May et al. 2007). Past protocols were not often founded on a statistically-based sampling design, leading to biased assessments of presence–absence, genetic integrity, or population abundance.

A variety of methods have been used to assess distributions of fish species in coldwater streams (Harig and Fausch 2002; Bateman et al. 2005; Young et al. 2005). The most appropriate method is related to study objectives and available resources, but maintaining comparability among studies is desirable. Systematic sampling of all available habitats and fish collection or observation techniques that provide a known probability of individual capture are important for establishing the extent of fish in a watershed (Bayley and Peterson 2001). If tissue samples for genetic analyses are also collected probabilistically, results can be used for statistical comparisons among sites and through time (Guy et al. 2008).

There is a rich literature concerning habitat assessment and estimation of habitat quality in streams at the local scale (i.e., individual study sections comprising transects and channel units; see Bauer and Ralph 2001 and Chapters 10 and 12). Focus on the local scale is a major shortcoming of much of the historical literature because changes in habitat use related to different life stages and movements are ignored at this scale (but see Petty et al. 2005). Recent efforts to use a nested approach with data from multiple spatial scales (that is, finer spatial scales, e.g., habitat units, combined over broader spatial scales, e.g., reaches) may prove useful (Frissell et al. 1986; Hankin and Reeves 1988; Gresswell et al. 2006). Moreover, newer statistical designs provide the means to expand estimates to the landscape scale (Urquhart et al. 1998; Larsen et al. 2001; Larsen et al. 2004). By incorporating these broad-scale techniques into habitat assessments, current and future resource conditions may be addressed (Petty et al. 2005).

Results from assessment of fish habitat in conjunction with species' distributions can be used to explain observed distributions (Steen et al. 2006). Such data are useful for identifying factors that may limit the occurrence of fish (i.e., presence or absence of a target species) and provide a basis for monitoring, rehabilitation, and management activities. Coordinated assessments require robust statistical sampling frameworks, but the additional information and predictive potential associated with these approaches provide justification for their costs. Emerging geographical information system (GIS) tools can be used to integrate information about hydrology, geomorphology, biology, connectivity, and water quality and to facilitate understanding of watershed function (see Annear et al. 2004). At the same time, it is important to recognize the importance of quality and consistency of data collection protocols and data management for any type of sampling.

18.3.2.3 Population and habitat monitoring

Population estimates of mature individuals are critical for species assessments, but they are especially useful for evaluating changes in population abundance through time. Mark–recapture and depletion techniques that provide estimates of precision are becoming more common (Budy et al. 2007). Many estimates lack inferential power, however, because evaluations are based on "happenstance samples" (sites originally chosen nonrandomly for a variety of purposes). Findings can be misleading if the sampling design is not statistically robust (Larsen et al. 2001). Recently developed protocols for evaluating changes in habitat quality can contribute to understanding population trends through time (Urquhart et al. 1998; Larsen et al. 2001, 2004).

At an individual site, measurement error for population estimates is important to consider, but variation among sample sites should be a major consideration when planning multiyear assessments (Olsen et al. 1999; Larsen et al. 2001). Sites selected using a probability-based sampling method assure substantial inferential power. Changes in habitat conditions that influence species' distributions may be detected using consistent annual monitoring of 30–50 sites (Larsen et al. 2004). Probability-based sample selection can be used for sampling at the watershed scale (Gresswell et al. 2004, 2006).

18.3.2.4 Habitat and population management

Habitat management. Habitat improvement has been, and will continue to be, critical to conservation efforts where habitat degradation resulted in decline and (or) extirpation of coldwater stream fishes. When habitat improvement is undertaken, it is important to focus on ecologically-based strategies at the watershed scale. Goals of an ecological strategy should include (1) sustaining diverse habitats and native aquatic biota that are supported in these areas, (2) securing existing populations and critical refugia that support historical ecosystem function, and (3) promoting recovery with the greatest probability of improving the status of native populations beginning from existing strongholds and incrementally extending the influences of these ecosystem processes (Frissell 1997).

A key concept of restoration of coldwater stream habitat incorporates both system capacity and development. Human decisions and actions influence ecosystem capacity across landscapes. Reducing or removing human land use pressures can facilitate restoration of structure and function to stream systems by natural processes. Rehabilitation thus involves identifying and relieving these stressors and allowing natural forces to proceed (Ebersole et al. 1997).

This approach requires a thorough watershed analysis identifying the factors that are negatively influencing habitat and the appropriate scale for improvements (Kershner 1997). Habitat improvement is often pursued where habitat loss is caused by local factors, such as streambank slumping related to cattle grazing. It is critical to consider the relationship of physical and climatic processes to stream habitat during planning and implementation of habitat improvements. These activities require close coordination among agencies, especially those agencies charged with landscape management, because activities affecting watershed vegetation can influence the hydrology of a stream system.

Although monitoring provides information necessary to evaluate success of restoration (Kershner 1997), this step is often ignored to avoid additional costs. However, despite expenditures of more than US$1 billion on habitat improvement annually, there is little information available about the results for most restoration efforts (Bernhardt et al. 2005, 2007). Only 10% of about 37,000 projects reviewed by Bernhardt et al. (2005) indicated that subsequent assessment or monitoring occurred. Ecological degradation typically motivated most of the restoration projects, but less than 50% of the projects had measurable objectives (Bernhardt et al. 2005, 2007).

Population isolation. Because nonnative species present a threat to persistence of native fishes in coldwater streams, curtailing the spread of nonnative species is critical. One strategy is the isolation of remaining genetically-unaltered populations of native fishes. In small headwater drainages, however, isolation and fragmentation can substantially increase the probability of demographic collapse (Kruse et al. 2001) associated with catastrophic disturbances (e.g., wildfire or flooding and debris flow events); furthermore, less mobile taxa may be at greater risk of local extirpation in isolated streams. Gradual reductions in habitat suitability related to climate change are more likely in small, isolated headwater streams. In addition, mobile life history types are more likely to be extirpated by curtailing upstream fish passage (see Box 18.2).

The minimum watershed size necessary for fish persistence is related to demographic characteristics and movement capacity of individual species. Wofford et al. (2005) reported that demographic isolation upstream from dispersal barriers can decrease genetic diversity of coastal cutthroat trout. At the regional scale, among-population genetic diversity of coastal cutthroat trout in headwater watersheds (500–1,000 ha) appears to be related to within-watershed complexity and connectivity (Guy et al. 2008). Despite variation in genetic diversity, coastal cutthroat trout have occupied these watersheds for thousands of years. In contrast, Gila trout populations have been extirpated from small headwater streams following wildfire and postfire floods (Rinne 1996). Moreover, Lahontan cutthroat trout have been documented in 89% of 47 networked systems, but in only 32% of 72 fragmented (isolated) watersheds (Dunham et al. 1997).

Much less is understood about the effects of fragmentation on non-salmonid fishes, but it may be reasonable to assume that although spatial scales may differ, connectivity in coldwater streams is important for persistence for these fishes as well. In western Oregon, for example, the occurrence of sculpins in small watersheds (<1,000 ha) above barriers to anadromous fishes appears to be related to complexity of the watershed (number of tributaries) and connectivity within the watershed (R. Gresswell and D. Bateman, unpublished data). In general, effects of disturbance in watersheds are greatest on those individuals and local populations of aquatic organisms that are least mobile, and reinvasion is most rapid by organisms with high mobility (Gresswell 1999).

Management decisions with respect to fragmentation in stream networks can be complex and each situation must be evaluated individually. In some cases, reconnection of networks that have been fragmented by anthropogenic activities (e.g., diversions, dams, or road culverts) may be desirable, but in other situations, fish passage may be intentionally blocked to prevent invasions by nonnative fishes (see Box 18.2).

Population removal. Where habitat has the capacity for supporting populations that are reproducing, removal of nonnative fishes and reintroduction of native fishes may be possible. The feasibility of this management alternative is limited by the size and complexity of the target drainage, but the probability of success can sometimes be increased by isolating an appropriately large portion of a watershed prior to the removal activities. Although removal of nonnative species is difficult and usually expensive, in many cases it may be the best option for restoring native coldwater fishes in their historic distribution (Finlayson et al. 2005). This type of management action may even be an appropriate alternative when installation of fish passage barriers is not feasible because of demographic risks to isolated populations of native fishes. Capturing native fish by electrofishing prior to piscicide application has been successful where native and nonnative fishes are sympatric.

The two most commonly used piscicides are rotenone and antimycin. Rotenone has been used more frequently because of its lower price and greater availability (Finlayson et al. 2005). Although removing nonnative fishes by means of electrofishing may be effective when habitat is simple and the target area is small, this technique seldom results in complete extirpation of target species (Thompson and Rahel 1998). On the other hand, repeated removals by electrofishing may increase short-term survival of native coldwater salmonids when hybridization is not a concern (Peterson et al. 2008a).

Box 18.2. Barriers, Invasion, and Conservation of Native Salmonids in Coldwater Streams

BRUCE RIEMAN[1], MICHAEL YOUNG[2], KURT FAUSCH[3], JASON DUNHAM[4], AND DOUGLAS PETERSON[5]

Habitat loss and fragmentation are threats to persistence of many native fish populations. Invading nonnative species that may restrict or displace native species are also important. These two issues are particularly relevant for native salmonids that are often limited to remnant habitats in cold, headwater streams. On the surface, reversing threats to native fishes would seem to be straightforward: focus all available resources on habitat restoration and control of invaders. However, there are trade-offs that make this a more complex problem. This is well illustrated by the installation or removal of barriers to fish movements because either action may simultaneously mitigate and exacerbate risks to native salmonid populations.

The size, distribution, and connectivity of suitable habitats are common issues in the conservation of native salmonid populations. The reason is that the size of stream habitat networks and connectivity among habitats are important to persistence of local populations. Loss of connectivity can lead to loss of genetic diversity (Wofford et al. 2005; Neville et al. 2006b; Guy et al. 2008), increased vulnerability to catastrophic events, loss of migratory life histories needed to access complementary habitats (Northcote 1997; Rieman and Dunham 2000), and loss of connectivity to other populations that historically facilitated demographic support, rescue, or even reinvasion (Rieman and Dunham 2000; Letcher et al. 2007). Declines in habitat size and connectivity have been caused by habitat degradation (e.g., streamflow diversion, increased water temperature, and decreased water quality) and habitat fragmentation by fish passage barriers (e.g., road culverts, hydroelectric dams, and diversion dams). Reversing habitat degradation can be a relatively complex process involving extensive watershed and streamside protection or restoration that can be expensive, controversial, and slow to take effect. In contrast, many fish passage barriers block access to relatively high-quality headwater habitat, and restoring access to these habitats would seem a simple matter of removing barriers. Most fish passage barriers are quite small, but there are thousands across the landscapes supporting native salmonids (GAO 2001). Restoration of fish passage thus offers an important opportunity to make rapid gains in restoring both size and connectivity of fish habitats and populations.

Nevertheless, even within the apparently simple arena of fish passage restoration involving smaller barriers, there are outstanding issues that require further consideration.

---

[1] U.S. Forest Service (Retired), Boise, Idaho.
[2] U.S. Forest Service, Missoula, Montana.
[3] Colorado State University, Fort Collins.
[4] U.S. Geological Survey, Corvallis, Oregon.
[5] U.S. Fish and Wildlife Service, Helena, Montana.

(Box continues)

Box 18.2. Continued.

First, it is clear that existing resources (people, money, time, and materials) are inadequate to restore fish passage in a timely manner for the vast majority of cases (GAO 2001). In this situation, it becomes important to justify the relevance of individual projects. Managers must prioritize limited resources effectively to make sure that projects actually gain the greatest benefits possible. Research on fish population persistence upstream of fish passage barriers (e.g., Morita and Yamamoto 2002) also has shown that the probability of extinction increases as a function of time. A process of triage by which the most urgent projects with the greatest chances of success are prioritized would be required.

A second major consideration is that restoring fish passage might allow invasions of nonnative fishes that could threaten native species and ecosystems. In many parts of the inland West, managers are actively installing passage barriers to protect upstream populations of native fishes from invasions by nonnative fishes. Some existing passage barriers may indeed be protecting upstream habitats from invasions, but in the long term, isolated populations of native fishes face an elevated risk of extinction. Thus, conflicts between management to reduce threats from nonnative fishes versus threats from habitat isolation highlight the real-world uncertainties and complexities in identifying priorities and use of fish passage barriers.

Trade-offs may be relatively clear to biologists with intimate knowledge of a particular system, and their efforts can be focused effectively. Elsewhere, where trade-offs may be more ambiguous or data and experience more limited, the result may be a decision that is influenced more by personal philosophy or public pressure than by knowledge. When differences in these choices cannot be clearly supported and articulated, the decision process can appear inconsistent and arbitrary to the public or administrators who fund these projects. A consistent decision process would include an analysis of the relative risks associated with either action.

Biologists can weigh risks and benefits of installing or removing migration barriers by articulating the biological processes and social values defining the problem. Fausch et al. (2006) suggested that the context for this particular problem can be defined by three key elements: (1) understanding conservation values at risk and recognizing that some (e.g., conservation of genetic purity) may require barriers, but others (e.g., reestablishment of main-stem fisheries supported by tributary spawning) may require barrier removal; (2) understanding how environmental conditions in a particular watershed favor or constrain nonnative fish invasion and displacement of the native species; and (3) understanding the likelihood of local extinction if a native population is isolated, with recognition that time, size, and quality of the isolated habitat, and the species in question can strongly influence that probability. By assembling this kind of information for streams and populations across a region of interest, biologists can begin to prioritize where to work and what to do more effectively. Formal decision models are now available to facilitate this process when the underlying biology is relatively well known (Peterson et al. 2008b); even when it is not, however, acknowledgment of the general gradients important to these trade-offs can help focus limited management resources.

Population redundancy. When most of the remaining genetically-unaltered populations of a species or subspecies of native salmonid are found in small, isolated headwater streams, expanding the number of populations is important because persistence of fishes in any single watershed cannot be assured. Although a strategy focused on population redundancy is often incorporated with removal of nonnative fishes in the watersheds, it is sometimes feasible to introduce native salmonids into watersheds where these fishes were not found historically. In many cases, management policies support replacement of introduced nonnative fishes with native salmonids; however, introduction of fish into waters that were historically uninhabited by fish is generally prohibited.

18.3.3 Collaborative Management Solutions

Building public and private partnerships is critical to restoration of coldwater streams systems. In the USA, many rehabilitation projects are spearheaded by state and federal natural resources agencies, but there is a significant amount of funding provided by nongovernmental organizations dedicated to conservation. For example, Trout Unlimited spent over US$11 million in 2006 on conservation, most of which targeted habitat rehabilitation projects. Sustained partnerships are founded on conservation needs. Success of these collaborative efforts requires that partners are treated equally and work is shared, nontraditional partners are encouraged, and partners remain flexible in the midst of unforeseen challenges (Tilt and Williams 1997).

Public participation in coldwater stream management reflects a widespread desire to be involved in natural resource decision making (Koontz and Johnson 2004). Individuals want to be involved in the management process, and this has created a shift from the historic expert-authority approach to management toward more inclusive collaborative methods that foster public involvement (see Chapters 5 and 6). There are numerous options for public involvement. For example, managers can encourage private landowners along coldwater streams to plant native vegetation for stabilizing streambanks. Farmers and ranchers can be involved in workshops designed to promote riparian zone restoration. Stakeholders can be invited to participate in agency-designed restoration projects. Watershed associations, conservation organizations, and angler groups constitute a valuable workforce for habitat rehabilitation in riparian corridors of coldwater streams.

Collaboration is a process in which diverse stakeholders work together to resolve a conflict or develop and advance a shared vision. Numerous agencies collaborate to address environmental issues. Many agencies promote collaborative relationships with citizens through creation of public involvement programs (Malone 2000). Widespread public involvement is vital to ensure collaborative environmental management is effective (Koontz and Johnson 2004) and that benefits translate to tangible results on the ground.

Neighborhood groups constitute a major type of collaborative management organization. By providing organizational support, agencies can facilitate evolution of a neighborhood group into a watershed association. Volunteers in such stakeholder-based collaborative groups set goals for watershed improvements and assist management agencies in development of scientifically-based action plans. In fact, groups that interact and receive support from agencies tend to exist longer, spend more effort per site, and participate more enthusiastically in monitoring efforts (Frost-Nerbourne and Nelson 2004). Perhaps most importantly, public involvement in the management process educates and informs participants in regards to local environmental issues.

18.4 CONCLUSIONS

Coldwater streams come in a variety of sizes, and although salmonids are generally the most highly-valued fishes inhabiting these systems, cottids, catostomids, and cyprinids are often abundant. Fish assemblages in coldwater streams are structured by a complex set of interacting abiotic and biotic factors. Coldwater assemblages are dependent on energy inputs from autotrophic production and organic matter that originate in adjacent riparian areas. Herbivorous and detritivorous invertebrates and fish use these energy sources and are in turn consumed by predators. Linkages among trophic levels in coldwater streams are complex and often vary through time in relation to changes in the physical environment. For example, climate and geologic structure shape the distribution and abundance of fishes in coldwater streams by influencing water chemistry, channel depth, temperature, discharge, substrate, and cover. Predation is often the dominant biological interaction influencing fish populations in coldwater streams, but both interspecific and intraspecific competition can influence fine-scale assemblage structure.

Management of coldwater streams has shifted from a focus on recreational fishing to a greater emphasis on native fish restoration and conservation. Assessments for management often focus on abundance and age structure of populations, habitat condition, and threats to population and habitats. One of the primary threats to native fishes has been the intentional introduction and subsequent spread of nonnative species. Habitat degradation and fragmentation of coldwater streams is ubiquitous throughout North America and it poses another major threat to the persistence of coldwater fish populations. Synergistic relationships among climate, invasive species, and habitat degradation make it difficult to predict the effects of climate change on persistence of fishes in coldwater streams. Management activities that promote the capacity of fish populations to adapt to changing environments will be important when addressing this complex issue. Collaborative partnerships will undoubtedly become more important for providing institutional and financial support for management actions and as sources of innovative solutions to problems at a variety of spatial and temporal scales.

18.5 REFERENCES

Allan, J. D. 1995. Stream ecology: structure and function of running waters. Chapman and Hall, London.

Anderson, D. K., R. B. Ditton, and K. M. Hunt. 2007. Measuring angler attitudes toward catch related aspects of fishing. Human Dimensions of Wildlife 12:181–191.

Annear, T., I. Chisholm, H. Beecher, A. Locke, and 12 other authors. 2004. Instream flows for riverine resource stewardship, revised edition. Instream Flow Council, Cheyenne, Wyoming.

Baltz, D. M., P. B. Moyle, and N. J. Knight. 1982. Competitive interactions between benthic stream fishes, riffle sculpin, *Cottus gulosus*, and speckled dace, *Rhinichthys osculus*. Canadian Journal of Fisheries and Aquatic Sciences 39:1502–1511.

Bardonnet, A. and M. Heland. 1994. The influence of potential predators on the habitat preferenda of emerging brown trout. Journal of Fish Biology 45 (Supplement A):131–142.

Bateman, D. S., R. E. Gresswell, and C. E. Torgersen. 2005. Evaluating single-pass catch as a tool for identifying spatial pattern in fish distribution. Freshwater Ecology 20:335–345.

Bateman, D. S., and H. W. Li. 2001. Nest site selection by reticulate sculpin in two streams of different geologies in the central Coast Range of Oregon. Transactions of the American Fisheries Society 130:823–832.

Bauer, S. B., and S. C. Ralph. 2001. Strengthening the use of aquatic habitat indicators in Clean Water Act programs. Fisheries 26(6):14–25.

Bayley, P. B., and J. T. Peterson. 2001. An approach to estimate probability of presence and richness of fish species. Transactions of the American Fisheries Society 130:620–633.

Belford, D. A., and W. R. Gould. 1989. An evaluation of trout passage through six highway culverts in Montana. Transactions of the American Fisheries Society 9:437–445.

Behnke, R. J. 1992. Native trout of western North America. American Fisheries Society, Monograph 6, Bethesda, Maryland.

Benson, N. G., O. B. Cope, and R. V. Bulkley. 1959. Fishery management studies on the Madison River system in Yellowstone National Park. U.S. Fish and Wildlife Service, Special Scientific Report: Fisheries 307.

Berger, A. M., and R. E. Gresswell. 2009. Factors influencing coastal cutthroat trout seasonal survival rates: a spatially continuous approach among stream network habitats. Canadian Journal of Fisheries and Aquatic Sciences 66:613–632.

Bernhardt, E. S., M. A. Palmer, J. D. Allan, G. Alexander, K. Barnas, S. Brooks, J. Carr, S. Clayton, C. Dahm, J. Follstad-Shah, D. Galat, S. Gloss, P. Goodwin, D. Hart, B. Hassett, R. Jenkinson, S. Katz, G. M. Kondolf, P. S. Lake, and R. Lave. 2005. Synthesizing U.S. river restoration efforts. Science 308:636–637.

Bernhardt, E. S., E. B. Sudduth, M. A. Palmer, J. D. Allan, J. L. Meyer, G. Alexander, J. Follastad-Shah, B. Hassett, R. Jenkinson, R. Lave, J. Rumps, and L. Pagano. 2007. Restoring rivers one reach at a time: results from a survey of U. S. river restoration practitioners. Restoration Ecology 15:482–493.

Bratovich, P., D. Olson, J. Cornell, A. Pitts, and A. Niggemyer. 2004. Evaluation of potential effects of fisheries management activities on ESA-listed fish species SP-F5/7 Task 1. State of California, Department of Water Resources, Final Report FERC (Federal Energy Regulation Commission) Project 2100.

Brown, A. V., M. M. Lyttle, and K. D. Brown. 1998. Impacts of gravel mining on gravel bed streams. Transactions of the American Fisheries Society 127:979–994.

Budy, P., G. P. Thiede, and P. McHugh. 2007. Quantification of the vital rates, abundance, and status of a critical, endemic population of Bonneville cutthroat trout. North American Journal of Fisheries Management 27:593–604.

Burrell, K. H., J. J. Isely, D. B. Bunnell Jr., D. H. Van Lear, and C. A. Dolloff. 2000. Seasonal movement of brown trout in a southern Appalachian River. Transactions of the American Fisheries Society 129:1373–1379.

Byers, J. E. 2002. Impact of nonindigenous species on natives enhanced by anthropogenic alteration of selection regimes. Oikos 97:449–458.

Chu, C., N. E. Jones, N. E. Mandrak, A. R. Piggott, and C. K. Minns. 2008. The influence of air temperature, groundwater discharge, and climate change on the thermal diversity of stream fishes in southern Ontario watersheds. Canadian Journal of Fisheries and Aquatic Sciences 65:297–308.

Clark, J. R. D., and G. R. Alexander. 1985. Effects of a slotted size limit on a multispecies trout fishery. Michigan Department of Natural Resources, Fisheries Research Report 1926, Ann Arbor.

Cohen, W. B., T. A. Spies, R. J. Alig, D. R. Oetter, T. K. Maiersperger, and M. Fiorella. 2002. Characterizing 23 years (1972–95) of stand replacement disturbance in western Oregon forests with LandSAT imagery. Ecosystems 5:122–137.

Coleman, M. A., and K. D. Fausch. 2007. Cold summer temperature limits recruitment of age-0 cutthroat trout in high-elevation Colorado streams. Transactions of the American Fisheries Society 136:1231–1244.

Colyer, W. T., R. H. Hilderbrand, and J. L. Kershner. 2005. Movements of fluvial Bonneville cutthroat trout in the Thomas Fork of the Bear River, Idaho–Wyoming. North American Journal of Fisheries Management 25:954–963.

Crisp, D. T. 1993. The environmental requirements of salmon and trout in freshwater. Freshwater Forum 3:176–202.

Crisp, D. T. 2000. Trout and salmon ecology, conservation and rehabilitation. Fishing News Books, Blackwell Science, Malden, Massachusetts.

Dewald, L., and M. A. Wilzbach. 1992. Interactions between native brook trout and hatchery brown trout: effects on habitat use, feeding, and growth. Transactions of the American Fisheries Society 121:287–296.

Dunham, J., B. Rieman, and G. Chandler. 2003a. Influences of temperature and environmental variables on the distribution of bull trout within streams at the southern margin of its range. North American Journal of Fisheries Management 23:894–904.

Dunham, J. B., G. L. Vinyard, and B. E. Rieman. 1997. Habitat fragmentation and extinction risk of Lahontan cutthroat trout. North American Journal of Fisheries Management 17:1126–1133.

Dunham, J. B., M. K. Young, R. E Gresswell, and B. E. Rieman. 2003b. Effects of fire on fish populations: landscape perspectives on persistence of native fishes and nonnative fish invasion. Forest Ecology and Management 178:183–196.

Dunson, W. A., and J. Travis. 1991. The role of abiotic factors in community organization. American Naturalist 138:1067–1091.

Ebersole, J. L., W. J. Liss, and C. A. Frissell. 1997. Restoration of stream habitats in the western United States: restoration as reexpression of habitat capacity. Environmental Management 21:1–14.

Elliott, J. M. 1994. Quantitative ecology and the brown trout. Oxford University Press, Oxford, UK.

Epifanio, J. 2000. The status of coldwater fishery management in the United States: an overview of state programs. Fisheries 25(7):13–27.

Farag, A. M., D. Skaar, D. A. Nimick, E. MacConnell, and C. Hogstrand. 2003. Characterizing aquatic health using salmonid mortality, physiology, and biomass estimates in streams with elevated concentrations of arsenic, cadmium, copper, lead, and zinc in the Boulder River Watershed, Montana. Transactions of the American Fisheries Society 132:450–467.

Fausch, K. D., B. E. Rieman, M. K. Young, and J. B. Dunham. 2006. Strategies for conserving native salmonid populations at risk from nonnative fish invasions: tradeoffs in using barriers to upstream movement. U.S. Depatment of Agriculture Forest Service, Rocky Mountain Research Station, General Technical Report RMRS-GTR-174, Fort Collins, Colorado.

Fausch, K. D., C. E. Torgersen, C. V. Baxter, and H. W. Li. 2002. Landscapes to riverscapes: bridging the gap between research and conservation of stream fishes. BioScience 52:483–498.

Field, B., L. D. Mortsch, M. Brklacich, D. L. Forbes, P. Kovacs, J. A. Patz, S. W. Running, and M. J. Scott. 2007. North America. Pages 617–652 in M. L. Parry, O. F. Canziani, J. P. Palutikof, P. J. Van der Linden, and C. E. Hanson, editors. Climate change 2007: impacts, adaptation, and vulnerability. Contribution of working group II to the fourth assessment report of the Intergovernmental Panel on Climate Change. Cambridge University Press, Cambridge, UK, and New York.

Finlayson, B., W. Somer, D. Duffield, D. Propst, C. Mellison, T. Pettengill, H. Sexauer, T. Nesler, S. Gurtin, J. Elliot, F. Partridge, and D. Skaar. 2005. Native inland trout restoration on national forests in the western United States: time for improvement. Fisheries 30(3):10–19.

Flebbe, P. A., and A. C. Dolloff. 1995. Trout use of woody debris and habitat in Appalachian wilderness streams of North Carolina. North American Journal of Fisheries Management 15:579–590.

Freeman, M. C., and G. D. Grossman. 1992. A field test for competitive interactions among foraging stream fishes. Copeia 1992:898–902.

Frissell, C. A. 1997. Ecological principles. Pages 96–115 in J. E. Williams, C. A. Wood, and M. P. Dombeck, editors. Watershed restoration: principles and practices. American Fisheries Society, Bethesda, Maryland.

Frissell, C. A., W. J. Liss, C. E. Warren, and M. D. Hurley. 1986. A hierarchical framework for stream habitat classification: viewing streams in a watershed context. Environmental Management 10:199–214.

Frost-Nerbourne, J., and K. C. Nelson. 2004. Volunteer macroinvertebrate monitoring in the United States: resource mobilization and comparative state structures. Society and Natural Resources 17:817–839.

Ganio, L. M., C. E. Torgersen, and R. E. Gresswell. 2005. Describing spatial pattern in stream networks: a practical approach. Frontiers in Ecology and the Environment 3:138–144.

GAO (General Accounting Office). 2001. Restoring fish passage through culverts on Forest Service and BLM lands in Oregon and Washington could take decades. U.S. General Accounting Office GAO-02-136, Washington, D.C.

Gilliam, J. F., and D. F. Fraser. 2001. Movement in corridors: enhancement by predation threat, disturbance, and habitat structure. Ecology 82:258–273.

Granek, E. F., E. M. P. Madin, M. A. Brown, W. Figueira, D. S. Cameron, Z. Hogan, G. Kristianson, P. de Villiers, J. E. Williams, J. Post, S. Zahn, and R. Arlinghaus. 2008. Engaging recreational fishers in management and conservation: global case studies. Conservation Biology 22:1125–1134.

Grant, J. W. A. 1990. Aggressiveness and the foraging behaviour of young-of-the-year brook char (*Salvelinus fontinalis*). Canadian Journal of Fisheries and Aquatic Sciences 47:915–920.

Grant, J. W. A., and D. L. Kramer. 1990. Territory size as a predictor of the upper limit to population density of juvenile salmonids in streams. Canadian Journal of Fisheries and Aquatic Science 47:1724–1737.

Gresswell, R. E., editor. 1988. Status and management of interior stocks of cutthroat trout. American Fisheries Society, Symposium 4, Bethesda, Maryland.

Gresswell, R. E. 1995. Yellowstone cutthroat trout. Pages 36–54 *in* M. Young, editor. Conservation assessment for inland cutthroat trout. U.S. Department of Agriculture Forest Service, Rocky Mountain Forest and Range Experiment Station, General Technical Report RM-GTR-256, Fort Collins, Colorado.

Gresswell, R. E. 1999. Fire and aquatic ecosystems in forested biomes of North America. Transactions of the American Fisheries Society 128:193–221.

Gresswell, R. E., D. S. Bateman, G. W. Lienkaemper, and T. J. Guy. 2004. Geospatial techniques for developing a sampling frame of watersheds across a region. Pages 517–530 *in* T. Nishida, P. J. Kailola, and C. E. Hollingworth, editors. GIS/Spatial Analyses in Fishery and Aquatic Sciences, volume 2. Fishery–Aquatic GIS Research Group, Saitama, Japan.

Gresswell, R. E., and R. D. Harding. 1997. The role of special angling regulations in management of coastal cutthroat trout. Pages 151–156 *in* J. D. Hall, P. A. Bisson, and R. E. Gresswell, editors. Sea-run cutthroat trout: biology, management, and future conservation. American Fisheries Society, Oregon Chapter, Corvallis.

Gresswell, R. E., and S. R. Hendricks. 2007. Population-scale movement of coastal cutthroat trout in a naturally isolated stream network. Transactions of the American Fisheries Society 136:238–253.

Gresswell, R. E., and W. J. Liss. 1995. Values associated with management of Yellowstone cutthroat trout in Yellowstone National Park. Conservation Biology 9:159–165.

Gresswell, R. E., W. J. Liss, and G. L. Larson. 1994. Life history organization of Yellowstone cutthroat trout (*Oncorhynchus clarki bouvieri*) in Yellowstone Lake. Canadian Journal of Fisheries and Aquatic Sciences 51 (Supplement 1):298–309.

Gresswell, R. E., W. J. Liss, G. L. Larson, and P. J. Bartlein. 1997. Influence of basin-scale physical variables on life history characteristics of cutthroat trout in Yellowstone Lake. North American Journal of Fisheries Management 17:1046–1064.

Gresswell, R. E., C. E. Torgersen, D. S. Bateman, T. J. Guy, S. R. Hendricks, and J. E. B. Wofford. 2006. A spatially explicit approach for evaluating relationships among coastal cutthroat trout, habitat, and disturbance in headwater streams. Pages 457–471 *in* R. Hughes, L. Wang, and P. Seelbach, editors. Influences of landscapes on stream habitats and biological assemblages. American Fisheries Society, Symposium 48, Bethesda, Maryland.

Griffith, J. S. 1993. Coldwater streams. Pages 481–504 *in* C. C. Kohler and W. A. Hubert, editors. Inland fisheries management in North America, 2nd edition. American Fisheries Society, Bethesda, Maryland.

Griffith, J. S., and D. A. Andrews. 1981. Effects of a small suction dredge on fishes and aquatic invertebrates in Idaho streams. North American Journal of Fisheries Management 1:21–28.

Grossman, G. D., R. E. Ratajczak Jr., M. Crawford, and M. C. Freeman. 1998. Assemblage organization in stream fishes: effects of environmental variation and interspecific interactions. Ecological Monographs 68:395–420.

Guy, T. J., R. E. Gresswell, and M. A. Banks. 2008. Landscape-scale evaluation of genetic structure among barrier-isolated populations of coastal cutthroat trout *Oncorhynchus clarkii clarkii*. Canadian Journal of Fisheries and Aquatic Sciences 165:1749–1762.

Hankin, D. G., and G. H. Reeves. 1988. Estimating total fish abundance and total habitat area in small streams based on visual estimation methods. Canadian Journal of Fisheries and Aquatic Sciences 45:834–844.

Harig, A. L., and K. D. Fausch. 2002. Minimum habitat requirements for establishing translocated cutthroat trout populations. Ecological Applications 12:535–551.

Harper, D. D., and A. M. Farag. 2004. Winter habitat use by cutthroat trout in the Snake River near Jackson, Wyoming. Transactions of the American Fisheries Society 133:15–25.

Harper, M. P., and B. L. Peckarsky. 2006. Emergence clues of a mayfly in a high-altitude stream ecosystem: potential response to climate change. Ecological Applications 16:612–621.

Harvey, B. C., and T. E. Lisle. 1998. Effects of suction dredging on streams: a review and an evaluation strategy. Fisheries 23(8):8–17.

Heggenes, J., J. L. Bagliniere, and R. A. Cunjak. 1999. Spatial niche variability for young Atlantic salmon (*Salmo salar*) and brown trout (*S. trutta*) in heterogeneous streams. Ecology of Freshwater Fish 8:1–21.

Henderson, R., J. L. Kershner, and C. A. Toline. 2000. Timing and location of spawning by nonnative wild rainbow trout and native cutthroat trout in the South Fork Snake River, Idaho, with implications. North American Journal of Fisheries Management 20:584–596.

Hicks, B. J., and J. D. Hall. 2003. Rock type and channel gradient structure salmonid populations in the Oregon Coast Range. Transactions of the American Fisheries Society 132:468–482.

Hilderbrand, R. H., and J. L. Kershner. 2000. Conserving inland cutthroat trout in small streams: how much stream is enough? North American Journal of Fisheries Management 20:513–520.

IPCC (Intergovernmental Panel on Climate Change). 2007. Climate change 2007: the physical science basis. Contribution of working group 1 to the fourth assessment report of the Intergovernmental Panel on Climate Change. Cambridge University Press, Cambridge, UK, and New York.

Jackson, D. A., P. R. Peres-Neto, and J. D. Olden. 2001. What controls who is where in freshwater fish communities—the roles of biotic, abiotic, and spatial factors. Canadian Journal of Fisheries and Aquatic Sciences 58:157–170.

Jager, H. I., W. Van Winkle, and B. D. Holcomb. 1999. Would hydrologic climate changes in Sierra Nevada streams influence trout persistence? Transactions of the American Fisheries Society 128:222–240.

Jenkins, T. M., S. Diehl, K. W. Kratz, and S. D. Cooper. 1999. Effects of population density on individual growth of brown trout in streams. Ecology 80:941–956.

Keeler, R. A., and R. Cunjak. 2007. Reproductive ecology of slimy sculpin in small New Brunswick streams. Transactions of the American Fisheries Society 136:1762–1768.

Keeley, E. R. 2003. An experimental analysis of self-thinning in juvenile steelhead trout. Oikos 102:543–550.

Keleher, C. J., and F. J. Rahel. 1996. Thermal limits to salmonid distributions in the Rocky Mountain region and potential habitat loss due to global warming: a geographic information system (GIS) approach. Transactions of the American Fisheries Society 125:1–13.

Kershner, J. L. 1997. Monitoring and adaptive management. Pages 96–115 in J. E. Williams, C. A. Wood, and M. P. Dombeck, editors. Watershed restoration: principles and practices. American Fisheries Society, Bethesda, Maryland.

Kondolf, G. M. 2000. Assessing salmonid gravel quality. Transactions of the American Fisheries Society 129:262–281.

Koontz, T. M., and E. M. Johnson. 2004. One size does not fit all: matching breadth of stakeholder participation to watershed group accomplishments. Policy Sciences 37:185–204.

Kruse, C. G., W. A. Hubert, and F. J. Rahel. 2000. Status of Yellowstone cutthroat trout in Wyoming waters. North American Journal of Fisheries Management 20:693–705.

Kruse, C. G., W. A. Hubert, and F. J. Rahel. 2001. An assessment of headwater isolation as a conservation strategy for cutthroat trout in the Absaroka Mountains of Wyoming. Northwest Science 75:1–11.

Kyle, G., W. Norman, L. Jodice, A. Graefe, and A. Marsinko. 2007. Segmenting anglers using their consumptive orientation profile. Human Dimensions of Wildlife 12:115–132.

Lake, P. S. 2000. Disturbance, patchiness, and diversity in streams. Journal of the North American Benthological Society 19:573–592.

Larsen, D. P., P. R. Kaufmann, T. M. Kincaid, and N. S. Urquhart. 2004. Detecting persistent change in the habitat of salmon-bearing streams in the Pacific Northwest. Canadian Journal of Fisheries and Aquatic Sciences 61:283–291.

Larsen, D. P., T. M. Kincaid, S. E. Jacobs, and N. S. Urquhart. 2001. Designs for evaluating local and regional scale trends. BioScience 51:1069–1078.

Letcher B. H., K. H. Nislow, J. A. Coombs, M. J. O'Donnell, and T. L. Dubreuil. 2007. Population response to habitat fragmentation in a stream-dwelling brook trout population. PLoS ONE 2(11):1139.

Lyons, J., B. M. Weigel, L. K. Paine, and D. J. Undersander. 2000. Influence of intensive rotational grazing on bank erosion, fish habitat quality, and fish communities in southwestern Wisconsin trout streams. Journal of Soil and Water Conservation 55:271–276.

Malone, C. R. 2000. State governments, ecosystem management, and the enlibra doctrine in the U.S. Ecological Economics 34:9–17.

May, B. E., S. E. Albeke, and T. Horton. 2007. Range-wide status of Yellowstone cutthroat trout (*Oncorhynchus clarkii bouvieri*): 2006. Montana Department of Fish, Wildlife and Parks, Helena.

McHugh, P., and P. Budy. 2005. An experimental evaluation of competitive and thermal effects on brown trout (*Salmo trutta*) and Bonneville cutthroat trout (*Oncorhynchus clarkii utah*) performance along an altitudinal gradient. Canadian Journal of Fisheries and Aquatic Science 62:2784–2795.

McIntosh, B. A., J. R. Sedell, J. E. Smith, R. C. Wissmar, S. E. Clarke, G. H. Reeves, and L. A. Brown. 1994. Historical changes in fish habitat for select river basins in eastern Oregon and Washington. Northwest Science 68:36–53.

McMahon, T. E., A. V. Zale, F. T. Barrows, J. H. Selong, and R. J. Danehy. 2007. Temperature and competition between bull trout and brook trout: a test of the elevation refuge hypothesis. Transactions of the American Fisheries Society 136:1313–1326.

Montgomery, D. R., and J. M. Buffington. 1998. Channel processes, classification, and response. Pages 13–42 in R. J. Naiman and R. E. Bilby, editors. River ecology and management: lessons from the Pacific Coastal Ecoregion. Springer-Verlag, New York.

Morita, K., and S. Yamamoto. 2002. Effects of habitat fragmentation by damming on the persistence of stream-dwelling char populations. Conservation Biology 16:1318–1323.

Morrill, J. C., R. C. Bales, and M. H. Conklin. 2005. Estimating stream temperature from air temperature: implications for future water quality. Journal of Environmental Engineering 131:139–146.

Moyle, P. B., and B. Vondracek. 1985. Persistence and structure of the fish assemblage in a small California stream. Ecology 66:1–13.

Mundahl, N. D., and R. A. Sagan. 2005. Spawning ecology of the American brook lamprey, *Lampetra appendix*. Environmental Biology of Fishes 73:283–292.

Nelson, R. L., M. L. Mchenry, and W. S. Platts. 1991. Mining. Pages 425–457 *in* W. R. Meehan, editor. Influences of forest and rangeland management on salmonid fishes and their habitats. American Fisheries Society, Special Publication 19, Bethesda, Maryland.

Neraas, L. P., and P. Spruell. 2001. Fragmentation of riverine systems: the genetic effects of dams on bull trout *Salvelinus confluentus* in the Clark Fork River system. Molecular Ecology 10:1153–1164.

Neville, H., J. Dunham, and M. Peacock. 2006a. Assessing connectivity in salmonid fishes with DNA microsatellite markers. Pages 318–342 *in* K. Crooks and M. A. Sanjayan, editors. Connectivity conservation. Cambridge University Press, Cambridge, UK.

Neville, H. M., J. B. Dunham, and M. M. Peacock. 2006b. Landscape attributes and life history variability shape genetic structure of trout populations in a stream network. Landscape Ecology 21:901–916

Nico, L. G., and P. L. Fuller. 1999. Spatial and temporal patterns of nonindigenous fish introductions in the United States. Fisheries 24(1):16–27.

Northcote, T. G. 1978. Migratory strategies and production in freshwater fishes. Pages 326–359 *in* S. D. Gerking, editor. Ecology of freshwater fish populations. John Wiley and Sons, New York.

Northcote, T. G. 1997. Potamodromy in Salmonidae—living and moving in the fast lane. North American Journal of Fisheries Management 17:1029–1045.

Olsen, A. R., J. E. D. Sedransk, C. A. Gotway, W. Liggett, S. Rathbun, K. H. Reckhow, and L. J. Young. 1999. Statistical issues for monitoring ecological and natural resources in the United States. Environmental Monitoring and Assessment 54:1–45.

Peterson, D. P., and K. D. Fausch. 2003. Testing population-level mechanisms of invasion by a mobile vertebrate: a simple conceptual framework for salmonids in streams. Biological Invasions 5:239–259.

Peterson, D. P., K. D. Fausch, J. Watmough, and R. A. Cunjak. 2008a. When eradication is not an option: modeling strategies for electrofishing suppression of nonnative brook trout to foster persistence of sympatric native cutthroat trout in small streams. North American Journal of Fisheries Management 28:1847–1867.

Peterson, D. P., K. D. Fausch, and G. C. White. 2004. Population ecology of an invasion: effects of brook trout on native cutthroat trout. Ecological Applications 14:754–772.

Peterson, D. P., B. E. Rieman, J. B. Dunham, K. D. Fausch, and M. K. Young. 2008b. Analysis of tradeoffs between threats of invasion by nonnative trout and intentional isolation for native westslope cutthroat trout. Canadian Journal of Fisheries and Aquatic Sciences 65:557–573.

Petty, J. T., and G. D. Grossman. 2007. Size-dependent territoriality of mottled sculpin in a southern Appalachian stream. Transactions of the American Fisheries Society 136:1750–1761.

Petty, J. T., P. J. Lamothe, and P. M. Mazik. 2005. Spatial and seasonal dynamics of brook trout populations inhabiting a central Appalachian watershed. Transactions of the American Fisheries Society 134: 572–587.

Pickett, S. T. A., and M. L. Cadenasso. 1995. Landscape ecology: spatial heterogeneity in ecological systems. Science 269:331–334.

Platts, W. S. 1991. Livestock grazing. Pages 389–423 *in* W. R. Meehan, editor. Influences of forest and rangeland management on salmonid fishes and their habitats. American Fisheries Society, Special Publication 19, Bethesda, Maryland.

Poff, N. L., J. D. Allan, M. B. Bain, J. R. Karr, K. L. Prestegaard, B. D. Richter, R. E. Sparks, and J. C. Stromberg. 1997. The natural discharge regime: a paradigm for river conservation and restoration. BioScience 47:769–784.

Poff, N. L., M. M. Brinson, and J. W. Day. 2002. Aquatic ecosystems and global climate change: potential impacts on inland freshwater and coastal wetland ecosystems in the United States. Pew Center on Global Climate Change, Arlington, Virginia.

Poff, N. L., and J. V. Ward. 1990. Physical habitat template of lotic systems: recovery in the context of historical pattern of spatiotemporal heterogeneity. Environmental Management 14:629–645.

Post, J. R., T. Rhodes, P. Askey, A. Paul, and B. T. VanPoorten. 2006. Fish entrainment into irrigation canals: an analytical approach and application to the Bow River, Alberta, Canada. North American Journal of Fisheries Management 26:875–887.

Post, J. R., M. Sullivan, S. Cox, N. P. Lester, C. J. Walters, E. A. Parkinson, A. J. Paul, L. Jackson, and B. J. Shuter. 2002. Canada's recreational fisheries: the invisible collapse? Fisheries 27(1):6–17.

Quist, M. C., and W. A. Hubert. 2004. Bioinvasive species and the preservation of cutthroat trout in the western United States: ecological, social, and economic issues. Environmental Science and Policy 7:303–313.

Quist, M. C., and W. A. Hubert. 2005. Relative effects of biotic and abiotic process: a test of the biotic–abiotic constraining hypothesis as applied to cutthroat trout. Transactions of the American Fisheries Society 134:676–686.

Rahel, F. J. 1997. From Johnny Appleseed to Dr. Frankenstein: changing values and the legacy of fisheries management. Fisheries 22(8):8–9.

Rahel, F. J., and J. D. Olden. 2008. Assessing the effects of climate change on aquatic invasive species. Conservation Biology 22:521–533.

Reeves, G. H., L. E. Benda, K. M. Burnett, P. A. Bisson, and J. R. Sedell. 1995. A disturbance-based ecosystem approach to maintaining and restoring freshwater habitats of evolutionarily significant units of anadromous salmonids in the Pacific Northwest. Pages 334–349 in J. L. Nielsen, editor. Evolution and the aquatic ecosystem: defining unique units in population conservation. American Fisheries Society, Symposium 17, Bethesda, Maryland.

Rieman, B. E., and Dunham, J. B. 2000. Metapopulation and salmonids: a synthesis of life history patterns and empirical observations. Ecology of Freshwater Fish 9:51–64.

Rieman, B. E., D. Isaak, S. Adams, D. Horan, D. Nagel, C. Luce, and D. Myers. 2007. Anticipated climate warming effects on bull trout habitats and populations across the interior Columbia River Basin. Transactions of the American Fisheries Society 136:1552–1565.

Rieman, B. E., and J. D. McIntyre. 1995. Occurrence of bull trout in naturally fragmented habitat patches of varied size. Transactions of the American Fisheries Society 124:285–296.

Rinne, J. N. 1996. Short-term effects of wildfire on fishes and aquatic macroinvertebrates in the southwestern United States. North American Journal of Fisheries Management 16:653–658.

Romero, N. R., R. E. Gresswell, and J. Li. 2005. Changing patterns in coastal cutthroat trout (*Oncorhynchus clarki clarki*) diet and prey in a gradient of deciduous canopies. Canadian Journal of Fisheries and Aquatic Sciences 62:1797–1807.

Schroeder, S. A., D. C. Fulton, L. Currie, and T. Goeman. 2006. He said, she said: gender and angling specialization, motivations, ethics, and behaviors. Human Dimensions of Wildlife 11:301–315.

Shetter, D. S., and G. R. Alexander. 1967. Angling and trout populations on the North Branch of the Au Sable River, Crawford and Otsego counties, Michigan, under special and normal regulations, 1958–63. Transactions of the American Fisheries Society 96:85–91.

Sinokrot, B. A., H. G. Stefan, J. H. McCormick, and J. G. Eaton. 1995. Modeling of climate change effects on stream temperatures and fish habitats below dams and near groundwater inputs. Climatic Change 30:181–200.

Smith, G. R., T. E. Dowling, K. W. Gobalet, T. Lugaski, D. K. Shiozawa, and R. P. Evans. 2002. Biogeography and timing of evolutionary events among Great Basin fishes. Pages 175–234 in R. Hershler, D. B. Madsen, and D. R. Currey, editors. Great Basin aquatic systems history. Smithsonian Contributions to the Earth Sciences 33, Washington, D.C.

Southwood, T. R. E. 1977. Habitat, the template for ecological strategies? Journal of Animal Ecology 46:337–365.

Steen, P. J., D. R. Passino-Reader, and M. J. Wiley. 2006. Modeling brook trout presence and absence from landscape variables using four different analytical methods. Pages 513–531 *in* R. Hughes, L. Wang, and P. Seelbach, editors. Influences of landscapes on stream habitats and biological assemblages. American Fisheries Society, Symposium 48, Bethesda, Maryland.

Taniguchi, Y., F. J. Rahel, D. C. Novinger, and K. G. Gerow. 1998. Temperature mediation of competitive interactions among three fish species that replace each other along longitudinal stream gradients. Canadian Journal of Fisheries and Aquatic Sciences 55:1894–1901.

Thompson, K. G., R. B. Nehring, D. C. Bowden, and T. Wygant. 1999. Field exposure of seven species or subspecies of salmonids to *Myxobolus cerebralis* in the Colorado River, Middle Park, Colorado. Journal of Aquatic Animal Health 11:312–329.

Thompson, P. D., and F. J. Rahel. 1998. Evaluation of artificial barriers in small Rocky Mountain streams for preventing the upstream movement of brook trout. North American Journal of Fisheries Management 18:206–210.

Thurow, R. F., D. C. Lee, and B. E. Rieman. 1997. Distribution and status of seven native salmonids in the interior Columbia basin and portions of the Klamath River and Great Basins. North American Journal of Fisheries Management 17:1094–1110.

Tilt, W., and C. A. Williams. 1997. Building public and private partnerships. Pages 145–157 *in* J. E. Williams, C. A. Wood, and M. P. Dombeck, editors. Watershed restoration: principles and practices. American Fisheries Society, Bethesda, Maryland.

Trenberth, K. E., A. Dai, R. M. Rasmussen, and D. B. Parsons. 2003. The changing character of precipitation. Bulletin of the American Meteorological Society 84:1205–1217.

Trombulak, S. C., and C. A. Frissell. 2000. Review of ecological effects of roads on terrestrial and aquatic communities. Conservation Biology 20:18–30.

Urquhart, N. S., S. G. Paulsen, and D. P. Larsen. 1998. Monitoring for policy-relevant regional trends over time. Ecological Applications 8:246–257.

Van Kirk, R. W., and L. Benjamin. 2001. Status and conservation of salmonids in relation to hydrologic integrity in the Greater Yellowstone Ecosystem. Western North American Naturalist 61:359–374.

Vannote, R. L., G. W. Minshall, K. W. Cummins, J. R. Sedell, and C. E. Cushing. 1980. The river continuum concept. Canadian Journal of Fisheries and Aquatic Sciences 37:130–137.

Varley, J. D., and R. E. Gresswell. 1988. Ecology, status, and management of the Yellowstone cutthroat trout. Pages 13–24 *in* R. E. Gresswell, editor. Status and management of interior stockis of cutthroat trout. American Fisheries Society, Symposium 4, Bethesda, Maryland.

Vashro, J. 1995. The "bucket brigade" is ruining our fisheries. Montana Outdoors 26:34–37.

Vincent, E. R. 1987. Effects of stocking catchable-size hatchery rainbow trout on two wild trout species in the Madison River and O'Dell Creek, Montana. North American Journal of Fisheries Management 7:91–105.

Vincent, E. R. 1996. Whirling disease and wild trout: the Montana experience. Fisheries 21(6):32–33.

Warren, C. E., and W. J. Liss. 1980. Adaptation to aquatic environments. Pages 15–40 *in* R. T. Lackey, and L. A. Nielsen, editors. Fisheries Management. Blackwell Scientific Publications, Oxford, UK.

Wehrly, K. E., L. Wang, and M. Mitro. 2007. Field-based estimates of thermal tolerance limits for trout: incorporating exposure time and temperature fluctuation. Transactions of the American Fisheries Society 136:365–374.

Westerling, A. L., H. G. Hidalgo, D. R. Ryan, and T. W. Swetnam. 2006. Warming and earlier spring increase western U.S. wildfire activity. Science 313:940–943.

Williams, J. E., A. L. Haak, H. M. Neville, W. T. Colyer, and N. G. Gillespie. 2007. Climate change and western trout: strategies for restoring resistance and resilience in native populations. Pages 236–246 *in* R. F. Carline and C. LoSapio, editors. Wild trout IX: sustaining wild trout in a changing world. Wild Trout Symposium, Bozeman, Montana.

Williams, J. E., C. A. Wood, and M. P. Dombeck. 1997. Understanding watershed-scale restoration. Pages 1–13 *in* J. E. Williams, C. A. Wood, and M. P. Dombeck, editors. Watershed restoration: principles and practices. American Fisheries Society, Bethesda, Maryland.

Wofford, J. E. B., R. E. Gresswell, and M. A. Banks. 2005. Influence of barriers to movement on within-watershed genetic variation of coastal cutthroat trout. Ecological Applications 15:628–637.

Woodward, D. F., A. M. Farag, M. E. Mueller, E. E. Little, and F. A. Vertucci. 1989. Sensitivity of endemic Snake River cutthroat trout to acidity and elevated aluminum. Transactions of the American Fisheries Society 118:630–643.

Woodward, D. F., J. N. Goldstein, A. M. Farag, and W. G. Brumbaugh. 1997. Cutthroat trout avoidance of metals and conditions characteristic of a mining waste site: Coeur d'Alene River, Idaho. Transactions of the American Fisheries Society 126:699–706.

Young, M. K. 1995. Conservation assessment for inland cutthroat trout. U.S. Deparment of Agriculture Forest Service, General Technical Report RM-GTR-256, Fort Collins, Colorado.

Young, M. K. 1996. Summer movements and habitat use by Colorado River cutthroat trout (*Oncorhynchus clarki pleuriticus*) in small, montane streams. Canadian Journal of Fisheries and Aquatic Sciences 53:1403–1408.

Young, M. K., P. M. Guenther-Gloss, and A. D. Ficke. 2005. Predicting cutthroat trout (*Oncorhynchus clarkii*) abundance in high-elevation streams: revisiting a model of translocation success. Canadian Journal of Fisheries and Aquatic Sciences 62:2399–2408.

Zimmerman, J. K. H., and B. Vondracek. 2006. Interactions of slimy sculpin *Cottus cognatus* with native and nonnative trout: consequences for growth. Canadian Journal of Fisheries and Aquatic Sciences 63:15261535.

Chapter 19

Coldwater Rivers

DARIN G. SIMPKINS AND JESSICA L. MISTAK

19.1 INTRODUCTION

Coldwater rivers are large streams that are most often sampled from a boat or raft and are occupied by coldwater species such as salmonids. Many coldwater rivers have been modified as a result of water storage and dam-regulated flows. Nevertheless, fisheries management for coldwater rivers should not be considered independent to that of coldwater streams. Many coldwater species have fluvial life histories that require connections between streams and rivers to accommodate migratory behaviors, growth, and survival of specific life stages. Physical and biological processes in cold headwater streams affect those in coldwater rivers, which, in turn, structure fish assemblages. Consequently, fisheries managers of coldwater rivers must establish goals and objectives relative to specific abiotic and biotic attributes of rivers, as well as to life history requirements of fishes and ecological processes and management practices within entire coldwater drainages.

Fisheries management for coldwater streams and rivers has focused predominantly on salmonids owing to their economic importance in North America. Forty-seven of 50 states manage recreational salmonid fisheries (Epifanio 2000). In 2006, approximately 6.8 million anglers in the USA fished for salmonids over 76 million days in streams and rivers, which represented 27% of all anglers and 18% of all days fished (USDI 2007). The average freshwater angler spent US$460 on travel, food, lodging, equipment, and licenses. In 2005, approximately 3.2 million Canadians fished for salmonids over 43 million days in streams and rivers, spending Can$2.5 billion (DFO 2007). Thus, salmonids in coldwater streams and rivers support important recreational fisheries and generate substantial revenues.

Historically, management of coldwater rivers focused on stocking native or nonnative salmonids, manipulating flows and habitat to improve growth and survival of salmonids, and creating fishing opportunities. Most of these efforts have focused on small spatial scales, and little attention has been given to the significance of connectivity among streams and rivers in a watershed. More recently, emphasis to conserve native fish assemblages in coldwater environments has evolved as knowledge has been gained about the effects of human disturbances on ecological interactions among species and processes that structure fish assemblages. The purpose of this chapter is to introduce management practices used to conserve, enhance, or create fisheries in coldwater rivers.

19.2 COLDWATER RIVER ECOLOGY

19.2.1 Structure and Function of Free-Flowing Coldwater Rivers

The ecological basis for understanding interactions between free-flowing coldwater streams and rivers is often presented within the context of the river continuum concept (Vannote et al. 1980). The premise of the river continuum concept is that downstream processes are linked to upstream processes, creating resource gradients that structure aquatic communities (Figure 19.1). Physical habitats in lotic environments are determined by interactions between hydrologic and geomorphic processes. Specifically, the morphology of streams and rivers is dependent on sediment transport, which is determined by discharge, channel width, water depth, channel slope, roughness of the bank and bed, shape of the bed, sediment size and supply, and riparian vegetation (Montgomery and Buffington 1998). A change in any of these factors results in compensatory responses of other factors as channels move toward a new dynamic equilibrium.

A hierarchical system can be used to describe the relative size of streams and rivers (Allan 1995). The smallest, permanently flowing stream is termed first order, and the union of two streams of the same order increases the downstream order by one. Small streams (i.e., stream orders 1–3; Allan 1995) may provide a relatively constant environment through most of the year because of groundwater inputs (Chapter 18). Channel slopes and substrate sizes are relatively large in small streams compared with larger rivers. Primary production in small streams is limited because of shading by riparian vegetation. As such, the primary energy input is from allochthonous organic matter washed or blown into the stream from the watershed. The aquatic invertebrate community in small streams consists primarily of shredders that process large debris and feed on coarse particulate organic matter and collectors that feed on fine particulate matter. Such aquatic invertebrates, along with terrestrial invertebrates that fall into streams, serve as the primary prey base for coldwater fishes residing in small streams (Baxter et al. 2004). Fishes that are adapted to a narrow range of coldwater temperatures, specific food resources, and maintaining themselves in swift currents of small streams are generally small, consisting primarily of salmonids, cottids, and some catostomids and cyprinids.

Intermediate-size streams and rivers (i.e., orders 4–6) are the most variable in hydrologic features and water temperatures, thereby exhibiting the greatest physical and biological diversity along the continuum. As the gradient of streams and rivers decline, currents slow and physical habitat characteristics become more variable than they are in small streams. Secondary channels and islands are formed during high-flow events, such as seasonal periods of snowmelt or precipitation. Increased channel widths open the forest canopy and allow sunlight penetration, thereby increasing the water temperature during summer. A decrease in stream slope, sediment size, and relative contribution of allochthonous inputs from the terrestrial landscape, along with an increase in light and water temperature, facilitates a shift from heterotrophy to autotrophy by which in-channel primary production becomes the more dominant energy source. Due to the lack of coarse particulate organic matter, the invertebrate community changes from shredders to collectors and grazers that feed on algae and rooted aquatic vegetation. Abrupt changes in channel morphology may occur at river confluences due to variation in tributary discharge and sediment transport (Figure 19.2). The variety of habitat conditions often results in patchy distributions of areas with relatively high biological productivity and diversity (network dynamic hypothesis; Benda et al. 2004). Fishes that oc-

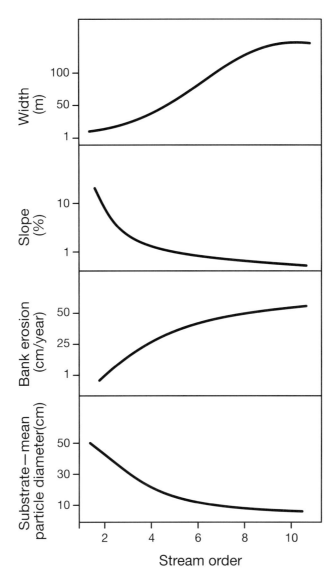

Figure 19.1. Gradual and continuous central tendencies of downstream (increasing stream order) changes in physical and biological processes predicted by the river continuum concept.

cupy intermediate-size rivers have life histories that are adapted to exploit dynamic environments and food resources.

Variability in hydrologic features and water temperature diminishes in systems with a stream order greater than six due to the volume of water within rivers (Johnson et al. 1995a). Shading from riparian vegetation has little effect on water temperature. The channel is wider with slower velocities than in smaller streams and rivers, causing fine sediments to accumulate. Main channels are generally unsuitable for macrophytes and periphyton owing to turbidity, water depth, and unstable substrates. Autochthonous production is primarily by phytoplankton, which is limited by turbidity and mixing (Allan 1995). Allochthonous inputs of fine particulate organic matter from upstream systems and the floodplain are the primary

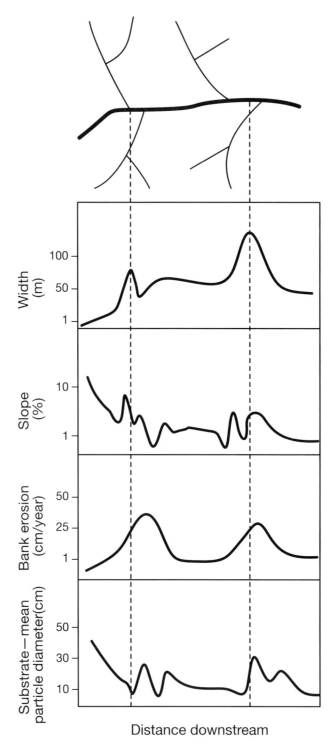

Figure 19.2. The network dynamic hypothesis suggests that tributary hydrology and sediment inputs cause deviations from central tendencies in physical and biological processes of coldwater rivers that are predicted by the river continuum concept and create habitat diversity.

energy inputs and are consumed by bottom-dwelling invertebrates characterized as collectors. Compared with small streams, terrestrial vegetation contributes relatively little allochthonous material in large rivers. Channels of large rivers are frequently characterized by repeating river bends and crossover regions and often lack distinctive pools, riffles, and runs (Trush et al. 2000). Channel morphology is dependent on geologic features and flow regimes, as well as erosional and depositional processes that result in channels meandering within the floodplain (Figure 19.3). The thalweg is the deepest part of the river; there water velocities are typically the swiftest, and the thalweg traverses the channel as sediments are eroded from outside bends and deposited on inside bends. Habitats of importance in large rivers include backwaters, islands, and floodplains (Trush et al. 2000). Aquatic species that are adapted to large rivers utilize these unique habitats during specific developmental periods of their life histories. For example, backwaters often support up to 90% of the fish biomass in rivers (Stalnaker et al. 1989) by serving as refugia, regions of nutrient enrichment, and larval fish habitat (Sheaffer and Nickum 1986; Scott and Nielsen 1989). Inundation of riparian vegetation within floodplains of coldwater rivers also provides important refuges for juvenile fish (Coutant 2004; Figure 19.4). The importance of the interaction between large rivers and their floodplain is emphasized by the flood pulse concept, which describes the advancement and retraction of the river onto the floodplain as the principal agent enhancing biological productivity and diversity (Bayley 1995). Many large coldwater rivers having these characteristics occur in Canada and northern costal regions of the USA (Cushing et al. 2006).

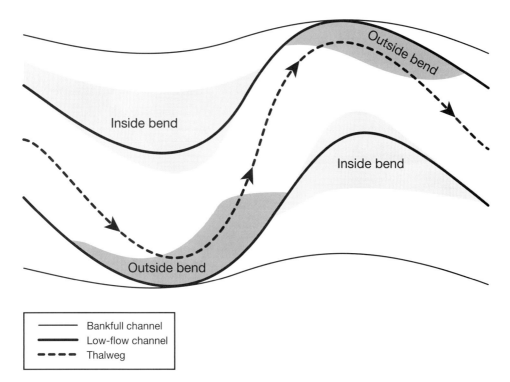

Figure 19.3. An example of a high-order river (six order or greater) with an idealized alternate bar sequence showing erosional and depositional processes that result in channels meandering within the floodplain.

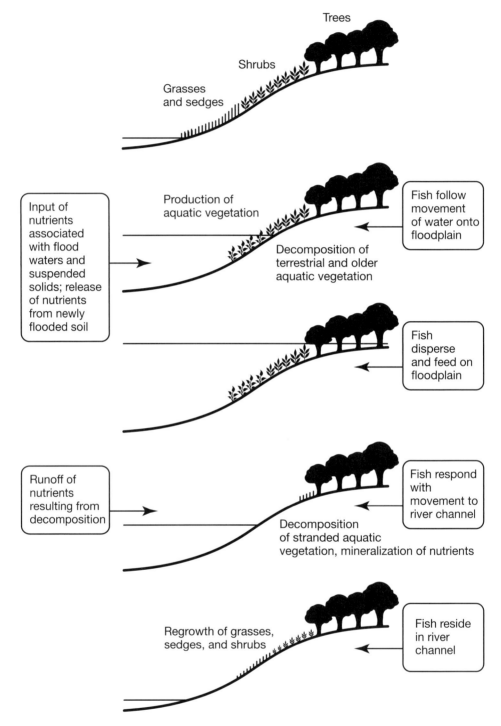

Figure 19.4. Schematic of the flood pulse concept showing a vertically exaggerated section of a floodplain in five snapshots of an annual hydrological cycle. Right hand column indicates typical responses of fish to flood pulse.

19.2.2 Factors Affecting Distribution and Abundance of Coldwater Fishes

19.2.2.1 Temperature and Hydrology

Although the river continuum concept suggests a shift from coldwater fish species in small streams to warmwater species in large rivers (Vannote et al. 1980), the sizes and diversity of coldwater species are often greatest in intermediate-size rivers (Platts 1979; Mackay 2006), whereas densities tend to be greater in large rivers (Gende et al. 2002). Nevertheless, water temperature is a primary factor restricting the spatial distribution of coldwater fishes. Most coldwater species inhabit rivers that do not exceed 22°C, but some salmonids occupy rivers having summer water temperatures of 24–29°C (Zoellick 1999; Schrank et al. 2003). Thus, coldwater species may at times co-exist in rivers with coolwater species (fishes with temperature preferences of 22–24°C, such as percids and esocids) or even warmwater species (fishes with temperature preferences >24°C, such as centrarchids and ictalurids), especially in systems that have been environmentally altered or degraded (Lyons et al. 1996). Though the spatial distributions of coldwater species in rivers are usually temperature limited, the distributions of coolwater and warmwater species in rivers can be limited by additional factors, such as water velocity (Faler et al. 1988).

Water temperature and hydrologic features vary on a seasonal basis depending on climate, geology, and water source. In systems where groundwater input is substantial, seasonal variation in water temperature and flow is nominal (Figure 19.5). For example, water temperatures and flows in free-flowing coldwater rivers in the northern portion of Michigan's Lower Peninsula are relatively stable due to the influence of groundwater (Wilhelm et al. 2005). At the opposite extreme, water temperatures can be highly variable in free-flowing rivers where flows are maintained by tributary inputs that are dependent on precipitation and snowmelt, such as rivers in the Rocky Mountains and Pacific Northwest (Ziemer and Lisle 1998). Such rivers may have hydrographs that depict peak flows that are ten to several hundred times that of average low flows. Coastal rivers, such as in the Pacific Northwest, may be characterized by hydrographs that reflect substantial rainfall from fall through winter. Water temperatures in coastal rivers are usually moderate in comparison to extremes exhibited by groundwater- and snowmelt-dependent rivers and generally follow seasonal air temperatures. Variability of water temperature and flow is dependent on the rate of overland runoff, which is determined by topographic and geologic features, vegetation, and soil permeability (Trush et al. 2000). Thus, changes in land or water use can lead to the alteration of water temperatures and flows. In urban areas, a relatively high proportion of impervious surfaces results in greater overland flows, higher peak flows, and greater variability in water temperatures (Allan 1995).

19.2.2.2 Downstream Effects of Dams

Numerous dams and reservoirs have been constructed on river systems throughout North America. The effects of dams on physical and biological attributes of downstream rivers have been documented (Allan 1995; Poff et al. 1997) and have been characterized by the serial discontinuity concept (Stanford and Ward 2001). The serial discontinuity concept is based on the river continuum concept but predicts changes and downstream shifts in physical and biological characteristics of rivers caused by impoundments. The premise of the serial discontinuity concept is that the ecological structure and function of rivers as naturally free-flowing

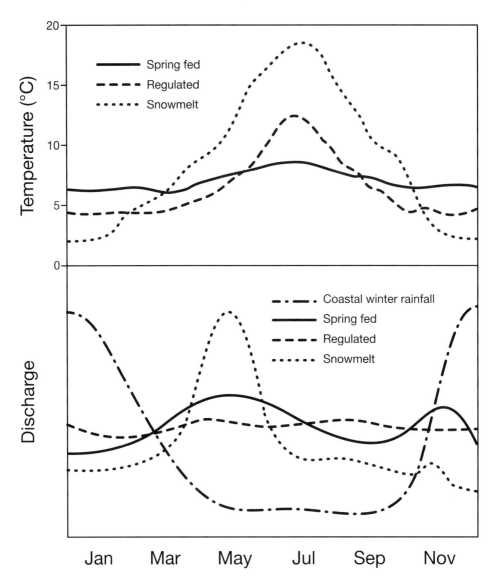

Figure 19.5. Typical seasonal temperature and flow patterns of coldwater rivers. An example of spring-fed profiles would be unregulated coldwater rivers in Michigan; regulated profiles would be impounded rivers with deep hypolimnetic water withdrawals that are not managed for power-peaking flows; snowmelt profiles would be unregulated coldwater rivers in the Rocky Mountains; and coastal winter rainfall profiles would be rivers along the coast in the Pacific Northwest.

and continuous systems are transformed or "reset" into smaller river segments by reservoirs. Consequently, the physical and biological characteristics of rivers immediately downstream from dams resemble lower-order streams and transition into characteristics of higher-order rivers with increasing distance downstream as allochthonous inputs are added.

Dams cause predictable changes in hydrologic and morphologic features of downstream rivers (Poff et al. 1997). Seasonal variation in peak-to-base flow common among large rivers is often reduced, resulting in more constant and uniform hydrographs (Figure 19.5). Because impoundments alter the natural flow regime and trap sediments, channel characteristics change downstream from reservoirs. Water released from dams scours fine sediments, causing bank erosion and armoring of the streambeds as streambeds downcut deep into channels. Streambed armoring is a process that tightly compacts bottom substrates due to the selective removal of small sediments from the stream bottom. Over time, channel lengths, widths, and sinuosity as well as the numbers of islands decrease. River characteristics downstream from dams change from diverse habitats to single, swift, deep, and narrow channels. Degradation of main channels leads to loss of lateral connectivity to the floodplain, dewatering of side-channel and backwater habitats, and decreases in the diversity and abundance of riparian vegetation.

As a result of thermal stratification in large reservoirs and hypolimnetic releases from most large dams, water temperatures are cooler during summer and warmer during winter in downstream tailwaters than in unregulated rivers (Figure 19.5; Allan 1995; Vinson 2001). Diel temperature fluctuations characteristic of unregulated rivers are also reduced. Water released during the winter is generally near 4°C, which is the temperature at which water reaches its maximum density. These relatively warm winter temperatures prevent surface ice formation for some distance downstream from the dam (Annear et al. 2002). The absence of surface ice allows for the formation of frazil ice during extremely cold weather. Frazil ice forms when flowing water is supercooled to less than 0°C by cold air temperatures. Simpkins et al. (2000a) found that nighttime air temperatures of less than −24°C resulted in frazil ice events in a Wyoming tailwater. After water is supercooled, small (0.1–5.0 mm) disk-shaped ice crystals form in the water column. Turbulent water keeps the frazil ice suspended, but it may adhere to underwater objects to create anchor ice. Frazil ice floats to the surface in slow-velocity habitats, creating surface ice along stream margins. Ice processes can affect fish in coldwater rivers. Trout have been observed to move long distances during periods of frazil and anchor ice formation (Simpkins et al. 2000a). Physical injury can cause winter mortality when surface ice collapses in shallow habitats (Annear et al. 2002) or frazil ice gets caught in mouths and gills (Brown 1999).

Downstream changes in water quality in coldwater rivers depend on limnological processes in reservoirs. Because dams trap inflowing sediments, water clarity typically increases in tailwaters. Initial responses to impoundment include nutrient enrichment of tailwaters, followed by nutrient reduction associated with aging of reservoirs (Vinson 2001; Chapter 17). Nutrient enrichment, coupled with enhanced water clarity and reduced variability of flow, allows the establishment of periphyton and aquatic macrophytes in tailwaters, which are generally rare elsewhere in unregulated coldwater rivers but are important for production of aquatic invertebrates that are shredders or collectors (Munn and Brusven 1991). However, production of aquatic plants in tailwaters diminishes over time as the channel degrades, the reservoir ages, and the availability of nutrients declines.

The initial changes in the physical and chemical environment caused by dams can result in dramatic decreases in species richness but increases in overall abundance of aquatic invertebrates (Blinn and Cole 1991). Reductions in habitat heterogeneity and large woody debris in tailwaters cause a reduction in invertebrate species diversity but greater abundances of species favored by altered conditions. Invertebrates that are often common in coolwater streams include mayflies, such as *Baetis* spp., *Rhithrogena* spp., and *Epeorus* spp., as well as blackflies and midges (Allan 1995; Vinson 2001). However, continued main-channel habitat degradation, entrainment of nutrients in reservoirs, reduced aquatic plant production, and lack of allochthonous inputs lead to long-term decreases in abundance of aquatic invertebrates.

Furthermore, moderation of temperature regimes in tailwaters has a significant impact on benthic fauna (Vinson 2001). Water temperatures in tailwaters that are warmer than temperatures in unregulated rivers during winter can eliminate thermal cues needed by many invertebrate species required to break egg diapause (Allan 1995). Cool summer temperatures reduce the growing period required to complete development of many aquatic invertebrates; thus tailwaters generally have smaller aquatic invertebrates than do unregulated rivers.

The physical, chemical, and biological changes that are caused by dams and reservoirs can have cascading effects on fish populations. Hypolimnetic water releases from most large dams maintain water temperatures in tailwaters that are close to optimal for metabolism and growth of salmonids during much of the year. Consequently, fisheries managers often stock salmonids in tailwaters to create sport fisheries. Salmonids flourish and grow rapidly due to aquatic invertebrate prey bases. It is not uncommon for salmonids that are stocked as fingerlings (10–15 cm) in the spring (April–June) to grow to 20–25 cm by the following fall (October–November) and eventually become trophy-size fish (Wright 1995). Initially, flow releases that scour fine sediments from coarser substrates (e.g., gravels and cobbles) may create habitats suitable for spawning and reproduction of many salmonids. However, in upstream reaches the quantity and quality of spawning habitats may degrade as channel armoring and downcutting processes occur. In downstream reaches, fine sediments may accumulate in spawning areas due to sediment contributions from tributaries and bank erosion (Figure 19.1) and lack of high spring flows (Kondolf and Wilcock 1996). Reductions in peak flows coupled with degradation of main channels can lead to the dewatering of side channel and riverbank habitats that are important for nursery, rearing, and overwinter habitat of young-of-the-year fish (Mitro et al. 2003).

The thermal influence of hypolimnetic water releases can create a mismatch between the metabolic demands of young salmonids and the life history of aquatic invertebrates in tailwaters during the winter. Metabolic rates of salmonids in tailwaters may be higher than those of fish in unregulated rivers during most of the winter owing to relatively warm water temperatures. Unlike salmonids in unregulated rivers, fish in tailwaters often must feed to offset metabolic costs during winter, when the availability of drifting aquatic insects is reduced (Simpkins and Hubert 2000). Though small salmonids in regulated rivers often lose a substantial amount of weight during winter (Cunjak 1988), starvation does not appear to be the sole cause for overwinter losses of small fish in tailwaters (Simpkins et al. 2003). Rather, overwinter losses in tailwaters appear to be due to starvation processes in combination with long-range and exhaustive movements associated with fall-to-winter changes in flows and habitat, occurrences of frazil ice during extreme cold periods, and predator avoidance (Mitro et al. 2003; Simpkins et al. 2004). As in many unregulated rivers during extremely cold periods, salmonids in tailwaters move from feeding habitats into slow-velocity areas with large substrates (Simpkins et al. 2000a). Such areas provide interstitial spaces and concealment cover that allow fish to conserve energy (Mitro et al. 2003; Roussel et al. 2004). Due to channel degradation in upstream reaches and siltation in downstream reaches of tailwaters, reduced availability of winter habitats providing concealment cover can limit overwinter survival of salmonids.

Flow variability can affect both quantity and quality of habitat available for use by coldwater fishes in tailwaters (Bunt et al. 1999). Changes in flow can result in changes in thermal conditions of tailwaters (Krause et al. 2005). Water depth and velocities and availability of riffles, side channels, and backwaters generally decline with decreasing flows (Dare et al. 2002). Since aquatic invertebrate production is typically highest in such habitats, reductions in invertebrate abundance are commonly associated with decreases in flow (Blinn et al. 1995).

Consequently, salmonids tend to move into deep pool habitats as flows decrease and become food limited (Dare et al. 2002). However, rapid flow reductions have been attributed to stranding of eggs and fish in dewatered habitats such as river margins, side channels, and backwaters (Saltveit et al. 2001; Pender and Kwak 2002).

High-flow releases can create diverse and abundant habitat and reconnect rivers with floodplains, thus increasing allochthonous inputs and terrestrial invertebrates in tailwaters (Figure 19.4). Increased water velocities dislodge aquatic invertebrates into the water column (i.e., drift), providing a short-term increase in abundance of food for fish (Lagarrigue et al. 2002). However, the magnitude and duration of sequential high-flow releases can reduce aquatic invertebrate populations because of armoring of the channel bed; instability in depths, velocities, and water temperatures; and lack of recolonization by aquatic invertebrates from an upstream source due to the dam (Shannon et al. 2001). Long-term decreases in aquatic invertebrates in tailwaters can limit food availability, decrease growth rates, reduce fecundity, and contribute to mortality of salmonids in tailwaters over time (Pender and Kwak 2002). Relatively small fishes that primarily feed on drifting aquatic invertebrates (e.g., cutthroat trout and rainbow trout) seem to be impacted by the physical and biological changes in tailwaters more than are larger, piscivorous fishes (e.g., brown trout), but over time even piscivorous fishes may decline in growth and abundance as the densities of prey species decline (McHugh et al. 2006).

Dams can interfere with upstream and downstream migrations of fish in coldwater rivers (Northcote 1998; Lichatowich 1999). Young fish migrating downstream can be killed or injured by hydropower turbines and spillways (Cada et al. 1997). Current velocities are reduced in reservoirs, forcing juveniles to expend more energy swimming during their downstream migration than they would in unregulated rivers. Such migration difficulties result in higher stress, reduced developmental rates, and increased mortality among small fish (Congleton et al. 2000). Impoundment of the Columbia River has led to an increase in the abundance of piscivorous fishes (e.g., northern pikeminnow, largemouth bass, smallmouth bass, and walleye) that pose additional risks to survival of young salmonids. Low-head dams, including those used to divert water into canals for irrigation or industrial purposes, also restrict fish movements but can additionally cause fish to become entrained and subsequently die when flows through diversion canals are reduced (Schrank and Rahel 2004; Carlson and Rahel 2007).

19.2.2.3 Biotic and Abiotic Interactions

In free-flowing river systems, the longitudinal continuum of biotic and abiotic variables structures the distribution of coldwater species. For example, segregation among salmonid species has been described in the Salmon River drainage in Idaho (Platts 1979; Figure 19.6). First-order streams were used only by bull trout, whereas fifth-order streams were not used by bull trout. Rainbow trout was the only salmonid species found in stream reaches with gradients over 16%. River reaches with gradients of 8–14% were favored by cutthroat trout, while reaches with gradients of 6–10% were favored by bull trout and reaches with gradients of 2–5% were favored by brook trout. Gradient has been shown to be negatively correlated with both brook trout and brown trout abundance (Quist and Hubert 2005). Furthermore, water temperature for optimal growth can differ among species (Selong et al. 2001) and contributes to the spatial structuring of populations (Paul and Post 2001). Coexistence of multiple species within a river depends on the species ability to avoid hybridization and specialize under conditions of variable habitat, food availability, and risks of predation. Biotic interactions

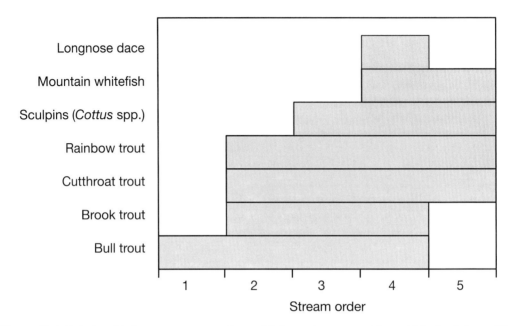

Figure 19.6. Relationship between stream order and fish species composition within the Salmon River watershed, Idaho (data are from Platts 1979).

among species, such as competition, predation, and hybridization, can have overriding negative influences on the distribution of fishes, even when abiotic conditions are suitable (Quist and Hubert 2005).

Competition can occur among fishes when they coexist in the same river and share limited common resources, such as food or suitable space. Competition can reduce growth, condition, and survival of one or more species and over time may lead to extirpation of one or more species. However, within- or among-year variability in environmental conditions, such as water temperature, may alternately favor one competitor over others (Dunham et al. 2002). Coexistence of species may result from interspecific differences in habitat use that reflect species differences in agonistic behavior, innate habitat preference, timing of emergence, morphologic features, or a combination of these factors. Introductions of nonnative salmonids often lead to competition with native salmonids for food and space. Examples include the decline of native cutthroat trout following the range expansion of brook trout (Dunham et al. 2002), decline of brook trout in the southeastern portion of its native range following invasion by rainbow trout (Flebbe 1994), and replacement of native brook trout by brown trout in streams of the midwestern USA (Zorn et al. 2002).

The effects of predation on fish populations can be substantial in coldwater rivers. In the Au Sable River of Michigan, predators of juvenile brook trout and brown trout included great blue herons, common mergansers, river otters, American mink, and adult brown trout (Alexander 1979). These predators were estimated to have consumed 79% of age-1 brook trout and 45% of age-1 brown trout. Derby and Lovvorn (1997) estimated that American white pelicans and cormorants consumed up to 80% of the rainbow trout that were stocked in a segment of the North Platte River in Wyoming. In the lower Columbia River of the Pacific Northwest, salmonids accounted for 74% of the mass of the diet for Caspian terns, 46% for double-crested cormorants, and 11% for glaucous-winged western gulls (Collis et al. 2002).

19.3 MANAGEMENT OF COLDWATER RIVERS

19.3.1 History

Early management efforts focused on propagation and stocking of salmonids to enhance or create fishing opportunities in coldwater rivers (Chapters 1 and 9). In Canada, Atlantic salmon were first cultured in 1866 to be stocked into coldwater streams and rivers entering Lake Ontario. As early as 1872, Pacific salmon eggs were cultured in the western USA and transported to the eastern USA to stock coldwater streams and rivers. Similarly, rainbow trout, brook trout, and brown trout were distributed throughout the USA by the late-19th century. Hatcheries were established on many coldwater rivers that were commercially fished. The emphasis on propagation and stocking of salmonids into coldwater rivers was due to the belief that sustainable fisheries could be maintained only by aquaculture. An emphasis on artificial propagation also contributed to poor understanding of ecological linkages among fish populations, habitat quality and connectedness, and land use practices. Difficulties assessing the status of fish populations and improving habitat conditions in rivers perpetuated the application of stocking as a primary management tool to offset anthropogenic factors affecting coldwater fish populations into the mid-20th century.

From the late 1960s through the 1970s, concern grew regarding the effects of stocked fishes and the role of habitat on native fish populations (Mahnken et al. 1998; Chapter 9). Fisheries managers began realizing that natural resource conservation and habitat improvement could provide unique angling opportunities for wild and native fishes. Awareness of habitat requirements in relation to life history strategies of various salmonid species in coldwater streams and rivers proliferated into the 1980s, and understanding of the significance of connectivity among streams and rivers to create habitat diversity to fulfill life history requirements of fishes has grown in parallel with principles of river ecology (Benda et al. 2004). Consequently, management alternatives now include maintaining or enhancing native species or subspecies of concern, applying special regulations (Chapter 7), mitigating for habitat impacted by development projects, and protecting and enhancing instream and riparian habitat needed to fulfill life history requirements of coldwater river fishes (Williams et al. 1997). Decision-making processes for implementing management actions now account for complex interactions among political, economic, and sociocultural entities (Chapter 5) that require managers to have skills in communications and conflict resolution (Chapter 6).

19.3.2 Agency Responsibilities and Competing Resource Use Issues

Management of coldwater rivers is complex due to competing uses of natural resources, interrelated responsibilities among agencies, and sampling difficulties. Ecological processes at the watershed scale affect fishes in coldwater rivers, so fisheries managers cannot only address fish and habitat within a river. Rather, they must coordinate their efforts with entities managing watershed issues, such as increased urbanization and reduction of floodplain areas, installation or removal of hydropower dams, or development of resource-extraction industries (e.g., agriculture, mining, logging, and grazing). Water development has been a substantial concern of fisheries managers responsible for coldwater rivers in western North America, where water diversions and withdrawal systems have been installed on most river systems to irrigate agricultural land. Rivers without legally-established minimum instream flows can be completely dewatered during periods of limited precipitation. The issue is not unique to western North America. High-capacity groundwater wells near streams and rivers can deplete aquifers and reduce inputs to streams, rivers, and wetlands. Reductions in flows caused by water diversions and groundwater withdrawals can result in elevated water temperatures in coldwater rivers. In addition, manufacturing industries and power plants use substantial quantities of water from rivers to cool various devices, including electrical generators, pumps, and reactors. Regulation of warmwater discharges from these cooling facilities is needed to avoid thermal barriers for migratory fishes and macroinvertebrates.

One of the biggest challenges in managing coldwater fisheries in tailwaters is accommodating multiple uses for water stored in reservoirs. Under the Rivers and Harbors Act and the Pick–Sloan Flood Control Act of 1944, dams were primarily built for flood control, hydroelectric power generation, and other municipal uses. Secondary benefits included water storage for irrigation and recreation. Consequently, most dam operations and maintenance responsibilities are managed by federal (i.e., Federal Energy Regulatory Commission, U.S. Army Corps of Engineers, and U.S. Bureau of Reclamation) or regional agencies (e.g., water councils or municipality districts) that are not responsible for managing fisheries resources. Competing demands for water in reservoirs and tailwaters require collaboration among water and fisheries managers to balance the interests of various stakeholders.

Responsibility of providing fisheries in public waters rests primarily with state or provincial fish and wildlife agencies under the public trust doctrine (Chapter 4), but management of water quality and aquatic habitat may be the responsibility of other agencies (Epifanio 2000). For example, multiple state agencies are often involved in decision-making processes when the U.S. Federal Energy Regulatory Commission considers issuing licenses for development and operation of hydroelectric dams (see Box 4.5). As part of the licensing effort, one agency may be responsible for monitoring to ensure operational compliance with standards for water temperature, dissolved oxygen, and minimum instream flows, while another agency may be responsible for ensuring that fish communities and fish passage are not affected by dam operations (Box 19.1). Multiple agencies representing water quality, agriculture, and fisheries management may have competing interests and must work together to develop recommendations. When management responsibilities of coldwater rivers are shared, close coordination is required to ensure that all resources are considered in management plans that may influence aquatic resources.

Management of endangered and threatened species, anadromous fishes, rivers on federal lands in the USA or Crown lands in Canada, or international fisheries generally fall within the responsibility of federal agencies (i.e., U.S. Fish and Wildlife Service [USFWS], U.S. National Park Service, U.S. Forest Service, and Fisheries and Oceans Canada). In the USA, state agencies are required by law to comply with decisions of federal agencies (Chapter 4). Nevertheless, numerous programs, such as the National Fish Passage Initiative of the USFWS, provide opportunities for collaboration among agencies. In addition, most biological opinions written by the USFWS under the Endangered Species Act require coordination among federal and state natural resource agencies. Recovery efforts that focus on a single species may not meet the needs of other imperiled fish, wildlife, and (or) plant species. Restoring ecosystem function and diversity at the watershed scale instead of focusing on single species can benefit imperiled species but is difficult due to conflicting interests among stakeholders (Williams et al. 1997). For example, erosion on outside riverbanks and deposition on inside riverbanks are important considerations when restoring natural ecosystem function and diversity but may

Box 19.1. Hydropower on Coldwater Rivers

KYLE KRUGER[1]

The Federal Energy Regulatory Commission (FERC) is responsible for licensing and inspecting nonfederal hydroelectric dams in the USA for safety conditions and for compliance with license terms and conditions. As part of the licensing effort, FERC considers recommendations by state and federal resource agencies, nongovernmental organizations, and tribes to offset negative environmental project effects. As of December 2007, FERC has issued 1,018 licenses for hydroelectric dams nationwide, many with substantial mitigation requirements based on resource agency recommendations.

Hydroelectric dams licensed by FERC on coldwater rivers pose particular problems for resource management agencies. Specific issues include water quality and quantity and how effects of changes in water quality and quantity are managed under the authorities of resource management agencies. For instance, both the increase in surface water temperatures caused by impounding a river and peaking operations have proven to have negative effects on trout populations in the state of Michigan, especially during the summer. In Michigan, instead of having one management authority for overall resource protection, the management authority for protection of water quality and fisheries resources is separated into two agencies, the Department of Environmental Quality and the Department of Natural Resources (MDNR), respectively. The Michigan Department of Environmental Quality is responsible for issuing U.S. Clean Water Act § 401 Water Quality Certifications to ensure discharges from hydroelectric dams into navigable waters are protective of natural resources. To ensure protection of aquatic habitat and reduce conflicting recommendations, the MDNR is granted a key advisory role in this process, especially related to reservoir elevations and flow monitoring.

To address further concerns related to water quality and quantity, FERC frequently requires that hydropower licensees establish measures to assure more naturalized flow regimes and mitigate temperature problems. Often, impoundments formed by dams allow increased contact of surface water with warm air and solar radiation. Dams that release water from near the surface of a reservoir cause water temperatures in downstream rivers to be warmer than they are in un-impounded rivers. Because of this, nonpeaking or run-of-river operation has become a standard license requirement for most dams. The intent of this standard is to maintain a hydrograph downstream of the dam that mimics natural flows as closely as possible. This improves natural river function in rivers downstream from dams and alleviates some problems associated with warming of the impounded water, bank erosion, and habitat dewatering often associated with power-peaking releases.

In some cases, licenses issued by FERC include provisions for the licensee to investigate the installation of downstream water temperature control devices. For example, on

(Box continues)

---

[1] Michigan Department of Natural Resources, Mio.

Box 19.1. Continued

the Manistee River in Michigan, a hydroelectric facility owner negotiated a settlement agreement with resource management agencies including the U.S. Fish and Wildlife Service, U.S. Forest Service, MDNR, and the National Park Service that involved studying water temperature effects of the project. The licensee collected field data in order to model the effects of the impoundment and recommended potential improvements that could be made through a water temperature control device. During meetings to discuss calibration of the model, the licensee indicated that accretion of water, which averaged 20% of the total flow being discharged between the upstream gauge and the tailrace gauge, was from tributaries that had the same temperature as the water in the impoundment. Under this assumption, the conclusion was there was not enough cool water to be manipulated at the powerhouse to cause a change downstream. The resource agencies challenged that assumption by claiming that the accreted water was most likely groundwater and was, therefore, significantly cooler than surface water, especially during the summer months. To validate their point, a multi-agency team measured flow from the tributaries to determine how much of the accretion was from surface water. The results of the field investigation showed that tributaries accounted for approximately 30% of the accreted flow, leaving the remaining 70% coming from groundwater. The U.S. Geological Survey verified that groundwater in the vicinity of the project was approximately 10°C, which was much lower than surface water temperatures. When the model was recalibrated to account for the temperature difference, it showed that there was enough cool water to enhance water temperatures downstream from the project.

Once the potential benefit of cooler water was established, the licensee developed a water temperature control device that delivered cold water from the lower portion of the impoundment to the intakes near the surface by means of a bubbler system. Initial tests of the system showed that downstream water temperatures in the tailrace were reduced by more than 2°C. Observations by the MDNR substantiated that the water temperature reduction improved survival of coldwater fishes, especially trout, during periods of prolonged warm temperatures and provided even greater temperature enhancements during cooler periods.

result in the loss of property by landowners and farmers and can disrupt sites held sacred by Native Americans or First Nations. River impoundment and high-flow releases can enhance erosive power of streams and result in loss of property. Thus, natural resource agencies must consider ecological, social, and economic factors when managing coldwater rivers.

Hunting and fishing rights in coldwater rivers were retained by Native Americans in treaties with the USA (Chapter 4). Federal courts, including the U.S. Supreme Court, have consistently held that time cannot erode the rights that were established when treaties were signed. Agencies representing tribal interest (e.g., U.S. Bureau of Indian Affairs and the Inter-Tribal Fish Commission) may manage fisheries on tribal lands and ensure that treaty rights are retained. Tribes play an important role in managing coldwater rivers by assisting with monitoring and assessment efforts in ceded territories and contributing harvest data that can be used to develop management plans. Managers of many coldwater river fisheries must consistently try to balance demands on fisheries resources among tribal fisheries, commercial fisheries, and various recreational user groups.

Watershed groups, often organized as nongovernmental organizations (NGOs), assist in management efforts by working with state, federal, and provincial agencies to collect information and conduct assessments and habitat improvement projects. These organizations are valuable in that they can mobilize funds and personnel quickly. Among other attributes, NGOs are able to build strength and diversity from local institutions, provide information and local knowledge regarding rivers, improve degraded watersheds, and promote market incentives for conservation (von Hagen et al. 1998). For instance, American Rivers is an NGO that advocates for selective removal of dams and protection of river habitat by highlighting threats to rivers in national and local media. Such actions led to the successful removal of Edwards Dam on the Kennebec River in Maine. Many of these organizations play an important role in fisheries management by contributing funds and personnel to habitat improvement of degraded coldwater rivers and watersheds. Such NGOs are particularly valuable fisheries management partners during times of agency budget and staffing constraints.

Collaboration among agencies and resource users is important for managing coldwater rivers efficiently and effectively. Sampling fish populations and habitat in coldwater rivers is difficult and expensive (Johnson et al. 1995a; Bonar et al. 2009). Given that agency budgets frequently dictate the effectiveness of monitoring and assessment programs, pooling resources and personnel among agencies often increases effectiveness. Monitoring activities may be allocated among agencies according to specific management objectives or geographic locations. However, standard procedures should be established to ensure that data are obtained and recorded consistently among partners. Various sampling techniques are available to assess fish populations and physical habitat characteristics in large rivers, but common practices need to be used to ensure data comparability (Bonar et al. 2009).

19.3.3 Management Goals

Effective management of coldwater rivers should follow the general process of fisheries management, where goals and objectives are well defined before implementing management actions, such as regulating harvest, controlling undesirable species, stocking, or undertaking habitat improvement, and before initiating monitoring and assessment activities (Chapter 5). Goals are long-term statements about what fisheries programs are to achieve. Objectives are measurable outcomes that indicate how and when achievement or progress toward goals will be accomplished. Goals should be developed specifically for individual coldwater river systems and are generally defined to improve or maintain fish populations and their habitats and to create, maintain, or enhance fishing experiences. Individual management goals require specific management activities. A relatively simple goal to create or enhance fishing opportunities for coldwater fishes requires a different set of objectives and management actions than does a more complex goal to rehabilitate self-sustaining fish populations. Management activities to create or enhance fishing opportunities may include specific harvest regulations, stocking, control of undesired species, or habitat improvements that are specific to a particular species. However, activities to improve or rehabilitate fish populations may include removing dams to enhance connectivity between tributaries and rivers, re-establishing natural flow regimes to facilitate lateral connectivity with floodplains and side channels, eliminating severely eroding riverbanks, minimizing competition between native and nonnative fishes, or improving water quality. In these cases, the goal is to provide adequate habitat to support natural reproduction and survival of multiple life stages of coldwater fishes. Establishing management goals can be difficult, especially when stakeholders value different aspects of coldwater river resources. Nevertheless, fisheries managers should avoid implementing specific actions or techniques that address short-term problems without establishing well-defined goals and objectives for the resource.

19.3.4 Regulation of Harvest

Most states and provinces issue general fishing licenses and establish harvest regulations for fishing in coldwater rivers, but some states and provinces may require an additional license or stamp to fish for salmonids (see Chapter 7). Sales of fishing licenses generate revenue for agencies responsible for managing fisheries resources and information provided on license applications can be used to design angler surveys. Other funding sources, such as salmonid-specific licenses or stamps, may also be available for management activities pertaining to coldwater river resources, such as habitat improvement or hatchery operations.

Some regulations are imposed to limit access to coldwater river resources, thereby reducing angling pressure and susceptibility of fish to capture and harvest. Species-specific seasons are commonly established in coldwater rivers to protect spawning fishes from harvest or redds from wading anglers. Access to rivers can also be regulated through permits to prevent overcrowding and improve recreational experiences of resource users, including anglers. For example, to reduce conflicts among users the number and type of watercrafts, as well the times that watercraft may be used, are regulated through the use of permits in the National Forest Wild and Scenic River section of the Pere Marquette River in Michigan. Restrictions can also be placed on the number of fishing guide permits issued for popular coldwater rivers.

Minimum length limits that prohibit the harvest of fish below a specified length, typically the length at which a given species becomes sexually mature, have been imposed to lower mortality rates in fish populations and to prevent excessive harvest. Consequently, it may be necessary to set minimum length limits relatively high for long-lived fishes that mature at large sizes or late in life, such as bull trout (Post et al. 2003). For brook trout and brown trout in the Au Sable River in Michigan, Clark et al. (1981) demonstrated that catch rates, release rates, and harvest of large trout by anglers increased as minimum length limits increased, whereas total harvest decreased. When coupled with reduced creel limits, increases in minimum length limits increased densities of brook trout and brown trout (Shetter and Alexander 1966). However, long-term effects of minimum length limits on salmonid populations in coldwater rivers are poorly understood. In some rivers, growth rates of trout have remained relatively constant over long periods in spite of high exploitation (Clark et al. 1980), but fishing quality has declined in other systems where faster-growing and larger individuals have been selectively removed (Favro et al. 1980). Average size and age of salmonids in coldwater rivers may be substantially reduced over time after minimum length limits are imposed, especially if the abundance of small fish substantially increases and food or habitat is limited (Barnhart and Engstrom-Heg 1984). Smaller fish may not be physically capable of migrating long distances upriver to spawning areas, especially in highly-modified systems (Ricker 1989). Consequently, minimum length limits may affect population abundance and length structure of salmonids. Long-term monitoring by fisheries managers is necessary to ensure that minimum length limits do not adversely affect coldwater fish populations (Box 19.2).

Box 19.2. Regulation of Harvest in Coldwater Rivers

Andrew Nuhfer and Todd Wills[1]

Coldwater rivers in Michigan support productive brook trout fisheries. Compared with smaller streams, brook trout in rivers have relatively fast growth but high mortality during the first 2 years of life. Brook trout will commonly reach 23 cm in total length (TL) by age 2, but larger and older fish are rarely found in rivers. Poor recruitment of brook trout to larger length-classes can be attributed to high fishing mortality when populations exhibit fast growth and low natural mortality rates. However, differentiating fishing mortality from natural mortality can be challenging without conducting extensive creel surveys, which are difficult and expensive for species with fluvial life histories. Consequently, research can be conducted to determine causes for poor survival of large fish by evaluating the effects of restrictive harvest regulations on population abundance of brook trout over long segments of coldwater rivers.

In the late 1990s, the Michigan Department of Natural Resources and university fisheries biologists convened with public stakeholders to develop acceptable management alternatives for enhancing angling experiences. A goal was established to create opportunities for anglers to catch large brook trout in various coldwater rivers throughout the state. Objectives were to increase the density and length structure of brook trout in rivers within 6 years by enacting new harvest regulations in 2000. Historical records on growth rates and professional opinion were used to develop combinations of specific restrictive regulations that were likely to achieve the desired goal for each river. Modification to existing regulations included reducing daily creel limits, increasing minimum length limits, and restricting terminal tackle. In selected Upper Peninsula rivers, the minimum length limit for brook trout was experimentally increased from 18 to 25 cm TL in 2000. Mark and recapture population estimates were conducted, and density and length structure of brook trout were compared during the 4 years before and 6 years after the minimum length limit was changed. Densities and length structure of brook trout in three out of four rivers studied did not significantly change after increasing minimum length limits. In the Iron River, brook trout 18–25 cm TL averaged 138 fish/ha before regulations were changed and 280 fish/ha after increasing the minimum length limit. However, the number of brook trout greater than 25 cm TL did not change. Though the abundance of brook trout increased, large fish were still rare. In most cases, the new regulation failed to meet the desired goal or accomplish the objectives to increase the density and length structure of brook trout because natural mortality, rather than fishing mortality, was the cause of poor recruitment to age-3, when fish became greater than 25 cm TL. When natural mortality is high, harvest restrictions do not appear to be effective for increasing length structure. In other rivers that were studied, catch-and-release regulations were not successful at increasing the length structure of brook trout where they were sympatric with predatory brown trout. Though the new regulations occasionally resulted in an increase in abundance of intermediate-length fish, trophy fish were not produced and harvest was substantially reduced or eliminated.

[1] Michigan Department of Natural Resources, Lewiston.

Protected slot length limits prohibit harvest in an intermediate length range and have been applied to salmonid populations with high levels of reproduction and recruitment in coldwater rivers. In theory, reduced abundance of small fish during summer should reduce competition for limited resources, increase growth rates, and allow midsize fish to recruit into the adult population (Clark et al. 1980). Success of slot length limits depends on exploitation rates and productivity of particular rivers, as well as vulnerability of the target species to angling. At high exploitation rates, a change from a minimum length limit to a slot length limit can increase the abundance and harvest of salmonids, but slot length limits coupled with lower exploitation rates can result in reduced abundance if the productivity of a river is limited (Nordwall and Lundberg 2000). To reduce the abundance of small fish and increase the number of trophy-length fish in the Au Sable River in Michigan, a slot length limit on brown trout between 33 and 41 cm was imposed (Clark and Alexander 1985). Although the number of small fish harvested increased, fish growth and length structure did not change for the population. Clark and Alexander (1985) attributed the lack of response to limited productivity of the river, but it is possible that harvest of small fish was not sufficient to achieve management objectives. Brown trout are generally more difficult for anglers to catch than are other salmonids species, such as rainbow trout, cutthroat trout, or brook trout (Anderson and Nehring 1984). Slot length limits may be better suited for productive coldwater rivers where fishing pressure is high and small fish are relatively easy to catch.

Maximum length limits that prohibit the harvest of large fish from coldwater rivers can be used to protect fish with the highest fecundity and reproductive potential while ensuring anglers the opportunity to catch trophy fish when fishing pressure is high. In the Rangitikei River in New Zealand, a maximum length limit of 55 cm in total length resulted in changes in angler behavior that led to a shift in exploitation from large rainbow trout to smaller fish (Barker et al. 2002). The number of fish harvested did not change, but the length and age distribution of the population became skewed toward older and larger fish. Survival increased with fish size, but survival of all fish improved after the maximum length limit was imposed.

In addition to length restrictions, creel limits have been imposed to reduce the rate of fishing mortality in coldwater rivers. Creel limits spread the allowable harvest among anglers over time and can be successful when catch or harvest rates are low. However, creel limits only restrict harvest by individual anglers and may not affect total fishing mortality if angling effort and catch rates are high. Catch-and-release regulations can be used to reduce fishing mortality further, maintain high catch rates, and increase catch of large fish. Achieving these objectives depends on angler compliance, but catch-and-release practices have been successful on coldwater rivers where long-lived fish populations exhibit good growth and low mortality. In the South Platte River in Colorado, the catch rate of trophy-size trout was 28 times greater in a catch-and-release area than in an area where harvest of 8 trout per day was permitted (Anderson and Nehring 1984). Rainbow trout dominated the population in the catch-and-release area; 500 kg/ha rainbow trout were present and 50% were longer than 30 cm. In the harvest area, brown trout were most abundant, total trout biomass was about one-third of that in the catch-and release area, and only 17% of the population exceeded 30 cm. Before implementing catch-and-release regulations, managers should determine if sufficient interest exists among anglers. Anglers are often supportive of catch-and-release fishing when it offers a unique opportunity to provide trophy-size fish, high catch rates, or protection of a species vulnerable to overfishing. Furthermore, managers should consider lower creel limits before implementing catch-and-release regulations. Eliminating harvest may not be necessary if most anglers voluntarily release fish already.

19.3.5 Use of Hatchery Fish

Salmonids are among the most popular groups of fishes used in stocking programs because they are relatively adaptable to hatchery environments. However, decisions to stock fish have sometimes been based on the ability of managers to produce and stock salmonids successfully rather than on clearly defined objectives. As with any management activity, specific objectives should be established prior to stocking fishes into coldwater rivers, and the success of stocking to meet management targets should be evaluated. Objectives often vary depending on angler opinions about quality fishing experiences and self-sustaining or native fish populations, but the main reasons for stocking fishes into coldwater rivers include the following.

Mitigation. stocking that is conducted to replace lost production of fishes caused by anthropogenic factors. For example, stocking of salmonids occurs in the Pacific Northwest to offset losses in natural reproduction and recruitment caused by hydropower development on the Columbia River.

Restoration. stocking that occurs after limiting factors to natural production have been identified and eliminated. The long-term objective for such a program would be to achieve a self-sustaining population. For example, restoration stocking of juvenile salmonids occurred on the Morrell River in Canada after improvements in water quality and habitat (Bielak et al. 1991). Within 8 years, salmon runs increased from 4–45 fish per year to 360–1,263 fish per year.

Enhancement. stocking that supplements an existing population where the production is less than the river can sustain due to limited spawning habitat or survival in early life history phases. Such stocking practices can compensate for effects of various natural and anthropogenic factors on salmonid production, such as periodic flooding, natural barriers to migration, or large- scale degradation of habitat due to urbanization, channelization, acidification, or changes in land use. Enhancement stocking can also be used to maintain artificially high capture rates and create put-and-take salmonid fisheries in coldwater rivers where fishing pressure is relatively high (Johnson et al. 1995b).

Creation of new fisheries. stocking that introduces coldwater fish into previously unoccupied rivers. For example, dams that release cold water from hypolimnetic portions of reservoirs create optimal thermal conditions for coldwater species in rivers that were once suitable for only coolwater and warmwater species. Consequently, fisheries managers have created popular coldwater fisheries by introducing various trout species, most frequently rainbow trout, into tailwaters. By stocking fish in tailwaters below hydroelectric dams, fishing opportunities have been created in rivers where they may not have otherwise existed.

Research. stocking that is aimed to address specific fisheries management issues. For example, stocking can be conducted to estimate potential carrying capacity of some coldwater rivers by stocking various densities of fish and surveying the population over time (Aprahamian et al. 2003). Growth and survival of stocked fish of various strains, stocking densities, and sizes can be determined.

Conservation. stocking that conserves populations of fish. Some endangered species plans require stocking to prevent extinction. Presently, there is a growing emphasis on stocking to aid in native fish recovery and a redirection of efforts toward maintaining self-sustaining coldwater river fisheries through habitat preservation or restoration (Brannon et al. 2004).

Stocking can benefit recreational fisheries in coldwater rivers by improving fishing experiences and increasing satisfaction among anglers, which, in turn, can increase recreational use and appreciation of coldwater river resources. It also can help maintain, improve, or conserve fish populations of concern. However, there are inherent risks associated with stocking salmonids in coldwater rivers. Stocking may not result in good growth or high survival of stocked fish owing to limitations in habitat availability, low river productivity, or excessive predation by birds, mammals, or piscivorous fishes (Alexander 1979; Derby and Lovvorn 1997; Walters et al. 1997). Excessive stocking densities can decrease prey bases, causing hatchery and wild fish to compete for limited prey resources (Fresh 1997). Reduced growth associated with competition can lead to increased vulnerability to predation, decreased survival, and changes in movements, feeding behavior, or habitat use by wild salmonids (Einum and Fleming 2001). Such factors have been implicated in the decline of native brook trout following brown trout stocking in coldwater rivers in the eastern USA (DeWald and Wilzbach 1992).

Since salmonids are often stocked at catchable sizes in the spring to optimize growth and survival during the growing season (Epifanio 2000), stocked fish are typically larger than are wild conspecifics and may outcompete wild fish for food and space (McMichael et al. 1999). After catchable-length rainbow trout were stocked in the Madison River of Montana, Vincent (1987) observed a decline in growth and abundance of wild brown trout and rainbow trout. After stocking ceased, the abundance and biomass of wild trout increased nearly threefold. Conversely, Marshall (1973) observed no difference in size or abundance of wild brown trout and rainbow trout after hatchery trout were stocked in the Cache la Poudre River of Colorado. Stocking hatchery rainbow trout had no effect on dispersion, abundance, growth, or survival of wild rainbow trout or cutthroat trout in two Idaho rivers (Petrosky and Bjornn 1988). Differences in population level effects due to stocking are attributable to different habitat characteristics, flows, stocking densities, and exploitation rates among rivers. If harvest of stocked fish is low, stocking could be detrimental to wild populations. Therefore, stocking should not occur if self-sustaining populations naturally exist and anglers are generally satisfied with their fishing experience.

Another risk of stocking is interbreeding of hatchery and wild fish that may reduce fitness of locally-adapted native fishes (Utter et al. 1993). Interbreeding of hatchery and wild salmonids may alter morphologic features, migration patterns, or feeding behaviors of locally-adapted fishes (Brannon et al. 2004). Consequently, the use of stocking as a management tool to recover threatened and endangered populations of salmonids has been controversial. Stocking has played a role in conservation of some threatened species by maintaining a balanced population structure, such as winter-run Chinook salmon in the Sacramento River (Hedrick et al. 2000). Nonetheless, stocking may not be effective to recover wild salmonid populations unless limiting factors are identified and managed. Stocking to maintain viable populations masked long-term problems causing poor natural reproduction and recruitment and resulted in near extinction of some wild populations of Atlantic salmon in coldwater rivers in eastern Canada (Myers et al. 2004). Stocking of nonnative salmonids into coldwater rivers may cause genetic alteration of wild populations. Native cutthroat trout can hybridize with rainbow trout, and bull trout can hybridize with brook trout. Reproduction by such hybrids has been linked to declines in native species (Rieman and McIntyre 1993). Stocking programs should be evaluated prior to implementation to minimize risk to resident salmonids through interbreeding and should be designed to preserve the genetic integrity of locally-adapted wild populations. One approach is to use local broodstock that encompasses the genetic variability in the population and release hatchery fish to rear or spawn near habitats used by wild fish (Reisenbichler and McIntyre 1986).

Stocking has also resulted in the spread of various parasites and diseases. Consequently, managers need to consider both the potential benefits and risks prior to stocking. Producing hatchery fish is expensive and generally accounts for substantial proportions of agency budgets (Johnson et al. 1995b). In some cases, management may be better directed at maintaining self-sustaining populations by improving habitat or applying appropriate angling regulations.

19.3.6 Management of Diseases and Undesired Species

Though some diseases and undesirable species are native to coldwater rivers, many become established in association with the translocation of fish. For example, whirling disease is endemic to coldwater rivers in Europe and was introduced into coldwater streams and rivers in the USA after frozen rainbow trout were imported from Scandinavia for market in grocery stores (Graff 1996). Water from residential areas, where fish were consumed, contained early life stages of the parasite *Myxobolus cerebralis* that is responsible for the disease and flowed into a river that was used as a water source for a hatchery producing rainbow trout. Before clinical signs of the disease were discovered, hatchery fish were shipped to other hatcheries and stocked in various coldwater systems throughout the USA (Bergersen and Anderson 1997). Piscivorous animals may also transmit the parasite. The disease has been implicated in the decline or elimination of salmonid year-classes in some coldwater rivers. Another example of a harmful translocation is the release of various live baitfishes by anglers into coldwater rivers that are outside the native distribution of the species (Litvak and Mandrak 1993). Baitfish can compete for resources with desired coldwater species or introduce disease, such as viral hemorrhagic septicemia, into coldwater river systems. Management agencies often establish regulations regarding fish importation, fish health assessments, and introductions of nonnative bait and sport fishes and strive to educate the public on the potential impacts that introduced fishes and other species can have on ecosystems.

Many nonnative fishes are now established in coldwater river systems as a result of human activities. The introduction of warmwater and coolwater fishes, such as channel catfish and walleye, into impoundments of the Columbia River has caused an increase in predation on salmonid smolts migrating toward the Pacific Ocean (Reeves et al. 1998). Competition with nonnative species may result in loss of native salmonids. In western North America, the introduction of brook trout has led to decline of native bull trout (Reeves et al. 1998), and introduced rainbow trout, brook trout, and brown trout have displaced Yellowstone cutthroat trout from native habitats through hybridization and competitive interactions (Behnke 2002).

In eastern North America, rainbow trout and brown trout displace native brook trout. Due to the size and nature of coldwater rivers, eradication of undesirable species, or the diseases that may be introduced with them, has been impossible after they become established.

Several fishing methods have been used in attempts to control the spread of undesirable fish species in coldwater rivers. Depending on management objectives, species to be controlled, and river size, success has been highly variable. Large rivers have diverse habitat conditions that reduce gear effectiveness. Substantial effort is needed to be effective, which often results in only short-term benefits if any. Most gears used to catch fish in large rivers, such as gill nets, fyke nets, seines, trawls, and electrofishing, are not species specific, and unintentional stress and mortality among fish species of concern may result (Bonar et al. 2009). Nevertheless, some methods may be effective at controlling undesirable species in some coldwater rivers. For example, Shetter and Alexander (1970) used electrofishing to reduce the abundance of large brown trout that were predators of brook trout in a coldwater river in Michigan. Despite reductions of 40–66%, it was concluded that higher reductions of piscivores were necessary to improve recruitment of young fish. On the Columbia and Snake rivers, recreational anglers were paid a bounty for each northern pikeminnow harvested, and managers fished near dams with hook and line and deployed gill nets near hatchery release sites to reduce adult northern pikeminnow predation on stocked juvenile salmonids (Beamesderfer et al. 1996). Annual exploitation rates using these methods have been 9–16%, but additional methods have been ineffective at further reducing adult northern pikeminnow populations due to the size of the river, patchy distribution of adults, and limited gear effectiveness. Some experimental gears resulted in catching more endangered white sturgeon than northern pikeminnow, so their use was discontinued. Nevertheless, effective methods to remove adult northern pikeminnow may be sufficient at reducing predation on juvenile salmonids in the Columbia River and Snake River without impacting nontarget species.

The use of toxicants to remove undesirable fish species from some coldwater rivers was common in past decades, primarily to depress nongame fishes and facilitate successful introductions of nonnative salmonids. In the 1950s, over 400 km of the Russian River watershed in California were treated with rotenone (Pintler and Johnson 1958). In the 1960s, over 700 km of the Green River were treated with rotenone prior to closing Flaming Gorge Dam in Utah (Holden 1991) and Fontenelle Reservoir in Wyoming (Zafft et al. 1995). In these examples, nonnative trout were stocked following the application of rotenone and resulted in popular fisheries. Although the use of toxicants in streams is still common, particularly in the western USA, its application to manage undesirable species residing in large coldwater rivers has declined in North America and is no longer permitted in several states and provinces. Treatment with toxicants is expensive and may be unpopular to anglers in coldwater rivers and tailwaters where nonnative salmonid fisheries may exist. Many reservoirs on coldwater rivers are used for municipal water supplies, thus managers are often unwilling or unable to use toxicants in these systems.

The use of pheromones to attract undesirable fish species into traps during spawning periods has not been widely applied in coldwater river systems but could be used to improve the effectiveness of fishing gears and reduce the abundance of undesirable fishes. Pheromones have been used to control sea lampreys that migrate into coldwater rivers from the Great Lakes, but they are generally selective toward males. Consequently, males may be sterilized and released back into the river under the premise that they will unsuccessfully attempt to mate with gravid females (Twohey et al. 2003). Sexually-mature brook trout have also been baited into traps in coldwater streams by use of ripe males (Young et al. 2003), so the use of pheromones has potential for controlling undesirable salmonids or other species in coldwater rivers.

19.3.7 Habitat Improvement

Since physical and biological conditions in coldwater rivers depend on the degree of connectedness with ecological processes in upstream tributaries and floodplains, managers should consider causes for habitat degradation prior to implementing improvement projects in coldwater rivers. Anthropogenic impacts in a watershed have cumulative effects on coldwater rivers. Habitat improvement strategies that re-establish some semblance of the natural hydrograph, sediment regime, and physical diversity in a river may be desirable for large-scale enhancement of aquatic and riparian habitats but are often limited by physical, socioeconomic, and political constraints. Most large rivers have been impounded, thus restoration of habitats in many coldwater rivers may be unrealistic. Because of the considerable complexity and expense associated with restoring large rivers, habitat management for coldwater fishes has been largely based on opportunities to repair or augment specific habitat conditions at small spatial scales rather than restoring landscape scale processes that form and sustain habitats for various life stages and species.

Millions of dollars are spent annually in many coldwater river basins to improve habitat for salmonids. This interest and funding has been partially due to listings of various salmonid stocks as threatened or endangered under the U.S. Endangered Species Act and efforts to restore native fish assemblages. Efforts such as the Eastern Brook Trout Joint Venture and Western Native Trout Initiative were implemented by the USFWS and state, federal, and local partners as part of the National Fish Habitat Action Plan to protect and improve fish habitat in coldwater rivers (see www.fishhabitat.org/; Higgins 2009; Box 10.2). The plan emphasizes the use of landscape scale, adaptive management to overcome strategically factors limiting the abundance and distribution of fishes throughout their native distributions. However, even with all of the attention and money being spent, few habitat improvement techniques have been evaluated, and their effectiveness is debatable (Kauffman et al. 1997). Monitoring the response of fish, invertebrates, and other biota to habitat improvement projects, which is the ultimate measure of effectiveness, has been neglected, and most monitoring has focused on short-term physical responses. While delivery of water, organic matter, and sediment are some of the major processes dictating channel morphology and formation of habitat in coldwater rivers, many processes that create habitat operate on long time scales, and decades of monitoring are often required to detect responses to habitat improvement activities (Bisson et al. 1992; Reeves et al. 1997). Consequently, limited information exists on the effectiveness of various habitat improvement techniques to coldwater fishes in rivers because many techniques have not been thoroughly evaluated.

Habitat improvement techniques for coldwater rivers include (1) restoring habitat connectivity, (2) constructing and improving roads, (3) restoring riparian vegetation, (4) modifying instream habitat, and (5) increasing nutrient enrichment.

19.3.7.1 Restoration of Habitat Connectivity

Reconnecting isolated habitats may be one of the most important components of restoring fish populations in coldwater rivers, but it is difficult in large rivers with impoundments.

Artificial barriers, such as culverts and dams, alter fish passage and reduce connectivity among river and tributary habitats that are critical to life history requirements of salmonids. For Pacific salmon, barriers may reduce the extent of upstream nutrient inputs by limiting the number of adults that move upstream to spawn and subsequently die. In coldwater rivers of the Pacific Northwest, impassable barriers have been estimated to have reduced coho salmon smolt production as much as 30–58% (Beechie et al. 1994; Pess et al. 1998). Scully et al. (1990) examined the relative benefit of barrier removal, off-channel habitat development, instream structure placement, and sediment reduction projects in the Salmon River basin, Idaho. Of the four types of projects, barrier removal improved steelhead and Chinook salmon parr production the most. Nevertheless, habitat quantity and quality along with fish presence above and below impassable barriers should be factors used to prioritize fish passage projects and to determine whether other improvement techniques might be cost-effective.

Off-channel habitats such as sloughs, small ponds adjacent to river channels, side channels, wetlands, and other permanently or seasonally flooded areas are important rearing habitats for juvenile salmonids. However, off-channel habitats normally associated with floodplains have been routinely isolated or altered by floodplain and (or) hillslope activities such as agriculture, urbanization, flood control, and transportation. Beechie et al. (1994) concluded that loss of side channels and sloughs was the major factor limiting smolt production of coho salmon in the Skagit River, Washington. Overwinter survival and growth of salmonids is often higher in off-channel habitats than in main-channel habitats (Swales and Levings 1989). Consequently, most off-channel improvement efforts have focused on providing overwinter habitat for juvenile salmonids. Reconnecting isolated natural off-channel habitats and excavating new habitats are management techniques used in coldwater rivers (Richards et al. 1992).

19.3.7.2 Road Construction and Improvement

The construction and improvement of roads, including forest logging roads and paved highways, can harm fish habitat in coldwater rivers through increased delivery of fine sediment, higher landslide frequencies, changes in hydrologic features, and reduced connectivity. Fine sediment produced by surface erosion and delivered to coldwater streams and rivers by roads can infiltrate spawning gravels, reduce survival of eggs (Reid et al. 1981), and decrease macroinvertebrate abundance (Lenat et al. 1981). Coarse sediments from road-related landslides can increase bed load supply, which fills pools and decreases bed and bank stability by causing bed aggregation or lateral migration (Tripp and Poulin 1986). Changes to hydrologic processes, such as flooding, can occur as a result of increased runoff and cause widening and downcutting of a river channel (Wheeler et al. 2005). Connectivity and downstream transport of organic material can be disrupted through a placement of bridges and, more frequently, culverts. Given the effects of road construction and improvement, and the secondary effects of road presence and associated urban development on the watershed, managers should carefully assess proposed road projects (Wheeler et al. 2005).

Avoiding or minimizing construction of new roads and closing existing roads may reduce effects associated with sedimentation, hydrology, and connectivity. When road improvements are needed, surface erosion and delivery of fine sediments into rivers may be reduced by proper road design and maintenance. For example, armoring the surface by paving or using crushed rock (7.6–15.2 cm diameter) can reduce surface erosion (Burroughs and King 1989) and the production of fine sediment. To avoid sediment inputs, ditches and drains should be directed away from rivers. Reduction in landslide hazards may require road removal or reconstruction (Waters 1995). Risks associated with new road projects can be minimized during the design phase or, for existing roads, mitigated to reduce impairments to the aquatic community.

19.3.7.3 Riparian Enhancement

Land use by humans can disrupt the processes that form and sustain aquatic habitats, such as the supply and movement of sediment from upland habitats, woody debris recruitment, shading by riparian cover, and water delivery. Timber harvest, urbanization, and agriculture (e.g., livestock grazing) have altered riparian areas and caused changes in both coldwater streams and rivers. For instance, extensive harvest of timber resources has resulted in loss of riparian vegetation, including a reduction in the contribution of large woody debris important for creating and maintaining habitat. Techniques to improve riparian areas affected by timber harvesting include thinning or harvesting to increase the growth rate of existing trees and planting to establish desired vegetation. Livestock permitted to graze along riverbanks can reduce diversity and abundance of riparian vegetation, which increases bank erosion, sediment inputs, and water temperatures in tributaries. These changes, which may reduce the quality and quantity of habitat for fishes in coldwater rivers, can be managed by developing grazing strategies and limiting access to riverbanks by livestock. Although bank stabilization, channel geometry, habitat complexity, and other channel characteristics can recover quickly after riparian enhancement in smaller rivers, recovery may take longer in large rivers, especially those with deeply incised channels (D'Aoust and Millar 2000). Furthermore, fish populations may not respond to changes in riparian cover over short time intervals owing to inherently large interannual variability in fish populations in coldwater rivers.

19.3.7.4 Modification of Instream Habitat

Habitat modification techniques to improve fish production, such as instream placement of woody debris or rocks, have been rarely applied in large rivers because these techniques are expensive and can fail to create expected habitat conditions (Beechie et al. 1996; Kauffman et al. 1997). Nonetheless, instream habitat modifications have occurred in rivers to stabilize bank erosion and increase habitat diversity to meet the life history requirements of coldwater fish species. For instance, bank stabilization structures have been used to maintain physical characteristics of existing or created off-channel habitats that are often important juvenile salmonids. Banks are typically stabilized to protect property from the natural meandering tendencies of rivers that have high sediment transport capabilities during periods of high flow. Rip-rap has been used effectively to stabilize eroding riverbanks, but tree revetments and log structures in combination with planting vegetation have been used when the needs of fish habitat are considered (Quinn and Kwak 2000). The placement of boulders and logs in river channels has been practiced to provide refuges from high current velocities for salmonids in rivers with poor habitat diversity (Saha et al. 2006). Such habitat modifications have resulted in increased biomass of salmonids due to immigration of large fish from unmodified areas rather than increased growth or survival (Gowan and Fausch 1996). Placement of instream structures can be successful if engineered properly and placed in locations where they would naturally occur.

Flow modification has probably been used the most by managers to enhance instream habitat for coldwater fishes in tailwaters. Application of a flow regime depends on specific management objectives but generally has been applied to improve spawning, recruitment, growth, and survival of salmonids. Flushing flows, designed to improve spawning and nursery habitats, are releases of water from dams to remove or flush sediments that accumulate during periods of low flow, remove fine sediments from gravels used for spawning, and enhance survival and recruitment of small salmonids (Box 19.3; Young et al. 1991). Though flushing flows can enhance spawning and nursery habitat in the short term, high-flow releases in the long term may enhance streambed degradation, channelization, and loss of side-channel habitats used as spawning and nursery habitats by small salmonids, particularly in reaches immediately downstream from dams (Simpkins et al. 2000a, 2000b). As a result, gravel augmentation has been used as a tool to enhance the availability of spawning substrates in degraded habitats downstream from dams (Merz and Setka 2004).

Flow releases have also managed to provide habitats that promote growth and survival of coldwater fish in regulated river systems. Using methods such as the instream flow incremental methodology, flow requirements can be established to provide seasonal habitats that are essential for various life stages and species of fish, as well as aquatic invertebrates used as prey (Annear et al. 2004). Magnitude of flow releases can be changed or dam outlets can be structurally modified to withdraw water from various reservoir depths to provide more optimal water temperatures for production of aquatic invertebrates and growth and survival of fish in coldwater tailwaters (Krause et al. 2005). Furthermore, flows can be managed to release warmer waters during the winter to minimize potential ice impacts on fish assemblages (Annear et al. 2002). Increased flows may also be necessary to permit spawning migrations of salmonids access to regulated rivers from lentic environments and may improve survival of juvenile anadromous salmonids by increasing rates of downstream transport through regulated river systems (Cada et al. 1997).

19.3.7.5 Nutrient Enrichment

Over the last decade, fisheries managers have realized that Pacific salmon are an important contributor to habitat where they spawn. Because Pacific salmon die after spawning, their bodies provide nutrient and organic material to vegetation bordering river channels and to various aquatic and terrestrial organisms (Larkin and Slaney 1997). Barriers, such as dams and culverts, limit access of Pacific salmon to natal spawning habitats and lead to reduced nutrients (inorganic nitrogen and phosphorous) and productivity in some coldwater systems. Artificially increasing nutrient availability by adding carcasses of hatchery-spawned salmon is practiced in much of the Pacific Northwest and has been found to increase primary productivity, invertebrate densities, and growth of juvenile salmonids (Bilby et al. 1998; Wipfli et al. 1999). However, nutrient augmentation projects should consider the nutrient status of the system not only at the location of the application but also downstream (Stockner et al. 2000). In some watersheds that flow through developed areas, excess nutrients may exist, and additional nutrients may create unsuitable conditions for coldwater fishes. The risks of degrading water quality in rivers should be weighed against potential benefits associated with adding fish carcasses.

Box 19.3. Use of Flushing Flows to Enhance Spawning Habitat in Coldwater Rivers

The construction and operation of dams has had a dramatic effect on natural processes in rivers. Since water is generally retained in reservoirs during wet times of the year for use during dry times of the year, flows released from dams tend to be lower during wet periods and higher during dry periods than they are in unregulated rivers. Compared with unregulated rivers, peak flows are moderate and occur later in regulated rivers. Such hydrologic changes modify sediment transport and substrate characteristics. Sediment accumulates on riverbeds during low flows, and high flows remove or flush deposited sediment downstream. The lack of naturally-high peak flows in regulated rivers can result in the accumulation of fine sediments (i.e., silt and sand) in habitats containing larger substrates (i.e., gravel, cobbles, and boulders).

Salmonids require gravel substrates for spawning. Juvenile salmonids use small eddies and interstitial spaces formed by cobble and boulder complexes during the winter and early spring in coldwater rivers. Fine sediments that are deposited and embed larger substrates can reduce spawning success, survival, and recruitment of salmonids in regulated rivers. Consequently, fisheries managers often prescribe flushing flows to remove fine sediments from regulated rivers. Flushing flows are controlled releases of high flows designed to mobilize gravel and small cobble substrates of a river in order to suspend fine sediment in the water column and flush it downstream.

During the 1990s, Wyoming Game and Fish Department, U.S. Bureau of Reclamation, and University of Wyoming biologists developed flushing flow recommendations and evaluated their effects on sediment transport, substrate characteristics, and spawning habitat for rainbow trout and brown trout in the Big Horn River downstream from Boysen Reservoir (Wiley 1995) and the North Platte River downstream from Gray Reef Reservoir (Wenzel 1993; Leonard 1995). The lack of high flows was thought to be responsible for limited availability of spawning habitat and recruitment. Prior to the flushing flows, substrate samples collected at spawning locations consisted of 22–28% fine sediment, which confirmed that spawning habitat was likely impaired. Sediment transport and substrate characteristics were assessed during a series of test flows, and spawning habitat condition was re-assessed after completion of the flushing flows. Though flushing flow recommendations were specific for each tailwater and habitat needs for specific trout species, the combined insight from these studies provided some general recommendations regarding the use of flushing flows to enhance and maintain spawning habitat for trout in regulated rivers. It was found that flushing flows should be of a magnitude approaching bank-full conditions to mobilize gravels and small cobbles and transport peak bed load and suspended sediment, should be of a duration equivalent to approximately 1.5 times the duration for water to travel over the prescribed reach of river, and should reoccur at 3.5–4.0 year intervals to be effective. After flushing flows, fines consisted of 7–16% of the substrates sampled at observed spawning locations. The availability of clean cobble substrate in the Big Horn River provided juvenile rainbow trout with refuge habitat during the flushing flows (Simpkins et al. 2000b). These results indicated that flushing flows can be useful tools for enhancing and maintaining spawning habitat and managing salmonid fisheries in regulated rivers without causing downstream displacement of larger fish.

19.4 CONCLUSIONS

Old challenges for the management of coldwater river fisheries still exist, such as determining appropriate fishing regulations and fish stocking strategies, but managers are increasingly faced with new challenges. New challenges vary from management with limited staff and operational budgets to control of nuisance species and fish diseases. Future managers will be expected to have the skills to communicate effectively with the public and function as a part of cross-disciplinary teams.

Today's fisheries managers understand the need for collaboration with individuals in different disciplines and across jurisdictional boundaries and the role of the public in managing coldwater rivers. Human population growth, increasing demands for water and natural resources, and climate change pose serious challenges to fisheries managers. Managers have to work with multiple partners to influence policy, develop effective regulations, and improve the structure and function of coldwater rivers. Habitat improvement efforts in coldwater rivers require collaboration with experts in fields such as engineering, water quality, wetlands ecology, hydrology, and fluvial geomorphology. The wave of the future is represented by initiatives such as the National Fish Habitat Action Plan, a coalition of anglers, conservation groups, scientists, state and federal agencies, and industry leaders working together to protect and improve fish habitat. Fisheries managers can also expect an increased emphasis on management of disease, nuisance species, and the conservation and restoration of native fish assemblages. The threat of undesirable species is expected to increase in conjunction with climate change and, like introduced diseases, will be difficult to control once spread. This challenge requires increased efforts to educate anglers on issues such as risks involved with transferring live fish, invertebrates, and disease-causing organisms among water bodies. Managers will use barriers, chemical control, hatchery procedures that reduce disease transmission, and restrictions on the sale of live bait to manage the risks of introduced species and diseases.

Difficulties with managing large river systems, such as accurate assessment, restoration, and management of habitat, will persist into the future. To address these challenges, managers need to use evolving methods and technologies along with shared knowledge of their success and failures. More than ever before, management of coldwater rivers requires innovative thinking and communication.

19.5 REFERENCES

Alexander, G. R. 1979. Predators of fish in coldwater streams. Pages 153–170 *in* H. Clepper, editor. Predator–prey systems in fisheries management. Sport Fishing Institute, Washington, D.C.

Allan, J. D. 1995. Stream ecology. Chapman and Hall, London.

Anderson, R. M., and R. B. Nehring. 1984. Effects of catch-and-release regulation on a wild trout population in Colorado and its acceptance by anglers. North American Journal of Fisheries Management 4:257–265.

Annear, T., I. Chisholm, H. Beecher, A. Locke, P. Aarrestad, C. Coomer, C. Estes, J. Hunt, R. Jacobson, G. Jobsis, J. Kauffman, J. Marshall, K. Mayes, G. Smith, R. Wentworth, and C. Stalnaker. 2004. Instream flows for riverine resource stewardship, revised edition. Instream Flow Council, Cheyenne, Wyoming.

Annear, T. A., W. Hubert, D. Simpkins, and L. Hebdon. 2002. Behavioral and physiological response of trout to winter habitat in tailwaters in Wyoming, USA. Hydrological Processes 16:915–925.

Aprahamian, M. W., K. Martin Smith, P. McGinnity, S. McKelvey, and J. Taylor. 2003. Restocking of salmonids: opportunities and limitations. Fisheries Research 62:211–227.

Barker, R. J., P. H. Taylor, and S. Smith. 2002. Effect of a change in fishing regulations on the survival and capture probabilities of rainbow trout in the upper Rangitikei River, New Zealand. North American Journal of Fisheries Management 22:465–473.

Barnhart, G. A., and R. Engstrom-Heg. 1984. A synopsis of some New York experiences with catch and release management of wild salmonids. Pages 91–101 in F. Richardson and R. H. Hamre, editors. Wild trout III. Trout Unlimited, Vienna, Virginia.

Baxter, C. V., K. D. Fausch, M. Murakami, and P. L. Chapman. 2004. Fish invasion restructures stream and forest food webs by interrupting reciprocal prey subsidies. Ecology 85:2656–2663.

Bayley, P. B. 1995. Understanding large river–floodplain ecosystems. BioScience 45:153–158.

Beamesderfer, R. C. P., D. L. Ward, and A. A. Nigro. 1996. Evaluation of the biological basis for a predator control program on northern squawfish (*Ptychocheilus oregonensis*) in the Columbia and Snake rivers. Canadian Journal of Fisheries and Aquatic Sciences 53:2898–2908.

Beechie, T., E. Beamer, B. Collins, and L. Benda. 1996. Restoration of habitat-forming processes in Pacific Northwest watersheds: a locally adaptable approach to salmonid habitat restoration. Pages 48–67 in D. L. Peterson and C. V. Klimas, editors. The role of restoration in ecosystem management. Society for Ecological Restoration, Madison, Wisconsin.

Beechie, T., E. Beamer, and L. Wasserman. 1994. Estimating coho salmon rearing habitat and smolt production losses in a large river basin, and implications for restoration. North American Journal of Fisheries Management 14:797–811.

Behnke, R. J. 2002. Trout and salmon of North America. The Free Press, New York.

Benda, L., N. L. Poff, D. Miller, T. Dunne, G. Reeves, G. Pess, and M. Pollock. 2004. The network dynamics hypothesis: how channel networks structure riverine habitats. BioScience 54:413–427.

Bergersen, E. P., and D. E. Anderson. 1997. The distribution and spread of *Myxobolus cerebralis* in the United States. Fisheries 22(8):6–7.

Bielak, A. T., R. W. Gray, T. G. Lutzak, M. G. Hambrook, and P. Cameron. 1991. Atlantic salmon restoration in the Morell River, P.E.I and the Nepisigutt, N. B., Canada. Pages 122–139 in D. Mills, editor. Strategies for the rehabilitation of salmon rivers. The Atlantic Salmon Trust, The Institute of Fisheries Management, and the Linnean Society of London, Pitlochry, Nottingham, and London, UK.

Bilby, R. E., B. R. Fransen, P. A. Bisson, and J. K. Walter. 1998. Response of juvenile coho (*Oncorhynchus kisutch*) and steelhead (*O. mykiss*) to the addition of salmon carcasses to two streams in southwestern Washington, USA. Canadian Journal of Fisheries and Aquatic Sciences 55:1909–1918.

Bisson, P. A., T. P. Quinn, G. H. Reeves, and S. V. Gregory. 1992. Best management practices, cumulative effects, and long-term trends in fish abundance in Pacific Northwest river systems. Pages 189–232 in R. J. Naiman, editor. Watershed management. Springer-Verlag, New York.

Blinn, D. W., and G. A. Cole. 1991. Algal and invertebrate biota in the Colorado River: comparison of pre- and post-dam conditions. Pages 85–123 in Committee on Glen Canyon Environmental Studies, editors. Colorado River ecology and dam management. National Academy Press, Washington, D.C.

Blinn, D. W., J. P. Shannon, L. E. Stevens, and J. P. Carder. 1995. Consequences of fluctuating discharge for lotic communities. Journal of the North American Benthological Society 14:233–248.

Bonar, S. A., W. A. Hubert, and D. W. Willis, editors. 2009. Standard methods for sampling North American freshwater fishes. American Fisheries Society, Bethesda, Maryland.

Brannon, E. L., D. F. Amend, M. A. Cronin, J. E. Lannan, S. LaPatra, W. J. McNeil, R. E. Nole, C. E. Smith, A. J. Talbot, G. A. Wedemeyer, and H. Westers. 2004. The controversy about salmon hatcheries. Fisheries 29(9):12–31.

Brown, R. S. 1999. Fall and early winter movements of cutthroat trout, *Oncorhynchus clarki*, in relation to water temperature and ice conditions in Dutch Creek, Alberta. Environmental Biology of Fishes 55:359–368.

Bunt, C. M., S. J. Cooke, C. Katopodis, and R. S. McKinley. 1999. Movement and summer habitat of brown trout (*Salmo trutta*) below a pulsed discharge hydroelectric generating station. Regulated Rivers: Research and Management 15:395–403.

Burroughs, E. R., Jr., and J. G. King. 1989. Reduction of soil erosion on forest roads. U.S. Forest Service General Technical Report INT-264, Ogden, Utah.

Cada, G. F., M. D. Deacon, S. V. Mitz, and M. S. Bevelhimer. 1997. Effects of water velocity on the survival of downstream-migrating juvenile salmon and steelhead: a review with emphasis on the Columbia River basin. Reviews in Fisheries Science 5:131–183

Carlson, A. J., and F. J. Rahel. 2007. A basinwide perspective on entrainment of fish in irrigation canals. Transactions of the American Fisheries Society 136:1335–1343.

Clark, R. D., Jr., and G. R. Alexander. 1985. Effects of a slotted size limit on a brown trout fishery, Au Sable River, Michigan. Pages 74–84 *in* F. Richardson and R. H. Hamre, editors. Wild trout III. Trout Unlimited, Vienna, Virgina.

Clark, R. D., Jr., G. R. Alexander, and H. Gowing. 1980. Mathematical description of trout stream fisheries. Transactions of the American Fisheries Society 109:587–602.

Clark, R. D., Jr., G. R. Alexander, and H. Gowing. 1981. A history and evaluation of regulations for brook trout and brown trout in Michigan streams. North American Journal of Fisheries Management 1:1–14.

Collis, K., D. D. Roby, D. P. Craig, S. Adamany, J. Y. Adkins, and D. E. Lyons. 2002. Colony size and diet composition of piscivorous waterbirds on the lower Columbia River: implications for losses of juvenile salmonids to avian predation. Transactions of the American Fisheries Society 131:537–550.

Congleton, J. L., W. J. LaVoie, C. B. Schreck, and L. E. Davis. 2000. Stress indices in migrating juvenile Chinook salmon and steelhead of wild and hatchery origin before and after barge transportation. Transactions of the American Fisheries Society 129:946–961.

Coutant, C. C. 2004. A riparian habitat hypothesis for successful reproduction of white sturgeon. Reviews in Fisheries Science 12:23–73.

Cunjak, R. A. 1988. Physiological consequences of overwintering in streams: the cost of acclimatization? Canadian Journal of Fisheries and Aquatic Sciences 45:443–452.

Cushing, C. E., K. W. Cummins, and G. W. Minshall. 2006. River and stream ecosystems of the world. University of California Press, Berkley.

D'Aoust, S. G. D., and R. G. Millar. 2000. Stability of ballasted woody debris habitat structures. Journal of Hydraulic Engineering 126:810–817.

Dare, M. R., W. A. Hubert, and K. G. Gerow. 2002. Changes in habitat availability and habitat use and movements by two trout species in response to declining discharge in a regulated river during winter. North American Journal of Fisheries Management 22:917–928.

Derby, C. E., and J. R. Lovvorn. 1997. Predation on fish by cormorants and pelicans in a cold-water river: a field and modeling study. Canadian Journal of Fisheries and Aquatic Sciences 54:1480–1493.

DeWald, L., and M. A. Wilzbach. 1992. Interactions between native brook trout and hatchery brown trout: effects on habitat use, feeding, and growth. Transactions of the American Fisheries Society 121:287–296.

DFO (Department of Fisheries and Oceans Canada). 2007. Survey of recreational fishing in Canada 2005. Fisheries and Oceans Canada, Report 2007–1303, Ottawa, Ontario.

Dunham, J. B., S. B. Adams, R. E. Schroeter, and D. C. Novinger. 2002. Alien invasions in aquatic ecosystems: toward an understanding of brook trout invasions and potential impacts on inland cutthroat trout in western North America. Reviews in Fish Biology and Fisheries 12:373–391.

Einum, S., and I. A. Fleming. 2001. Implications of stocking: ecological interactions between wild and released salmonids. Nordic Journal of Freshwater Research 75:56–70.

Epifanio, J. 2000. The status of coldwater fishery management in the United States: an overview of state programs. Fisheries 25(7):13–27.

Favro, L. D., P. K. Kuo, and J. F. McDonald. 1980. Effects of unconventional size limits on the growth rate of trout. Canadian Journal of Fisheries and Aquatic Sciences 37:873–876.

Faler, M. P., L. M. Miller, and K. I. Welke. 1988. Effects of variation in flow on distributions of northern squawfish in the Columbia River below McNary Dam. North American Journal of Fisheries Management 8:30–35.

Flebbe, P. A. 1994. A regional view of the margin: salmonid abundance and distribution in the southern Appalachian Mountains of North Carolina and Virginia. Transactions of the American Fisheries Society 123:657–667.

Fresh, K. L. 1997. The role of competition and predation in the decline of Pacific salmon and steelhead. Pages 245–275 in D. J. Stouder, P. A. Bisson, R. J. Naiman, and M. G. Duke, editors. Pacific salmon and their ecosystems: status and future options. Chapman and Hall, New York.

Gende, S. M., R. T. Edwards, M. F. Willson, and M. S. Wipfli. 2002. Pacific salmon in aquatic and terrestrial ecosystems. BioScience 52:917–928.

Gowan, C., and K. D. Fausch. 1996. Long-term demographic responses of trout populations to habitat manipulation in six Colorado streams. Ecological Applications 6:931–946.

Graff, D. R. 1996. Whirling disease: the Midwest/Eastern experience. Pages 43–53 in E. P. Bergerson and B. A. Knoph, editors. Proceedings whirling disease workshop—where do we go from here? Colorado Cooperative Fish and Wildlife Research Unit, Colorado State University, Fort Collins.

Hedrick, P. W., V. K. Rashbrook, and D. Hedgecock. 2000. Effective population size of winter-run Chinook salmon based on microsatellite analysis of returning spawners. Canadian Journal of Fisheries and Aquatic Sciences 57:2368–2373.

Higgins, J. 2009. The national fish habitat action plan: a partnership approach to protect and restore fish populations. Journal of the American Water Works Association 101:20–22.

Holden, P. B. 1991. Ghosts of the Green River: impacts of Green River poisoning on management of native fishes. Pages 43–54 in W. L. Minckley and J. E. Deacon, editors. Battle against extinction: native fish management in the American West. University of Arizona Press, Tucson.

Johnson, B. L., W. B. Richardson, and T. J. Naimo. 1995a. Past, present, and future concepts in large river ecology. BioScience 45:134–141.

Johnson, D. M., R. J. Behnke, D. A. Haprman, and R. G. Walsh. 1995b. Economic benefits and cost of stocking catchable rainbow trout: a synthesis of economic analysis in Colorado. North American Journal of Fisheries Management 15:26–32.

Kauffman, J. B., R. L. Beschta, N. Otting, and D. Lytijen. 1997. An ecological perspective of riparian and stream restoration in the western United States. Fisheries 22(5):12–25.

Kondolf, G. M., and P. R. Wilcock. 1996. The flushing flow problem: defining and evaluating objectives. Water Resource Research 32:2589–2600.

Krause, C. W., T. J. Newcomb, and D. J. Orth. 2005. Thermal habitat assessment of alternative flow scenarios in a tailwater fishery. River Research and Applications 21:581–593.

Lagarrigue T., R. Cereghino, P. Lim, P. Reyes-Marchant, R. Chappaz, P. Lavandier, and A. Belaud. 2002. Diel and seasonal variations in brown trout (*Salmo trutta*) feeding patterns and relationship with invertebrate drift under natural and hydropeaking conditions in a mountain stream. Aquatic Living Resources 15:129–137.

Larkin, G., and P. A. Slaney. 1997. Implications of trends in marine-derived nutrient influx to south coastal British Columbia salmonids production. Fisheries 22(11):16–24.

Lenat, D. R., D. L. Penrose, and K. W. Eagleson. 1981. Variable effects of sediment addition on stream benthos. Hydrobiologia 79:187–194.

Leonard, D. S. 1995. An evaluation of the North Platte River flushing flow releases. Master's thesis. University of Wyoming, Laramie.

Lichatowich, J. 1999. Salmon without rivers: a history of the Pacific salmon crisis. Island Press, Washington, D.C.

Litvak, M. K., and N. E. Mandrak. 1993. Ecology of freshwater baitfish use in Canada and the United States. Fisheries 18(12):6–13.

Lyons, J., L. Wang, and T. D. Simonson. 1996. Development and validation of an index of biotic integrity for coldwater streams in Wisconsin. North American Journal of Fisheries Management 16:241–256.

Mackay, R. J. 2006. River and stream ecosystems of Canada. Pages 33–60 in C. E. Cushing, K. W. Cummins, and G. W. Minshall, editors. River and stream ecosystems of the world. University of California Press, Berkley.

Mahnken, C., G. Ruggerone, W. Waknitz, and T. Flagg. 1998. A historical perspective on salmonid production from Pacific rim hatcheries. North Pacific Anadromous Fish Commission Bulletin 1:38–53.

Marshall, T. L. 1973. Trout populations, angler harvest and value of stocked and unstocked fisheries in the Cache la Poudre River, Colorado. Doctoral dissertation. Colorado State University, Fort Collins.

McHugh, P., P. Budy, G. Thiede, and Erin VanDyke. 2006. Trophic relationships of nonnative brown trout, Salmo trutta, and native Bonneville cutthroat trout, *Oncorhynchus clarkii utah*, in a northern Utah, USA river. Environmental Biology of Fishes 81:63–75.

McMichael, G. A., T. N. Pearsons, and S. A. Leider. 1999. Minimizing ecological impacts of hatchery-reared juvenile steelhead on wild salmonids in a Yakima basin tributary. Pages 365–380 in E. Knudson, C. R. Steward, D. D. MacDonald, J. E. Williams, and D. W. Reiser, editors. Sustainable fisheries management: Pacific salmon. CRC Press, Boca Raton, Florida.

Merz, J. E., and J. D. Setka. 2004. Evaluation of spawning habitat enhancement site for Chinook salmon in a regulated California River. North American Journal of Fisheries Management 24:397–407.

Mitro, M. G., A. V. Zale, and B. A. Rich. 2003. The relation between age-0 rainbow trout (*Oncorhynchus mykiss*) abundance and winter discharge in a regulated river. Canadian Journal of Fisheries and Aquatic Sciences 60:135–139.

Montgomery, D. R., and J. M. Buffington. 1998. Channel processes, classification, and response. Pages 13–42 in R. J. Naiman and R. E. Bilby, editors. River ecology and management: lessons from the Pacific Coastal ecoregion. Springer, New York.

Munn, M. D., and M. A. Brusven. 1991. Benthic macroinvertebrate communities in nonregulated and regulated waters of the Clearwater River, Idaho, USA. Regulated Rivers: Research and Management 6:1–11.

Myers, R. A., S. A. Levin, R. Lande, F. C. James, W. W. Murdoch, and R. T. Paine. 2004. Hatcheries and endangered salmon. Policy Forum Ecology, Science 303:1980.

Northcote, T. G. 1998. Migratory behaviour of fish and its significance to movement through riverine fish passage facilities. Pages 3–18 in M. Jungwirth, S. Schmutz, and S. Weiss, editors. Fish migration and fish bypasses. Blackwell Scientific Publications, Malden, Massachusetts.

Nordwall, F., and P. Lundberg. 2000. Simulated harvesting of stream salmonids with a seasonal life history. North American Journal of Fisheries Management 20:481–492.

Paul, A. J., and J. R. Post. 2001. Spatial distribution of native and nonnative salmonids in streams of the eastern slopes of the Canadian Rocky Mountains. Transactions of the American Fisheries Society 130:417–430.

Pender, D. R., and T. J. Kwak. 2002. Factors influencing brown trout reproductive success in Ozark tailwater rivers. Transactions of the American Fisheries Society 131:698–717.

Pess, G. R., M. E. McHugh, D. Fagen, P. Stevenson, and J. Drotts. 1998. Stillaguamish salmonids barrier evaluation and elimination project (Phase III). National Marine Fisheries Service, Final Report to the Tulalip Tribes, Marysville, Washington.

Petrosky, C. E., and T. C. Bjornn. 1988. Response of wild rainbow (*Salmo gairdweri*) and cutthroat trout (*S. clarki*) to stocked rainbow trout in fertile and infertile streams. Canadian Journal of Fisheries and Aquatic Sciences 45:2087–2105.

Pintler, H. E., and W. C. Johnson. 1958. Chemical control of rough fish in the Russian River drainage, California. California Fish and Game 44:91–124.

Platts, W. S. 1979. Relationships among stream order, fish populations, and aquatic geomorphology in an Idaho River drainage. Fisheries 4(2):5–9.

Poff, N. L., J. D. Allan, M. B. Bain, J. R. Karr, K. L. Prestegaard, B. D. Richter, R. E. Sparks, and J. C. Stromberg. 1997. The natural flow regime: a paradigm for river conservation and restoration. BioScience 47:769–784.

Post, J. R., C. Mushiens, A. Paul, and M. Sullivan. 2003. Assessment of alternative harvest regulations for sustaining fisheries: model development and application to bull trout. North American Journal of Fisheries Management 23:22–34.

Quinn, J. W., and T. J. Kwak. 2000. Use of rehabilitated habitat by brown trout and rainbow trout in an Ozark tailwater river. Fisheries 20(3):737–751.

Quist, M. C., and W. A. Hubert. 2005. Relative effects of biotic and abiotic processes: a test of the biotic–abiotic constraining hypothesis as applied to cutthroat trout. Transactions of the American Fisheries Society 134:676–686.

Reeves, G. H., P. A. Bisson, and J. M. Dambacher. 1998. Fish communities. Pages 200–234 *in* R. J. Naiman and R. E. Bilby, editors. River ecology and management: lessons from the Pacific coastal ecoregion. Springer, New York.

Reeves, G. H., D. B. Hohler, B. E. Hansen, F. H. Everest, J. R. Sedall, T. L. Hickman, and D. Shively. 1997. Fish habitat restoration in the Pacific Northwest: Fish Creek of Oregon. Pages 335–359 *in* J. E. Williams, C. A. Wood, and M. P. Dombeck, editors. Watershed restoration: principals and practices. American Fisheries Society, Bethesda, Maryland.

Reid, L. M., T. Dunne, and C. J. Cederholm. 1981. Application of sediment budget studies to the evaluation of logging road impact. Journal of Hydrology (New Zealand) 20:49–62.

Reisenbichler, R. R., and J. D. McIntyre. 1986. Requirements for integrating natural and artificial production of anadromous salmonids in the Pacific Northwest. Pages 365–374 *in* R. H. Stroud, editor. Fish culture in fisheries management. American Fisheries Society, Fish Culture Section and Fisheries Management Section, Bethesda, Maryland.

Richards, C. P., J. Cernera, M. P. Ramey, and D. W. Reiser. 1992. Development of off-channel habitats for use by juvenile Chinook salmon. North American Journal of Fisheries Management 12:721–727.

Ricker, W. E. 1989. History and present state of the odd-year pink salmon runs of the Fraser River region. Canadian Technical Report of Fisheries and Aquatic Sciences 1702.

Rieman, B. E., and J. D. McIntyre. 1993. Demographic and habitat requirements for conservation of bull trout. U.S. Forest Service Intermountain Research Station, General Technical Report INT-302, Ogden, Utah.

Roussel, J. M., R. A. Cunjak, R. Newbury, D. Caissie, and A. Haro. 2004. Movements and habitat use by PIT-tagged Atlantic salmon parr in early winter: influence of anchor ice. Freshwater Biology 49:1026–1035.

Saha, B., V. S. Neary, and P. W. Bettoli. 2006. Feasibility study of trout habitat enhancement in the Caney Fork River using boulder clusters. Tennessee Wildlife Resources Agency, Fisheries Report 06–10, Nashville.

Saltveit, S. J., J. H. Halleraker, J. V. Arnekleiv, and A. Harby. 2001. Field experiments on stranding in juvenile Atlantic salmon (*Salmo salar*) and brown trout (*Salmo trutta*) during rapid flow decreases caused by hydropeaking. Regulated Rivers: Research and Management 17:609–622.

Schrank, A. J., and F. J. Rahel. 2004. Movement patterns in inland cutthroat trout (*Oncorhynchus clarki utah*): management and conservation implications. Canadian Journal of Fisheries and Aquatic Sciences 61:1528–1537.

Schrank, A. J., F. J. Rahel, and H. C. Johnstone. 2003. Evaluating laboratory-derived thermal criteria in the field: an example involving Bonneville cutthroat trout. Transactions of the American Fisheries Society 132:100–109.

Scott, M. T., and L. A. Nielsen. 1989. Young fish distribution in backwaters and main-channel borders of the Kanawha River, West Virginia. Journal of Fish Biology 35:21–27.

Scully, R. J., E. J. Leitzinger, and C. E. Petrosky. 1990. Idaho habitat evaluation for off-site mitigation record. Idaho Department of Fish and Game, 1988 Annual Report to Bonneville Power Administration. Contract Report DE-179–84BP13381, Portland, Oregon.

Selong, J. H., T. E. McMahon, A. V. Zale, and F. T. Barrows. 2001. Effect of temperature on growth and survival of bull trout, with application of an improved method for determining thermal tolerance in fishes. Transactions of the American Fisheries Society 130:1026–1037.

Shannon, J. P., D. W. Blinn, T. McKinney, E. P. Benenati, K. P. Wilson, and C. O'Brien. 2001. Aquatic food base response to the 1996 test flood below Glen Canyon Dam, Colorado River, Arizona. Ecological Applications 11:672–685.

Sheaffer, W. A., and J. G. Nickum. 1986. Backwater areas as nursery habitats for fishes in Pool 13 of the upper Mississippi River. Hydrobiologia 136:131–139.

Shetter, D. S., and G. R. Alexander. 1966. Angling and trout populations on the North Branch of the Au Sable River, Crawford and Otsego counties, Michigan, under special and normal regulations, 1958–63. Transactions of the American Fisheries Society 95:85–91.

Shetter, D. S., and G. R. Alexander. 1970. Results of predator reduction on brook trout and brown trout in 4.2 mi (6.76 km) of the North Branch of the Au Sable River. Transactions of the American Fisheries Society 99:312–319.

Simpkins, D. G., and W. A. Hubert. 2000. Drifting invertebrates, stomach contents, and body condition of juvenile rainbow trout from fall through winter in a Wyoming tailwater. Transactions of the American Fisheries Society 129:1187–1195.

Simpkins, D. G., W. A. Hubert, and C. Martinez del Rio. 2004. Factors affecting the swimming performance of fasted rainbow trout with implications of exhaustive exercise on overwinter mortality. Journal of Freshwater Ecology 19:557–566.

Simpkins, D. G., W. A. Hubert, C. Martinez del Rio, and D. C. Rule. 2003. Interacting effects of water temperature and swimming activity on body composition and mortality of fasted juvenile rainbow trout. Journal of Fish Biology 81:1641–1649.

Simpkins, D. G., W. A. Hubert, and T. A Wesche. 2000a. Effects of fall-to-winter changes in habitat and frazil ice on the movements and habitat use of juvenile rainbow trout in a Wyoming tailwater. Transactions of the American Fisheries Society 129:101–118.

Simpkins, D. G., W. A. Hubert, and T. A. Wesche. 2000b. Effects of a spring flushing flow on the distribution of radio-tagged juvenile rainbow trout in a Wyoming tailwater. North American Journal of Fisheries Management 20:546–551.

Stalnaker, C. B., R. T. Milhous, and K. D. Bovee. 1989. Hydrology and hydraulics applied to fishery management in large rivers. Canadian Special Publication of Fisheries and Aquatic Sciences 106:13–40.

Stanford, J. A., and J. V. Ward. 2001. Revisiting the serial discontinuity concept. Regulated Rivers: Research and Management 17:303–310.

Stockner, J. G., F. Ryden, and P. Hyenstrand. 2000. Cultural oligotrophication: causes and consequences for fisheries resources. Fisheries 25(5):7–14.

Swales, S., and C. D. Levings. 1989. Role of off-channel ponds in the life cycle of coho salmon (*Oncorhynchus kisutch*) and other juvenile salmonids in the Coldwater River, British Columbia. Canadian Journal of Fisheries and Aquatic Sciences 46:232–242.

Tripp, D. B., and V. A. Poulin. 1986. The effects of logging and mass wasting on juvenile fish habitats in streams on the Queen Charlotte Islands. British Columbia Ministry of Forests and Lands, Land Management Report 45, Victoria.

Trush, W. J., S. M. McBain, and L. B. Leopold. 2000. Attributes of an alluvial river and their relation to water policy and management. Proceedings of the National Academy of Science 97:11858–11863.

Twohey, M. B., J. W. Heinrich, J. G. Seelye, K. T. Fredricks, R. A. Bergstedt, C. A. Kaye, R. J. Scholefield, R. B. McDonald, and G. C. Christie. 2003. The sterile-male-release technique in Great Lakes sea lamprey management. Journal of Great Lakes Research 29 (Supplement 1):410–423.

USDI (U.S. Department of the Interior). 2007. 2006 National survey of fishing, hunting, and wildlife-associated recreation. U.S. Department of the Interior, Fish and Wildlife Service and U.S. Department of Commerce, Census Bureau, Washington, D.C.

Utter, F. M., J. E. Seeb, and L. W. Seeb. 1993. Complementary uses of ecological and biochemical genetic data in identifying and conserving salmon populations. Fisheries Research 18:59–76.

Vannote, R. L., G. W. Minshall, K. W. Cummins, J. R. Sedell, and C. E. Cushing. 1980. The river continuum concept. Canadian Journal of Fisheries and Aquatic Sciences 37:130–137.

Vincent, E. R. 1987. Effects of stocking catchable-size rainbow trout on two wild trout species in the Madison River and O'Dell Creek, Montana. North American Journal of Fisheries Management 7:91–105.

Vinson, M. 2001. Long-term dynamics of an invertebrate assemblage downstream from a large dam. Ecological Applications 11:711–730.

von Hagen, B., S. Beebe, P. Schoonmaker, and R. Kellogg. 1998. Nonprofit organizations and watershed management. Pages 625–641 in R. J. Naiman and R. E. Bilby, editors. River ecology and management: lessons from the Pacific coastal ecoregion. Springer, New York.

Walters, J. P., T. D. Fresques, and S. D. Bryan. 1997. Comparison of creel returns from rainbow trout stocked at two sizes. North American Journal of Fisheries Management 17:474–476.

Waters, T. F. 1995. Sediment in streams: sources, biological effects, and control. American Fisheries Society, Monograph 7, Bethesda, Maryland.

Wenzel, C. R. 1993. Flushing flow requirements for a large, regulated, Wyoming river to maintain trout spawning habitat quality. Master's thesis. University of Wyoming, Laramie.

Wheeler, A. P., P. L. Angermeier, and A. E. Rosenberger. 2005. Impacts of new highways and subsequent landscape urbanization on stream habitat and biota. Reviews in Fisheries Science 13:141–164.

Wiley, D. 1995. Development and evaluation of flushing flow recommendations for the Big Horn River. Master's thesis. University of Wyoming, Laramie.

Wilhelm, J. G. O., J. D. Allan, K. J. Wessell, R. W. Merritt, and K. W. Cummins. 2005. Habitat assessment of nonwadeable rivers in Michigan. Environmental Management 36:592–609.

Williams, J. E., C. A. Wood, and M. P. Dombeck. 1997. Watershed restoration: principals and practices. American Fisheries Society, Bethesda, Maryland.

Wipfli, M. S., J. P. Hudson, D. T. Chaloner, and J. P. Couette. 1999. Influence of salmon spawner densities on stream productivity in southeast Alaska. Canadian Journal of Fisheries and Aquatic Sciences 56:1600–1611.

Wright, S. A. 1995. Ozark trout tales: a fishing guide for the White River system. White River Chronicle, Fayetteville, Arkansas.

Young, M. K., W. A. Hubert, and T. A. Wesche. 1991. Selection of measures of substrate composition to estimate survival to emergence of salmonids and to detect changes in stream substrates. North American Journal of Fisheries Management 11:339–346.

Young, M. K., B. K. Micek, and M. Rathbun. 2003. Probable pheromonal attraction of sexually mature brook trout to mature male conspecifics. North American Journal of Fisheries Management 23:276–282.

Zafft, D. J., P. J. Braaten, K. M. Johnson, and T. C. Annear. 1995. Comprehensive study of the Green River fishery between the New Fork River confluence and Flaming Gorge Reservoir: 1991–1994. Wyoming Game and Fish Department, Cheyenne, Wyoming.

Ziemer, R. R., and T. E. Lisle. 1998. Hydrology. Pages 43–68 in R. J. Naiman and R. E. Bilby, editors. River ecology and management: lessons from the Pacific coastal ecoregion. Springer, New York.

Zoellick, B. W. 1999. Stream temperatures and the elevational distribution of redband trout in southwestern Idaho. Great Basin Naturalist 59:136–143.

Zorn, T. G., P. W. Seelbach, and M. J. Wiley. 2002. Distributions of stream fishes and their relationship to stream size and hydrology in Michigan's Lower Peninsula. Transactions of the American Fisheries Society 131:70–85.

Chapter 20

Warmwater Streams

Daniel C. Dauwalter, William L. Fisher, and Frank J. Rahel

20.1 INTRODUCTION

Warmwater streams are those streams and rivers with warm temperatures and support diverse fish assemblages including populations of basses, sunfishes, and catfishes. Warmwater streams are distinguished from coldwater streams because they lack salmonid populations, typically occur at lower elevations, and have cool to warm water in summer, medium to high streamflows, clear to turbid water, diverse substrates, and low gradients (Winger 1981). Warmwater streams occur throughout the United Mexican States (Mexico), the USA, and central Canada, except in mountainous regions in the west and north. Fishing in warmwater streams occurs in the entire USA except Alaska and is the predominant type of fishing in over half of those states (Funk 1970). Not surprisingly, the criteria for classifying a stream as warmwater differ among individuals and management agencies. One criterion is the presence of trout: if trout are present, then the stream is considered to be a coldwater stream; if they are absent, then it is classified as a warmwater stream. Other criteria use water temperature statistics to classify streams (e.g., instantaneous maximum, daily mean, or monthly mean); an average daily summer water temperature 20°C or more is often used as general rule to define warmwater streams (Winger 1981). Adding to the confusion is the fact that coldwater streams can become warmwater streams when anthropogenic disturbances increase stream temperatures, and classifications can be based on either current conditions or potential conditions in the absence of disturbance. Regardless, criteria for classifying streams as coldwater or warmwater are based on management goals.

Because warmwater streams occur throughout North America, their physical and chemical characteristics vary in relation to their environmental setting. Most of the USA has a temperate climate with moderate air temperatures and rainfall amounts, but warmwater streams also occur in the hot, dry climate of Mexico and the southwestern USA; the cool, moist, continental climate of the northern USA and Canada; and the warm, wet, subtropical climate of southern Florida.

Climate, geology, land use, and physiography play a role in controlling the hydrologic and sediment regimes of warmwater streams (Knighton 1998) and ultimately the local habitats of fishes. Warmwater streams flow through forests, grasslands, and deserts. These land cover types affect the amount of stream shading and consequently affect stream temperatures. Some warmwater streams arise in high-gradient, cold headwater mountain streams and become warm as they flow downstream, but many originate in lower-gradient prairies and coastal areas and have warm headwaters. Warmwater stream temperatures can vary from 0°C during winter months to 40°C during summer in the southwestern USA (Matthews and Zimmerman 1990).

Warmwater streams contain diverse assemblages of fishes and many important sport and commercial fisheries. The diversity of fishes is highest in eastern North American temperate streams, particularly in the Mississippi–Missouri–Ohio river basins that contain 375 species in 31 families (Burr and Mayden 1992). Nearly 50 species may be found at a site (Matthews 1998), most of which belong to a few families such as the catfishes, suckers, minnows, sunfishes, and perches. State and provincial agencies recognize at least 32 species of sport fish in warmwater streams that range geographically from "catfish streams" in Oklahoma to "smallmouth bass streams" in Virginia (Rabeni and Jacobson 1999).

Warmwater streams also are rich in invertebrates including crayfishes and mussels. Both crayfishes and mussels reach their highest levels of diversity in the warmwater streams of North America. Approximately 77% of the more than 500 crayfish species worldwide occur in North America (Taylor et al. 2007). However, nearly half (48%) of these North American species are imperiled, primarily because of limited natural distributions, introductions of nonnative crayfishes, and habitat alterations. Freshwater mussels also reach their greatest diversity in North American warmwater streams with 297 species and subspecies known to occur (Williams et al. 1993). Mussels are the most imperiled faunal group in North America with 60% of described species considered threatened or endangered and 12% presumed extinct. Threats to mussels include widespread habitat degradation from pollution, overharvest, impoundments and channel alterations, and recent introductions and invasions of the Eurasian zebra mussel (Ricciardi et al. 1998).

Because of their diversity, productivity, and beauty, warmwater streams are valued by anglers and the general public. Results of a 1991 national survey showed that around 5 million of 13 million anglers in the southeastern USA fished in streams and rivers (Fisher et al. 1998), and a survey of Mississippi anglers revealed that most preferred to fish in streams (Jackson and Jackson 1989). Stream angling can provide substantial benefit to a regional economy (Fisher et al. 2002). As early as the 1950s, fisheries managers in the southeastern USA decried the effects of large water resource projects (e.g., reservoirs) on southeastern warmwater streams, recognized their recreational uses and esthetic values, and called for their preservation (Alexander 1959). Management of warmwater stream fisheries has not kept pace with the intensity of stream-fishing activity, particularly in the southeastern USA (Fisher et al. 1998). However, all state agencies in the USA devote resources to managing stream fisheries (Fisher and Burroughs 2003). Based on a survey of state agencies, one of the top goals for warmwater streams management is improving ecosystem integrity in the face of declining water quality (Fisher and Burroughs 2003). Accomplishing this goal requires developing techniques to assess the status of both fish assemblages and fish habitat (Quist et al. 2006).

20.2 FACTORS INFLUENCING WARMWATER STREAM FISHES

Factors that influence the composition of warmwater stream fish assemblages and fish populations can be grouped into two types: habitat (abiotic factors) or biological (biotic factors; Figure 20.1). The biogeographic processes of speciation, extinction, and extirpation influence modern-day distribution patterns of fishes. A warmwater fish assemblage at any stream

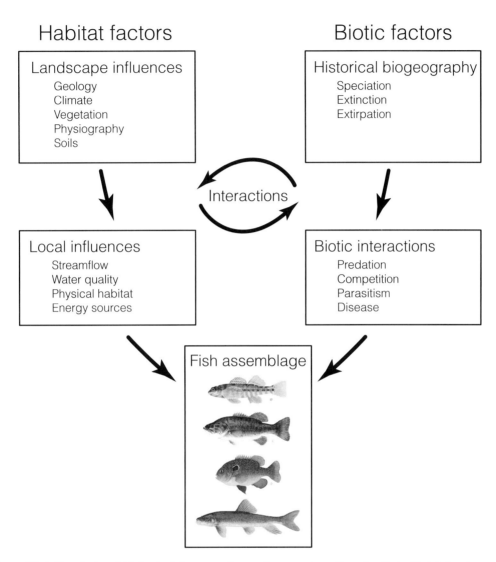

Figure 20.1. Physical and biological factors acting on large and small scales that affect fishes in warmwater streams.

site is conceptually the result of species passing through a series of nested, successively finer habitat filters that select species by their traits (Poff 1997). For example, a species that occurs in a particular microhabitat, such as a darter living on the bottom of shallow riffle, possesses traits (e.g., small body size, no swim bladder, and large pectoral fins) that are suitable for that channel unit type (riffle), stream reach (headwaters), and watershed (forested). However, biogeographic history will determine whether the species occurs within the watershed (Dauwalter et al. 2008), and biotic interactions can prevent a species occurrence in habitats that are otherwise suitable (Quist et al. 2005).

20.2.1 Habitat Factors

Habitat factors that influence warmwater fish assemblages vary from small-scale local influences to large-scale landscape influences (Figure 20.1). These habitat factors are hierarchically nested from microhabitats to channel units, stream reaches, valley segments, watersheds, and geomorphic provinces (Figure 20.2). Landscape influences are the result of climate and geology that affect soils and vegetation within a geomorphic province.

Ecosystem processes in warmwater streams are influenced by land cover types and land use activities in a watershed (Figure 20.1). Land cover in North American watersheds ranges widely, from shrublands in the Southwest, to grasslands or croplands in the Midwest and Great Plains, to broadleaf deciduous forests in the East, and to needleleaf conifer forests in the Northwest, Northeast, and Southeast. Stream ecosystems derive their energy from dead and living organic matter. Dead organic matter falls into a stream as leaves, grasses, and wood or enters as dissolved organic matter from groundwater sources (Cummins 1974; Brunke and Gonser 1997). Live organic matter is produced in streams from primary production by algae and macrophytes (Cummins 1974; Baxter et al. 2005). Dead and live organic matter is consumed by microbes and invertebrates. In turn, aquatic invertebrates, and terrestrial invertebrates that fall into a stream, are eaten by fishes and other vertebrates. Longitudinally, much of the organic matter comes from the headwaters of forested streams (Vannote et al. 1980); however, lowland forested streams in the southeastern USA have extensive floodplains with wetlands throughout their length that contribute organic matter (Meyer and Edwards 1990). In contrast, some warmwater streams in prairie and arid regions of the central and southwestern USA derive much of their energy from primary production in headwater regions that is then transported downstream (Fisher et al. 1982; Wiley et al. 1990; Gray 1997).

An important local habitat factor influencing warmwater stream fish assemblages is streamflow (Figure 20.1). Streamflow exerts control over many stream attributes including channel structure and form, substrate composition, and instream habitat (e.g., wood and vegetation) available for aquatic organisms (Poff and Ward 1989). Streamflows can be characterized by five components—magnitude, frequency, duration, timing, and rate of change (Poff et al. 1997) —and these components vary widely across North American streams (Benke and Cushing 2005). These flow patterns have a strong effect on the availability of habitat for fish populations and fish assemblages in warmwater streams (Poff and Allan 1995; Remshardt and Fisher 2009).

Water quality can strongly influence the structure of fish assemblages and individual fish populations in warmwater streams (Figure 20.1). Water quality includes the biological, chemical, and physical characteristics of a water body in relation to water uses (Armantrout 1998). Warmwater stream temperatures vary widely throughout North America and can fluctuate as much as 20°C a day in headwater streams (Matthews 1998). Stream temperature strongly affects the distributiona and health of warmwater fishes. For example, smallmouth bass and largemouth bass co-occur in the Ozark streams of Missouri, but smallmouth bass dominate at cooler temperatures and grow optimally at around 22°C whereas largemouth bass dominate and grow optimally at warmer temperatures. In addition to stream temperatures, warmwater fishes are sensitive to low levels of dissolved oxygen (DO), although tolerance varies widely among species. Acute lethal DO concentrations for warmwater fishes generally occur below 3 mg/L. Low DO concentrations are associated with eutrophication caused by high levels of phosphorus and nitrogen in streams (Mallin et al. 2006). Smale and Rabeni (1995a, 1995b) found that DO minima varied from 0.8

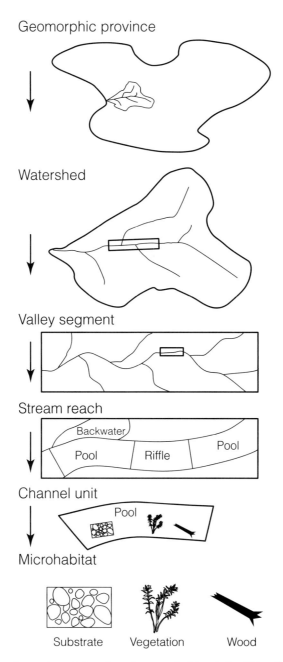

Figure 20.2. The hierarchy and spatial scales of stream habitat (after Frissell et al. 1986; Montgomery and Bolton 2003).

to 6.0 mg/L and temperature maxima varied from 19.6°C to 30.7°C among 35 species of stream fishes in Missouri. In general, species from prairie streams tolerated lower DO and higher temperatures than did species in upland Ozark streams, reflecting the high occurrence of intermittent flows in prairie streams. Low DO concentrations had a stronger effect on fish assemblage composition, and high water temperatures only affected composition at sites with sufficient DO.

Physical habitat in streams is considered the template for ecological interactions (Southwood 1977). Habitat is the result of hydrologic, geomorphic, and vegetation (particularly wood) transport processes that interact to form pools, riffles, and bars in meandering streams (Montgomery and Bolton 2003). These features, in turn, influence water depths and velocities, substrates, presence of wood, and aquatic vegetation important to fishes (Fore et al. 2007). Habitat features change over a time scale depending on the spatial scale of interest, from tens of thousands of years for a watershed to months or days for a microhabitat (Frissell et al. 1986). Hydrogeomorphic processes, including floods, droughts, and landslides, create, modify, and destroy habitat features used by stream organisms and shape the dynamics of stream ecosystems (Montgomery 1999; Montgomery and Bolton 2003).

Scientific studies of streams and their watersheds have helped shape warmwater stream management (Table 20.1). Hynes (1970) completed the first comprehensive review of the physical and biological components of streams. Since then, the science of flowing waters has continued to advance. Gorman and Karr (1978) identified a positive relationship between local habitat diversity and fish species diversity in warmwater streams in central Indiana, which focused attention on maintaining habitat diversity in streams. By the 1980s, stream ecologists began viewing stream habitat at different spatial scales within a watershed context (Frissell et al. 1986) and realizing the importance of fluvial geomorphology. For example, Dauwalter et al. (2007) showed that channel unit size and stream size, geomorphic factors representing two different spatial scales, were the primary determinants of smallmouth bass density in streams in eastern Oklahoma. Currently there is increased focus on how longitudinal and lateral connectivity of stream habitats allows fish to move among different habitats needed for spawning, feeding, or refugia from harsh environmental conditions (Schlosser and Angermeier 1995; Belica and Rahel 2008; Dauwalter and Fisher 2008). An area of a stream that contains all needed habitats has been defined as a functional habitat unit (Figure 20.3), and the arrangement of habitats units has been shown to affect fish population dynamics (Kocik and Ferreri 1998; Le Pichon et al. 2006). Headwater streams make up over two-thirds of the total stream length in a typical watershed (Freeman et al. 2007) and supply invertebrate food resources and detritus to downstream food webs (Cummins and Wilzbach 2005), highlighting the importance of longitudinal connectivity in stream networks. Incorporating a spatial perspective in stream management has been facilitated by the rapid development of technological tools, such as geographic information systems (GIS) and global positioning systems (GPS), that allow biologists to view streams at multiple spatial scales and to evaluate the spatial relationships among features of the landscape, watershed, and stream reach (Fausch et al. 2002; Fisher and Rahel 2004a).

20.2.2 Biotic Factors

An assemblage consists of many fish species that interact with one another. These interactions can be either beneficial or detrimental (Hildrew 1996). Predation can reduce or eliminate a species and is a powerful selective force and a primary biological determinant of

Table 20.1. Major concepts in stream and river science that have influenced warmwater stream management.

Concept	Overview	Source
River continuum concept	Physical, energetic, and biological attributes of streams change along a continuous longitudinal gradient from small streams to large rivers	Vannote et al. (1980)
Nutrient spiraling concept	Nutrients move downstream in a spiraling manner between abiotic and biotic components of streams	Newbold et al. (1981)
Serial discontinuity concept	The continuums and nutrient spirals of streams are disrupted by dams and reservoirs	Ward and Stanford (1983)
Predation	Predation can strongly affect stream assemblage structure during stable environmental conditions	Power et al. (1985)
Stream habitat hierarchy	Stream habitats are organized within a nested set of spatial and temporal scales ranging from large watersheds that are affected across long time periods to microhabitats that are affected across short time periods	Frissell et al. (1986)
Disturbance	The frequency, intensity, and severity of unpredictable disturbance events has a strong influence on the structure and function of stream ecosystems	Resh et al. (1988)
Flood pulse concept	Flood pulses drive the existence, productivity, and interactions of biota in river–floodplain ecosystems	Junk et al. (1989)
Four dimensions	Stream ecosystems have four dimensions: longitudinal, lateral, vertical, and temporal	Ward (1989)
Riverine productivity model	Local energy production and organic inputs from the riparian zone strongly affect ecosystems of streams with constricted channels and infrequent floodplain interactions	Thorp and DeLong (1994)
Natural flow regime paradigm	The natural magnitude, frequency, duration, timing, and rate of change in streamflows is required to maintain the ecological integrity of stream ecosystems	Poff et al. (1997)

Table 20.1. Continued.

Concept	Overview	Source
Network dynamics hypothesis	Abrupt changes in stream habitat, biota, and ecosystem function can occur at the junctions of tributaries recently disturbed or set in a unique environmental setting	Benda et al. (2004)
Riverine ecosystem synthesis	Incorporates many previous concepts into an integrated, heuristic model of lotic biocomplexity across spatiotemporal scales	Thorp et al. (2006)
Land cover cascade	Disturbances to natural land cover cascade through various ecosystem components and affect stream ecosystems	Burcher et al. (2007)

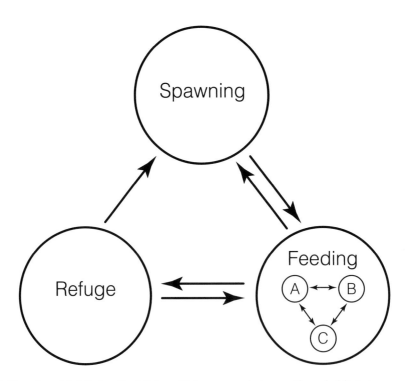

Figure 20.3. Functional habitat unit of fishes. Fishes move between these habitats to reproduce, seek refuge from predators, and feed and can even use different areas within a habitat (A, B, and C). All habitats are essential to completing their life cycle (modified from Schlosser and Angermeier 1995).

local fish assemblage structure (Figure 20.1). Apex predators in many warmwater streams are black bass, catfishes, and some sunfishes. For instance, largemouth bass and spotted bass can strongly influence the distribution and abundance of prey fishes (Power et al. 1985), and introduced flathead catfish can suppress native fish assemblage biomass through predation (Pine et al. 2007). Competition is the joint utilization of a limited resource by multiple species that reduces the fitness of one or more species. Although competition has been described among some co-existing fishes (e.g., darters) in warmwater streams (Greenberg 1988), it is generally considered to be minimized by partitioning of available food and habitat resources among species (Fisher and Pearson 1987; Gray et al. 1997). Other biotic factors, such as disease and parasitism, can negatively affect the health of warmwater fishes, particularly in streams where pollution from sewage or pesticides and high temperature cause stress in fishes (Snieszko 1974). Hybridization can also be detrimental. For example, the white sucker was introduced into the Colorado River Basin where it hybridizes with native suckers that are now threatened by extinction through hybridization (McDonald et al. 2008). Positive interactions also occur in streams, such as when different species of minnows school together to feed or avoid predators. Although biotic interactions can predominate under low-flow conditions (Power et al. 1985), abiotic factors, such as streamflow variability, are thought to be most important in controlling fish assemblage structure (Horwitz 1978; Schlosser 1982).

20.3 WARMWATER STREAMS: ISSUES AND MANAGEMENT

Human activities in watersheds and along stream corridors have vastly altered both physical and biological components of warmwater streams. There is a long history of human influences on stream fish assemblages through agriculture, dams, discharge of oxygen-demanding wastes and toxic chemicals, overconsumption of water, and exotic species introductions (Karr et al. 1985). This history of disturbance, and the fact that warmwater streams are species rich and often contain multispecies fisheries, makes these streams one of the most challenging aquatic systems for a fisheries biologist to manage.

20.3.1 Management Goals

Management of warmwater streams can be focused on habitat, aquatic organisms, people, or any combination of these fishery components. As in other aquatic systems, management of warmwater streams is done through a process that involves a constituent base, goal setting, plan development, management implementation, monitoring, and reevaluation (Figure 20.4). Management goals for warmwater streams can often be placed into one of four categories.

Restoration. Reestablishment of reference stream conditions and natural processes that created those conditions. Agricultural land in a watershed may be converted back to native vegetation to promote natural levels of sediment, water, and wood transport that create natural stream habitats. An invasive species may be removed from a stream system to restore natural fish assemblage interactions.

Rehabilitation. Improvement of stream conditions to near original condition but not restoring the natural processes that created those conditions. A streambank may be artificially stabilized to reduce erosion and fine sediment production, but the cause of streambank erosion is not addressed. A fish population may be maintained with supplemental stocking, but the cause of recruitment failure remains.

Conservation. Maintenance of existing stream conditions and fish populations. Watersheds with little human impact may be given a special protection status to conserve current conditions and prohibit future disturbances to the stream ecosystem. Legislation may be enacted that prohibits the transport of fishes to prevent fish species from being introduced into drainage basins where they are not native.

Enhancement. Improvement of stream conditions that benefit stream habitat or stream fishes. Log weirs and rock vanes might be placed into a stream to create additional habitat that improves fish population structure and increases angling opportunities. Large sport fish can be stocked to supplement the natural population and increase angler catch rates of large fish.

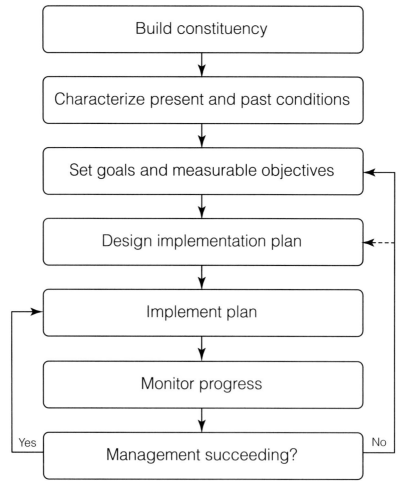

Figure 20.4. Conceptual framework for stream habitat management that begins by building a constituency base and then setting goals, developing and implementing a management plan, and monitoring progress. Monitoring may reveal the need to change goals and revise plans or to continue plan implementation.

20.3.2 Habitat Issues and Management

Warmwater streams have a long history of habitat degradation that is a major concern for fisheries managers. Point source discharges, dams, agricultural practices, and road and bridge construction are among the many human activities that influence habitat in warmwater streams (Table 20.2).

Humans can change the sources of energy in streams by altering the size and source of particulate organic matter. Agriculture and wastewater effluents add fine organic matter or dissolved organic carbon into streams. Loss of riparian canopy cover and increased solar radiation in small streams can lead to enhanced algal production. In contrast, medium-sized streams have naturally high levels of algal production that results from more sunlight through the open riparian canopy. However, agricultural and silvicultural activities that increase suspended sediments and sedimentation can reduce benthic algal production by shading the stream bottom and smothering hard substrates that are habitats for periphyton. The timing of energy inputs, like leaf litter, can also be important to certain life stages of microbes and aquatic invertebrates. Changes to the source and timing of energy inputs can alter the structure of invertebrate assemblages and thus affect food resources for fishes.

Fishes in warmwater streams are adapted to natural streamflow conditions that are often altered directly by dams, channelization, or water withdrawal and indirectly by land use (Figure 20.5). Of the 5.3 million kilometers of rivers in the U.S. coterminous 48 states, only 42 free-flowing rivers greater than 200 km in length have not been influenced by dams (Benke 1990). Dams have been constructed on streams and rivers across North America for several reasons: hydroelectric power, flood control, navigation, water supply, irrigation, and recreation. Although dams block fish movements, create reservoirs with nonnative sport fishes, and alter downstream habitats in ways more suitable to nonnative coldwater trout (Quinn and Kwak 2003), they most notably alter natural streamflows. Their effects on streamflows are dependent on the size and purpose of the dam (Hart et al. 2002), but they often alter the magnitude, frequency, duration, timing, and rate of change of streamflows. This is particularly true when dams are operated for hydroelectric power (Figure 20.6). Dams are often operated to maintain certain water levels in reservoirs and release little water during low-flow periods. This results in extremely low streamflows or dewatering below the dam. Water withdrawal for irrigation or municipal use can dewater stream channels and entrain fish in irrigation canals where mortality is usually high (Jaeger et al. 2005; Roberts and Rahel 2008). Groundwater pumping also lowers water tables, decreases base flows, and reduces or eliminates springs important to some fishes (Hargrave and Johnson 2003). Both dams and water diversions can reduce the magnitude of floods. Smaller floods may not inundate floodplain habitats that are a source of nutrients and important spawning areas for some fishes. In contrast, channelization of streams often expedites the transportation of water from the watershed and can increase the frequency and magnitude of floods. Unnaturally large floods alter physical habitat in streams by increasing streambank erosion and channel incision, disconnecting streams from the riparian zone, and reducing the amount of pool habitat and instream cover. These factors that increase runoff and expedite water transport in streams also reduce the duration of flood events. Decreased flood duration may limit access time to floodplain spawning habitats and nursery habitats. It also decreases the time for inundated organic matter to be processed and incorporated into the food web by microbes and macroinvertebrates.

Table 20.2. Human activities and their impacts to warmwater streams in North America (adapted from Bryan and Rutherford 1993).

Activity and factor impacted	Impact
Point source discharges	
Water quality	Thermal, chemical, and biological pollution; excessive nutrients; dissolved oxygen depletion and hypoxia
Energy sources	Increased fine organic matter inputs
Dams	
Physical habitat	Lentic habitat above dam; channel scour; altered channel morphology; disrupted sediment transport; disrupted habitat connectivity
Streamflow	Desiccation; reduced seasonal streamflow variation; reduced flood peaks; increased daily variation during hydropeaking
Water quality	Volatile temperature; decreased turbidity; low dissolved oxygen and hypoxia
Energy	Decreased primary production with hydropeaking; organic inputs from reservoir
Channelization	
Physical habitat	Increased channel gradient; altered sediment transport and increased bank erosion; widened or deepened channel; homogenization of channel morphology; sedimentation; wood removal
Streamflow	Change and homogenization of water velocity; increased runoff; increased storm peaks; lower base flows; increased flood peaks downstream
Water quality	Increased temperature if riparian vegetation removed; increased suspended sediments
Clearing and snagging	
Physical habitat	Reduced cover; altered channel morphology; decreased habitat diversity
Streamflows	Reduced flood peaks and flood durations
Energy sources	Increased coarse and fine organic matter transport; decreased macroinvertebrate habitat and production
Instream gravel mining	
Physical habitat	Altered channel morphology; widened channel; changed sediment dynamics; sedimentation; loss of instream cover and riparian vegetation
Water quality	Increased suspended sediments
Mining	
Physical habitat	Sedimentation
Water quality	Acidification; metal contamination

Table 20.2. Continued.

Activity and factor impacted	Impact
Road and bridge construction	
Physical habitat	Increased sedimentation; armored streambanks; disrupted sediment transport; interrupted connectivity
Water quality	Pollutants from roads
Timber harvest	
Physical habitat	Altered wood recruitment; increased sedimentation; altered channel morphology
Streamflows	Higher peak flows; lower base flows
Water quality	Increased suspended sediments; herbicide and pesticide pollution; increased temperature; increased nutrients
Energy sources	Increased organic inputs
Land drainage	
Physical habitat	Altered channel morphology; sedimentation
Streamflows	Increased peak flows; decreased flood duration; decreased base flows
Water quality	Increased suspended sediments; increased temperature from removal of riparian vegetation
Energy sources	Increased fine organic matter
Agriculture	
Physical habitat	Increase sedimentation; decreased streambank stability
Streamflows	Increased peak flows; decreased base flows
Water quality	Increased nutrients; dissolved oxygen depletion; pesticides and herbicides
Energy sources	Increased fine organic matter
Water withdrawal	
Physical habitat	Altered channel morphology; decreased wetted area
Streamflows	Decreased streamflows below diversion
Water quality	Increased turbidity, temperature, nutrients and salinity from return flows from agricultural areas

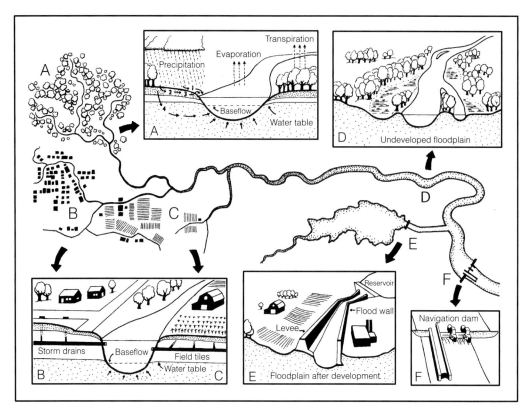

Figure 20.5. The hydrologic cycle. Streams originate in headwater reaches (**A**) through overland, subsurface, and groundwater flow. Urban (**B**) and agricultural (**C**) land uses reduce infiltration and increase surface flows by increasing impermeable surfaces, reducing vegetative cover, and installing drainage systems. Natural floodplain habitats (**D**) are disconnected from streams and rivers when levees are constructed (**E**) or water levels are controlled for hydroelectric power or navigation (**F**) (from Poff et al. 1997; Copyright, American Institute of Biological Sciences).

Water quality degradation historically has plagued warmwater streams and rivers. Pollutants from industrial and municipal activities were often discharged directly into streams. The discharge of pollutants by industry is what led the Cuyahoga River near Cleveland, Ohio, to catch fire many times from 1936 to 1969 (Box 20.1). Degradation of the Cuyahoga River was not an isolated incident, however. The water quality of many streams and rivers in the USA was a threat to public health, and many fisheries declined and fish kills became common. These types of events led the U.S. Congress to pass laws, such as the Clean Water Act of 1972, to clean up national waters (Adler et al. 1993). Fortunately, many of the toxic effluents originating from point sources have been eliminated and water quality problems associated with them have subsided because of laws like the Clean Water Act. Even the Cuyahoga River has improved tremendously (Box 20.1). However, water quality issues still exist for many warmwater streams due to non-point-source pollutants that the Clean Water Act was not designed to address. During periods of naturally low streamflow, nutrient-laden discharges from sewage treatment plants may constitute the majority or all of a stream's flow. In such situations, fish assemblages can be negatively affected by excessive algal production and low DO

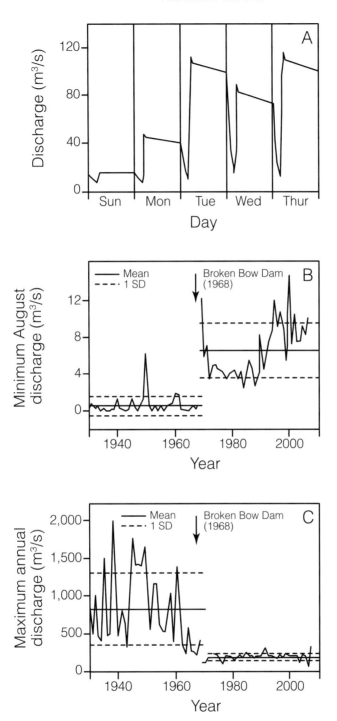

Figure 20.6. Daily streamflows (**A**), minimum August streamflows (**B**), and maximum annual streamflows (**C**) below Broken Bow Dam on the Mountain Fork River in eastern Oklahoma. Daily streamflows reflect water releases to generate hydroelectric power during peak use periods. Minimum August flows increased after dam construction due to constant summer releases for hydropeaking. Maximum flows decreased after dam construction because the reservoir stores flood waters and dampens downstream flood peaks.

concentrations, with species composition often shifting toward a few tolerant species, such as common carp and bullheads. A recent phenomenon is the "feminization" of fishes as a result of estrogenic compounds found in sewage effluents. The estrogenic compounds originate from various human pharmaceutical compounds (e.g., birth control pills) and cause male fish to develop female characteristics (Jobling and Tyler 2003).

Deterioration of physical habitat is a common problem in warmwater streams in North America. Sediment pollution is most often the cause of physical habitat deterioration (Waters 1995). All streams contain sediment from natural erosion processes, but sediment produced from roads, agricultural lands, and logged forests increases the amount of fine sediments in streams. Increased streamflows from human modifications to the landscape exacerbate these effects. Fine sediments embed coarser substrates that are required by many fishes for spawning and that are habitat for invertebrate prey populations. Fine sediments may also remain suspended, increasing turbidity and thereby reducing primary productivity and the foraging efficiency of fishes. Land uses that increase streambank erosion and sediment production also change stream geomorphology, resulting in fewer riffles and pools and more shallow-run habitats. This change in morphology is typically paralleled by a reduction in the diversity of water depths and velocities important to stream fishes. Channelization and instream sand and gravel mining alter channel morphology directly. A study of warmwater streams in Indiana showed that channelized streams had more fine substrates, less riffles and pools, and less cover than did nonchannelized streams (Lau et al. 2006). Consequently, channelized reaches did not have darters that require riffles for spawning and large-bodied fishes that require deep pools. Many land use practices adversely affect the recruitment of wood into streams. Riparian trees may be cleared to increase pasture or row-crop acreage, or livestock may consume or trample riparian vegetation resulting in fewer large trees available for recruitment into streams. In addition, wood is often removed from streams to reduce the retention time of floodwaters or increase boating safety.

Land use also varies widely in North America, with urban areas interspersed throughout the landscape, agriculture and grazing dominating throughout much of central and western North America, and silviculture occurring throughout forested regions. Land cover disturbances cascade through stream ecosystems and affect stream biota (Burcher et al. 2007). Urbanization can alter local physical habitat, streamflow, water quality, and energy sources in warmwater streams (Bernhardt and Palmer 2007). As mentioned earlier, urban areas historically have discharged pollutants that deteriorate water quality directly into streams and rivers. Pollutants such as leaking fluids from motor vehicles can directly enter streams as they are washed from roads and highways by stormwater runoff. Pollutants from landfills can contaminate groundwater and enter streams through diffuse groundwater pathways. Urban areas have a high percentage of impervious surface area that can alter streamflows by impeding infiltration during periods of high precipitation and by funneling runoff directly into streams (Figure 20.5). Expedited transport of water results in more frequent and larger floods that increase streambank erosion, incise stream channels, destroy habitat used by aquatic organisms, and reduce the retention of organic matter. Expedited water transport also decreases infiltration and reduces base flows that buffer warm summer temperatures. In Wisconsin streams, watersheds with 10–20% urban land use had impoverished fish assemblages as measured using the index of biotic integrity (see Chapter 12; Wang et al. 1997), and fish species richness, fish diversity, and fish density decreased as urban land cover in watersheds increased (Wang et al. 2001).

Box 20.1. The Cuyahoga—Recovery of a Burning River

Steve Tuckerman[1]

Brief history of the Cuyahoga River

Throughout the 20th century, the Cuyahoga River in Ohio was used for disposal of a wide variety of wastes. Sewage and industrial sludge covered the stream bottom, pipes delivered stinking multicolored discharges, oil slicks were common, and anglers did not venture near the river. It was not uncommon for the river to catch fire near its mouth in Cleveland, a highly-visible testament to its extremely polluted condition (see Figure). The most infamous fire on June 22, 1969 played a pivotal role in galvanizing the environmental movement of the 1970s and led to the passage of the U.S. Federal Clean Water Act of 1972 and establishment of the U.S. Environmental Protection Agency (EPA). "The Fire" was also the catalyst cited by Gaylord Nelson for the formation of the first Earth Day in 1970.

Figure A. The Cuyahoga River has come a long way from its days as a symbol of America's degraded waterways (left; photograph provided by the Ohio EPA). Massive clean-up efforts have restored water quality and allowed fish populations to recover. A recreational fishery has developed between Akron and Cleveland, Ohio (right; photograph provided by Ohio EPA).

Prior to clean-up efforts, concentrations of ammonia, heavy metals, and fecal coliform bacteria in the Cuyahoga River normally exceeded today's water quality standards. Dissolved oxygen (DO) was often absent due to high nutrient loads from inadequately treated municipal sewage. Benthic macroinvertebrates, if present, consisted primarily of pollution-tolerant taxa such as sludge worms (Tubificidae) and air-breathing snails. Fish were absent from many parts of the river, and the few fish present had visible tumors or deformities. As late as 1984, 1 h of electrofishing in the river between Akron and Cleveland resulted in the capture of only 27 individuals (3 white suckers, 1 bluegill, 1 bluntnose minnow, 1 fathead minnow, and 21 gizzard shad). Something had to be done.

[1] Ohio Environmental Protection Agency, Twinsburg.

(Box continues)

Box 20.1. Continued

Approaches to improve the water resource

River pollution was the spark that helped establish the programs needed to restore the Cuyahoga River and other rivers in the USA. Local industries and municipal leaders in Ohio acknowledged the deplorable state of the river and formed the Cuyahoga River Basin Water Quality Committee. A few years later the Clean Water Act was passed, resulting in increased regulation of discharges to the nation's waters. Initial steps in the restoration of the Cuyahoga River focused on the most obvious problems: inadequate treatment of point sources from factories and municipal wastewater treatment plants. Control of point source pollution greatly improved water quality, but other problems remained. In 1999, two dams were identified as contributing to water quality problems. Removal of these dams and their upstream impoundments restored river habitat, improved DO concentrations, and removed barriers to fish movement within the river.

Not all changes that benefited the river have been regulatory. The region has shifted toward a service economy due to the decline of the rubber, steel, automobile, and other manufacturing industries. As the number of large factories situated along the river declined, so did industrial discharges into the river. Municipal, business, citizen, and regulatory agency leaders have joined forces to improve further the quality of life in and along the river. A committee of stakeholders appointed by the Ohio Environmental Protection Agency (EPA) in 1988 started a remedial action plan (RAP) to restore the Cuyahoga River. Most remaining causes of impairment identified by the Cuyahoga River RAP involved nonpoint sources. Stewardship groups such as the Friends of the Crooked River and the Cuyahoga River RAP helped fix some of the problems associated with poor habitat and nonpoint sources of pollution. Projects that have aided in the river's recovery include creation of wetlands, streambank restoration, construction of stormwater basins, and implementation of setback ordinances that prevent development in riparian areas

Cuyahoga River response to restoration efforts

In 2000, the Cuyahoga River for the first time was in full attainment of the Ohio EPA's Aquatic Life Water Quality Standards between Akron and Cleveland. By 2008, almost 70 species of fish, including walleye, smallmouth bass, northern pike, and rainbow trout were found in the river. Extirpated species such as rainbow darters, mimic shiners, and golden redhorse have returned to the river near Cleveland, indicating a healthy fish assemblage (see Figure B on next page). Most of the lower half of the river supports good or excellent benthic invertebrate assemblages, and pollution-intolerant taxa such as hellgrammites, mayflies, and caddisflies have returned. Water quality standards are now seldom exceeded. However, high bacteria levels still occur during rainfall events because of combined sewer overflows and contaminated stormwater runoff. A recreational fishery has been reestablished (see above Figure A), although there are fish consumption warnings for portions of the river due to persistent toxicants. Problems remain where the

(Box continues)

Box 20.1. Continued

Cuyahoga River meets Lake Erie. There, the river has been modified to form a ship navigation channel. The deep, U-shaped channel has low DO concentrations and provides little habitat for fishes (see Figure below). Adult fish migrate from Lake Erie upstream to suitable spawning habitat but there is significant mortality of larval fish that drift into the navigation channel.

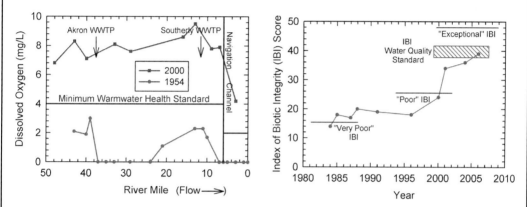

Figure B. Left panel: Dissolved oxygen concentrations in the Cuyahoga River increased to meet warmwater health standards owing to better treatment of municipal sewage from wastewater treatment plants (WWTPs). However, low DO remains a problem in the navigation channel portion of the river near Lake Erie. Right panel: Improving water quality allowed recovery of fish assemblages as evidenced by increasing scores of the index of biotic integrity (IBI), a measure of the well-being of fish assemblages. The shaded box represents the IBI score needed to meet water quality standards for Ohio.

The future

Management agencies are continuing to focus on nonpoint sources of sediments, nutrients, and toxicants. There is increased recognition that a river is more than just the stream channel. Local communities have passed riparian protection ordinances, and some are contemplating a no net increase or even a reduction of surface runoff from impervious surfaces. Additional dam removals are planned. The Cuyahoga River RAP has initiated a "Green Bulkhead" program to create shoreline habitat in lieu of the sheet steel bulkheads that line the Cuyahoga River near Lake Erie. Watershed stewardship organizations are being established in major tributary streams. All of these efforts are contributing to restoration of the once burning river.

Converting natural land cover to agricultural use can have a major impact on stream habitat conditions. Agriculture is a major source of fine sediments (Waters 1995). Sediment is easily eroded from plowed fields when vegetation, which promotes infiltration and reduces runoff, is removed. Agricultural fields can also change watershed hydrology. Oftentimes drainage ditches are constructed and drain tiles are installed in fields to remove excess water (Figure 20.5). These practices increase runoff and flood frequency and magnitude, which increase suspended sediments in streams. For instance, the Minnesota River has a watershed of 4 million hectares that is primarily row-crop agriculture and discharges 635,000 metric tons of suspended sediments per year into the Mississippi River—that is equal to 86 large dump trucks per day! Suspended sediments increase turbidity and cause problems for fishes that rely on visual cues to avoid predators, locate prey while feeding, or find mates for reproduction. Converting riparian vegetation to agricultural fields can also reduce channel shading and increase water temperatures. In Missouri, increased stream temperature due to land use changes negatively affected smallmouth bass populations and favored largemouth bass populations (Sowa and Rabeni 1995; Whitledge et al. 2006). Finally, excess nutrients from fertilization of agricultural fields and livestock waste represent diffuse, non-point-source pollutants that can increase algal production, cause hypoxia (low DO), and alter the composition of invertebrate prey assemblages. These changes, in turn, favor algivorous, omnivorous, and pollution-tolerant fishes (Dauwalter et al. 2003; Weigel and Robertson 2007). Excess nutrients alter stream ecosystem processes (e.g., primary production) and are exported downstream to large-river ecosystems. For example, nitrate levels in the Mississippi River have more than doubled since 1950 due to inputs from tributary streams draining agricultural lands (Alexander et al. 2000).

Forestry is another land use that alters streamflows, increases sediment production, and alters water temperatures. Streamflows are affected when roads, landings, and skid trails compact soils and reduce infiltration. Forest roads channel runoff directly into streams, while forestry activities expose soils, increase erosion, and increase sediment levels in streams (Miller et al. 1988; Eaglin and Hubert 1993). Forest activities also result in higher peak flows and lower base flows. These changes to stream habitats can adversely affect stream fish assemblages by favoring opportunistic species and adversely affecting sensitive species (Rutherford et al. 1992; Hlass et al. 1998).

Of the 5.3 million kilometers of rivers in the conterminous USA, only 2% remains relatively unaffected by human activities such as urbanization, agriculture, road building, or impoundment (Palmer et al. 2007). Thus, it is easy to see why there is much interest in managing streams through restoration, rehabilitation, conservation, and enhancement. Management of stream habitat in North America began in the 1920s and was traditionally focused in cold-water trout streams (White 1996). Management of warmwater stream habitat first occurred in the 1940s on Sugar Creek, Indiana (Lyons and Courtney 1990). Twenty-two improvement structures (e.g., rock bulkheads, current deflectors, and low-head dams) were used in Sugar Creek to control streambank erosion and deepen pools to improve bass fishing. Many agencies now manage stream habitat at some level (Fisher and Burroughs 2003). Historically, habitat management was focused on correcting physical habitat and water quality problems. Today, it is also concerned with the alteration of streamflows and energy sources. Likewise, early habitat management focused on small-scale problems that were a result of large-scale issues. For example, local streambank erosion was controlled by armoring the streambank even though the erosion was caused by altered streamflows due to land use change. Managers are more cognizant of how watershed changes can influence stream habitat and are often concerned with restoring the watershed processes of sediment, water, and wood transport that create natural habitat conditions (Fisher and Burroughs 2003). Working at the watershed scale is difficult, slow, and costly, but it is becoming more popular as an alternative to local habitat work that is often a short-term solution to large-scale watershed problems (Williams et al. 1997; Roni et al. 2002; Wissmar and Bisson 2003).

Energy sources in warmwater streams can also be managed. Regulation and treatment of effluents not only improves water quality but also reduces the amount of fine particulate organic matter associated with some effluents (e.g., sewage treatment plants). Protecting riparian areas maintains stream shading and prevents excessive algal production. Retention of energy is also important. Wood not only provides habitat for fishes but also traps leaves and other organic materials that are processed by biota and incorporated into stream food webs. However, managers also remove wood from rivers when it threatens infrastructure, such as when it aggregates around bridge pilings, and makes recreational boating unsafe.

Management of streamflows is often focused on maintaining or restoring natural streamflow patterns important to fishes—termed environmental flows. Fisheries managers occasionally work with dam operators to have water released in a way that improves fish habitat downstream, such as maintaining minimum streamflows. After a minimum-flow policy was established for a hydroelectric dam on the Tallapoosa River, Alabama, fish species richness more than doubled, and the fish assemblage shifted from generalist species to species that required a fluvial environment (Travnichek et al. 1995). Fisheries management agencies sometimes buy water rights to conserve fish habitat (see Chapter 4). Dam releases have been implemented to mimic natural flood events that create habitats needed by fishes. Dam removal is also a viable streamflow management option. Removal can restore natural streamflows and removes barriers to fish movement (Stanley and Doyle 2003).

Historic management of water quality was done through regulation of point source dischargers. For example, the Clean Water Act and its amendments require dischargers to obtain permits to release effluents into streams. Total maximum daily loads (TMDLs) are the maximum amount of a given pollutant that can be discharged into a stream by all sources and are set by environmental protection agencies that often employ fisheries biologists. Today, a majority of water quality problems result from diffuse non-point-source pollutants. Elevated nutrient concentrations, excessive sediments, and chemicals are water quality problems related to land use activities. Non-point pollutants often require fisheries managers to work with other land management agencies to identify source areas (e.g., feedlots and intensive agriculture) and implement sound management practices to protect or improve water quality. Sometimes fisheries biologists are involved with development of water quality standards. For instance, it is often recommended that fish surveys be conducted when determining TMDLs for streams (Yoder 1995). In fact, fish assemblages are often used to assess and monitor the quality of stream resources because fish species richness, diversity, and composition change when water quality conditions deteriorate. This reflection of stream conditions by fish assemblages has driven the development of bioassessment tools, such as the index of biotic integrity, that are used by state and provincial agencies to monitor and report water quality conditions (see Chapter 12; Kwak and Peterson 2007).

The physical habitat of streams is often the primary focus of management efforts. As previously mentioned, excessive sediment from streambank erosion is a typical cause of poor physical habitat, but there are many methods for its control (Table 20.3). Hard materials such

Table 20.3. Methods for reducing streambank erosion (after Waters 1995).

Treatment zone and method	Comments
Lower bank contacted by streamflow	
Reduce water energy by means of instream structures	Option may be incompatible with recreation, esthetics, or fishery goals
Reduce bank angle	Lower bank angle needed with higher flood peak, gradient, and flood frequency
Protect bank by means of rocks or trees	Most commonly used option, preferably executed with natural materials
Revegetate	Grass turf, shrubs, and trees with strong root systems should be used
Upper bank above streamflow maximum	
Reduce bank angle	Upper bank angle should be gentler than lower bank angle
Terrace at toe of slope	Terracing reduces runoff velocity
Revegetate	Grass turf, shrubs, and trees with strong root systems should be used
Riparian zone	
Maintain vegetation	Roots stabilize banks and soils
Install fencing	Fencing eliminates livestock trampling and grazing
Watershed	
Promote infiltration	Greater infiltration prevents unnaturally high flood peaks that can cause erosion

as rock rip-rap are used to protect streambanks from high streamflows. Structures made of boulders or wood are used to deflect streamflows away from streambanks. Protection or establishment of riparian vegetation is commonly used to stabilize streambanks and reduce erosion. Artificial structures made of wood and boulders are also used to control streambank erosion, recreate channel morphology, and improve habitat for fishes (Figure 20.7). When such improvements were made in a Mississippi stream, large-bodied fish and piscivorous fishes such as basses and sunfishes increased in a fish assemblage previously dominated by small-bodied and opportunistic fishes such as minnows (Shields et al. 2007). In the Wabash River basin in Indiana, restoring riparian vegetation was more cost-effective than was installing logs for improving warmwater stream habitat and fish assemblages (Frimpong et al. 2006).

It is evident that many problems associated with warmwater streams are closely tied to land use. Consequently, fisheries managers work with landowners and land management agencies to promote "stream friendly" land management. Many states and provinces have guidelines or laws that restrict the disturbance of riparian areas. For example, the Minnesota Forest Resources Council has guidelines for minimum riparian widths and riparian tree harvest that are meant to ensure streambank stability and recruitment of wood into streams after logging occurs. Conservation farming, such as the use of conservation tillage, terraces, grass

Figure 20.7. Cedar tree revetment (top) used to control streambank erosion and conserve fish habitat in Spring Creek, Oklahoma. J-hook rock vanes (bottom) installed in Honey Creek, Oklahoma, to stabilize streambank erosion and create fish habitat. Erosion control blankets were placed on the streambanks after each project to limit erosion until natural vegetation became reestablished. (Photos by Oklahoma Department of Wildlife Conservation).

waterways, and filter strips, can also be used to enhance or restore water quality in streams. The U.S. Farm Bill is legislation that promotes soil conservation and implements improved farming techniques on private lands, but recent reauthorizations have additional implications for aquatic conservation (Garvey 2007). The bill rewards landowners for adopting best management practices (BMPs) on their lands to reduce soil erosion. The bill also authorizes cost-share provisions and rental payments to landowners that enroll croplands into the Conservation Reserve Program or Wetland Reserve Program (Gray and Teels 2006). The Conservation Reserve Program is designed to maintain permanent cover on marginal and erodible lands. The Wetland Reserve Program is designed to restore wetlands that were drained for agriculture. Over 50% of the land in the USA is in agricultural production, and warmwater streams drain much of this land. Often marginal croplands and wetlands are on the floodplains of streams and rivers. Clearly, programs that take highly erodible and floodplain lands out of agricultural production will be beneficial by reducing fine sediment and nutrient inputs into streams. Riparian buffer zones also trap sediments and absorb nutrients derived from agricultural fields. However, even when agricultural land is reverted back to natural land cover, the effects of past land use on aquatic biota may persist for long time periods (Harding et al. 1998). Consequently, fisheries managers must be aware that several decades may be required before stream fisheries respond to management efforts focused on land use changes throughout watersheds.

20.3.3 Fish Issues and Management

In addition to management of stream habitat, management of warmwater streams has focused on altering fish populations. Historical management of fish populations was done through fish stocking with varied success. More recently, however, there has been interest in protecting native species, including nongame taxa. Native fish conservation may include supplementing populations of imperiled species with individuals from hatcheries, but it often requires management of entire communities. Because warmwater fish assemblages often contain many species, sometimes up to 70 species, new approaches are needed to manage such diverse systems for multiple uses. For example, management goals often include identifying streams or watersheds with high species diversity so that they can be prioritized for conservation.

Some fish species found in warmwater streams are threatened or endangered because they naturally have a restricted geographic range. However, habitat degradation, introduced species, hybridization, and overharvest are human-caused reasons for fish species being listed as endangered, vulnerable, or threatened. For example, the Pecos gambusia is listed as endangered by the U.S. Fish and Wildlife Service (USFWS) because the species' distribution has declined considerably due to loss of habitat and introduced fish species; natural populations are now restricted to a few springs and sinkholes in the Pecos River basin in New Mexico and Texas (USFWS 1982). In the southern USA, there are 662 fish species and 28% are classified as vulnerable, threatened, endangered, or extinct (Warren et al. 2000), and 39% of all species in North America are imperiled to some degree (Jelks et al. 2008). Many of these species spend all or part of their life in warmwater streams. In Mexico, 169 of the 506 known fish species are at some level of risk, and 25 are now extinct (Contreras-Balderas et al. 2003). Many of these species reside in streams in arid regions in northern Mexico and are impacted by habitat degradation, water development, and introduced species.

Introduced species are a major concern for managers of warmwater streams. Historic introductions were often done to expand sportfishing opportunities. Black bass, sunfishes, crappies, and catfishes were commonly introduced to create sportfishing opportunities. In many places, the introduction of predaceous largemouth bass has decimated populations of native fishes (Jackson 2002). Red shiners, fathead minnows, and white suckers were widely introduced as prey for sport fishes or as a result of bait bucket introductions, and they can compete or hybridize with native fishes. Unauthorized introductions result from knowingly illegal introductions such as bait bucket releases, unintended colonization from other aquatic systems due to water diversions or removal of migration barriers, or inadvertent introductions by contaminated fish stockings (Rahel 2004b). Species introductions, often for fisheries management purposes, have been listed as an important cause of the extinction of 61 North American fishes (Jelks et al. 2008). In Canada, 68% of at-risk species are threatened by introduced species (Dextrase and Mandrak 2006). The effect of introduced species on native fishes and stream ecosystem function varies geographically and is often more pronounced in areas with fewer native fish species, such as the western USA and the Atlantic Coast. Authorized introductions by fisheries management agencies has declined over the last century, likely because early introductions satisfied public demand for specific fisheries and because managers are more aware of the negative effects of such introductions (Rahel 2004b).

The introduction of nonnative species can result in many problems (see Chapter 8). Introductions homogenize fish assemblages, decrease biodiversity, reduce the abundance of desired sport fishes and native species, and may have negative economic impacts (Rahel 2002). Warmwater streams are more widespread geographically and have higher diversities of fishes than do coldwater streams in North America. As a result, warmwater streams are more likely to have introduced species because there are more potential species available for introduction and there are more habitats in which they can be introduced. Human alteration of warmwater streams also facilitates establishment of introduced species that fill an ecological void left by a native species that was extirpated.

Although management of warmwater streams is typically focused on stream habitat, there are also ways to manage fishes. Single-species management in warmwater streams can be focused on sport fishes or nongame species. Enhancement and restoration of sport fish populations can be done by stocking. Although not a widely-used management option, stocking has been used with varied success to supplement populations adversely affected by reduced habitat quality, reestablish extirpated populations, or establish new populations to increase angling opportunities (see Chapter 9). For example, walleye fry were stocked into some Iowa rivers to improve angling opportunities in populations with poor natural reproduction (Paragamian and Kingery 1992). In Wisconsin, introductions of muskellunge increased the length of streams and rivers managed for this highly-valued sport fish from 1,145 km in 1970 to 2,708 km in 1996 (Simonson and Hewett 1999). Recovery plans for fishes listed as threatened or endangered by the USFWS under the Endangered Species Act (see Chapter 4) often call for introducing new populations or supplementing existing populations with individuals from hatcheries (Williams et al. 1988). However, stocking is not always cost-effective, and plans to stock fishes should always consider the impacts to existing fish populations and communities and the stream's capacity to support stocked fish.

Oftentimes conservation of stream fishes focuses on control of harmful nonnative species. Flathead catfish have been introduced into many streams and rivers on the Atlantic Coast to promote recreational fisheries, but they can have a detrimental impact on native fishes through predation (Pine et al. 2005, 2007). Management options that promote exploitation of flathead catfish, such as bounties and subsidies to commercial fishers and allowing unlimited harvest by anglers, may keep populations in check to allow persistence of native fish populations (Pine et al. 2007).

Management options are also available to conserve the genetic integrity of native stream fishes. The Oklahoma Department of Wildlife Conservation ceased stocking nonnative strains of smallmouth bass in reservoirs to conserve the genetic integrity of native smallmouth bass in eastern Oklahoma steams (Stark and Echelle 1998). In Texas, a dam was built to restrict contact between Clear Creek gambusia, an endemic fish limited to the headwaters of a single tributary stream, and the widespread western mosquitofish in attempt to limit hybridization and genetic introgression between the two species (Davis et al. 2006).

Because warmwater streams can be some of the most speciose aquatic systems in North America, there has been increasing interest in multiple-species management approaches. One approach places species with similar morphology, reproduction or feeding strategies, or habitat use into groups, called guilds, that are expected to respond to environmental change or management in a similar manner (Austen et al. 1994). For example, fish species that use coarse rocky substrates for spawning are often placed into the lithophilic spawning guild. Collectively, the abundance of this guild may show a stronger negative response to increased sedimentation than the response shown by any individual species in the guild. Another multispecies approach is the recognition of fish assemblage types. Often different physiographic regions or ecoregions have different fish assemblages, as do drainage basins that have different evolutionary histories (Dauwalter et al. 2008). Angermeier and Winston (1999) found distinct fish assemblage types among physiographic regions and major drainage basins in Virginia. They recommended that representative assemblages from each physiographic–drainage combination be identified as a conservation goal. This type of approach has resulted in a shift in management focus from individual streams to watersheds and regions, especially when highly-mobile species are considered (Wishart and Davies 2003). After areas with unique assemblage types or highly-diverse aquatic assemblages have been identified, they can be managed as freshwater protected areas. Freshwater protected areas are portions of freshwater environments that are protected from disturbance to allow natural processes to govern ecosystems, communities, and populations (Suski and Cooke 2007). Because streams are tightly linked with their watersheds, protected streams must include substantial portions of the watershed to be effective. Geographic information systems (GISs) are a powerful tool for identifying where species of conservation concern occur or where species diversity is high (Sowa et al. 2007; Dauwalter and Rahel 2008). Spatial information on species occurrences can be incorporated with spatial information on human impacts and protected lands to identify areas or watersheds that should be given priority for protection (Box 20.2) (Wall et al. 2004).

20.3.4 People Issues and Management

Fisheries managers must balance many competing uses of warmwater streams. Recreational fishing and bait harvest are major fisheries activities on warmwater streams in North America although commercial fishing can be important in large rivers (see Chapter 21). In addition to fishing, warmwater streams are used for other recreational activities such as sightseeing, canoeing, hunting, swimming, camping, and picnicking (Hess and Ober 1981). Warmwater streams are also sources of water for irrigation and livestock watering and sites for the discharge of industrial and municipal effluents. Fisheries management agencies often collaborate with other governmental agencies to regulate the many ways that humans can affect warmwater stream ecosystems.

Important sport fishes in warmwater streams include walleye, black bass, and catfishes. Because these species are highly sought after by anglers, they are subject to overexploitation, and thus fishing regulations are usually needed to maintain a quality fishery. Historically, seasonal closures were used to protect walleye and black bass during the spring spawning season when these species are aggregated in shallow waters and vulnerable to angling. Seasonal closures have become less common for black bass, especially in the southern USA (Paukert

Box 20.2. Utility of Geographic Information Systems to Fisheries Management

Scott P. Sowa[1]

Many issues facing stream resource managers are spatially oriented. In fact, it is hard to identify instances in which some form of spatial analysis would not improve the fisheries management process. It was not very long ago that spatial analyses were a monumental or impossible task because spatially-explicit (i.e., map-based) information on much or all of the ecological, political, economic, and sociocultural factors pertinent to fisheries management was lacking or not easily integrated.

Fortunately, in recent decades fisheries managers have embraced the use of a geographic information system (GIS) for addressing spatial issues. A GIS is a collection of computer hardware, software, data, and personnel designed to collect, store, update, manipulate, analyze, and display georeferenced information (i.e., information referenced to a particular place on the earth; Rahel 2004a). A GIS can be used to generate spatially-explicit inventories, devise sampling designs for monitoring or research, identify and prioritize locations in need of conservation, or conduct complex spatial analyses dealing with issues of habitat juxtaposition, connectivity, patch size, or habitat fragmentation. Fisher and Rahel (2004b) discuss in detail the use of GIS in fisheries management.

A common question facing resource management agencies is, Where should we focus our management efforts in order to…? The complexity of the spatial analyses required to answer the question depends upon what follows the word "to" and the amount of GIS data available. For this example, the question of interest is, Where should we focus our management efforts in order to conserve fish species that are listed as rare, threatened, or endangered in the state of Nebraska?

Nebraska has 22 fish species listed as rare, threatened, or endangered and nearly 130,000 km of stream. To maximize efficiency, conservation efforts should focus on streams that harbor a high number of these species. This can be accomplished by developing GIS-based predicted distribution maps for each listed species and identifying areas of distributional overlap (see Figure A; Sowa et al. 2006; Sowa et al. 2007). The simple but powerful maps depicted below can be used by management agencies to direct necessary resources to conserve habitat in specific regions of the state. In this example three regions of the state stand out as having a high number of listed species, (1) the Missouri River main stem along the eastern border of the state, (2) the lower main stem of the Platte River in east-central Nebraska, and (3) the headwater streams draining the northern slope of the Nebraska Sandhills in the northwest.

---

[1] Missouri Resource Assessment Partnership, School of Natural Resources, University of Missouri, Columbia. Current contact information: The Nature Conservancy, Lansing, Michigan.

(Box continues)

Box 20.2. Continued.

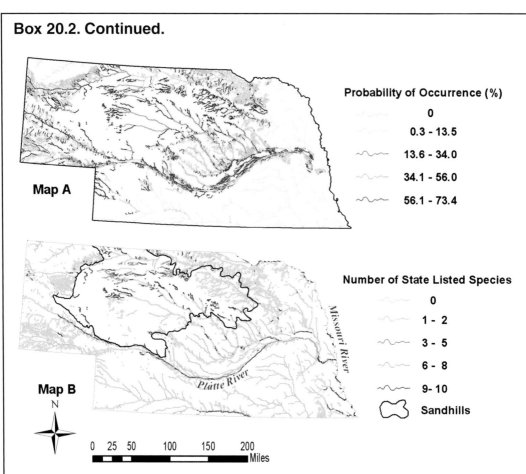

Figure A. Map A: Probability of occurrence (%) for the plains topminnow throughout Nebraska. Map B: Species richness map for the 22 state-listed rare, threatened, and endangered fish species in Nebraska.

Knowing where concentrations of state-listed species occur in the state is powerful information for decision makers. However, even more helpful would be to know the management options and management issues for each region of interest. A GIS can be used to address these tasks as well. For instance, map A in Figure B shows a land cover map of Nebraska overlaid with a map of public lands.

Collectively maps A and B in Figure B illustrate that there are relatively few human disturbances affecting headwater streams in the Nebraska Sandhills and therefore proactive protection measures will likely be a key to conserving stream habitats in this region. However, the Missouri River main stem and lower Platte River are large rivers that are influenced by an extensive and diverse suite of disturbances throughout their watersheds in Nebraska and other states, which suggests that more intensive restoration efforts will likely be needed. Regardless, these maps can help resource managers identify and prioritize disturbances and management issues facing each region. Each spatial data layer provides a critical piece of information to improve the decision-making process.

(Box continues)

Box 20.2. Continued.

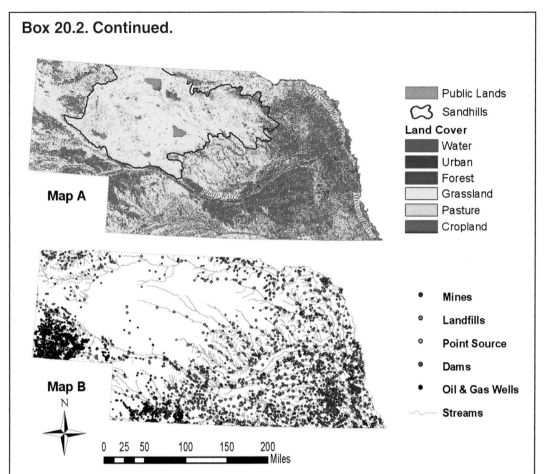

Figure B. Map A: Land cover and public lands of Nebraska. Map B: Location and spatial distribution of human disturbances potentially affecting stream habitat conditions throughout Nebraska.

The old cliché that "a picture is worth a thousand words" is certainly true and is fully appreciated by those who use GIS as a tool for fisheries management. The simple example provided above only scratches the surface of what can be accomplished as more GIS data become available and spatial modeling techniques continue to improve. Hopefully, in the not too distant future all fisheries managers will understand what a powerful tool GIS can be for addressing many resource management issues.

et al. 2007). In part, this reflects a belief that recruitment of black bass is determined more by environmental conditions than by the number of spawning fish (Kubacki et al. 2002). Also, seasonal closures are difficult to enforce because anglers can usually pursue other species in areas with spawning black bass, and illegal harvest in such areas can be substantial (Kubacki et al. 2002). Creel limits have long been used to regulate black bass fisheries, and the general trend has been toward reduced daily limits.

Because of the premium placed on large fish by anglers, length limits have become a popular way to regulate the length structure of fish populations (see Chapter 7). Interestingly, length limits have had varied success depending on the angling constituency. In Elkhorn Creek, Kentucky, where a slot limit of 305–405 mm was implemented for smallmouth bass, 50% of anglers said they would not keep fish under 305 mm even if it would benefit the fishery (Buynak and Mitchell 2002). Hence, the slot limit would not have the desired effect of thinning out an overabundance of small bass. In contrast, catfish anglers in Texas appear to be more harvest oriented than are black bass anglers, making length limits a useful management tool for catfish populations (Wilde and Ditton 1999). A recent survey showed that catfish anglers across the Midwest and Great Plains supported more restrictive regulations for catfishes (Arterburn et al. 2002).

Historically, most angling regulations on warmwater streams were implemented at the state or provincial level. This largely reflected a lack of information about regional differences in fish production and a belief that regulations should be easily understood by the angling public. However, our understanding of how climate and habitat factors influence the response of fish populations to exploitation has increased greatly (Beamesderfer and North 1995; Paukert et al. 2007). As a result, angling regulations are increasingly being tailored to local conditions. For example, in Arkansas the minimum length limit for smallmouth bass is higher for streams in the Ozarks, reflecting the higher productivity of these streams and thus the faster growth of smallmouth bass. In Mississippi, the growth of channel catfish is higher in southern streams flowing through fertile, agricultural landscapes than in northern streams flowing through less fertile, forested landscapes (Shepard and Jackson 2006). This suggests that basin-specific regulations would help to maximize angler satisfaction in the channel catfish fishery.

Baitfish collecting is another important activity in warmwater streams. Species harvested for bait are primarily minnows and suckers, but a variety of other taxa are also collected including sculpins, topminnows, and crayfishes. Collection of fish and crayfish for bait poses two main concerns for fisheries managers. The first concern involves overexploitation of wild populations. Most states and provinces allow anglers to harvest a small number of baitfish for personal use (typically 50–100 individual baitfish per day), but they require a commercial license if fish are sold to the public. The second concern is the potential for fish to be transported and released into waterways other than where they were collected, leading to the spread of nonnative species and transfer of diseases (e.g., viral hemorrhagic septicemia in the Great Lakes region). A high proportion of anglers release their unused bait at the end of the day, despite the fact that such releases are increasingly illegal (Litvak and Mandrak 1999). Also, collection of wild baitfish typically results in nontarget species being captured. In Maine, 10 fish species not legal for use as bait were found mixed with 23 legal baitfish species during bait shop inspections (Kircheis 1998). As a result of bait bucket releases, over 100 fish species are thought to have been introduced outside their native distributions in North America, and such releases now constitute a major mechanism by which illegal fish introductions continue to occur (Litvak and Mandrak 1999; Rahel 2004b). Although species introductions due to bait bucket releases or other means are likely to continue into the future, managers can slow the rate of introductions by educating the public (Figure 20.8), implementing legislation that prohibits transfer of fishes among water bodies, or attempting to control introduced species through eradication (Rahel 2004b).

In addition to angling, warmwater streams are used for a variety of other purposes. There are generally few conflicts among anglers and other recreational users such as canoeists, float tubers, or wildlife watchers. In fact, most recreational users have the same concerns for warmwater streams as do fisheries managers—poor water quality, erosion, sedimentation, and litter—because the factors that adversely affect fisheries also affect esthetics and recreational

Figure 20.8. Educating the public is one way fisheries managers can slow the introduction of nonnative fishes and conserve native fishes.

experiences (Pardee et al. 1981). Conflicts do arise when streams are used for livestock watering and irrigation withdrawal. Livestock grazing not only impacts stream habitats used by fishes, but it also reduces the esthetic value of streams and can pose a health risk to swimmers (Rinne 1999). Fisheries managers often work with land management agencies and local landowners to fence riparian areas and remove livestock impacts to streams. Managers often work with government agencies charged with determining the environmental flows necessary to maintain fish habitat and conserve fish populations. Managers are also consulted by environmental protection agencies that develop and set biotic criteria for streams, determine TMDLs, and issue permits to industrial and municipal entities that discharge effluents into streams. Thus, fisheries managers often play a crucial role in balancing the public's water needs versus conservation of fisheries resources. Oftentimes stream rehabilitation and restoration decision making includes private citizens, public interest groups, public officials, and economic interests to ensure that all community interests are considered and the project is not undermined by a constituency that is not represented in the management process (Figure 20.4). Such multi-agency coordination and public participation in management activities have become common themes of fisheries management in the 21st century (Fisher and Burroughs 2003).

20.4 CONCLUSIONS

Historically, fisheries managers focused their efforts on maintaining or enhancing sportfishing opportunities. Recently, however, there has been a shift to a more holistic approach to stream management. A survey of state agency programs in the USA in 2000 showed that maintenance and improvement of ecosystem integrity was a management goal for 35% of states working in warmwater streams, whereas increasing angling quality and opportunities was a goal for only 27% of states (Fisher and Burroughs 2003). This represents a major management shift from directly restoring instream habitat toward restoration of the watershed processes that create natural habitat conditions (White 1996; Williams et al. 1997). It also demonstrates a shift from management at small spatial scales to consideration of spatial scales from individual habitat units to entire watersheds (Quist et al. 2006). Managers must now not only consider how stocking fishes influences angling opportunities but also how it might affect native fish assemblages both where stocking occurs and throughout the watershed.

Future managers of warmwater streams will encounter new issues that will exacerbate or interact with old problems. Urbanization, agricultural production, and extraction of natural resources such as oil and gas will intensify to meet the construction and energy demands of the growing human population. Water consumption will also increase, and water withdrawals will further reduce fish habitat. And there will be increased pressure to build dams and create reservoirs to store surface waters for human use. Continued nonnative fish introductions and accelerated climate change are other factors associated with the frontiers of fisheries management. Climate warming allows the spread of tropical nonnative species such as cichlids that are currently limited by cold winter temperatures (Rahel and Olden 2008). Native species at the edge of their upper thermal tolerances will need to adapt or become locally extinct as streams become warmer, more saline, and more intermittent (Box 20.3). The future managers of warmwater streams will have to tackle the problems that managers have dealt with in the past, but they will also be confronted with these new problems associated with continued human population growth and climate change.

Fortunately, there are many groups already working to maintain and restore the health of streams and their fisheries. Advocacy groups, such as American Rivers, promote and work to maintain healthy rivers that are vital to human health, safety, and quality of life. The Izaak Walton League's Save Our Streams program and Iowa's IOWATER program promote stream and watershed education and organize citizen groups to monitor stream water quality. The U.S. Environmental Protection Agency provides information and education for watershed planning and restoration through its Watershed Academy program, and multiple agencies are engaged in the National Wild and Scenic Rivers program that affords protection to river resources. The U.S. Department of Agriculture administers the Conservation Reserve Program and Wetland Reserve Program that also benefit stream health. All of these organizations help to protect and restore stream fisheries and are often partners on stream management projects that require interdisciplinary expertise and have a large constituency base. The widespread interest in protecting and restoring warmwater streams, when coupled with some of the most speciose aquatic ecosystems in North America, promise to make management of warmwater streams one of the most exciting jobs for a fisheries biologist!

Box 20.3. Effects of Climate Warming on Warmwater Streams

Concentrations of atmospheric carbon dioxide (CO_2) have increased since the industrial revolution and have lead to an increase in global air temperatures. Doubling of CO_2 concentrations is expected to increase global temperatures by 3°C to 4°C. Concurrent with the increase in air temperatures is an increase in stream temperatures and increased stream temperatures have clear implications for fishes. Mohseni et al. (2003) estimated that the amount of suitable stream habitat for coolwater fishes like smallmouth bass, walleye, and northern pike is expected to decrease by 15% in the USA. Habitat for warmwater fishes is estimated to increase by 31%, but in some cases streams might become too warm for some warmwater species. Although increased stream temperatures have a direct effect on temperature-sensitive fishes, there are other changes to warmwater streams that are expected to accompany climate warming. The expected changes are

- increased habitat for warmwater fishes as coldwater streams become warmwater streams;
- shifted distributions of thermally-suitable habitat for fishes with narrow temperature tolerances;
- increased primary production, organic decomposition, and nutrient cycling due to higher temperatures and longer growing seasons;
- altered streamflow patterns due to altered precipitation regimes;
- decreased water quality and suitable habitat in summer due to lower base flows, reduced DO concentrations, and higher salinities;
- altered invertebrate assemblages due to loss of species that have life histories dependent on specific environment cues (e.g., temperature and streamflow);
- shifts in predator–prey balances due to changes in growth, feeding behavior, and timing of reproduction; and
- expansion of habitat for nonnative fishes, invertebrates, and diseases.

Fisheries managers will have to adapt to changing stream conditions as a result of climate warming (Ficke et al. 2007). Endemic fishes with restricted geographic distributions are prone to extinction, leading to a loss of biodiversity, as stream habitats become unsuitable. Streams that historically supported high-quality fisheries may fail to do so as temperatures warm and water quality deteriorates, whereas other streams may increase in their ability to support an abundance of large fish. Changes in stream habitat may also cause changes in species composition. For example, increasing temperatures may change a smallmouth bass fishery to a largemouth bass fishery because largemouth bass are better adapted to warmer temperatures. Managers will be expected to maintain fisheries in the face of habitat changes due to climate warming. The changes to a fishery can have a cascading effect on local economies supported by those fisheries. Consequently, communication among managers, anglers, and the general public will be imperative as warmwater stream fisheries change in response to a changing climate.

20.5 REFERENCES

Adler, R. W., J. C. Landman, and D. M. Cameron. 1993. The Clean Water Act 20 years later. Island Press, Washington, D.C.

Alexander, H. E. 1959. Stream values, recreational use, and preservation in the Southeast. Proceedings of the Annual Conference of Southeastern Association of Game and Fish Commissioners 13(1959):338–343.

Alexander, R. B., R. A. Smith, and G. E. Schwarz. 2000. Effect of stream channel size on the delivery of nitrogen to the Gulf of Mexico. Nature (London) 403:758–761.

Angermeier, P. L., and M. R. Winston. 1999. Characterizing fish community diversity across Virginia landscapes: prerequisite for conservation. Ecological Applications 9:335–349.

Armantrout, N. B. 1998. Glossary of aquatic habitat inventory terminology. American Fisheries Society, Bethesda, Maryland.

Arterburn, J. E., D. J. Kirby, and C. R. Berry Jr. 2002. A survey of angler attitudes and biologist opinions regarding trophy catfish and their management. Fisheries 27(5):10–21.

Austen, D. J., P. B. Bayley, and B. W. Menzel. 1994. Importance of the guild concept to fisheries research and management. Fisheries 19(6):12–19.

Baxter, C. V., K. D. Fausch, and W. C. Saunders. 2005. Tangled webs: reciprocal flows of invertebrate prey link streams and riparian zones. Freshwater Biology 50:201–220.

Beamesderfer, R. C. P., and J. A. North. 1995. Growth, natural mortality, and predicted response to fishing for largemouth bass and smallmouth bass populations in North America. North American Journal of Fisheries Management 15:688–704.

Belica, L. A. T., and F. J. Rahel. 2008. Movements of creek chubs, *Semotilus atromaculatus*, among habitat patches in a plains stream. Ecology of Freshwater Fish 17:258–272.

Benda, L., N. L. Poff, D. Miller, T. Dunne, G. Reeves, G. Pess, and M. Pollock. 2004. The network dynamics hypothesis: how channel networks structure riverine habitat. BioScience 54:413–427.

Benke, A. C. 1990. A perspective on America's vanishing streams. Journal of the North American Benthological Society 9:77–88.

Benke, A. C., and C. E. Cushing. 2005. Rivers of North America. Elsevier Academic Press, Burlington, Massachusetts.

Bernhardt, E. S., and M. A. Palmer. 2007. Restoring streams in an urbanizing world. Freshwater Biology 52:738–751.

Brunke, M., and T. Gonser. 1997. The ecological significance of exchange processes between rivers and groundwater. Freshwater Biology 37:1–33.

Bryan, C. F., and D. A. Rutherford. 1993. Impacts on warmwater streams: guidelines for evaluation. American Fisheries Society, Southern Division, Little Rock, Arkansas.

Burcher, C. L., H. M. Valett, and E. F. Benfield. 2007. The land-cover cascade: relationships coupling land and water. Ecology 88:228–242.

Burr, B. M., and R. L. Mayden. 1992. Phylogenetics and North American freshwater fishes. Pages 18–75 *in* R. L. Mayden, editor. Systematics, historical ecology, and North American freshwater fishes. Stanford University Press, Stanford, California.

Buynak, G. L., and B. Mitchell. 2002. Response of smallmouth bass to regulatory and environmental changes in Elkhorn Creek, Kentucky. North American Journal of Fisheries Management 22:500–508.

Contreras-Balderas, S., P. Almada-Villela, M. d. L. Lozano-Vilano, and M. E. García-Ramírez. 2003. Freshwater fish at risk or extinct in México. Reviews in Fish Biology and Fisheries 12:241–251.

Cummins, K. W. 1974. Structure and function of stream ecosystems. BioScience 24:631–641.

Cummins, K. W., and M. A. Wilzbach. 2005. The inadequacy of the fish-bearing criterion for stream management. Aquatic Sciences 67:486–491.

Dauwalter, D. C., and W. L. Fisher. 2008. Spatial and temporal patterns in stream habitat and smallmouth bass populations in eastern Oklahoma. Transactions of the American Fisheries Society 137:1072–1088.

Dauwalter, D. C., E. J. Pert, and W. E. Keith. 2003. An index of biotic integrity for fish assemblages in Ozark Highland streams of Arkansas. Southeastern Naturalist 2:447–468.

Dauwalter, D. C., and F. J. Rahel. 2008. Distribution modeling to guide stream fish conservation: an example using the mountain sucker in the Black Hills National Forest, USA. Aquatic Conservation: Marine and Freshwater Ecosystems 18:1263–1276.

Dauwalter, D. C., D. K. Splinter, W. L. Fisher, and R. A. Marston. 2007. Geomorphology and stream habitat relationships with smallmouth bass abundance at multiple spatial scales in eastern Oklahoma. Canadian Journal of Fisheries and Aquatic Sciences 64:1116–1129.

Dauwalter, D. C., D. K. Splinter, W. L. Fisher, and R. A. Marston. 2008. Biogeography, ecoregions, and geomorphology affect fish species composition in streams of eastern Oklahoma, USA. Environmental Biology of Fishes 82:237–249.

Davis, S. K., A. A. Echelle, and R. A. Van Den Bussche. 2006. Lack of cytonuclear genetic introgression despite long-term hybridization and backcrossing between two poeciliid fishes (*Gambusia heterochir* and *G. affinis*). Copeia 2006:351–359.

Dextrase, A. J., and N. E. Mandrak. 2006. Impacts of alien invasive species on freshwater fauna at risk in Canada. Biological Invasions 8:13–24.

Eaglin, G. S., and W. A. Hubert. 1993. Effects of logging and roads on substrate and trout in streams of the Medicine Bow National Forest, Wyoming. North American Journal of Fisheries Management 13:844–846.

Fausch, K. D., C. E. Torgersen, C. V. Baxter, and H. W. Li. 2002. Landscapes to riverscapes: bridging the gap between research and conservation of stream fishes. BioScience 52:483–498.

Ficke, A. D., C. A. Myrick, and L. J. Hansen. 2007. Potential impacts of global climate change on freshwater fisheries. Reviews in Fish Biology and Fisheries 17:581–613.

Fisher, S. G., L. J. Gray, N. B. Grimm, and D. E. Busch. 1982. Temporal succession in a desert stream ecosystem following flash flooding. Ecological Monographs 52:93–110.

Fisher, W. L., and J. P. Burroughs. 2003. Stream fisheries management in the United States: a survey of state agency programs. Fisheries 28(2):10–18.

Fisher, W. L., and W. D. Pearson. 1987. Patterns of resource utilization among four species of darters in three central Kentucky streams. Pages 69–76 *in* W. J. Matthews and D. C. Heins, editors. Community and evolutionary ecology of North American stream fishes. University of Oklahoma Press, Norman.

Fisher, W. L., and F. J. Rahel. 2004a. Geographic information systems applications in stream and river fisheries. Pages 49–84 *in* W. L. Fisher and F. J. Rahel, editors. Geographic information systems in fisheries. American Fisheries Society, Bethesda, Maryland.

Fisher, W. L., and F. J. Rahel, editors. 2004b. Geographic information systems in fisheries. American Fisheries Society, Bethesda, Maryland.

Fisher, W. L., D. F. Schreiner, C. D. Martin, Y. A. Negash, and E. Kessler. 2002. Recreational fishing and socioeconomic characteristics of eastern Oklahoma stream anglers. Proceedings of the Oklahoma Academy of Science 82:79–87.

Fisher, W. L., A. F. Surmont, and C. D. Martin. 1998. Warmwater stream and river fisheries in the southeastern United States: are we managing them in proportion to their values? Fisheries 23(12):16–24.

Fore, J. D., D. C. Dauwalter, and W. L. Fisher. 2007. Microhabitat use by smallmouth bass in an Ozark stream. Journal of Freshwater Ecology 22:189–199.

Freeman, M. C., C. M. Pringle, and C. R. Jackson. 2007. Hydrologic connectivity and the contribution of stream headwaters to ecological integrity at regional scales. Journal of the American Water Resources Association 43:5–14.

Frimpong, E. A., J. G. Lee, and T. M. Sutton. 2006. Cost effectiveness of vegetative filter strips and instream half-logs for ecological restoration. Journal of the American Water Resources Association 42:1349–1361.

Frissell, C. A., W. J. Liss, C. E. Warren, and M. D. Hurley. 1986. A hierarchical framework for stream habitat classification: viewing streams in a watershed context. Environmental Management 10:199–214.

Funk, J. L. 1970. Warm-water streams. Pages 141–152 *in* N. G. Benson, editor. A century of fisheries in North America. American Fisheries Society, Special Publication 7, Bethesda, Maryland.

Garvey, J. E. 2007. Farm bill 2007: placing fisheries upstream of conservation provisions. Fisheries 32(8):399–404.

Gorman, O. T., and J. R. Karr. 1978. Habitat structure and stream fish communities. Ecology 59:507–515.

Gray, E. V., J. M. Boltz, K. A. Kellogg, and J. R. Stauffer. 1997. Food resource partitioning by nine sympatric darter species. Transactions of the American Fisheries Society 126:822–840.

Gray, L. J. 1997. Organic matter dynamics in Kings Creek, Konza Prairie, Kansas, USA. Journal of the North American Benthological Society 16:50–54.

Gray, R. L., and B. M. Teels. 2006. Wildlife and fish conservation through the Farm Bill. Wildlife Society Bulletin 34:906–913.

Greenberg, L. A. 1988. Interactive segregation between the stream fishes *Etheostoma simoterum* and *E. rufilineatum*. Oikos 51:193–202.

Harding, J. S., E. F. Benfield, P. V. Bolstad, G. S. Helfman, and E. B. D. Jones III. 1998. Stream biodiversity: the ghost of land use past. Proceedings of the National Academy of Sciences of the USA 95:14843–14847.

Hargrave, C. W., and J. E. Johnson. 2003. Status of Arkansas darter, *Etheostoma cragini*, and least darter, *E. microperca*, in Arkansas. Southwestern Naturalist 48:89–92.

Hart, D. D., T. E. Johnson, K. L. Bushaw-Newton, R. J. Horwitz, A. T. Bednarek, D. F. Charles, D. A. Kreeger, and D. J. Velinsky. 2002. Dam removal: challenges and opportunities for ecological research and river restoration. BioScience 52:669–681.

Hess, T. B., and R. D. Ober. 1981. Recreational use surveys on two Georgia rivers. Pages 14–20 *in* L. A. Krumholz, editor. Warmwater streams symposium. American Fisheries Society, Southern Division, Bethesda, Maryland.

Hildrew, A. G. 1996. Food webs and species interactions. Pages 123–144 *in* G. Petts and P. Calow, editors. River biota: diversity and dynamics. Blackwell Scientific Publications, Oxford, UK.

Hlass, L. J., W. L. Fisher, and D. J. Turton. 1998. Use of the index of biotic integrity to assess water quality in forested streams in the Ouachita Mountains ecoregion, Arkansas. Journal of Freshwater Ecology 13:181–192.

Horwitz, R. J. 1978. Temporal variability patterns and the distributional patterns of stream fishes. Ecological Monographs 48:307–321.

Hynes, H. B. N. 1970. The ecology of running waters. University of Toronto Press, Toronto.

Jackson, D. A. 2002. Ecological effects of *Micropterus* introductions: the dark side of black bass. Pages 221–232 *in* D. P. Philipp and M. S. Ridgway, editors. Black bass: ecology, conservation, and management. American Fisheries Society, Symposium 31, Bethesda, Maryland.

Jackson, D. C., and J. R. Jackson. 1989. A glimmer of hope for stream fisheries in Mississippi. Fisheries 14(3):4–9.

Jaeger, M. E., A. V. Zale, T. E. McMahon, and B. J. Schmitz. 2005. Seasonal movements, habitat use, aggregation, exploitation, and entrainment of saugers in the lower Yellowstone River: an empirical assessment of factors affecting population recovery. North American Journal of Fisheries Management 25:1550–1568.

Jelks, H. L., S. J. Walsh, S. Contreras-Balderas, E. Díaz-Pardo, N. M. Burkhead, D. A. Hendrickson, J.

Lyons, N. E. Mandrak, F. McCormick, J. S. Nelson, S. P. Platania, B. A. Porter, C. B. Renaud, J. J. Schmitter-Soto, E. B. Taylor, and M. L. Warren. 2008. Conservation status of imperiled North American freshwater and diadromous fishes. Fisheries 33(8):372–407.

Jobling, S., and C. R. Tyler. 2003. Topic 4.3 Endocrine disruption in wild freshwater fish. Pure and Applied Chemistry 75:2219–2234.

Junk, W. J., P. B. Bayley, and R. E. Sparks. 1989. The flood pulse concept in river–floodplain systems. Pages 110–127 in D. P. Dodge, editor. Proceedings of the international large river symposium. Canadian Special Publication in Fisheries and Aquatic Sciences 106.

Karr, J. R., L. A. Toth, and D. R. Dudley. 1985. Fish communities of midwestern rivers: a history of degradation. BioScience 35:90–95.

Kircheis, F. W. 1998. Species composition and economic value of Maine's winter baitfish industry. North American Journal of Fisheries Management 18:175–180.

Knighton, D. 1998. Fluvial forms and processes: a new perspective. Arnold, London.

Kocik, J. F., and C. P. Ferreri. 1998. Juvenile production variation in salmonids: population dynamics, habitat, and the role of spatial relationships. Canadian Journal of Fisheries and Aquatic Sciences 55:191–200.

Kubacki, M. F., F. J. S. Phelan, J. E. Claussen, and D. B. Philipp. 2002. How well does a closed season protect spawning bass in Ontario? Pages 379–386 in D. P. Philipp and M. S. Ridgway, editors. Black bass: ecology, conservation, and management. American Fisheries Society, Symposium 31, Bethesda, Maryland.

Kwak, T. J., and J. T. Peterson. 2007. Community indices, parameters, and comparisons. Pages 677–763 in C. S. Guy and M. L. Brown, editors. Analysis and interpretation of freshwater fisheries data. American Fisheries Society, Bethesda, Maryland.

Lau, J. K., T. E. Lauer, and M. L. Weinman. 2006. Impacts of channelization on stream habitats and associated fish assemblages in east central Indiana. American Midland Naturalist 156:319–330.

Le Pichon, C., G. Gorges, P. Boët, J. Baudry, F. Goreaud, and T. Faure. 2006. A spatially explicit resource-based approach for managing stream fishes in riverscapes. Environmental Management 37:322–335.

Litvak, M. K., and N. E. Mandrak. 1999. Baitfish trade as a vector of aquatic introductions. Pages 163–180 in R. Claudi and J. H. Leach, editors. Nonindigenous freshwater organisms: vectors, biology, and impacts. Lewis Publishers, Boca Raton, Florida.

Lyons, J., and C. C. Courtney. 1990. A review of fisheries habitat improvement projects in warmwater streams, with recommendations for Wisconsin. Wisconsin Department of Natural Resources, Technical Bulletin 169, Madison.

Mallin, M. A., V. L. Johnson, S. H. Ensign, and T. A. MacPherson. 2006. Factors contributing to hypoxia in rivers, lakes, and streams. Limnology and Oceanography 51:690–701.

Matthews, W. J. 1998. Patterns in freshwater fish ecology. Chapman and Hall, New York.

Matthews, W. J., and E. G. Zimmerman. 1990. Potential effects of global warming on native fishes of the southern Great Plains and the Southwest. Fisheries 15(6):26–32.

McDonald, D. B., T. L. Parchman, M. R. Bower, W. A. Hubert, and F. J. Rahel. 2008. An introduced and a native vertebrate hybridize to form a genetic bridge to a second native species. Proceedings of the National Academy of Sciences of the USA 105:10842–10847.

Meyer, J. L., and R. T. Edwards. 1990. Ecosystem metabolism and turnover of organic carbon along a blackwater river continuum. Ecology 71:668–677.

Miller, E. L., R. S. Beasley, and E. R. Lawson. 1988. Forest harvest and site preparation effects on erosion and sedimentation in the Ouachita Mountains. Journal of Environmental Quality 17:219–225.

Mohseni, O., H. G. Stefan, and J. G. Eaton. 2003. Global warming and potential changes in fish habitat in U.S. streams. Climate Change 59:389–409.

Montgomery, D. R. 1999. Process domains and the river continuum. Journal of the American Water Resources Association 36:397–410.

Montgomery, D. R., and S. M. Bolton. 2003. Hydrogeomorphic variability and river restoration. Pages 39–80 in R. C. Wissmar and P. A. Bisson, editors. Strategies for restoring river ecosystems: sources of variability and uncertainty in natural and managed systems. American Fisheries Society, Bethesda, Maryland.

Newbold, J. D., J. W. Elwood, R. V. O'Neill, and W. Van Winkle. 1981. Nutrient spiraling in streams: the concept and its field measurement. Canadian Journal of Fisheries and Aquatic Sciences 38:860–863.

Palmer, M., J. D. Allan, J. Meyer, and E. Bernhardt. 2007. River restoration in the twenty-first century: data and experiential knowledge to inform future efforts. Restoration Ecology 15:472–481.

Paragamian, V. L., and R. Kingery. 1992. A comparison of walleye fry and fingerling stockings in three rivers in Iowa. North American Journal of Fisheries Management 12:313–320.

Pardee, L., D. D. Tarbet, and J. D. Gregory. 1981. Management practices related to recreational use and the responses of recreationists to those practices along the French Broad River in North Carolina. Pages 21–30 in L. A. Krumholz, editor. Warmwater streams symposium. American Fisheries Society, Southern Division, Bethesda, Maryland.

Paukert, C. P., M. C. McInerny, and R. D. Schultz. 2007. Historical trends in creel limits, length-based limits, and season restrictions for black basses in the United States and Canada. Fisheries 32(2):62–72.

Pine, W. E., T. J. Kwak, and J. A. Rice. 2007. Modeling management scenarios and the effects of an introduced apex predator on a coastal riverine fish community. Transactions of the American Fisheries Society 136:105–120.

Pine, W. E., T. J. Kwak, D. S. Waters, and J. A. Rice. 2005. Diet selectivity of introduced flathead catfish in coastal rivers. Transactions of the American Fisheries Society 134:901–909.

Poff, N. L. 1997. Landscape filters and species traits: towards mechanistic understanding and prediction in stream ecology. Journal of the North American Benthological Society 16:391–409.

Poff, N. L., and J. D. Allan. 1995. Functional organization of stream fish assemblages in relation to hydrological variability. Ecology 76:606–627.

Poff, N. L., J. D. Allan, M. B. Bain, J. R. Karr, K. L. Prestegaard, B. D. Richter, R. E. Sparks, and J. C. Stromberg. 1997. The natural flow regime. BioScience 47:769–784.

Poff, N. L., and J. V. Ward. 1989. Implications of streamflow variability and predictability for lotic community structure: a regional analysis of streamflow patterns. Canadian Journal of Fisheries and Aquatic Sciences 46:1805–1818.

Power, M. E., W. J. Matthews, and A. J. Stewart. 1985. Grazing minnows, piscivorous bass, and stream algae: dynamics of a strong interaction. Ecology 66:1448–1456.

Quinn, J. W., and T. J. Kwak. 2003. Fish assemblage changes in an Ozark river after impoundment: a long-term perspective. Transactions of the American Fisheries Society 132:110–119.

Quist, M. C., W. A. Hubert, M. Fowden, S. W. Wolff, and M. R. Bower. 2006. The Wyoming habitat assessment methodology (WHAM): a systematic approach to evaluating watershed conditions and stream habitat. Fisheries 31(2):75–81.

Quist, M. C., F. J. Rahel, and W. A. Hubert. 2005. Hierarchical faunal filters: an approach to assessing effects of habitat and nonnative species on native fishes. Ecology of Freshwater Fish 14:24–39.

Rabeni, C. F., and R. B. Jacobson. 1999. Warmwater streams. Pages 505–528 in C. C. Kohler and W. A. Hubert, editors. Inland fisheries management in North America, 2nd edition. American Fisheries Society, Bethesda, Maryland.

Rahel, F. J. 2002. Homogenization of freshwater faunas. Annual Review of Ecology and Systematics 33:291–315.

Rahel, F. J. 2004a. Introduction to geographic information systems in fisheries. Pages 1–12 in W. L.

Fisher and F. J. Rahel, editors. Geographic information systems in fisheries. American Fisheries Society, Bethesda, Maryland.

Rahel, F. J. 2004b. Unauthorized fish introductions: fisheries management of the people, for the people, or by the people. Pages 431–443 in M. J. Nickum, P. M. Mazik, J. G. Nickum, and D. D. MacKinlay, editors. Propagated fish in resource management. American Fisheries Society, Symposium 44, Bethesda, Maryland.

Rahel, F. J., and J. D. Olden. 2008. Effects of climate change on aquatic invasive species. Conservation Biology 22:521–533.

Remshardt, W. J., and W. L. Fisher. 2009. Effects of variation in streamflow and channel structure on smallmouth bass habitat in an alluvial stream. River Research and Applications 25:661–674.

Resh, V. H., A. V. Brown, A. P. Covich, M. E. Gurtz, H. W. Li, G. W. Minshall, S. R. Reice, A. L. Sheldon, J. B. Wallace, and R. C. Wissmar. 1988. The role of disturbance in stream ecology. Journal of the North American Benthological Society 7:433–455.

Ricciardi, A., R. J. Neves, and J. B. Rasmussen. 1998. Impending extinctions of North American freshwater mussels (Unionoida) following the zebra mussel (*Dressena polymorpha*) invasion. Journal of Animal Ecology 67:613–619.

Rinne, J. N. 1999. Fish and grazing relationships: the facts and some pleas. Fisheries 24(8):12–21.

Roberts, J. J., and F. J. Rahel. 2008. Irrigation canals as sink habitat for trout and other fishes in a Wyoming drainage. Transactions of the American Fisheries Society 137:951–961.

Roni, P., T. J. Beechie, R. E. Bilby, F. E. Leonetti, M. M. Pollock, and G. R. Pess. 2002. A review of stream restoration techniques and a hierarchical strategy for prioritizing restoration in Pacific Northwest watersheds. North American Journal of Fisheries Management 22:1–20.

Rutherford, D. A., A. A. Echelle, and O. E. Maughan. 1992. Drainage-wide effects of timber harvesting on the structure of stream fish assemblages in southeastern Oklahoma. Transactions of the American Fisheries Society 121:716–728.

Schlosser, I. J. 1982. Fish community structure and function along two habitat gradients in a headwater stream. Ecological Monographs 52:395–414.

Schlosser, I. J., and P. L. Angermeier. 1995. Spatial variation in demographic processes of lotic fishes: conceptual models, empirical evidence, and implications for conservation. Pages 392–401 in J. L. Nielsen, editor. Evolution and the aquatic ecosystem: defining unique units in population conservation. American Fisheries Society, Symposium 17, Bethesda, Maryland.

Shepard, S., and D. C. Jackson. 2006. Difference in channel catfish growth among Mississippi stream basins. Transactions of the American Fisheries Society 135:1224–1229.

Shields, F. D., S. S. Knight, and C. M. Cooper. 2007. Can warmwater streams be rehabilitated using watershed standard erosion control measures alone? Environmental Management 40:62–79.

Simonson, T. D., and S. W. Hewett. 1999. Trends in Wisconsin's muskellunge fishery. North American Journal of Fisheries Management 19:291–299.

Smale, M. A., and C. F. Rabeni. 1995a. Hypoxia and hyperthermia tolerances of headwater stream fishes. Transactions of the American Fisheries Society 124:698–710.

Smale, M. A., and C. F. Rabeni. 1995b. Influences of hypoxia and hyperthermia on fish species composition in headwater streams. Transactions of the American Fisheries Society 124:711–725.

Snieszko, S. F. 1974. The effects of environmental stress on outbreaks of infectious diseases of fishes. Journal of Fish Biology 6:197–208.

Southwood, T. R. E. 1977. Habitat, the template for ecological strategies? Journal of Animal Ecology 46:336–365.

Sowa, S. P., G. Annis, M. E. Morey, and D. D. Diamond. 2007. A GAP analysis and comprehensive conservation strategy for riverine ecosystems of Missouri. Ecological Monographs 77:301–334.

Sowa, S. P., G. Annis, M. E. Morey, and A. Garringer. 2006. Developing predicted distribution models for fish species in Nebraska. Final Report submitted to the USGS (U.S. Geological Survey) National Gap Analysis Program, Moscow, Idaho.

Sowa, S. P., and C. F. Rabeni. 1995. Regional evaluation of the relation of habitat to distribution and abundance of smallmouth bass and largemouth bass in Missouri streams. Transactions of the American Fisheries Society 124:240–251.

Stanley, E. H., and M. W. Doyle. 2003. Trading off: the ecological effects of dam removal. Frontiers in Ecology and the Environment 1:15–22.

Stark, W. J., and A. A. Echelle. 1998. Genetic structure and systematics of smallmouth bass, with emphasis on Interior Highlands populations. Transactions of the American Fisheries Society 127:393–416.

Suski, C. D., and S. J. Cooke. 2007. Conservation of aquatic resources through the use of freshwater protected areas: opportunities and challenges. Biodiversity and Conservation 16:2015–2029.

Taylor, C. A., G. A. Schuster, J. E. Cooper, R. J. DiStefano, A. G. Eversole, P. Hamr, H. H. I. Hobbs, H. W. Robison, C. E. Skelton, and R. F. Thoma. 2007. A reassessment of the conservation status of crayfishes of the United States and Canada after 10+ years of increased awareness. Fisheries 32(8):372–389.

Thorp, J. H., and M. D. DeLong. 1994. The riverine productivity model: an heuristic view of carbon sources and organic processing in large river ecosystems. Oikos 70:305–308.

Thorp, J. H., M. C. Thoms, and M. D. DeLong. 2006. The riverine ecosystem synthesis: biocomplexity in river networks across space and time. River Research and Applications 22:123–147.

Travnichek, V. H., M. B. Bain, and M. J. Maceina. 1995. Recovery of a warmwater fish assemblage after the initiation of a minimum-flow release downstream from a hydroelectric dam. Transactions of the American Fisheries Society 124:836–844.

USFWS (U.S. Fish and Wildlife Service). 1982. Pecos gambusia (*Gambusia nobilis*) recovery plan. U.S. Fish and Wildlife Service, Albuquerque, New Mexico.

Vannote, R. L., G. W. Minshall, K. W. Cummins, and C. E. Cushing. 1980. The river continuum concept. Canadian Journal of Fisheries and Aquatic Sciences 37:130–137.

Wall, S. S., C. R. Berry Jr., C. M. Blausey, J. A. Jenks, and C. J. Kopplin. 2004. Fish-habitat modeling for gap analysis to conserve the endangered Topeka shiner (*Notropis topeka*). Canadian Journal of Fisheries and Aquatic Sciences 61:954–973.

Wang, L., J. Lyons, and R. Gatti. 1997. Influences of watershed land use on habitat quality and biotic integrity in Wisconsin streams. Fisheries 22(6):6–12.

Wang, L., J. Lyons, P. Kanehl, and R. Bannerman. 2001. Impacts of urbanization on stream habitat and fish across multiple spatial scales. Environmental Management 28:255–266.

Ward, J. V. 1989. The four-dimensional nature of lotic ecosystems. Journal of the North American Benthological Society 8:2–8.

Ward, J. V., and J. A. Stanford. 1983. The serial discontinuity concept of lotic ecosystems. Pages 29–42 *in* T. D. I. Fontaine and S. M. Bartell, editors. Dynamics of lotic ecosystems. Ann Arbor Science Publishers, Ann Arbor, Michigan.

Warren, M. L., Jr., B. M. Burr, S. J. Walsh, H. L. Bart Jr., R. C. Cashner, D. A. Etnier, B. J. Freeman, B. R. Kuhajda, R. L. Mayden, H. W. Robison, S. T. Ross, and W. C. Starnes. 2000. Diversity, distribution, and conservation status of the native freshwater fishes of the southern United States. Fisheries 25(10):7–31.

Waters, T. F. 1995. Sediment in streams: sources, biological effects, and control. American Fisheries Society, Monograph 7, Bethesda, Maryland.

Weigel, B. M., and D. M. Robertson. 2007. Identifying biotic integrity and water chemistry relations in nonwadeable rivers of Wisconsin: toward the development of nutrient criteria. Environmental Management 40:691–708.

White, R. J. 1996. Growth and development of North American stream habitat management for fish. Canadian Journal of Fisheries and Aquatic Sciences 53 (Supplement 1):342–363.

Whitledge, G. W., C. F. Rabeni, G. Annis, and S. P. Sowa. 2006. Riparian shading and groundwater enhance growth potential for smallmouth bass in Ozark streams. Ecological Applications 16:1461–1473.

Wilde, G. R., and R. B. Ditton. 1999. Differences in attitudes and fishing motives among Texas catfish anglers. Pages 395–405 *in* E. R. Irwin, W. A. Hubert, C. F. Rabeni, H. L. Schramm Jr., and T. Coon, editors. Catfish 2000: proceedings of the international ictalurid symposium. American Fisheries Society, Symposium 24, Bethesda, Maryland.

Wiley, M. J., L. L. Osborne, and R. W. Larimore. 1990. Longitudinal structure of an agricultural prairie river system and its relationship to current stream ecosystem theory. Canadian Journal of Fisheries and Aquatic Sciences 47:373–384.

Williams, J. D., M. L. Warren, K. S. Cummings, J. L. Harris, and R. J. Neves. 1993. Conservation status of fresh water mussels of the United States and Canada. Fisheries 18(9):6–22.

Williams, J. E., D. W. Sada, C. D. Williams, J. R. Bennett, J. E. Johnson, P. C. Marsh, D. E. McAllister, E. P. Pister, R. D. Radant, J. N. Rinne, M. D. Stone, L. Ulmer, and D. L. Withers. 1988. American Fisheries Society guidelines for introductions of threatened and endangered fishes. Fisheries 13(5):5–11.

Williams, J. E., C. A. Wood, and M. P. Dombeck. 1997. Watershed restoration: principles and practices. American Fisheries Society, Bethesda, Maryland.

Winger, P. V. 1981. Physical and chemical characteristics of warmwater streams: a review. Pages 32–44 *in* L. A. Krumholz, editor. The warmwater streams symposium. Southern Division, American Fisheries Society, Southern Division, Bethesda, Maryland.

Wishart, M. J., and B. R. Davies. 2003. Beyond catchment considerations in the conservation of lotic diversity. Aquatic Conservation: Marine and Freshwater Ecosystems 13:429–437.

Wissmar, R. C., and P. A. Bisson. 2003. Strategies for restoring river ecosystems: sources of variability and uncertainty in natural and managed systems. American Fisheries Society, Bethesda, Maryland.

Yoder, C. O. 1995. Policy issues and management applications for biological criteria. Pages 327–344 *in* J. R. Davis and T. P. Simon, editors. Biological assessment and criteria: tools for water resource planning and decision making. CRC Press, Boca Raton, Florida.

Chapter 21

Warmwater Rivers

CRAIG P. PAUKERT AND DAVID L. GALAT

21.1 INTRODUCTION

Warmwater rivers are diverse ecosystems with substantial spatial variability, both longitudinal and lateral. Management of warmwater rivers is challenging and requires creative and diverse solutions to management problems. In addition to the biological and physical complexity of warmwater rivers, the human component is also complex. Warmwater rivers often traverse many political boundaries (e.g., states, provinces, and countries) and have multiple stakeholder interests that affect management actions. Warmwater rivers provide transportation corridors and serve as centers of human settlement. While a fisheries manager may be concerned with how river modification may affect fish abundance, growth, or movements, other stakeholders may be interested in commercial development in riparian areas, hydroelectric power generation, flood control, or commerce. Therefore, management of warmwater rivers involves input from multiple users. In fact, fisheries are a small portion of the management focus and economic benefits of warmwater rivers.

Warmwater rivers have been altered throughout North America and the world; only about 23% of the discharge of the world's large rivers remaining unaltered (Dynesius and Nilsson 1994). In North America, unaltered rivers are typically in northern regions, such as the Yukon River (Benke and Cushing 2005). Most warmwater rivers occur in areas to the south, where human populations have altered riparian zones, vegetative cover across watersheds, hydrologic regimes, and floodplains.

Fisheries management in warmwater rivers has lagged behind management of lakes, ponds, and reservoirs for several reasons. Warmwater rivers are one of the more demanding aquatic habitats to sample because of their size, diversity of habitats, and variation in flow. Because most warmwater rivers have been altered, there is little preregulation information on reference conditions (Emery et al. 2003), which is critical for determining the "natural" state of rivers to aid management. Similarly, warmwater rivers are typically unique within a region, so unaltered spatial references for management or restoration are also lacking.

Warmwater rivers are generally managed by a diversity of resource agencies for multiple uses, making fisheries management complex. Diverse stakeholder groups have competing interests for services that warmwater rivers provide. For example, users may need specific flows for agriculture, hydropower, or boating that do not necessarily reflect the best interests of fisheries or fish habitat. Often, these competing interests have varied political powers that also play into decisions affecting river management. Although there may be certain flows, temperatures, or habitat requirements for fishes, these needs may not override those for municipal water supplies, flood protection for cities, or dredging of sand and gravel for construction materials

21.2 CHARACTERISTICS OF LARGE WARMWATER RIVERS

This chapter focuses on large warmwater rivers, but there is no universally-recognized definition of a large river. Large rivers have been distinguished from small rivers and streams as being "those large enough to intimidate research workers" (Hynes 1989). What technically distinguishes rivers from streams, and sizes of rivers, is often based on drainage basin size, Strahler order (Strahler 1957), discharge, and (or) length (Benke and Cushing 2005). Benke (1990) considered rivers greater than 1,000 km long to be large rivers. Large rivers have been variously defined as those with drainage areas exceeding 1,600 km^2 (Ohio EPA 1989), 2,590–5,180 km^2 (Simon and Lyons 1995), or greater than 20,000 km^2 (Reash 1999). Rivers with wetted widths greater than 50 m were classified as large by Simonson et al. (1994) and those with a mean depth greater than 1 m were considered large by Stalnaker et al. (1989). Rivers with Strahler stream orders of more than 6 have also been defined as large rivers (Sheehan and Rasmussen 1999). Dynesius and Nilsson (1994) provide another definition pertinent to the northern one-third of the world based on the discharge before any significant direct human manipulation, or the virgin mean annual discharge (VMAD); rivers having a VMAD greater than or equal to 350 m^3/s are considered large rivers. In this chapter a large river is defined similar to Benke and Cushing (2005) to include rivers with a mean annual discharge of 100 m^3/s or greater; a basin area of 217 km^2 or greater; and a Strahler order of at least 3, but almost always 5 or greater (Appendix 21.1). We do not distinguish between large and "great" rivers (e.g., Simon and Emery 1995); we include both in this chapter. Appendix 21.1 summarizes the diversity of features that contribute to management challenges of warmwater rivers.

Temperature affects nearly all biological rates of fishes, particularly individual growth and reproduction, and therefore temperature is a primary factor structuring riverine fish assemblages (Wolter 2007). What makes a "warmwater" river can be best represented by the composition of its fish fauna. Magnuson et al. (1979) defined the thermal niches for families of freshwater temperate fishes as temperature centers based on laboratory preferences and temperatures occupied by fishes in rivers during summer. Using this approach they grouped temperate freshwater fishes in the USA and Canada into coldwater, coolwater, and warmwater thermal guilds. Fisheries managers have adopted these general ranges to classify warmwater streams and rivers as those where the temperature becomes too warm to support a self-sustaining salmonid population. More specifically, the families Salmonidae and Cotttidae dominate coldwater fishes, Percidae and Esocidae compose coolwater fishes, and Centrarchidae, Ictaluridae, and Moronidae are widely recognized as warmwater fishes. However, classification of rivers solely by summer thermal preferences of families of fishes is not entirely reasonable for all families as Cyprinidae and Catostomidae have species in all three thermal guilds.

Warmwater rivers are defined in this chapter as those rivers dominated by non-salmonid fishes and with annual mean water temperatures greater than 11°C (Magnuson et al. 1979). However, it can be misleading to classify an entire river as being warmwater because upper segments may support either coldwater or coolwater fishes. This chapter focuses on fisheries management in segments of large rivers dominated by warmwater fishes and excludes tailwater fisheries dominated by salmonids (see Chapter 19). Warmwater segments are predominately lowland, floodplain segments in temperate and subtropical regions of North America. Using these criteria, 108 of the 218 rivers covered in Benke and Cushing (2005) provide a representative sample of warmwater rivers in North America (Appendix 21.1).

21.3 RIVER ECOLOGY CONCEPTS

Effective management of riverine fishes is based on management of river–floodplain systems and an understanding of river ecology is an essential precursor to managing riverine fishes. Below are ecological concepts that are commonly used in management of fisheries in rivers.

21.3.1 Longitudinal Zonation of Fishes

Rivers have been viewed as linear systems and managed accordingly. Early ideas of river management classified rivers into relatively distinct zones based on fish species composition (e.g., Sheldon 1968). For example, upstream, montane reaches in western U.S. streams and small rivers were dominated by coldwater fishes and abruptly changed to domination by warmwater fishes as rivers left the mountains and became lower gradient (Rahel and Hubert 1991). Management of rivers based on zonation of fish species has its critics because the concept is based on rivers that are unchanged by human activities, which are rare. Zonation also ignores variation in fish species composition associated with the vertical and lateral dimensions of rivers (Ward 1989) and variation in fish species composition among regions and watersheds with differing hydrologic and geomorphic features (e.g., Poff and Ward 1989). The fish zonation concept is not completely applicable to large rivers because alluvial reaches exhibit multiple biophysical gradients and nodes of environmental heterogeneity (Stanford et al. 1996). However, the realization that rivers are connected systems with longitudinal gradients led to other concepts in river ecology.

21.3.2 River Continuum Concept

A more integrated perspective of longitudinal changes is the river continuum concept (RCC; Vannote et al. 1980). The RCC was developed for temperate river systems in North America and postulates that riverine biota adapt to structural and functional patterns of the abiotic environment along a continuous gradient from a river's headwaters to its mouth. The RCC emphasizes gradients. River networks are viewed as longitudinally-linked systems wherein biotic assemblages progressively change (Figure 21.1a).

The direct influence of riparian vegetation on physical (e.g., water temperature) and biological (e.g., dominance of shredding invertebrates) processes decreases with movement down the river continuum to warmwater segments. The RCC predicts that the effect of riparian vegetation is low in large rivers due to the greater channel width. In large rivers, energy sources that become fish food are dominated by fine particulate matter derived from upstream processing of allochthonous material (e.g., leaves) and phytoplankton production until a river channel becomes so large (e.g., Strahler order ≥ 8) that light attenuation due

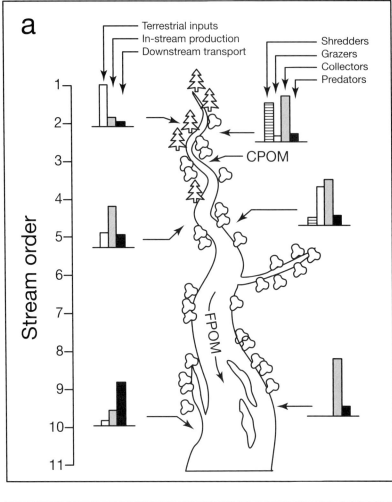
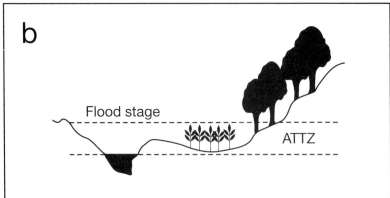

Figure 21.1. Conceptual diagram of the river continuum concept (**a**) and the flood pulse concept (**b**). In diagram (a), CPOM is coarse particulate organic matter and FPOM is fine particulate organic matter. In diagram (b) the dashed lines represent water levels at low flow (bottom dashed line) and flood pulse (top dashed line), with the aquatic–terrestrial-transition zone represented by ATTZ. (Figure modified from Johnson et al. 1995)

to increasing river depth and turbidity reduces phytoplankton primary production; then the system becomes heterotrophic (production:respiration is less than 1).

The RCC postulates that total biotic diversity increases downstream to a point where it then decreases further downstream. This RCC concept has been supported for fishes in several European rivers (e.g., Aarts and Nienhuis 2003) and may be related to cumulative anthropogenic impacts being more significant downriver.

21.3.3 Flood Pulse Concept

Challenges to the universal relevance of the RCC arose when applied to both large coldwater and warmwater rivers (Sedell et al. 1989). Recognition of the importance of floodplains in lowland, warmwater rivers resulted in a complementary concept, the flood pulse concept (FPC). The FPC describes interactions between the biota and environment of a river channel and its adjacent floodplain and emphasizes the role of periodic flooding of unmodified river–floodplain systems (Junk et al. 1989; Figure 21.1b). The FPC recognized a periodic interaction between the flood pulse and floodplain that is critical to recruitment of riverine fishes (Welcomme 1985). The FPC related life cycles of riverine biota, particularly fishes, to an annual flood and recognized its role in food supply, spawning, and refugia (Lorenz et al. 1997). Strong year-classes of fishes resulting from increased river water levels and a high-amplitude flood of long duration in lowland temperate and tropical rivers became known as the "flood pulse advantage" (Bayley 1991). The FPC acknowledges the importance of a lateral component (i.e., floodplain) and the role of periodic flooding to fish production.

Studies have evaluated the FPC in large rivers and have related it to growth of fishes, suggesting that increased growth of fishes occurs following floods (Jones and Noltie 2007). Gutreuter et al. (1999) hypothesized that growth of fishes that use floodplains would differ among years with varying flow regimes, whereas fishes that solely use main-channel habitats would not show a response in the upper Mississippi River. Their study found that littoral zone fishes (e.g., largemouth bass and bluegill) had increased growth in years when floods inundated the floodplain and that growth of main-channel fishes (e.g., white bass) did not differ among years, which is consistent with the FPC. In the lower Mississippi River, research has indicated that the FPC may apply, but only if inundation of the floodplain occurs when water temperature is suitable for feeding by littoral zone fishes (Schramm and Eggleton 2006).

21.3.4 Riverine Ecosystem Synthesis Model

The riverine ecosystem synthesis (RES) model (Thorp et al. 2008) provides a hierarchal framework to explain discontinuous and broad longitudinal and lateral patterns at multiple spatial scales. The RES model builds on two areas of study critical to river management: ecology and geomorphology. A novel difference in the RES model compared with earlier concepts (i.e., RCC) is that a river network is not viewed as a continuous longitudinal gradient but as a mosaic of patches, termed functional process zones (FPZs). Functional process zones are based on hydrologic and geomorphic features and are distinguished by climate and vegetation. These FPZs may have different nutrient inputs, runoff, sediment loads, and floodplain vegetation. The unique habitats characteristic of each FPZ contributes to the potential fish species pool. The distribution and diversity of fish species from the headwaters to the river's mouth reflects the matrix of FPZs rather than a clinal position along the river continuum.

The critical concept in the RES model is that geomorphic patches form the basis of interactions of biota with surrounding areas. Ecosystem function (productivity and movement of energy) varies among distinct FPZs. The RES model suggests that FPZs may serve as a useful tool to assess river modification or management actions. Therefore, assessing river condition should be conducted at a large spatial scale based on geomorphology and hydrology (Thorp et al. 2008). Because FPZs have different river structures and functions they also have distinct fish assemblages and likely respond differently to management actions. In addition, at a larger spatial scale the number or diversity of FPZs (sensu species richness and species diversity) may also be a useful metric to identify river reaches that have had less human disturbance and that have maintained a more natural flow and geomorphic regime (Thoms et al. 2008; Thorp et al. 2008)

21.3.5 Serial Discontinuity Concept

The previously described concepts (i.e., RCC and FPC) largely were developed for undisturbed systems that have continuous longitudinal gradients of connectivity between the main channel and floodplains. However, many rivers are modified and disconnected longitudinally by dams and diversions and laterally by levees and dikes. The serial discontinuity concept (SDC; Ward and Stanford 1983) addressed the effects of dams on rivers. Dams disrupt the natural processes of rivers, but the severity of the discontinuities depends on the position of the dam in the watershed and the type of dam release (e.g., hypolimnetic releases or releases from throughout the water column). The SDC views the river system as a longitudinal gradient with alteration of the gradient by dams. Ward and Stanford (1995) built on this framework to include floodplain habitat in different types of river segments. According to the SDC, where artificial longitudinal discontinuities (e.g., dams) predominate, headwater segments have stable, confined channels. Lower segments with braided and meandering channel reaches have higher channel instability and thermal heterogeneity, which affect fish assemblages. In these lower segments, regulation by dams will reset the system by stabilizing channels downstream of reservoirs, but thermal variability will remain relatively high. Connectivity of floodplains and backwaters, which are used by many fishes, will be more strongly affected by reservoirs in the lower meandering segments because these segments originally had the most diverse habitats.

21.3.6 Natural Flow Regime

The importance of flow as the master variable sustaining riverine ecological integrity (the ability to maintain a balanced community of organisms with the species composition, diversity, and functional organization comparable with natural habitat of that region; Karr 1981) is encompassed by the river's "natural flow regime" (Poff et al. 1997). The magnitude, frequency, timing, duration, and rate of rise and fall of both high- and low-flow pulses are critical components of the flow regime that regulate ecological structure and function in river–floodplain systems (Richter et al. 1997; Bunn and Arthington 2002). These flow dynamics are strongly linked to ecological integrity (Poff et al. 1997; Figure 21.2; also see Chapter 12 for a discussion of ecological integrity) and river modifications, such as dams and channelization, caused by humans. A river with a relatively natural flow regime often has high native fish recruitment because these fishes have evolved in that natural flow regime. Natural flow regimes

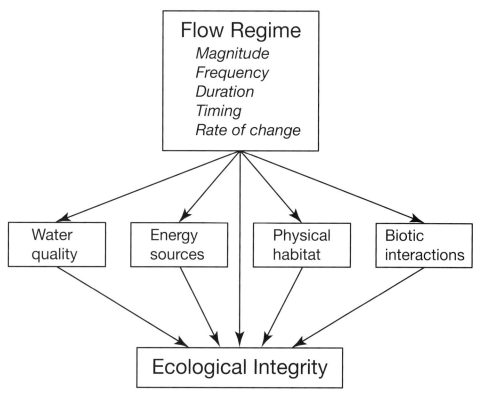

Figure 21.2. Conceptual diagram of how water flows affects ecological integrity through a cascading process of the five components of flow regime (from Poff et al. 1997, from concepts in Karr 1991).

can be relatively predictable and trigger fish to migrate, spawn, and access floodplain habitat (Welcomme 1992; Junk et al. 1989; Paukert and Fisher 2001). In addition, natural flows allow for more suitable habitat for native fishes and other aquatic biota, including freshwater mussels (Di Maio and Corkum 1995). High flows transport fine sediments downstream and clear interstitial spaces in channel substrates that are needed for spawning by many fishes. Natural high flows also transport large woody debris, a critical habitat component for riverine fishes (Angermeier and Karr 1984; Benke et al. 1985; Lehtinen et al. 1997).

Successful recruitment of riverine fishes can depend on high flows, but this is species dependent. Harris and Gehrke (1994) proposed a flood recruitment model to describe how some species respond to appropriately timed rises in flow and flooding. The model implies that flow initiates spawning and places emphasis on inundation of the floodplain to provide spawning and rearing habitat (Welcomme 1985; Humphries et al. 1999; Winemiller 2005). However, other species may show successful recruitment during low flows (Humphries et al. 1999), including species in North American warmwater rivers such as the Brazos River, Texas (Zeug and Winemiller 2008). Humphries et al. (1999) proposed the low-flow recruitment hypothesis to explain how some native riverine fishes can spawn during periods of reduced flow, suggesting some species may be advantaged by low flows for spawning because prey are concentrated. However, some riverine fishes are less responsive to flow and depend largely on photoperiod and thermal cues to initiate spawning (Winemiller 2005). For example, several species of cyprinids in rivers of the Great Plains have protracted spawning periods with no relationship between discharge and reproductive success (Durham and Wilde 2006). Persistence of flow in the channel throughout the reproductive season appears to be important to reproductive success for many cyprinid species in arid regions.

21.3.7 A Unifying Theme

Underlying the array of river ecology concepts is the theory that physical habitat and spatiotemporal heterogeneity greatly influence the richness, distribution, and abundance of lotic biota, including fishes (Vannote et al. 1980; Ward 1989; Thorp et al. 2008). It is theorized that the "habitat template" (Southwood 1977) largely structures riverine fish assemblages, and greater spatial heterogeneity in physical habitat results in more microhabitats and hydraulic diversity that yields great biotic diversity (Poff and Ward 1990). Riverine habitats, largely created by water flows and developed within a spatial hierarchy, is a valuable perspective from which to understand fish assemblage structure, factors that affect riverine fishes, and, consequently, management options.

The RCC and FPC hypotheses help managers to understand life history adaptations of riverine fishes, identify functional groups, and identify mechanisms operating over multiple spatiotemporal scales (Bayley and Li 1992). There have been numerous reviews, evaluations, and critiques of river ecology concepts (Johnson et al. 1995; Lorenz et al. 1997; Thorp et al. 2006, 2008) that collectively provide a unifying foundation to aid managers in addressing human impacts and management alternatives to conserve and restore warmwater rivers and their fishes. Although no single concept is unilaterally accepted, the ecology and management of river fisheries can be explained by an integration of these concepts as well as other concepts in fluvial geomorphology and natural flow regimes (Poff et al. 1997; Dettmers et al. 2001; Thorp et al. 2006). The most unifying feature of these concepts is that rivers are, longitudinally, laterally, and temporally dynamic and are influenced by hydrologic and geomorphic processes and also by dams and other structures built by humans.

21.4 MAJOR ISSUES

Management of fishes in warmwater rivers is dictated by abiotic and biotic processes that affect river systems. Abiotic processes provide a foundation for managing riverine fishes that is refined by biotic factors. The primary abiotic factors affecting warmwater riverine fisheries are fragmentation, flow regulation, habitat alterations, water quality, and climate change. Primary biotic factors affecting riverine fishes are overfishing and invasive species.

21.4.1 Abiotic Factors

Many North American warmwater rivers are highly fragmented and regulated. For example, of 15 warmwater rivers discussed in Dynesius and Nilsson (1994), 8 (53%) were considered strongly affected. In contrast, only one river was not affected by fragmentation and flow regulations.

21.4.1.1 Fragmentation

There are many sources of fragmentation, but dams and water diversion structures are the most obvious (Figure 21.3). These structures can block longitudinal movements of migratory fishes, isolate populations that were previously connected, reduce spawning habitat, decrease sediment transport, alter water temperatures, and change substrate composition. In addition, reservoirs can be sources of nonnative fishes that may consume, hybridize, or compete with native fishes. Longitudinal fragmentation of rivers has occurred throughout North America and has affected native riverine fish and freshwater mussel assemblages. Razorback suckers, Colorado pikeminnow, and humpback chub in the Colorado River have declined owing, in part, to changes in thermal regime and hydrology of rivers after dam construction (Minckley et al. 2003). Paddlefish in the Mississippi and Missouri rivers have declined because dams have inundated spawning habitat and blocked spawning migrations (Jennings and Zigler 2009). American shad in the Susquehanna River have declined due to blockage of spawning migrations (St. Pierre 2003). Shoal bass in the Apalachicola–Chattahoochee–Flint river drainage have declined in part from destruction of spawning and rearing habitat caused by dams (Williams and Burgess 1999).

Entrainment of fish by diversions and dams is also a concern for managers of altered river systems. Diversion of water through irrigation canals, municipal water supplies, or power-generating turbines may allow fish to be flushed through these structures and killed. In the Yellowstone River, Montana, up to 78% of the nonfishing mortality of saugers was attributed to entrainment though a water diversion canal (Jaeger et al. 2005), indicating that mortality of fishes can be substantial. However, management actions, such as screens over release structures, are often in place to minimize fish entrainment (Moyle and Israel 2005). Strobe lights and sound or bubble-producing structures in reservoirs have been used to minimize fish entrainment through dams (e.g., Patrick et al. 1985; Popper and Carlson 1998; Hamel et al. 2008). Electrical barriers have also been used to minimize introduction of undesirable fishes from canals (Clarkson 2004).

Lateral fragmentation caused by the decoupling of the floodplain and main channel has also altered habitat and affected recruitment of fishes in warmwater rivers. Lateral fragmentation occurs when levees and other structures are constructed to reduce flooding and bank erosion and to keep the river in the main channel. These structures are often coupled with maintaining a deepwater channel for navigation. However, restoration of large warmwater rivers by reconnecting main-channel and floodplain habitats is being attempted. For example, the Kissimmee River in Florida was channelized to reduce flooding in the 1960s and turned a meandering 167-km river into a 90-km canal. Efforts are underway to restore the natural hydrologic regime and physical form of the river by reconstructing meanders and reconnecting the main channel with the floodplain (Whalen et al. 2002). Similarly, reconnecting backwaters to the main channel has been undertaken in the Illinois River, and use of restored backwaters by larval and adult fishes is occurring (but is dependent upon habitat quality and hydrologic factors; Schultz et al. 2007; Csoboth and Garvey 2008).

21.4.1.2 Flow regulation

Flow regulation is another important driver of fish population dynamics and assemblage structure in large warmwater rivers (Figure 21.4). Flow regulation is caused by dams and

Figure 21.3. Major anthropogenic effects in large warmwater rivers. From the top: water withdrawal for irrigation, dams (Glen Canyon Dam on the Colorado River, Arizona), wing dike, and instream sand dredging.

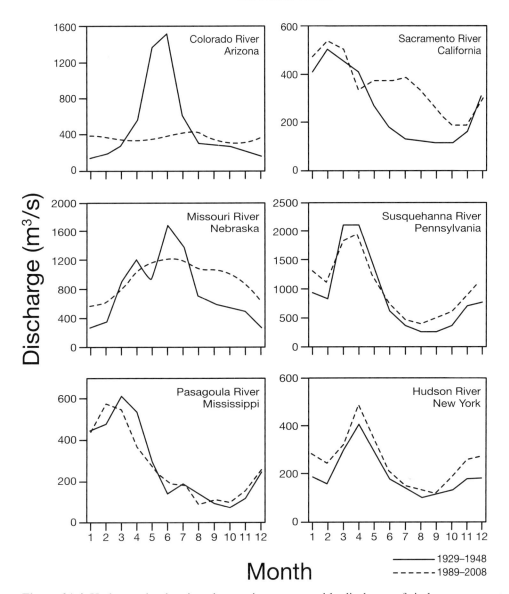

Figure 21.4. Hydrographs showing changes in mean monthly discharge of six large warmwater rivers. Values represent monthly means from a 20-year current time period (1989–2008) and historical time period (1929–1948).

water diversions that manipulate river flows to benefit hydropower, irrigation, navigation, recreation, and other economic activities not related to management of fishes. These multiple uses of water commonly lead to flows that are not favorable to ecological processes in rivers. Reductions in native biodiversity and establishment of nonnative species are linked to hydrologic alteration (Stanford et al. 1996; Bunn and Arthington 2002). Many riverine fishes synchronize life history traits, especially reproduction, with specific flow events, and altered flow regimes put these species at risk (Bunn and Arthington 2002). Additionally, nonnative fishes that can adapt to regulated flows often thrive and outcompete native fishes. Therefore, rivers with relatively natural flow regimes often have increased abundance and diversity of native fishes, whereas rivers with altered flow regimes often have increased abundance of nonnative fishes (Marchetti and Moyle 2001).

Flow regulation has several meanings. The magnitude, duration, timing, and rate of change of flow events are all important factors to spawning and recruitment of fishes (Poff et al. 1997). Management of flows in warmwater rivers has focused on restoring aspects of natural flow regimes. However, completely-restoring natural flow regimes is often not possible because of multiple competing water interests. For example, dams used for power generation often have highly-fluctuating flow regimes (hydropeaking) that alter the natural hydrograph to accommodate municipal power needs, which, in turn, alters spawning, recruitment, and available habitat for warmwater fishes (Freeman et al. 2001). Efforts to design more natural flows are often a management goal (Richter et al. 2003; Jacobson and Galat 2008). Efforts are underway in the Green River, Kentucky, to change high reservoir releases from early fall (September and October) to later in the fall to mimic natural flows associated with fall rains (Richter et al. 2003). More natural flows in the Missouri River for the endangered pallid sturgeon are being attempted, but specific flow requirements for pallid sturgeon life history needs are unknown. As such, the "best guess" of the most appropriate flow is to mimic the natural flow (Jacobson and Galat 2008). The Colorado River through Grand Canyon has had numerous large-scale experiments related to changing flows to aid recovery of endangered fishes, including controlled floods in spring and steady low flows during summer (e.g., Valdez et al. 2001).

Characterizing the flow regime of a river is complex as the flow regime comprises multiple components that may influence many life history aspects of a species. Because magnitude, duration, frequency, timing, and rate of change define a river's flow pattern, a composite metric linked to these attributes has become a useful tool to assess hydrologic alteration (Olden and Poff 2003). The index of hydrologic alteration (IHA) calculates 32 individual metrics related to the five features of flow patterns mentioned above (Richter et al. 1996) and has received wide use to assess hydrologic alterations in warmwater rivers. The range of variability approach (RVA) expanded the IHA to help river managers design a range of ecologically relevant flow targets (Richter et al. 1997, 1998). Together, the IHA–RVA have been applied to numerous warmwater rivers including the Colorado (Richter et al. 1998), Missouri (Galat and Lipkin 2000), Tallapoosa (Irwin and Freeman 2002), and Illinois (Koel and Sparks 2002). Recommendations for more naturalized flows downstream of dams to improve conditions for age-0 fishes in the Illinois River (Koel and Sparks 2002) and to benefit recovery of the federally -threatened Neosho madtom (Wildhaber et al. 2000) have been proposed based on the IHA–RVA. However, we are unaware of cases (including these) in which, following IHA–RVA recommendations, more naturalized flows have resulted in improvement of native warmwater fishes. This seems more a consequence of failure to implement IHA–RVA-based recommendations than a lack of documenting responses by fishes to a more naturalized flow regime.

More recently, Poff et al. (2010) proposed ecological limits of hydrological alteration (ELOHA) as a framework for restoring natural flows in rivers. The advantage of ELOHA is that it links flows to regionally-based ecological processes (Poff et al. 2010). As such, rivers in different regions or with different hydrologic regimes are managed differently. The five-step ELOHA process involves (1) defining historical and current streamflow conditions, (2) classifying the stream type based on ecologically-relevant streamflow characteristics, (3) assessing flow alteration from historic conditions, (4) generating flow alteration–ecological response relationships (e.g., responses of species richness or larval fish abundance to flows), and (5) implementing environmental flows in cooperation with stakeholders. The ELOHA process has been implemented for several projects throughout the USA and beyond.

21.4.1.3 Habitat Alteration

Riparian and instream habitat alteration in warmwater rivers has been linked to fish and freshwater mussel abundance and assemblage structure. Channelization and dredging of rivers for navigation or dredging sand and gravel for construction materials has modified large river habitats and altered sediment transport, substrate, and water velocity, all linked to fishes. Natural habitats (e.g., sand bars, islands, connected backwaters, and wetlands) have been eliminated or highly altered due to channel modification practices (Pflieger and Grace 1987; Whalen et al. 2002; Galat et al. 2005; Paukert et al. 2008).

Structures used to create and maintain navigation channels and reduce flooding include wing dikes, revetments, and levees. Rock structures built perpendicular to (or nearly so) the main current, wing dikes, or in an "L" shape with the long arm pointed downstream, L-dikes, direct flow to the center of the thalweg to maximize scouring of sediment and removal of large woody debris. Low-velocity habitats used by large river fishes often form immediately downstream from wing dikes, but sediment deposition can eliminate the value of these habitats (Sheehan and Rasmussen 1999). Rock revetments are commonly used along riverbanks to reduce lateral bank erosion. Levees are constructed landward along the bank to reduce overland flooding during high flows and reduce connectivity between the channel and its floodplain. Effects of these structures are varied and scale dependent. At larger spatial scales, they may reduce biotic integrity because of loss of natural river habitat; at local scales they may increase fish biomass or abundance by providing large rock substrate, which creates habitat analogous to that provided by large woody debris and boulders in unchannelized rivers (White et al. 2009). Hartman and Titus (2009) found that artificial dike structures were important habitats for centrarchid species in the Kanawha River, West Virginia. In addition, modification of these structures by removing a section of rock (i.e., notching) in the lower Missouri and middle Mississippi rivers allowed water flow behind the structures, reduced sedimentation, and diversified backwater habitats to benefit native fishes (Schloesser 2008).

Removal of large woody debris from rivers is commonly conducted to benefit navigation (Angradi et al. 2004) or as part of riparian vegetation removal, but the practice can have a detrimental influence on riverine fishes. In the upper Missouri River, North Dakota, 59% of the original floodplain forest was eliminated by 1979, primarily due to clearing for settlement and agriculture (Johnson 1992). Collectively, these human activities have reduced occurrence of large woody debris in warmwater rivers and have contributed to lowered habitat heterogeneity, resulting in lower invertebrate diversity and less habitat and food for fishes (Gurnell et al. 2002). Large woody debris accounted for only 4% of the surface area, but supported 60% of the total biomass of invertebrate production in the Satilla River, Georgia (Benke et al. 1985). In addition, fishes used woody habitat for cover, and their diets were primarily composed of invertebrates originating from woody debris. Similar links between large woody debris and fishes have been shown in other warmwater rivers (Lehtinen et al. 1997).

21.4.1.4 Water quality

Warmwater rivers have highly variable water quality due to their large size, dams and diversions, diversity of urban and agricultural land use in their floodplains and watersheds, and instream and riparian zone habitat alterations. One measure that can be substantially altered by these anthropogenic disturbances is water temperature, which is a critical factor regulating fish reproduction, recruitment, and growth. For example, creation of Flaming Gorge Dam on the Green River, Utah, changed midsummer water temperatures from 15–21°C to about 7°C and increased winter water temperatures from near 0°C to 3–8°C due to hypolimnetic releases from the dam (Vanicek et al. 1970). Similar changes have occurred in the Colorado River, where water temperatures changed from a peak of over 25°C in summer to 10–12°C following construction of Glen Canyon Dam (Petersen and Paukert 2005). Nonnative rainbow trout have been introduced in the tailwaters of both the Green and Colorado rivers and have thrived, whereas native warmwater fishes have declined. Other warmwater rivers have been thermally altered through hypolimnetic water releases from dams (Quinn and Kwak 2000). Often these tailwaters are intensively managed for nonnative sport fishes, and the temperature changes are perceived by some stakeholder groups as beneficial.

Excess fine sediment caused by land uses such as row-crop agriculture in the watershed of warmwater rivers fills interstitial spaces of coarse bed substrates that may be used for spawning. Excess fine sediment also reduces visibility, which limits the foraging ability of fishes and causes declines in primary production due to low sunlight penetration. Increased sediment has been linked to low proportions of specialist fish species and high proportions of generalist fish species (Berkman and Rabeni 1987). Increased turbidity can reduce foraging efficiency of fishes (Sweka and Hartman 2003), leading to decreased growth and survival. Increased sedimentation has also been declared one of the most detrimental factors influencing freshwater mussel populations in rivers (Hopkins 2009).

However, increased sediment is not necessary detrimental to fishes. Dams have reduced sediment transport in formerly turbid warmwater rivers and have caused channel scouring and bed degradation downstream from dams. Sediment transport in the Colorado River below Glen Canyon Dam, Arizona (Stevens et al. 1997), and the Missouri River below Gavins Point Dam, South Dakota (Galat et al. 1996), is less than 1% of preimpoundment levels. Reduced sediment transport can increase water clarity, which can alter the benthic ecology of rivers and the food web of fishes (Stevens et al. 1997) and may increase predation on fishes by sight-feeding predators.

Contaminants are relatively common in warmwater rivers and can be detrimental to fishes. Selenium and dichlorodiphenyltrichloroethane (DDT) have caused decreased survival of adults and eggs of several fish species (Jarvinen and Ankley 1999), and mercury has suppressed hormone levels, reproduction, and female gonad development (Jarvinen and Ankley 1999; Hink et al. 2006). Agriculture and urbanization in the floodplain and watershed have led to increased chemical use and both point and nonpoint discharges into rivers. Common contaminants found in fishes from warmwater rivers include organochloride pesticides (e.g., DDT, chlordane, and atrazine) linked to agricultural and urban land uses. Heavy metals (e.g., mercury and selenium) from mining and urbanization also are sources of contaminants in fishes. Increased environmental regulation of these chemicals has reduced their concentrations in many watersheds. However, contaminants often remain in the water and substrate leading to bioaccumulation of these contaminants in fishes and subsequent fish consumption advisories in many rivers. Increased levels of polychlorinated biphenyls (PCBs) related to urban development remain in the upper Ohio River and other urbanized areas of the Mississippi River basin. However, there appears to be a general decline in contaminants in rivers (Schmitt 2002). Although natural erosion and weather can release selenium and other heavy metals into water, much contamination has been linked to agriculture, mining activities, and municipal and industrial water discharges (Hink et al. 2006). Fishes from nearly all sampling sites in the Colorado River basin had elevated selenium levels and many sites had elevated mercury concentrations (Hink et al. 2006).

Endocrine-disrupting compounds are an emerging threat to warmwater fishes in rivers. These anthropogenic chemical compounds disrupt normal endocrine function in fishes and can cause reduced fertility, hatching success, and survival as well as occurrence of fish with both male and female sex organs (Schmitt 2002; Hink et al. 2006). Organochlorines (e.g., chlordane and PCBs) are linked to urban areas, and major rivers flow through large cities. These compounds commonly originate from effluent discharges of sewage. The highest levels of PCBs and chlordane in the Mississippi River were found near St. Louis, Missouri, where there has been an increased proportion of intersex shovelnose sturgeon (Harshbarger et al. 2000). However, fish population level effects of endocrine-disrupting compounds remain unknown. Nonetheless, continued efforts to reduce these compounds in rivers will likely result in increased health of fish populations and normal reproductive activity.

Contaminants in waters are increasingly scrutinized and environmental regulations (e.g., Clean Water Act) curb discharges into rivers. Watershed practices that minimize surface runoff (e.g., riparian buffers, grass waterways, and wetland restoration) help to control runoff into rivers and its associated contaminants. However, point sources of contaminants such as mining and industrial operations are ongoing, and careful regulation of these activities is needed to minimize release of contaminants to the environment.

21.4.1.5 Climate change

Climate change will likely have a profound effect on fishes in warmwater rivers. Recent predictions indicate that ambient air temperature may increase by 1.5–6.0°C in the next century (Houghton et al. 2001), which may cause changes in fish species composition due to altered water temperatures. Rivers at higher latitudes will be more affected than will be rivers at lower latitudes (Ficke et al. 2007). However, precipitation patterns will likely be more extreme, culminating in more severe floods and droughts (Dai et al. 2009). Increased temperatures in northern latitudes may create more warmwater rivers and shift warmwater species to more northern latitudes. In North America, climate change could result in the decrease of coolwater and coldwater streams by as much as 50% as these streams become warmwater streams (Eaton and Scheller 1996). In addition, fishes adapted to specific flow and temperature requirements will likely be reduced or extirpated, and more generalist species may increase as more unpredictable hydrologic regimes occur (Ficke et al. 2007). Warmwater rivers that already have extreme environmental conditions (e.g., Great Plains) currently have fish assemblages dominated by generalist species (Bramblett and Fausch 1991; Eitzmann and Paukert 2010) and these regions may be less affected than are regions with currently more stable hydrology.

The effects of climate change on warmwater rivers may be mediated because many of these systems already have highly-regulated flows to meet agricultural, hydroelectric, and other demands. The future effects of climate change will be strongly affected by how fish management competes with other water allocations in warmwater rivers as hydrologic conditions in these systems become more variable.

21.4.2 Biotic Factors

21.4.2.1 Overfishing

Overfishing is a concern in warmwater rivers, but both commercial and sport harvests have been increasingly regulated. Commercial fisheries are important economic resources in many warmwater rivers. In the Mississippi River commercial catches of buffaloes, catfishes, and freshwater drum have doubled from 1945 to 1999, but there is little evidence to suggest these stocks are overfished (Schramm 2003). Commercial fishing for paddlefish averaged 420 metric tons for the Ohio, Tennessee, and Mississippi rivers from 2001 to 2006, and paddlefish harvests increased 32-fold in the Ohio River from 1965–1975 to 2001–2006 (Quinn 2009). The effects of harvest may be species specific. Detrimental effects from overharvest have been shown for channel catfish (Pitlo 1997), shovelnose sturgeon (Koch et al. 2009), and paddlefish (Scholten and Bettoli 2005). Commercial harvest of warmwater river fishes is ongoing, particularly in the midwestern and southeastern USA. Harvest is more regulated now than in the past, and better stock assessment methods are available to evaluate harvest effects.

Sturgeons throughout warmwater rivers in North America have been overfished (e.g., Boreman 1997; Quist et al. 2002; Peterson et al. 2008). Shovelnose sturgeon in the Mississippi River basin provides an excellent example of the potential effects of harvest on warmwater river fishes. Commercial fishing for this species, primarily for caviar, has increased while populations have decreased (Colombo et al. 2007). Minimum length limits to protect the adult stock are needed to maintain sustainable harvest, but even low exploitation levels (e.g., 20%) may affect the long-term viability of this species (Koch et al. 2009). Efforts are underway in the Mississippi River basin to reduce or eliminate harvest of shovelnose sturgeon, in part because of the similarity of its appearance to the federally-endangered pallid sturgeon. Although regulating harvest of large warmwater river fishes has been historically neglected, there is a growing emphasis on fishes in these systems and this increased concern about overfishing has led to more monitoring of populations.

Several commercially-important fishes are rebounding due to changes in harvest regulations. A change from a 330-mm to a 381-mm minimum length limit for channel catfish in the Mississippi River led to increased yield of harvested fish and abundance of age-0 fish (Pitlo 1997). Therefore, subtle changes in regulations may allow for sustained harvest in some commercial fisheries in large rivers. Overharvest adversely affected riverine fishes historically, but more intensive management and regulations have led to better information on harvest levels and efforts to promote sustained fisheries of many species (e.g., Pitlo 1997; Scholten and Bettoli 2005; Koch et al. 2009). Additionally, highly cost-effective aquaculture has diminished consumer demand for wild fishes.

Sport fishes are typically not overharvested in many warmwater rivers, but often there are no reliable estimates of harvest or exploitation (Schramm 2003). Recreational fisheries are common in large rivers and may vary seasonally for migratory species (e.g., paddlefish). Recreational fishing for resident fishes is popular in impoundments on warmwater rivers, but many of these fishes also reside and provide angling opportunities in upstream and downstream river channels and backwaters. For example, recreational centrarchid fisheries are common in backwater reaches of the upper Mississippi River in the pools created by locks and dams (e.g., Gent et al. 1995), and recreational catfish fisheries are common in rivers throughout the Mississippi River basin (e.g., Makinster and Paukert 2008).

21.4.2.2 Invasive Species

Introduction of nonnative (or nonindigenous, which are species introduced beyond their native distribution by human activity; Kolar and Lodge 2001) aquatic organisms is one of the most important conservation and management concerns in warmwater rivers. Details about invasive species (a nonnative species that spreads from the point of introduction and becomes abundant; Kolar and Lodge 2001) can be found in Chapter 8, but a summary of the issues regarding invasive species in warmwater rivers is warranted. Nonnative fishes occur throughout warmwater rivers in the USA. For example, bighead carp and silver carp were introduced into the USA in the 1970s and have become established in many rivers throughout the Mississippi River basin (Kolar et al. 2007). The primary concerns with bighead carp and silver carp in warmwater rivers in North America are competition with native fishes for food resources and changes in nutrient concentrations due to consumption of phytoplankton and excretion of nutrients in feces. Direct competition with planktivorous fishes (e.g., gizzard shad, bigmouth buffalo, and paddlefish) may occur with the establishment of bighead carp or silver carp (Sampson et al. 2009); reduced condition of gizzard shad and bigmouth buffalo has been observed following establishment of these fishes (Irons et al. 2007). Another example of an invasive species in warmwater rivers is the flathead catfish. This species is native to the midwestern USA but has been introduced into southeastern U.S. rivers where it competes with and consumes native fishes. Modeling results by Pine et al. (2007) indicated that flathead catfish may reduce the native fish biomass by as much as 50%. In the southwestern USA, red shiner, a tolerant cyprinid native to the Mississippi River basin, may outcompete native fishes for preferred habitat and has been implicated in the decline of many native species (e.g., Douglas et al. 1994).

The zebra mussel is another important nonnative species that may have management implications for large warmwater rivers. The zebra mussel is a bivalve native to the Caspian, Black, and Azov seas of Eurasia and was introduced into North America via bilge water from ocean-going ships in the late 1980s (Lockwood et al. 2007). The zebra mussel has become established throughout the Great Lakes and in several river systems in the Great Plains and eastern USA. Zebra mussels may reduce microzooplankton abundance and result in food limitations for zooplanktivorous fishes (Caraco et al. 1997). These reductions, in turn, may lead to decreased growth and (or) abundance of those fishes (Strayer et al. 1999). Although the impacts of zebra mussels on fishes in warmwater rivers are not fully realized, these invasive bivalves spread rapidly though river systems and may impact growth and abundance of river and reservoir fishes. Control of zebra mussels is difficult, and there are substantial efforts to inform and educate the public to prevent transport of zebra mussels into new water bodies. Biological control by use of native fishes as predators on zebra mussel has been considered (Eggleton et al. 2004). However, consumption of zebra mussels by fishes is unlikely to limit population densities because of the zebra mussel's high reproductive capacity (Magoulick and Lewis 2002).

Introduced fishes have been implicated in changes in fish assemblages of 13 U.S. rivers

(Rinne et al. 2005). Changes associated with nonnative fishes are particularly noteworthy in the western USA (Schade and Bonar 2005). For example, over 60% of the fish species currently found in the lower Colorado River are nonnative (Minckley et al. 2003). In the Sacramento River, 43 invasive species have increasing or stable population size, whereas there were historically only 28 native species, and 14 of those species extirpated, threatened, or endangered (Rinne et al. 2005). Over 50 nonnative fish species have been introduced into the upper Colorado River basin, where only 14 native species occurred (Valdez and Muth 2005). Of the 15 most common fishes introduced into the western USA, 13 were introduced as sport fish or prey fish for sport fishes (Schade and Bonar 2005). Although introduced fishes include salmonids in cold, tailwater systems, most introduced species are warmwater fishes that thrive in river channels or backwater habitats (e.g., red shiner, fathead minnow, centrarchids, and common carp). However, nonnative fish introductions are not unique to the western USA. Flathead catfish, native to the Mississippi River basin, have been introduced into at least 13 Atlantic and southeastern states, and they have had detrimental effects on native river fishes (Pine et al. 2007). The introduction of flathead catfish is an example of current challenges when managing both native and nonnative warmwater riverine species and is also an example of how management strategies differ throughout North America. Whereas there is increased management of flathead catfish in the Mississippi River basin to protect and enhance sport fisheries for this native species (e.g., Makinster and Paukert 2008), enhancing harvest and removal of this species is encouraged in basins outside its native distribution because of its adverse effects on native species (e.g., Pine et al. 2007).

21.5 ECOLOGICAL INTEGRITY

Ideal indicators to provide ecological assessments of warmwater rivers should meet several criteria. They should (1) characterize current river health (status), (2) monitor changes in river health at multiple spatial and temporal scales (trends); (3) identify and respond to major stressors and rehabilitation programs; and (4) interact across ecological, economic, and social realms (see Chapter 12). Fish are particularly suitable ecological indicators for large rivers because their various guilds integrate a wide range of riverine conditions, from properties of bed sediments for egg development at the microscale to longitudinal integrity for spawning migrations at the landscape scale (Scheimer 2000). As migratory organisms, fishes are indicators of longitudinal and lateral connectivity across the riverscape (Fausch et al. 2002) because many fishes use clearly-defined habitats at specific life stages. In addition, the longevity of fishes allows for the integration of environmental alterations across long time periods.

Managers often evaluate abundance, distribution, spawning condition, and (or) growth of individual fish species to determine river health. This can be particularly relevant if a fish species is a keystone species (one that is critical in maintaining the organization and diversity of their ecological communities; Mills et al. 1993; Richter et al. 2006). River segments considered suitable for native warmwater fishes tend to have fish with higher body condition (e.g., flannelmouth sucker, Paukert and Rogers 2004), abundance (e.g., blue sucker, Eitzmann et al. 2007), and (or) growth (e.g., flathead catfish, Paukert and Makinster 2009) than do other river segments. In addition, restoration efforts to recover species often use indices of abundance in certain river reaches to assess responses to actions such as dam removal (Box 21.1).

Box 21.1. The Response of Native Anadromous Fishes to Dam Removal on the Neuse River, North Carolina

Joseph E. Hightower

The removal of Quaker Neck Dam on the Neuse River, North Carolina, provided an opportunity to measure the response of anadromous fishes to an increase in accessible spawning habitat. This low-head dam 225 km upstream from the mouth of the river was built in 1952 to supply cooling water for a steam-electric plant. The dam structure included a fish ladder, but the ladder was ineffective. Several studies demonstrated that the dam restricted the migration of striped bass and American shad, the two predominant anadromous species in the basin. Concerns about biological impacts led to a redesign of the cooling water intake, which eliminated the need for the dam, and the dam was removed in 1998.

Prior to the dam's removal, Beasley and Hightower (2000) used sonic telemetry to monitor the upstream migrations of striped bass and American shad. Low passage rates by these fishes occurred even though the low-head dam was completely submerged during some periods of high flow. Of 13 striped bass and 8 American shad that had been fitted with transmitters and that had migrated to the base of the dam, only 3 striped bass passed the structure, demonstrating that the dam was an impediment to migration. The telemetry results and observations of spawning activity indicated that striped bass spawning was concentrated within 1.5 km downstream of the dam. Some American shad were able to move upstream of the dam during periods of high flow, and American shad spawning was observed from the base of the dam to 1.5 km downstream as well as 3 km above the dam.

Information about spawning habitat indicated that both species would benefit from removal of the dam. Striped bass spawned at sites with higher water velocities and larger substrates, and American shad spawned at sites that were shallower and had larger substrates. Spawning habitat with these characteristics was relatively rare downstream from the dam but more available upstream of the dam.

Striped bass and American shad appeared to benefit from the dam removal. Radio telemetry revealed that 15 of 23 striped bass and 12 of 22 American shad migrated past the former dam site (Bowman 2001). American shad spawning was concentrated about 14 km upstream of the former dam site near relatively coarse substrates (gravel, cobble, and boulder) and intermediate current velocities (0.20–0.60 m/s) and water depths (50–125 cm). These results indicated that the fish selected the same type of spawning habitat as downstream from the dam, habitat that was relatively rare downstream from the dam.

Another study used ichthyoplankton sampling to identify spawning locations (Burdick and Hightower 2006) and indicated a substantial upstream expansion in spawning activity following dam removal (Burdick and Hightower 2006). Eggs and larvae of American shad, striped bass, and hickory shad were collected at sites up to 120 km upstream of the former dam. However, the use of the newly accessible habitat varied among species. American shad and striped bass spawning was concentrated in the main stem of the river, whereas hickory shad made greater use of tributaries. Use of the newly available habitat varied between years, with greater use of the newly available habitat in a year with higher spring flows.

(Box continues)

Box 21.1. Continued.

The benefits to anadromous fishes gained by dam removal likely depend on several factors. An obvious one is the distance upstream to the next barrier. It was easy to justify the removal of Quaker Neck Dam on biological grounds because it provided access to 120 km of main-stem and 1,488 km of tributary spawning habitat. The response of anadromous fishes to dam removal will also depend on whether spawning or nursery habitat is a limiting factor. Other factors, such as overfishing or bycatch mortality, could regulate population size, so a response to increased habitat may not be observed. These other factors would not negate the long-term value of removing a dam, but they should factor into expectations about how fish populations might respond. Nonetheless, the removal of the Quaker Neck Dam on the Neuse River was an important conservation and management action that provided additional habitat to native anadromous fishes.

21.6 CONCLUSIONS

Managing fisheries in warmwater rivers shares similarities with other aquatic systems discussed in this book, including traditional regulations (e.g., length limits and seasonal restrictions) to manage sport and commercial fisheries. However, management of large rivers also needs to consider that these are open systems which allow immigration and emigration of fishes that are often highly migratory. Many warmwater rivers cross political boundaries (e.g., states, provinces, and countries), so interjurisdictional collaboration and cooperation are needed. A good example of interjurisdictional management is sport harvest of paddlefish from the Missouri River. Paddlefish are highly migratory and may travel hundreds to thousands of kilometers (Stancill et al. 2002). The Missouri River is a boundary water between South Dakota and Nebraska, and a popular paddlefish sport fishery exists below Gavins Point Dam, South Dakota. A harvest quota system is jointly managed by the states of South Dakota and Nebraska (Mestl and Sorensen 2009; see also Box 7.1). Paddlefish are managed further through season closures and other region-specific regulations as agreed upon by other states (Hansen and Paukert 2009).

Another fundamental management aspect of warmwater rivers is the need to accommodate multiple uses because rivers commonly have numerous stakeholders with varied interests. Not only are natural resource conservation and management agencies involved, but also municipalities, recreational users, agricultural producers, navigation interests, power companies, Native American tribes, and potentially other stakeholders. An approach that is gaining in popularity to meet the diverse needs of river interests is adaptive management.

Adaptive management is a systematic process that uses lessons learned from management outcomes to improve resource management (Walters 2001). The central tenets of adaptive management are that a management problem needs to be bound by specific objectives or hypotheses, use exiting knowledge to hypothesize outcomes from a management action, recognize uncertainty in the response to the management action, and adjust the actions as more is learned about the repose to the system from these actions (Walters 2001; Chapter 5, this volume). Box 21.2 illustrates the application of adaptive management to address complex, multi-stakeholder resource issues for the Colorado River in Grand Canyon, Arizona, where

Box 21.2. Adaptive Management of the Colorado River, Grand Canyon: Meeting the Needs of Multiple Stakeholders

Lew Coggins, Jr.[1]

Following construction of Glen Canyon Dam near the upstream boundary of Grand Canyon National Park in 1963, the character of the Colorado River in the Grand Canyon was altered, severely affecting physical processes and biota of the river. Discharge patterns were altered (Figure 21.4) and approximately 90% of the sediment supply was trapped behind Glen Canyon Dam. Hypolimnetic releases from the dam changed summer water temperatures in the river from a maximum of 29°C to less than 12°C and caused the river to change from being turbid to predominantly clear. The postdam river is inhabited by a host of introduced nonnative fishes and four of the eight native fish species have been extirpated from the Grand Canyon reach. The most numerous nonnative fish is the rainbow trout, which has been stocked below Glen Canyon Dam to provide a recreational fishery. The river has also changed from a sediment-rich system with sand beaches and little riparian vegetation to a river that is sediment starved and has much-reduced sand beaches and lush, predominantly nonnative, riparian vegetation. These changes, coupled with use of the region by tourists to Grand Canyon National Park, hikers, and recreational rafters, dictate that management of the Colorado River in Grand Canyon needs to address multiple stakeholder groups who have multiple interests.

The Glen Canyon Dam Adaptive Management Program (GCDAMP) was initiated in 1996 to address the degradation of the river and multiple stakeholder interests in the river. Although the overarching goal of the GCDAMP is to assist the U.S. Secretary of the Interior to comply with the management of Colorado River water resources and Grand Canyon National Park and Glen Canyon National Recreation Area, the program has a significant river restoration intent based on in the Grand Canyon Protection Act of 1992. The GCDAMP is responsible for making recommendations to the Secretary of the Interior regarding the operation of Glen Canyon Dam and other management actions to achieve a diverse set of program goals, including conserving native and endangered species, providing recreational opportunities, preserving cultural and archeological resources, and providing hydropower generation capabilities. The program is composed of stakeholders representing seven states spanning the Colorado River basin, six Native American tribes, seven federal and state agencies, two environmental groups, two recreation groups, and two groups that market electrical power from Glen Canyon Dam.

The GCDAMP and the Secretary of the Interior have the difficult task of achieving a broad suite of program goals simultaneously. Although the GCDAMP has attempted to develop specific and measurable objectives shared by all of the participating stakeholders, the GCDAMP makes no prioritization of program goals. Nevertheless, fishes are a focal resource. The primary goals are to conserve the federally endangered humpback chub and other native fishes and to maintain a quality tailwater fishery for nonnative rainbow trout in

[1] U.S. Fish and Wildlife Service, National Conservation Training Center, Shepherdstown, West Virginia.

(Box continues)

Box 21.2. Continued.

the Lees Ferry reach just below Glen Canyon Dam. Conserving native fishes and providing a nonnative trout fishery may seem mutually exclusive. However, these goals may be attainable because the Lees Ferry reach is approximately 100 km upstream from the primary humpback chub spawning and rearing area in the Little Colorado River and the main-stem Colorado River near the confluence.

Current knowledge indicates that dominant drivers of the population dynamics of native fishes in Grand Canyon include interactions with nonnative fishes and the suboptimal water temperatures and unstable flow that hinder growth and survival of native fishes. As such, improving rearing conditions for native fishes in the Colorado River will likely provide benefits to those fishes. Therefore, controlled field experiments were proposed by the GCDAMP to manipulate water temperatures, water flows, and nonnative fishes to help conserve native fishes.

After months of discussions among stakeholders, an experiment was begun in 2003 to manipulate dam operations and suppress the abundance of nonnative fishes in a stretch of the Colorado River that was home to humpback chub and other native fish species. However, an additional challenge was that 2003 marked the beginning of unplanned release of warmer water from Glen Canyon Dam owing to decreased reservoir levels associated with drought conditions. Nonetheless, there were increasing trends in the abundance of humpback chub and other native fish species in the Colorado River following the planned activities. While these trends would be predicted given the hypothesized positive effects of nonnative fish suppression and increased water temperatures on native fishes, it was not clear which of these factors (or possibly others) might be responsible for trends.

One strategy to determine the relative influence of these factors is to manipulate the system so that one of the factors (either temperature or nonnative abundance) remained stable while the other was varied. Perhaps the most tenable option to achieve this circumstance would be for the GCDAMP to continue nonnative fish control and wait for a period with greater precipitation to raise reservoir levels and decrease release temperatures. If under these conditions native fish abundance and recruitment remained high, the conclusion would be that interactions with nonnative fishes was the dominant factor. In contrast, if native fish populations declined, the data would indicate that water temperature was the dominant factor and lend support to construction of a proposed selective withdrawal structure on Glen Canyon Dam to provide control of release-water temperature.

Experimental manipulation of the system to understand the relative effects of these various factors is appealing from a scientific standpoint, particularly in such a large and dynamic system. However, adaptive management is primarily concerned with discovering successful management practices, and greater scientific understanding is a secondary concern. This view was apparently shared by participants in a workshop of scientific experts sponsored by the GCDAMP to advise on the design of future experimentation for native fish conservation. While participants debated the merits of design alternatives, they favored an approach to continue nonnative fish suppression and advocated construction and testing of

(Box continues)

Box 21.2. Continued.

a selective withdrawal device capable of providing both warmer and colder water temperatures in dam releases. They justified their advice by pointing out that past policies appeared to be supporting native fish management objectives and that dramatic changes should be considered only if trends in native fish populations began to reverse. Committee members were concerned about the lack of control of release-water temperatures: continued drought could result in sustained periods of warmwater discharge and highly predatory warmwater fishes could expand in the system and have more of an impact on native fishes.

Application of adaptive management principles to achieve resource goals in Grand Canyon shows promise, but many hurdles remain. Of particular concern is the lack of well-defined and prioritized management objectives to provide program guidance and to arbitrate among potentially conflicting resource goals. Additionally, the program would benefit from more rigorously-defined decision rules to integrate formally research and monitoring findings into future policy choices. Nonetheless, the program does provide a process for science to be considered in the decision-making process and allows all stakeholders to have a voice in that process.

---

management interests include protection of the federally-endangered humpback chub, maintenance of sport fisheries, recreation, sediment transport, power generation, preservation of cultural integrity, and other uses. Adaptive management is also being used to aid large-scale ecosystem restoration programs in the Sacramento–San Joaquin Delta, Platte River, and upper Mississippi River (Doyle and Drew 2008). Conservation and management of warmwater fishes are components of each of these programs: the federally-threatened delta smelt (Sacramento–San Joaquin Delta), federally endangered pallid sturgeon (Platte River), and multiple recreational species (upper Mississippi River).

Fisheries management in warmwater rivers requires broad, watershed scale approaches that include multiple stakeholder inputs and recognize the challenges and uncertainty associated with these dynamic and diverse systems. Effective management of warmwater rivers requires consideration of multiple species and taxa (both game and nongame, native and nonnative) and the need to recognize that single-species management is seldom appropriate. When harvest is a concern, regulating anglers or commercial harvesters may be necessary. Management of habitat is a primary focus or warmwater river management. Restoration of ecologically-relevant flows suitable for fishes, protection and rehabilitation of main channel, floodplain, and estuarine habitats, and improvement of water quality not only benefits fishes but the broader ecological function of warmwater rivers. Nonnative species will likely play a critical role in the future management of warmwater rivers. Unfortunately, nonnative species occur in waterways throughout North America and rivers serve as conduits for these invasions. The future of river management will need to address the biological and social effects of these species.

Fisheries managers are one part of the mix when it comes to managing warmwater rivers. These systems often cross several political boundaries and input from numerous regulatory agencies and stakeholder groups will be the norm. Fisheries managers that understand multiple aspects of river management (e.g., ecology, hydrology, geomorphology, political science, and resource economics) and realize that these systems will be managed in collaboration with an array of stakeholder groups will likely be most effective. Involvement of stakeholders in an adaptive framework from initial planning, through goal setting, to evaluating of management actions, and to using results for improving management and learning is essential.

21.7 REFERENCES

Aarts, B. G. W., and P. H. Nienhuis. 2003. Fish zonations and guilds as the basis for assessment of ecological integrity of large rivers. Hydrobiologia 500:157–178.

Angermeier P. L., and J. R. Karr. 1984. Relationships between woody debris and fish habitat in a small warmwater stream. Transactions of the American Fisheries Society 113:716–726.

Angradi, T. R., E. W. Schweiger, D. W. Bolgrien, P. Ismert, and T. Selle. 2004. Banks stabilization, riparian land use, and the distribution of large woody debris in a regulated reach of the Upper Missouri River, North Dakota, USA. River Research and Applications 20:829–846.

Bayley, P. 1991. The flood pulse advantage and the restoration of river–floodplain systems. Regulated Rivers: Research and Management 6:75–86.

Bayley, P., and H. Li. 1992. Riverine fishes. Pages 251–281 *in* P. Calow, and G. Petts, editors. The rivers handbook: hydrological and ecological principles, volume 1. Blackwell Scientific Publications, Oxford, UK.

Beasley, C. A., and J. E. Hightower. 2000. Effects of a low-head dam on the distribution and characteristics of spawning habitat used by striped bass and American shad. Transactions of the American Fisheries Society 129:1372–1386.

Benke, A. 1990. A perspective on America's vanishing streams. Journal of the North American Benthological Society 9:77–88.

Benke, A., and C. Cushing. 2005. Rivers of North America. Elsevier Academic Press, Burlington, Massachusetts.

Benke, A. C., R. L. Henry III, D. M. Gillespie, and R. J. Hunter. 1985. Importance of snag habitat for animal production in southeastern streams. Fisheries 10(5):8–13.

Berkman, H. E., and C. F. Rabeni. 1987. Effects of siltation on stream fish communities. Environmental Biology of Fishes 18:285–294.

Boreman, J. 1997. Sensitivity of North American sturgeons and paddlefish to fishing mortality. Environmental Biology of Fishes 48:399–405.

Bowman, S. W. 2001. American shad and striped bass spawning migration and habitat selection in the Neuse River, North Carolina. Master's thesis. North Carolina State University, Raleigh.

Bramblett, R. G., and K. D. Fausch. 1991. Variable fish communities and the index of biotic integrity in a western Great Plains river. Transactions of the American Fisheries Society 120:752–769.

Bunn, S., and A. Arthington. 2002. Basic principles and ecological consequences of altered flow regimes for aquatic biodiversity. Environmental Management 4:492–507.

Burdick, S. M., and J. E. Hightower. 2006. Distribution of spawning activity by anadromous fishes in an Atlantic slope drainage after removal of a low-head dam. Transactions of the American Fisheries Society 135:1290–1300.

Caraco, N. F., J. J. Cole, P. A. Raymond, D. L. Strayer, M. L. Pace, S. E. Findlay, and D. T. Fischer. 1997. Zebra mussel invasion in a large turbid river: phytoplankton response to increased grazing. Ecology 78:588–602.

Clarkson, R. W. 2004. Effectiveness of electrical fish barriers associated with the central Arizona project. North American Journal of Fisheries Management 24:94–105.

Colombo, R. E., J. E. Garvey, N. D. Jackson. R. Brooks, D. P. Herzog, R. A. Hrabik, and T. W. Spier. 2007. Harvest of Mississippi River sturgeon drives abundance and reproductive success: a harbinger of collapse? Journal of Applied Ichthyology 23:444–451.

Csoboth, L. A., and J. E. Garvey. 2008. Lateral exchange of larval fish between a restored backwater and a large river in the east-central USA. Transactions of the American Fisheries Society 137:33–44.

Dai, A., T. Qian, K. E. Trenberth, and J. D. Milliman. 2009. Changes in continental freshwater discharge from 1948 to 2004. Journal of Climate 22:2773–2792.

Dettmers, J. M., S. Gutreuter, D. H. Wahl, and D. A. Soluk. 2001. Patterns in abundance of fishes in main channels of the upper Mississippi River system. Canadian Journal of Fisheries and Aquatic Sciences 58:933–942.

Di Maio, J., and L. D. Corkum. 1995. Relationship between the spatial distribution of freshwater mussels and the hydraulic variability of rivers. Canadian Journal of Zoology 73:663–671.

Douglas, M. E., P. C. Marsh, and W. L. Minckley. 1994. Indigenous fishes of western North America and the hypothesis of competitive displacement: *Meda fulgida* (Cyprinidae) as a case study. Copeia 1994:9–19.

Doyle, M., and C. A. Drew, editors. 2008. Large-scale ecosystem restoration: five case studies from the United States. Island Press, Washington, D.C.

Durham, B. W., and G. R. Wilde. 2006. Influence of stream discharge on reproduction of a prairie stream fish assemblage. Transactions of the American Fisheries Society 135:1644–1653.

Dynesius, M., and C. Nilsson. 1994. Fragmentation and flow regulation of river systems in the northern third of the world. Science 266:753–762.

Eaton, J. G., and R. M. Scheller. 1996. Effects of climate warming on fish thermal habitat in streams of the United States. Limnology and Oceanography 41:1109–1115.

Eggleton, M. A., L. E. Miranda, and J. P. Kirk. 2004. Assessing the potential for fish predation to impact zebra mussels: insight from bioenergetics models. Ecology of Freshwater Fish 13:85–95.

Eitzmann, J. L., A. S. Makinster, and C. P. Paukert. 2007. Distribution and growth of blue suckers in a Great Plains river, USA. Fisheries Management and Ecology 14:255–262.

Eitzmann, J. L., and C. P. Paukert. 2010. Longitudinal differences in habitat complexity and fish assemblage structure of a Great Plains River. American Midland Naturalist 163:14–32.

Emery, E. B., T. P. Simon, F. H. McCormick, P. L. Angermeier, J. E. Deshon, C. O. Yoder, R. E. Sanders, W. D. Pearson, G. D. Hickman, R. J. Reash, and J. A. Thomas. 2003. Development of a multimetric index for assessing the biological condition of the Ohio River. Transactions of the American Fisheries Society 132:791–808.

Fausch, K. D., C. E. Torgerson, C. E. Baxter, and H. W. Li. 2002. Landscapes to riverscapes: bridging the gap between research and conservation of stream fishes. BioScience 52:483–498.

Ficke, A. D., C. A. Myrick, and L. J. Hansen. 2007. Potential impacts of global climate change on freshwater fisheries. Reviews in Fish Biology and Fisheries 17:581–613.

Freeman, M. C., Z. H. Bowen, K. D. Bovee, and E. R. Irwin. 2001. Flow and habitat effects on juvenile fish abundance in natural and altered flow regimes. Ecological Applications 11:179–190.

Galat, D. L., C. R. Berry, W. M. Gardner, J. C. Hendrickson, G. E. Mestl, G. J. Power, C. Stone, and M. R. Winston. 2005. Spatiotemporal patterns and changes in Missouri River fishes. Pages 249–291 *in* J. Rinne, R. M. Hughes, and R. Calamusso, editors. Historical changes in large river fish assemblages of the Americas. American Fisheries Society, Symposium 45, Bethesda, Maryland.

Galat, D. L. and R. Lipkin. 2000. Restoring ecological integrity of great rivers: historic hydrographs aid in defining reference conditions for the Missouri River. Hydrobiologia 422/423:29–48.

Galat, D. L., J. W. Robinson, and L. W. Hesse. 1996. Restoring aquatic resources to the lower Missouri River: issues and initiatives. Pages 49–72 *in* D. L. Galat and A. G. Frazier, editors. Overview of river–floodplain ecology in the upper Mississippi River basin. U.S. Government Printing Office, Washington, DC.

Gent, R., J. Pitlo, and T. Boland. 1995. Largemouth bass response to habitat and water quality rehabilitation in a backwater of the upper Mississippi River. North American Journal of Fisheries Management 15:784–793.

Gurnell, A. M., H. S. Piegay, and S. V. Gregory. 2002. Large wood and fluvial processes. Freshwater Biology 47:601–619.

Gutreuter, S., A. D. Bartells, K. Irons, and M. B. Sandheinrich. 1999. Evaluation of the flood pulse concept based on statistical models of growth of selected fishes of the upper Mississippi River system. Canadian Journal of Fisheries and Aquatic Sciences 56:2282–2291.

Hamel, M. J., M. L. Brown, and S. R. Chipps. 2008. Behavioral responses of rainbow smelt to in situ strobe lights. North American Journal of Fisheries Management 28:394–401.

Hansen, K., and C. P. Paukert. 2009. Current management of paddlefish sport fisheries. Pages 277–290 in C. P. Paukert, C. P. and G. D. Scholten, editors. Paddlefish management, propagation, and conservation in the 21st century: building from 20 years of research and management. American Fisheries Society, Symposium 66, Bethesda, Maryland.

Harris, J. H., and P. C. Gehrke. 1994. Modeling the relationship between streamflow and population recruitment to manage freshwater fisheries. Australian Fisheries 6:28–30.

Harshbarger, J. C., M. J. Coffey, and M. Y. Young. 2000. Intersexes in Mississippi River shovelnose sturgeon sampled below Saint Louis, Missouri, USA. Marine Environmental. Research 50:247–250.

Hartman, K. J., and J. L. Titus. 2009. Fish use of artificial dike structures in a navigable river. River Research and Applications. Wiley InterScience Online DOI 10.1002/rra.1329.

Hink, J. E., V. S. Blazer, N. D. Denslow, T. S. Gross, K. R. Echols, A. P. Davis. T. W. May, C. E. Orazio, J. J. Coyle, and D. E. Tillitt. 2006. Biomonitoring of environmental status and trends (BEST) program: environmental contaminants, health indicators, and reproductive biomarkers in fish from the Colorado River basin. U.S. Geological Survey, Scientific Investigations Report 2006–5163, Washington, D.C.

Hopkins, R. L. 2009. Use of landscape pattern metrics and multiscale data in aquatic species distribution models: a case study of a freshwater mussel. Landscape Ecology 24:943–955.

Houghton, J. T., Y. Ding, D. J. Griggs, M. Noguer, P. J. van der Linden, X. Dai, K. Maskell, and C. A. Johnson, editors. 2001. Climate change 2001: the scientific basis. Contribution of working group I to the third assessment report of the Intergovernmental Panel on Climate Change. Cambridge University Press, Cambridge, UK, and New York. Available: http://www.grida.no/climate/ipcc_tar. (February 2010).

Humphries, P., A. King, and J. Koehn. 1999. Fish, flows and floodplains: links between freshwater fish and their environment in the Murray–Darling River system, Australia. Environmental Biology of Fishes 59:129–151.

Hynes, H. 1989. Keynote address. Pages 5–10 in D. Dodge, editor. Proceedings of the international large river symposium. Canadian Special Publications of Fisheries and Aquatic Sciences 106.

Irons, K. S., G. G. Sass, M. A. McClalland, and J. D. Stafford. 2007. Reduced condition factor of two native fish species with invasion of nonnative Asian carps in the Illinois River USA: is this evidence of competition and reduced fitness? Journal of Fish Biology 71:258–273.

Irwin, E. R. and M. C. Freeman. 2002. Proposal for adaptive management to conserve biotic integrity in a regulated segment of the Tallapoosa River, Alabama, USA. Conservation Biology 16:1212–1222.

Jacobson, R. B. and D. L. Galat. 2008. Design of a naturalized flow regime—an example from the lower Missouri River, USA. Ecohydrology 1:81–104.

Jaeger, M. E., A. V. Zale, T. E. McMahon, and B. J. Schmitz. 2005. Seasonal movements, habitat use, aggregation, exploitation, and entrainment of saugers in the lower Yellowstone River: an empirical assessment of factors affecting population recovery. North American Journal of Fisheries Management 25:1550–1568.

Jarvinen, A. W., and G. T. Ankley. 1999. Linkages to effects of tissue residues: development of a comprehensive database for aquatic organisms exposed to inorganic and organic chemicals. SETAC Press, Pensacola, Florida.

Jennings, C. A., and S. J. Zigler. 2009. Biology and life history of the paddlefish: an update. Pages 1–22 *in* C. P. Paukert and G. D. Scholten, editors. Paddlefish management, propagation, and conservation in the 21st century: building from 20 years of research and management. American Fisheries Society, Symposium 66, Bethesda, Maryland.

Johnson, B. L., W. B. Richardson, and T. J. Naimo. 1995. Past, present, and future concepts in large river ecology. BioScience 45:134–141.

Johnson, W. C. 1992. Dams and riparian forests: a case study from the upper Missouri River. Rivers 3:229–242.

Jones, B. D., and D. B. Noltie 2007. Flooded flatheads: evidence of increased growth in Mississippi River *Pylodictis olivaris* (Pisces: Ictaluridae) following the great Midwest flood of 1993. Hydrobiologia 592:183–209.

Junk, W., P. Bayley, and R. Sparks. 1989. The flood pulse concept in river–floodplain systems. Pages 110–127 *in* D. Dodge, editor. Proceedings of the international large river symposium. Canadian Special Publication of Fisheries and Aquatic Sciences 106.

Karr, J. R. 1981. Assessment of biotic integrity using fish communities. Fisheries 6(6):21–27.

Karr, J. R. 1991. Biotic integrity: a long neglected aspect of water resource management. Ecological Applications 1:66–84.

Koch, J. D., M. C. Quist, C. L. Pierce, K. A. Hanson, and M. J. Steuck. 2009. Effects of commercial harvest on shovelnose sturgeon populations in the upper Mississippi River. North American Journal of Fisheries Management 29:84–100.

Koel, T. M. and R. E. Sparks. 2002. Historical patterns of river stage and fish communities as criteria for operations of dams on the Illinois River. River Research and Applications 18:3–19.

Kolar, C. S., D. C. Chapman, W. R. Courtenay Jr., C. M. Housel, J. D. Williams, and D. P. Jennings. 2007. Bigheaded carps: a biological synopsis and environmental risk assessment. American Fisheries Society, Special Publication 33, Bethesda, Maryland.

Kolar, C. S., and D. M. Lodge. 2001. Progress in invasion biology: predicting invaders. Trends in Ecology and Evolution 16:199–2004.

Lehtinen, R. M., N. D. Mundahl, and J. C. Madejczyk. 1997. Autumn use of woody snags by fishes in backwater and channel border habitats of a large river. Environmental Biology of Fishes 49:7–19.

Lockwood, J. L., M. F. Hoopes, and M. P. Marchetti. 2007. Invasion ecology. Blackwell Publishing, Malden, Massachusetts.

Lorenz, C. M., G. M. Van Dijk, A. G. M. Van Hattum, and W. P. Cofino. 1997. Concepts in river ecology: implications for indicator development. Regulated Rivers: Research and Management 13:501–516.

Magnuson, J. J., L. B. Crowder, and P. A. Medvick. 1979. Temperature as an ecological resource. American Zoologist 19:331–343.

Magoulick, D. D., and L. C. Lewis. 2002. Predation on exotic zebra mussels by native fish: effects on predator and prey. Freshwater Biology 47:1908–1918.

Makinster, A. S., and C. P. Paukert. 2008. Effects and utility of minimum length limits and mortality caps for flathead catfish in discrete reaches of a large prairie river. North American Journal of Fisheries Management 28:97–108.

Marchetti, M. P., and P. B. Moyle. 2001. Effects of flow regime on fish assemblages in a regulated California stream. Ecological Applications 11:530–539.

Mestl, G., and J. Sorensen. 2009. Joint management of an interjurisdictional paddlefish snag fishery in the Missouri River below Gavins Point Dam, South Dakota and Nebraska. Pages 235–260 *in* C. P. Paukert and G. D. Scholten, editors. Paddlefish management, propagation, and conservation in the 21st century: building from 20 years of research and management. American Fisheries Society, Symposium 66, Bethesda, Maryland.

Mills, L. S., M. E. Soule, and D. F. Doak. 1993. The keystone species concept in ecology and conservation. BioScience 43:219–224.

Minckley, W. L., P. C. Marsh, J. E. Deacon, T. E. Dowling, P. W. Hedrick, W. J. Matthews, and G. Mueller. 2003. A conservation plan for native fishes of the lower Colorado River. BioScience 53:219–234.

Moyle, P. B., and J. A. Israel. 2005. Untested assumptions: effectiveness of screening diversions for conservation of fish populations. Fisheries 30(5):20–28.

Ohio EPA (Environmental Protection Agency). 1989. Biological criteria for the protection of aquatic life. Volume III: Standardized field sampling and laboratory methods for assessing fish sampling and macroinvertebrate communities. Ohio Environmental Protection Agency, Division of Water Quality Monitoring and Assessment, Columbus, Ohio.

Olden, J. D., and N. L. Poff. 2003. Redundancy and the choice of hydrologic indices for characterizing streamflow regimes. River Research and Applications 19:101–121.

Patrick, P. H., A. E. Christie, D. Sager, C. Hocutt, and J. Stauffer Jr. 1985. Responses of fish to a strobe light/air-bubble barrier. Fisheries Research 3:157–172.

Paukert, C. P., and W. L. Fisher. 2001. Spring movements of paddlefish in a prairie reservoir system. Journal of Freshwater Ecology 16:113–124.

Paukert, C. P., and A. S. Makinster. 2009. Longitudinal patterns in flathead catfish abundance and growth within a large river: effects of an urban gradient. River Research and Applications 25:861–873.

Paukert, C. P., and R. S. Rogers. 2004. Factors affecting condition of flannelmouth suckers in the Colorado River, Grand Canyon, Arizona. North American Journal of Fisheries Management 24:648–653.

Paukert, C. P., J. Schloesser, J. Eitzmann, J. Fischer, K. Pitts, and D. Thornbrugh. 2008. Effect of instream sand dredging on fish communities in the Kansas River USA: current and historical perspectives. Journal of Freshwater Ecology 23:623–633.

Petersen, J. H., and C. P. Paukert. 2005. Development of a bioenergetics model for humpback chub and evaluation of water temperature changes in the Grand Canyon, Colorado River. Transactions of the American Fisheries Society 134:960–974.

Peterson, D. L., P. Schueller, R. DeVries, J. Fleming, C. Grunwald, and I. Wirgin. 2008. Annual run size and genetic characteristics of Atlantic sturgeon in the Altamaha River, Georgia. Transactions of the American Fisheries Society 137:393–401.

Pflieger, W. L., and T. B. Grace. 1987. Changes in the fish fauna of the lower Missouri River, 1940–1983. Pages 166–177 in W. J. Matthews and D. C. Heins, editors. Community and evolutionary ecology of North American stream fishes. University of Oklahoma Press, Norman.

Pine, W. E., III, T. J. Kwak, and J. A. Rice. 2007. Modeling management scenarios and the effects of an introduced apex predator on a coastal riverine fish community. Transactions of the American Fisheries Society 135:105–120.

Pitlo, J., Jr. 1997. Response of upper Mississippi River channel catfish populations to changes in commercial harvest regulations. North American Journal of Fisheries Management 17:848–859.

Poff, N. L., J. D. Allan, M. D. Bain, J. L. Karr, K. L. Prestegaard, B. D. Richter, R. E. Sparks, and J. C. Stromberg. 1997. The natural flow regime. BioScience 47:769–784.

Poff, N. L., B. Richter, A. H. Arthington, S. E. Bunn, R. Naiman, E. Kendy, M. Acreman, C. Apse, B. P. Bledsoe, M. Freeman, J. Henriksen, R. B. Jacobson, J. Kennen, D. R. Merritt, J. O'Keeffe, J. D. Olden, K. Rogers, R. E. Tharme, and A. Warner. 2010. The ecological limits of hydrologic alteration (ELOHA): a new framework for developing regional environmental flow standards. Freshwater Biology 55:147–170.

Poff, N. L., and J. V. Ward. 1989. Implications of streamflow variability and predictability for lotic community structure: a regional analysis of streamflow patterns. Canadian Journal of Fisheries and Aquatic Sciences 45:1805–1818.

Poff, N. L., and J. V. Ward. 1990. Physical habitat template of lotic systems: recovery in the context of historical pattern of spatiotemporal heterogeneity. Environmental Management 14:629–645.

Popper, A. N., and T. J. Carlson. 1998. Application of sound and other stimuli to control fish behavior. Transactions of the American Fisheries Society 127:673–707.

Quinn, J. R. 2009. Harvest of paddlefish in North America. Pages 203–222 in C. P. Paukert and G. D. Scholten, editors. Paddlefish management, propagation, and conservation in the 21st century: building from 20 years of research and management. American Fisheries Society, Symposium 66, Bethesda, Maryland.

Quinn, J. W. and T. J. Kwak. 2000. Use of rehabilitated habitat by brown trout and rainbow trout in an Ozark tailwater river. North American Journal of Fisheries Management 20:737–751.

Quist, M. C., C. S. Guy, M. A. Pegg, C. L. Pierce, and V. H. Travnichek. 2002. Potential influence of harvest of shovelnose sturgeon populations in the Missouri River system. North American Journal of Fisheries Management 22:537–549.

Rahel, F. J., and W. A. Hubert. 1991. Fish assemblage and habitat gradients in a Rocky Mountain–Great Plains stream: biotic zonation and additive patterns of community change. Transactions of the American Fisheries Society 120:319–332.

Reash, R. 1999. Considerations for characterizing midwestern large river habitats. Pages 463–474 in T. Simon, editor. Assessing the sustainability and biological integrity of water resources using fish communities. CRC Press, Boca Raton, Florida.

Richter, B., J. Baumgartner, J. Powell, and D. Braun. 1997. How much water does a river need? Freshwater Biology 37:231–249.

Richter, B. D., J. V. Baumgartner, D. P. Braun, and J. Powell. 1998. A spatial assessment of hydrologic alteration within a river network. Regulated Rivers: Research and Management 14:329–340.

Richter, B. D., J. V. Baumgartner, J. Powell, and D. P. Braun. 1996. A method of assessing hydrologic alteration within ecosystems. Conservation Biology 10:1163–1174.

Richter, B. D., R. Matthews, D. L. Harrison, and R. Wigington. 2003. Ecologically sustainable water management: managing river flows for ecological integrity. Ecological Applications 13:206–224.

Richter, B. D., A. T. Warner, J. L. Meyer, and K. Lutz. 2006. A collaborative and adaptive process for developing environmental flow recommendations. River Research and Applications 22:297–318.

Rinne, J. N., R. M. Hughes, and B. Calamusso. 2005. Historical changes in large river fish assemblages of the Americas. American Fisheries Society, Symposium 45, Bethesda, Maryland.

Sampson, S. J., J. H. Chick, and M. A. Pegg. 2009. Diet overlap among two Asian carp and three native fishes in backwater lakes on the Illinois and Mississippi rivers. Biological Invasions 11:783–496.

Schade, C. B., and S. A. Bonar. 2005. Distribution and abundance of nonnative fishes in streams of the western Unites States. North American Journal of Fisheries Management 25:1386–1394.

Scheimer, F. 2000. Fish as indicators for the assessment of ecological integrity of large rivers. Hydrobiologia 422/423:271–278.

Schloesser, J. T. 2008. Large river fish community sampling strategies and fish associations to engineered and natural river channel structures. Master's thesis. Kansas State University, Manhattan.

Schmitt, C. J. 2002. Biomonitoring of environmental status and trends (BEST) program: environmental contaminants and their effects on fish in the Mississippi River basin. U.S. Geological Survey, Biological Science Report USGS/BRD/BSR 2002–0004, Washington, D.C.

Scholten, G. D., and P. B. Bettoli. 2005. Population characteristics and assessment of overfishing for an exploited paddlefish population in the lower Tennessee River. Transactions of the American Fisheries Society 134:1285–1298.

Schramm, H. L., Jr. 2003. Stats and management of Mississippi River fisheries. Pages 301–333 in R. L. Welcomme and T. Petr, editors. Proceedings of the second international symposium on management of large river fish fisheries, volume 1. FAO (Food and Agriculture Operations of the United Nations) Regional Office for Asia and the Pacific, Publication 2004/16, Bangkok, Thailand.

Schramm, H. L., Jr., and M. Eggleton. 2006. Applicability of the flood-pulse concept in a temperate floodplain river ecosystem: thermal and temporal components. River Research and Applications 22:543–553.

Schultz, D. W., J. E. Garvey, and R. C. Brooks. 2007. Backwater immigration by fishes through a water control structure: implications for connectivity and restoration. North American Journal of Fisheries Management 27:172–180.

Sedell, J. R., J. E. Richey, and F. J. Swanson. 1989. The river continuum concept: a basis for the expected ecosystem behavior of very large rivers? Pages 49-55 *in* D. Dodge, editor. Proceedings of the international large river symposium. Canadian Journal of Fisheries and Aquatic Sciences Special Publication 106.

Sheehan, R. J., and J. L. Rasmussen. 1999. Large rivers. Pages 529–559 *in* C. C. Kohler and W. A. Hubert, editors. Inland fisheries management in North America, 2nd edition. American Fisheries Society, Bethesda, Maryland.

Sheldon, A. L. 1968. Species diversity and longitudinal succession in stream fishes. Ecology 49:194–198.

Simon, T. P., and E. B. Emery. 1995. Modification and assessment of an index of biotic integrity to quantify water resource quality in great rivers. Regulated Rivers: Research and Management 11:283–298.

Simon, T. P., and J. Lyons. 1995. Application of the index of biotic integrity to evaluate water resource integrity in freshwater ecosystems. Pages 245–262 *in* T. Simon, editor. Assessing the sustainability and biological integrity of water resources using fish communities. CRC Press, Boca Raton, Florida.

Simonson, T., J. Lyons, and P. Kanehl. 1994. Guidelines for evaluation of fish habitat in Wisconsin streams. U.S. Department of Agriculture, Forest Service, North Central Forest Experiment Station, General Technical Report NC-164, St. Paul, Minnesota.

Southwood, T. 1977. Habitat, the template for ecological strategies. Journal of Animal Ecology 46:337–365.

Stalnaker, C., R. Milhous, and K. Bovee. 1989. Hydrology and hydraulics applied to fisheries management in large rivers. Pages 13–30 *in* D. Dodge, editor. Proceedings of the international large river symposium. Canadian Special Publication of Fisheries and Aquatic Sciences 106.

Stancill, W., G. R. Jordan, and C. P. Paukert. 2002. Seasonal migration patterns and site fidelity of adult paddlefish in Lake Francis Case, Missouri River. North American Journal of Fisheries Management 22:815–824.

Stanford, J. A., J. V. Ward, W. J. Liss, C. A. Frissell, R. N. Williams, J. A. Lichatowich, and C. C. Coutant. 1996. A general protocol for restoration of regulated rivers. Regulated Rivers: Research and Management 12:391–414.

Stevens, L. E., J. P. Shannon, and D. W. Blinn. 1997. Colorado River benthic ecology in Grand Canyon, Arizona USA: dam, tributary, and geomorphological influences. Regulated Rivers: Research and Management 13:129–149.

St. Pierre, R. A. 2003. A case history: American shad restoration on the Susquehanna River. Pages 315–321 *in* K. E. Limburg and J. R. Waldman, editors. Biodiversity, status, and conservation of the world's shads. American Fisheries Society, Symposium 35, Bethesda, Maryland.

Strahler, A. N. 1957. Quantitative analysis of watershed geomorphology. American Geophysical Union Transactions 38:913–920.

Strayer, D. L., N. F. Caraco, J. J. Cole, S. Findlay, and M. L. Pace. 1999. Transformation of freshwater ecosystems by bivalves: a case study of zebra mussels in the Hudson River. Bioscience 49:19–27.

Sweka, J. A., and K. J. Hartman. 2003. Reduction of reactive distance and relative foraging success in smallmouth bass exposed to elevated turbidity levels. Environmental Biology of Fishes 67:341–347.

Thoms, M. C., S. Rayberg, and B. Neave. 2008. The physical diversity and assessment of a large river system: the Murray–Darling Basin, Australia. Pages 587–608 *in* A. Gupta, editor. Large rivers. John Wiley and Sons, West Sussex, UK.

Thorp, J. H., M. C. Thomas, and M. D. Delong. 2006. The riverine ecosystem synthesis: biocomplexity in river networks across space and time. River Research and Applications 22:123–147.

Thorp, J. H., M. C. Thomas, and M. D. Delong. 2008. The riverine ecosystem synthesis: towards conceptual cohesiveness in river science. Academic Press, Amsterdam.

Valdez, R. A., T. L. Hoffnagle, C. C. McIvor, T. McKinney, and W. C. Leibfried. 2001. Effects of a test flood on fishes of the Colorado River, Grand Canyon, Arizona. Ecological Applications 11:686–700.

Valdez, R.A., and R. M. Muth. 2005. Ecology and conservation of native fishes in the upper Colorado River basin. Pages 157–205 *in* J. Rinne, R. M. Hughes, and B. Calamusso (editors). Historical changes on large river fish assemblages of the Americas. American Fisheries Society, Symposium 45, Bethesda, Maryland.

Vanicek, C. D., R. H. Kramer, and D. R. Fraklin. 1970. Distribution of Green River fishes in Utah and Colorado following closure of Flaming Gorge Dam. Southwestern Naturalist 14:297–315.

Vannote, R. L., G. W. Minshall, K. W. Cummins, J. R. Sedell, and C. E. Cushing. 1980. The river continuum concept. Canadian Journal of Fisheries and Aquatic Sciences 37:130–137.

Walters, C. 2001. Adaptive management of renewable resources. Blackburn Press, Caldwell, New Jersey.

Ward, J. V. 1989. The four-dimensional nature of lotic ecosystems. Journal of the North American Benthological Society 8:2–8.

Ward, J. V., and J. A. Stanford. 1983. The serial discontinuity concept of lotic ecosystems. Pages 29–42 *in* D. Fontaine and S. M. Bartell, editors. Dynamics of lotic ecosystems. Ann Arbor Science, Ann Arbor, Michigan.

Ward, J. V., and J. A. Stanford. 1995. The serial discontinuity concept: extending the model to floodplain rivers. Regulated Rivers: Research and Management 10:159–168.

Welcomme, R. 1985. River fisheries. FAO (Food and Agricultural Organization of the United Nations) Fisheries Technical Paper 262, Rome.

Welcomme, R. 1992. River conservation—future prospects. Pages 454–462 *in* P. J. Boone, P. Calow, and G. E. Petts, editors. River conservation and management. John Wiley and Sons, New York.

Whalen, P. J., L. A. Toth, J. W. Koebel, and P. K. Strayer. 2002. Kissimmee River restoration; a case study. Water Science and Technology 45:55–62.

White, K., J. Gerken, C. P. Paukert, and A. S. Makinster. 2009. Fish community structure in natural and engineered habitats in the Kansas River. River Research and Applications. Wiley InterScience Online DOI 10.1002/rra.1287.

Wildhaber, M. L., V. M. Tabor, J. E. Whitaker, A. L. Albert, D. W. Mulhern, P. J. Lamberson, and K. L. Powell. 2000. Ictalurid populations in relation to the presence of a mainstem reservoir in a midwestern warmwater stream with emphasis on the threatened Neosho madtom. Transactions of the American Fisheries Society 129:1320–1336.

Williams, J. D., and G. H. Burgess. 1999. A new species of bass, *Micropterus cataractae* (Teleostei: Centrarchidae), from the Apalachicola River basin in Alabama, Florida, and Georgia. Bulletin of the Florida Museum of Natural History 42:80–114.

Winemiller, K. 2005. Floodplain river food webs: generalizations and implications for fisheries management. Pages 285–312 in R. Welcomme and T. Petr, editors. Proceedings of the second international symposium on the management of large rivers for fisheries, volume 2. FAO (Food and Agriculture Operations of the United Nations) Regional Office for Asia and the Pacific, Publication 2004/17, Bangkok, Thailand.

Wolter, C. 2007. Temperature influence on the fish assemblage structure in a large lowland river, the lower Oder River, Germany. Ecology of Freshwater Fish 16:493–503.

Zeug, S. C., and K. O. Winemiller. 2008. Relationships between hydrology, spatial heterogeneity, and river fish recruitment dynamics in a temperature floodplain river. River Research and Applications 24:90–102.

Appendix: Summary of Features for Selected Warmwater Rivers

Table. Summary of features for selected warmwater rivers (mean water temperature > 11.0°C or non-salmonid dominated) reported in Benke and Cushing (2005). Degree of main-stem fragmentation: 0 = no dams on main channel; 1 = moderately fragmented (1–3 dams on main channel); 2 = strongly fragmented (>4 dams on main channel). Number of main-stem dams was obtained from one-page summaries including regional maps. Number of all native fish species, when reported, are in parentheses. Asterisk indicates endangered fishes were not identified.

Major basin or region and river name	Basin area (km²)	Mean discharge (m³/s)	River order	Annual runoff (cm)	Mean water temperature (°C)	Number physiographic provinces	Number terrestrial ecoregions	Number fish species (native)	Number endangered fish species	Degree of fragmentation	Percent basin area Agricultural	Percent basin area Urban	Population density (people/km²)
Atlantic and northeastern USA													
Susquehanna	71,432	1153	7	51.0	14.0	4	4	103	0	2	27	9	56
Delaware	33,041	422	7	59.0	13.8	5	3	105	1	0	24	9	214
Hudson	34,615	592	7	59.0	12.4	5	3	165	1	2	25	8	92
Connecticut	29,160	445	7	58.0		1	2	64	1	2	11		69
Potomac	37,995	320	7	33.0	14.0	5	5	95 (65)	1	1	32	5	138
Raritan	2,862	34	5	52.0	13.6	3	1	88	0	0	19	36	419
Atlantic and southeastern USA													
York–North Anna	6,892	45	6	30.0	15.0	2	2	75	2	1	19	8	54
James	26,164	227	7	39.0	16.0	4	3	109	3	1	22	6	96
Roanoke	25,326	232	6	33.0	16.0	4	3	119	9	2	25	3	31
Cape Fear	24,150	217	6	37.0	17.0	2	2	95	8	1	24	9	69
Great Pee Dee	27,560	371	7	39.0	17.0	3	3	101	6	2	28	8	49
Santee	39,500	434	7	41.0	18.0	3	3	125	5	2	26	6	65
Savannah	27,414	319	7	39.0	17.0	3	4	106	7	2	22	4	35
Ogeechee	13,500	115	6	31.0	19.0	2	3	80	6	0	18	1	30
Altamaha	37,600	393	7	35.0	20.0	2	2	93	12	0	26	3	28
Satilla	9,143	65	6	22.0	20.0	1	1	52	2	0	26	1	11
St. Johns	22,539	222	5	30.0	22.0	1	2	190	4	0	25	6	78

Table. Continued.

Major basin or region and river name	Basin area (km²)	Mean discharge (m³/s)	River order	Annual runoff (cm)	Mean water temperature (°C)	Number physiographic provinces	Number terrestrial ecoregions	Number fish species (native)	Number endangered fish species	Degree of fragmentation	Percent basin area Agricultural	Percent basin area Urban	Population density (people/km²)
Eastern Gulf Coast													
Mobile–Alabama	111,369	1914	8	51.0	19.9	5	4	236	12	2	18	<2	44
Pearl	21,999	373	6	53.0	19.0	1	2	119	1	1	24	<2	42
Apalachicola–Chattahoochee													
Flint	50,688	759	8	45.0	20.6	3	4	104	1	2	25	2	51
Suwannee	24,967	294	7	37.0	19.7	1	1	81	1	0	30	1	22
Cahaba	4,730	80	6	56.0	18.1	2	2	135	3	0	11	2	33
Pascagoula–Black	24,599	432	6	55.0	19.7	1	2	114	1	0	17	<1	29
Flint	22,377	283	7	40.0	19.8	2	2	71	1	1	34	2	27
Upper Tombigbee	18,800	336	6	56.0	18.8	1	1	122	1	2	20	2	18
Escambia–Conecuh	10,963	196	6	56.0	20.4	1	2	102	0	1	15	<1	33
Sipsey	2,044	34	5	53.0	17.1	2	1	83	0	0	10	<1	11
Choctawhatchee	12,033	212	6	55.0	20.0	1	2	80	1	0	25	1	18
Western Gulf Coast													
Rio Grande	870,000	37	7	0.13	14.0	7	8	86	≥16	2	5	7	16
San Antonio–Guadalupe	26,231	79	7	11.7	23.0	2	4	88 (60)	≥7	2	75	25	85
Colorado	103,341	75	7	2.8	22.0	3	7	98 (72)	≥4	2	85	15	35
Brazos	115,566	249	8	4.0	21.0	3	6	93 (72)	≥4	1	24	16	20
Sabine	25,268	238	8	25.0	21.0	1	4	104 (88)	≥4	1	10	8	18
Pecos	113,960	2	5	0.08	21.0	3	4	70	≥12	2	10	2	3
Nueces–Frio	43,512	20	7	1.6	23.0	2	4	66	≥3	1	40	5	16
Trinity	46,540	222	7	16.0	21.0	3	5	99	3	1	15	30	98
Neches–								96	≥4	1	15	5	11

Table. Continued.

Major basin or region and river name	Basin area (km^2)	Mean discharge (m^3/s)	River order	Annual runoff (cm)	Mean water temperature (°C)	Number physiographic provinces	Number terrestrial ecoregions	Number fish species (native)	Number endangered fish species	Degree of fragmentation	Percent basin area Agricultural	Percent basin area Urban	Population density (people/km^2)
Lower Mississippi													
Lower Mississippi	3,270,000	18400	10	17.5	16.0	3	5	375	3	0	57	14	10
White	72,189	979	7	43.0	19.0	2	3	163	10	2	~55	1	18
Yazoo	35,000	523	6	48.0	21.5	1	1	119	2	0	80		16
Buffalo	3,465	48	4	44.0	13.0	1	1	66	0	0	10	0.5	7
Big Black	8,770	107	4	46.0	17.2	1	3	112	1	0	35	0	25
Saline	5,465	89	5	42.0	17.0	2	1	85	1	0	>50		17
Ouachita	64,454	843	6	50.0	16.0	2	3	80	2	2	13	6	16
Current	6,776	77	6	36.0	17.0	2	1	112	2	0	17	<1	6
Atchafalaya	8,345	3761	NA	NA	22.0	1	1	181	3	0	63	3.5	18
Cache	5,227	68	4	40.0	17.0	1	1	32	0	0	83	1.5	23
Southern Plains													
Arkansas	414,910	1004	7	7.5	18.0	6	5	171 (141)	1	2	~40	<1	14.6
Red	169,890	852	7	15.0	19.3	4	4	171 (152)	0	2	~50	<1	9.1
Canadian	122,070	174	6	4.0	18.0	4	3	63	1	2	~50	0	9.1
Little	10,720	183	6	47.0	16.5	2	3	110	1	1	15	0	3.2
Kiamichi	4,650	48	5	41.0	16.7	2	2	86	0	1	25	0	5.6
Blue	1,650	9	4	23.0	17.0	2	1	85	0	0	70	0	18.9
Illinois	4,260	54	5	40.0	16.5	1	1	101	0	1	30	0	18.3
Washita	20,230	44	4	7.0	18.4	3	2	51	0	1	85	0	8.7
Neosho–Grand	32,427	254	6	25.0	15.4	2	1	94	1	2	~50		13.5
Poteau	4,840	68	4	46.0	17.0	1	1	95	0	1	27	0	9.4
Cimarron	50,540	42	4	2.6	18.4	2	1	48	1	1	85	0	6.7

Table. Continued.

Major basin or region and river name	Basin area (km^2)	Mean discharge (m^3/s)	River order	Annual runoff (cm)	Mean water temperature (°C)	Number physiographic provinces	Number terrestrial ecoregions	Number fish species (native)	Number endangered fish species	Degree of fragmentation	Percent basin area Agricultural	Percent basin area Urban	Population density (people/km^2)
Upper Mississippi													
Upper Mississippi	489,510	3576	10	22.0	14.3	3	5	145	7	2	70	5	54
Illinois	75,136	649	9	29.0	16.0	1	1	127	0	2	87	5	97
Wisconsin	30,000	261	8	29.0	11.8	2	2	119	0	2	88	2	14
Wapsipinicon	6,050	47	5	32.0	12.0	1	1	74	1	0	88	2	14
Des Moines	31,127	182	6	21.0	11.6	1	2	84	3	1	86	5	22
Rock	28,101	184	7	23.0	12.0	1	2	115	0	2	71	14	52
Kaskaskia	15,025	107	6	29.0	15.2	1	1	112	0	1	80	3	30
Ohio													
Ohio	529,000	8733	9	52.0	14.0	6	8	240–250	0	2	48	4	49
Tennessee	105,870	2000	8	59.0	19.0	5	4	225–240	9	2	36	4	19
Cumberland	46,430	862	7	59.0	16.0	3	2	172–186	3	2	40	5	16
Wabash	85,340	1001	7	37.0	15.0	2	3	95	0	1	65	5	62.7
Kanawha	31,690	537	6	54.0	14.0	3	2	126	0	2	23	3	29
Green	23,850	420	7	55.0	16.0	1	2	151	1	2	55	3	34
Kentucky	18,025	285	6	50.0	15.0	2	2	110–115	1	2	42	4	39
Great Miami	39,915	152	6	33.0	15.0	1	1	120–125	0	1	80	5	134
Scioto	16,882	189	6	35.0	15.0	2	2	120–130	1	1	69	9	108
Licking	9,600	145	6	48.0	14.0	2	2	110	1	1	42	4	36
Monongahela	19,110	377	7	62.0	14.0	2	2	>120	0	2	29	4	65

Table. Continued.

Major basin or region and river name	Basin area (km²)	Mean discharge (m³/s)	River order	Annual runoff (cm)	Mean water temperature (°C)	Number physiographic provinces	Number terrestrial ecoregions	Number fish species (native)	Number endangered fish species	Degree of fragmentation	Percent basin area Agricultural	Percent basin area Urban	Population density (people/km²)
Missouri													
Missouri	1,371,017	1956	9	6.0	12.0	7	13	183 (145)	1	2	37	9	8
White	26,418	16	6	2.0	12.7	1	2	49	0	0	57	0	1.3
North–South Platte	230,362	203	5	4.0	11.8	4	6	100	2	2	>90	3	16
Gasconade	9,256	87	6	33.0	15.2	1	2	103	0	0	23	4	90
Big Sioux	23,325	35	5	5.0	12.0	1	1	70	1	1	75	1	20
Cheyenne–Belle Fourche	63,455	25	4	6.0	12.7	1	2	52	0	1	42	1	5.7
Niobrara	32,600	49	4	5.0	12.3	1	3	67	1	1	95	0	1.1
Kansas–Republican	159,171	214	7	4.0	11.6	2	3	99	1	2	>90	3	30
Grand	20,461	117	7	21.0	13.0	1	1	60	2	0	73	0	43
Colorado													
Colorado	642,000	550	6	3.0	16.0	6	7	80 (42)	16	2	67	8	7
Little Colorado	69,000	7	4	0.3	18.0	2	2	33 (9)	2	1			1.5
Gila	149,832	10	6	0.2	21.0	2	4	36 (14)	2	1	70	10	25
San Juan	59,600	65	4	4.0	12.0	2	2	26 (7)	2	1	75	5	0.6
Virgin	13,200	7	3	2.0	17.0	2	2	21 (6)	5	0	60	10	7
Bill Williams	13,950	4	3	1.0	20.0	2	1	13 (1)	1	1	75	0	1.5
Verde	16,190	17	4	3.0	16.5	2	1	27 (10)	8	1	70	5	3
Black	3,400	12	4	11.0	15.0	2	1	13 (5)	2	0	15	5	1
Salt	35,480	25	5	7.0	19.0	2	2	16 (9)	7	2	75	5	90

Table. Continued.

Major basin or region and river name	Basin area (km^2)	Mean discharge (m^3/s)	River order	Annual runoff (cm)	Mean water temperature (°C)	Number physiographic provinces	Number terrestrial ecoregions	Number fish species (native)	Number endangered fish species	Degree of fragmentation	Percent basin area Agricultural	Percent basin area Urban	Population density (people/km^2)
Pacific USA													
Sacramento	72,132	657	8	34.0		3	5	69 (29)	*	1	15	1.7	23.6
San Joaquin	83,409	132	8	5.0		2	3	63 (23)	*	1	30	1.9	29.3
Salinas	10,983	12.7	6	4.0				36 (16)	*	1	13	0.7	10.1
Santa Margarita	1,896	1	5	2.0		2	1	17 (6)	*	1	11.6	3.2	51.5
Santa Ana	6,314	1.7	6	1.0		1	2	45 (9)	*	1	11	32	334
Mexico													
Yaqui	73,000	79	6	4.3	18	3	5	107	7	1	2	3	7
Conchos	68,386	20.5		2.7		2	2	53 (>38)	5	1			>16
Pánuco	98,227	473		15.0		4	6	>88 (>80)	7	0			
Usumacinta–Grijalva	112,550	2678		77.0		3	7	>112 (103)	11	1	31	3	28
Tamesí	19,127	64.6	3	11.0		2	3	93	4	1			
Lacanjá	800					1	1	44	2	0	15		<1
Salado	60,000	10				2	4	52	0	1			
Candelaria	10,755	46		13.0		2	4	>65	0	0			
Armería–Ayuquila	9,803	30.4	6	10.0	21.8	2	3	38 (32)	3	1	30	10	56
Fuerte	34,247	31		2.8		2	3	51	6	1			

Symbols and Abbreviations

The following symbols and abbreviations may be found in this book without definition. Also undefined are standard mathematical and statistical symbols given in most dictionaries.

A	ampere	G	giga (10^9, as a prefix)
AC	alternating current	gal	gallon (3.79 L)
Bq	becquerel	Gy	gray
C	coulomb	h	hour
°C	degrees Celsius	ha	hectare (2.47 acres)
cal	calorie	hp	horsepower (746 W)
cd	candela	Hz	hertz
cm	centimeter	in	inch (2.54 cm)
Co.	Company	Inc.	Incorporated
Corp.	Corporation	i.e.	(id est) that is
cov	covariance	IU	international unit
DC	direct current; District of Columbia	J	joule
D	dextro (as a prefix)	K	Kelvin (degrees above absolute zero)
d	day	k	kilo (10^3, as a prefix)
d	dextrorotatory	kg	kilogram
df	degrees of freedom	km	kilometer
dL	deciliter	l	levorotatory
E	east	L	levo (as a prefix)
E	expected value	L	liter (0.264 gal, 1.06 qt)
e	base of natural logarithm (2.71828…)	lb	pound (0.454 kg, 454g)
		lm	lumen
e.g.	(exempli gratia) for example	log	logarithm
eq	equivalent	Ltd.	Limited
et al.	(et alii) and others	M	mega (10^6, as a prefix); molar (as a suffix or by itself)
etc.	et cetera		
eV	electron volt	m	meter (as a suffix or by itself); milli (10^{-3}, as a prefix)
F	filial generation; Farad		
°F	degrees Fahrenheit	mi	mile (1.61 km)
fc	footcandle (0.0929 lx)	min	minute
ft	foot (30.5 cm)	mol	mole
ft³/s	cubic feet per second (0.0283 m³/s)	N	normal (for chemistry); north (for geography); newton
g	gram	N	sample size

NS	not significant	tris	tris(hydroxymethyl)-aminomethane (a buffer)
n	ploidy; nanno (10^{-9}, as a prefix)	UK	United Kingdom
o	ortho (as a chemical prefix)	U.S.	United States (adjective)
oz	ounce (28.4 g)	USA	United States of America (noun)
P	probability	V	volt
p	para (as a chemical prefix)	V, Var	variance (population)
p	pico (10^{-12}, as a prefix)	var	variance (sample)
Pa	pascal	W	watt (for power); west (for geography)
pH	negative log of hydrogen ion activity	Wb	weber
ppm	parts per million	yd	yard (0.914 m, 91.4 cm)
qt	quart (0.946 L)	α	probability of type I error (false rejection of null hypothesis)
R	multiple correlation or regression coefficient	β	probability of type II error (false acceptance of null hypothesis)
r	simple correlation or regression coefficient	Ω	ohm
rad	radian	μ	micro (10^{-6}, as a prefix)
S	siemens (for electrical conductance); south (for geography)	$'$	minute (angular)
		$''$	second (angular)
SD	standard deviation	$°$	degree (temperature as a prefix, angular as a suffix)
SE	standard error		
s	second	%	per cent (per hundred)
T	tesla	‰	(per thousand)

Index

A

abiotic processes
    in coldwater rivers, 629
    in coldwater streams, 588–590
    population dynamics of, 332
    in warmwater streams, 658
abundance. *See also* population dynamics; population size
    absolute, 331, 336, 339–340
    average, 44
    of coldwater fishes, 625–630
    estimation, 68, 339–340
    genetic variability-related, 597
    indices, 533, 535
    juvenile, 333
    larval, 333
    natural flow regime effects, 709
    overfishing and, 470
    predation effects, 409–410
    relative, 331, 334, 340, 369
    removal model, 326, 328
    slot length limit effects, 685–686
abundance-size relationship, 475
Academy of Natural Sciences of Philadelphia, 4
acidification, 484–485
acidity, of ponds, 513–514, 516
acid neutralization capacity (ANC), 484–485
acid rain, 484–485
active management, 476–481
Adams Lake, British Columbia, 452
adaptation
    of coldwater fishes, 620
    to food availability, 591
adaptive management, 100–101, 137
    definition, 718
    of natural lakes, 479–480
    population monitoring in, 344
    process, 142–149
    of warmwater rivers, 718–721
advocacy, 107, 158, 161, 635, 688
aeration, 483, 502, 514
Agassiz, Louis, 4–5, 158

age-at-maturity, 331, 342, 342 461, 461, 475
age estimation, 329–330, 341
agencies. *See also* fish and wildlife agencies; nongovernmental organizations (NGOs); *names of specific agencies*
    in coldwater river fisheries management, 631–636
    in coldwater stream fisheries management, 594–595
    in endangered and threatened species management, 632
    habitat improvement involvement, 300, 302–304
    history, 33–34
    human dimensions data use, 428, 429
    interjurisdictional conflict, 33–34
    in invasive species management, 222–226
agriculture
    conservation, 384, 678–679
    drainage, 670, 676
    ecological integrity effects, 374
    habitat effects, 595, 667, 669, 676
    irrigation, 597, 667, 708
    stream ecosystem effects, 669, 672, 676
    warmwater river effects, 712
    water quality effects, 669, 712
    in watersheds, 570–571, 573
air pollution, 485
air temperature, yield effects, 459–460
Alaska
    Afognak Island nature preserve, 14
    ocean ranching in, 262
    subsistence fisheries, 456–457, 471
alevins, 592–593
alewives
    biological control, 217
    diet, 408–409
    as introduced species, 557
    as prey species, 557
    in reservoirs, 577–578
    stocking, 577–578

algae. See also phytoplankton
    abundance, 410
    competition with macrophytes, 405
    filamentous, 518
    in natural lakes, 453
    in oligotrophic lakes, 403
    in warmwater streams, 667
algal biomass
    benthivorous fishes and, 410, 415
    chlorophyll-*a* indicator, 396, 399, 413, 452, 458, 552, 553, 557
    in food web, 399, 400, 403, 407, 410
    productivity index, 396
    in reservoirs, 552, 553
algal blooms
    as fish kill cause, 558
    increase in, 31
    management, 412–413
    in reservoirs, 558
    in urban lakes, 502
    in warmwater streams, 670, 672
alkalinity, of ponds, 513–514
allelic diversity, 337
allelopathy, 405
alternate stable state hypothesis, 412
altitude
    assemblage structure effects, 454
    phytoplankton productivity effects, 457
aluminum sulfate (alum), 412–413, 483, 516
American Association for the Advancement of Science, 9
American Fish Cultural Association, 8–9
American Fisheries Society (AFS)
    ecological integrity statements, 356, 386
    eradication project training courses, 231
    first woman president, 5, 6
    Fisheries History Section, 34
    Fish Management Chemicals Subcommittee, 236
    history, 5, 6, 9
    influence techniques use, 161, 162–163
    introduced species statement, 228, 229
    North American Fish Policy, 20
    propagated fish symposium, 284
    resource policy statements, 354, 356
    scientific writing guidelines, 178

American Rivers, 385, 635
American shad. *See* shad
American Society of Limnology and Oceanography, 386
American water willow, 565–566
amphibians, endangered, 135, 136
anadromous fishes management. *See also* salmonids
    fish passage, 569, 717–718
    interagency involvement, 632
    stocking, 263–264, 282
anadromy, 592
analysis of variance (ANOVA), 93, 96, 97
angler data
    on anglers, 619
    on fish populations, 536–537
anglers
    attitudes, 201, 204, 432–433
    beliefs, 432–433
    characteristics, 430–431
    coldwater stream angling, 601–602
    expectations, 433–434
    expenditures, 427, 428
    influence on decision making, 143–144
    introduction of fish by, 134, 137
    motivations, 433
    national-level studies of, 442
    participation patterns, 431
    perceptions, 432–433
    preferences, 432, 455
    resource-level studies of, 443–444
    satisfaction, 144, 433–434, 465, 466, 602, 640, 686
    species-specific, 443
    as stakeholders, 427–428
    state- or province-level studies of, 442–443
    value orientations, 434–435
angling
    as mortality cause, 591–592
    types of, 186
    warmwater streams, 658
angling effort, 481, 482
anomalies, proportion of population with, 331, 343
antibiotics, 278–279
antimycin, 231, 232, 236, 239, 488, 605
Antiquities Act, 27
Apalachicola River, Florida, 707

aphotic (profundal) zone, 453
aquaculture. See also fish culture
    conservation, 280
    criticisms of, 284
    definition, 261
    history, 18, 20
    mussels, 11
    Regional Centers, 537
    regulations, 223
aquaculture facilities, fish escapes from, 218–219
aquaculture industry, 223
aquarium fishes
    as nuisance species, 217
    release, 227
Aquatic Resources Trust Fund, 118
aquatic vegetation. See also phytoplankton; submerged vegetation
    benthic, 411–412
    biological control, 217, 218, 484, 518, 563, 565
    categories, 517
    chemical control, 484, 518, 563, 565
    density, 484
    ecological assessment, 362
    in ecosystem repair, 566
    epiphytes, 405–406
    eradication, 484
    functions, 517
    as habitat improvement indicator, 316
    macrophytes, 405–406, 414, 518
    mechanical control, 484, 518, 563
    of natural lakes, 452–453, 483–484
    of ponds, 511, 517–518
    of reservoirs, 563, 565–566
archaeological data, 31–32
archives, of fisheries management information, 34
Arctic char, 454
Arctic grayling, 454
Arizona Cooperative Fish and Wildlife Research Unit, 159
Arizona Fish and Game Commission, 114–116, 117
Arizona Fish and Game Department, 228
Army Corps of Engineers (ACE), 18, 112, 125, 309–310, 548
artificial structures
    for habitat enhancement, 561–563
    for streambank erosion control, 677, 678
    in warmwater rivers, 711. 708
Asian carp, 225
Asian clams, 359
Asian fish species, introduction of, 220
Asian swamp eels, 229–230, 238
assemblages
    of coldwater streams, 589–594
    community indicator, 474
    in Everglades, 314–315
    hierarchical organization, 87, 89–90
    of large reservoirs, 551, 552
    manipulation, 488
    of natural lakes, 487–488
    physiographic-drainage combination, 682
    predation effects on, 659, 662, 663, 665
    of reservoirs, 308, 549, 567–568, 571–572, 575–576, 577–578
    of river basins, 575–576
    riverine fishes, 706
    of warmwater rivers, 715–716
    of warmwater streams, 658–665, 682
    as water quality indicator, 677
assessment
    of coldwater stream fisheries, 602–603
    differentiated from monitoring, 344
    of ecological integrity, 361–373
    of economic information, 435–441
    of fish populations. See population estimation
    of habitat, 296–298, 602–603
    of habitat improvement, 312–316, 317
    social and economic information use in, 425–447
    strategies, 149, 150–151
    of strategies, 149, 150–151
Association of Fish and Wildlife Agencies, 303–304
Atherinidae, 281
Atlantic salmon, harvest regulations, 6
Atlantic States Marine Fisheries Commission, 278
Atlantic Striped Bass Conservation Act, 278
Atlas of North American Freshwater Fishes (Lee et al.), 4

ATLSS project, 100
Atomic Energy Act, 28
A_r value, 530, 531, 533
Audubon, John James, 4
Au Sable River, Michigan, 630, 636
authority figures, 161, 162

B

backwater lakes, habitat restoration, 309–311
backwaters
    of coldwater rivers, 627
    connectivity, 707
    of reservoirs, 704
    of rivers, 623, 627, 707
bacterial load, 331, 344
bag limits. *See* creel limits
Baird, Spencer Fullerton, 4, 5, 7, 158, 215
bait-bucket releases, 213, 223, 680–681, 686, 687
baitfishes
    as disease vectors, 641
    harvest regulations, 686
    tapeworm infections, 218
    unauthorized release, 213, 223, 680–681, 686, 687
    from warmwater streams, 686
bait restrictions, 197
ballast-water introductions, 31, 219–220, 223–224
Baranov's equation, 477
barriers
    to fish passage, 18–20
    to habitat connectivity, 644
    installation, 606
    to migration, 593, 597, 629, 707
    for nuisance fish control, 239–240
    removal, 606–607
Bayesian belief networks (BBNs), 98, 99, 100
Bayluscide, 231, 237
Bear River, 599–601
before-after-control-impact (BACI) design, 372
behavior, of hatchery fish, 269–270, 277
benchmark values, 473
benefits, optimum, 474
benthic-pelagic links, in food webs, 411–412
benthic zone, of natural lakes, 453
benthivorous fishes, algal biomass effects, 415
best management practices, 226, 573, 679

bias
    in abundance estimates, 328
    in population sampling, 325, 326–327
Big Horn River, Wyoming, 647
bioaccumulation, 485, 712–713
biocides, 231
biodiversity
    of coldwater rivers, 620, 623
    invasive species as threats to, 223
    river continuum concept of, 703
    of warmwater rivers, 709–710
biodiversity conservation
    conflict with hatchery programs, 271–273
    in stocking, 227–228
bioenergetics models, 403–404, 416, 555
biological control
    of aquatic vegetation, 217, 218, 484, 518, 563, 565
    diseases as, 240–241
    exotic species as, 285
    fish as, 213
    introduced species as, 216, 217
    of invasive species, 240–241
    of mosquitoes, 264
biological data. *See also* biological statistics
    in population dynamics, 330–336
biological reference points, 200
biological statistics. *See also* indices
    for ecological integrity assessment, 372–373
    for population estimation, 330
    precision (repeatability), 325, 326–327
    sampling bias, 326
    sampling design, 329
biologic data
    analysis, 327
    collection, 142, 145
    presentation of, 177
    relationship to models, 98
biologic integrity index, 677
biomanipulation, 264, 414–415
biomass. *See also* algal biomass
    availability, 475
    of catch-and-release populations, 638
    equilibrium, 473
    factors affecting, 44
    indices, 530–532

at maximum sustainable yield, 473
morphoedaphic index (MEI), 399–400, 557
of phytoplankton, 458
precautionary, 470–471
as productivity index, 400
of stocked fish, 287
of unexploited populations, 461, 462–463
water turbidity and, 412
biomass:production ratio, 460–461
biotic indices, 365–372
community structure parameters, 366, 367–369
fish, 366–372
macroinvertebrates, 366, 367
biotic integrity. *See also* ecological integrity
criticism of, 366
definition, 353
index, 357–358
biotic processes
in coldwater streams, 590–594
in lentic ecosystems, 308
in lotic ecosystems, 308
population dynamics of, 332
in warmwater rivers, 706, 714–716
in warmwater streams, 658–659
biotic resistance hypothesis, 360
birds
migratory, 20, 122, 123
piscivorous, 21, 630
birth rate, 330–331
Bisazic, 241
black bass
anglers' preference for, 455
as apex predator, 665
catch-and-release regulations, 190
closed season, 205
creel limits, 685
latitudinal gradient, 454
length restrictions, 185
maximum length limit, 196
pond management, 523–524
seasonal closures, 682, 686
total catch model, 64
black carp, 225
black crappie, 281, 327
catch-per-unit effort measurement, 61
total-length-at-ages estimates, 59–60
yield-per-recruit model, 64, 65–67
blackflies, 627
blood analysis, 331, 343–344
bluegill
coppernose, 526–527
edge habitats, 414
largemouth bass predation on, 190, 196
pond management, 512, 516, 520, 521, 525, 526–527, 536
stocking, 281, 487, 525
stock-length catch rate, 340
supplemental feeding, 516
in urban lakes, 502
warmwater assemblages, 454
blue tilapia, 239
boating regulations, 486, 636
body size, of fishes. *See* size, of fishes
Boundary Waters Treaty, 21
bounties, 681
Brando, Marlon, 161
broodstock
in genetically-segregated hatcheries, 272
genetic drift, 267–268
genetic integrity, 277
population size, 268
selection, 261
spawning behavior, 269
in supplemental hatcheries, 265
brook trout
climate change effects on, 598
displacement by rainbow trout, 219
Eastern Brook Trout Joint Venture, 643
eradication programs, 232, 233, 234
harvest regulations, 637
hybrid, 337, 641
as introduced species, 641
minimum length limits, 637
in reservoirs, 549
stocking, 12, 14
stream order-related distribution, 629, 630
brown trout
climate change effects on, 598
competition with stocked fish, 640
hatchery-reared, 270
hybrid, 219
as introduced species, 215, 641–642

in reservoirs, 549
stocking, 12
stream order-related distribution, 629
bubble curtains, 240
buffalo
commercial fisheries, 11
in reservoirs, 552, 571
buffer strips, 572
buffer zones, 396, 578, 680
bullheads
anglers' preference for, 455
pond management, 521
as turbidity cause, 518–519
bull trout
climate change effects on, 598
heavy metal toxicity in, 598
hybrid, 337, 641
recovery plan, 91–92
stream order-related distribution, 629, 630
bulrushes, 518
Bureau of Biological Survey, 8, 20
Bureau of Fisheries, 8, 15, 17, 18, 20
Bureau of Indian Affairs, 124
Bureau of Land Management, 112, 130, 141
Bureau of Reclamation, 126, 548
bycatch, 475

C

Cache la Poudre River, Colorado, 640
calcium
as fertilizer, 512
as nutrient, 452
CALFED Bay-Delta Program, 26
California
eradication programs, 242
native species management, 135–137
northern pike eradication program, 152–154
California Central Valley Project, 26
California Department of Fish and Game, 135–137, 152–154, 181
California Environmental Quality Act, 136, 153
California Water Code, 128
Canada. *See also* First Nations peoples
Crown lands, 20–21, 130, 632
early fisheries research, 8
Environmental Assessment Agency, 561
eradication programs, 238
fisheries agencies, 7
Fisheries and Oceans Canada, 26, 122, 222, 225
fish habitat program, 302
fish hatcheries, 7, 286
Freshwater Fish Marketing Corporation, 455–456
Great Lakes habitat improvement participation, 301
harvest regulations, 187
international treaties, 121, 122
introduced species, 359–360
invasive species regulations, 222, 223, 224, 225–226
large reservoirs, 546, 547, 551
marine fisheries regulations, 139–140
natal lake management, 478–481
national parks, 14
natural lake access, 489
piscicide regulations, 231
political system, 108–109, 110
public trust principle, 119
recreational fishing surveys, 442, 455
Salmon Enhancement Program, 262
salmon stocking, 631
stocking regulations, 228
water rights, 22, 128
Canadian laws
Canada Water Act, 558
Constitution Act, 20–21, 121
Environmental Assessment Act, 561
Environmental Protection Act, 22
Federal Water Policy Act, 380, 382
Fisheries Act, 6, 26, 112, 124, 125, 225, 302
Imperial Fisheries Judgment, 14
international treaties, 121, 122
Migratory Birds Convention Act, 122
Natural Resource Transfer Act, 20–21
Rocky Mountain Park Act, 14
Shipping Act, 223
Species at Risk Act, 22, 380, 382
canneries, 18
canning, of fishery products, 11

capture-and-release, 203
capture-recapture, 68
carbon
    as productivity-limiting nutrient, 396
    release from macrophytes, 405
carbon dioxide, as global warming cause, 689
carcasses
    genetic analysis, 339
    as nutrient source, 417, 646
carp. *See also* common carp
    commercial fisheries, 456
    diet, 409
    eradication programs, 233, 237, 240, 241
    feeding behavior, 409
    as "injurious wildlife," 225
    as introduced/invasive species, 215, 225, 715
    stocking, 280, 281
    as turbidity cause, 518–519
    in warmwater rivers, 715
carrying capacity
    intraspecific competition and, 594
    life cycle approach, 463
    of unexploited populations, 462–463
Carson, Rachel, 158
Caspian Sea, 450, 451
catadromous fishes, fish passage, 569
catch
    median length, 475
    in recreational fisheries, 465, 466
    catch-and-release, 186, 190, 198
    for coldwater fishes, 638–639
    cultural attitudes toward, 434
    management objectives, 467
    for trout, 148
    in urban lakes, 502
catch-at-age matrix, 68
catch-at-age models, 64, 68–77, 477
catch curve analysis, 51, 52–53
catch limits. *See* creel limits
catch per unit effort (C/f) (CPUE), 332–335, 535
    in fishery-independent surveys, 473
    gill netting-based, 480–481
    in natural lakes, 465–466, 473
    for overfishing estimation, 481, 482
    in recreational fisheries, 465–466
    as recruitment measure, 60–62
    as relative abundance index, 340
    variation among anglers, 465–466
catfish. *See also* channel catfish
    anglers' preference for, 455
    as apex predator, 665
    commercial fisheries, 11
    flathead, 681
    harvest regulations, 686
    as introduced/nuisance species, 219, 549–550, 716
    latitudinal gradient, 454
    length-based creel limits, 195
    overfishing, 714
    pond management, 523
    recreational fisheries, 715
    in reservoirs, 549–550, 552
    sailfin, 219
    stocking, 263, 280
    transportation, 279
    walking, 225
    in warmwater rivers, 716
Catostomidae, 700
cattails, 518
caviar, 456
Center for Climate Strategies, 180
Centrarchidae, 281, 454, 700
Centre for Strategic Management, 172
Centre of Expertise for Aquatic Risk Assessment (CEARA), 223
change, through community participation, 180–183
channel catfish
    commercial fisheries, 456
    diet, 409
    growth, 686
    pond management, 516, 521, 523, 525
    stocking, 502, 525
    supplemental feeding, 516
    in urban lakes, 502
channelization
    flood magnitude effects, 667
    habitat effects, 668, 711
    of rivers, 704
    streamflow effects, 667, 668
    of warmwater streams, 667, 668, 672
    water quality, 668

channels
    of coldwater rivers and streams, 620, 621, 623
    downstream from reservoirs, 626–627
    navigation, 707, 711
    of warmwater rivers, 707
charal, 456
charts, use in project management, 169, 171
Chattahoochee River, Georgia, 707
chemical analysis, for population estimation, 330
chemical control, 231, 237
    antimycin, 231, 232, 236, 239, 488, 605
    for aquatic vegetation, 484, 518, 563, 565
    Fintrol, 236
    in natural lakes, 488
    Niclosamide, 246–247
    rotenone, 152–154, 231, 232–233, 235–237, 242, 417, 488, 519, 605, 642
    for sea lampreys, 488
chemicals
    bioaccumulation, 485, 712–713
    as water contaminants, 356, 358, 712–713
Chesapeake Bay, striped bass restoration program, 278
Chesapeake Bay Program, 305
Chinook language, 21
Chinook salmon, 18, 640
    genetic analysis, 337
    genetic drift, 268
    habitat, 644
    homing behavior, 339
chlordane, 713
chlorophyll a, 396, 399, 413, 452, 458, 552, 553, 557
chub, commercial fisheries, 456
Cialdini, Robert, 159–161, 162–163
cisco, 454, 456
civil law, 113, 116
clay
    as pond sealant, 504, 507, 509
    as pond turbidity cause, 516–517
Clean Air Act, 22, 28
Clean Water Act, 26, 28, 110, 125, 127, 140–141, 296–297, 353–354, 380, 381–382, 558, 602, 670, 674, 677
clear-water state, 412, 413, 417
climate change, 180, 182
    coldwater stream effects, 598–601, 604
    ecological integrity effects, 379–380
    genetic monitoring, 338
    invasive fish management effects, 248
    warmwater river effects, 713–714
    warmwater stream effects, 689
climatic-morphoedaphic index model, 459–460
cline, 568
Clinton, Bill, 214
closed areas, 196–197
closed seasons, 186, 196, 682, 686
closure, of fisheries, 197
Clupeidae, 281
Coastal Barrier Resources Act, 30
Coastal Zone Management Act, 22, 127, 128
Coast Guard, 226
Coe, Ernest, 313
coefficient of determination (r^2), 333–334
coexistence, of fish species, 629–630
    grain analysis, 83
    of native and nonnative species, 83
cohorts. See also year classes
    biomass, 44
    optimal harvest periods, 467
coho salmon, 18
    hatchery-reared, 270
    stocking, 263–264
"coldwater disease," 591
coldwater fishes
    distribution and abundance, 625–630
    hierarchical filters, 454
    in ponds, 520–524
    thermal guilds, 700
coldwater ponds, 526
coldwater rivers, 619–656
    abiotic and biotic processes in, 629–630
    definition, 619
    ecology, 620–630
    management, 631–647
    structure and function, 620–624
coldwater streams, 587–618
    abiotic processes in, 588–590
    biotic integrity index, 357

biotic processes in, 590–594
characteristics, 588–594
fish assemblages, 589–594
fisheries management, 594–610
invasive species, 604–607
restoration, 604
salmonid life histories and behavioral patterns, 590
trout in, 657
water temperature, 578, 588–589, 593, 594, 595–596
collaborative management, 608, 609
Colorado River, 121, 332, 575, 708, 710, 712, 716, 718–721
Colorado River Basin Restoration, 26
Columbia River, 18–19, 23, 33, 318, 577, 629, 639, 641, 642
Columbia River Basin Fish and Wildlife Program, 26
commercial fisheries
    comparison with recreational fisheries, 474
    fishing license caps, 489
    fish sizes, 460
    in Great Lakes, 14–15, 455–456
    history, 11
    management goals and objectives, 467–471
    in natural lakes, 455–456, 467–471, 489
    overfishing, 18
    public trust doctrine and, 14–15
    for sport fish, 186
Commission for Environmental Cooperation, 122
Commission on Fish and Fisheries, 4, 7, 8, 215, 216, 219
    publications, 9–10
commitment, consistency with, 160, 162
committees, 168
common carp
    commercial fisheries, 456
    control and eradication programs, 233, 488
    stocking, 281
common law, 5–6, 113, 119, 126
common ownership principle, 116, 119
communication techniques, 133, 157–184
    community participation, 180–183

conflict resolution, 161, 163–166
for effective meetings, 178–180
for group project management, 168–173
importance of, 157–158
influence techniques, 158–161, 162–163
negotiation, 166–168
presentations, 173–174, 175, 176–177
scientific publications, 174–175, 177–178
community indicators, 474
community manipulations, 488
community participation, 180–183
community structure parameters, 366, 367–369
competition
    among coexisting species, 630, 665
    "apparent," 419
    in cold water streams, 593–594
    definition, 665
    between hatchery and wild fish, 640
    interspecific, 593–594
    intraspecific, 593, 594
    between native and nonnative fishes, 594, 641–642, 709
    in warmwater streams, 665
competitive or tournament fishing, 186, 198
complex phenomena, 26, 31
Comprehensive Environmental Response, Compensation and Liability Act, 30
Concho River, Texas, 332
conflict resolution, 161, 163–166, 631
connectivity
    of backwaters, 707
    of floodplains, 711
    of habitat, 643–644
    of natural lakes, 453–454
    of streams, 662
conservation
    beginning, 215
    fisheries management versus, 120
    units, 92
    of warmwater streams, 666
conservation agriculture, 678–679
Conservation Reserve Program, 679, 688
contingent valuation method (CVM), 440
continuous fisheries, mortality rate, 49–50

Contreras-Balderas, Salvador, 231
Convention on Nature Protection and Wildlife Preservation in the Western Hemisphere, 121
convict cichlid, 235
coolwater fishes, 554, 625
    climate change effects on, 689
    hierarchical filters, 454
    thermal guilds, 700
Cooperative Fishery Research Units, 25
cost-benefit analysis, 440
cost-effectiveness analysis, of stocking programs, 280, 282
cottids, in coldwater streams, 587
Council for Environmental Cooperation, 224
crabs, mitten, 225
crappie. *See also* black crappie
    anglers' preference for, 455
    pond management, 521–522, 528
    predators, 528
Crater Lake, Oregon, 451
crayfish
    as bait, 686
    fish population effects, 406
    as turbidity cause, 517
    in warmwater streams, 658
creel limits, 186, 192, 193–194, 432–433
    for black bass, 685
    for coldwater fishes, 638–639
    harvest regulations and, 205
    length-based, 194–195
    in marine fisheries, 140
creel surveys, 68, 202
    cost, 467
    disadvantages, 204
    for exploitation estimation, 336
    functions, 467
    of stocked fish, 283
    of urban lakes, 503
cryptic species, 336
culling, 193
cultural eutrophication, 125, 483, 558–559
cultural patterns, 434–435
culverts, 597
cumulative impacts, 360
current status evaluation, of fisheries, 142, 145
cutthroat trout
    climate change effects on, 598, 599–601
    density, 97
    heavy metal toxicity in, 598
    hybrid, 641
    Lahontan, genetic analysis, 338–339
    population isolation, 605
    in reservoirs, 549
    stocking, 14
    stream order-related distribution, 629, 630
Cuyahoga River, Ohio, 670, 673–675
cyanobacteria, 398, 403, 405, 558
Cyprinidae, 281, 700
    in coldwater streams, 587

D

dams. *See also* hydroelectric dams/facilities
    downstream effects, 625–629
    ecological integrity effects, 374, 383
    erosion, 517
    fish passage, 5, 18–20, 376, 569–570, 717–718
    flow regimes, 710
    functions, 359
    Gavins Point, South Dakota, 190, 191–192
    history, 18–20
    as migration barriers, 593, 629
    nonnative species transport and, 360
    on ponds, 504, 505, 506, 507, 509, 517
    regulations, 632
    removal, 297–298, 359, 376, 677, 716, 717–718
    river effects, 704
    serial discontinuity concept, 704
    stream flow effects, 667, 671
    as stream fragmentation cause, 597
    in warmwater rivers, 707
    warmwater stream effects, 667, 668, 670
    water releases, 677, 710, 711, 719
Daphnia, 407, 408, 409
Darwin, Charles, 174–175
DDT, 712
decision making, in fisheries management, 138
    reference point framework, 473–474
demographic data, on anglers, 430

denitrification, 398, 405
density
    definition, 340
    effect of harvest regulations on, 204–205
    estimation, 331, 340
    hierarchical model, 97
    influence on recruitment, 58, 60
    of lake fish species, 463–464
    relationship to growth, 204–205
depletion technique, 603
desert fishes, patterns of rarity, 93
Devils Hole pupfish, 129
dewatering
    below dams, 667
    of coldwater rivers, 631–632
    of salmonid habitats, 628–629
diatoms, 403, 453, 590
diffusers, 514
digital record systems, 34
dimorphism, sexual, 342
Dingell-Johnson Act, 25, 118
dinoflagellates, 403
discrete fisheries, mortality rate, 48–49, 50
disease
    as biological control, 240–241
    in coldwater stream fishes, 591
    proportion of population with, 331
    in warmwater fishes, 665
disease transmission
    in baitfish, 686
    in hatchery fish, 285
    in introduced/invasive species, 224–225
    stocking-related, 641
    in transported fish, 524–525
dispersal, gene flow, 85
dissolved oxygen (DO) concentration
    clinograde profiles, 554, 555
    fluctuation, 414
    heterograde profiles, 554, 555
    in lakes, 463
    lethal, 660
    low, 559, 676
    macrophytes' effects on, 405
    in natural lakes, 452
    phosphorus loading effects, 396–397
    in ponds, 514, 515, 516
    in reservoirs, 552, 554–556
    standards, 364, 555
    tolerance to, 555–556
    in warmwater streams, 660, 662, 670, 672
Disston, Hamilton, 313
distribution
    assessment, 603
    of coldwater fishes, 625–630
    in coldwater rivers, 629–630
    estimation, 331
    historic data, 32
    of North American freshwater fishes, 4
    scale-area curves, 84
disturbance events, 595–596, 604, 663
diversity biologic indices, 365, 367–369
DNA analysis, 338–339
domestication, of hatchery fish, 268–269, 277
dominance biologic index, 369
dominance hierarchy, 590
"do no harm." *See* precautionary management
downstream, effect of dams on, 625–629
drainage
    agricultural, 670, 676
    as eradication method, 239
    of Everglades, 313
    as nuisance fish control method, 519
    into warmwater streams, 669
dredging, 483, 598, 708, 711
dried fish, as food, 456
drift, 590
droughts, 595

E

Eastern Brook Trout Joint Venture, 643
ecological conditions, definition, 353
ecological efficiency, 458
ecological integrity, 353–394, 361
    assessment, 361–373
    definition, 353
    factors influencing, 354–360
    fisheries management for, 361, 373–380
    indices, 357–358, 364–372
    legislation affecting, 380–386
    long-term, large-scale actions for, 378–380

near-term, local-scale actions for, 375–378
quantitative assessment, 372–373
reference conditions, 363–364
of rivers, 704–705
of warmwater rivers, 716
ecological limits of hydrological alteration (ELOHA), 710–711
Ecological Society of America, 386
economic impact analysis, 435–438
economic information and research
definition, 426
general uses, 428
input-output analysis, 435–438
economic value, of fish, 12, 20, 21
economic yield, 467, 468–469
ecosystem
four dimensions, 663
health, 354
historic reconstruction, 319–320
of rivers, 703–704, 706
structure and function, 298
ecosystem-based management, 124–125
ecosystem integrity. See also biotic integrity; ecological integrity
as stream management goal, 688
of warmwater streams, 658
ecosystem restoration
adaptive management, 721
interagency involvement, 632, 635
educational programs, 688
in integrated pest management (IPM), 245
in invasive species management, 226–227
electrofishing
in abundance estimation, 533, 535
catch rate, 327, 340
as eradication or control method, 232, 233–234, 238, 242, 605, 642
in mark-recapture programs, 477
in population assessment, 326
electromechanical barriers, 40, 239
Elkhorn River, Nebraska, 216
Emergency Planning and Community Right-to-Know Act, 30
emergent plants, 518
emigration rate, 330–331
empathy, 164, 165
employment, in fisheries management, 25
endangered species. See also Threatened species
agencies responsible for, 632
"ark" populations, 378
definition, 111
flow regimes for, 710
genetic integrity, 269
increased number, 354
Kootenai River populations, 45
population models, 46
stocking, 264, 280
in warmwater streams, 680, 681
Endangered Species Act, 22, 29, 33, 110, 111–112, 127, 131, 135, 136, 140–141, 271, 380, 382–383, 471–472, 602, 632, 643
Endangered Species Act (Ontario), 382
endocrine-disrupting compounds, 713
energy acquisition, 342–343
energy allocation indices, 343
energy sources. See also food webs; nutrients; trophic levels
in warmwater streams, 667, 677
energy storage, 342–343
enhancement, definition, 297
entrainment, 597, 707
environmental drivers, 375
environmental filters, 364
environmental impact reports, 153
environmental impact statements, 153, 560
environmental orientation, 434
Environmental Protection Act, 560
Environmental Protection Agency (EPA), 112, 518
National Pollutant Discharge Elimination System, 125
water quality standards, 364, 365
Watershed Academy, 688
epilimnion, 453, 556–557
epiphytes, 405–406
EPT, 366
eradication. See also biological control; chemical control; mechanical control; physical control
examples, 232–235, 241–243
of introduced/invasive species, 152–154, 227, 228–243

methods, 642
in ponds, 518–519
of undesirable coldwater fishes, 642–643
water quality and, 415
erosion
control, 677, 679
near ponds, 508, 509, 517
near reservoirs, 551
riverbank, 632, 634, 711
from roads, 644–645
shoreline, 485–486
streambank, 672, 677
in watersheds, 573, 574
Esocidae, 700
estrogenic compounds, 342, 672
ethics, 158–159
ethnic/minority groups, as anglers, 430, 431–432, 434
European fishes, importation of, 215, 216
eutrophication, 356, 358
causes, 483
control, 483
cultural, 125, 483, 558–559
dissolved oxygen concentrations, 660
in the Everglades, 313
invasive species establishment and, 596–597
in lakes, 396
in natural lakes, 452, 458
reduction techniques, 308
trophic state index (TSI) of, 399, 400
in watersheds, 570–571
evaluation. See assessment
evaporation, from lakes, 449
evenness, species, 369
Everglades, 313–315
exclusive economic zones (EEZ), 139, 140
excretion rate, 411
executive branch, of government, 110, 112–113
exotic fishes
definition, 263
economic benefits, 285
stocking, 284
expert judgment modeling, 98–100
exploitation. See also mortality rate, fishing
age-at-maturity size and, 461
definition, 336
juvenile:adult ratio, 341–342
of nonnative fish, 681
size-specific, 341
exploitation rate, 642
slot length limits and, 638
exploited species
population models, 46
proportional size distribution, 340–341
explosives, use in eradication, 232, 238–239
extinction. See also endangered species; threatened species
human activities-related, 354
species introduction-related, 681
extirpation, 597

F

Farm Bill, 679
farm ponds. See ponds
FAST (Fisheries Analysis and Simulation Tools), 64, 330
fathead minnow, 281
$F{:}C$ ratio, 530, 531
fecundity
endocrine-disrupting compounds and, 713
length-specific, 331, 343
relationship to weight, 55
weight-specific, 331, 343
Federal Aid in Sport Fish Restoration Act, 25, 118
Federal Aid in Sport Fish Restoration Program, 442
Federal Aid in Wildlife Restoration Act, 25
Federal Energy Regulatory Commission (FERC), 126, 127–128, 383, 560, 632, 633–634
Federal Food, Drug, and Cosmetic Act, 27
Federal Highway Agency, 112
Federal Insecticide, Fungicide, and Rodenticide Act, 27
Federal Land Policy and Management Act, 29, 130
Federal Register, 112
Federal Water Pollution Control Act Amendments, 22
feeding, supplemental, 526–527
feeding behavior. See also foraging
of coldwater stream fishes, 590–591
feed-trained fish, 527–528

female-only stocking, 528, 529
feminization, of male fish, 672
fertility. *See* fecundity
fertilization, intentional
    of lakes, 414, 417–419
    of ponds, 511–513, 521
Field and Stream (magazine), 9
fingerlings, stocking, 282
Finlayson, Brian, 231
Fintrol, 236. *See also* antimycin
fires
    on rivers, 673
    wildfires, 595–596, 599
First Nations peoples
    fishing rights, 24, 124
    legislative process, 109, 110
    reserve lands, 20, 124
    subsistance fisheries, 457
fish and wildlife agencies
    funding, 25, 427
    history, 20
    primary responsibility, 631
    regulatory authority and function,
        20, 112–113, 117, 118–119, 146
fish and wildlife commissions, 146–147
Fish and Wildlife Coordination Act, 20, 112, 127
Fish and Wildlife Service
    archives, 34
    budget, 25
    Bureau of Commercial Fisheries, 8
    Bureau of Sport Fisheries, 8
    creation, 110
    Endangered Species Act and, 111,
        382–383, 623
    inland water management role, 141
    invasive species management role,
        218, 224, 226
    National Fish Passage Initiative
        program, 623
    National Instream Flow Program
        Assessment Project, 128
    organization, 20
    statutory authority, 115
fish and wildlife state commissioners,
    qualifications, 139
fish biologic indices, 366–372
fish carcasses
    genetic analysis, 339
    as nutrient source, 417, 646

"fish cars," 10, 216
Fish Commission. *See* Commission on Fish
    and Fisheries
fish consumption
    among Native Americans, 2
    anglers' attitudes toward, 433, 434
    of contaminated fish, 485
    of genetically-modified fish, 241
fish culture. *See also* hatchery fishes
    in Canada, 7
    criticism of, 283–286
    definition, 261
    future, 286–287
    history, 7
    production-oriented, 265
fish feeds
    fish meal- or fish oil-based, 285–286
    floating, 516
fisheries
    closure of, 197
    types of, 186–187
Fisheries Analysis and Simulation Tools
    (FAST), 64, 330
fisheries biologists, training, 25
Fisheries Conservation and Management
    Act, 22, 29
fisheries management
    changing concepts of, 5–31
    complexity, 26, 31
    conservation versus, 120
    definition, 425
    dual focus of, 457
    employment, 25
    funding, 118–119
    holistic approach, 26
    new approaches, 26, 31
fisheries managers, 137–138
fishery, definition, 426
fishery commissioners, 123
fishery commissions, 12
    fisheries resource reports, 31–32
    history, 31–32
fishes
    biotic indices, 366–372
    economic and social value, 12, 20, 21
    as food. *See* fish consumption
    as food web component, 408
    as habitat improvement indicator, 316
    plant-eating, 484

fishing effect indicators, 473–475
fishing effort, effect of harvest regulations on, 204
fishing in balance, 475
fishing licenses, 189–190
    in Canada, 14
    for coldwater rivers, 636
    for commercial fisheries, 489
    fees, 427, 440, 441
    as fisheries management funding source, 118
    history, 23–24
    species-specific, 443
fishing rights
    of First Nations peoples, 24, 124
    interjurisdictional disputes regarding, 21
    of Native Americans, 23–24, 150, 161, 634–635
    public trust doctrine of, 14–18, 20–21
fishing techniques. *See also* electrofishing; nets
    restrictions on, 197
fish kills
    algal bloom-related, 558
    aquatic vegetation-related, 517, 558
    stocking following, 264
    summer, 483, 512
    winter, 407, 483, 512, 514, 515
"fishless" waters, 12, 14, 587
fish passage, 379
    barriers, 18–20
    at dams, 5, 376, 569–570
    of reservoir fishes, 569–570
    restoration, 606–607
fish rescue, 12, 13–14
flathead catfish, 681
Flint River, Georgia, 707
flood control, 547, 560, 569, 711
floodplains, 560, 623
    connectivity with channels, 711
    decoupling from channels, 707
    legal restrictions on use, 116
    serial discontinuity concept of, 704
    of warmwater streams, 660, 670
flood pools, fish rescue from, 12, 13–14
flood pulse concept, 623, 624, 663, 702, 703
floods, channelization effects, 667

Florida
    aquarium fish culture, 217
    invasive species, 229–230
    invasive species eradication programs, 238, 240
flow dynamics
    of coldwater rivers, 626, 632
    ecological integrity effects, 359, 377, 379
    hydroelectric power facility effects, 667, 668, 671
    natural flow regime, 663, 704–706, 709–710
    of rivers, 704–706
    of spawning, 705–706
    of tail waters, 628–629
    of warmwater streams, 659, 660, 667, 668, 669, 671, 677
flow regulation, in warmwater rivers, 706, 707, 708, 709–711
flow restoration, 239
fluridone, 565
flushing flows, 646, 647
fly fishing, 197
food. *See also* fish feeds
    availability, 408
    consumption, bioenergetics models of, 403–404
Food and Drug Administration (FDA), 518, 528
Food Quality Protection Act, 30
food webs
    of coldwater rivers and streams, 590, 621–622, 623
    components, 402–410
    descriptive, 401, 402
    dynamics, 415–419
    energy flow, 401, 402
    fish production in, 399–401
    floods and, 667
    interaction, 401, 402
    of lakes, 395, 453
    management, 412–415
    of reservoirs, 551
    theories, 410–412
foraging
    by coldwater stream fishes, 590
    in large reservoirs, 552
    turgidity effects, 712

forestry. *See also* logging
    stream habitat effects, 676
Forest Service, 112, 133, 141, 153
fossil fuels, as acid rain cause, 485
four dimensions, of ecosystems, 663
fragmentation, 706–707
    migration effects, 593
    of warmwater rivers, 708
    of watersheds, 595
Frazier River, 20
frazil ice, 627
freezing, of fishery products, 11
Freshwater Fish Marketing Corporation, 455–456
freshwater protected areas, 682
frogs, endangered, 135, 136, 137
frozen fish products, 456
fruit-eating fishes, 409
fry
    mortality rate, 591
    stocking, 282
    territorialism, 590
functional habitat units, 662, 664
functional process zone (FPZ), 703–704, 706

G

gambusia, 681–682
Gantt charts, 169, 171
gear. *See also* nets
    of Native Americans, 2
    restrictions, 197
gene flow, dispersal-related, 85
genetic analysis, in population estimation, 330, 331, 336–339
genetic diversity
    assessment, 337
    in hatchery-raised fishes, 378
    in species restoration, 472
genetic drift, 267–268
genetic integrity
    habitat fragmentation effects, 597
    of hatchery fish, 284
    stocking effects, 488
    of warmwater fishes, 681–682
genetic markers, 336–337
geographic information systems (GIS), 603, 662, 682, 683–685

geomorphology, historic data, 32
German trout. See brown trout
Girard, Charles Frederic, 4–5
gizzard shad
    growth, 523
    pond management, 522–523
    as prey species, 334, 488, 522–523
    in reservoirs, 552, 571
    stocking, 281
    trophic cascade, 410–411
glacial lakes, 449–452
Glen Canyon Dam Adaptive Management Program (GCDAMP), 718–721
global positioning systems (GPS), 662
global warming. See climate change
glycogen storage, 343
goals, in fisheries management, 21–22, 149, 150–151
    in coldwater river management, 635–636
    definition, 464–465
    differentiated from objectives, 464–465
    for habitat improvement, 298, 300
    identification, 145, 146–147
    in species restoration, 472
    in urban fisheries management, 503
goby, 220
golden shiner, 281
goldfish, as introduced species, 215
gonadosomatic index (GSI), 331, 343
goods and services concept, 297
governmental authority, in fisheries management. *See also* specific agencies, laws, and regulations
    fisheries management policy directives and, 16–18
    government structure and, 108–113
    history, 5–7, 12, 14, 18–21, 113–120
    land use issues, 129–131
    national-level, 121–122
    primary management structure, 121–124
    sub-national-level, 122–124
    watershed or ecosystem management framework, 124–129
    water use issues, 125–129
graphs, as a data presentation technique, 177
grass carp
    as biological control agent, 518, 563

stocking, 281
triploid, 281
grassroots environmental organizations, 386
gravel, as spawning substrate, 589–590, 592–593
gravel mining, instream, 668, 671
gravel pits, 504, 509
grazing, habitat effects, 595, 597, 672, 687
Great Bear Lake, Canada, 450, 451, 457
Great Lakes. *See also* Lake Erie; Lake Huron; Lake Michigan; Lake Ontario; Lake Superior
    commercial fisheries, 14–15, 455–456
    interjurisdictional management, 21
    invasive species, 220
    lake trout monitoring programs, 476–477
    lake trout restoration programs, 471, 487
    maintenance stocking programs, 487–488
    public trust doctrine and, 14–15
    recreational fisheries, 455
    sea lamprey eradication programs, 245, 246–247, 488
    stocking programs, 487
    subsistence fisheries, 457
    zebra mussels in, 715
Great Lakes Fishery Commission, 24, 121, 187, 246–247
Great Lakes Fishery Convention, 121
Great Lakes Regional Collaboration, 301–302
Great Slave Lake, Canada, 450, 451, 457
Green, Seth, 7
Green River, 237, 712
grizzly bears, 137
groundwater
    as coldwater stream input, 587, 598
    as small stream input, 620
    as stream input, 589
group project management, 168–173
growth
    definition, 44
    density-dependent decline, 204–205
    energy utilization for, 343
    estimation, 55–58
    flood pulse concept, 703
    health status and, 344
    indices, 336
    life stage-related variation, 56
    models, 334, 336
    as rate function, 332
growth curves
    allometric, 56–57
    fitted, 57–58, 59–60
    isometric, 56
    models, 55–58
growth rate
    aquatic vegetation effects, 484
    average, 475
    estimates, 330
    importance, 55
    length-at-age, 341
    in population estimation, 334, 336
    sex differences, 336
    specific, 336
guide curves, 548, 549, 560–562
gypsum, 516

H

habitat
    assessment, 296–298, 602–603
    carrying capacity and, 463
    climate change effects, 689
    of coldwater rivers, 620, 622, 623
    of coldwater streams, 589–590
    concept of, 296
    connectivity, 643–644
    definition, 296
    fragmentation, 593, 597, 599
    historic conditions and data, 31, 32–33
    interspecific use, 630
    of lakes, 463–464
    land acquisition for, 122
    of ponds, 504–519
    quality indices, 364, 365
    of riverbanks, 628
    of riverine fishes, 706
    shoreline alterations, 485–486
    spawning, 486–487
    thermal, 463, 464. *See also* water temperature
    time-series photographs, 34
    of warmwater rivers, 711

of warmwater streams, 659,
  660–662, 667–680, 668
 water flow alteration in, 359
habitat-conservation plans (HCP), 383
habitat filters, 658–659
habitat management, 133, 134
 agencies' involved in, 300, 302–304
 in altered systems, 295–324
 with artificial structures, 561–563
 in coldwater rivers, 643–647
 in coldwater streams, 604
 for ecological integrity, 375–378
 ecological principles, 317, 318–320
 economic cost, 604
 evaluation, 312–316, 317
 as fisheries management focus, 26
 funding, 302, 304, 306
 instream, 645–646
 in lentic ecosystems, 308
 in lotic ecosystems, 307–308
 in natural lakes, 482–487, 486–487
 process, 298–300
 public acceptance, 302, 304
 relationship to stocking, 273, 277, 278
 in reservoirs, 308, 311–312
 Upper Mississippi River System
  Environmental Management Plan,
  308, 309–311
 in warmwater streams, 666, 676–677
 in watersheds, 304–307
half-log structures, 561, 562
hand grabbing, 186
Harvard Negotiations Project, 166
harvest
 historic data, 33
 illegal, 202
 for invasive fish control, 240
 in ponds, 527
 in recreational fisheries, 465, 466
 selective, 520, 529
harvest regulations, 185–212
 adverse effects, 205
 in coldwater river management,
  636–639
 for coldwater stream fishes, 591
 enforcement, 187, 202
 evaluation, 201–202
 factors affecting effectiveness,
  202–205

 history, 185–186
 implementation, 187–189
 input controls, 489
 justification, 189
 length-based, 55, 205
 models of response to, 201
 for natural lakes, 489
 noncompliance with, 202
 output controls, 489
 for ponds, 520–524
 for predator-prey balance,
  415–416
 selection, 198, 200–201
 terminology, 296, 297
 types, 189–198
 for warmwater fishes, 714
 weight-based, 55
harvest-tag systems, 190, 191–192, 193
hatcheries
 categories, 264–267
 conservation-oriented, 265–267,
  275–276
 federally-operated, 10, 286
 first, 10
 fish escapes from, 217, 218–219
 floating, 8, 9
 goals and objectives, 275–276
 history, 7
 husbandry techniques, 274–277
 private, 7
 production-oriented, 265, 275–276
 supplemental-oriented, 265
hatchery fishes, 261–293
 behavior, 269–270, 277
 comparison with wild fish, 269–270,
  271–273
 criticisms of, 283–286
 domestication, 268–269, 277
 escapes, 217, 218–219
 genetic management, 267–269,
  271–273
 inbreeding, 267
Hatchery Scientific Review Group, 266, 271
hazard and critical control points (HACCP),
 226
headwaters
 hydrologic cycle, 670
 population genetic analysis,
  338–339

headwater streams, 587, 662. *See also* coldwater streams
    drift, 590
health certificates, 525
health status estimation, 343–344
heavy metals, 331, 343–344, 485, 598
hepatosomatic index, 329, 331, 343
herbicides, 484, 518, 563, 565
heterocysts, 398
heterozygosity, 337
hierarchical filters, 454
hierarchical models/theory, of scale, 87, 89–90, 93, 95–97
Hilsenhoff, William, 357
Hilsenhoff biotic index (HBI), 357–358
historic conditions, as habitat improvement basis, 298, 319–320
historic data
    in contemporary management, 31–34
    sources, 31
history, of North American inland fisheries management, 1–41
    changing concepts of management, 5–31
    coldwater river management, 631
    early ichthyological surveys, 4–5
    introduction of fish species, 214–221
    prior to European settlements, 2–3
    scale and complexity changes, 26, 31
holistic approach, 26
homing, natal, 592
    sex-biased, 339
hook restrictions, 197
Hubbs, Carl Leavitt, 5
Hudson River, New York, 215
Hudson's Bay Company, 33
human dimensions, in fisheries management, 26, 133-134, 152-154. *See also* anglers; stakeholders
    definition, 425
    ecological integrity effects, 354–356, 358–360
    economic and social research, 426, 428–438, 442–444
    of natural lakes management, 489
    of warmwater stream management, 682, 685–686
hunting licenses, 24

hybridization
    genetic analysis, 337
    between native and nonnative fishes, 593, 596
    prevention, 681–682
    of warmwater fishes, 665
hydrilla, 483, 565
hydroelectric dams or facilities
    adverse effects, 597
    in Canada, 546
    on coldwater rivers, 632, 633–634
    history, 18–20
    in Mexico, 546–547
    minimum flow policy, 677
    warmwater stream effects, 667, 670, 671
    water rights and, 126, 127–128
hydrogeomorphic processes, of stream ecosystems, 662
hydrologic alteration
    ecological limits (ELOHA), 710–711
    of warmwater rivers, 709, 710–711
hydrologic cycle, of headwater streams, 670
hydrology. *See also* channelization; channels
    of coldwater rivers, 620, 621, 622, 625
hydropeaking, 710
hypolimnion
    of natural lakes, 452–453, 453, 483
    nutrient enrichment, 483
    of reservoirs, 554, 555, 556–557
    of stratified ponds, 514
    water temperature, 598
hypotheses, 32, 97–98, 101

I

ice, 627
ichthyocides. *See* piscicides
ichthyological surveys, 4–5, 7
ichythology, North American, "father" of, 5
Ictaluridae, 700
ide, 216
identification, of invasive and nonnative species, 230
I_F value, 531
Illinois
    farm pond stocking programs, 263
    fish rescue operations, 13
Illinois River, 707

immigration rate, 330–331
impact analysis for planning (IMPLAN) software, 435, 436–437
importation, of invasive species, 224
impoundments. *See also* ponds; large reservoirs; reservoirs
    nonnative species eradication, 239
inbreeding, of hatchery fish, 267, 277
inbreeding depression, 337
index of biotic integrity (IBI), 370–373
index of vulnerability (IOV), 334
indicator organisms
    biotic indices of, 365–372
    of ecological integrity, 716
indicators
    definition, 473
    of fishing effects, 473–475
    reference point relationship, 473–474
indices
    definition, 327
    ecological integrity, 357–358, 364–372
    ecological integrity assessment, 364–372
    energy acquisition and storage, 342–343
    growth, 336
    habitat assessment, 364, 365
    length-at-age, 341, 342
    maturity status-based, 341–342
    population dynamics, 341–342
    population parameters, 327
    population structure, 340–342
    sex identification-based, 341–342
    size structure, 332–335
    water pollution, 357–358
    water quality, 364–365
influence techniques, 158–161, 162–163
inland fisheries management. *See* fisheries management
inland silverside, 281
input controls, 489
input-output analysis, 435–438
insects
    mayflies, 599, 627
    midges, 453, 627
    mosquitoes, 264
    plant-eating, 484
Instream Flow Council, 128
integrated pest management (IPM), 243–245, 246–247
interbreeding, between hatchery and wild fishes, 640–641
interjurisdictional management, 21, 33–34
international cooperation, for invasive species control, 224
International Joint Commission, 21
International Waterways Commission, 301
intersexing, 713
interspecies hybrids, stocking, 263, 264
introduced fishes. *See also* names of individual fish species, e.g., striped bass
    as biological control agents, 216, 217
    definition, 263
    history, 214–221
    illegal or unauthorized, 216, 686, 687. *See also* invasive/nuisance fishes
    new species, 134–137
    in ponds, 524–525
    in reservoirs, 577–578
    "tens rule," 360
    unauthorized, 216
Inuit, 109
invasive/nuisance fishes. *See also* names of individual fish species, e.g., snakeheads
    adverse environmental effects, 137
    agencies involved in control of, 222
    bait-bucket releases, 213, 223, 680–681, 686, 687
    as biodiversity threat, 223
    biological control, 240–241
    in coldwater streams, 596–597, 604–607
    definition, 214
    ecological integrity effects, 359–360
    economic costs, 359
    eradication and control programs, 152–154, 228
    in Everglades, 313–314
    as extinction cause, 681
    identification, 230
    impact on native species, 134–137
    integrated pest management (IPM), 243–245, 246–247
    management, 221–227
    quarantine laws, 224–225
    in reservoirs, 577–578
    risk reduction, 227–228

"tens rule," 360
transportation restrictions, 279–280
vectors, 215–221
in warmwater streams, 680–681
inventory, 145
inventory, in adaptive management, 142, 145
invertebrates. *See also* macroinvertebrates
of coldwater rivers, 621, 623, 626
of coldwater streams, 590
of coolwater streams, 627–628
of natural lakes, 452–453
as stream prey base, 620
of tailwaters, 628, 629
of warmwater streams, 658
Iowa
fish rescue operations, 12, 13
IOWATER program, 688
lake and watershed management programs, 573–574
iron, as phosphorus binding agent, 396–397
irrigation, 597, 667, 708
islands, 623
Izaak Walton League, 688

J

Jordan, David Starr, 5, 158
journal articles, how to write, 178
judicial branch, of government, 113
jug lines, 197
juvenile:adult ratio, 331, 341–342
juvenile fish
competition with adult fish, 594
diet, 408
habitats, 644
stocking, 282

K

Kanawha River, West Virginia, 711
Karr, James, 357
Kennebec River, Maine, 635
keystone species concept, 401
Kissimmee River, Florida, 308, 313, 707
Klamath River, 26, 33
kokanee
competition with opossum shrimp, 137, 406, 410
diet, 408

as prey species, 416, 488
stocking, 487

L

Lacey Act, 122, 224, 225
lake(s). *See also* natural lakes; names of specific lakes
dimictic, 413
eutrophic, 412–413
"fishless," 12, 14
karst, 452
oligotrophic, 414, 418–419
polymictic, 413
restoration, 375–377, 414–415
Lake as a Microcosm, The (Forbes), 419
Lake Baikal, Russia, 452
Lake Chelan, Washington, 452
Lake Davis, California, 153–154, 242–243
Lake Erie, 450, 451, 674–675
Lake George, New York, 5
Lake Gogebic, Michigan, 33
Lake Guerrero, Mexico, 455
Lake Huron, 24, 33, 450, 451, 487
Lake Mead, Nevada, 575
Lake Merced, California, 240
Lake Michigan, 24, 450, 451, 472, 488
Lake Okeechobee, Florida, 313
Lake Ontario, 450, 451
Lake Powell, Utah, 575
lake sturgeon, 21, 33
Lake Superior, 24, 240, 450, 451, 452, 472, 473, 476–477, 487
Lake Tahoe, California and Nevada, 452
Lake Tanganyika, 452
lake trout
anglers' preference for, 455
coldwater assemblages, 454
commercial fisheries, 456
control programs, 488
optimal habitat, 463
potential yield assessment, 463
predator-prey relationship, 415–416
restoration program, 471, 472
sea lamprey predation, 476–477
stocking, 12
sustained yield, 461
Lake Washington, Washington, 483

lake whitefish
	commercial fisheries, 456
	sustained yield, 461
	thermal habitat area versus lake surface area, 464
Lake Winnipeg, 450, 451
land ownership, 129–131
land use management, near coldwater rivers, 645
land use practices. See also agriculture; logging; mining; urbanization
	aquatic ecosystem effects, 355–356
	ecological integrity effects, 373–374
	warmwater stream effects, 660, 678–680
	in watersheds, 396
largemouth bass
	as apex predator, 665
	as bluegill predator, 190, 196
	catch-and-release regulations, 190
	diet, 408
	feed-trained, 527–528
	female-only stocking, 528
	as introduced species, 680
	minimum length limits, 204–205, 520
	overfishing, 519–520
	pond management, 512, 516, 519–521, 522–524, 525, 527–528, 536, 537
	population estimation, 332–335
	proportional size distribution, 340–341
	quotas, 521
	stocking, 263, 281, 487, 488, 525, 527–528
	stock-length catch rate, 340
	supplemental feeding, 516
	trophy size, 522–523, 527
	in urban lakes, 502
	warmwater assemblages, 454
large reservoirs, 545–586
	definition, 545, 570
	diversity, 546–550
	flood control, 547
	guide curves, 548, 549
	hydroelectric, 546, 547
	longitudinal gradients or patterns, 566–568, 574–576
	main-stem, 547, 548
	navigation, 547
	nutrients and water quality, 552–559
	in river basins, 574–576
	run-of-river, 547, 548
	submerged structures and vegetation, 561–566
	suspended sediments and sedimentation, 550–552
	tributary storage, 547–548
	water retention and water level fluctuation, 559–561
	water supply, 547
large warmwater rivers. See warmwater rivers
Larkin, Peter, 133
larval fish
	diet, 408
	as prey, 409
	stocking, 282
latitude
	assemblage structure effects, 454
	phytoplankton productivity effects, 457, 458
least-disturbed conditions, 363–364
Leech Lake, 456
legal processes, in fisheries management. See also governmental authority, in fisheries management; specific laws and regulations
	government structure and, 108–113
legislation. See also specific legislation
	affecting ecological integrity, 380–386
	for invasive species management, 222, 224–226
legislative branch, of government, 108–110
legislative process, 108–110
length
	average, 466
	average maximum, 475
	growth rate relationship, 56
	mean distribution, 475
	median, at maturity, 475
	structure indices, 533
length-at-age growth curve, 56, 57
length-at-age index, 331, 341, 342
length limits, 194–196. See also maximum length limits; minimum length limits
	for coldwater species, 636, 637, 638
	for predator-prey balance, 415–416

restrictive, 185, 186
for trophy fishes, 467
lentic ecosystems, 308
Leopold, Aldo, 133, 158
lesions, 343
LeSueur, Charles Alexandre, 4
levees, 670, 711
licenses. *See also* fishing licenses
for hydroelectric facilities, 127–128, 560–561
life cycle assessment, 463, 464
life history
of coldwater fishes, 592–593
habitat improvement and, 319
migratory, 90–91
as potential yield determinant, 460
relationship to natural mortality rate, 461–462
liking and similarity principle, 159–160, 162
liming
for acidification control, 485
of lakes, 483
of ponds, 513–514, 516
limnology, 395
linear expression analysis, 93, 95–97
lipids, tissue levels, 331, 343
littoral law, 113, 116
littoral zones
fish assemblages, 568
fish growth in, 703
of natural lakes, 453
of reservoirs, 551, 560–561, 562, 563, 568, 571
submerged vegetation in, 483–484
live food fish industry, 223
liver weight. *See* hepatosomatic index
livestock
access to ponds, 517
grazing, 595, 597, 672, 687
logging
clear-cutting, 596
coldwater river effects, 645
coldwater stream effects, 595, 596
ecological integrity effects, 374
stream ecosystem effects, 672
warmwater stream effects, 669
logging roads, 644
logistic population growth model, 44, 45
longnose gar, 232, 238

lotic ecosystems, 307–308
Louisiana, fish rescue operations, 13

M

Mackenzie River, 307
macroinvertebrates
benthic, 406–407, 408
biotic indices, 366, 367
ecological assessment, 362
as habitat improvement indicator, 316
as lake food web component, 406–407, 408
macrophytes
definition, 518
as lake food web component, 405–406
management, 414
Madison River, Montana, 640
magnesium, as fertilizer, 512
Management Pro, 172
Manistee River, Michigan, 633–634
marine fisheries management, 139–142
ecosystem effects, 474
holistic approach, 26
National Marine Fisheries Service, 127
Marine Mammal Protection Act, 29
mark-recapture, 477–478, 603
Maslow, Abraham, 159, 161
maturity
age at, 331, 342, 342 461, 475
status, 341
maximum economic yield (MEY)
in commercial fisheries, 469–469
in lakes, 469–469
maximum length limits, 196, 638
maximum sustainable yield (MSY), 185, 458–459
of commercial fisheries, 468
estimation, 460–461
as exploitation limiting point, 474
in lakes, 468
of natural lakes, 457
as overfishing indicator, 481, 482
reference point framework, 473–474
mayflies, 599, 627
mean length, 331

mean length of population, 475
mechanical control. *See also* barriers
 of aquatic vegetation, 484, 518
mechanical methods, for control and
 eradication, 232, 233–234, 238–240
median length of catch, 475
meetings
    effective, 172, 178–180
    presentations at, 173–174, 175,
      176–177
mercury, 485, 712
metalimnion, 453, 554
methyltestosterone, 528
Mexico
    commercial fisheries, 456
    Constitution, 121
    eradication programs, 238, 242
    indigenous peoples, 124
    international treaties, 121, 122
    introduced species, 359–360
    invasive fish regulations, 222,
      223–224, 225, 226
    legal system, 113
    marine fisheries regulations, 140
    National Commission for The
      Knowledge and Use of
      Biodiversity, 226
    National Water Law, 380, 382
    political system, 108–109, 110
    recreational fisheries, 455
    reservoirs, 546–547, 551
    water quality legislation, 558
Michigan Department of Environmental
 Quality, 633
Michigan Department of Natural Resources,
 637
Michigan Fish Commission, 33
Microsoft® Office PowerPoint, 173
midges, 453, 627
migration
    barriers to, 593, 597, 629
    of birds, 20, 122, 123
    international agreements regarding,
      121
    of reservoir fishes, 569–570
    spawning, 646
    through lake inlets and outlets, 454
    of zooplankton, 408
Migratory Bird Conservation Act, 122

milfoil, 518
minimum length limits, 55, 194, 196, 204–205
    for coldwater species, 626, 637
    for largemouth bass, 520
    in pond management, 520
    stunting and, 341
    for warmwater river fishes, 714
mining, 713
    coldwater stream effects, 598
    historic data, 32
    warmwater stream effects, 668, 671
Minnesota, fish rescue operations, 13
Minnesota Forest Resources Council, 678
Minnesota River, 676
Miramichi River, New Brunswick, Canada, 3
Mississippi, fish rescue operations, 13
Mississippi Interstate Cooperative Resource
 Agreement (MICRA), 26
Mississippi River, 8, 12, 13–14, 225, 308,
 676, 703, 707, 711, 713, 714, 715
Missouri, fish rescue operations, 13
Missouri River, 707, 710, 711
Mitchell, Peter, 6
Mitchell, Samuel Latham, 4
Mitchell Act, 18
mitigation, 297
mixed fisheries, 186, 592
models, relationship to data, 98
molecular genetic data, 338–339
molluscides, 23, 231
mollusks. *See also* mussels, freshwater
    as introduced species, 359
monitoring
    in adaptive management, 149,
      150–151
    differentiated from assessment, 344
    of habitat improvement programs,
      312, 315–316
    lake trout, 476–477
    walleye, 477–478
monosex fish, 528
Montana Department of Fish, Wildlife, and
 Parks, 150–151
Moore, Emmeline, 5, 6
Moronidae, 700
morphoedaphic index (MEI), 399–400, 459,
 557
    calculation, 459
    climatic model, 459–460

of lake fishery production, 459
in large reservoirs, 557
in natural lakes, 452
Morrell River, 639
mortality rate, 46–51
    among coldwater fishes, 591–592
    catch-at-age analysis, 68, 77
    in continuous fisheries, 49–50
    in discrete fisheries, 48–49, 50
    estimation, 330
    estimation of, 51–55
    expressions, 46–50
    finite, 47, 48, 49
    fishing, 44, 46, 48–49, 49–50, 51, 53–55, 336, 470, 473, 475
    health status and, 344
    importance, 330–331
    instantaneous, 47, 48, 49
    instantaneous natural, 462
    larval, 46, 58
    natural, 44, 46–47, 48–49, 53, 55, 68, 77, 203, 336, 460–461, 461–462
    post-release, 203, 336
    precautionary, 470–471
    as rate function, 332
    relationship to harvest regulations, 203, 204
    total, 51, 52–53, 475
    total annual, 51, 336
    total instantaneous, 336
mosquitoes, biological control, 264
mosquitofish, 681–682
    eradication programs, 232
    prey-species consumption estimates, 404
multiple-species management, 682
multiple-use land management, 130
Museum of Comparative Zoology, 5
muskellunge
    as prey species, 409
    in reservoirs, 549
    stocking, 681
    trophy fisheries, 467
muskgrass, 518
mussels, freshwater, 11, 316, 658, 707, 711
    commercial fisheries, 11
    culture, 11
    as endangered species, 658
    as habitat improvement indicator, 316
    invasive, 220, 224
    zebra, 225, 359, 658, 715

N

National Administrative Procedure Act, 110, 112
National Atmosphric and Oceanic Administration (NOAA), 34
National Environmental Policy Act (NEPA), 22, 28, 112, 127, 140–141, 153
National Fish Habitat Action Plan, 300, 302, 303–304, 643, 648
National Fish Passage Initiative, 623
National Forest Management Act, 602
National Historic Preservation Act, 127
National Instream Flow Program Assessment Project, 128
National Invasive Species Council, 222
National Invasive Species Management Plan, 214, 222
National Marine Fisheries Service, 127, 382–383
National Oceanic and Atmospheric Administration (NOAA), 8
National Park Service, 12, 110, 141
National Park Service Organic Act, 27
national park system, 12, 14. *See also* names of specific national parks
National Pollutant Discharge Elimination System, 125
National River Restoration Science Synthesis database, 295
National Salmon Park, 14
National Science Foundation, Partnerships for Enhancing Expertise in Taxonomy, 230
National Survey of Fishing, Hunting, and Wildlife-Associated Recreation, 438, 442, 501
National Wetlands Program, 118
National Wild and Scenic Rivers program, 688
Native Americans. *See also* First Nations peoples; tribal fisheries
    coldwater river management involvement, 634–635
    commercial fisheries, 456
    fisheries management involvement, 23, 24

fishing and hunting management
  involvement, 123, 124
fishing rights, 23–24, 150, 161,
  634–635
government interactions with,
  123–124
habitat improvement involvement,
  301
harvest restrictions, 187
hunting rights, 634
land rights, 22–23
legislative process, 109
subsistence fisheries, 457
water management authority, 142
water rights, 129
native species, protection from introduced
  species, 134–137
natural flow regime, 663, 704–706, 709–710
natural lakes, 449–500
  active management, 478–481
  basins, 449
  classification, 450–452, 452
  commercial fisheries, 455–456,
    467–471
  connectivity, 453–454
  depth, 463
  distribution, 450–452
  drainage areas, 451, 453, 454
  drained, 450
  fishery potential, 457–464
  formation processes, 449–450, 452
  glacial, 449–452
  habitat zones, 453
  inlets and outlets, 449–450, 454
  management goals and objectives,
    464–472
  management strategies, 481–489,
    490
  passive management, 478–479
  recreational fisheries, 455, 464–467,
    474, 476
  reference point framework, 473–476
  seepage, 449–450
  spatial scale, 479
  subsistence fisheries, 456–457, 471
  trophic state continuum, 452–453
natural resources, as public trust, 138
Natural Resources Conservation Service,
  112

Nature Conservancy, 385
navigation, right of, 119
Nebraska
  Aquatic Habitat Plan, 306
  geographic information systems
    (GIS) use in, 683–685
needs, hierarchy of, 159, 161
negotiation, 166–168
  for fishing rights, 21
Neosho madtom, 710
nests (redds), 592, 593
net pens, 283
nets
  fyke, 327, 477, 642
  gill, 234, 238, 480–481, 564, 642
  mesh-size restrictions, 197
  use in eradication, 231, 238
  use in urban lakes, 503
network dynamic hypothesis, 620–621, 622,
  664
Neuse River, North Carolina, 717–718
new species, intentional introduction,
  134–137
New York State Department of Conservation,
  5, 6
Niclosamide, 246–247
nitrates, 676
nitrification, 398
nitrogen
  as fertilizer, 512
  in natural lakes, 452
  as productivity-limiting nutrient, 396,
    398
  release from macrophytes, 405
  in reservoirs, 552, 553
  in streams, 660
nitrogen cycle, 398
nitrogen:phosphorus ratio, 398, 414, 483
no-kill regulations, See catch-and-release
nongame fishes
  conservation, 135–137, 142
  eradication programs, 21
  hatchery programs, 377
  population assessment, 326
nongovernmental organizations (NGOs)
  coldwater river management
    involvement, 635
  ecological integrity management
    involvement, 384–385

habitat improvement involvement, 299, 300, 301–302
natural resource management involvement, 141
Nonindigenous Aquatic Nuisance Prevention and Control Act, 222
nonnative fishes. *See also* introduced fishes; invasive/nuisance fishes
    competition with native species, 630
    fish passage, 607
    identification, 230
    stocking, 227–228
Norris Lake, Tennessee, 561–563, 564
Norris Reservoir Fishery, Tennessee, 142, 143–144
North American Benthological Society, 386
North American Fish Policy, 20
North American Free Trade Agreement (NAFTA), 122, 224
North American Lake Management Society, 386
northern pike
    eradication programs, 152–154, 233, 235, 242–243
    female-only stocking, 529
    intraguild predation, 409
    latitudinal gradient, 454
    pond management, 529
    in reservoirs, 549
    sustained yield, 461
    thermal habitat area versus lake surface area, 464
North Platte River, 647
Novak, Vanna, 176–177
nuisance fishes. *See* invasive/nuisance fishes
nutrients. *See also* carbon; nitrogen; phosphorus
    abatement programs, 483
    cycling, in lakes, 459
    in downstream environments, 627
    as fertilizer component, 512
    from fish carcasses, 417, 646
    high concentration, 412–413
    in lakes, 396–399
    loading, in reservoirs, 552–553, 558
    low concentration, 414
    management, 412–414
    in natural lakes, 457–458
    productivity-limiting, 396–398, 400
    in salmonid ecosystems, 416–417, 418
nutrient spiraling concept, 663

O

objectives, in fisheries management, 149
    definition, 464–465
    differentiated from goals, 464–465
    for ecological integrity management, 374
    ecosystem-related, 474
    for habitat improvement projects, 300
    identification, 145, 148
    reference point framework, 473–474
    for species restoration, 472
ocean ranching, 262, 264
off-channel habitats, 644
Ohio Environmental Procession Agency, 674
Ohio River, 4
Oil Pollution Act, 30
Oklahoma Department of Wildlife Conservation, 681
oligotrophication, 417–419, 558–559
omnivorous fishes, 409
Oneida Lake, New York
On the Origin of Species (Darwin), 174–175
opossum shrimp, 137, 406, 410
optimum sustainable yield (OSY), 186
organic matter, as turbidity control, 517
organization, hierarchical, 87, 89–90
Organization of Wildlife Planners, 168–169
organochlorines, 485
ornamental fishes
    culture, 261
    shipment, 216
outbreeding depression, 488
output controls, 489
outreach programs, 9
    for invasive species management, 226–227
ovaries, egg content, 343
overexploitation, of prey species, 484
overfishing
    calculation, 481, 482

in commercial fisheries, 18, 470
definition, 46
growth, 46, 58, 60, 62–64, 65–67, 489
harvest regulations, 200
identification, 43
in lakes, 470
in marine fisheries, 140
in natural lakes, 481, 482, 489
in ponds, 519–520
quality, 489
recruitment, 46, 58, 60, 200, 470, 489
in urban lakes, 504
in warmwater rivers, 714–715
yield-per-recruit models, 62–64, 65–67
Overman, Barton Warren, 5, 215
overwintering, in coldwater pools, 590
oxygen requirements. See also dissolved oxygen (DO) concentration
    of transported fish, 279

P

Pacific Northwest
    salmon hatchery programs, 271–273
    salmon restoration, 267
Pacific Railway Act, 22–23
Pacific salmon
    habitat, 644
    Native American fisheries, 2
paddlefish
    harvest regulations, 190, 191–192, 196, 718
    overfishing, 714
palmetto bass, 528
panfishes
    anglers' preference for, 455
    latitudinal gradient, 454
    stocking, 520, 521–522
parasites
    in coldwater stream fishes, 591
    proportion of population with, 331, 343
    stocking-related transmission, 641
    transmission in transported fish, 524–525
    in warmwater fishes, 665
parental care, 592, 593
passive management, 474, 476, 478–479
patches, 92
patch networks, 92
pathology laboratories, 344
PCBs (polychlorinated biphenyls), 713
peer review process, 320
pelagic-to-demersal ratio, 475
Percidae, 700
Pere Marquette River, Michigan, 636
performance criteria, for habitat improvement, 320
permits, for fishing, 190, 193. *See also* fishing licenses
pesticides, 712–713
Peterson, Tom, 180, 182–183
pH, 484–485
pheromones, 241, 488, 642–643
phobic zone, 453
phosphorus
    algal blooms and, 412–413
    as fertilizer, 512
    inactivation, 483
    in lakes, 458
    in natural lakes, 452
    as productivity-limiting nutrient, 396–397, 398, 400, 458, 552, 553, 660
    in reservoirs, 552, 553
    in streams, 660
photographs, 34
physical methods, for control and eradication. *See* mechanical methods, for control and eradication
phytoplankton
    biomass, 458
    definition, 518
    in eutrophic reservoirs, 558
    as food web component, 402–403
    in natural lakes, 452–453
    primary productivity, 457, 458
Pinchot, Gifford, 133
piranha, black, 232
piscicides, 231, 232–233, 235–237
    in coldwater rivers, 642
    in combination with drainage, 239
    estimation of presence, 331
    regulations, 231, 236

screening for, 343–344
TFM, 246–247
piscivorous fishes
    as biological control agents, 264
    in coldwater streams, 590, 591
    diet, 409–410
    fishing effects on, 475
    prey species, 408
    stocking, 415
    in tailwaters, 629
    trophic cascade of, 410
Pittman-Robertson Act, 25
plankton. *See also* phytoplankton; zooplankton
    as habitat improvement indicator, 316
plastic bags, for fish transportation, 216
point source discharges, 668
policymakers, in fisheries management, 137–138, 139. *See also* governmental authority, in fisheries management
    in adaptive management, 148–149
    goal-setting role, 145, 146–147
ponds, 501–543
    coldwater, 526
    coldwater fishes in, 520–524
    construction methods, 504–509, 518
    definition, 501
    embankment (hill), 504, 505
    excavated, 504, 509
    fertilization, 511–513, 521
    habitat, 504–519
    habitat development, 504–511
    innovative management strategies, 526–529
    leakage prevention, 507, 508, 509, 510
    liming, 513–514, 516
    outlet systems, 506, 508
    overfishing, 519–520
    population balance, 521, 530–537
    predator-prey relationship, 520
    stocking, 520–524, 520–526
    supplemental feeding i, 516
    traditional management strategies, 519–526
    warmwater, 525
    warmwater fishes in, 520–524
pondweed, 518

pools
    habitat availability, 98, 99
    as salmonid habitat, 590
    warmwater, 672
population
    definition, 325
    geographic boundaries, 329
    monitoring programs, 344
population assessment. *See* population estimation
population balance. *See also* predator-prey balance
    in ponds, 521, 530–537
population density. *See* density
population dynamics, 43-49, 330-336. *See also* growth; mortality rate; recruitment
    definition, 325
    estimation, 341–342
    models in, 44, 45–46
    overview, 44–46
    population parameters, 46–62
population estimation, 325-351, 603-604. *See also* abundance
    bias and precision, 326–327
    fish characteristics-based, 329–330
    genetic assessment, 336–339
    geographic delineation, 329
    models, 330
    need for, 325–326
    nonprobability sampling, 329
    parameters, 330–344
    probability sampling, 329, 603–604
    removal model, 328
    standardized sampling, 327
population growth, exponential, 44, 45
population isolation, 604–605, 606–607
population models, 44, 45–46
    catch-at-age, 64, 68–77
    yield-per-recruit, 62–64, 65–67
population redundancy, 608
population sampling. *See* sampling
population size, historic data, 33
population structure, 331
    indices, 340–342
possession limits, 192–193
potamodromous fishes, in coldwater streams, 587
potamodromy, 592
potassium (potash), as fertilizer, 512

potential yield
    habitat factors in, 463
    of lake fish species, 457–464,
        463–464
Potomac River, Washington, D.C., 216, 220
Powers, Bill, 181
precautionary management, 467, 468,
    469–470
precipitation
    acid rain, 484–485
    climate change-related changes,
        598, 601
    into lakes, 449
    into ponds, 505
    into streams, 589
precision (repeatability), 325, 326–327
predation
    aquatic vegetation effects, 484
    on coldwater fish populations, 591,
        593, 630
    fish assemblage effects, 659, 662,
        663, 665
    intraguild, 409
    in natural lakes, 488
    by nonnative species, 227
    relationship to abundance, 409–410
    in warmwater streams, 659, 662,
        663, 665
predator-prey balance
    ecological efficiency, 458
    in natural lakes, 487
    in ponds and small impoundments,
        501, 511, 520
    as trophic economics, 415–416
predators
    apex, 596–597, 665
    eradication programs, 21–22
    piscivorous fishes, 415, 590, 591,
        629, 410
    prey overexploitation, 484
presentations, 173–174, 175, 176–177
preservation, of historic fisheries
    information, 33–34
press disturbances, 595–596
prey, of fishes
    availability estimation, 342, 343
    benthic, 411–412
prey fishes
    index of vulnerability (IOV), 334

introduced, 488
    in reservoirs, 552, 577–578
    stocking, 557
    supplemental stocking, 264
primary productivity
    correlation with secondary
        productivity, 458
    of phytoplankton, 457, 458
    in reservoirs, 552, 557, 566–567
    in small streams, 620
    in warmwater streams, 660
Prince, Edward E., 8
private property principle, 113, 116, 130–131
private rights, versus public trust, 14–18
probability sampling, 329, 603–604
process, of fisheries management, 133-155.
    See also human dimensions, of fisheries
    management
    adaptive management, 142–149
    design and implementation of
        strategies, 148–149
    habitat improvement, 298–300
    identification of goals and objectives,
        145–148, 150–151
process-based simulations, 100
production:biomass ratio, 460–461
productivity. See also primary productivity;
secondary productivity
    benthic-pelagic, 411–412
    of coldwater rivers and streams,
        587, 620, 623
    index, 399
    of lakes, 452
    limiting factors, 361, 457
    limiting nutrients, 396–398, 400
    maximum sustainable yield (MSY),
        185, 457, 458–459, 460–461,
        468, 473–474, 481, 482
    morphoedaphic index (MEI),
        399–400, 459–460, 557
    of natural lakes, 457–460
    of reservoirs, 575
    riverine model, 663
profundal (aphotic) zone, 453
prohibition, of fishing, 197
project management, 168–173
propagated fishes. See hatchery fishes
proportional size distribution (PSD), 331,
    332–335

of exploited populations, 340–341
fishing effects on, 475
of lake fishes, 466–467
subdivisions, 533, 534, 535
proportional stock density. *See* proportional size distribution
publications, scientific, 174–175, 177–178
public interest, 138–139
public opinion
toward fisheries management, 149, 152–154
toward habitat improvement, 302, 304
toward harvest regulations, 188–189, 201
public participation
in coldwater stream fisheries management, 608, 609
in warmwater stream management, 687–688
public property principle, 113, 116
public trust doctrine, 5, 14–18, 116, 119
in Canada, 20–21
public waters principle, 116
Puget Sound, 23
pulp and paper mill effluents, 342
pulse disturbances, 595–596
"put-and-take" fisheries, 263
Puyallup River, Washington, 3

Q

qualitative analysis, of ecological integrity, 361
quantitative analysis
of ecological integrity, 361, 372–373
of population dynamics, 43–79
quarantine laws, 224–225
quarries, as pond sites, 504
Quesnel Lake, British Columbia, 452
quotas, 197–198, 489, 521

R

Rafinesque, Constantine Samuel, 4
rainbow smelt
commercial fisheries, 456
diet, 408–409
as prey species, 557

rainbow trout
brook trout displacement by, 219
eradication programs, 232, 234
hybridization, 641
as introduced species, 641–642
pond management, 516, 523, 526
as prey species, 416, 523
removal estimates, 328
in reservoirs, 549
stocking, 12, 14, 263, 280, 487, 502, 526, 640
stream order-related distribution, 629, 630
supplemental feeding, 516
in urban lakes, 502
ramp disturbances, 595
rare species. *See also* endangered species; threatened species
genetic integrity, 269
population models, 46
stocking, 264, 280
rarity, patterns, 93–94
rate functions, 332
razorback sucker, 377–378
reciprocation/reciprocity, 21, 160, 162
reclamation, 297
recreational activities, invasive species introduction through, 223
recreational fisheries. *See also* anglers
active management, 476–481
comparison with commercial fisheries, 474
economic impact analysis, 435–438
economic valuation, 427, 438–441
management goals and objectives, 464–467
of natural lakes, 455, 464–467, 474, 476, 489
passive management, 474, 476
spatial definitions, 465
special definitions, 465
surveys, 438, 442, 455
target fish sizes, 460
taxonomic definitions, 465
temporal definitions, 465
types, 186
in warmwater rivers, 714–715
recruitment
definition, 44, 58

estimation, 58, 60–62, 330
health status and, 344
low-flow hypothesis, 705–706
poor/failed, 341, 342
population dynamics effects, 332, 333–334
as rate function, 332
of riverine fishes, 703, 705–706
variability, 50, 58, 61, 62, 203–204, 332, 333–334, 341
redds (nests), 592, 593
reference conditions, 363–364
reference points
definition, 473
in natural lake management, 473–476, 481
for spearfishing exploitation, 478
reference populations, 202
regulations. *See also* governmental authority, in fisheries management; specific regulations
affecting marine fisheries, 139–140
creation of, 112–113, 117, 118–119, 145, 146–147
enforcement, 110, 112–113
federal, 141
history, 12, 14
for pond construction, 504–505
states, 141
rehabilitation. *See also* habitat improvement
of ecosystems, 297
historic data in, 33
of warmwater streams, 665–666
rehabilitation. *See also* habitat management
definition, 297
Reindeer Lake, Saskatchewan-Manitoba, Canada, 452
relative growth index (RGI), 336
relative stock density. *See* proportional size distribution
relative weight (W_r), 331, 332, 535, 537
definition, 342
as energy acquisition and storage index, 342–343
removal model, of abundance, 326, 328
renovation, 517, 518–519
reports, 31–32
reproduction. *See also* fecundity; spawning
endocrine-disrupting compounds and, 713

reproductive capacity, 468
reproductive fitness, of hatchery fish, 269, 270
reproductive output, lifetime, 342
reptiles, piscivorous, 21
research
American Fisheries Society policy document on, 20
funding, 25
history, 25
as nuisance fish introduction cause, 223
reservoir fishes
assemblages, 308
spawning and migration, 568–570
in watersheds, 571–572
reservoirs. *See also* large reservoirs
economic analysis, 436–437, 439–440
effect on channels, 626–627
fish assemblages, 308
flood control, 569
flow release, 647
guide curves, 560–562
habitat improvement, 308, 311–312
morphoedaphic index (MEI), 400
nonnative fishes control and eradication, 239, 707
sediment/sedimentation, 311–312
thermal stratification, 627
toxicant use prohibition in, 642
tributary streams, 568–570
water level, 311, 312
in watersheds, 576, 578
Resource Conservation and Recovery Act, 29
resource level studies, 443–444
resource management, North American model, 138
resource utilization, in coldwater river management, 631–636
restoration. *See also* habitat management
of coldwater streams, 604
definition, 297
of ecosystems, 297
of lakes, 375–377, 414–415
of native habitats, 471–472
of streams, 375–376
of warmwater rivers, 673–675
of warmwater streams, 665

of watersheds, 680
of wetlands, 679–680
reticulate sculpin, 593
revetments, 679, 711
riffles, 589, 672
Rio Grande River, 121
riparian law, 113
riparian vegetation
    buffer strip, 572
    of floodplains, 560
    habitat effects, 678
    management, 565–566, 645
    near ponds, 508, 511
    removal, 711
riparian water rights, 126, 128–129
riparian zones, 571–572, 574, 578
    disconnection, 667
    ownership, 6
risk assessment
    for habitat improvement, 297–298
    of stocking programs, 273
    techniques, 297
risk management, 170–171
riverbank. *See also* shoreline
    erosion, 632, 634
    habitats, 628
river basins, 574–576
river continuum concept, 574–576
    of coldwater rivers, 620–624, 625
    of coldwater streams, 588
    of warmwater rivers, 701–703, 706
    of warmwater streams, 663
river ecosystem synthesis (RES) model, 664, 703–704
riverine fishes
    habitat, 705
    overfishing, 714–715
    recruitment, 703, 705–706
    in reservoirs, 575
riverine productivity model, 663
riverkeeper organizations, 385
rivers. *See also* names of specific rivers
    historic data, 32–33, 33
    international treaties regarding, 121
    most endangered, 385
    size, 620–621
Rivers and Harbors Act, 27
roads, 644–645
    environmental effects, 595
    forestry, 676
    stream habitat effects, 676
    warmwater stream effects, 669
Romans, ancient, legal system, 113, 116
rotenone, 152–154, 232–233, 235, 237, 242
    use in coldwater rivers, 642
    use in coldwater streams, 605
    use in lakes, 417, 488
    use in ponds, 519
runoff
    into ponds, 505, 506
    into reservoirs, 308
    into streams, 589
    into urban ponds, 502
    into warmwater streams, 672

S

Sacramento River, California, 640, 716
Safe Drinking Water Act, 29
Saginaw Bay, 33
salmon. *See also* Chinook salmon; coho salmon; sockeye salmon
    anglers' preference for, 455
    diseases, 285
    as endangered species, 271–273
    feeds for, 286
    hatchery strategies, 275–276
    historical reports, 2, 3
    as "injurious wildlife," 225
    in reservoirs, 549
    stable isotope analysis, 416–417
    stocking, 263–264, 577
salmon canneries, 18
salmon carcasses
    genetic analysis, 339
    as nutrient source, 417, 646
salmonids
    as biological control agents, 217
    in coldwater rivers and streams, 587, 590, 619
    diseases and parasites, 591
    distribution, 599
    as endangered species, 643
    fish passage restoration, 606–607
    hatchery-reared, 266, 267, 269
    pool habitats, 590
    spawning behavior, 592–593
    spawning migrations, 592

stocking, 628, 639–641
in tailwaters, 628
Salmon River, Idaho, 629
sampling, 325, 326–329
    bias in, 325, 326–327
    for density estimation, 340
    for distribution estimation, 340
    in ecological integrity assessment, 362–363
    with electrofishing, 340
    hydroacoustic, 68
    nonprobability, 329, 340
    for population assessment, 325, 326–329
    precision, 325, 326–327
    probability, 329, 603–604
    standardized, 327
sanctuaries, 196–197
sand pits, 504, 509
saponins, as fish toxicants, 237
saprobien system, 365–366
sauger, 455, 549
Save Our Streams program, 688
scale, 31, 81–105
    conceptual models, 86, 88–89, 91–92
    definition and importance, 81–86
    in ecological integrity assessment, 362, 365
    empirical approach, 86–87, 91, 93, 95–97
    expert judgment models, 98–100
    extent of observation, 81, 82
    grain of observation, 81–85
    grain-to-extent ratio, 83
    hierarchical models, 93, 95–97
    identification of appropriate scales, 86–97, 101–102
    incorporation into fishery management, 97–101
    limnological, 395
    multiple, 85
    process-based simulations, 100
    qualitative, 91
    quantitative, 91, 93–97
    of species-specific characteristics, 90–91
    theoretical approach, 86, 87, 89–91
scale-area curves, 83–84

scarcity, 160–161, 162
Schmidt, Terry, 172–173
schooling behavior, 590, 665
Scientific Style and Format (Council of Science Editors), 177
sculpin, 593, 594
sea lamprey, 596
    control and eradication programs, 121, 237, 241, 245, 246–247, 488, 642
    mortality rate, 477
    sterile-male release programs, 488
    wounding rate, 476
sea lions, 240
seasons
    closed, 186, 196, 682, 686
    species-specific, 636
Secchi disk depth, 399, 400, 552, 553
secondary impacts, 360
secondary productivity, 458
sediment/sedimentation
    agriculture-related, 676
    flushing flow removal, 646, 647
    in large reservoirs, 311–312
    as phosphorus buffer, 396–397
    in ponds, 506
    in reservoirs, 311–312, 550–552, 566, 567, 570, 571, 574
    spawning substrate effects, 316–317
    transport, 620
    in warmwater rivers, 712
    in warmwater streams, 672, 677–678
    in watersheds, 570, 571, 572, 573
seed dispersal, 409
seining method, of pond analysis, 530–532, 533, 535
selection pressure, 342
selenium, 712, 713
serial discontinuity concept, 625–626, 663, 704
sewage treatment plant discharge, 670, 672, 673, 677
    estrogenic compounds in, 342
sex identification, 341, 342
sex ratio, 331, 342
S_F value, 531
shad
    biological control, 217
    migration barriers, 707

in reservoirs, 567–568
spawning, 707, 717
stocking, 8, 281
Shannon's index of diversity, 368–369
sheepshead minnow, 232, 238
Shepherd, John, 43
shoreline
erosion, 485–486, 517
of natural lakes, 485–486
of ponds, 517
protection, 486
stabilization, 574
of urban lakes, 503
shoreline development index, 365
shoreline vegetation. See riparian vegetation
side channels, 627, 628, 644
Sierra Club, 385
Sierra Nevada lakes, 135–137
silver carp, 225
silviculture. See forestry
simulation models, 100
single-species management, 26, 681
site selection, subjectively-fixed, 329
site surveys, of ponds, 506
size, of fishes
average, at maturity, 475
in coldwater environments, 620
edible, 460
in recreational fisheries, 460
relationship to yield, 462
spectrum, 475
structure indices, 332–335
size-abundance relationship, 475
slides, 173–174, 175
slot length limits
abundance effects, 685–686
for black bass, 523–524
for lake trout, 415–416
in pond management, 520
protected, 195, 196, 415–416, 638
sloughs, 644
small impoundments. See also ponds
definition, 501
smallmouth bass
eradication programs, 234
latitudinal gradient, 454
length limits, 685–686
pond management, 522, 529

in reservoirs, 568
stocking, 281
SMART statements, 145, 148
Smithsonian Institution, 4–5, 7
smoky madtom, 378
snagging, 668
harvest-tag regulation, 190, 191–192
snag removal, 374
snakeheads, 220–221, 225
northern, 235
Snake River, 18–19, 642
social accounting matrix (SAM) multipliers, 436–437
social information and research
definition, 426
general uses, 428
specific types and uses, 428–435
social proof, 161, 162
social value, of fish, 12, 20
Society of Wetland Scientists, 386
sockeye salmon, 18
software
Fisheries Analysis and Simulation Tools (FAST), 64, 330
impact analysis for planning (IMPLAN), 435, 436–437
soil conservation, 679
soluble reactive phosphorus (SRP), 396, 397
Southeast Rivers and Streams program, 385
South Platte River, Colorado, 638
Spangler, A.M., 15, 17
spatial autocorrelation, 91, 93
spatial scale, 84–85, 86, 101
of artificial structures, 711
in ecological integrity management, 378–379
watersheds, 304–305, 307
spawning
habitat manipulations for, 486–487
migrations, 646
by reservoir fishes, 568–569
in reservoirs, 557
by riverine fishes, 705–706
substrates for, 316–317, 486, 589–590, 592–593, 646, 647, 712
in tailwaters, 628
types, 592
spawning benches, 561, 562

spawning guilds, 682
spawning migrations, barriers to, 593, 597, 629, 646, 707, 717
spawning reefs, 486
spearfishing, 477–478
    harvest regulations, 199–200
speciation, 658–659
species councils/committees, 222
species diversity
    of coldwater fishes, 589
    in coldwater streams, 587
    drainage areas, 454
    in reservoirs, 568
    in urban lakes, 504
    in warmwater streams, 658
species dominance biologic index, 369
species evenness, 369
species richness, 367, 371, 372–373
    correlation with pH, 484–485
    of natural lakes, 454
    in reservoirs, 568
species traits, ordination of, 475
spillways, for ponds, 505, 506, 508, 509
sport fisheries
    history, 23–24
    population assessment, 326
    stocking, 280
    of warmwater rivers, 714
    of warmwater streams, 682, 685
Sport Fish Restoration Act, 118
Sport Fish Restoration Program, 118–119
spotted bass, as apex predator, 665
spotted jewelfish, 234
squoxin, 237
stable isotope analysis, 416–417
stake beds, 561, 562
stakeholders
    in adaptive management, 148–149, 150–151
    in coldwater stream fisheries management, 608, 609
    decision-making involvement, 140–142, 143–144, 427, 428
    definition, 425, 426
    fisheries management role, 137–139
    in habitat improvement, 298–300
    in marine fisheries management, 139
    in native species management, 136–137
    in recreational fisheries, 465
    self-determination of, 180, 182–183
    types, 426–427
stamps, for fishing, 190
    for coldwater rivers, 636
    habitat, 306
standard weight (W_s), 535, 536
starvation, 591
states. *See also* specific state fish and wildlife agencies
    fisheries management authority, 123
State Wildlife Grants Program, 120
statistical analysis. *See* biological statistics
steelhead, 18, 455, 644
sterile-fish stocking, for nuisance fish control, 228, 241, 246–247, 488, 642
Stevens, Isaac Ingalls, 23
stock assessment, 472–481
stocking
    adverse effects, 640
    after piscicide use, 237
    biodiversity conservation in, 227–228
    for biological control, 217–218
    choice of fish taxa, 280, 281–282
    in coldwater rivers, 631, 639–641
    in coldwater streams, 602
    conservation, 640
    cost-effectiveness analysis, 280, 282
    criticisms of, 284–285
    early policies and programs, 12, 14, 15, 17–18
    evaluation, 283
    female-only, 528, 529
    fish sizes and numbers, 280, 282, 287
    history, 12, 14, 15, 17–18, 33
    illegal, 519
    introductory rates, 525
    justification, 227
    maintenance-type, 487–488
    management practices, 270, 273–283
    in natural lakes, 487–488
    of nonnative species, 22, 219, 227–228
    philosophy of, 263–264
    of piscivorous fishes, 415
    in ponds, 520–524, 520–526
    of prey species, 557

private sector-sponsored, 262
public relations aspect, 262
public sector-sponsored, 262
put-and-take, 283
regulations, 223
relationship to habitat restoration, 273, 277, 278
in reservoirs, 577–578
restoration, 487
of self-sustaining populations, 487
"soft" versus "hard" releases in, 283
of sterile fish, as nuisance fish control, 228, 241, 246–247, 488, 642
stocking site, 282–283
successive/split, 525
supplemental, 264, 265, 283, 639
in tailwaters, 628
techniques, 277–283
timing, 282–283
unauthorized, 223
in urban lakes, 502
in warmwater streams, 681
stock-length, 340, 533, 535
"stockpiling," 204–205, 335
stock-recruitment relationship, 26, 58, 60
stomach content analysis, 330, 331, 342
Stone, Livingstone, 10, 14
stonewort, 518
storm water retention ponds, 396
Strategic Action Planning (Schmidt), 172–173
strategies, design and implementation, 148–149
stream(s). *See also* coldwater streams; warmwater streams
connectivity, 662
fishless, 587
hierarchical concept, 588, 589
restoration, 375–376
size, 620, 621
streambed armoring, 627
stream discharge, 590
stream habitat hierarchy concept, 663
stream networks, ecological integrity assessment, 362
stream order
fish species composition and, 603, 629
of large warmwater rivers, 700

stress effects, 343
striped bass
as biological control agent, 217
hybrid (wipers), 264, 487, 528
as introduced species, 564
pond management, 516
in reservoirs, 577
restoration programs, 278
spawning, 717
stocking, 143–144, 263, 264, 487, 577
supplemental feeding, 516
trophy-size, 528
structural uncertainty, 101
stunting, 341
sturgeon
lake, 21, 33
overfishing of, 714
white, 45
sub-basins, 92
submerged vegetation
definition, 518
in natural lakes, 483–484
in ponds, 511, 512
subsistence fisheries
management goals and objectives, 471
in natural lakes, 456–457, 471
of natural lakes, 471
Sugar Creek, Indiana, 676
sulfur, as fertilizer, 512
summerkill, in ponds, 483, 512
sunfish
as apex predator, 665
green, 281
hybrid, 281, 522
pond management, 516, 522, 525
redear, 54, 281
stocking, 280, 281, 525
supplemental feeding, 516
sunshine bass, 528
Surface Mining Control and Reclamation Act, 29
survey maps, 31
surveys
descriptive, 5
fisheries, 24
fishery-dependent, 472, 473, 490
fishery-independent, 472–473, 490

as historic data source, 33
ichthyological, 4–5, 7
large-fish netting, 480–481
of recreational fisheries, 438, 442, 455
survival
of hatchery fish, 269, 277
of larval fish, 408
in tailwaters, 628
Susquehanna River, 707
sustainability
of fish resources, 12
of Native American fisheries, 2
Sustainable Rivers Project, 385
sustainable yield, 467, 468
sustenance use, of fisheries, 432-433. *See also* subsistence fisheries
Swan Lake Habitat and Restoration Enhancement Project, 309–311
Swindle, H.S., 530–532, 533, 535
Systematics, Historical Ecology and North American Freshwater Fishes (Mayden), 4

T

tagging methods
for abundance estimation, 68
for growth estimation, 56
for mortality estimation, 51, 53–55
tag loss, evaluation, 51, 53
tag-reward studies, 331, 336
tailwaters, 627–628, 632, 712
Tallapoosa River, Alabama, 710
talweg, 623
tapeworms, 218
taxa richness, 366, 367
taxonomic definitions, 465
taxonomic identification, 363
tectonic activity, as lake formation process, 452
telemetry, use in active tagging, 53, 55
telephone interviews, 428
temperature
air. *See* air temperature
water. *See* water temperature
temperature-oxygen squeeze, 554–556, 559
temporal autocorrelation, 91, 93
temporal scale, 81–85, 86, 101
in natural lakes management, 481

tench, 216
Tennessee River, 575–576
Tennessee Valley Authority (TVA), 20, 143–144, 548, 559
"tens rule," 360
TFM (3-trifluoromethyl-4-nitrophenol), 231, 237, 241, 246–247
thermal acclimation state, 554–556
thermal enrichment, 125
thermal guilds, 700
thermal pollution, 356, 358
thermal stratification, in reservoirs, 627
thermal tolerance, 239
thermocline, 554
thermograph, 82–83
threadfin shad, 281, 488, 523
Threatened and Endangered Species Database System, 112
threatened species
definition, 111
stocking, 640
in warmwater streams, 680, 681
"three-story" fisheries, 557
Tijuana River, 121
tilapia
blue, 233, 239, 240
commercial fisheries, 456
tissue analysis, 331, 343–344
Titcomb, J.W., 15, 17–18
total dissolved solids (TDS), 459
total maximum daily load (TMDL), 381, 382, 677, 687
total phosphorus (TP), 396, 399
correlation with fish production, 458
as fish yield predictor, 459
of lakes, 458
morphoedaphic index and, 459
toxicants. *See* piscicides
Toxic Substances Control Act, 29
translocation
as benthic process, 397
of wild-caught introduced species, 377–378
transportation, of live fish
of hatchery fish, 277–280
history, 215, 216
illegal, 524
of introduced species, 215, 221
parasite transmission in, 524–525

plastic bags for, 216
  by railroad, 10, 216
trapping
  of crayfish, 517
  as nuisance fish control method, 234, 242, 642
"trash fish," 21, 285–286
trees, riparian, 672, 678
tribal fisheries
  in coldwater rivers, 643–635
  harvest policies, 187
  prior to European settlements, 2–3
  spearfishing, 199–200, 477–478
tributary streams
  management, 570
  of reservoirs, 568–570
triploidy, 281
trophic cascade hypothesis, 410–411
trophic composition metrics, 370–371
trophic depression, 564
trophic economics, 415–416
trophic guilds, 550, 576
trophic levels. *See also* primary productivity; secondary productivity
  mean, 475
trophic state, of reservoirs, 567, 572
trophic state continuum, 452–453
trophic state index (TSI), 399, 400
trophic upsurge, 564
trophy fish, 638
trophy fisheries
  in lakes, 467
  management objectives, 467
  stocking, 283
trophy-size fish
  largemouth bass, 522–523
  striped bass, 528
trotlines, 197
trout. *See also* brook trout; brown trout; bull trout; cutthroat trout; lake trout; rainbow trout; steelhead trout
  adaptive management, 150–151
  anglers' preference for, 455
  catch-and-release regulations, 148
  as coldwater stream indicator, 657
  as introduced species, 135–137
  pond management, 524
  in reservoirs, 549
  restocking, 154
  stocking, 577
  total catch model, 64
$TSI_{chl\text{-}a}$ index, 399, 400
Tui chub, 235
turbidity
  adverse effects, 676
  fish as cause of, 518–519
  in ponds, 511, 512, 516–517
  in reservoirs, 552
  in warmwater rivers, 712
turbid-water state, 412, 413, 417
"two-story" fisheries, 263, 556–557

U

uncertainty management, 100–101
unexploited populations, 461
United Nations Convention on Biodiversity, 223
United States Code, 22
United States Department of Agriculture (USDA), 8
  Animal and Plant Health Inspection Service, 224–225
  Conservation Reserve Program, 688
  Forest Service, 112, 133, 141, 153
  Wetland Reserve Program, 688
United States Department of Commerce, 8, 127
United States Department of the Interior, 8, 127
University of Wisconsin Sea Grant College Program, 403
Upper Mississippi River System Environmental Management Plan, 308, 309–311
urban fisheries
  management, 467
  ponds and small impoundments, 501, 502–504
urban fishing programs, 280
urbanization
  as pollution source, 672
  warmwater stream effects, 670, 672
  watershed effects, 672
Utah chub, 233
Utah sucker, 233

V

value
        of fisheries resources, 427
        of fishes, 12, 20, 21
        nonmarket, 438
value choices, 138, 140, 142, 143–144, 147, 154
value orientations, 434–435
variability range, 710
vegetation
        aquatic. *See* aquatic vegetation
        riparian. *See* riparian vegetation
"verbal judo," 161, 163–166
video data, 34
viral load, 331
virgin mean annual discharge (VMAD), 700
virtual population analysis (VPA), 68, 69–76
vital load, 344
volcanic activity, as lake formation process, 452
von Bertalaffy growth curve, 56, 57, 334, 336
        fitted in Microsoft® Office Excel, 59–60

W

Wabash River, 678
walleye
        anglers' preference for, 455
        commercial fisheries, 456
        coolwater assemblages, 454
        diet, 408
        harvest regulations, 200, 205
        historic data, 33
        latitudinal gradient, 454
        in natural lakes, 477–478, 487, 488
        pond management, 529
        in reservoirs, 549, 568
        seasonal closures, 682
        stocking, 33, 280, 487, 488, 681
        sustained yield, 461
        thermal habitat area versus lake surface area, 464
        tournament fishing, 198
        virtual population analysis, 68, 69–76
Wallop-Breaux Act, 25
warmwater discharge, into coldwater rivers, 632
warmwater fishes, 625
        abiotic processes affecting, 706–714
        biotic processes affecting, 706, 714–716
        hierarchical filters, 454
        management issues, 706–716
        in ponds, 520–524
        thermal guilds, 700
warmwater ponds, 525
warmwater rivers, 699–736
        adaptive management, 718–721
        characteristics, 700–701
        definition, 700
        ecological integrity, 716
        ecology, 701–706
        multiple-use function, 718, 719–721
        river continuum concept, 701–703
warmwater streams, 657–697
        characteristics, 731–736
        climate change effects, 689
        contemporary management issues, 665–688
        definition, 657
        fish assemblages, 658–665
        future management issues, 688
        goals, 665–666
        index of biotic integrity, 357
        introduced species, 680–681
        multiple-species management, 682
        multiple-use, 686–687
        single-species management, 681
Washington Treaty, 21
water clarity, 405-406, 412, 413. *See also* turbidity
        of coldwater rivers, 627
        of natural lakes, 452
        potential yield effects, 460
        of reservoirs, 552, 567
        species abundance effects, 460
        of warmwater rivers, 712
water depth/level
        of ponds, 505, 506
        of reservoirs, 311, 312, 559–561
water development, history of, 32
Waterkeeper Alliance, 385
water pollution
        federal regulations, 358
        indices, 357–358
        of natural lakes, 485

non-point sources, 125–126, 670
point-sources, 125, 670
pulp and paper mill effluents, 342
regulations, 125–126, 713
sources, 356, 358–359
testing for, 344
total maximum daily load (TMDL), 381, 382, 677
of warmwater rivers, 673–675, 712–713
of warmwater streams, 670, 672–675

water quality
biomanipulation improvement, 264
of coldwater rivers, 627
eradication programs and, 415
historic management, 677
indices, 364–365, 677
regulations, 558
regulatory authority for, 125–126
sampling, 362–363
standards, 22, 364, 365, 555
at stocking sites, 282–283
of urban ponds, 502
of warmwater rivers and streams, 662, 668, 669, 670, 712–713

water quantity, 126–129
Water Resources Development Act, 296–297
water retention time, 559, 566
water rights, 126–129, 677
Watershed Academy, 688
watersheds, 570–574
definition, 304, 570
fragmentation, 595, 605
habitat alteration, 304–307, 662
habitat improvement, 304–307
of lakes, 449–450
management, 572–574
of natural lakes, 449–450, 453–454
of ponds, 511
population isolation, 605
of reservoirs, 576, 578
restoration, 680
spatial scale, 304–305, 307
urban land use effects on, 672

water temperature
of coldwater rivers and streams, 578, 588–589, 593, 594, 595–596, 620, 621, 627

downstream changes, 627
effects of, 598
habitat segregation effects, 594
land use effects, 676
natural mortality rate effects, 461, 462
thermal acclimation to, 554–556
thermograph measurement, 82–83
of warmwater rivers, 700, 712
of warmwater streams, 657–658

water temperature regulation, as nuisance fish control method, 233, 239
Water Utilization Treaty, 121
water withdrawal, 134, 667, 669
weight
correlation with maturity status, 342
as growth indicator, 55
at maturity, 331
relative (W_r), 331, 332, 342–343, 535, 537
standard (W_s), 535, 536

weight-length relationship models, 56–57
Western Hemisphere Convention, 121
Western Native Trout Initiative, 643
Wetland Reserve Program, 679–680, 688
wetlands
Everglades, 313–315
legal restrictions on use, 116
permit programs, 125
as public trust, 119
restoration, 679–680
of warmwater streams, 660

whirling disease, 285, 591
white bass, in reservoirs, 550
white crappie, 281
whitefish, 454
white sturgeon, 45
Wild and Scenic Rivers Act, 28, 127
Wild and Scenic Rivers Program, 688
Wild Animal and Plant Protection and Regulation of International and Interprovincial Trade Act, 122
Wilderness Act, 28
wilderness fisheries, management objectives, 467
wildfires, 595–596, 599
wild fishes, competition with hatchery fishes, 640
wildlife, definition, 20

wildlife management. *See also* fish and wildlife agencies
    "father" of, 133
    history, 25
wildlife management plans, 120
Wildlife Research Units, 25
Wildlife Restoration Program, 25
willingness-to-pay (WTP), 438, 439, 440, 441
windbreaks, 486, 517
wing dikes, 708, 711
winterkill, in ponds, 512, 514, 515
winterkills, 407, 512, 514, 515
wipers (hybrid striped bass), 264, 487, 528
Wisconsin
    fish rescue operations, 14
        walleye population monitoring, 477–478
Wisconsin Department of Natural Resources, 357–358
Woods Hole, Massachusetts, 7
woody debris, as habitat, 486, 677, 711
World Wildlife Fund, 385
writing skills, 174–175, 177–178

Y

Y:C ratio, 530, 531
year classes
    age at maturity estimation, 342
    age estimation, 341
    per sample, 331
    recruitment, 341
    relative abundance, 344
    size, 333
    in warmwater rivers, 703
yellow bass, in reservoirs, 550
yellow perch
    anglers' preference for, 455
    commercial fisheries, 456
    coolwater assemblages, 454
    diet of, 408
    harvest regulations, 186
    latitudinal gradient, 454
    pond management, 521
    in reservoirs, 549
Yellowstone Lake, 488
Yellowstone National Park, 12, 232, 587
Yellowstone River, Montana, 707
yield
    annual air temperature and, 459–460
    definition, 458
    economic, 467, 468
    equilibrium, 468, 473
    long-term average, 458–459
    morphoedaphic index (MEI), 399–400, 459–460, 557
    potential, 463–464
    as production index (kg/ha/year), 458–459
    safe, 469–470
yield (harvest) models, 330
yield isopleths, 67
yield-per-recruit models, 62–64, 65–67
yolk sac fry, 592–593
Yukon River, 337

Z

zonation, longitudinal, 701
zoning regulations, 130–131
Zoogeography of North American Freshwater Fishes (Hocutt and Wiley), 4
zooplankton, 403, 408–409
    migration, 408
    pelagic, 407–408
    in reservoirs, 551, 558
    trophic cascade, 410
zooplanktivorous fish, 408–409